D0078970

PROBABILITY, RANDOM VARIABLES, AND STOCHASTIC PROCESSES

McGraw-Hill Series in Electrical Engineering

Consulting Editor

Stephen W. Director, *Carnegie-Mellon University*

Circuits and Systems
Communications and Signal Processing
Control Theory
Electronics and Electronic Circuits
Power and Energy
Electromagnetics
Computer Engineering
Introductory
Radar and Antennas
VLSI

Previous Consulting Editors

Ronald N. Bracewell, Colin Cherry, James F. Gibbons, Willis W. Harman, Hubert Heffner, Edward W. Herold, John G. Linvill, Simon Ramo, Ronald A. Rohrer, Anthony E. Siegman, Charles Susskind, Frederick E. Terman, John G. Truxal, Ernst Weber, and John R. Whinnery

Communications and Signal Processing

Consulting Editor

Stephen W. Director, *Carnegie-Mellon University*

Antoniou: *Digital Filters: Analysis and Design*
Candy: *Signal Processing: The Model-Based Approach*
Candy: *Signal Processing: The Modern Approach*
Carlson: *Communications Systems: An Introduction to Signals and Noise in Electrical Communication*
Cherin: *An Introduction to Optical Fibers*
Collin: *Antennas and Radiowave Propagation*
Collin: *Foundations for Microwave Engineering*
Cooper and McGillem: *Modern Communications and Spread Spectrum*
Davenport: *Probability and Random Processes: An Introduction for Applied Scientists and Engineers*
Drake: *Fundamentals of Applied Probability Theory*
Huelsman and Allen: *Introduction to the Theory and Design of Active Filters*
Jong: *Method of Discrete Signal and System Analysis*
Keiser: *Local Area Networks*
Keiser: *Optical Fiber Communications*
Kraus: *Antennas*
Kuc: *Introduction to Digital Signal Processing*
Papoulis: *Probability, Random Variables, and Stochastic Processes*
Papoulis: *Signal Analysis*
Papoulis: *The Fourier Integral and Its Applications*
Peebles: *Probability, Random Variables, and Random Signal Principles*
Proakis: *Digital Communications*
Schwartz: *Information Transmission, Modulation, and Noise*
Schwartz and Shaw: *Signal Processing*
Smith: *Modern Communication Circuits*
Taub and Schilling: *Principles of Communication Systems*

PROBABILITY, RANDOM VARIABLES, AND STOCHASTIC PROCESSES

Third Edition

Athanasios Papoulis

University Professor

Polytechnic University

McGraw-Hill, Inc.

New York St. Louis San Francisco Auckland Bogotá
Caracas Lisbon London Madrid Mexico City Milan
Montreal New Delhi San Juan Singapore
Sydney Tokyo Toronto

This book was set in Times Roman by Science Typographers, Inc.
The editors were Roger L. Howell and John M. Morriss;
the production supervisor was Richard A. Ausburn.
The cover was designed by Joseph Gillians.
Project supervision was done by Science Typographers, Inc.
R. R. Donnelley & Sons Company was printer and binder.

PROBABILITY, RANDOM VARIABLES, AND STOCHASTIC PROCESSES

4 5 6 7 8 9 0 DOC DOC 9 0 9 8 7 6 5 4

ISBN 0-07-048477-5

Library of Congress Cataloging-in-Publication Data

Papoulis, Athanasios, (date).
Probability, random variables, and stochastic processes/
Athanasios Papoulis.—3rd ed.
p. cm.—(McGraw-Hill series in electrical engineering. Communications and signal processing)
Includes bibliographical references and index.
ISBN 0-07-048477-5
1. Probabilities. 2. Random variables. 3. Stochastic processes.
I. Title. II. Series.
QA273.P2 1991
519.2—dc20 90-23127

CONTENTS

PREFACE TO THE THIRD EDITION

In this edition, about a third of the text is either new or substantially revised. The new topics include the following:

A chapter on statistics. With this addition, the first nine chapters of the book could form the basis for a senior–graduate course in probability and statistics.

A chapter on spectral estimation. This chapter starts with an expanded treatment of ergodicity and it covers the fundamentals of parametric and nonparametric estimation in the context of system identification.

A section on the meaning and generation of random numbers. This material is essential for the understanding of computer simulation of random phenomena and the use of statistics in the solution of deterministic problems (Monte Carlo techniques).

Other topics include bispectra, state variables and vector processes, factorization, and spectral representation.

I wrote the first edition of this book long ago. My objective was to develop the subject of probability and stochastic processes as a deductive discipline and to illustrate the theory with basic applications of general interest. I tried to stress clarity and economy, avoiding sophisticated mathematics or, at the other extreme, detailed discussion of practical applications. It appears that this approach met with some success. For over a quarter of a century, the book has been used as a basic text and standard reference not only in this country but throughout the world. I am deeply grateful.

McGraw-Hill and I would like to thank the following reviewers for their many helpful comments and suggestions: John Adams, Lehigh University; David Anderson, University of Michigan; V. Krishnan, University of Lowell; Robert J. Mulholland, University of Oklahoma; Stephen Sebo, Ohio State University; and Samir S. Soliman, Southern Methodist University.

Athanasios Papoulis

PREFACE TO THE SECOND EDITION

This is an extensively revised edition reflecting the developments of the last two decades. Several new topics are added, important areas are strengthened, and sections of limited interest are eliminated. Most additions, however, deal with applications; the first ten chapters are essentially unchanged.

In the selection of the new material I have attempted to concentrate on subjects that not only are of current interest, but also contribute to a better understanding of the basic properties of stochastic processes. The new material includes the following:

Discrete-time processes with applications in system theory

Innovations, factorization, spectral representation

Queueing theory, level crossings, spectra of FM signals, sampling theory

Mean square estimation, orthonormal expansions, Levinson's algorithm, Wold's decomposition, Wiener, lattice, and Kalman filters

Spectral estimation, windows, extrapolation, Burg's method, detection of line spectra

This book concludes with a self-contained chapter on entropy developed axiomatically from first principles. It is presented in the context of earlier chapters, and it includes the method of maximum entropy in parameter estimation and elements of coding theory.

As in the first edition, I made a special effort to stress the conceptual difference between mental constructs and physical reality. This difference is summarized in the following paragraph, taken from the first edition:

Scientific theories deal with concepts, not with reality. All theoretical results are derived from certain axioms by deductive logic. In physical sciences the theories are so formulated as to correspond in some useful sense to the real world, whatever that may mean. However, this correspondence is approxi-

mate, and the physical justification of all theoretical conclusions is based on some form of inductive reasoning.

Responding to comments by a number of readers over the years, I would like to emphasize that this passage in no way questions the existence of natural laws (patterns). It is merely a reminder of the fundamental difference between concepts and reality.

During the preparation of the manuscript I had the benefit of lengthy discussions with a number of colleagues and friends. I thank in particular Hans Schreiber of Grumman, William Shanahan of Norden Systems, and my colleagues Frank Cassara and Basil Maglaris for their valuable suggestions. I wish also to express my appreciation to Mrs. Nina Adamo for her expert typing of the manuscript.

Athanasios Papoulis

PREFACE TO THE FIRST EDITION

Several years ago I reached the conclusion that the theory of probability should no longer be treated as adjunct to statistics or noise or any other terminal topic, but should be included in the basic training of all engineers and physicists as a separate course. I made then a number of observations concerning the teaching of such a course, and it occurs to me that the following excerpts from my early notes might give you some insight into the factors that guided me in the planning of this book:

"Most students, brought up with a deterministic outlook of physics, find the subject unreliable, vague, difficult. The difficulties persist because of inadequate definition of the first principles, resulting in a constant confusion between assumptions and logical conclusions. Conceptual ambiguities can be removed only if the theory is developed axiomatically. They say that this approach would require measure theory, would reduce the subject to a branch of mathematics, would force the student to doubt his intuition leaving him without convincing alternatives, but I don't think so. I believe that most concepts needed in the applications can be explained with simple mathematics, that probability, like any other theory, should be viewed as a conceptual structure and its conclusions should rely not on intuition but on logic. The various concepts must, of course, be related to the physical world, but such motivating sections should be separated from the deductive part of the theory. Intuition will thus be strengthened, but not at the expense of logical rigor.

"There is an obvious lack of continuity between the elements of probability as presented in introductory courses, and the sophisticated concepts needed in today's applications. How can the average student, equipped only with the probability of cards and dice, understand prediction theory or harmonic analysis? The applied books give at most a brief discussion of background material; their objective is not the use of the applications to strengthen the student's understanding of basic concepts, but rather a detailed discussion of special topics.

"Random variables, transformations, expected values, conditional densities, characteristic functions cannot be mastered with mere exposure. These concepts must be clearly defined and must be developed, one at a time, with sufficient elaboration. Special topics should be used to illustrate the theory, but they must be so presented as to minimize peripheral, descriptive material and to concentrate on probabilistic content. Only then the student can learn a variety of applications with economy and perspective."

I realized that to teach a convincing course, a course that is not a mere presentation of results but a connected theory, I would have to reexamine not only the development of special topics, but also the proofs of many results and the method of introducing the first principles.

"The theory must be mathematical (deductive) in form but without the generality or rigor of mathematics. The philosophical meaning of probability most somehow be discussed. This is necessary to remove the mystery associated with probability and to convince the student of the need for an axiomatic approach and a clear distinction between assumptions and logical conclusions. The axiomatic foundation should not be a mere appendix but should be recognized throughout the theory.

"Random variables must be defined as functions with domain an abstract set of experimental outcomes and not as points on the real line. Only then infinitely dimensional spaces are avoided and the extension to stochastic processes is simplified.

"The inadequacy of averages as definitions and the value of an underlying space is most obvious in the treatment of stochastic processes. Time averages must be introduced as stochastic integrals, and their relationship to the statistical parameters of the process must be established only in the form of ergodicity.

"The emphasis on second-order moments and spectra, utilizing the student's familiarity with systems and transform techniques, is justified by the current needs.

"Mean-square estimation (prediction and filtering), a topic of considerable importance, needs a basic reexamination. It is best understood if it is divorced from the details of integral equations or the calculus of variations, and is presented as an application of the orthogonality principle (linear regression), simply explained in terms of random variables.

"To preserve conceptual order, one must sacrifice continuity of special topics, introducing them as illustrations of the general theory."

These ideas formed the framework of a course that I taught at the Polytechnic Institute of Brooklyn. Encouraged by the students' reaction, I decided to make it into a book. I should point out that I did not view my task as an impersonal presentation of a complete theory, but rather as an effort to explain the essence of this theory to a particular group of students. The book is written neither for the handbook-oriented students nor for the sophisticated few who can learn the subject from advanced mathematical texts. It is written for the majority of engineers and physicists who have sufficient maturity to appreciate and follow a logical presentation, but, because of their limited mathematical background, would find a book such as Doob's too difficult for a beginning text.

Although I have included many useful results, some of them new, my hope is that the book will be judged not for completeness but for organization and clarity. In this context I would like to anticipate a criticism and explain my approach. Some readers will find the proofs of many important theorems lacking in rigor. I emphasize that it was not out of negligence, but after considerable thought, that I decided to give, in several instances, only plausibility arguments. I realize too well that "a proof is a proof or it is not." However, a rigorous proof must be preceded by a clarification of the new idea and by a plausible explanation of its validity. I felt that, for the purpose of this book, the emphasis should be placed on explanation, facility, and economy. I hope that this approach will give you not only a working knowledge, but also an incentive for a deeper study of this fascinating subject.

Although I have tried to develop a personal point of view in practically every topic, I recognize that I owe much to other authors. In particular, the books "Stochastic Processes" by J. L. Doob and "Théorie des Functions Aléatoires" by A. Blanc-Lapierre and R. Forter influenced greatly my planning of the chapters on stochastic processes.

Finally, it is my pleasant duty to express my sincere gratitude to Mischa Schwartz for his encouragement and valuable comments, to Ray Pickholtz for his many ideas and constructive suggestions, and to all my colleagues and students who guided my efforts and shared my enthusiasm in this challenging project.

Athanasios Papoulis

PROBABILITY, RANDOM VARIABLES, AND STOCHASTIC PROCESSES

PART I

PROBABILITY AND RANDOM VARIABLES

CHAPTER

1

THE MEANING OF PROBABILITY

1-1 INTRODUCTION

The theory of probability deals with averages of mass phenomena occurring sequentially or simultaneously: electron emission, telephone calls, radar detection, quality control, system failure, games of chance, statistical mechanics, turbulence, noise, birth and death rates, and queueing theory, among many others.

It has been *observed* that in these and other fields certain averages approach a constant value as the number of observations increases and this value remains the same if the averages are evaluated over any subsequence specified before the experiment is performed. In the coin experiment, for example, the percentage of heads approaches 0.5 or some other constant, and the same average is obtained if we consider every fourth, say, toss (no betting system can beat the roulette).

The purpose of the theory is to describe and predict such averages in terms of probabilities of events. The probability of an event $\mathscr{A}$ is a number $P(\mathscr{A})$ assigned to this event. This number could be interpreted as follows:

> If the experiment is performed n times and the event $\mathscr{A}$ occurs $n_{\mathscr{A}}$ times, then, *with a high degree of certainty*, the relative frequency $n_{\mathscr{A}}/n$ of the occurrence of $\mathscr{A}$ is *close* to $P(\mathscr{A})$:
>
> $$P(\mathscr{A}) \simeq n_{\mathscr{A}}/n \qquad (1\text{-}1)$$
>
> provided that n is *sufficiently large*.

This interpretation is imprecise: The terms "with a high degree of certainty," "close," and "sufficiently large" have no clear meaning. However, this lack of precision cannot be avoided. If we attempt to define in probabilistic terms the "high degree of certainty" we shall only postpone the inevitable conclusion that probability, like any physical theory, is related to physical phenomena only in inexact terms. Nevertheless, the theory is an exact discipline developed logically from clearly defined axioms, and when it is applied to real problems, *it works*.

OBSERVATION, DEDUCTION, PREDICTION. In the applications of probability to real problems, the following steps must be clearly distinguished:

Step 1 (physical) We determine by an inexact process the probabilities $P(\mathscr{A}_i)$ of certain events $\mathscr{A}_i$.

This process could be based on the relationship (1-1) between probability and observation: The probabilistic data $P(\mathscr{A}_i)$ equal the observed ratios $n_{\mathscr{A}_i}/n$. It could also be based on "reasoning" making use of certain symmetries: If, out of a total of N outcomes, there are $N_{\mathscr{A}}$ outcomes favorable to the event $\mathscr{A}$, then $P(\mathscr{A}) = N_{\mathscr{A}}/N$.

For example, if a loaded die is rolled 1000 times and *five* shows 203 times, then the probability of *five* equals 0.2. If the die is fair, then, because of its symmetry, the probability of *five* equals 1/6.

Step 2 (conceptual) We assume that probabilities satisfy certain axioms, and by deductive reasoning we determine from the probabilities $P(\mathscr{A}_i)$ of certain events $\mathscr{A}_i$ the probabilities $P(\mathscr{B}_j)$ of other events $\mathscr{B}_j$.

For example, in the game with a fair die we deduce that the probability of the event *even* equals 3/6. Our reasoning is of the following form:

$$\text{If} \quad P(1) = \cdots = P(6) = \tfrac{1}{6} \qquad \text{then} \quad P(\textit{even}) = \tfrac{3}{6}$$

Step 3 (physical) We make a physical *prediction* based on the numbers $P(\mathscr{B}_j)$ so obtained.

This step could rely on (1-1) applied in reverse: If we perform the experiment n times and an event $\mathscr{B}$ occurs $n_{\mathscr{B}}$ times, then $n_{\mathscr{B}} \simeq nP(\mathscr{B})$.

If, for example, we roll a fair die 1000 times, our prediction is that *even* will show about 500 times.

We could not emphasize too strongly the need for separating the above three steps in the solution of a problem. We must make a clear distinction between the data that are determined empirically and the results that are deduced logically.

Steps 1 and 3 are based on *inductive reasoning*. Suppose, for example, that we wish to determine the probability of *heads* of a given coin. Should we toss the coin 100 or 1000 times? If we toss it 1000 times and the average number of heads equals 0.48 what kind of prediction can we make on the basis of this observation? Can we deduce that at the next 1000 tosses the number of heads will be about 480? Such questions can be answered only inductively.

In this book, we consider mainly step 2, that is, from certain probabilities we derive *deductively* other probabilities. One might argue that such derivations are mere tautologies because the results are contained in the assumptions. This is true in the same sense that the intricate equations of motion of a satellite are included in Newton's laws.

To conclude, we repeat that the probability $P(\mathscr{A})$ of an event $\mathscr{A}$ will be interpreted as a number assigned to this event as mass is assigned to a body or resistance to a resistor. In the development of the theory, we will not be concerned about the "physical meaning" of this number. This is what is done in circuit analysis, in electromagnetic theory, in classical mechanics, or in any other scientific discipline. These theories are, of course, of no value to physics unless they help us solve real problems. We must assign specific, if only approximate, resistances to real resistors and probabilities to real events (step 1); we must also give physical meaning to all conclusions that are derived from the theory (step 3). But this link between concepts and observation must be separated from the purely logical structure of each theory (step 2).

As an illustration, we discuss in the next example the interpretation of the meaning of resistance in circuit theory.

Example 1-1. A resistor is commonly viewed as a two-terminal device whose voltage is proportional to the current

$$R = \frac{v(t)}{i(t)} \tag{1-2}$$

This, however, is only a convenient abstraction. A real resistor is a complex device with distributed inductance and capacitance having no clearly specified terminals. A relationship of the form (1-2) can, therefore, be claimed only within certain errors, in certain frequency ranges, and with a variety of other qualifications. Nevertheless, in the development of circuit theory we ignore all these uncertainties. We assume that the resistance R is a precise number satisfying (1-2) and we develop a theory based on (1-2) and on Kirchhoff's laws. It would not be wise, we all agree, if at each stage of the development of the theory we were concerned with the *true* meaning of R.

1-2 THE DEFINITIONS

In this section, we discuss various definitions of probability and their roles in our investigation.

Axiomatic Definition

We shall use the following concepts from set theory (for details see Chap. 2): The certain event $\mathscr{S}$ is the event that occurs in every trial. The union $\mathscr{A} + \mathscr{B}$ of two events $\mathscr{A}$ and $\mathscr{B}$ is the event that occurs when $\mathscr{A}$ or $\mathscr{B}$ or both occur. The intersection $\mathscr{A}\mathscr{B}$ of the events $\mathscr{A}$ and $\mathscr{B}$ is the event that occurs when both events $\mathscr{A}$ and $\mathscr{B}$ occur. The events $\mathscr{A}$ and $\mathscr{B}$ are mutually exclusive if the occurrence of one of them excludes the occurrence of the other.

We shall illustrate with the die experiment: The certain event is the event that occurs whenever any one of the six faces shows. The union of the events *even* and *less than 3* is the event *1 or 2 or 4 or 6* and their intersection is the event *2*. The events *even* and *odd* are mutually exclusive.

The axiomatic approach to probability is based on the following three postulates and on nothing else: The probability $P(\mathscr{A})$ of an event $\mathscr{A}$ is a non-negative number assigned to this event:

$$P(\mathscr{A}) \geq 0 \tag{1-3}$$

The probability of the certain event equals 1:

$$P(\mathscr{S}) = 1 \tag{1-4}$$

If the events $\mathscr{A}$ and $\mathscr{B}$ are mutually exclusive, then

$$P(\mathscr{A} + \mathscr{B}) = P(\mathscr{A}) + P(\mathscr{B}) \tag{1-5}$$

This approach to probability is relatively recent (A. Kolmogoroff,†1933). However, in our view, it is the best way to introduce a probability even in elementary courses. It emphasizes the deductive character of the theory, it avoids conceptual ambiguities, it provides a solid preparation for sophisticated applications, and it offers at least a beginning for a deeper study of this important subject.

The axiomatic development of probability might appear overly mathematical. However, as we hope to show, this is not so. The elements of the theory can be adequately explained with basic calculus.

Relative Frequency Definition

The relative frequency approach is based on the following definition: The probability $P(\mathscr{A})$ of an event $\mathscr{A}$ is the limit

$$P(\mathscr{A}) = \lim_{n \to \infty} \frac{n_{\mathscr{A}}}{n} \tag{1-6}$$

where $n_{\mathscr{A}}$ is the number of occurrences of $\mathscr{A}$ and n is the number of trials.

This definition appears reasonable. Since probabilities are used to describe relative frequencies, it is natural to define them as limits of such frequencies. The problem associated with a priori definitions are eliminated, one might think, and the theory is founded on observation.

However, although the relative frequency concept is fundamental in the applications of probability (steps 1 and 3), its use as the basis of a deductive theory (step 2) must be challenged. Indeed, in a physical experiment, the numbers $n_{\mathscr{A}}$ and n might be large but they are only finite; their ratio cannot, therefore, be equated, even approximately, to a limit. If (1-6) is used to define

†A. Kolmogoroff: Grundbegriffe der Wahrscheinlichkeits Rechnung, *Ergeb. Math und ihrer Grensg.* vol. 2, 1933.

$P(\mathscr{A})$, the limit must be accepted as a *hypothesis*, not as a number that can be determined experimentally.

Early in the century, Von Mises† used (1-6) as the foundation for a new theory. At that time, the prevailing point of view was still the classical and his work offered a welcome alternative to the a priori concept of probability, challenging its metaphysical implications and demonstrating that it leads to useful conclusions mainly because it makes implicit use of relative frequencies based on our collective experience. The use of (1-6) as the basis for deductive theory has not, however, enjoyed wide acceptance even though (1-6) relates $P(\mathscr{A})$ to observed frequencies. It has generally been recognized that the axiomatic approach (Kolmogoroff) is superior.

We shall venture a comparison between the two approaches using as illustration the definition of the resistance R of an ideal resistor. We can define R as a limit

$$R = \lim_{n \to \infty} \frac{e(t)}{i_n(t)}$$

where $e(t)$ is a voltage source and $i_n(t)$ are the currents of a sequence of real resistors that tend in some sense to an ideal two-terminal element. This definition might show the relationship between real resistors and ideal elements but the resulting theory is complicated. An axiomatic definition of R based on Kirchhoff's laws is, of course, preferable.

Classical Definition

For several centuries, the theory of probability was based on the classical definition. This concept is used today to determine probabilistic data and as a working hypothesis. In the following, we explain its significance.

According to the classical definition, the probability $P(\mathscr{A})$ of an event $\mathscr{A}$ is determined a priori without actual experimentation: It is given by the ratio

$$P(\mathscr{A}) = \frac{N_{\mathscr{A}}}{N} \tag{1-7}$$

where N is the number of *possible* outcomes and $N_{\mathscr{A}}$ is the number of outcomes that are *favorable* to the event $\mathscr{A}$.

In the die experiment, the possible outcomes are six and the outcomes favorable to the event *even* are three; hence $P(\textit{even}) = 3/6$.

It is important to note, however, that the significance of the numbers N and $N_{\mathscr{A}}$ is not always clear. We shall demonstrate the underlying ambiguities with the following example.

†Richard Von Mises: *Probability, Statistics and Truth*, English edition, H. Geiringer, ed., G. Allen and Unwin Ltd., London, 1957.

Example 1-2. We roll two dice and we want to find the probability p that the sum of the numbers that show equals 7.

To solve this problem using (1-7), we must determine the numbers N and $N_{\mathscr{A}}$. (*a*) We could consider as possible outcomes the 11 sums $2, 3, \ldots, 12$. Of these, only one, namely the sum 7, is favorable; hence $p = 1/11$. This result is of course wrong. (*b*) We could count as possible outcomes all pairs of numbers not distinguishing between the first and the second die. We have now 21 outcomes of which the pairs (3, 4), (5, 2), and (6, 1) are favorable. In this case, $N_{\mathscr{A}} = 3$ and $N = 21$; hence $p = 3/21$. This result is also wrong. (*c*) We now reason that the above solutions are wrong because the outcomes in (*a*) and (*b*) are not *equally likely*. To solve the problem "correctly," we must count all pairs of numbers distinguishing between the first and the second die. The total number of outcomes is now 36 and the favorable outcomes are the six pairs (3, 4), (4, 3), (5, 2), (2, 5), (6, 1), and (1, 6); hence $p = 6/36$.

The above example shows the need for refining definition (1-7). The improved version reads as follows:

> The probability of an event equals the ratio of its favorable outcomes to the total number of outcomes provided that all outcomes are *equally likely*.

As we shall presently see, this refinement does not eliminate the problems associated with the classical definition.

Notes 1. The classical definition was introduced as a consequence of the *principle of insufficient reason*†: "In the absence of any prior knowledge, we *must* assume that the events $\mathscr{A}_i$ have equal probabilities." This conclusion is based on the subjective interpretation of probability as a *measure of our state of knowledge* about the events $\mathscr{A}_i$. Indeed, if it were not true that the events $\mathscr{A}_i$ have the same probability, then changing their indices we would obtain different probabilities without a change in the state of our knowledge.

2. As we explain in Chap. 15, the principle of insufficient reason is equivalent to the *principle of maximum entropy*.

CRITIQUE. The classical definition can be questioned on several grounds.

A. The term *equally likely* used in the improved version of (1-7) means, actually, *equally probable*. Thus, in the definition, use is made of the concept to be defined. As we have seen in Example 1-2, this often leads to difficulties in determining N and $N_{\mathscr{A}}$.

B. The definition can be applied only to a limited class of problems. In the die experiment, for example, it is applicable only if the six faces have the same probability. If the die is loaded and the probability of *four* equals 0.2, say, the number 0.2 cannot be derived from (1-7).

---

†H. Bernoulli, *Ars Conjectandi*, 1713.

C. It appears from (1-7) that the classical definition is a consequence of logical imperatives divorced from experience. This, however, is not so. We accept certain alternatives as equally likely because of our collective experience. The probabilities of the outcomes of a fair die equal 1/6 not only because the die is symmetrical but also because it was observed in the long history of rolling dice that the ratio $n_{\mathscr{A}}/n$ in (1-1) is close to 1/6. The next illustration is, perhaps, more convincing:

We wish to determine the probability p that a newborn baby is a boy. It is generally assumed that $p = 1/2$; however, this is not the result of pure reasoning. In the first place, it is only approximately true that $p = 1/2$. Furthermore, without access to long records we would not know that the boy–girl alternatives are equally likely regardless of the sex history of the baby's family, the season or place of its birth, or other conceivable factors. It is only after long accumulation of records that such factors become irrelevant and the two alternatives are accepted as equally likely.

D. If the number of possible outcomes is infinite, then to apply the classical definition we must use length, area, or some other measure of infinity for determining the ratio $N_{\mathscr{A}}/N$ in (1-7). We illustrate the resulting difficulties with the following example known as the *Bertrand paradox*.

Example 1-3. We are given a circle C of radius r and we wish to determine the probability p that the length l of a "randomly selected" cord AB is greater than the length $r\sqrt{3}$ of the inscribed equilateral triangle.

We shall show that this problem can be given at least three reasonable solutions.

I. If the center M of the cord AB lies inside the circle C_1 of radius $r/2$ shown in Fig. 1-1a, then $l > r\sqrt{3}$. It is reasonable, therefore, to consider as favorable outcomes all points inside the circle C_1 and as possible outcomes all points inside the circle C. Using as measure of their numbers the corresponding areas $\pi r^2/4$ and πr^2, we conclude that

$$p = \frac{\pi r^2/4}{\pi r^2} = \frac{1}{4}$$

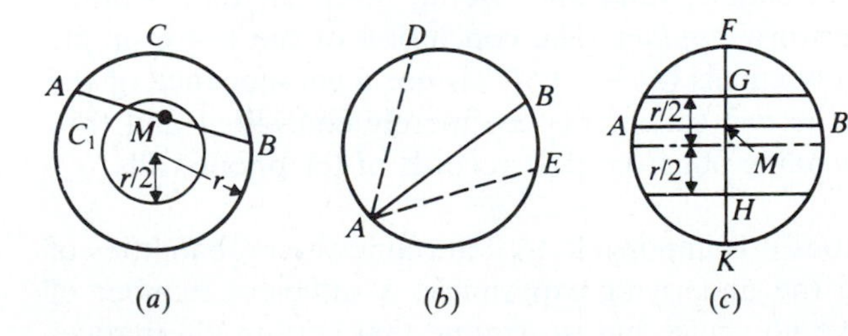

FIGURE 1-1

II. We now assume that the end A of the cord AB is fixed. This reduces the number of possibilities but it has no effect on the value of p because the number of favorable locations of B is reduced proportionately. If B is on the $120°$ arc DBE of Fig. 1-1b, then $l > r\sqrt{3}$. The favorable outcomes are now the points on this arc and the total outcomes all points on the circumference of the circle C. Using as their measurements the corresponding lengths $2\pi r/3$ and $2\pi r$, we obtain

$$p = \frac{2\pi r/3}{2\pi r} = \frac{1}{3}$$

III. We assume finally that the direction of AB is perpendicular to the line FK of Fig. 1-1c. As in II this restriction has no effect on the value of p. If the center M of AB is between G and H, then $l > r\sqrt{3}$. Favorable outcomes are now the points on GH and possible outcomes all points on FK. Using as their measures the respective lengths r and $2r$, we obtain

$$p = \frac{r}{2r} = \frac{1}{2}$$

We have thus found not one but three different solutions for the same problem! One might remark that these solutions correspond to three different experiments. This is true but not obvious and, in any case, it demonstrates the ambiguities associated with the classical definition, and the need for a clear specification of the outcomes of an experiment and the meaning of the terms "possible" and "favorable."

VALIDITY. We shall now discuss the value of the classical definition in the determination of probabilistic data and as a working hypothesis.

A. In many applications, the assumption that there are N equally likely alternatives is well established through long experience. Equation (1-7) is then accepted as self-evident. For example, "If a ball is selected at random from a box containing m black and n white balls, the probability that it is white equals $n/(m+n)$," or, "If a call occurs at random in the time interval $(0, T)$, the probability that it occurs in the interval (t_1, t_2) equals $(t_2 - t_1)/T$."

Such conclusions are of course, valid and useful; however, their validity rests on the meaning of the word *random*. The conclusion of the last example that "the unknown probability equals $(t_2 - t_1)/T$" is not a consequence of the "randomness" of the call. The two statements are merely equivalent and they follow not from a priori reasoning but from past records of telephone calls.

B. In a number of applications it is impossible to determine the probabilities of various events by repeating the underlying experiment a sufficient number of times. In such cases, we have no choice but to *assume* that certain alternatives are equally likely and to determine the desired probabilities from (1-7). This means that we use the classical definition as a *working* hypothesis. The hypothesis is accepted if its observable consequences agree with experience, otherwise it is rejected. We illustrate with an important example from statistical mechanics.

Example 1-4. Given n particles and $m > n$ boxes, we place at random each particle in one of the boxes. We wish to find the probability p that in n preselected boxes, one and only one particle will be found.

Since we are interested only in the underlying assumptions, we shall only state the results (the proof is assigned as Prob. 3-15). We also verify the solution for $n = 2$ and $m = 6$. For this special case, the problem can be stated in terms of a pair of dice: The $m = 6$ faces correspond to the m boxes and the $n = 2$ dice to the n particles. We assume that the preselected faces (boxes) are 3 and 4.

The solution to this problem depends on the choice of possible and favorable outcomes. We shall consider the following three celebrated cases:

Maxwell–Boltzmann statistics. If we accept as outcomes all possible ways of placing n particles in m boxes distinguishing the identity of each particle, then

$$p = \frac{n!}{m^n}$$

For $n = 2$ and $m = 6$ the above yields $p = 2/36$. This is the probability for getting 3, 4 in the game of two dice.

Bose–Einstein statistics. If we assume that the particles are not distinguishable, that is, if all their permutations count as one, then

$$p = \frac{(m-1)!n!}{(n+m-1)!}$$

For $n = 2$ and $m = 6$ this yields $p = 1/21$. Indeed, if we do not distinguish between the two dice, then $N = 21$ and $N_{\mathscr{A}} = 1$ because the outcomes 3, 4 and 4, 3 are counted as one.

Fermi–Dirac statistics. If we do not distinguish between the particles and also we assume that in each box we are allowed to place at most one particle, then

$$p = \frac{n!(m-n)!}{m!}$$

For $n = 2$ and $m = 6$ we obtain $p = 1/15$. This is the probability for 3, 4 if we do not distinguish between the dice and also we ignore the outcomes in which the two numbers that show are equal.

One might argue, as indeed it was in the early years of statistical mechanics, that only the first of these solutions is logical. The fact is that in the absence of direct or indirect experimental evidence this argument cannot be supported. The three models proposed are actually only *hypotheses* and the physicist accepts the one whose consequences agree with experience.

C. Suppose that we know the probability $P(\mathscr{A})$ of an event $\mathscr{A}$ in experiment 1 and the probability $P(\mathscr{B})$ of an event $\mathscr{B}$ in experiment 2. In general, from this information we cannot determine the probability $P(\mathscr{AB})$ that both events $\mathscr{A}$ and $\mathscr{B}$ will occur. However, if we know that the two experiments are *independent*, then

$$P(\mathscr{AB}) = P(\mathscr{A})P(\mathscr{B}) \tag{1-8}$$

In many cases, this independence can be established a priori by reasoning that the outcomes of experiment 1 have no effect on the outcomes of experiment 2. For example, if in the coin experiment the probability of *heads* equals $1/2$ and in the die experiment the probability of *even* equals $1/2$, then, we conclude "logically" that if both experiments are performed, the probability that we get *heads* on the coin and *even* on the die equals $1/2 \times 1/2$. Thus, as in (1-7), we accept the validity of (1-8) as a logical necessity without recourse to (1-1) or to any other direct evidence.

D. The classical definition can be used as the basis of a deductive theory if we accept (1-7) as an *assumption*. In this theory, no other assumptions are used and postulates (1-3) to (1-5) become theorems. Indeed, the first two postulates are obvious and the third follows from (1-7) because, if the events $\mathscr{A}$ and $\mathscr{B}$ are mutually exclusive, then $N_{\mathscr{A}+\mathscr{B}} = N_{\mathscr{A}} + N_{\mathscr{B}}$; hence

$$P(\mathscr{A} + \mathscr{B}) = \frac{N_{\mathscr{A}+\mathscr{B}}}{N} = \frac{N_{\mathscr{A}}}{N} + \frac{N_{\mathscr{B}}}{N} = P(\mathscr{A}) + P(\mathscr{B})$$

As we show in (2-25), however, this is only a very special case of the axiomatic approach to probability.

1-3 PROBABILITY AND INDUCTION

In the applications of the theory of probability we are faced with the following question: Suppose that we know somehow from past observations the probability $P(\mathscr{A})$ of an event $\mathscr{A}$ in a given experiment. What conclusion can we draw about the occurrence of this event in a *single* future performance of this experiment? (See also Sec. 9-1.)

We shall answer this question in two ways depending on the size of $P(\mathscr{A})$: We shall give one kind of an answer if $P(\mathscr{A})$ is a number distinctly different from 0 or 1, for example 0.6, and a different kind of an answer if $P(\mathscr{A})$ is close to 0 or 1, for example 0.999. Although the boundary between these two cases is not sharply defined, the corresponding answers are fundamentally different.

Case 1 Suppose that $P(\mathscr{A}) = 0.6$. In this case, the number 0.6 gives us only a "certain degree of confidence that the event $\mathscr{A}$ will occur." The known probability is thus used as a "measure of our belief" about the occurrence of $\mathscr{A}$ in a single trial. This interpretation of $P(\mathscr{A})$ is subjective in the sense that it cannot be verified experimentally. In a single trial, the event $\mathscr{A}$ will either occur or will not occur. If it does not, this will not be a reason for questioning the validity of the assumption that $P(\mathscr{A}) = 0.6$.

Case 2 Suppose, however, that $P(\mathscr{A}) = 0.999$. We can now state with practical certainty that at the next trial the event $\mathscr{A}$ will occur. This conclusion is objective in the sense that it can be verified experimentally. At the next trial the event $\mathscr{A}$ must occur. If it does not, we must seriously doubt, if not outright reject, the assumption that $P(\mathscr{A}) = 0.999$.

The boundary between these two cases, arbitrary though it is (0.9 or 0.99999?), establishes in a sense the line separating "soft" from "hard" scientific conclusions. The theory of probability gives us the analytic tools (step 2) for transforming the "subjective" statements of case 1 to the "objective" statements of case 2. In the following, we explain briefly the underlying reasoning.

As we show in Chap. 3, the information that $P(\mathscr{A}) = 0.6$ leads to the conclusion that if the experiment is performed 1000 times, then "almost certainly" the number of times the event $\mathscr{A}$ will occur is between 550 and 650. This is shown by considering the repetition of the original experiment 1000 times as a *single* outcome of a new experiment. In this experiment the probability of the event

$$\mathscr{A}_1 = \{\text{the number of times } \mathscr{A} \text{ occurs is between 550 and 650}\}$$

equals 0.999 (see Prob. 3-6). We must, therefore, conclude that (case 2) the event $\mathscr{A}_1$ will occur with practical certainty.

We have thus succeeded, using the theory of probability, to transform the "subjective" conclusion about $\mathscr{A}$ based on the *given* information that $P(\mathscr{A}) = 0.6$, to the "objective" conclusion about $\mathscr{A}_1$ based on the *derived* conclusion that $P(\mathscr{A}_1) = 0.999$. We should emphasize, however, that both conclusions rely on inductive reasoning. Their difference, although significant, is only quantitative. As in case 1, the "objective" conclusion of case 2 is not a certainty but only an inference. This, however, should not surprise us; after all, no prediction about future events based on past experience can be accepted as logical certainty.

Our inability to make categorical statements about future events is not limited to probability but applies to all sciences. Consider, for example, the development of classical mechanics. It was *observed* that bodies fall according to certain patterns, and on this evidence Newton formulated the laws of mechanics and used them to *predict* future events. His predictions, however, are not logical certainties but only plausible inferences. To "prove" that the future will evolve in the predicted manner we must invoke metaphysical causes.

1-4 CAUSALITY VERSUS RANDOMNESS

We conclude with a brief comment on the apparent controversy between causality and randomness. There is no conflict between causality and randomness or between determinism and probability if we agree, as we must, that scientific theories are not *discoveries* of the laws of nature but rather *inventions* of the human mind. Their consequences are presented in deterministic form if we examine the results of a single trial; they are presented as probabilistic statements if we are interested in averages of many trials. In both cases, all statements are qualified. In the first case, the uncertainties are of the form "with certain errors and in certain ranges of the relevant parameters"; in the

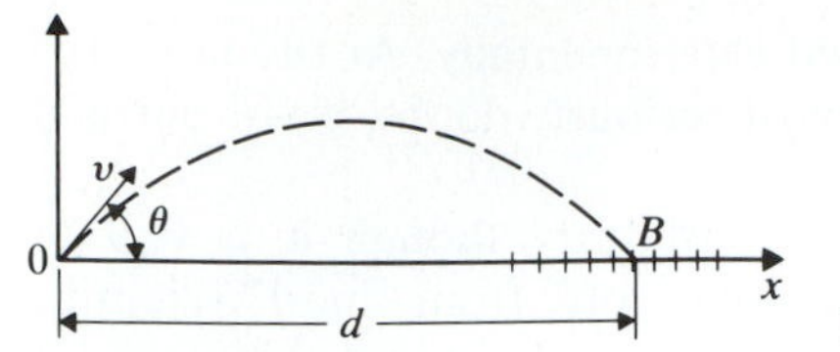

FIGURE 1-2

second, "with a high degree of certainty if the number of trials is large enough." In the next example, we illustrate these two approaches.

Example 1-5. A rocket leaves the ground with an initial velocity v forming an angle θ with the horizontal axis (Fig. 1-2). We shall determine the distance $d = OB$ from the origin to the reentry point B.

From Newton's law it follows that

$$d = \frac{v^2}{g} \sin 2\theta \tag{1-9}$$

The above seems to be an unqualified consequence of a causal law; however, this is not so. The result is approximate and it can be given a probabilistic interpretation.

Indeed, (1-9) is not the solution of a real problem but of an idealized model in which we have neglected air friction, air pressure, variation of g, and other uncertainties in the values of v and θ. We must, therefore, accept (1-9) only with qualifications. It holds within an error ε provided that the neglected factors are smaller than δ.

Suppose now that the reentry area consists of numbered holes and we want to find the reentry hole. Because of the uncertainties in v and θ, we are in no position to give a deterministic answer to our problem. We can, however, ask a different question: If many rockets, nominally with the same velocity, are launched, what percentage will enter the nth hole? This question no longer has a causal answer; it can only be given a random interpretation.

Thus the same physical problem can be subjected either to a deterministic or to a probabilistic analysis. One might argue that the problem is inherently deterministic because the rocket has a precise velocity even if we do not know it. If we did, we would know exactly the reentry hole. Probabilistic interpretations are, therefore, necessary because of our ignorance.

Such arguments can be answered with the statement that the physicists are not concerned with what *is true* but only with what *they can observe*.

Concluding Remarks

In this book, we present a deductive theory (step 2) based on the axiomatic definition of probability. Occasionally, we use the classical definition but only to determine probabilistic data (step 1).

To show the link between theory and applications (step 3), we give also a relative frequency interpretation of the important results. This part of the book, written in small print under the title *Frequency interpretation*, does not obey the rules of deductive reasoning on which the theory is based.

CHAPTER 2

THE AXIOMS OF PROBABILITY

2-1 SET THEORY

A *set* is a collection of objects called *elements*. For example, "car, apple, pencil" is a set whose elements are a car, an apple, and a pencil. The set "heads, tails" has two elements. The set "1, 2, 3, 5" has four elements.

A *subset* $\mathscr{B}$ of a set $\mathscr{A}$ is another set whose elements are also elements of $\mathscr{A}$. All sets under consideration will be subsets of a set $\mathscr{S}$ which we shall call *space*.

The elements of a set will be identified mostly by the Greek letter ζ. Thus

$$\mathscr{A} = \{\zeta_1, \ldots, \zeta_n\} \tag{2-1}$$

will mean that the set $\mathscr{A}$ consists of the elements $\zeta_1, \ldots, \zeta_n$. We shall also identify sets by the properties of their elements. Thus

$$\mathscr{A} = \{\text{all positive integers}\} \tag{2-2}$$

will mean the set whose elements are the numbers $1, 2, 3, \ldots$.

The notation

$$\zeta_i \in \mathscr{A} \qquad \zeta_i \notin \mathscr{A}$$

will mean that ζ_i is or is not an element of $\mathscr{A}$.

The *empty* or *null* set is by definition the set that contains no elements. This set will be denoted by $\{\emptyset\}$.

If a set consists of n elements, then the total number of its subsets equals 2^n.

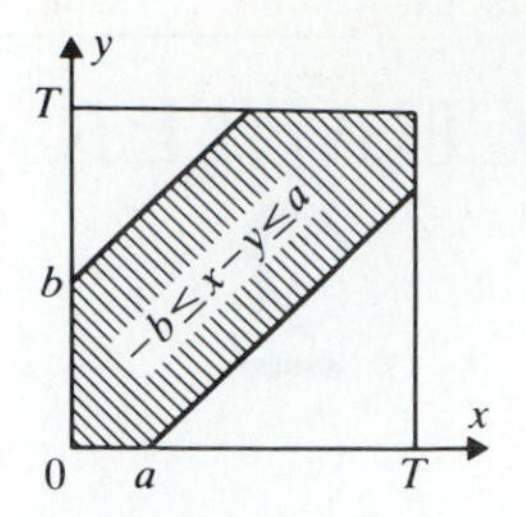

FIGURE 2-1

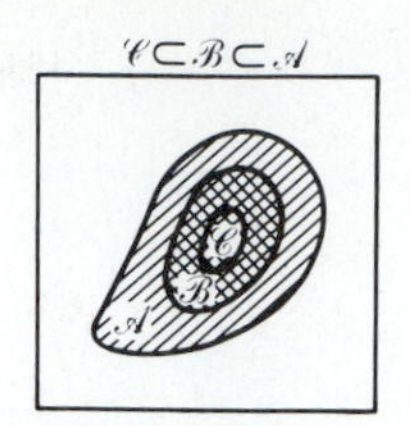

FIGURE 2-2

Note In probability theory, we assign probabilities to the subsets (events) of $\mathscr{S}$ and we define various functions (random variables) whose domain consists of the elements of $\mathscr{S}$. We must be careful, therefore, to distinguish between the element ζ and the set $\{\zeta\}$ consisting of the single element ζ.

Example 2-1. We shall denote by f_i the faces of a die. These faces are the elements of the set $\mathscr{S} = \{f_1, \ldots, f_6\}$. In this case, $n = 6$; hence $\mathscr{S}$ has $2^6 = 64$ subsets:

$$\{\emptyset\}, \{f_1\}, \ldots, \{f_1 f_2\}, \ldots, \{f_1 f_2 f_3\}, \ldots, \mathscr{S}$$

In general, the elements of a set are arbitrary objects. For example, the 64 subsets of the set $\mathscr{S}$ in the above example can be considered as the elements of another set. In Example 2-2, the elements of $\mathscr{S}$ are pairs of objects. In Example 2-3, $\mathscr{S}$ is the set of points in the square of Fig. 2-1.

Example 2-2. Suppose that a coin is tossed twice. The resulting outcomes are the four objects hh, ht, th, tt forming the set

$$\mathscr{S} = \{hh, ht, th, tt\}$$

where hh is an abbreviation for the element "heads–heads." The set $\mathscr{S}$ has $2^4 = 16$ subsets. For example,

$$\mathscr{A} = \{\text{heads at the first toss}\} = \{hh, ht\}$$

$$\mathscr{B} = \{\text{only one head showed}\} = \{ht, th\}$$

$$\mathscr{C} = \{\text{heads shows at least once}\} = \{hh, ht, th\}$$

In the first equality, the sets $\mathscr{A}$, $\mathscr{B}$, and $\mathscr{C}$ are represented by their properties as in (2-2); in the second, in terms of their elements as in (2-1).

Example 2-3. In this example, $\mathscr{S}$ is the set of all points in the square of Fig. 2-1. Its elements are all ordered pairs of numbers (x, y) where

$$0 \le x \le T \qquad 0 \le y \le T$$

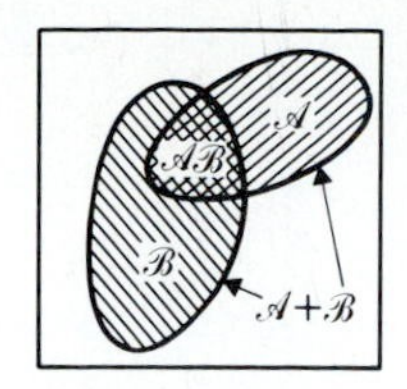

FIGURE 2-3

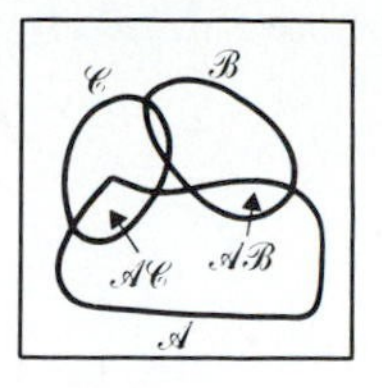

FIGURE 2-4

The shaded area is a subset $\mathscr{A}$ of $\mathscr{S}$ consisting of all points (x, y) such that $-b \le x - y \le a$. The notation

$$\mathscr{A} = \{-b \le x - y \le a\}$$

describes $\mathscr{A}$ in terms of the properties of x and y as in (2-2).

Set Operations

In the following, we shall represent a set $\mathscr{S}$ and its subsets by plane figures as in Fig. 2-2 (*Venn diagrams*).

The notation $\mathscr{B} \subset \mathscr{A}$ or $\mathscr{A} \supset \mathscr{B}$ will mean that $\mathscr{B}$ is a subset of $\mathscr{A}$, that is, that every element of $\mathscr{B}$ is an element of $\mathscr{A}$. Thus, for any $\mathscr{A}$,

$$\{\emptyset\} \subset \mathscr{A} \subset \mathscr{A} \subset \mathscr{S}$$

Transitivity If $\mathscr{C} \subset \mathscr{B}$ and $\mathscr{B} \subset \mathscr{A}$ then $\mathscr{C} \subset \mathscr{A}$

Equality $\mathscr{A} = \mathscr{B}$ iff† $\mathscr{A} \subset \mathscr{B}$ and $\mathscr{B} \subset \mathscr{A}$

Unions and intersections The *sum* or *union* of two sets $\mathscr{A}$ and $\mathscr{B}$ is a set whose elements are all elements of $\mathscr{A}$ or of $\mathscr{B}$ or of both (Fig. 2-3). This set will be written in the form

$$\mathscr{A} + \mathscr{B} \qquad \text{or} \qquad \mathscr{A} \cup \mathscr{B}$$

The above operation is commutative and associative:

$$\mathscr{A} + \mathscr{B} = \mathscr{B} + \mathscr{A} \qquad (\mathscr{A} + \mathscr{B}) + \mathscr{C} = \mathscr{A} + (\mathscr{B} + \mathscr{C})$$

We note that, if $\mathscr{B} \subset \mathscr{A}$, then $\mathscr{A} + \mathscr{B} = \mathscr{A}$. From this it follows that

$$\mathscr{A} + \mathscr{A} = \mathscr{A} \qquad \mathscr{A} + \{\emptyset\} = \mathscr{A} \qquad \mathscr{S} + \mathscr{A} = \mathscr{S}$$

The *product* or *intersection* of two sets $\mathscr{A}$ and $\mathscr{B}$ is a set consisting of all elements that are common to the set $\mathscr{A}$ and $\mathscr{B}$ (Fig. 2-3). This set is written in the form

$$\mathscr{AB} \qquad \text{or} \qquad \mathscr{A} \cap \mathscr{B}$$

The above operation is commutative, associative, and distributive (Fig. 2-4):

$$\mathscr{AB} = \mathscr{BA} \qquad (\mathscr{AB})\mathscr{C} = \mathscr{A}(\mathscr{BC}) \qquad \mathscr{A}(\mathscr{B} + \mathscr{C}) = \mathscr{AB} + \mathscr{AC}$$

†Iff is an abbreviation for *if and only if*.

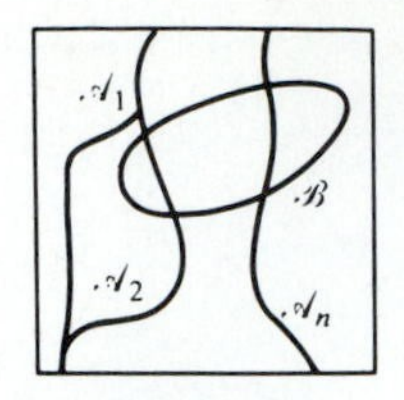

FIGURE 2-5

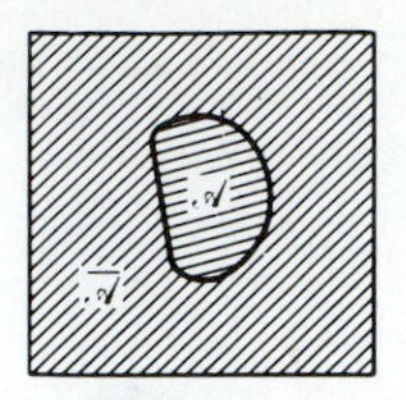

FIGURE 2-6

We note that if $\mathscr{A} \subset \mathscr{B}$, then $\mathscr{A}\mathscr{B} = \mathscr{A}$. Hence

$$\mathscr{A}\mathscr{A} = \mathscr{A} \qquad \{\emptyset\}\mathscr{A} = \{\emptyset\} \qquad \mathscr{A}\mathscr{S} = \mathscr{A}$$

Note If two sets $\mathscr{A}$ and $\mathscr{B}$ are described by the properties of their elements as in (2-2), then their intersection $\mathscr{A}\mathscr{B}$ will be specified by including these properties in braces. For example, if

$$\mathscr{S} = \{1,2,3,4,5,6\} \qquad \mathscr{A} = \{\text{even}\} \qquad \mathscr{B} = \{\text{less than 5}\}$$

then†

$$\mathscr{A}\mathscr{B} = \{\text{even, less than 5}\} = \{2,4\} \tag{2-3}$$

Mutually exclusive sets Two sets $\mathscr{A}$ and $\mathscr{B}$ are called *mutually exclusive* or *disjoint* if they have no common elements, that is, if

$$\mathscr{A}\mathscr{B} = \{\emptyset\}$$

Several sets $\mathscr{A}_1, \mathscr{A}_2, \ldots$ are called mutually exclusive if

$$\mathscr{A}_i\mathscr{A}_j = \{\emptyset\} \qquad \text{for every} \quad i \text{ and } j \neq i$$

Partitions A partition $\mathfrak{A}$ of a set $\mathscr{S}$ is a collection of mutually exclusive subsets $\mathscr{A}_i$ of $\mathscr{S}$ whose union equals $\mathscr{S}$ (Fig. 2-5).

$$\mathscr{A}_1 + \cdots + \mathscr{A}_n = \mathscr{S} \qquad \mathscr{A}_i\mathscr{A}_j = \{\emptyset\} \qquad i \neq j \tag{2-4}$$

All partitions will be denoted by boldface German script (Fraktur) letters. Thus

$$\mathfrak{A} = [\mathscr{A}_1, \ldots, \mathscr{A}_n]$$

†We should stress the difference in the meaning of commas in (2-1) and (2-3). In (2-1) the braces include all elements ζ_i and

$$\{\zeta_1, \ldots, \zeta_n\} = \{\zeta_1\} \cup \cdots \cup \{\zeta_n\}$$

is the union of the sets $\{\zeta_i\}$. In (2-3) the braces include the properties of the sets {even} and {less than 5}, and

$$\{\text{even, less than 5}\} = \{\text{even}\} \cap \{\text{less than 5}\}$$

is the intersection of the sets {even} and {less than 5}.

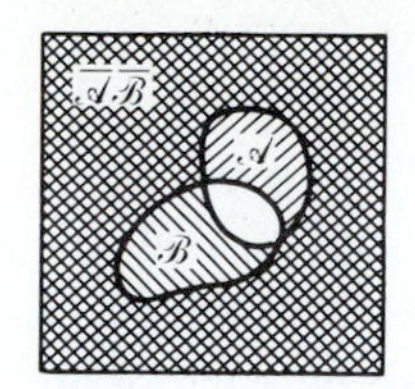

FIGURE 2-7

Complements The complement $\overline{\mathscr{A}}$ of a set $\mathscr{A}$ is the set consisting of all elements of $\mathscr{S}$ that are not in $\mathscr{A}$ (Fig. 2-6). From the definition it follows that

$$\mathscr{A} + \overline{\mathscr{A}} = \mathscr{S} \qquad \mathscr{A}\overline{\mathscr{A}} = \{\emptyset\} \qquad \overline{\overline{\mathscr{A}}} = \mathscr{A} \qquad \overline{\mathscr{S}} = \{\emptyset\} \qquad \{\overline{\emptyset}\} = \mathscr{S}$$

If $\mathscr{B} \subset \mathscr{A}$ then $\overline{\mathscr{B}} \supset \overline{\mathscr{A}}$; if $\mathscr{A} = \mathscr{B}$ then $\overline{\mathscr{A}} = \overline{\mathscr{B}}$.

De Morgan's law Clearly (see Fig. 2-7)

$$\overline{\mathscr{A} + \mathscr{B}} = \overline{\mathscr{A}}\,\overline{\mathscr{B}} \qquad \overline{\mathscr{A}\mathscr{B}} = \overline{\mathscr{A}} + \overline{\mathscr{B}} \tag{2-5}$$

Repeated application of (2-5) leads to the following:

If in a set identity we replace all sets by their complements, all unions by intersections, and all intersections by unions, the identity is preserved.

We shall demonstrate the above using as example the identity

$$\mathscr{A}(\mathscr{B} + \mathscr{C}) = \mathscr{A}\mathscr{B} + \mathscr{A}\mathscr{C} \tag{2-6}$$

From (2-5) it follows that

$$\overline{\mathscr{A}(\mathscr{B} + \mathscr{C})} = \overline{\mathscr{A}} + \overline{\mathscr{B} + \mathscr{C}} = \overline{\mathscr{A}} + \overline{\mathscr{B}}\,\overline{\mathscr{C}}$$

Similarly,

$$\overline{\mathscr{A}\mathscr{B} + \mathscr{A}\mathscr{C}} = \left(\overline{\mathscr{A}\mathscr{B}}\right)\left(\overline{\mathscr{A}\mathscr{C}}\right) = (\overline{\mathscr{A}} + \overline{\mathscr{B}})(\overline{\mathscr{A}} + \overline{\mathscr{C}})$$

and since the two sides of (2-6) are equal, their complements are also equal. Hence

$$\overline{\mathscr{A}} + \overline{\mathscr{B}}\,\overline{\mathscr{C}} = (\overline{\mathscr{A}} + \overline{\mathscr{B}})(\overline{\mathscr{A}} + \overline{\mathscr{C}}) \tag{2-7}$$

Duality principle As we know, $\overline{\mathscr{S}} = \{\emptyset\}$ and $\{\overline{\emptyset}\} = \mathscr{S}$. Furthermore, if in an identity like (2-7) all overbars are removed, the identity is preserved. This leads to the following version of De Morgan's law:

If in a set identity we replace all unions by intersections, all intersections by unions, and the sets $\mathscr{S}$ and $\{\emptyset\}$ by the sets $\{\emptyset\}$ and $\mathscr{S}$, the identity is preserved.

Applying the above to the identities

$$\mathscr{A}(\mathscr{B} + \mathscr{C}) = \mathscr{A}\mathscr{B} + \mathscr{A}\mathscr{C} \qquad \mathscr{S} + \mathscr{A} = \mathscr{S}$$

we obtain the identities

$$\mathscr{A} + \mathscr{B}\mathscr{C} = (\mathscr{A} + \mathscr{B})(\mathscr{A} + \mathscr{C}) \qquad \{\emptyset\}\mathscr{A} = \{\emptyset\}$$

2-2 PROBABILITY SPACE

In probability theory, the following set terminology is used: The space $\mathscr{S}$ is called the *certain event*, its elements *experimental outcomes*, and its subsets *events*. The empty set $\{\emptyset\}$ is the *impossible event*, and the event $\{\zeta_i\}$ consisting of a single element ζ_i is an *elementary event*. All events will be identified by script letters.

In the applications of probability theory to physical problems, the identification of experimental outcomes is not always unique. We shall illustrate this ambiguity with the die experiment as might be interpreted by players X, Y, and Z.

X says that the outcomes of this experiment are the six faces of the die forming the space $\mathscr{S} = \{f_1, \ldots, f_6\}$. This space has $2^6 = 64$ subsets and the event {even} consists of the three outcomes f_2, f_4, and f_6.

Y wants to bet on *even* or *odd* only. He argues, therefore that the experiment has only the two outcomes *even* and *odd* forming the space $\mathscr{S} = \{\text{even}, \text{odd}\}$. This space has only $2^2 = 4$ subsets and the event {even} consists of a single outcome.

Z bets that *one* will show and the die will rest on the left side of the table. He maintains, therefore, that the experiment has infinitely many outcomes specified by the coordinates of its center and by the six faces. The event {even} consists not of one or of three outcomes but of infinitely many.

In the following, when we talk about an experiment, we shall assume that its outcomes are clearly identified. In the die experiment, for example, $\mathscr{S}$ will be the set consisting of the six faces $f_1, \ldots, f_6$.

In the relative frequency interpretation of various results, we shall use the following terminology.

Trial A single performance of an experiment will be called a *trial*. At each trial we observe a single outcome ζ_i. We say that an event $\mathscr{A}$ *occurs* during this trial if it contains the element ζ_i. The certain event occurs at every trial and the impossible event never occurs. The event $\mathscr{A} + \mathscr{B}$ occurs when $\mathscr{A}$ or $\mathscr{B}$ or both occur. The event $\mathscr{A}\mathscr{B}$ occurs when both events $\mathscr{A}$ and $\mathscr{B}$ occur. If the events $\mathscr{A}$ and $\mathscr{B}$ are mutually exclusive and $\mathscr{A}$ occurs, then $\mathscr{B}$ does not occur. If $\mathscr{A} \subset \mathscr{B}$ and $\mathscr{A}$ occurs, then $\mathscr{B}$ occurs. At each trial, either $\mathscr{A}$ or $\overline{\mathscr{A}}$ occurs.

If, for example, in the die experiment we observe the outcome f_5, then the event $\{f_5\}$, the event {odd}, and 30 other events occur.

The Axioms

We assign to each event $\mathscr{A}$ a number $P(\mathscr{A})$ which we call *the probability of the event* $\mathscr{A}$. This number is so chosen as to satisfy the following three conditions:

I $$P(\mathscr{A}) \geq 0 \tag{2-8}$$

II $$P(\mathscr{S}) = 1 \tag{2-9}$$

III $$\text{if} \quad \mathscr{A}\mathscr{B} = \{\emptyset\} \quad \text{then} \quad P(\mathscr{A} + \mathscr{B}) = P(\mathscr{A}) + P(\mathscr{B}) \tag{2-10}$$

These conditions are the axioms of the theory of probability. In the development of the theory, all conclusions are based directly or indirectly on the axioms and only on the axioms. The following are simple consequences.

Properties. The probability of the impossible event is 0:

$$P\{\emptyset\} = 0 \tag{2-11}$$

Indeed, $\mathscr{A}\{\emptyset\} = \{\emptyset\}$ and $\mathscr{A} + \{\emptyset\} = \mathscr{A}$; therefore [see (2-10)]

$$P(\mathscr{A}) = P(\mathscr{A} + \emptyset) = P(\mathscr{A}) + P\{\emptyset\}$$

For any $\mathscr{A}$,

$$P(\mathscr{A}) = 1 - P(\overline{\mathscr{A}}) \le 1 \tag{2-12}$$

because $\mathscr{A} + \overline{\mathscr{A}} = \mathscr{S}$ and $\mathscr{A}\overline{\mathscr{A}} = \{\emptyset\}$; hence

$$1 = P(\mathscr{S}) = P(\mathscr{A} + \overline{\mathscr{A}}) = P(\mathscr{A}) + P(\overline{\mathscr{A}})$$

For any $\mathscr{A}$ and $\mathscr{B}$,

$$P(\mathscr{A} + \mathscr{B}) = P(\mathscr{A}) + P(\mathscr{B}) - P(\mathscr{A}\mathscr{B}) \le P(\mathscr{A}) + P(\mathscr{B}) \tag{2-13}$$

To prove the above, we write the events $\mathscr{A} + \mathscr{B}$ and $\mathscr{B}$ as unions of two mutually exclusive events:

$$\mathscr{A} + \mathscr{B} = \mathscr{A} + \overline{\mathscr{A}}\mathscr{B} \qquad \mathscr{B} = \mathscr{A}\mathscr{B} + \overline{\mathscr{A}}\mathscr{B}$$

Therefore [see (2-10)]

$$P(\mathscr{A} + \mathscr{B}) = P(\mathscr{A}) + P(\overline{\mathscr{A}}\mathscr{B}) \qquad P(\mathscr{B}) = P(\mathscr{A}\mathscr{B}) + P(\overline{\mathscr{A}}\mathscr{B})$$

Eliminating $P(\overline{\mathscr{A}}\mathscr{B})$, we obtain (2-13).

Finally, if $\mathscr{B} \subset \mathscr{A}$, then

$$P(\mathscr{A}) = P(\mathscr{B}) + P(\mathscr{A}\overline{\mathscr{B}}) \ge P(\mathscr{B}) \tag{2-14}$$

because $\mathscr{A} = \mathscr{B} + \mathscr{A}\overline{\mathscr{B}}$ and $\mathscr{B}(\mathscr{A}\overline{\mathscr{B}}) = \{\emptyset\}$.

Frequency interpretation The axioms of probability are so chosen that the resulting theory gives a satisfactory representation of the physical world. Probabilities as used in real problems must, therefore, be compatible with the axioms. Using the frequency interpretation

$$P(\mathscr{A}) \simeq \frac{n_{\mathscr{A}}}{n}$$

of probability, we shall show that they do.

I. Clearly, $P(\mathscr{A}) \ge 0$ because $n_{\mathscr{A}} \ge 0$ and $n > 0$.

II. $P(\mathscr{S}) = 1$ because $\mathscr{S}$ occurs at every trial; hence $n_{\mathscr{S}} = n$.

III. If $\mathscr{A}\mathscr{B} = \{\emptyset\}$, then $n_{\mathscr{A}+\mathscr{B}} = n_{\mathscr{A}} + n_{\mathscr{B}}$ because if $\mathscr{A} + \mathscr{B}$ occurs then $\mathscr{A}$ or $\mathscr{B}$ occurs but not both. Hence

$$P(\mathscr{A} + \mathscr{B}) \simeq \frac{n_{\mathscr{A}+\mathscr{B}}}{n} = \frac{n_{\mathscr{A}}}{n} + \frac{n_{\mathscr{B}}}{n} \simeq P(\mathscr{A}) + P(\mathscr{B})$$

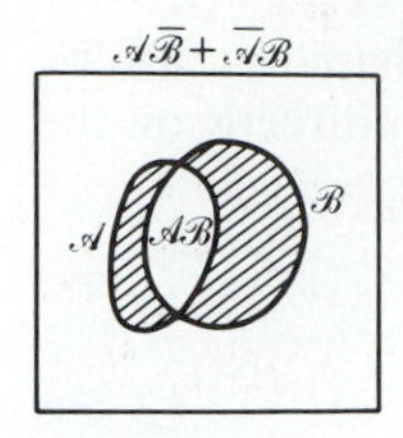

FIGURE 2-8

Equality of events. Two events $\mathscr{A}$ and $\mathscr{B}$ are called *equal* if they consist of the same elements. They are called *equal with probability 1* if the set

$$(\mathscr{A} + \mathscr{B})(\overline{\mathscr{A}\mathscr{B}}) = \mathscr{A}\overline{\mathscr{B}} + \overline{\mathscr{A}}\mathscr{B}$$

consisting of all outcomes that are in $\mathscr{A}$ or in $\mathscr{B}$ but not in $\mathscr{A}\mathscr{B}$ (shaded area in Fig. 2-8) has zero probability.

From the definition it follows that (see Prob. 2-4) the events $\mathscr{A}$ and $\mathscr{B}$ are equal with probability 1 iff

$$P(\mathscr{A}) = P(\mathscr{B}) = P(\mathscr{A}\mathscr{B}) \tag{2-15}$$

If $P(\mathscr{A}) = P(\mathscr{B})$ then we say that $\mathscr{A}$ and $\mathscr{B}$ are *equal in probability*. In this case, no conclusion can be drawn about the probability of $\mathscr{A}\mathscr{B}$. In fact, the events $\mathscr{A}$ and $\mathscr{B}$ might be mutually exclusive.

From (2-15) it follows that, if an event $\mathscr{N}$ equals the impossible event with probability 1 then $P(\mathscr{N}) = 0$. This does not, of course, mean that $\mathscr{N} = \{\emptyset\}$.

The Class $\mathfrak{F}$ of Events

Events are subsets of $\mathscr{S}$ to which we have assigned probabilities. As we shall presently explain, we shall not consider as events all subsets of $\mathscr{S}$ but only a class $\mathfrak{F}$ of subsets.

One reason for this might be the nature of the application. In the die experiment, for example, we might want to bet only on *even* or *odd*. In this case, it suffices to consider as events only the four sets $\{\emptyset\}$, {even}, {odd}, and $\mathscr{S}$.

The main reason, however, for not including all subsets of $\mathscr{S}$ in the class $\mathfrak{F}$ of events is of a mathematical nature: In certain cases involving sets with infinitely many outcomes, it is impossible to assign probabilities to all subsets satisfying all the axioms including the generalized form (2-21) of axiom III.

The class $\mathfrak{F}$ of events will not be an arbitrary collection of subsets of $\mathscr{S}$. We shall assume that, if $\mathscr{A}$ and $\mathscr{B}$ are events, then $\mathscr{A} + \mathscr{B}$ and $\mathscr{A}\mathscr{B}$ are also events. We do so because we will want to know not only the probabilities of various events, but also the probabilities of their unions and intersections. This leads to the concept of a field.

FIELDS. A field $\mathfrak{F}$ is a nonempty class of sets such that:

$$\text{If} \quad \mathscr{A} \in \mathfrak{F} \qquad \text{then} \quad \overline{\mathscr{A}} \in \mathfrak{F} \tag{2-16}$$

$$\text{If} \quad \mathscr{A} \in \mathfrak{F} \quad \text{and} \quad \mathscr{B} \in \mathfrak{F} \qquad \text{then} \quad \mathscr{A} + \mathscr{B} \in \mathfrak{F} \tag{2-17}$$

These two properties give a minimum set of conditions for $\mathfrak{F}$ to be a field. All other properties follow:

$$\text{If} \quad \mathscr{A} \in \mathfrak{F} \quad \text{and} \quad \mathscr{B} \in \mathfrak{F} \qquad \text{then} \quad \mathscr{A}\mathscr{B} \in \mathfrak{F} \tag{2-18}$$

Indeed, from (2-16) it follows that $\overline{\mathscr{A}} \in \mathfrak{F}$ and $\overline{\mathscr{B}} \in \mathfrak{F}$. Applying (2-17) and (2-16) to the sets $\overline{\mathscr{A}}$ and $\overline{\mathscr{B}}$, we conclude that

$$\overline{\mathscr{A}} + \overline{\mathscr{B}} \in \mathfrak{F} \qquad \overline{\overline{\mathscr{A}} + \overline{\mathscr{B}}} = \mathscr{A}\mathscr{B} \in \mathfrak{F}$$

A field contains the certain event and the impossible event:

$$\mathscr{S} \in \mathfrak{F} \qquad \{\emptyset\} \in \mathfrak{F} \tag{2-19}$$

Indeed, since $\mathfrak{F}$ is not empty, it contains at least one element $\mathscr{A}$; therefore [see (2-16)] it also contains $\overline{\mathscr{A}}$. Hence

$$\mathscr{A} + \overline{\mathscr{A}} = \mathscr{S} \in \mathfrak{F} \qquad \mathscr{A}\overline{\mathscr{A}} = \{\emptyset\} \in \mathfrak{F}$$

From the above it follows that all sets that can be written as unions or intersections of *finitely many* sets in $\mathfrak{F}$ are also in $\mathfrak{F}$. This is not, however, necessarily the case for infinitely many sets.

Borel fields. Suppose that $\mathscr{A}_1, \ldots, \mathscr{A}_n, \ldots$ is an infinite sequence of sets in $\mathfrak{F}$. If the union and intersection of these sets also belongs to $\mathfrak{F}$, then $\mathfrak{F}$ is called a Borel field.

The class of all subsets of a set $\mathscr{S}$ is a Borel field. Suppose that $\mathfrak{C}$ is a class of subsets of $\mathscr{S}$ that is not a field. Attaching to it other subsets of $\mathscr{S}$, all subsets if necessary, we can form a field with $\mathfrak{C}$ as its subset. It can be shown that there exists a smallest Borel field containing all the elements of $\mathfrak{C}$.

Example 2-4. Suppose that $\mathscr{S}$ consists of the four elements a, b, c, d and $\mathfrak{C}$ consists of the sets $\{a\}$ and $\{b\}$. Attaching to $\mathfrak{C}$ the complements of $\{a\}$ and $\{b\}$ and their unions and intersections, we conclude that the smallest field containing $\{a\}$ and $\{b\}$ consists of the sets

$$\{\emptyset\} \quad \{a\} \quad \{b\} \quad \{a,b\} \quad \{c,d\} \quad \{b,c,d\} \quad \{a,c,d\} \quad \mathscr{S}$$

Events. In probability theory, events are certain subsets of $\mathscr{S}$ forming a Borel field. This permits us to assign probabilities not only to finite unions and intersections of events, but also to their limits.

For the determination of probabilities of sets that can be expressed as limits, the following extension of axiom III is necessary.

Repeated application of (2-10) leads to the conclusion that, if the events $\mathscr{A}_1, \ldots, \mathscr{A}_n$ are mutually exclusive, then

$$P(\mathscr{A}_1 + \cdots + \mathscr{A}_n) = P(\mathscr{A}_1) + \cdots + P(\mathscr{A}_n) \tag{2-20}$$

The extension of the above to infinitely many sets does not follow from (2-10). It is an additional condition known as the *axiom of infinite additivity*:

III*a*. If the events $\mathscr{A}_1, \mathscr{A}_2, \ldots$ are mutually exclusive, then

$$P(\mathscr{A}_1 + \mathscr{A}_2 + \cdots) = P(\mathscr{A}_1) + P(\mathscr{A}_2) + \cdots \qquad (2\text{-}21)$$

We shall assume that all probabilities satisfy axioms I, II, III, and III*a*.

Axiomatic Definition of an Experiment

In the theory of probability, an experiment is specified in terms of the following concepts:

1. The set $\mathscr{S}$ of all experimental outcomes.
2. The Borel field of all events of $\mathscr{S}$.
3. The probabilities of these events.

The letter $\mathscr{S}$ will be used to identify not only the certain event, but also the entire experiment.

We discuss next the determination of probabilities in experiments with finitely many and infinitely many elements.

Countable spaces. If the space $\mathscr{S}$ consists of N outcomes and N is a finite number, then the probabilities of all events can be expressed in terms of the probabilities

$$P\{\zeta_i\} = p_i$$

of the elementary events $\{\zeta_i\}$. From the axioms it follows, of course, that the numbers p_i must be nonnegative and their sum must equal 1:

$$p_i \ge 0 \qquad p_1 + \cdots + p_N = 1 \qquad (2\text{-}22)$$

Suppose that $\mathscr{A}$ is an event consisting of the r elements ζ_{k_i}. In this case, $\mathscr{A}$ can be written as the union of the elementary events $\{\zeta_{k_i}\}$. Hence [see (2-20)]

$$P(\mathscr{A}) = P\{\zeta_{k_1}\} + \cdots + P\{\zeta_{k_r}\} = p_{k_1} + \cdots + p_{k_r} \qquad (2\text{-}23)$$

The above is true even if $\mathscr{S}$ consists of an infinite but countable number of elements $\zeta_1, \zeta_2, \ldots$ [see (2-21)].

Classical definition If $\mathscr{S}$ consists of N outcomes and the probabilities p_i of the elementary events are all equal, then

$$p_i = \frac{1}{N} \qquad (2\text{-}24)$$

In this case, the probability of an event $\mathscr{A}$ consisting of r elements equals r/N:

$$P(\mathscr{A}) = \frac{r}{N} \tag{2-25}$$

This very special but important case is equivalent to the classical definition (1-7), with one important difference, however: In the classical definition, (2-25) is deduced as a logical necessity; in the axiomatic development of probability, (2-24), on which (2-25) is based, is a mere assumption.

Example 2-5. (*a*) In the coin experiment, the space $\mathscr{S}$ consists of the outcomes h and t:

$$\mathscr{S} = \{h, t\}$$

and its events are the four sets $\{\emptyset\}, \{t\}, \{h\}, \mathscr{S}$. If $P\{h\} = p$ and $P\{t\} = q$, then $p + q = 1$.

(*b*) We consider now the experiment of the toss of a coin three times. The possible outcomes of this experiment are:

$$hhh, hht, hth, htt, thh, tht, tth, ttt$$

We shall assume that all elementary events have the same probability as in (2-24) (fair coin). In this case, the probability of each elementary event equals $1/8$. Thus the probability $P\{hhh\}$ that we get three heads equals $1/8$. The event

$$\{\text{heads at the first two tosses}\} = \{hhh, hht\}$$

consists of the two outcomes hhh and hht; hence its probability equals $2/8$.

The real line. If $\mathscr{S}$ consists of a noncountable infinity of elements, then its probabilities cannot be determined in terms of the probabilities of the elementary events. This is the case if $\mathscr{S}$ is the set of points in an n-dimensional space. In fact, most applications can be presented in terms of events in such a space. We shall discuss the determination of probabilities using as illustration the real line.

Suppose that $\mathscr{S}$ is the set of all real numbers. Its subsets can be considered as sets of points on the real line. It can be shown that it is impossible to assign probabilities to all subsets of $\mathscr{S}$ so as to satisfy the axioms. To construct a probability space on the real line, we shall consider as events all intervals $x_1 \le x \le x_2$ and their countable unions and intersections. These events form a field $\mathfrak{F}$ that can be specified as follows:

It is the smallest Borel field that includes all half-lines $x \le x_i$ where x_i is any number.

This field contains all open and closed intervals, all points, and, in fact, every set of points on the real line that is of interest in the applications. One might wonder whether $\mathfrak{F}$ does not include *all* subsets of $\mathscr{S}$. Actually, it is possible to show that there exist sets of points on the real line that are not countable unions and intersections of intervals. Such sets, however, are of no interest in most applications. To complete the specification of $\mathscr{S}$, it suffices to

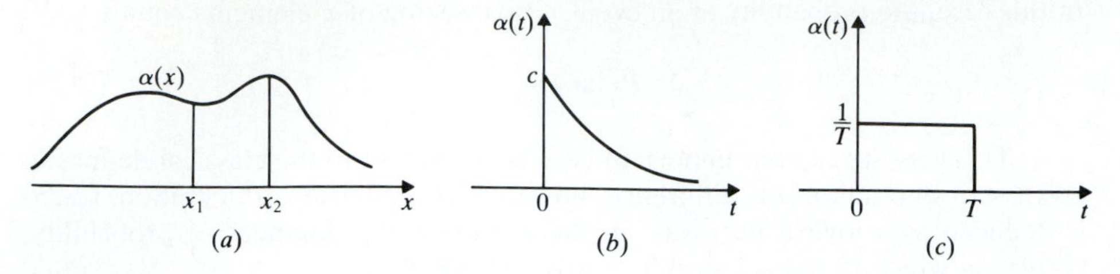

FIGURE 2-9

assign probabilities to the events $\{x \le x_i\}$. All other probabilities can then be determined from the axioms.

Suppose that $\alpha(x)$ is a function such that (Fig. 2-9*a*)

$$\int_{-\infty}^{\infty} \alpha(x)\,dx = 1 \qquad \alpha(x) \ge 0 \tag{2-26}$$

We define the probability of the event $\{x \le x_i\}$ by the integral

$$P\{x \le x_i\} = \int_{-\infty}^{x_i} \alpha(x)\,dx \tag{2-27}$$

This specifies the probabilities of all events of $\mathscr{S}$. We maintain for example, that the probability of the event $\{x_1 < x \le x_2\}$ consisting of all points in the interval (x_1, x_2) is given by

$$P\{x_1 < x \le x_2\} = \int_{x_1}^{x_2} \alpha(x)\,dx \tag{2-28}$$

Indeed, the events $\{x \le x_1\}$ and $\{x_1 < x \le x_2\}$ are mutually exclusive and their union equals $\{x \le x_2\}$. Hence [see (2-10)]

$$P\{x \le x_1\} + P\{x_1 < x \le x_2\} = P\{x \le x_2\}$$

and (2-28) follows from (2-27).

We note that, if the function $\alpha(x)$ is bounded, then the integral in (2-28) tends to 0 as $x_1 \to x_2$. This leads to the conclusion that the probability of the event $\{x_2\}$ consisting of the single outcome x_2 is 0 for every x_2. In this case, the probability of all elementary events of $\mathscr{S}$ equals 0, although the probability of their unions equals 1. This is not in conflict with (2-21) because the total number of elements of $\mathscr{S}$ is not countable.

Example 2-6. A radioactive substance is selected at $t = 0$ and the time t of emission of a particle is observed. This process defines an experiment whose outcomes are all points on the positive t axis. This experiment can be considered as a special case of the real line experiment if we assume that $\mathscr{S}$ is the entire t axis and all events on the negative axis have zero probability.

Suppose then that the function $\alpha(t)$ in (2-26) is given by (Fig. 2-9b)

$$\alpha(t) = ce^{-ct}U(t) \qquad U(t) = \begin{cases} 1 & t \geq 0 \\ 0 & t < 0 \end{cases}$$

Inserting into (2-28), we conclude that the probability that a particle will be emitted in the time interval $(0, t_0)$ equals

$$c\int_0^{t_0} e^{-ct}\,dt = 1 - e^{-ct_0}$$

Example 2-7. A telephone call occurs at *random* in the interval $(0, T)$. This means that the probability that it will occur in the interval $0 \leq t \leq t_0$ equals t_0/T. Thus the outcomes of this experiment are all points in the interval $(0, T)$ and the probability of the event {the call will occur in the interval (t_1, t_2)} equals

$$P\{t_1 \leq t \leq t_2\} = \frac{t_2 - t_1}{T}$$

This is again a special case of (2-28) with $\alpha(t) = 1/T$ for $0 \leq t \leq T$ and 0 otherwise (Fig. 2-9c).

Probability masses. The probability $P(\mathscr{A})$ of an event $\mathscr{A}$ can be interpreted as the mass of the corresponding figure in its Venn diagram representation. Various identities have similar interpretations. Consider, for example, the identity $P(\mathscr{A} + \mathscr{B}) = P(\mathscr{A}) + P(\mathscr{B}) - P(\mathscr{A}\mathscr{B})$. The left side equals the mass of the event $\mathscr{A} + \mathscr{B}$. In the sum $P(\mathscr{A}) + P(\mathscr{B})$, the mass of $\mathscr{A}\mathscr{B}$ is counted twice (Fig. 2-3). To equate this sum with $P(\mathscr{A} + \mathscr{B})$, we must, therefore, subtract $P(\mathscr{A}\mathscr{B})$.

2-3 CONDITIONAL PROBABILITY

The *conditional probability* of an event $\mathscr{A}$ assuming $\mathscr{M}$, denoted by $P(\mathscr{A}|\mathscr{M})$, is by definition the ratio

$$P(\mathscr{A}|\mathscr{M}) = \frac{P(\mathscr{A}\mathscr{M})}{P(\mathscr{M})} \tag{2-29}$$

where we assume that $P(\mathscr{M})$ is not 0.

The following properties follow readily from the definition:

$$\text{If} \quad \mathscr{M} \subset \mathscr{A} \quad \text{then} \quad P(\mathscr{A}|\mathscr{M}) = 1 \tag{2-30}$$

because then $\mathscr{A}\mathscr{M} = \mathscr{M}$. Similarly,

$$\text{if} \quad \mathscr{A} \subset \mathscr{M} \quad \text{then} \quad P(\mathscr{A}|\mathscr{M}) = \frac{P(\mathscr{A})}{P(\mathscr{M})} \geq P(\mathscr{A}) \tag{2.31}$$

Frequency interpretation Denoting by $n_{\mathscr{A}}$, $n_{\mathscr{M}}$, and $n_{\mathscr{A}\mathscr{M}}$ the number of occurrences of the events $\mathscr{A}$, $\mathscr{M}$, and $\mathscr{A}\mathscr{M}$ respectively, we conclude from (1-1) that

$$P(\mathscr{A}) \simeq \frac{n_{\mathscr{A}}}{n} \qquad P(\mathscr{M}) \simeq \frac{n_{\mathscr{M}}}{n} \qquad P(\mathscr{A}\mathscr{M}) \simeq \frac{n_{\mathscr{A}\mathscr{M}}}{n}$$

Hence

$$P(\mathscr{A}|\mathscr{M}) = \frac{P(\mathscr{A}\mathscr{M})}{P(\mathscr{M})} \simeq \frac{n_{\mathscr{A}\mathscr{M}}/n}{n_{\mathscr{M}}/n} = \frac{n_{\mathscr{A}\mathscr{M}}}{n_{\mathscr{M}}} \tag{2-32}$$

This result can be phrased as follows: If we discard all trials in which the event $\mathscr{M}$ did not occur and we retain only the subsequence of trials in which $\mathscr{M}$ occurred, then $P(\mathscr{A}|\mathscr{M})$ equals the relative frequency of occurrence $n_{\mathscr{A}\mathscr{M}}/n_{\mathscr{M}}$ of the event $\mathscr{A}$ in that subsequence.

Fundamental remark. We shall show that, for a specific $\mathscr{M}$, the conditional probabilities are indeed probabilities; that is, they satisfy the axioms.

The first axiom is obviously satisfied because $P(\mathscr{A}\mathscr{M}) \geq 0$ and $P(\mathscr{M}) > 0$:

$$P(\mathscr{A}|\mathscr{M}) \geq 0 \tag{2-33}$$

The second follows from (2-30) because $\mathscr{M} \subset \mathscr{S}$:

$$P(\mathscr{S}|\mathscr{M}) = 1 \tag{2-34}$$

To prove the third, we observe that if the events $\mathscr{A}$ and $\mathscr{B}$ are mutually exclusive, then (Fig. 2-10) the events $\mathscr{A}\mathscr{M}$ and $\mathscr{B}\mathscr{M}$ are also mutually exclusive. Hence

$$P(\mathscr{A} + \mathscr{B}|\mathscr{M}) = \frac{P[(\mathscr{A} + \mathscr{B})\mathscr{M}]}{P(\mathscr{M})} = \frac{P(\mathscr{A}\mathscr{M}) + P(\mathscr{B}\mathscr{M})}{P(\mathscr{M})}$$

This yields the third axiom:

$$P(\mathscr{A} + \mathscr{B}|\mathscr{M}) = P(\mathscr{A}|\mathscr{M}) + P(\mathscr{B}|\mathscr{M}) \tag{2-35}$$

From the above it follows that all results involving probabilities holds also for conditional probabilities. The significance of this conclusion will be appreciated later.

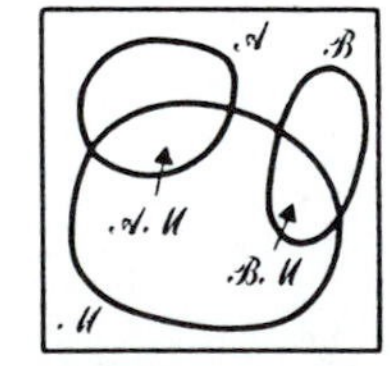

FIGURE 2-10

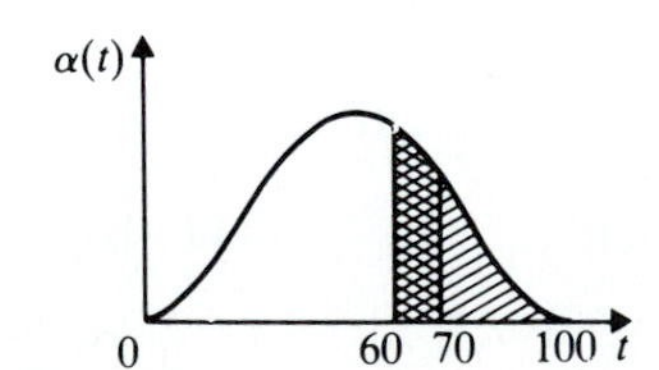

FIGURE 2-11

Example 2-8. In the fair-die experiment, we shall determine the conditional probability of the event $\{f_2\}$ assuming that the event *even* occurred. With

$$\mathscr{A} = \{f_2\} \qquad \mathscr{M} = \{\text{even}\} = \{f_2, f_4, f_6\}$$

we have $P(\mathscr{A}) = 1/6$ and $P(\mathscr{M}) = 3/6$. And since $\mathscr{A}\mathscr{M} = \mathscr{A}$, (2-29) yields

$$P\{f_2|\text{even}\} = \frac{P\{f_2\}}{P\{\text{even}\}} = \frac{1}{3}$$

This equals the relative frequency of the occurrence of the event {two} in the subsequence whose outcomes are even numbers.

Example 2-9. We denote by t the age of a person when he dies. The probability that $t \le t_o$ is given by

$$P\{t \le t_o\} = \int_0^{t_o} \alpha(t)\,dt$$

where $\alpha(t)$ is a function determined from mortality records. We shall assume that

$$\alpha(t) = 3 \times 10^{-9} t^2 (100 - t)^2 \qquad 0 \le t \le 100 \text{ years}$$

and 0 otherwise (Fig. 2-11).

From (2-28) it follows that the probability that a person will die between the ages of 60 and 70 equals

$$P\{60 \le t \le 70\} = \int_{60}^{70} \alpha(t)\,dt = 0.154$$

This equals the number of people who die between the ages of 60 and 70 divided by the total population.

With

$$\mathscr{A} = \{60 \le t \le 70\} \qquad \mathscr{M} = \{t \ge 60\} \qquad \mathscr{A}\mathscr{M} = \mathscr{A}$$

it follows from (2-29) that the probability that a person will die between the ages of 60 and 70 assuming that he was alive at 60 equals

$$P\{60 \le t \le 70 | t \ge 60\} = \frac{\int_{60}^{70} \alpha(t)\,dt}{\int_{60}^{100} \alpha(t)\,dt} = 0.486$$

This equals the number of people who die between the ages 60 and 70 divided by the number of people that are alive at age 60.

Example 2-10. A box contains three white balls w_1, w_2, w_3 and two red balls r_1, r_2. We remove at random two balls in succession. What is the probability that the first removed ball is white and the second is red?

We shall give two solutions to this problem. In the first, we apply (2-25); in the second, we use conditional probabilities.

First solution. The space of our experiment consists of all *ordered* pairs that we can form with the five balls:

$$w_1w_2 \quad w_1w_3 \quad w_1r_1 \quad w_1r_2 \quad \cdots \quad r_2w_1 \quad r_2w_2 \quad r_2w_3 \quad r_2r_1$$

The number of such pairs equals $5 \times 4 = 20$. The event {white first, red second} consists of the six outcomes

$$w_1r_1 \quad w_1r_2 \quad w_2r_1 \quad w_2r_2 \quad w_3r_1 \quad w_3r_2$$

Hence [see (2-25)] its probability equals 6/20.

Second solution. Since the box contains three white and two red balls, the probability of the event $\mathscr{W}_1$ = {white first} equals 3/5. If a white ball is removed, there remain two white and two red balls; hence the conditional probability $P(\mathscr{R}_2 | \mathscr{W}_1)$ of the event $\mathscr{R}_2$ = {red second} assuming {white first} equals 2/4. From this and (2-29) it follows that

$$P(\mathscr{W}_1\mathscr{R}_2) = P(\mathscr{R}_2 | \mathscr{W}_1)P(\mathscr{W}_1) = \frac{2}{4} \times \frac{3}{5} = \frac{6}{20}$$

where $\mathscr{W}_1\mathscr{R}_2$ is the event {white first, red second}.

Total Probability and Bayes' Theorem

If $\mathfrak{A} = [\mathscr{A}_1, \ldots, \mathscr{A}_n]$ is a partition of $\mathscr{S}$ and $\mathscr{B}$ is an arbitrary event (Fig. 2-5), then

$$P(\mathscr{B}) = P(\mathscr{B}|\mathscr{A}_1)P(\mathscr{A}_1) + \cdots + P(\mathscr{B}|\mathscr{A}_n)P(\mathscr{A}_n) \qquad (2\text{-}36)$$

Proof. Clearly,

$$\mathscr{B} = \mathscr{B}\mathscr{S} = \mathscr{B}(\mathscr{A}_1 + \cdots + \mathscr{A}_n) = \mathscr{B}\mathscr{A}_1 + \cdots + \mathscr{B}\mathscr{A}_n$$

But the events $\mathscr{B}\mathscr{A}_i$ and $\mathscr{B}\mathscr{A}_j$ are mutually exclusive because the events $\mathscr{A}_i$ and $\mathscr{A}_j$ are mutually exclusive [see (2-4)]. Hence

$$P(\mathscr{B}) = P(\mathscr{B}\mathscr{A}_1) + \cdots + P(\mathscr{B}\mathscr{A}_n)$$

and (2-36) follows because [see (2-29)]

$$P(\mathscr{B}\mathscr{A}_i) = P(\mathscr{B}|\mathscr{A}_i)P(\mathscr{A}_i) \qquad (2\text{-}37)$$

This result is known as the *total probability theorem*.

Since $P(\mathscr{B}\mathscr{A}_i) = P(\mathscr{A}_i|\mathscr{B})P(\mathscr{B})$ we conclude with (2-37) that

$$P(\mathscr{A}_i|\mathscr{B}) = P(\mathscr{B}|\mathscr{A}_i)\frac{P(\mathscr{A}_i)}{P(\mathscr{B})} \qquad (2\text{-}38)$$

Inserting (2-36) into (2-38), we obtain *Bayes' theorem*†:

$$P(\mathscr{A}_i|\mathscr{B}) = \frac{P(\mathscr{B}|\mathscr{A}_i)P(\mathscr{A}_i)}{P(\mathscr{B}|\mathscr{A}_1)P(\mathscr{A}_1) + \cdots + P(\mathscr{B}|\mathscr{A}_n)P(\mathscr{A}_n)} \qquad (2\text{-}39)$$

†The main idea of this theorem is due to Thomas Bayes (1763). However, its final form (2-39) was given by Laplace several years later.

Note The terms *a priori* and *a posteriori* are often used for the probabilities $P(\mathscr{A}_i)$ and $P(\mathscr{A}_i | \mathscr{B})$.

Example 2-11. We have four boxes. Box 1 contains 2000 components of which 5 percent are defective. Box 2 contains 500 components of which 40 percent are defective. Boxes 3 and 4 contain 1000 each with 10 percent defective. We select *at random* one of the boxes and we remove *at random* a single component.

(a) What is the probability that the selected component is defective?

The space of this experiment consists of 4000 good (g) components and 500 defective (d) components arranged as follows:

$$\begin{array}{llll} \text{Box 1:} & 1900g,\ 100d & \text{Box 2:} & 300g,\ 200d \\ \text{Box 3:} & 900g,\ 100d & \text{Box 4:} & 900g,\ 100d \end{array}$$

We denote by $\mathscr{B}_i$ the event consisting of all components in the ith box and by $\mathscr{D}$ the event consisting of all defective components. Clearly,

$$P(\mathscr{B}_1) = P(\mathscr{B}_2) = P(\mathscr{B}_3) = P(\mathscr{B}_4) = \tfrac{1}{4} \tag{2-40}$$

because the boxes are selected at random. The probability that a component taken from a specific box is defective equals the ratio of the defective to the total number of components in that box. This means that

$$\begin{aligned} P(\mathscr{D}|\mathscr{B}_1) &= \frac{100}{2000} = 0.05 \qquad & P(\mathscr{D}|\mathscr{B}_2) &= \frac{200}{500} = 0.4 \\ P(\mathscr{D}|\mathscr{B}_3) &= \frac{100}{1000} = 0.1 \qquad & P(\mathscr{D}|\mathscr{B}_4) &= \frac{100}{1000} = 0.1 \end{aligned} \tag{2-41}$$

And since the events $\mathscr{B}_1, \mathscr{B}_2, \mathscr{B}_3, \mathscr{B}_4$ form a partition of $\mathscr{S}$, we conclude from (2-36) that

$$P(\mathscr{D}) = 0.05 \times \tfrac{1}{4} + 0.4 \times \tfrac{1}{4} + 0.1 \times \tfrac{1}{4} + 0.1 \times \tfrac{1}{4} = 0.1625$$

This is the probability that the selected component is defective.

(b) We examine the selected component and we find it defective. On the basis of this evidence, we want to determine the probability that it came from box 2.

We now want the conditional probability $P(\mathscr{B}_2 | \mathscr{D})$. Since

$$P(\mathscr{D}) = 0.1625 \qquad P(\mathscr{D}|\mathscr{B}_2) = 0.4 \qquad P(\mathscr{B}_2) = 0.25$$

(2-38) yields

$$P(\mathscr{B}_2|\mathscr{D}) = 0.4 \times \frac{0.25}{0.1625} = 0.615$$

Thus the a priori probability of selecting box 2 equals 0.25 and the a posteriori probability assuming that the selected component is defective equals 0.615. These probabilities have the following frequency interpretation: If the experiment is performed n times, then box 2 is selected $0.25n$ times. If we consider only the $n_{\mathscr{D}}$ experiments in which the removed part is defective, then the number of times the part is taken from box 2 equals $0.615n_{\mathscr{D}}$.

We conclude with a comment on the distinction between assumptions and deductions: Equations (2-40) and (2-41) are not derived; they are merely reasonable *assumptions*. Based on these assumptions and on the axioms, we *deduce* that $P(\mathscr{D}) = 0.1625$ and $P(\mathscr{B}_2 | \mathscr{D}) = 0.615$.

Independence

Two events $\mathscr{A}$ and $\mathscr{B}$ are called *independent* if

$$P(\mathscr{A}\mathscr{B}) = P(\mathscr{A})P(\mathscr{B}) \tag{2-42}$$

The concept of independence is fundamental. In fact, it is this concept that justifies the mathematical development of probability, not merely as a topic in measure theory, but as a separate discipline. The significance of independence will be appreciated later in the context of repeated trials. We discuss here only various simple properties.

Frequency interpretation Denoting by $n_{\mathscr{A}}$, $n_{\mathscr{B}}$, and $n_{\mathscr{A}\mathscr{B}}$ the number of occurrences of the events $\mathscr{A}$, $\mathscr{B}$, and $\mathscr{A}\mathscr{B}$ respectively, we have

$$P(\mathscr{A}) \simeq \frac{n_{\mathscr{A}}}{n} \qquad P(\mathscr{B}) \simeq \frac{n_{\mathscr{B}}}{n} \qquad P(\mathscr{A}\mathscr{B}) \simeq \frac{n_{\mathscr{A}\mathscr{B}}}{n}$$

If the events $\mathscr{A}$ and $\mathscr{B}$ are independent, then

$$\frac{n_{\mathscr{A}}}{n} \simeq P(\mathscr{A}) = \frac{P(\mathscr{A}\mathscr{B})}{P(\mathscr{B})} \simeq \frac{n_{\mathscr{A}\mathscr{B}}/n}{n_{\mathscr{B}}/n} = \frac{n_{\mathscr{A}\mathscr{B}}}{n_{\mathscr{B}}}$$

Thus, if $\mathscr{A}$ and $\mathscr{B}$ are independent, then the relative frequency $n_{\mathscr{A}}/n$ of the occurrence of $\mathscr{A}$ in the original sequence of n trials equals the relative frequency $n_{\mathscr{A}\mathscr{B}}/n_{\mathscr{B}}$ of the occurrence of $\mathscr{A}$ in the subsequence in which $\mathscr{B}$ occurs.

We show next that if the events $\mathscr{A}$ and $\mathscr{B}$ are independent, then the events $\overline{\mathscr{A}}$ and $\mathscr{B}$ and the events $\overline{\mathscr{A}}$ and $\overline{\mathscr{B}}$ are also independent.

As we know, the events $\mathscr{A}\mathscr{B}$ and $\overline{\mathscr{A}}\mathscr{B}$ are mutually exclusive and

$$\mathscr{B} = \mathscr{A}\mathscr{B} + \overline{\mathscr{A}}\mathscr{B} \qquad P(\overline{\mathscr{A}}) = 1 - P(\mathscr{A})$$

From this and (2-42) it follows that

$$P(\overline{\mathscr{A}}\mathscr{B}) = P(\mathscr{B}) - P(\mathscr{A}\mathscr{B}) = [1 - P(\mathscr{A})]P(\mathscr{B}) = P(\overline{\mathscr{A}})P(\mathscr{B})$$

This establishes the independence of $\overline{\mathscr{A}}$ and $\mathscr{B}$. Repeating the argument, we conclude that $\overline{\mathscr{A}}$ and $\overline{\mathscr{B}}$ are also independent.

In the next two examples, we illustrate the concept of independence. In Example 2-12a, we start with a known experiment and we show that two of its events are independent. In Examples 2-12b and 2-13 we use the concept of independence to complete the specification of each experiment. This idea is developed further in the next chapter.

Example 2-12. If we toss a coin twice, we generate the four outcomes hh, ht, th, and tt.

(a) To construct an experiment with these outcomes, it suffices to assign probabilities to its elementary events. With a and b two positive numbers such that $a + b = 1$, we assume that

$$P\{hh\} = a^2 \qquad P\{ht\} = P\{th\} = ab \qquad P\{tt\} = b^2$$

These probabilities are consistent with the axioms because

$$a^2 + ab + ab + b^2 = (a + b)^2 = 1$$

In the experiment so constructed, the events

$$\mathscr{H}_1 = \{\text{heads at first toss}\} = \{hh, ht\}$$

$$\mathscr{H}_2 = \{\text{heads at second toss}\} = \{hh, th\}$$

consist of two elements each, and their probabilities are [see (2-23)]

$$P(\mathscr{H}_1) = P\{hh\} + P\{ht\} = a^2 + ab = a$$

$$P(\mathscr{H}_2) = P\{hh\} + P\{th\} = a^2 + ab = a$$

The intersection $\mathscr{H}_1\mathscr{H}_2$ of these two events consists of the single outcome $\{hh\}$. Hence

$$P(\mathscr{H}_1\mathscr{H}_2) = P\{hh\} = a^2 = P(\mathscr{H}_1)P(\mathscr{H}_2)$$

This shows that the events $\mathscr{H}_1$ and $\mathscr{H}_2$ are independent.

(b) The above experiment can be specified in terms of the probabilities $P(\mathscr{H}_1) = P(\mathscr{H}_2) = a$ of the events $\mathscr{H}_1$ and $\mathscr{H}_2$, and the information that these events are independent.

Indeed, as we have shown, the events $\overline{\mathscr{H}}_1$ and $\mathscr{H}_2$ and the events $\overline{\mathscr{H}}_1$ and $\overline{\mathscr{H}}_2$ are also independent. Furthermore,

$$\mathscr{H}_1\mathscr{H}_2 = \{hh\} \qquad \mathscr{H}_1\overline{\mathscr{H}}_2 = \{ht\} \qquad \overline{\mathscr{H}}_1\mathscr{H}_2 = \{th\} \qquad \overline{\mathscr{H}}_1\overline{\mathscr{H}}_2 = \{tt\}$$

and $P(\overline{\mathscr{H}}_1) = 1 - P(\mathscr{H}_1) = 1 - a$, $P(\overline{\mathscr{H}}_2) = 1 - P(\mathscr{H}_2) = 1 - a$. Hence

$$P\{hh\} = a^2 \qquad P\{ht\} = a(1-a) \qquad P\{th\} = (1-a)a \qquad P\{tt\} = (1-a)^2$$

Example 2-13. Trains X and Y arrive at a station at random between 8 A.M. and 8.20 A.M. Train X stops for four minutes and train Y stops for five minutes. Assuming that the trains arrive independently of each other, we shall determine various probabilities related to the times x and y of their respective arrivals. To do so, we must first specify the underlying experiment.

The outcomes of this experiment are all points (x, y) in the square of Fig. 2-12. The event

$$\mathscr{A} = \{X \text{ arrives in the interval } (t_1, t_2)\} = \{t_1 \le x \le t_2\}$$

is a vertical strip as in Fig. 2-12a and its probability equals $(t_2 - t_1)/20$. This is

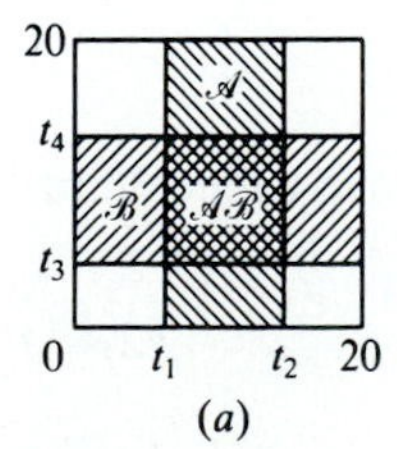

(a)

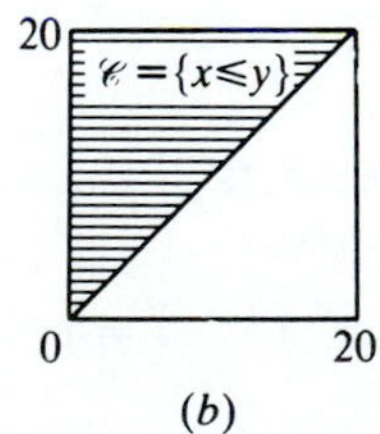

(b)

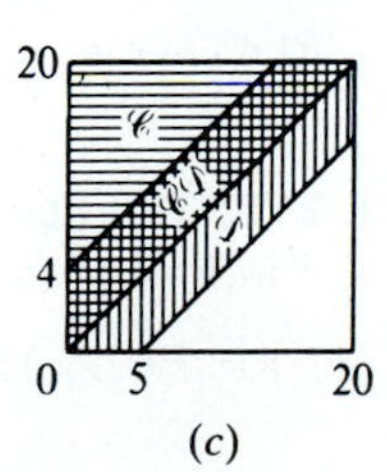

(c)

FIGURE 2-12

our interpretation of the information that the train arrives at random. Similarly, the event

$$\mathscr{B} = \{Y \text{ arrives in the interval } (t_3, t_4)\} = \{t_3 \le y \le t_4\}$$

is a horizontal strip and its probability equals $(t_4 - t_3)/20$.

Proceeding similarly, we can determine the probabilities of any horizontal or vertical sets of points. To complete the specification of the experiment, we must determine also the probabilities of their intersections. Interpreting the independence of the arrival times as independence of the events $\mathscr{A}$ and $\mathscr{B}$, we obtain

$$P(\mathscr{A}\mathscr{B}) = P(\mathscr{A})P(\mathscr{B}) = \frac{(t_2 - t_1)(t_4 - t_3)}{20 \times 20}$$

The event $\mathscr{A}\mathscr{B}$ is the rectangle shown in the figure. Since the coordinates of this rectangle are arbitrary, we conclude that the probability of any rectangle equals its area divided by 400. In the plane, all events are unions and intersections of rectangles forming a Borel field. This shows that the probability that the point (x, y) will be in an arbitrary region R of the plane equals the area of R divided by 400. This completes the specification of the experiment.

(*a*) We shall determine the probability that train X arrives before train Y. This is the probability of the event

$$\mathscr{C} = \{x \le y\}$$

shown in Fig. 2-12*b*. This event is a triangle with area 200. Hence

$$P(\mathscr{C}) = \frac{200}{400}$$

(*b*) We shall determine the probability that the trains meet at the station. For the trains to meet, x must be less than $y + 5$ and y must be less than $x + 4$. This is the event

$$\mathscr{D} = \{-4 \le x - y \le 5\}$$

of Fig. 2-12*c*. As we see from the figure, the region $\mathscr{D}$ consists of two trapezoids with common base, and its area equals 159.5. Hence

$$P(\mathscr{D}) = \frac{159.5}{400}$$

(*c*) Assuming that the trains met, we shall determine the probability that train X arrived before train Y. We wish to find the conditional probability $P(\mathscr{C}|\mathscr{D})$. The event $\mathscr{C}\mathscr{D}$ is a trapezoid as shown and its area equals 72. Hence

$$P(\mathscr{C}|\mathscr{D}) = \frac{P(\mathscr{C}\mathscr{D})}{P(\mathscr{D})} = \frac{72}{159.5}$$

INDEPENDENCE OF THREE EVENTS. The events $\mathscr{A}_1, \mathscr{A}_2, \mathscr{A}_3$ are called (mutually) independent if they are independent in pairs:

$$P(\mathscr{A}_i\mathscr{A}_j) = P(\mathscr{A}_i)P(\mathscr{A}_j) \qquad i \ne j \tag{2-43}$$

and

$$P(\mathscr{A}_1\mathscr{A}_2\mathscr{A}_3) = P(\mathscr{A}_1)P(\mathscr{A}_2)P(\mathscr{A}_3) \tag{2-44}$$

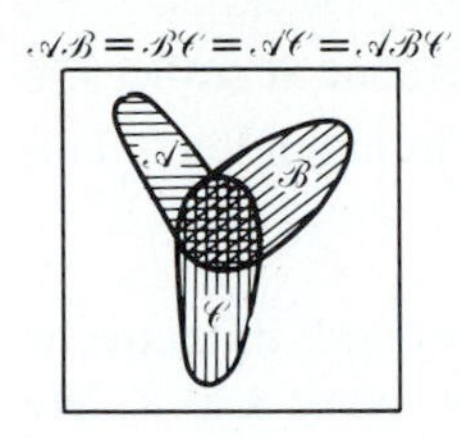

FIGURE 2-13

We should emphasize that three events might be independent in pairs but not independent. The next example is an illustration.

Example 2-14. Suppose that the events $\mathscr{A}$, $\mathscr{B}$, and $\mathscr{C}$ of Fig. 2-13 have the same probability

$$P(\mathscr{A}) = P(\mathscr{B}) = P(\mathscr{C}) = \tfrac{1}{5}$$

and the intersections $\mathscr{A}\mathscr{B}$, $\mathscr{A}\mathscr{C}$, $\mathscr{B}\mathscr{C}$, and $\mathscr{A}\mathscr{B}\mathscr{C}$ also have the same probability

$$p = P(\mathscr{A}\mathscr{B}) = P(\mathscr{A}\mathscr{C}) = P(\mathscr{B}\mathscr{C}) = P(\mathscr{A}\mathscr{B}\mathscr{C})$$

(*a*) If $p = 1/25$, then these events are independent in pairs but they are not independent because

$$P(\mathscr{A}\mathscr{B}\mathscr{C}) \neq P(\mathscr{A})P(\mathscr{B})P(\mathscr{C})$$

(*b*) If $p = 1/125$, then $P(\mathscr{A}\mathscr{B}\mathscr{C}) = P(\mathscr{A})P(\mathscr{B})P(\mathscr{C})$ but the events are not independent because

$$P(\mathscr{A}\mathscr{B}) \neq P(\mathscr{A})P(\mathscr{B})$$

From the independence of the events $\mathscr{A}$, $\mathscr{B}$, and $\mathscr{C}$ it follows that:

1. Any one of them is independent of the intersection of the other two.

Indeed, from (2-43) and (2-44) it follows that

$$P(\mathscr{A}_1\mathscr{A}_2\mathscr{A}_3) = P(\mathscr{A}_1)P(\mathscr{A}_2)P(\mathscr{A}_3) = P(\mathscr{A}_1)P(\mathscr{A}_2\mathscr{A}_3) \quad (2\text{-}45)$$

Hence the events $\mathscr{A}_1$ and $\mathscr{A}_2\mathscr{A}_3$ are independent.

2. If we replace one or more of these events with their complements, the resulting events are also independent.

Indeed, since

$$\mathscr{A}_1\mathscr{A}_2 = \mathscr{A}_1\mathscr{A}_2\mathscr{A}_3 + \mathscr{A}_1\mathscr{A}_2\overline{\mathscr{A}}_3 \qquad P(\overline{\mathscr{A}}_3) = 1 - P(\mathscr{A}_3)$$

we conclude with (2-45) that

$$P(\mathscr{A}_1\mathscr{A}_2\overline{\mathscr{A}}_3) = P(\mathscr{A}_1\mathscr{A}_2) - P(\mathscr{A}_1\mathscr{A}_2)P(\mathscr{A}_3) = P(\mathscr{A}_1)P(\mathscr{A}_2)P(\overline{\mathscr{A}}_3)$$

Hence the events $\mathscr{A}_1$, $\mathscr{A}_2$, and $\overline{\mathscr{A}}_3$ are independent because they satisfy (2-44) and, as we have shown earlier in the section, they are also independent in pairs.

3. Any one of them is independent of the union of the other two.

To show that the events $\mathscr{A}_1$ and $\mathscr{A}_2 + \mathscr{A}_3$ are independent, it suffices to show that the events $\mathscr{A}_1$ and $\overline{\mathscr{A}_2 + \mathscr{A}_3} = \overline{\mathscr{A}}_2\overline{\mathscr{A}}_3$ are independent. This follows from 1 and 2.

Generalization. The independence of n events can be defined inductively: Suppose that we have defined independence of k events for every $k < n$. We then say that the events $\mathscr{A}_1, \ldots, \mathscr{A}_n$ are independent if any $k < n$ of them are independent and

$$P(\mathscr{A}_1 \cdots \mathscr{A}_n) = P(\mathscr{A}_1) \cdots P(\mathscr{A}_n) \tag{2-46}$$

This completes the definition for any n because we have defined independence for $n = 2$.

PROBLEMS

2-1. Show that (a) $\overline{\overline{\mathscr{A}} + \overline{\mathscr{B}}} + \overline{\overline{\mathscr{A}} + \mathscr{B}} = \mathscr{A}$; (b) $(\mathscr{A} + \mathscr{B})(\overline{\mathscr{A}\mathscr{B}}) = \mathscr{A}\overline{\mathscr{B}} + \mathscr{B}\overline{\mathscr{A}}$.

2-2. If $\mathscr{A} = \{2 \le x \le 5\}$ and $\mathscr{B} = \{3 \le x \le 6\}$, find $\mathscr{A} + \mathscr{B}$, $\mathscr{A}\mathscr{B}$, and $(\mathscr{A} + \mathscr{B})(\overline{\mathscr{A}\mathscr{B}})$.

2-3. Show that if $\mathscr{A}\mathscr{B} = \{\emptyset\}$, then $P(\mathscr{A}) \le P(\overline{\mathscr{B}})$.

2-4. Show that (a) if $P(\mathscr{A}) = P(\mathscr{B}) = P(\mathscr{A}\mathscr{B})$, then $P(\mathscr{A}\overline{\mathscr{B}} + \mathscr{B}\overline{\mathscr{A}}) = 0$; (b) if $P(\mathscr{A}) = P(\mathscr{B}) = 1$, then $P(\mathscr{A}\mathscr{B}) = 1$.

2-5. Prove and generalize the following identity

$$P(\mathscr{A} + \mathscr{B} + \mathscr{C}) = P(\mathscr{A}) + P(\mathscr{B}) + P(\mathscr{C}) - P(\mathscr{A}\mathscr{B}) - P(\mathscr{A}\mathscr{C}) - P(\mathscr{B}\mathscr{C}) + P(\mathscr{A}\mathscr{B}\mathscr{C})$$

2-6. Show that if $\mathscr{S}$ consists of a countable number of elements ζ_i and each subset $\{\zeta_i\}$ is an event, then all subsets of $\mathscr{S}$ are events.

2-7. If $\mathscr{S} = \{1, 2, 3, 4\}$, find the smallest field that contains the sets $\{1\}$ and $\{2, 3\}$.

2-8. If $\mathscr{A} \subset \mathscr{B}$, $P(\mathscr{A}) = 1/4$, and $P(\mathscr{B}) = 1/3$, find $P(\mathscr{A}|\mathscr{B})$ and $P(\mathscr{B}|\mathscr{A})$.

2-9. Show that $P(\mathscr{A}\mathscr{B}|\mathscr{C}) = P(\mathscr{A}|\mathscr{B}\mathscr{C})P(\mathscr{B}|\mathscr{C})$ and $P(\mathscr{A}\mathscr{B}\mathscr{C}) = P(\mathscr{A}|\mathscr{B}\mathscr{C})$ $P(\mathscr{B}|\mathscr{C})P(\mathscr{C})$.

2-10. (*Chain rule*) Show that

$$P(\mathscr{A}_n \cdots \mathscr{A}_1) = P(\mathscr{A}_n|\mathscr{A}_{n-1} \cdots \mathscr{A}_1) \cdots P(\mathscr{A}_2|\mathscr{A}_1)\mathscr{P}(\mathscr{A}_1)$$

2-11. We select at random m objects from a set $\mathscr{S}$ of n objects and we denote by $\mathscr{A}_m$ the set of the selected objects. Show that the probability p that a particular element ζ_0 of $\mathscr{S}$ is in $\mathscr{A}_m$ equals m/n.

Hint: p equals the probability that a randomly selected element of $\mathscr{S}$ is in $\mathscr{A}_m$.

2-12. A call occurs at time t where t is a random point in the interval $(0, 10)$. (a) Find $P\{6 \le t \le 8\}$. (b) Find $P\{6 \le t \le 8 | t > 5\}$.

2-13. The space $\mathscr{S}$ is the set of all positive numbers t. Show that if $P\{t_0 \le t \le t_0 + t_1 | t \ge t_0\} = P\{t \le t_1\}$ for every t_0 and t_1, then $P\{t \le t_1\} = 1 - e^{-ct_1}$ where c is a constant.

2-14. The events $\mathscr{A}$ and $\mathscr{B}$ are mutually exclusive. Can they be independent?

2-15. Show that if the events $\mathscr{A}_1, \ldots, \mathscr{A}_n$ are independent and $\mathscr{B}_i$ equals $\mathscr{A}_i$ or $\overline{\mathscr{A}}_i$ or $\mathscr{S}$, then the events $\mathscr{B}_1, \ldots, \mathscr{B}_n$ are also independent.

2-16. Show that $2^n - (n + 1)$ equations are needed to establish the independence of n events.

2-17. Box 1 contains 1 white and 999 red balls. Box 2 contains 1 red and 999 white balls. A ball is picked from a randomly selected box. If the ball is red what is the probability that it came from box 1?

2-18. Box 1 contains 1000 bulbs of which 10 percent are defective. Box 2 contains 2000 bulbs of which 5 percent are defective. Two bulbs are picked from a randomly selected box. (*a*) Find the probability that both bulbs are defective. (*b*) Assuming that both are defective, find the probability that they came from box 1.

2-19. A train and a bus arrive at the station at random between 9 A.M. and 10 A.M. The train stops for 10 minutes and the bus for x minutes. Find x so that the probability that the bus and the train will meet equals 0.5.

2-20. Show that a set $\mathscr{S}$ with n elements has

$$\frac{n(n-1)\cdots(n-k+1)}{1\cdot 2\cdots k} = \frac{n!}{k!(n-k)!}$$

k-element subsets.

2-21. We have two coins; the first is fair and the second two-headed. We pick one of the coins at random, we toss it twice and heads shows both times. Find the probability that the coin picked is fair.

CHAPTER 3

REPEATED TRIALS

3-1 COMBINED EXPERIMENTS

We are given two experiments: The first experiment is the rolling of a fair die

$$\mathscr{S}_1 = \{f_1, \dots, f_6\} \qquad P_1\{f_i\} = \tfrac{1}{6}$$

The second experiment is the tossing of a fair coin

$$\mathscr{S}_2 = \{h, t\} \qquad P_2\{h\} = P_2\{t\} = \tfrac{1}{2}$$

We perform both experiments and we want to find the probability that we get "two" on the die and "heads" on the coin.

If we make the reasonable assumption that the outcomes of the first experiment are independent of the outcomes of the second, we conclude that the unknown probability equals $1/6 \times 1/2$.

The above conclusion is reasonable; however, the notion of independence used in its derivation does not agree with the definition given in (2-42). In that definition, the events $\mathscr{A}$ and $\mathscr{B}$ were subsets of the *same* space. In order to fit the above conclusion into our theory, we must, therefore, construct a space $\mathscr{S}$ having as subsets the events "two" and "heads." This is done as follows:

The two experiments are viewed as a single experiment whose outcomes are pairs $\zeta_1\zeta_2$ where ζ_1 is one of the six faces of the die and ζ_2 is heads or tails.† The resulting space consists of the 12 elements

$$f_1h, \dots, f_6h, f_1t, \dots, f_6t$$

In this space, {two} is not an elementary event but a subset consisting of two elements

$$\{\text{two}\} = \{f_2h, f_2t\}$$

Similarly, {heads} is an event with six elements

$$\{\text{heads}\} = \{f_1h, \dots, f_6h\}$$

To complete the experiment, we must assign probabilities to all subsets of $\mathscr{S}$. Clearly, the event {two} occurs if the die shows "two" no matter what shows on the coin. Hence

$$P\{\text{two}\} = P_1\{f_2\} = \tfrac{1}{6}$$

Similarly,

$$P\{\text{heads}\} = P_2\{h\} = \tfrac{1}{2}$$

The intersection of the events {two} and {heads} is the elementary event $\{f_2h\}$. Assuming that the events {two} and {heads} are independent in the sense of (2-42), we conclude that $P\{f_2h\} = 1/6 \times 1/2$ in agreement with our earlier conclusion.

CARTESIAN PRODUCTS. Given two sets $\mathscr{S}_1$ and $\mathscr{S}_2$ with elements ζ_1 and ζ_2 respectively, we form all *ordered* pairs $\zeta_1\zeta_2$ where ζ_1 is any element of $\mathscr{S}_1$ and ζ_2 is any element of $\mathscr{S}_2$. The *cartesian product* of the sets $\mathscr{S}_1$ and $\mathscr{S}_2$ is a set $\mathscr{S}$ whose elements are all such pairs. This set is written in the form

$$\mathscr{S} = \mathscr{S}_1 \times \mathscr{S}_2$$

Example 3-1. The cartesian product of the sets

$$\mathscr{S}_1 = \{\text{car}, \text{apple}, \text{bird}\} \qquad \mathscr{S}_2 = \{h, t\}$$

has six elements

$$\mathscr{S}_1 \times \mathscr{S}_2 = \{\text{car-}h, \text{car-}t, \text{apple-}h, \text{apple-}t, \text{bird-}h, \text{bird-}t\}$$

Example 3-2. If $\mathscr{S}_1 = \{h, t\}$, $\mathscr{S}_2 = \{h, t\}$. Then

$$\mathscr{S}_1 \times \mathscr{S}_2 = \{hh, ht, th, tt\}$$

In this example, the sets $\mathscr{S}_1$ and $\mathscr{S}_2$ are identical. We note also that the element ht is different from the element th.

---

†In the earlier discussion, the symbol ζ_i represented a *single* element of a set $\mathscr{S}$. In the following, ζ_i will also represent an arbitrary element of a set $\mathscr{S}_i$. It will be understood from the context whether ζ_i is one particular element or any element of $\mathscr{S}_i$.

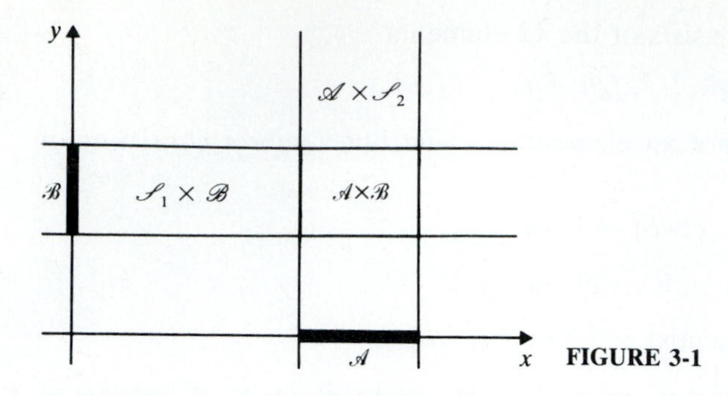

FIGURE 3-1

If $\mathscr{A}$ is a subset of $\mathscr{S}_1$ and $\mathscr{B}$ is a subset of $\mathscr{S}_2$, then the set

$$\mathscr{C} = \mathscr{A} \times \mathscr{B}$$

consisting of all pairs $\zeta_1\zeta_2$ where $\zeta_1 \in \mathscr{A}$ and $\zeta_2 \in \mathscr{B}$, is a subset of $\mathscr{S}$.

Forming similarly the sets $\mathscr{A} \times \mathscr{S}_2$ and $\mathscr{S}_1 \times \mathscr{B}$, we conclude that their intersection is the set $\mathscr{A} \times \mathscr{B}$:

$$\mathscr{A} \times \mathscr{B} = (\mathscr{A} \times \mathscr{S}_2) \cap (\mathscr{S}_1 \times \mathscr{B}) \tag{3-1}$$

Note Suppose that $\mathscr{S}_1$ is the x axis, $\mathscr{S}_2$ is the y axis, and $\mathscr{A}$ and B are two intervals:

$$\mathscr{A} = \{x_1 \le x \le x_2\} \qquad \mathscr{B} = \{y_1 \le y \le y_2\}$$

In this case, $\mathscr{A} \times \mathscr{B}$ is a rectangle, $\mathscr{A} \times \mathscr{S}_2$ is a vertical strip, and $\mathscr{S}_1 \times \mathscr{B}$ is a horizontal strip (Fig. 3-1).

We can thus interpret the cartesian product $\mathscr{A} \times \mathscr{B}$ of two arbitrary sets as a generalized rectangle.

Cartesian product of two experiments. The cartesian product of two experiments $\mathscr{S}_1$ and $\mathscr{S}_2$ is a new experiment $\mathscr{S} = \mathscr{S}_1 \times \mathscr{S}_2$ whose events are all cartesian products of the form

$$\mathscr{A} \times \mathscr{B} \tag{3-2}$$

where $\mathscr{A}$ is an event of $\mathscr{S}_1$ and $\mathscr{B}$ is an event of $\mathscr{S}_2$, and their unions and intersections.

In this experiment, the probabilities of the events $\mathscr{A} \times \mathscr{S}_2$ and $\mathscr{S}_1 \times \mathscr{B}$ are such that

$$P(\mathscr{A} \times \mathscr{S}_2) = P_1(\mathscr{A}) \qquad P(\mathscr{S}_1 \times \mathscr{B}) = P_2(\mathscr{B}) \tag{3-3}$$

where $P_1(\mathscr{A})$ is the probability of the event $\mathscr{A}$ in the experiments $\mathscr{S}_1$ and $P_2(\mathscr{B})$ is the probability of the event $\mathscr{B}$ in the experiments $\mathscr{S}_2$. The above is motivated by the interpretation of $\mathscr{S}$ as a combined experiment. Indeed, the event $\mathscr{A} \times \mathscr{S}_2$ of the experiment $\mathscr{S}$ occurs if the event $\mathscr{A}$ of the experiment $\mathscr{S}_1$ occurs no matter what the outcome of $\mathscr{S}_2$ is. Similarly, the event $\mathscr{S}_1 \times \mathscr{B}$ of the experiment $\mathscr{S}$ occurs if the event $\mathscr{B}$ of the experiment $\mathscr{S}_2$ occurs no matter what the outcome of $\mathscr{S}_1$ is. This justifies the two equations in (3-3).

These equations determine only the probabilities of the events $\mathscr{A} \times \mathscr{S}_2$ and $\mathscr{S}_1 \times \mathscr{B}$. The probabilities of events of the form $\mathscr{A} \times \mathscr{B}$ and of their unions and intersections cannot in general be expressed in terms of P_1 and P_2. To determine them, we need additional information about the experiments $\mathscr{S}_1$ and $\mathscr{S}_2$.

Independent experiments. In many applications, the events $\mathscr{A} \times \mathscr{S}_2$ and $\mathscr{S}_1 \times \mathscr{B}$ of the combined experiment $\mathscr{S}$ are independent for any $\mathscr{A}$ and $\mathscr{B}$. Since the intersection of these events equals $\mathscr{A} \times \mathscr{B}$ [see (3-1)], we conclude from (2-42) and (3-3) that

$$P(\mathscr{A} \times \mathscr{B}) = P(\mathscr{A} \times \mathscr{S}_2)P(\mathscr{S}_1 \times \mathscr{B}) = P_1(\mathscr{A})P_2(\mathscr{B}) \qquad (3\text{-}4)$$

This completes the specification of the experiment $\mathscr{S}$ because all its events are unions and intersections of events of the form $\mathscr{A} \times \mathscr{B}$.

We note in particular that the elementary event $\{\zeta_1\zeta_2\}$ can be written as a cartesian product $\{\zeta_1\} \times \{\zeta_2\}$ of the elementary events $\{\zeta_1\}$ and $\{\zeta_2\}$ of $\mathscr{S}_1$ and $\mathscr{S}_2$. Hence

$$P\{\zeta_1\zeta_2\} = P_1\{\zeta_1\}P_2\{\zeta_2\} \qquad (3\text{-}5)$$

Example 3-3. A box B_1 contains 10 white and 5 red balls and a box B_2 contains 20 white and 20 red balls. A ball is drawn from each box. What is the probability that the ball from B_1 will be white and the ball from B_2 red?

The above operation can be considered as a combined experiment. Experiment $\mathscr{S}_1$ is the drawing from B_1 and experiment $\mathscr{S}_2$ is the drawing from B_2. The space $\mathscr{S}_1$ has 15 elements: 10 white and 5 red balls. The event

$$\mathscr{W}_1 = \{\text{all white balls in } B_1\}$$

has 10 elements and its probability equals 10/15. The space $\mathscr{S}_2$ has 40 elements: 20 white and 20 red balls. The event

$$\mathscr{R}_2 = \{\text{all red balls in } B_2\}$$

has 20 elements and its probability equals 20/40. The space $\mathscr{S}_1 \times \mathscr{S}_2$ has 40×15 elements: all possible pairs that can be drawn.

We want the probability of the event

$$\mathscr{W}_1 \times \mathscr{R}_2 = \{\text{white from } B_1 \text{ and red from } B_2\}$$

Assuming independence of the two experiments, we conclude from (3-4) that

$$P(\mathscr{W}_1 \times \mathscr{R}_2) = P_1(\mathscr{W}_1)P_2(\mathscr{R}_2) = \frac{10}{15} \times \frac{20}{40}$$

Example 3-4. Consider the coin experiment where the probability of "heads" equals p and the probability of "tails" equals $q = 1 - p$. If we toss the coin twice, we obtain the space

$$\mathscr{S} = \mathscr{S}_1 \times \mathscr{S}_2 \qquad \mathscr{S}_1 = \mathscr{S}_2 = \{h, t\}$$

Thus $\mathscr{S}$ consists of the four outcomes hh, ht, th, and tt. Assuming that the nts $\mathscr{S}_1$ and $\mathscr{S}_2$ are independent, we obtain

$$P\{hh\} = P_1\{h\}P_2\{h\} = p^2$$

$$P\{ht\} = pq \qquad P\{th\} = qp \qquad P\{tt\} = q^2$$

We shall use the above to find the probability of the event

$$\mathscr{H}_1 = \{\text{heads at the first toss}\} = \{hh, ht\}$$

Since $\mathscr{H}_1$ consists of the two outcomes hh and ht, (2-23) yields

$$P(\mathscr{H}_1) = P\{hh\} + P\{ht\} = p^2 + pq = p$$

This follows also from (3-4) because $\mathscr{H}_1 = \{h\} \times \mathscr{S}_2$.

Generalization. Given n experiments $\mathscr{S}_1, \dots, \mathscr{S}_n$, we define as their cartesian product

$$\mathscr{S} = \mathscr{S}_1 \times \cdots \times \mathscr{S}_n \tag{3-6}$$

the experiment whose elements are all ordered n tuplets $\zeta_1 \cdots \zeta_n$ where ζ_i is an element of the set $\mathscr{S}_i$. Events in this space are all sets of the form

$$\mathscr{A}_1 \times \cdots \times \mathscr{A}_n$$

where $\mathscr{A}_i \subset \mathscr{S}_i$, and their unions and intersections. If the experiments are independent and $P_i(\mathscr{A}_i)$ is the probability of the event $\mathscr{A}_i$ in the experiment $\mathscr{S}_i$, then

$$P(\mathscr{A}_1 \times \cdots \times \mathscr{A}_n) = P_1(\mathscr{A}_1) \cdots P_n(\mathscr{A}_n) \tag{3-7}$$

Example 3-5. If we toss the coin of Example 3-4 n times, we obtain the space $\mathscr{S} = \mathscr{S}_1 \times \cdots \times \mathscr{S}_n$ consisting of the 2^n elements $\zeta_1 \cdots \zeta_n$ where $\zeta_i = h$ or t. Clearly,

$$P\{\zeta_1 \cdots \zeta_n\} = P_1\{\zeta_1\} \cdots P_n\{\zeta_n\} \qquad P_i\{\zeta_i\} = \begin{cases} p & \zeta_i = h \\ q & \zeta_i = t \end{cases} \tag{3-8}$$

If, in particular $p = q = 1/2$, then

$$P\{\zeta_1 \cdots \zeta_n\} = \frac{1}{2^n} \tag{3-9}$$

From (3-8) it follows that, if the elementary event $\{\zeta_1 \cdots \zeta_n\}$ consists of k heads and $n - k$ tails (in a specific order), then

$$P\{\zeta_1 \cdots \zeta_n\} = p^k q^{n-k} \tag{3-10}$$

We note that the event $\mathscr{H}_1 = \{\text{heads at the first toss}\}$ consists of 2^{n-1} outcomes $\zeta_1 \cdots \zeta_n$ where $\zeta_1 = h$ and $\zeta_i = t$ or h for $i > 1$. The event $\mathscr{H}_1$ can be written as a cartesian product

$$\mathscr{H}_1 = \{h\} \times \mathscr{S}_2 \times \cdots \times \mathscr{S}_n$$

Hence [see (3-7)]

$$P(\mathscr{H}_1) = P_1\{h\}P_2(\mathscr{S}_2) \cdots P_n(\mathscr{S}_n) = p$$

because $P_i(\mathscr{S}_i) = 1$. We can similarly show that if

$$\mathscr{H}_i = \{\text{heads at the } i\text{th toss}\} \qquad \mathscr{T}_i = \{\text{tails at the } i\text{th toss}\}$$

then

$$P(\mathscr{H}_i) = p \qquad P(\mathscr{T}_i) = q$$

Dual meaning of repeated trials. In the theory of probability, the notion of repeated trials has two fundamentally different meanings. The first is the approximate relationship (1-1) between the probability $P(\mathscr{A})$ of an event $\mathscr{A}$ in an experiment $\mathscr{S}$ and the relative frequency of the occurrence of $\mathscr{A}$. The second is the creation of the experiment $\mathscr{S} \times \cdots \times \mathscr{S}$.

For example, the repeated tossings of a coin can be given the following two interpretations:

First interpretation (physical) Our experiment is the *single* toss of a fair coin. Its space has two elements and the probability of each elementary event equals 1/2. A trial is the toss of the coin *once*.

If we toss the coin n times and heads shows n_h times, then almost certainly $n_h/n \simeq 1/2$ provided that n is sufficiently large. Thus the first interpretation of repeated trials is the above inprecise statement relating probabilities with observed frequencies.

Second interpretation (conceptual) Our experiment is now the toss of the coin *n times* where n is any number large or small. Its space has 2^n elements and the probability of each elementary event equals $1/2^n$. A trial is the toss of the coin *n times*. All statements concerning the number of heads are precise and in the form of probabilities.

We can, of course, give a relative frequency interpretation to these statements. However, to do so, we must repeat the *n tosses of the coin* a large number of times.

3-2 BERNOULLI TRIALS

It is well known from combinatorial analysis that, if a set has n elements, then the total number of its subsets consisting of k elements each equals

$$\binom{n}{k} = \frac{n(n-1)\cdots(n-k+1)}{1\cdot 2\cdots k} = \frac{n!}{k!(n-k)!} \tag{3-11}$$

For example, if $n = 4$ and $k = 2$, then

$$\binom{4}{2} = \frac{4\cdot 3}{1\cdot 2} = 6$$

Indeed, the two-element subsets of the four-element set *abcd* are

$$ab \qquad ac \qquad ad \qquad bc \qquad bd \qquad cd$$

The above result will be used to find the probability that an event occurs k times in n independent trials of an experiment $\mathscr{S}$. This problem is essentially the same as the problem of obtaining k heads in n tossings of a coin. We start, therefore, with the coin experiment.

Example 3-6. A coin with $P\{h\} = p$ is tossed n times. We maintain that the probability $p_n(k)$ that heads shows k times is given by

$$p_n(k) = \binom{n}{k} p^k q^{n-k} \qquad q = 1 - p \tag{3-12}$$

Proof. The experiment under consideration is the n-tossing of a coin. A single outcome is a particular sequence of heads and tails. The event $\{k$ heads in any order$\}$ consists of all sequences containing k heads and $n - k$ tails. The k heads of each such sequence form a k-element subset of a set of n heads. As we noted, there are $\binom{n}{k}$ such subsets. Hence the event $\{k$ heads in any order$\}$ consists of $\binom{n}{k}$ elementary events containing k heads and $n - k$ tails in a specific order. Since the probability of each of these elementary events equals $p^k q^{n-k}$, we conclude that

$$P\{k \text{ heads in any order}\} = \binom{n}{k} p^k q^{n-k}$$

Special Case. If $n = 3$ and $k = 2$, then there are three ways of getting two heads, namely, *hht*, *hth*, and *thh*. Hence $p_3(2) = 3p^2 q$ in agreement with (3-12).

Success or Failure of an Event $\mathscr{A}$ in n Independent Trials

We consider now our main problem. We are given an experiment $\mathscr{S}$ and an event $\mathscr{A}$ with

$$P(\mathscr{A}) = p \qquad P(\overline{\mathscr{A}}) = q \qquad p + q = 1$$

We repeat the experiment n times and the resulting product space we denote by $\mathscr{S}^n$. Thus

$$\mathscr{S}^n = \mathscr{S} \times \cdots \times \mathscr{S}$$

We shall determine the probability $p_n(k)$ that the event $\mathscr{A}$ occurs exactly k times.

FUNDAMENTAL THEOREM

$$p_n(k) = P\{\mathscr{A} \text{ occurs } k \text{ times in any order}\} = \binom{n}{k} p^k q^{n-k} \tag{3-13}$$

Proof. The event $\{\mathscr{A}$ occurs k times in a specific order$\}$ is a cartesian product $\mathscr{B}_1 \times \cdots \times \mathscr{B}_n$ where k of the events $\mathscr{B}_i$ equal $\mathscr{A}$ and the remaining $n - k$ equal $\overline{\mathscr{A}}$. As we know from (3-7), the probability of this event equals

$$P(\mathscr{B}_1) \cdots P(\mathscr{B}_n) = p^k q^{n-k}$$

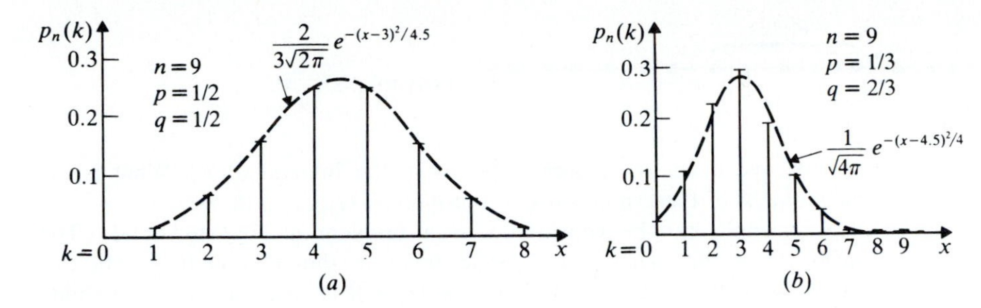

FIGURE 3-2

because

$$P(\mathscr{B}_i) = \begin{cases} p & \text{if } \mathscr{B}_i = \mathscr{A} \\ q & \text{if } \mathscr{B}_i = \overline{\mathscr{A}} \end{cases}$$

In other words,

$$P\{\mathscr{A} \text{ occurs } k \text{ times in a specific order}\} = p^k q^{n-k} \tag{3-14}$$

The event $\{\mathscr{A}$ occurs k times in any order$\}$ is the union of the $\binom{n}{k}$ events $\{\mathscr{A}$ occurs k times in a specific order$\}$ and since these events are mutually exclusive, we conclude from (2-20) that $p_n(k)$ is given by (3-13).
In Fig. 3-2, we plot $p_n(k)$ for $n = 9$. The meaning of the dashed curves will be explained later.

Example 3-7. A fair die is rolled five times. We shall find the probability $p_5(2)$ that "six" will show twice.

In the single roll of a die, $\mathscr{A} = \{\text{six}\}$ is an event with probability 1/6. Setting

$$P(\mathscr{A}) = \tfrac{1}{6} \qquad P(\overline{\mathscr{A}}) = \tfrac{5}{6} \qquad n = 5 \qquad k = 2$$

in (3-13), we obtain

$$P_5(2) = \frac{5!}{2!3!}\left(\frac{1}{6}\right)^2\left(\frac{5}{6}\right)^3$$

Example 3-8. A pair of fair dice is rolled four times. We shall find the probability $p_4(0)$ that "seven" will not show at all.

The space of the single roll of the two dice consists of the 36 elements $f_i f_j$. The event $\mathscr{A} = \{\text{seven}\}$ consists of the six elements

$$f_1 f_6, f_6 f_1, f_2 f_5, f_5 f_2, f_3 f_4, f_4 f_3$$

Therefore $P(\mathscr{A}) = 6/36$ and $P(\overline{\mathscr{A}}) = 5/6$. With $n = 4$ and $k = 0$, (3-13) yields

$$p_4(0) = \left(\tfrac{5}{6}\right)^4$$

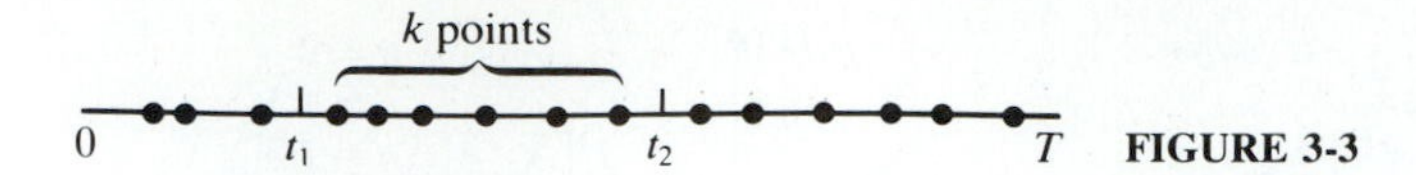

FIGURE 3-3

Example 3-9. We place at random n points in the interval $(0, T)$. What is the probability that k of these points are in the interval (t_1, t_2) (Fig. 3-3)?

This example can be considered as a problem in repeated trials. The experiment $\mathscr{S}$ is the placing of a *single* point in the interval $(0, T)$. In this experiment, $\mathscr{A} = \{$the point is in the interval $(t_1, t_2)\}$ is an event with probability

$$P(\mathscr{A}) = p = \frac{t_2 - t_1}{T}$$

In the space $\mathscr{S}^n$, the event $\{\mathscr{A}$ occurs k times$\}$ means that k of the n points are in the interval (t_1, t_2). Hence [see (3-13)]

$$P\{k \text{ points are in the interval } (t_1, t_2)\} = \binom{n}{k} p^k q^{n-k} \tag{3-15}$$

Example 3-10. A system containing n components is put into operation at $t = 0$. The probability that a particular component will fail in the interval $(0, t)$ equals

$$p = \int_0^t \alpha(\tau)\, d\tau \qquad \text{where} \quad \alpha(t) \ge 0 \qquad \int_0^\infty \alpha(t)\, dt = 1 \tag{3-16}$$

What is the probability that k of these components will fail prior to time t?

This example can also be considered as a problem in repeated trials. Reasoning as above, we conclude that the unknown probability is given by (3-15).

Most likely number of successes. We shall now examine the behavior of $p_n(k)$ as a function of k for a fixed n. We maintain that as k increases, $p_n(k)$ increases reaching a maximum for

$$k = k_{\max} = [(n+1)p] \tag{3-17}$$

where the brackets mean the largest integer that does not exceed $(n+1)p$. If $(n+1)p$ is an integer, then $p_n(k)$ is maximum for two consecutive values of k:

$$k = k_1 = (n+1)p \qquad \text{and} \quad k = k_2 = k_1 - 1 = np - q$$

Proof. We form the ratio

$$\frac{p_n(k-1)}{p_n(k)} = \frac{kq}{(n-k+1)p}$$

If this ratio is less than 1, that is, if $k < (n+1)p$, then $p_n(k-1)$ is less than $p_n(k)$. This shows that as k increases, $p_n(k)$ increases reaching its maximum for $k = [(n+1)p]$. For $k > (n+1)p$, the above ratio is greater than 1: hence $p_n(k)$ decreases.

If $k_1 = (n+1)p$ is an integer, then

$$\frac{p_n(k_1 - 1)}{p_n(k_1)} = \frac{k_1 q}{(n - k_1 + 1)p} = \frac{(n+1)pq}{[n - (n-1)p + 1]p} = 1$$

This shows that $p_n(k)$ is maximum for $k = k_1$ and $k = k_1 - 1$.

Example 3-11. (*a*) If $n = 10$ and $p = 1/3$, then $(n+1)p = 11/3$; hence $k_{\max} = [11/3] = 3$.

(*b*) If $n = 11$ and $p = 1/2$, then $(n+1)p = 6$; hence $k_1 = 6$, $k_2 = 5$.

We shall, finally, find the probability

$$P\{k_1 \le k \le k_2\}$$

that the number k of occurrences of $\mathscr{A}$ is between k_1 and k_2. Clearly, the events {$\mathscr{A}$ occurs k times}, where k takes all values from k_1 to k_2, are mutually exclusive and their union is the event $\{k_1 \le k \le k_2\}$. Hence [see (3-13)]

$$P\{k_1 \le k \le k_2\} = \sum_{k=k_1}^{k_2} p_n(k) = \sum_{k=k_1}^{k_2} \binom{n}{k} p^k q^{n-k} \tag{3-18}$$

Example 3-12. An order of 10^4 parts is received. The probability that a part is defective equals 0.1. What is the probability that the total number of defective parts does not exceed 1100?

The experiment $\mathscr{S}$ is the selection of a single part. The probability of the event $\mathscr{A}$ = {the part is defective} equals 0.1. We want the probability that in 10^4 trials, $\mathscr{A}$ will occur at most 1100 times. With

$$p = 0.1 \qquad n = 10^4 \qquad k_1 = 0 \qquad k_2 = 1100$$

(3-18) yields

$$P\{0 \le k \le 1100\} = \sum_{k=0}^{1100} \binom{10^4}{k} (0.1)^k (0.9)^{10^4 - k} \tag{3-19}$$

3-3 ASYMPTOTIC THEOREMS

In the preceding section, we showed that if the probability $P(\mathscr{A})$ of an event $\mathscr{A}$ of a certain experiment equals p and the experiment is repeated n times, then the probability that $\mathscr{A}$ occurs k times in any order is given by (3-13) and the probability that k is between k_1 and k_2 by (3-18). In this section, we develop simple approximate formulas for evaluating these probabilities.

Gaussian functions. In the following and throughout the book we use extensively the *normal* or *gaussian* function

$$\mathfrak{g}(x) = \frac{1}{\sqrt{2\pi}} e^{-x^2/2} \tag{3-20}$$

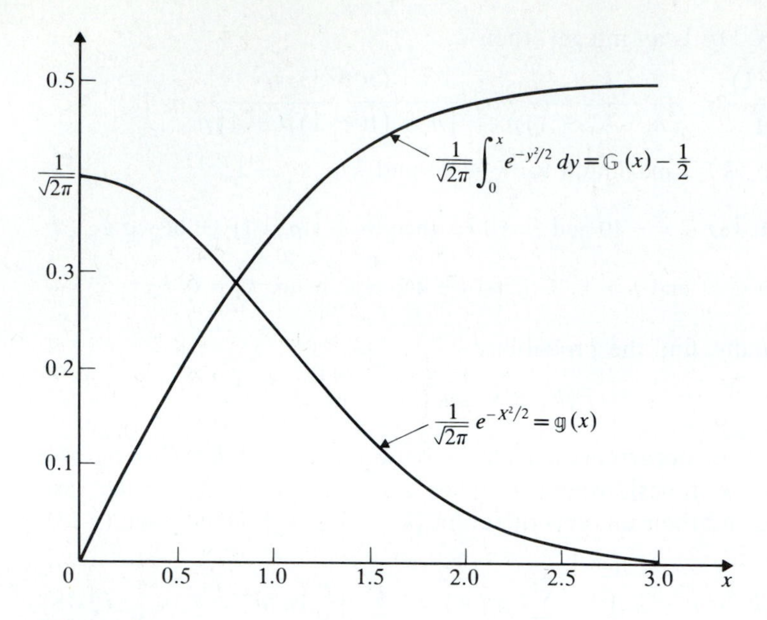

FIGURE 3-4

and its integral (see Fig. 3-4 and Table 3-1).

$$\mathbb{G}(x) = \int_{-\infty}^{x} \mathbb{g}(y)\, dy = \frac{1}{\sqrt{2\pi}} \int_{-\infty}^{x} e^{-y^2/2}\, dy \tag{3-21}$$

As is well known

$$\int_{-\infty}^{\infty} e^{-\alpha x^2}\, dx = \sqrt{\frac{\pi}{\alpha}} \tag{3-22}$$

From this it follows that

$$\mathbb{G}(\infty) = \frac{1}{\sqrt{2\pi}} \int_{-\infty}^{\infty} e^{-x^2/2}\, dx = 1 \tag{3-23}$$

Since $\mathbb{g}(-x) = \mathbb{g}(x)$, we conclude that

$$\mathbb{G}(-x) = 1 - \mathbb{G}(x) \tag{3-24}$$

With a change of variables, (3-21) yields

$$\frac{1}{a\sqrt{2\pi}} \int_{x_1}^{x_2} e^{-(x-b)^2/2a^2}\, dx = \mathbb{G}\left(\frac{x_2 - b}{a}\right) - \mathbb{G}\left(\frac{x_1 - b}{a}\right) \tag{3-25}$$

for any a and b.

TABLE 3-1

$$\text{erf } x = \frac{1}{\sqrt{2\pi}}\int_0^x e^{-y^2/2}\,dy = \mathbb{G}(x) - \frac{1}{2}$$

x	erf x	x	erf x	x	erf x	x	erf x
0.05	0.01994	0.80	0.28814	1.55	0.43943	2.30	0.48928
0.10	0.03983	0.85	0.30234	1.60	0.44520	2.35	0.49061
0.15	0.05962	0.90	0.31594	1.65	0.45053	2.40	0.49180
0.20	0.07926	0.95	0.32894	1.70	0.45543	2.45	0.49286
0.25	0.09871	1.00	0.34134	1.75	0.45994	2.50	0.49379
0.30	0.11791	1.05	0.35314	1.80	0.46407	2.55	0.49461
0.35	0.13683	1.10	0.36433	1.85	0.46784	2.60	0.49534
0.40	0.15542	1.15	0.37493	1.90	0.47128	2.65	0.49597
0.45	0.17364	1.20	0.38493	1.95	0.47441	2.70	0.49653
0.50	0.19146	1.25	0.39435	2.00	0.47726	2.75	0.49702
0.55	0.20884	1.30	0.40320	2.05	0.47982	2.80	0.49744
0.60	0.22575	1.35	0.41149	2.10	0.48214	2.85	0.49781
0.65	0.24215	1.40	0.41924	2.15	0.48422	2.90	0.49813
0.70	0.25804	1.45	0.42647	2.20	0.48610	2.95	0.49841
0.75	0.27337	1.50	0.43319	2.25	0.48778	3.00	0.49865

For large x, $\mathbb{G}(x)$ is given approximately by (see Prob. 3-9)

$$\mathbb{G}(x) \simeq 1 - \frac{1}{x}\mathbf{g}(x) \tag{3-26}$$

We note, finally, that $\mathbb{G}(x)$ is often expressed in terms of the *error function*

$$\text{erf } x = \frac{1}{\sqrt{2\pi}}\int_0^x e^{-y^2/2}\,dy = \mathbb{G}(x) - \frac{1}{2}$$

DeMoivre–Laplace Theorem

It can be shown that, if $npq \gg 1$, then

$$\binom{n}{k}p^k q^{n-k} \simeq \frac{1}{\sqrt{2\pi npq}}e^{-(k-np)^2/2npq} \tag{3-27}$$

for k in $\sqrt{npq}$ neighborhood of np. This important approximation, known as the DeMoivre–Laplace theorem, can be stated as an equality in the limit: The ratio of the two sides tends to 1 as $n \to \infty$. The proof is based on *Stirling's formula*

$$n! \simeq n^n e^{-n}\sqrt{2\pi n} \qquad n \to \infty \tag{3-28}$$

The details, however, will be omitted.†

†The proof can be found in Feller, 1957 (see references at the end of the book).

Thus the evaluation of the probability of k successes in n trials, given exactly by (3-13), is reduced to the evaluation of the normal curve

$$\frac{1}{\sqrt{2\pi npq}} e^{-(x-np)^2/2npq} \tag{3-29}$$

for $x = k$.

Example 3-13. A fair coin is tossed 1000 times. Find the probability p_a that heads will show 500 times and the probability p_b that heads will show 510 times.

In this example

$$p = q = 0.5 \qquad n = 1000 \qquad \sqrt{npq} = 5\sqrt{10}$$

(a) If $k = 500$ then $k - np = 0$ and (3-27) yields

$$p_a \simeq \frac{1}{\sqrt{2\pi npq}} = \frac{1}{10\sqrt{5\pi}} = 0.0252$$

(b) If $k = 510$ then $k - np = 10$ and (3-27) yields

$$p_b \simeq \frac{e^{-0.2}}{10\sqrt{5\pi}} = 0.0207$$

As the next example indicates, the approximation (3-27) is satisfactory even for moderate values of n.

Example 3-14. We shall determine $p_n(k)$ for $p = 0.5$, $n = 10$, and $k = 5$.

(a) Exactly from (3-13)

$$p_n(k) = \binom{n}{k} p^k q^{n-k} = \frac{10!}{5!5!} \frac{1}{2^{10}} = 0.246$$

(b) Approximately from (3-27)

$$p_n(k) \simeq \frac{1}{\sqrt{2\pi npq}} e^{-(k-np)^2/2npq} = \frac{1}{\sqrt{5\pi}} = 0.252$$

APPROXIMATE EVALUATION OF $P\{k_1 \le k \le k_2\}$. Using the approximation (3-27), we shall show that

$$\sum_{k=k_1}^{k_2} \binom{n}{k} p^k q^{n-k} \simeq \mathbb{G}\left(\frac{k_2 - np}{\sqrt{npq}}\right) - \mathbb{G}\left(\frac{k_1 - np}{\sqrt{npq}}\right) \tag{3-30}$$

Thus, to find the probability that in n trials the number of occurrences of an event $\mathscr{A}$ is between k_1 and k_2, it suffices to evaluate the tabulated normal function $\mathbb{G}(x)$. The approximation is satisfactory if $npq \gg 1$ and the differences $k_1 - np$ and $k_2 - np$ are of the order of $\sqrt{npq}$.

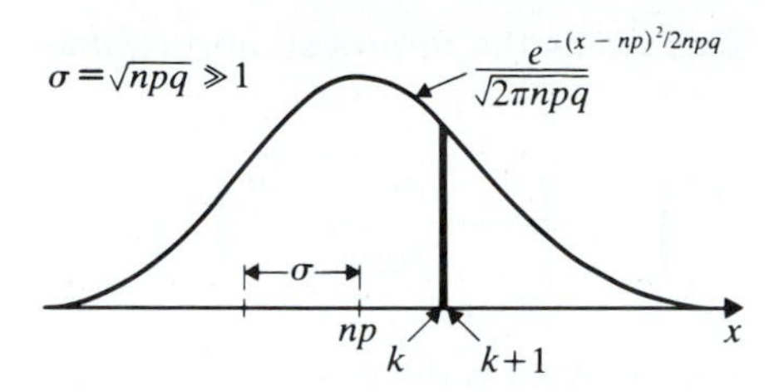

FIGURE 3-5

Proof. Inserting (3-27) into (3-18), we obtain

$$\sum_{k=k_1}^{k_2} \binom{n}{k} p^k q^{n-k} \simeq \frac{1}{\sigma\sqrt{2\pi}} \sum_{k=k_1}^{k_2} e^{-(k-np)^2/2\sigma^2} \tag{3-31}$$

The normal curve is nearly constant in any interval of length 1 because $\sigma^2 = npq \gg 1$ by assumption; hence its area in such an interval equals approximately its ordinate (Fig. 3-5). From this it follows that the right side of (3-31) can be approximated by the integral of the normal curve in the interval (k_1, k_2). This yields

$$\frac{1}{\sigma\sqrt{2\pi}} \sum_{k=k_1}^{k_2} e^{-(k-np)^2/2\sigma^2} \simeq \frac{1}{\sigma\sqrt{2\pi}} \int_{k_1}^{k_2} e^{-(x-np)^2/2\sigma^2}\,dx \tag{3-32}$$

and (3-30) results [see (3-25)].

Error correction. The sum on the left of (3-31) consists of $k_2 - k_1 + 1$ terms. The integral in (3-32) is an approximation of the shaded area of Fig. 3-6*a*, consisting of $k_2 - k_1$ rectangles. If $k_2 - k_1 \gg 1$ the resulting error can be neglected. For moderate values of $k_2 - k_1$, however, the error is no longer negligible. To reduce it, we replace in (3-30) the limits k_1 and k_2 by $k_1 - 1/2$

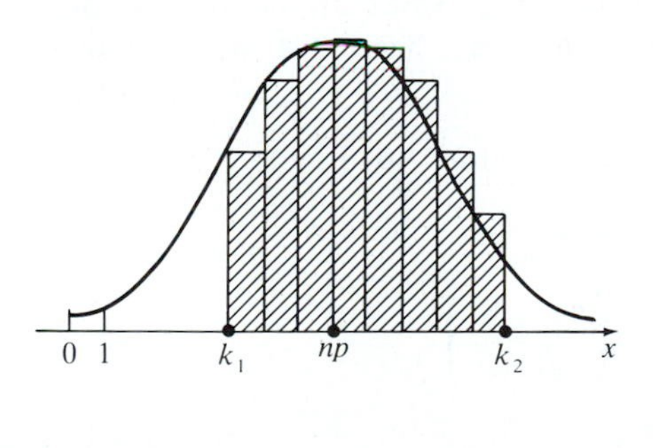

(*a*)

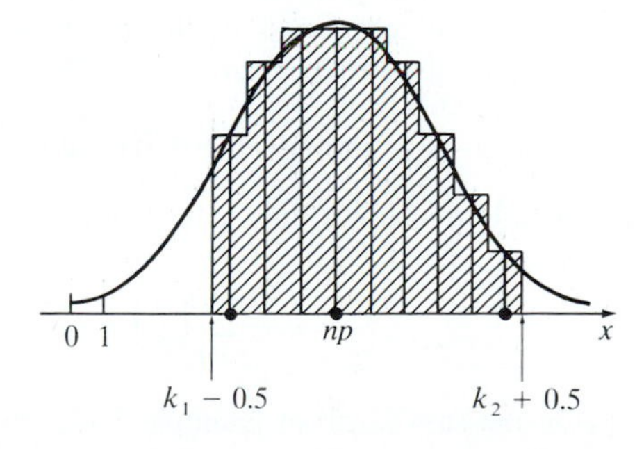

(*b*)

FIGURE 3-6

and $k_2 + 1/2$ respectively (see Fig. 3-6*b*). This yields the improved approximation

$$\sum_{k=k_1}^{k_2} \binom{n}{k} p^k q^{n-k} \simeq \mathbb{G}\left(\frac{k_2 + 0.5 - np}{\sqrt{npq}}\right) - \mathbb{G}\left(\frac{k_1 - 0.5 - np}{\sqrt{npq}}\right) \tag{3-33}$$

Example 3-15. A fair coin is tossed 10 000 times. What is the probability that the number of heads is between 4900 and 5100?

In this problem

$$n = 10\,000 \qquad p = q = 0.5 \qquad k_1 = 4900 \qquad k_2 = 5100$$

Since $(k_2 - np)/\sqrt{npq} = 100/50$ and $(k_1 - np)/\sqrt{npq} = -100/50$, we conclude from (3-30) that the unknown probability equals

$$\mathbb{G}(2) - \mathbb{G}(-2) = 2\mathbb{G}(2) - 1 = 0.9545$$

Example 3-16. Over a period of 12 hours 180 calls are made at random. What is the probability that in a four-hour interval the number of calls is between 50 and 70?

The above can be considered as a problem in repeated trials with $p = 4/12$ the probability that a particular call will occur in the four-hour interval. The probability that k calls will occur in this interval equals [see (3-27)]

$$\binom{180}{k}\left(\frac{1}{3}\right)^k\left(\frac{2}{3}\right)^{180-k} \simeq \frac{1}{4\sqrt{5\pi}} e^{-(k-60)^2/80}$$

and the probability that the number of calls is between 50 and 70 equals [see (3-30)]

$$\sum_{k=50}^{70} \binom{180}{k}\left(\frac{1}{3}\right)^k\left(\frac{2}{3}\right)^{180-k} \simeq \mathbb{G}(\sqrt{2.5}) - \mathbb{G}(-\sqrt{2.5}) \simeq 0.886$$

Note It seems that we cannot use the approximation (3-30) if $k_1 = 0$ because the sum contains values of k that are not in the $\sqrt{npq}$ vicinity of np. However, the corresponding terms are small compared to the terms with k near np; hence the errors of their estimates are also small. Since

$$\mathbb{G}(-np/\sqrt{npq}) = \mathbb{G}(-\sqrt{np/q}) \simeq 0 \qquad \text{for} \quad np/q \gg 1$$

we conclude that if not only $n \gg 1$ but also $np \gg 1$, then

$$\sum_{k=0}^{k_2} \binom{n}{k} p^k q^{n-k} \simeq \mathbb{G}\left(\frac{k_2 - np}{\sqrt{npq}}\right) \tag{3-34}$$

In the sum (3-19) of Example 3-12,

$$np = 1000 \qquad npq = 900 \qquad \frac{k_2 - np}{\sqrt{npq}} = \frac{10}{3}$$

Using (3-34), we obtain

$$\sum_{k=0}^{1100} \binom{10^4}{k}(0.1)^k(0.9)^{10^4-k} \simeq \mathbb{G}\left(\frac{10}{3}\right) = 0.99936$$

We note that the sum of the terms of the above sum from 900 to 1100 equals $2\mathbb{G}(10/3) - 1 \simeq 0.99872$.

The Law of Large Numbers

According to the relative frequency interpretation of probability, if an event $\mathscr{A}$ with $P(\mathscr{A}) = p$ occurs k times in n trials, then $k \simeq np$. In the following, we rephrase this heuristic statement as a limit theorem.

We start with the observation that $k \simeq np$ does not mean that k will be close to np. In fact [(see (3-27)]

$$P\{k = np\} \simeq \frac{1}{\sqrt{2\pi npq}} \to 0 \qquad \text{as} \quad n \to \infty \tag{3-35}$$

As we show in the next theorem, the approximation $k \simeq np$ means that the ratio k/n is close to p in the sense that, for any $\varepsilon > 0$, the probability that $|k/n - p| < \varepsilon$ tends to 1 as $n \to \infty$.

THEOREM. For any $\varepsilon > 0$,

$$P\left\{\left|\frac{k}{n} - p\right| \le \varepsilon\right\} \to 1 \qquad \text{as} \quad n \to \infty \tag{3-36}$$

Proof. The inequality $|k/n - p| < \varepsilon$ means that

$$n(p - \varepsilon) \le k \le n(p + \varepsilon)$$

With $k_1 = n(p - \varepsilon)$ and $k_2 = n(p + \varepsilon)$ we have

$$P\left\{\left|\frac{k}{n} - p\right| \le \varepsilon\right\} = P\{k_1 \le k \le k_2\} = \sum_{k=k_1}^{k_2} \binom{n}{k} p^k q^{n-k}$$

Inserting into (3-30), we obtain

$$P\{k_1 \le k \le k_2\} \simeq \mathbb{G}\left(\frac{k_2 - np}{\sqrt{npq}}\right) - \mathbb{G}\left(\frac{k_1 - np}{\sqrt{npq}}\right) = 2\mathbb{G}\left(\frac{n\varepsilon}{\sqrt{npq}}\right) - 1$$

But $\varepsilon\sqrt{n/pq} \to \infty$ as $n \to \infty$ for any ε. Hence

$$P\left\{\left|\frac{k}{n} - p\right| \le \varepsilon\right\} \simeq 2\mathbb{G}\left(\varepsilon\sqrt{\frac{n}{pq}}\right) - 1 \to 1 \qquad \text{as} \quad n \to \infty \tag{3-37}$$

Example 3-17. Suppose that $p = q = 0.5$ and $\varepsilon = 0.05$. In this case

$$n(p - \varepsilon) = 0.45n \qquad n(p + \varepsilon) = 0.55n \qquad \varepsilon\sqrt{n/pq} = 0.1\sqrt{n}$$

In the table below we show the probability $2\mathbb{G}(0.1\sqrt{n}) - 1$ that k is between $0.45n$ and $0.55n$ for various values of n.

n	100	400	900
$0.1\sqrt{n}$	1	2	3
$2\mathbb{G}(0.1\sqrt{n}) - 1$	0.682	0.954	0.997

Example 3-18. We now assume that $p = 0.6$ and we wish to find n such that the probability that k is between $0.59n$ and $0.61n$ is at least 0.98.

In this case, $p = 0.6$, $q = 0.4$, and $\varepsilon = 0.01$. Hence

$$P\{0.59n \le k \le 0.61n\} \simeq 2\mathbb{G}\left(0.01\sqrt{n/0.24}\right) - 1$$

Thus n must be such that

$$2\mathbb{G}\left(0.01\sqrt{n/0.24}\right) - 1 \ge 0.98$$

From Table 3-1 we see that $\mathbb{G}(x) > 0.99$ if $x > 2.35$. Hence $0.01\sqrt{n/0.24} > 2.35$ yielding $n > 13\,254$.

GENERALIZATION OF BERNOULLI TRIALS. The experiment of repeated trials can be phrased in the following form: The events $\mathscr{A}_1 = \mathscr{A}$ and $\mathscr{A}_2 = \overline{\mathscr{A}}$ of the space $\mathscr{S}$ form a partition and their respective probabilities equal $p_1 = p$ and $p_2 = 1 - p$. In the space $\mathscr{S}^n$, the probability of the event $\{\mathscr{A}_1$ occurs $k_1 = k$ times and $\mathscr{A}_2$ occurs $k_2 = n - k$ times in any order$\}$ equals $p_n(k)$ as in (3-13). We shall now generalize.

Suppose that

$$\mathfrak{A} = [\mathscr{A}_1, \ldots, \mathscr{A}_r]$$

is a partition of $\mathscr{S}$ consisting of the r events $\mathscr{A}_i$ with

$$P(\mathscr{A}_i) = p_i \qquad p_1 + \cdots + p_r = 1$$

We repeat the experiment n times and we denote by $p_n(k_1, \ldots, k_r)$ the probability of the event $\{\mathscr{A}_1$ occurs k_1 times, $\ldots, \mathscr{A}_r$ occurs k_r times in any order$\}$ where

$$k_1 + \cdots + k_r = n$$

We maintain that

$$p_n(k_1, \ldots, k_r) = \frac{n!}{k_1! \cdots k_r!} p_1^{k_1} \cdots p_r^{k_r} \tag{3-38}$$

Proof. Repeated application of (3-11) leads to the conclusion that the number of events of the form $\{\mathscr{A}_1$ occurs k_1 times, $\ldots, \mathscr{A}_r$ occurs k_r times in a specific order$\}$ equals

$$\frac{n!}{k_1! \cdots k_r!}$$

Since the trials are independent, the probability of each such event equals

$$p^{k_1} \cdots p_r^{k_r}$$

and (3-38) results.

Example 3-19. A fair die is rolled 10 times. We shall determine the probability that f_1 shows three times, and "even" shows six times.

In this case

$$\mathscr{A}_1 = \{f_1\} \qquad \mathscr{A}_2 = \{f_2, f_4, f_6\} \qquad \mathscr{A}_3 = \{f_3, f_5\}$$

Clearly,

$$p_1 = \tfrac{1}{6} \qquad p_2 = \tfrac{3}{6} \qquad p_3 = \tfrac{2}{6} \qquad k_1 = 3 \qquad k_2 = 6 \qquad k_3 = 1$$

and (3-38) yields

$$p_{10}(3, 6, 1) = \frac{10!}{3!6!1!}\left(\frac{1}{6}\right)^3\left(\frac{1}{2}\right)^6\frac{1}{3} = 0.002$$

DeMoivre–Laplace theorem. We can show as in (3-27) that, if k_i is in the $\sqrt{n}$ vicinity of np_i and n is sufficiently large, then

$$\frac{n!}{k_1! \cdots k_r!}p_1^{k_1} \cdots p_1^{k_r} \simeq \frac{\exp\left\{-\dfrac{1}{2}\left[\dfrac{(k_1 - np_1)^2}{np_1} + \cdots + \dfrac{(k_r - np_r)^2}{np_r}\right]\right\}}{\sqrt{(2\pi n)^{r-1}p_1 \cdots p_r}} \tag{3-39}$$

Equation (3-27) is a special case.

3-4 POISSON THEOREM AND RANDOM POINTS

We have shown in (3-13) that the probability that an event $\mathscr{A}$ occurs k times in n trials equals

$$\frac{n(n-1)\cdots(n-k+1)}{1 \cdot 2 \cdots k}p^k q^{n-k} \tag{3-40}$$

In the following, we obtain an approximate expression for this probability under the assumption that $p \ll 1$. If n is so large that $np \simeq npq \gg 1$, then we can use the DeMoivre–Laplace theorem (3-27). If, however, np is of order of one, (3-27) is no longer valid. In this case, the following approximation can be used: For k of the order of np,

$$\frac{n!}{k!(n-k)!}p^k q^{n-k} \simeq e^{-np}\frac{(np)^k}{k!} \tag{3-41}$$

Indeed, if k is of the order of np, then $k \ll n$ and $kp \ll 1$. Hence

$$n(n-1)\cdots(n-k+1) \simeq n \cdot n \cdots n = n^k$$

$$q = 1 - p \simeq e^{-p} \qquad q^{n-k} \simeq e^{-(n-k)p} \simeq e^{-np}$$

Inserting into (3-40), we obtain (3-41).

The above approximation can be stated as a limit theorem (see Feller, 1957):

POISSON THEOREM. If

$$n \to \infty \qquad p \to 0 \qquad np \to a$$

then

$$\frac{n!}{k!(n-k)!} p^k q^{n-k} \xrightarrow[n\to\infty]{} e^{-a} \frac{a^k}{k!} \tag{3-42}$$

Example 3-20. A system contains 1000 components. Each component fails independently of the others and the probability of its failure in one month equals 10^{-3}. We shall find the probability that the system will function (i.e., no component will fail) at the end of one month.

This can be considered as a problem in repeated trials with $p = 10^{-3}$, $n = 10^3$, and $k = 0$. Hence [see (3-15)]

$$P\{k = 0\} = q^n = 0.999^{1000}$$

Since $np = 1$, the approximation (3-41) yields

$$P\{k = 0\} \simeq e^{-np} = e^{-1} = 0.368$$

Applying (3-41) to the sum in (3-18), we obtain the following approximation for the probability that the number k of occurrences of $\mathscr{A}$ is between k_1 and k_2:

$$P\{k_1 \le k \le k_2\} \simeq e^{-np} \sum_{k=k_1}^{k_2} \frac{(np)^k}{k!} \tag{3-43}$$

Example 3-21. An order of 3000 parts is received. The probability that a part is defective equals 10^{-3}. We wish to find the probability $P\{k > 5\}$ that there will be more than five defective parts.

Clearly,

$$P\{k > 5\} = 1 - P\{k \le 5\}$$

With $np = 3$, (3-43) yields

$$P\{k \le 5\} = e^{-3} \sum_{k=0}^{5} \frac{3^k}{k!} = 0.916$$

Hence

$$P\{k > 5\} = 0.084$$

Generalization of Poisson theorem. Suppose that $\mathscr{A}_1, \ldots, \mathscr{A}_{m+1}$ are the $m + 1$ events of a partition with $P\{\mathscr{A}_i\} = p_i$. Reasoning as in (3-42), we can show that if $np_i \to a_i$ for $i \le m$, then

$$\frac{n!}{k_1! \cdots k_{m+1}!} p_1^{k_1} \cdots p_{m+1}^{k_{m+1}} \xrightarrow[n \to \infty]{} \frac{e^{-a_1} a_1^{k_1}}{k_1!} \cdots \frac{e^{-a_m} a_m^{k_m}}{k_m!} \tag{3-44}$$

Random Poisson Points

An important application of Poisson's theorem is the approximate evaluation of (3-15) as T and n tend to ∞. We repeat the problem: We place at random n points in the interval $(-T/2, T/2)$ and we denote by $P\{k \text{ in } t_a\}$ the probability that k of these points will lie in an interval (t_1, t_2) of length $t_2 - t_1 = t_a$. As we have shown in (3-15)

$$P\{k \text{ in } t_a\} = \binom{n}{k} p^k q^{n-k} \qquad \text{where} \quad p = \frac{t_a}{T} \tag{3-45}$$

We now assume that $n \gg 1$ and $t_a \ll T$. Applying (3-41), we conclude that

$$P\{k \text{ in } t_a\} \simeq e^{-nt_a/T} \frac{(nt_a/T)^k}{k!} \tag{3-46}$$

for k of the order of nt_a/T.

Suppose, next, that n and T increase indefinitely but the ratio

$$\lambda = n/T$$

remains constant. The result is an infinite set of points covering the entire t axis from $-\infty$ to $+\infty$. As we see from (3-46), the probability that k of these points are in an interval of length t_a is given by

$$P\{k \text{ in } t_a\} = e^{-\lambda t_a} \frac{(\lambda t_a)^k}{k!} \tag{3-47}$$

POINTS IN NONOVERLAPPING INTERVALS. Returning for a moment to the original interval $(-T/2, T/2)$ containing n points, we consider two nonoverlapping subintervals t_a and t_b (Fig. 3-7).

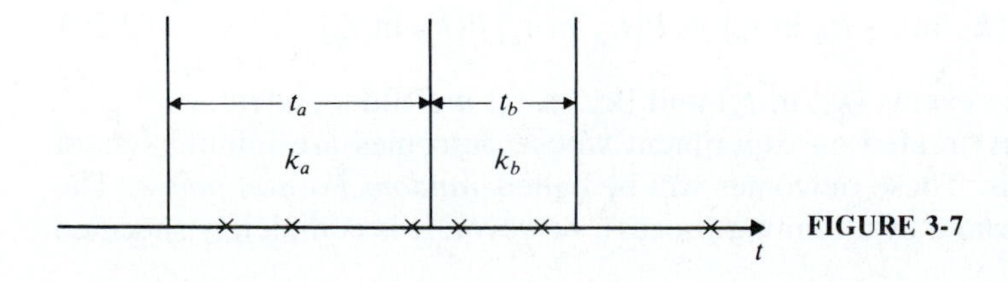

FIGURE 3-7

We wish to determine the probability

$$P\{k_a \text{ in } t_a, k_b \text{ in } t_b\}$$

that k_a of the n points are in interval t_a and k_b in the interval t_b. We maintain that

$$P\{k_a \text{ in } t_a, k_b \text{ in } t_b\} = \frac{n!}{k_a!k_b!k_3!}\left(\frac{t_a}{T}\right)^{k_a}\left(\frac{t_b}{T}\right)^{k_b}\left(1-\frac{t_a}{T}-\frac{t_b}{T}\right)^{k_3} \qquad (3\text{-}48)$$

where $k_3 = n - k_a - k_b$.

Proof. The above can be considered as a generalized Bernoulli trial. The original experiment $\mathscr{S}$ is the random selection of a single point in the interval $(-T/2, T/2)$. In this experiment, the events $\mathscr{A}_1 = \{\text{the point is in } t_a\}$, $\mathscr{A}_2 = \{\text{the point is in } t_b\}$, and $\mathscr{A}_3 = \{\text{the point is outside the intervals } t_a \text{ and } t_a\}$ form a partition and

$$P(\mathscr{A}_1) = \frac{t_a}{T} \qquad P(\mathscr{A}_2) = \frac{t_b}{T} \qquad P(\mathscr{A}_3) = 1 - \frac{t_a}{T} - \frac{t_b}{T}$$

If the experiment $\mathscr{S}$ is performed n times, then the event $\{k_a$ in t_a and k_b in $t_b\}$ will equal the event $\{\mathscr{A}_1$ occurs $k_1 = k_a$ times, $\mathscr{A}_2$ occurs $k_2 = k_b$ times, and $\mathscr{A}_3$ occurs $k_3 = n - k_1 - k_2$ times$\}$. Hence (3-48) follows from (3-38) with $r = 3$.

We note that the events $\{k_a$ in $t_a\}$ and $\{k_b$ in $t_b\}$ *are not* independent because the probability (3-48) of their intersection $\{k_a$ in t_a, k_b in $t_b\}$ does not equal $P\{k_a$ in $t_a\}P\{\{k_b$ in $t_b\}$.

Suppose now that

$$\frac{n}{T} = \lambda \qquad n \to \infty \qquad T \to \infty$$

Since $nt_a/T = \lambda t_a$ and $nt_b/T = \lambda t_b$, we conclude from (3-48) and Prob. 3-16 that

$$P\{k_a \text{ in } t_a, k_b \text{ in } t_b\} = e^{-\lambda t_a}\frac{(\lambda t_a)^{k_a}}{k_a!}e^{-\lambda t_b}\frac{(\lambda t_b)^{k_b}}{k_b!} \qquad (3\text{-}49)$$

From (3-47) and (3-49) it follows that

$$P\{k_a \text{ in } t_a, k_b \text{ in } t_b\} = P\{k_a \text{ in } t_a\}P\{k_b \text{ in } t_b\} \qquad (3\text{-}50)$$

This shows that the events $\{k_a$ in $t_a\}$ and $\{k_b$ in $t_b\}$ *are* independent.

We have thus created an experiment whose outcomes are infinite sets of points on the t axis. These outcomes will be called *random Poisson points*. The experiment was formed by a limiting process; however, it is completely specified in terms of the following two properties:

1. The probability $P\{k_a \text{ in } t_a\}$ that the number of points in an interval (t_1, t_2) equals k_a is given by (3-47).
2. If two intervals (t_1, t_2) and (t_3, t_4) are nonoverlapping, then the events $\{k_a \text{ in } (t_1, t_2)\}$ and $\{k_b \text{ in } (t_3, t_4)\}$ are independent.

The experiment of random Poisson points is fundamental in the theory and the applications of probability. As illustrations we mention electron emission, telephone calls, cars crossing a bridge, and shot noise, among many others.

Example 3-22. Consider two consecutive intervals (t_1, t_2) and (t_2, t_3) with respective lengths t_a and t_b. Clearly, (t_1, t_3) is an interval with length $t_c = t_a + t_b$. We denote by k_a, k_b, and $k_c = k_a + k_b$ the number of points in these intervals. We assume that the number of points k_c in the interval (t_1, t_3) is specified. We wish to find the probability that k_a of these points are in the interval (t_1, t_2). In other words, we wish to find the conditional probability

$$P\{k_a \text{ in } t_a | k_c \text{ in } t_c\}$$

With $k_b = k_c - k_a$, we observe that

$$\{k_a \text{ in } t_a, k_c \text{ in } t_c\} = \{k_a \text{ in } t_a, k_b \text{ in } t_b\}$$

Hence

$$P\{k_a \text{ in } t_a | k_c \text{ in } t_c\} = \frac{P\{k_a \text{ in } t_a, k_b \text{ in } t_b\}}{P\{k_c \text{ in } t_c\}}$$

From (3-47) and (3-49) it follows that the above fraction equals

$$\frac{e^{-\lambda t_a}\left[(\lambda t_a)^{k_a}/k_a!\right] e^{-\lambda t_b}\left[(\lambda t_b)^{k_b}/k_b!\right]}{e^{-\lambda t_c}\left[(\lambda t_c)^{k_c}/k_c!\right]}$$

Since $t_c = t_a + t_b$ and $k_c = k_a + k_b$, the above yields

$$P\{k_a \text{ in } t_a | k_c \text{ in } t_c\} = \frac{k_c!}{k_a! k_b!} \left(\frac{t_a}{t_c}\right)^{k_a} \left(\frac{t_b}{t_c}\right)^{k_b} \tag{3-51}$$

This result has the following useful interpretation: Suppose that we place at random k_c points in the interval (t_1, t_3). As we see from (3-15), the probability that k_a of these points are in the interval (t_1, t_2) equals the right side of (3-51).

Density of Poisson points. The experiment of Poisson points is specified in terms of the parameter λ. We show next that this parameter can be interpreted as the density of the points. Indeed, if the interval $\Delta t = t_2 - t_1$ is sufficiently small, then

$$\lambda \Delta t e^{-\lambda \Delta t} \simeq \lambda \Delta t$$

From this and (3-47) it follows that

$$P\{\text{one point in } (t, t + \Delta t)\} \simeq \lambda \Delta t \tag{3-52}$$

Hence

$$\lambda = \lim_{\Delta t \to 0} \frac{P\{\text{one point in } (t, t + \Delta t)\}}{\Delta t} \tag{3-53}$$

Nonuniform density Using a nonlinear transformation of the t axis, we shall define an experiment whose outcomes are Poisson points specified by a minor modification of property 1.

Suppose that $\lambda(t)$ is a function such that $\lambda(t) \geq 0$ but otherwise arbitrary. We define the experiment of the nonuniform Poisson points as follows:

1. The probability that the number of points in the interval (t_1, t_2) equals k is given by

$$P\{k \text{ in } (t_1, t_2)\} = \exp\left[-\int_{t_1}^{t_2} \lambda(t)\, dt\right] \frac{\left[\int_{t_1}^{t_2} \lambda(t)\, dt\right]^k}{k!} \tag{3-54}$$

2. The same as in the uniform case.

The significance of $\lambda(t)$ as density remains the same. Indeed, with $t_2 - t_1 = \Delta t$ and $k = 1$, (3-54) yields

$$P\{\text{one point in } (t, t + \Delta t)\} \simeq \lambda(t)\, \Delta t \tag{3-55}$$

as in (3-52).

PROBLEMS

3-1. A pair of fair dice is rolled 10 times. Find the probability that "seven" will show at least once.

Answer: $1 - (5/6)^{10}$.

3-2. A coin with $p\{h\} = p = 1 - q$ is tossed n times. Show that the probability that the number of heads is even equals $0.5[1 + (q - p)^n]$.

3-3. (*Hypergeometric series*) A shipment contains K good and $N - K$ defective components. We pick at random $n \leq K$ components and test them. Show that the probability p that k of the tested components are good equals

$$p = \binom{K}{k}\binom{N - K}{n - k} \Big/ \binom{N}{n}$$

3-4. A fair coin is tossed 900 times. Find the probability that the number of heads is between 420 and 465.

Answer: $\mathbb{G}(2) + \mathbb{G}(1) - 1 \simeq 0.819$.

3-5. A fair coin is tossed n times. Find n such that the probability that the number of heads is between $0.49n$ and $0.52n$ is at least 0.9.

Answer: $\mathbb{G}(0.04\sqrt{n}) + \mathbb{G}(0.02\sqrt{n}) \geq 1.9$; hence $n > 4556$.

3-6. If $P(\mathscr{A}) = 0.6$ and k is the number of successes of $\mathscr{A}$ in n trials (a) show that $P\{550 \leq k \leq 650\} = 0.999$, for $n = 1000$. (b) Find n such that $P\{0.59n \leq k \leq 0.61n\} = 0.95$.

3-7. A system has 100 components. The probability that a specific component will fail in the interval (a, b) equals $e^{-a/T} - e^{-b/T}$. Find the probability that in the interval $(0, T/4)$, no more than 100 components will fail.

3-8. A coin is tossed an infinite number of times. Show that the probability that k heads are observed at the nth toss but not earlier equals $\binom{n-1}{k-1}p^k q^{n-k}$.

3-9. Show that

$$\frac{1}{x}\left(1 - \frac{1}{x^2}\right)\mathbb{g}(x) < 1 - \mathbb{G}(x) < \frac{1}{x}\mathbb{g}(x) \qquad x > 0$$

Hint: Prove the following inequalities and integrate from x to ∞:

$$-\frac{d}{dx}\left(\frac{1}{x}e^{-x^2/2}\right) > e^{-x^2/2} \qquad -\frac{d}{dx}\left[\left(\frac{1}{x} - \frac{1}{x^3}\right)e^{-x^2/2}\right] > e^{-x^2/2}$$

3-10. Suppose that in n trials, the probability that an event $\mathscr{A}$ occurs at least once equals P_1. Show that, if $P(\mathscr{A}) = p$ and $pn \ll 1$, then $P_1 \simeq np$.

3-11. The probability that a driver will have an accident in 1 month equals 0.02. Find the probability that in 100 months he will have three accidents.

Answer: About $4e^{-2}/3$.

3-12. A fair die is rolled five times. Find the probability that *one* shows twice, *three* shows twice, and *six* shows once.

3-13. Show that (3-27) is a special case of (3-39) obtained with $r = 2$, $k_1 = k$, $k_2 = n - k$, $p_1 = p$, $p_2 = 1 - p$.

3-14. Players X and Y roll dice alternately starting with X. The player that rolls *eleven* wins. Show that the probability p that X wins equals $18/35$.

Outline: Show that

$$P(\mathscr{A}) = P(\mathscr{A}|\mathscr{M})P(\mathscr{M}) + P(\mathscr{A}|\overline{\mathscr{M}})P(\overline{\mathscr{M}})$$

Set $\mathscr{A} = \{X \text{ wins}\}$, $\mathscr{M} = \{\textit{eleven} \text{ shows at first try}\}$. Note that $P(\mathscr{A}) = p$, $P(\mathscr{A}|\mathscr{M}) = 1$, $P(\mathscr{M}) = 2/36$, $P(\mathscr{A}|\overline{M}) = 1 - p$.

3-15. We place at random n particles in $m > n$ boxes. Find the probability p that the particles will be found in n preselected boxes (one in each box). Consider the following cases: (a) M–B (Maxwell–Boltzmann)—the particles are distinct; all alternatives are possible, (b) B–E (Bose–Einstein)—the particles cannot be distinguished; all alternatives are possible, (c) F–D (Fermi–Dirac)—the particles cannot be distinguished; at most one particle is allowed in a box.

Answer:

	M–B	B–E	F–D
$p =$	$\dfrac{n!}{m^n}$	$\dfrac{n!(m-1)!}{(m+n-1)!}$	$\dfrac{n!(m-n)!}{m!}$

Outline: (a) The number N of all alternatives equals m^n. The number $N_{\mathscr{A}}$ of favorable alternatives equals the $n!$ permutations of the particles in the preselected boxes. (b) Place the $m - 1$ walls separating the boxes in line ending with the n particles. This corresponds to one alternative where all particles are in the last box. All other possibilities are obtained by a permutation of the $n + m - 1$ objects consisting of the $m - 1$ walls and the n particles. All the $(m - 1)!$ permutations of the walls and the $n!$ permutations of the particles count as one alternative. Hence $N = (m + n - 1)!/(m - 1)!n!$ and $N_{\mathscr{A}} = 1$. (*c*) Since the particles are not distinguishable, N equals the number of ways of selecting n out of m objects: $N = \binom{m}{n}$ and $N_{\mathscr{A}} = 1$.

3-16. Reasoning as in (3-41), show that, if

$$k_1 + k_2 + k_3 = n \qquad p_1 + p_2 + p_3 = 1 \qquad k_1 p_1 \ll 1 \qquad k_2 p_2 \ll 1$$

then

$$\frac{n!}{k_1!k_2!k_3!} \simeq \frac{n^{k_1+k_2}}{k_1!k_2!} \qquad p_3^{k_3} \simeq e^{-n(p_1+p_2)}$$

Use the above to justify (3-49).

3-17. We place at random 200 points in the interval (0, 100). Find the probability that in the interval (0, 2) there will be one and only one point (*a*) exactly and (*b*) using the Poisson approximation.

CHAPTER 4

THE CONCEPT OF A RANDOM VARIABLE

4-1 INTRODUCTION

A random variable (abbreviation: RV) is a number $\mathbf{x}(\zeta)$ assigned to every outcome ζ of an experiment. This number could be the gain in a game of chance, the voltage of a random source, the cost of a random component, or any other numerical quantity that is of interest in the performance of the experiment.

Example 4-1. (*a*) In the die experiment, we assign to the six outcomes f_i the numbers $\mathbf{x}(f_i) = 10i$. Thus

$$\mathbf{x}(f_1) = 10, \ldots, \mathbf{x}(f_6) = 60$$

(*b*) In the same experiment, we assign the number 1 to every even outcome and the number 0 to every odd outcome. Thus

$$\mathbf{x}(f_1) = \mathbf{x}(f_3) = \mathbf{x}(f_5) = 0 \qquad \mathbf{x}(f_2) = \mathbf{x}(f_4) = \mathbf{x}(f_6) = 1$$

THE MEANING OF A FUNCTION. An RV is a function whose domain is the set $\mathscr{S}$ of experimental outcomes. To clarify further this important concept, we review briefly the notion of a function. As we know, a function $x(t)$ is a rule of correspondence between values of t and x. The values of the independent variable t form a set $\mathscr{S}_t$ on the t axis called the *domain* of the function and the values of the dependent variable x form a set $\mathscr{S}_x$ on the x axis called the *range* of the function. The rule of correspondence between t and x could be a curve, a table, or a formula, for example, $x(t) = t^2$.

The notation $x(t)$ used to represent a function is ambiguous: It might mean either the particular number $x(t)$ corresponding to a specific t, or the function $x(t)$, namely, the rule of correspondence between any t in $\mathscr{S}_t$ and the corresponding x in $\mathscr{S}_x$. To distinguish between these two interpretations, we shall denote the latter by x, leaving its dependence on t understood.

The definition of a function can be phrased as follows: We are given two sets of numbers $\mathscr{S}_t$ and $\mathscr{S}_x$. To every $t \in \mathscr{S}_t$ we assign a number $x(t)$ belonging to the set $\mathscr{S}_x$. This leads to the following generalization: We are given two sets of objects $\mathscr{S}_\alpha$ and $\mathscr{S}_\beta$ consisting of the elements α and β respectively. We say that β is a function of α if to every element of the set $\mathscr{S}_\alpha$ we make correspond an element β of the set $\mathscr{S}_\beta$. The set $\mathscr{S}_\alpha$ is the domain of the function and the set $\mathscr{S}_\beta$ its range.

Suppose, for example, that $\mathscr{S}_\alpha$ is the set of children in a community and $\mathscr{S}_\beta$ the set of their fathers. The pairing of a child with his or her father is a function.

We note that to a given α there corresponds a single $\beta(\alpha)$. However, more than one element from $\mathscr{S}_\alpha$ might be paired with the same β (a child has only one father but a father might have more than one child). In Example 4-1*b*, the domain of the function consists of the six faces of the die. Its range, however, has only two elements, namely, the numbers 0 and 1.

The Random Variable

We are given an experiment specified by the space $\mathscr{S}$, the field of subsets of $\mathscr{S}$ called events, and the probability assigned to these events. To every outcome ζ of this experiment, we assign a number $\mathbf{x}(\zeta)$. We have thus created a function $\mathbf{x}$ with domain the set $\mathscr{S}$ and range a set of numbers. This function is called random-variable if it satisfies certain mild conditions to be soon given.

All random variables will be written in boldface letters. The symbol $\mathbf{x}(\zeta)$ will indicate the number assigned to the specific outcome ζ and the symbol $\mathbf{x}$ will indicate the rule of correspondence between any element of $\mathscr{S}$ and the number assigned to it. Example 4-1*a*, $\mathbf{x}$ is the table pairing the six faces of the die with the six numbers $10, \ldots, 60$. The domain of this function is the set $\mathscr{S} = \{f_1, \ldots, f_6\}$ and its range is the set of the above six numbers. The expression $\mathbf{x}(f_2)$ is the number 20.

Events generated by random variables. In the study of RVs, questions of the following form arise: What is the probability that the RV $\mathbf{x}$ is less than a given number x, or what is the probability that $\mathbf{x}$ is between the numbers x_1 and x_2. If, for example, the RV is the height of a person, we might want the probability that it will not exceed certain bounds. As we know, probabilities are assigned only to events; therefore, in order to answer such questions, we should be able to express the various conditions imposed on $\mathbf{x}$ as events.

We start with the meaning of the notation

$$\{\mathbf{x} \leq x\}$$

This notation represents a subset of $\mathscr{S}$ consisting of all outcomes ζ such that $\mathbf{x}(\zeta) \le x$. We elaborate on its meaning: Suppose that the RV $\mathbf{x}$ is specified by a table. At the left column we list all elements ζ_i of $\mathscr{S}$ and at the right the corresponding values (numbers) $\mathbf{x}(\zeta_i)$ of $\mathbf{x}$. Given an arbitrary number x, we find all numbers $\mathbf{x}(\zeta_i)$ that do not exceed x. The corresponding elements ζ_i on the left column form the set $\{\mathbf{x} \le x\}$. Thus $\{\mathbf{x} \le x\}$ is not a set of numbers but *a set of experimental outcomes*.

The meaning of

$$\{x_1 \le \mathbf{x} \le x_2\}$$

is similar. It represents a subset of $\mathscr{S}$ consisting of all outcomes ζ such that $x_1 \le \mathbf{x}(\zeta) \le x_2$ where x_1 and x_2 are two given numbers.

The notation

$$\{\mathbf{x} = x\}$$

is a subset of $\mathscr{S}$ consisting of all outcomes ζ such that $\mathbf{x}(\zeta) = x$.

Finally, if R is a set of numbers on the x axis, then

$$\{\mathbf{x} \in R\}$$

represents the subset of $\mathscr{S}$ consisting of all outcomes ζ such that $\mathbf{x}(\zeta) \in R$.

Example 4-2. We shall illustrate the above with the RV $\mathbf{x}(f_i) = 10i$ of the die experiment (Fig. 4-1).

The set $\{\mathbf{x} \le 35\}$ consists of the elements f_1, f_2, f_3 because $\mathbf{x}(f_i) \le 35$ only if $i = 1, 2$, or 3.

The set $\{\mathbf{x} \le 5\}$ is empty because there is no outcome such that $\mathbf{x}(f_i) \le 5$.

The set $\{20 \le \mathbf{x} \le 35\}$ consists of the elements f_2 and f_3 because $20 \le \mathbf{x}(f_i) \le 35$ only if $i = 2$ or 3.

The set $\{\mathbf{x} = 40\}$ consists of the element f_4 because $\mathbf{x}(f_i) = 40$ only if $i = 4$.

Finally, $\{\mathbf{x} = 35\}$ is the empty set because there is no experimental outcome such that $\mathbf{x}(f_i) = 35$.

Note In the applications, we are interested in the probability that an RV $\mathbf{x}$ takes values in a certain region R of the x axis. This requires that the set $\{\mathbf{x} \in R\}$ be an event. As we noted in Sec. 2-2, that is not always possible. However, if $\{\mathbf{x} \le x\}$ is an event for every x and R is a countable union and intersection of intervals, then $\{\mathbf{x} \in R\}$ is also an event. In

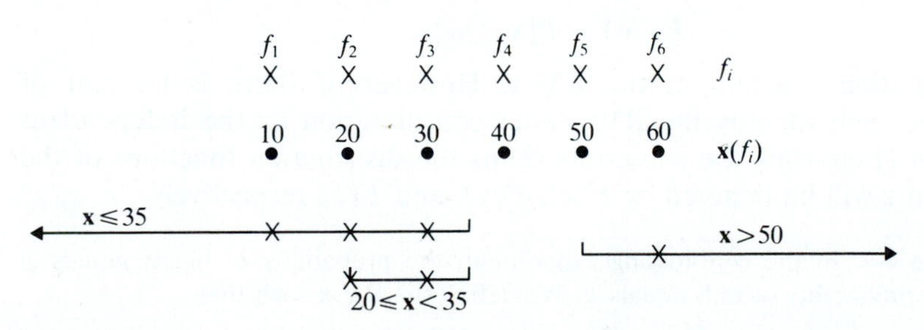

FIGURE 4-1

the definition of RVs we shall assume, therefore, that the set $\{\mathbf{x} \le x\}$ is an event. This mild restriction is mainly of mathematical interest.

We conclude with a formal definition of an RV.

DEFINITION. An RV $\mathbf{x}$ is a process of assigning a number $\mathbf{x}(\zeta)$ to every outcome ζ. The resulting function must satisfy the following two conditions but is otherwise arbitrary:

I. The set $\{\mathbf{x} \le x\}$ is an event for every x.
II. The probabilities of the events $\{\mathbf{x} = \infty\}$ and $\{\mathbf{x} = -\infty\}$ equal 0:

$$P\{\mathbf{x} = \infty\} = 0 \qquad P\{\mathbf{x} = -\infty\} = 0$$

The second condition states that, although we allow $\mathbf{x}$ to be $+\infty$ or $-\infty$ for some outcomes, we demand that these outcomes form a set with zero probability.

A *complex* RV $\mathbf{z}$ is a sum

$$\mathbf{z} = \mathbf{x} + j\mathbf{y}$$

where $\mathbf{x}$ and $\mathbf{y}$ are real RVs. Unless otherwise stated, it will be assumed that all RVs are real.

4-2 DISTRIBUTION AND DENSITY FUNCTIONS

The elements of the set $\mathscr{S}$ that are contained in the event $\{\mathbf{x} \le x\}$ change as the number x takes various values. The probability $P\{\mathbf{x} \le x\}$ of the event $\{\mathbf{x} \le x\}$ is, therefore, a number that depends on x. This number is denoted by $F_x(x)$ and is called the (*cumulative*) *distribution function* of the RV $\mathbf{x}$.

DEFINITION. The distribution function of the RV $\mathbf{x}$ is the function

$$F_x(x) = P\{\mathbf{x} \le x\} \tag{4-1}$$

defined for every x from $-\infty$ to ∞.

The distribution functions of the RVs $\mathbf{x}$, $\mathbf{y}$, and $\mathbf{z}$ are denoted by $F_x(x)$, $F_y(y)$, and $F_z(z)$ respectively. In this notation, the variables x, y, and z can be identified by any letter. We could, for example, use the notation $F_x(w)$, $F_y(w)$, and $F_z(w)$ to represent the above functions. Specifically,

$$F_x(w) = P\{\mathbf{x} \le w\}$$

is the distribution function of the RV $\mathbf{x}$. However, if there is no fear of ambiguity, we shall identify the RVs under consideration by the independent variable in (4-1) omitting the subscripts. Thus the distribution functions of the RVs $\mathbf{x}$, $\mathbf{y}$, and $\mathbf{z}$ will be denoted by $F(x)$, $F(y)$, and $F(z)$ respectively.

Example 4-3. In the coin-tossing experiment, the probability of heads equals p and the probability of tails equals q. We define the RV $\mathbf{x}$ such that

$$\mathbf{x}(h) = 1 \qquad \mathbf{x}(t) = 0$$

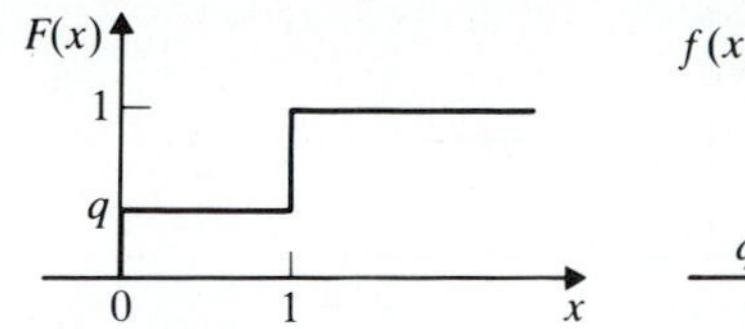

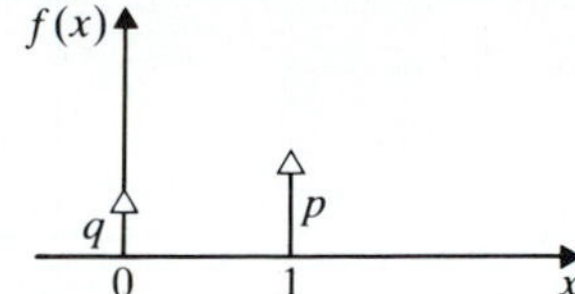

FIGURE 4-2

We shall find its distribution function $F(x)$ for every x from $-\infty$ to ∞.

If $x \geq 1$, then $\mathbf{x}(h) = 1 \leq x$ and $\mathbf{x}(t) = 0 \leq x$. Hence (Fig. 4-2)

$$F(x) = P\{\mathbf{x} \leq x\} = P\{h, t\} = 1 \qquad x \geq 1$$

If $0 \leq x < 1$, then $\mathbf{x}(h) = 1 > x$ and $\mathbf{x}(t) = 0 \leq x$. Hence

$$F(x) = P\{\mathbf{x} \leq x\} = P\{t\} = q \qquad 0 \leq x < 1$$

If $x < 0$, then $\mathbf{x}(h) = 1 > x$ and $\mathbf{x}(t) = 0 > x$. Hence

$$F(x) = P\{\mathbf{x} \leq x\} = P\{\emptyset\} = 0 \qquad x < 0$$

Example 4-4. In the die experiment of Example 4-2, the RV $\mathbf{x}$ is such that $\mathbf{x}(f_i) = 10i$. If the die is fair, then the distribution function of $\mathbf{x}$ is a staircase function as in Fig. 4-3.

We note, in particular, that

$$F(100) = P\{\mathbf{x} \leq 100\} = P(\mathscr{S}) = 1$$

$$F(35) = P\{\mathbf{x} \leq 35\} = P\{f_1, f_2, f_3\} = \tfrac{3}{6}$$

$$F(30.01) = P\{\mathbf{x} \leq 30.01\} = P\{f_1, f_2, f_3\} = \tfrac{3}{6}$$

$$F(30) = P\{\mathbf{x} \leq 30\} = P\{f_1, f_2, f_3\} = \tfrac{3}{6}$$

$$F(29.99) = P\{\mathbf{x} \leq 29.99\} = P\{f_1, f_2\} = \tfrac{2}{6}$$

Example 4-5. A telephone call occurs at random in the interval (0, 1). In this experiment, the outcomes are time distances t between 0 and 1 and the probability that t is between t_1 and t_2 is given by

$$P\{t_1 \leq t \leq t_2\} = t_2 - t_1$$

We define the RV $\mathbf{x}$ such that

$$\mathbf{x}(t) = t \qquad 0 \leq t \leq 1$$

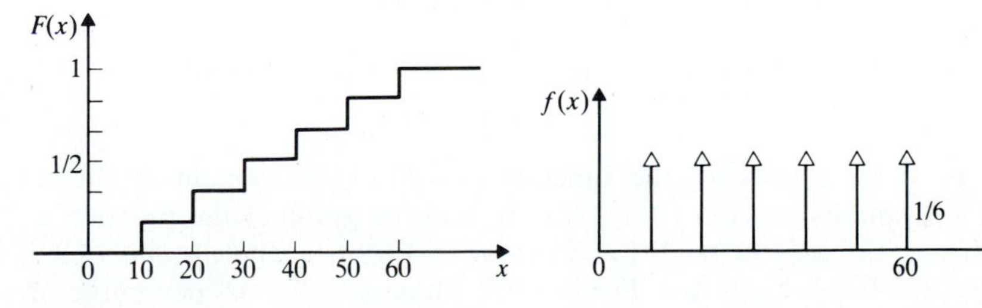

FIGURE 4-3

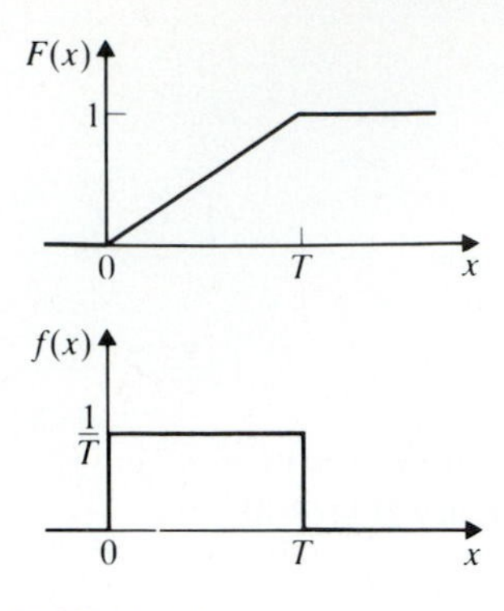

FIGURE 4-4

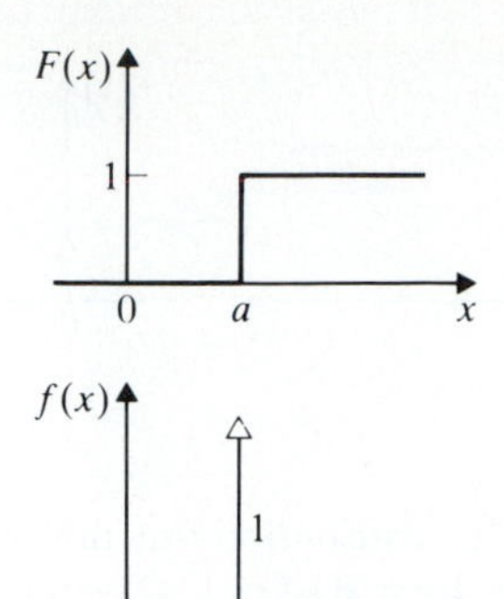

FIGURE 4-5

Thus the variable t has a double meaning: It is the outcome of the experiment and the corresponding value $\mathbf{x}(t)$ of the RV $\mathbf{x}$. We shall show that the distribution function $F(x)$ of $\mathbf{x}$ is a ramp as in Fig. 4-4.

If $x > 1$, then $\mathbf{x}(t) \leq x$ for every outcome. Hence

$$F(x) = P\{\mathbf{x} \leq x\} = P\{0 \leq t \leq 1\} = P(\mathscr{S}) = 1 \qquad x > 1$$

If $0 \leq x \leq 1$, then $\mathbf{x}(t) \leq x$ for every t in the interval $(0, x)$. Hence

$$F(x) = P\{\mathbf{x} \leq x\} = P\{0 \leq t \leq x\} = x \qquad 0 \leq x \leq 1$$

If $x < 0$, then $\{\mathbf{x} \leq x\}$ is the impossible event because $\mathbf{x}(t) \geq 0$ for every t. Hence

$$F(x) = P\{\mathbf{x} \leq x\} = P\{\emptyset\} = 0 \qquad x < 0$$

Example 4-6. Suppose that an RV $\mathbf{x}$ is such that $\mathbf{x}(\zeta) = a$ for every ζ in $\mathscr{S}$. We shall find its distribution function.

If $x \geq a$, then $\mathbf{x}(\zeta) = a \leq x$ for every ζ. Hence

$$F(x) = P\{\mathbf{x} \leq x\} = P(\mathscr{S}) = 1 \qquad x \geq a$$

If $x < a$, then $\{\mathbf{x} \leq x\}$ is the impossible event because $\mathbf{x}(\zeta) = a$. Hence

$$F(x) = P\{\mathbf{x} \leq x\} = P\{\emptyset\} = 0 \qquad x < a$$

Thus a constant can be interpreted as an RV with distribution function a delayed step $U(x - a)$ as in Fig. 4-5.

Note A complex RV $\mathbf{z} = \mathbf{x} + j\mathbf{y}$ has no distribution function because the inequality $\mathbf{x} + j\mathbf{y} \leq x + jy$ has no meaning. The statistical properties of $\mathbf{z}$ are specified in terms of the *joint distribution* of the RVs $\mathbf{x}$ and $\mathbf{y}$ (see Chap. 6).

Percentiles. The u percentile of an RV $\mathbf{x}$ is the smallest number x_u such that

$$u = P\{\mathbf{x} \leq x_u\} = F(x_u) \tag{4-2}$$

Thus x_u is the inverse of the function $u = F(x)$. Its domain is the interval $0 \leq u \leq 1$, and its range is the x axis. To find the graph of the function x_u, we interchange the axes of the $F(x)$ curve as in Fig. 4-6. The *median* of $\mathbf{x}$ is the smallest number m such that $F(m) = 0.5$. Thus m is the 0.5 percentile of $\mathbf{x}$.

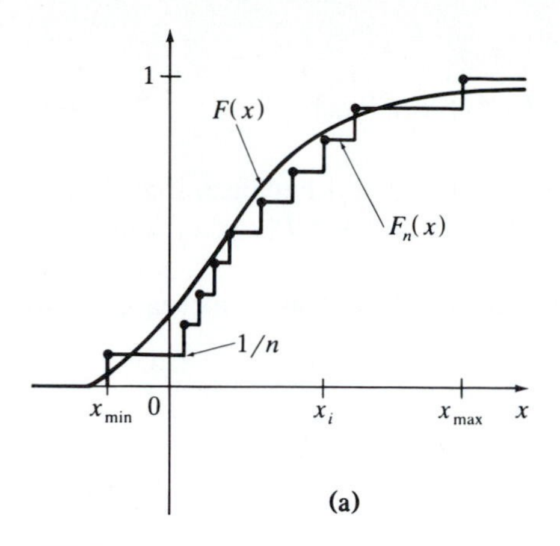

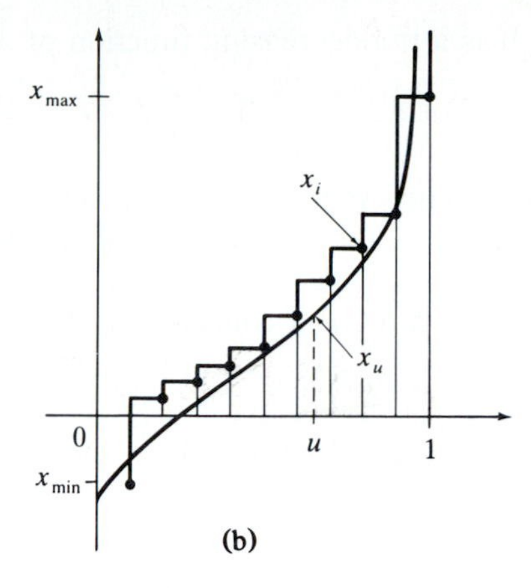

FIGURE 4-6

Frequency interpretation of $F(x)$ **and** x_u. We perform the experiment n times and we observe n values $x_1, \dots, x_n$ of the RV $\mathbf{x}$. We place these numbers on the x axis and we form a staircase function $F_n(x)$ as in Fig 4-6*a*. The steps are located at the points x_i and their height equals $1/n$. They start at the smallest value $x_{\min}$ of x_i, and $F_n(x) = 0$ for $x < x_{\min}$. The function $F_n(x)$ so constructed is called the *empirical distribution* of the RV $\mathbf{x}$.

For a specific x, the number of steps of $F_n(x)$ equals the number n_x of x_is that are smaller than x; thus $F_n(x) = n_x/n$. And since $n_x/n \simeq P\{\mathbf{x} \le x\}$ for large n, we conclude that

$$F_n(x) = \frac{n_x}{n} \to P\{\mathbf{x} \le x\} = F(x) \qquad \text{as} \quad n \to \infty \tag{4-3}$$

The empirical interpretation of the u percentile x_u is the *Quetelet curve* defined as follows: We form n line segments of length x_i and place them vertically in order of increasing length, distance $1/n$ apart. We then form a staircase function with corners at the endpoints of these segments as in Fig. 4-6*b*. The curve so obtained is the empirical interpretation of x_u and it equals the empirical distribution $F_n(x)$ if its axes are interchanged.

Properties of Distribution Functions

In the following, the expressions $F(x^+)$ and $F(x^-)$ will mean the limits

$$F(x^+) = \lim F(x + \varepsilon) \qquad F(x^-) = \lim F(x - \varepsilon) \qquad 0 < \varepsilon \to 0$$

The distribution function has the following properties

1.
$$F(+\infty) = 1 \qquad F(-\infty) = 0$$

Proof.

$$F(+\infty) = P\{\mathbf{x} \le \infty\} = P(\mathscr{S}) = 1 \qquad F(-\infty) = P\{\mathbf{x} = -\infty\} = 0$$

2. It is a nondecreasing function of x:

$$\text{if} \quad x_1 < x_2 \qquad \text{then} \quad F(x_1) \le F(x_2) \tag{4-4}$$

Proof. The event $\{\mathbf{x} \le x_1\}$ is a subset of the event $\{\mathbf{x} \le x_2\}$ because, if $\mathbf{x}(\zeta) \le x_1$ for some ζ, then $\mathbf{x}(\zeta) \le x_2$. Hence [see (2-14)] $P\{\mathbf{x} \le x_1\} \le P\{\mathbf{x} \le x_2\}$ and (4-4) results.

From (4-4) it follows that $F(x)$ increases from 0 to 1 as x increases from $-\infty$ to ∞.

3. $$\text{if} \quad F(x_0) = 0 \qquad \text{then} \quad F(x) = 0 \qquad \text{for every} \quad x \le x_0 \tag{4-5}$$

Proof. It follows from (4-4) because $F(-\infty) = 0$. The above leads to the following conclusion: Suppose that $\mathbf{x}(\zeta) \ge 0$ for every ζ. In this case, $F(0) = P\{\mathbf{x} \le 0\} = 0$ because $\{\mathbf{x} \le 0\}$ is the impossible event. Hence $F(x) = 0$ for every $x \le 0$.

4. $$P\{\mathbf{x} > x\} = 1 - F(x) \tag{4-6}$$

Proof. The events $\{\mathbf{x} \le x\}$ and $\{\mathbf{x} > x\}$ are mutually exclusive and

$$\{\mathbf{x} \le x\} + \{\mathbf{x} > x\} = \mathscr{S}$$

Hence $P\{\mathbf{x} \le x\} + P\{\mathbf{x} > x\} = P(\mathscr{S}) = 1$ and (4-6) results.

5. The function $F(x)$ is continuous from the right:

$$F(x^+) = F(x) \tag{4-7}$$

Proof. It suffices to show that $P\{\mathbf{x} \le x + \varepsilon\} \to F(x)$ as $\varepsilon \to 0$ because $P\{\mathbf{x} \le x + \varepsilon\} = F(x + \varepsilon)$ and $F(x + \varepsilon) \to F(x^+)$ by definition. To prove the above, we must show that the sets $\{\mathbf{x} \le x + \varepsilon\}$ tend to the set $\{\mathbf{x} \le x\}$ as $\varepsilon \to 0$ and to use the axiom IIIa of finite additivity. We omit, however, the details of the proof because we have not introduced limits of sets.

6. $$P\{x_1 < \mathbf{x} \le x_2\} = F(x_2) - F(x_1) \tag{4-8}$$

Proof. The events $\{\mathbf{x} \le x_1\}$ and $\{x_1 < \mathbf{x} \le x_2\}$ are mutually exclusive because $\mathbf{x}(\zeta)$ cannot be less than x_1 and between x_1 and x_2. Furthermore,

$$\{\mathbf{x} \le x_2\} = \{\mathbf{x} \le x_1\} + \{x_1 < \mathbf{x} \le x_2\}$$

Hence

$$P\{\mathbf{x} \le x_2\} = P\{\mathbf{x} \le x_1\} + P\{x_1 < \mathbf{x} \le x_2\}$$

and (4-8) results.

7. $$P\{\mathbf{x} = x\} = F(x) - F(x^-)$$

Proof. Setting $x_1 = x - \varepsilon$ and $x_2 = x$ in (4-8), we obtain

$$P\{x - \varepsilon < \mathbf{x} \le x\} = F(x) - F(x - \varepsilon)$$

and with $\varepsilon \to 0$, (4-9) results.

8. $$P\{x_1 \le \mathbf{x} \le x_2\} = F(x_2) - F(x_1^-) \tag{4-10}$$

Proof. It follows from (4-8) and (4-9) because

$$\{x_1 \le \mathbf{x} \le x_2\} = \{x_1 < \mathbf{x} \le x_2\} + \{\mathbf{x} = x_1\}$$

and the last two events are mutually exclusive.

Statistics We shall say that the statistics of an RV $\mathbf{x}$ are known if we can determine the probability $P\{\mathbf{x} \in R\}$ that $\mathbf{x}$ is in a set R of the x axis consisting of countable unions or intersections of intervals. From (4-1) and the axioms it follows that the statistics of $\mathbf{x}$ are determined in terms of its distribution function.

Continuous, discrete, and mixed types. We shall say that an RV $\mathbf{x}$ is of *continuous type* if its distribution function $F(x)$ is continuous. In this case, $F(x^-) = F(x)$; hence

$$P\{\mathbf{x} = x\} = 0 \tag{4-11}$$

for every x.

We shall say that $\mathbf{x}$ is of *discrete type* if $F(x)$ is a staircase function as in Fig. 4-7. Denoting by x_i by discontinuity points of $F(x)$, we have

$$F(x_i) - F(x_i^-) = P\{\mathbf{x} = x_i\} = p_i \tag{4-12}$$

In this case, the statistics of $\mathbf{x}$ are determined in terms of x_i and p_i. If the points x_i are equidistant, that is, if $x_i = a + bi$, then the RV $\mathbf{x}$ is of *lattice type*.

We shall say that $\mathbf{x}$ is of *mixed type* if $F(x)$ is discontinuous but not a staircase.

Note that if the set $\mathscr{S}$ has finitely many elements, then any RV defined on $\mathscr{S}$ is of discrete type. However, an RV $\mathbf{x}$ might be of discrete type even if $\mathscr{S}$ has infinitely many elements.

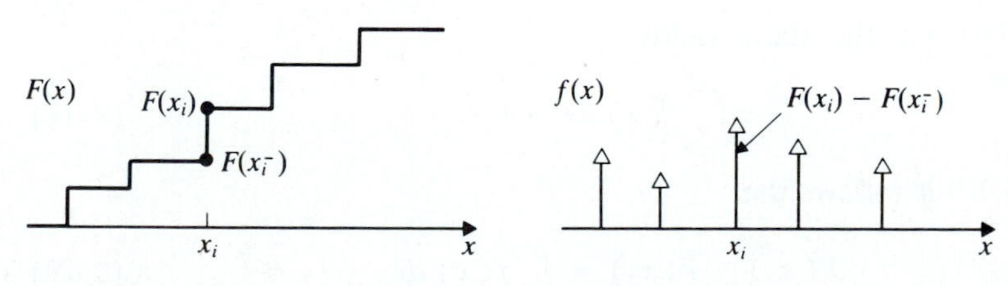

FIGURE 4-7

Example 4-7. If $\mathscr{A}$ is an arbitrary event of $\mathscr{S}$ and $\mathbf{x}_{\mathscr{A}}$ is an RV such that

$$\mathbf{x}_{\mathscr{A}}(\zeta) = \begin{cases} 1 & \zeta \in \mathscr{A} \\ 0 & \zeta \notin \mathscr{A} \end{cases} \tag{4-13}$$

then $\mathbf{x}_{\mathscr{A}}$ is called the *zero–one* RV associated with the event $\mathscr{A}$. Thus

$$\{\mathbf{x}_{\mathscr{A}} = 1\} = \mathscr{A} \qquad \{\mathbf{x}_{\mathscr{A}} = 0\} = \overline{\mathscr{A}}$$

Hence $\mathbf{x}_{\mathscr{A}}$ is of discrete type taking only the two values 0 and 1 with

$$P\{\mathbf{x}_{\mathscr{A}} = 1\} = P(\mathscr{A}) \qquad P\{\mathbf{x}_{\mathscr{A}} = 0\} = 1 - P(\mathscr{A})$$

The space $\mathscr{S}$, however, might have infinitely many elements.

The Density Function

The derivative

$$f(x) = \frac{dF(x)}{dx} \tag{4-14}$$

of $F(x)$ is called the *density function* (known also as the *frequency function*) of the RV $\mathbf{x}$.

If the RV $\mathbf{x}$ is of discrete type taking the values x_i with probabilities p_i, then

$$f(x) = \sum_i p_i \delta(x - x_i) \qquad p_i = P\{\mathbf{x} = x_i\} \tag{4-15}$$

where $\delta(x)$ is the impulse function (Fig. 4-7). The term $p_i \delta(x - x_i)$ is shown as a vertical arrow at $x = x_i$ with length equal to p_i.

In Example 4-2, the RV $\mathbf{x}$ is of discrete type taking the six values $x_1 = 10, \ldots, x_6 = 60$ with $p_i = 1/6$. Hence

$$f(x) = \tfrac{1}{6}[\delta(x - 10) + \delta(x - 20) + \cdots + \delta(x - 60)]$$

PROPERTIES. From the monotonicity of $F(x)$ it follows that

$$f(x) \geq 0 \tag{4-16}$$

Integrating (4-14) from $-\infty$ to x and using the fact that $F(-\infty) = 0$, we obtain

$$F(x) = \int_{-\infty}^{x} f(\xi)\, d\xi \tag{4-17}$$

Since $F(\infty) = 1$, the above yields

$$\int_{-\infty}^{\infty} f(x)\, dx = 1 \tag{4-18}$$

From (4-17) it follows that

$$F(x_2) - F(x_1) = \int_{x_1}^{x_2} f(x)\, dx \tag{4-19}$$

Hence [see (4-8)]

$$P\{x_1 < \mathbf{x} \le x_2\} = \int_{x_1}^{x_2} f(x)\, dx \tag{4-20}$$

If the RV $\mathbf{x}$ is of continuous type, then the set on the left might be replaced by the set $\{x_1 \le \mathbf{x} \le x_2\}$. However, if $F(x)$ is discontinuous at x_1 or x_2, then the integration must include the corresponding impulses of $f(x)$.

With $x_1 = x$ and $x_2 = x + \Delta x$ it follows from (4-20) that, if $\mathbf{x}$ is of continuous type, then

$$P\{x \le \mathbf{x} \le x + \Delta x\} \simeq f(x)\, \Delta x \tag{4-21}$$

provided that Δx is sufficiently small. This shows that $f(x)$ can be defined directly as a limit

$$f(x) = \lim_{\Delta x \to 0} \frac{P\{x \le \mathbf{x} \le x + \Delta x\}}{\Delta x} \tag{4-22}$$

Note As we can see from (4-21), the probability that $\mathbf{x}$ is in a small interval of specified length Δx is proportional to $f(x)$ and it is maximum if that interval contains the point x_m where $f(x)$ is maximum. This point is called the *mode* or the *most likely value* of $\mathbf{x}$. An RV is called *unimodal* if it has a single mode.

Frequency interpretation We denote by Δn_x the number of trials such that

$$x \le \mathbf{x}(\zeta) \le x + \Delta x$$

From (1-1) and (4-21) it follows that

$$f(x)\, \Delta x \simeq \frac{\Delta n_x}{n} \tag{4-23}$$

4-3 SPECIAL CASES

In the preceding sections, we defined RVs starting from known experiments. In this section and throughout the book, we shall often consider RVs having specific distribution or density functions without any reference to a particular probability space.

Existence theorem. To do so, we must show that given a function $f(x)$ or its integral

$$F(x) = \int_{-\infty}^{x} f(\xi)\, d\xi$$

we can construct an experiment and an RV $\mathbf{x}$ with distribution $F(x)$ or density $f(x)$. As we know, these functions must have the following properties:

The function $f(x)$ must be nonnegative and its area must be 1. The function $F(x)$ must be continuous from the right and, as x increases from $-\infty$ to ∞, it must increase monotonically from 0 to 1.

Proof. We consider as our space $\mathscr{S}$ the set of all real numbers, and as its events all intervals on the real line and their unions and intersections. We define the probability of the event $\{x \le x_1\}$ by

$$P\{x \le x_1\} = F(x_1) \tag{4-24}$$

where $F(x)$ is the given function. This specifies the experiment completely (see Sec. 2-2).

The outcomes of our experiment are the real numbers. To define an RV $\mathbf{x}$ on this experiment, we must know its value $\mathbf{x}(x)$ for every x. We define $\mathbf{x}$ such that

$$\mathbf{x}(x) = x \tag{4-25}$$

Thus x is the outcome of the experiment and the corresponding value of the RV $\mathbf{x}$ (see also Example 4-5).

We maintain that the distribution function of $\mathbf{x}$ equals the given $F(x)$. Indeed, the event $\{x \le x_1\}$ consists of all outcomes x such that $\mathbf{x}(x) \le x_1$. Hence

$$P\{\mathbf{x} \le x_1\} = P\{x \le x_1\} = F(x_1) \tag{4-26}$$

and since this is true for every x_1, the theorem is proved.

In the following, we discuss briefly a number of common densities.

Normal. An RV $\mathbf{x}$ is called *normal* or *gaussian* if its density is the normal curve $\mathbb{g}(x)$ [see (3-20)], shifted and scaled

$$f(x) = \frac{1}{\sigma}\,\mathbb{g}\left(\frac{x-\eta}{\sigma}\right) = \frac{1}{\sigma\sqrt{2\pi}}e^{-(x-\eta)^2/2\sigma^2} \tag{4-27}$$

This is a bell-shaped curve, symmetrical about the line $x = \eta$ (Fig. 4-8) and its area equals 1 as it should [see (3-22)]. The corresponding distribution function is given by

$$F(x) = \mathbb{G}\left(\frac{x-\eta}{\sigma}\right) \tag{4-28}$$

where $\mathbb{G}(x)$ is the tabulated integral of $\mathbb{g}(x)$ [see (3-21)].

We shall use the notation

$$N(\eta;\sigma)$$

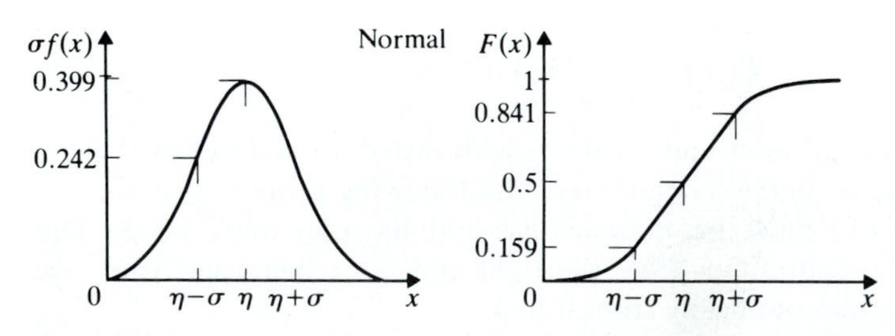

FIGURE 4-8

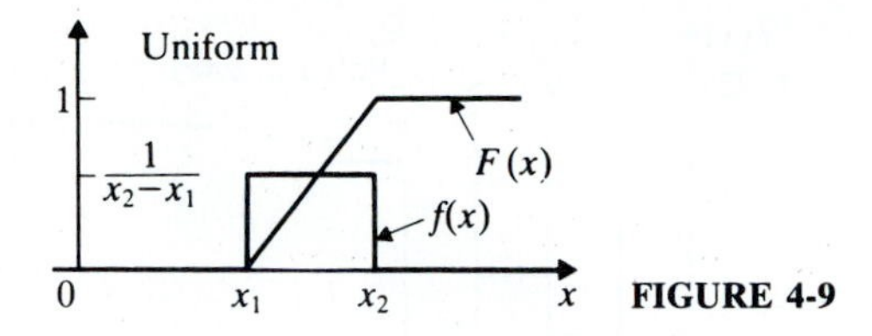

FIGURE 4-9

to indicate that an RV **x** is normal as in (4-27). The significance of the constants η and σ will be given in Sec. 5-4 (η: mean, σ: standard deviation).

Example 4-8. An RV **x** is $N(1000; 50)$. We shall find the probability that **x** is between 900 and 1050. Clearly,

$$P\{900 \le \mathbf{x} \le 1050\} = F(1050) - F(900) = \mathbb{G}(1) - \mathbb{G}(-2)$$

Since

$$\mathbb{G}(-x) = 1 - \mathbb{G}(x) \tag{4-29}$$

we conclude from Table 3-1 that

$$P\{900 \le \mathbf{x} \le 1050\} = \mathbb{G}(1) + \mathbb{G}(2) - 1 = 0.819$$

Uniform. An RV **x** is called uniform between x_1 and x_2 if its density is constant in the interval (x_1, x_2) and 0 elsewhere

$$f(x) = \begin{cases} \dfrac{1}{x_2 - x_1} & x_1 \le x \le x_2 \\ 0 & \text{otherwise} \end{cases} \tag{4-30}$$

The corresponding distribution function is a ramp as in Fig. 4-9.

Example 4-9. A resistor **r** is an RV uniform between 900 and 1100 Ω. We shall find the probability that **r** is between 950 and 1050 Ω.

Since $f(r) = 1/200$ in the interval (900, 1100), (4-20) yields

$$P\{950 \le \mathbf{r} \le 1050\} = \frac{1}{200}\int_{950}^{1050} dr = 0.5$$

Binomial. We say that an RV **x** has a *binomial* distribution of order n if it takes the values $0, 1, \ldots, n$ with

$$P\{\mathbf{x} = k\} = \binom{n}{k} p^k q^{n-k} \qquad p + q = 1 \tag{4-31}$$

Thus **x** is of lattice type and its density is a sum of impulses (Fig. 4-10a)

$$f(x) = \sum_{k=0}^{n} \binom{n}{k} p^k q^{n-k} \delta(x - k) \tag{4-32}$$

The corresponding distribution is a staircase function and in the interval $(0, n)$ it is given by

$$F(x) = \sum_{k=0}^{m} \binom{n}{k} p^k q^{n-k} \qquad m \le x < m + 1 \tag{4-33}$$

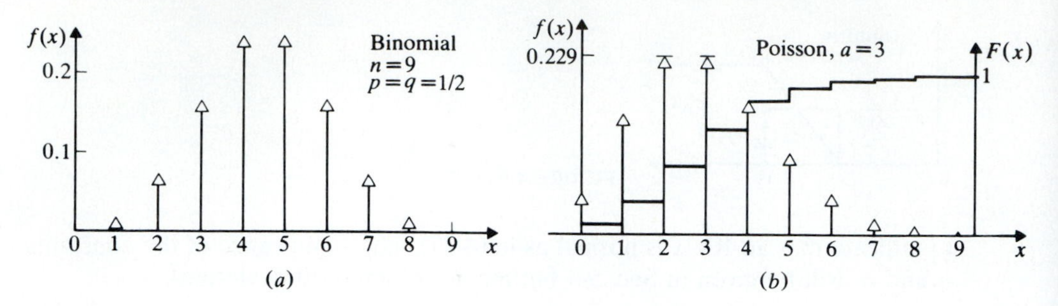

FIGURE 4-10

We note that, if n is large, then [see (3-34)] $F(x)$ is close to an $N(np, \sqrt{npq})$ distribution. In other words,

$$F(x) \simeq \mathbb{G}\left(\frac{x - np}{\sqrt{npq}}\right) \tag{4-34}$$

Example 4-10 Bernoulli trials. In the experiment of the n tosses of a coin, an outcome is a sequence $\zeta_1 \cdots \zeta_n$ of k heads and $n - k$ tails where $k = 0, \ldots, n$. We define the RV $\mathbf{x}$ such that

$$\mathbf{x}(\zeta_1 \cdots \zeta_n) = k$$

Thus $\mathbf{x}$ equals the number of heads. As we know [see (3-13)], the probability that $\mathbf{x} = k$ equals the right side of (4-31). Hence $\mathbf{x}$ has a binomial distribution.

Suppose that the coin is fair and it is tossed $n = 100$ times. We shall find the probability that $\mathbf{x}$ is between 40 and 60. In this case

$$p = q = 0.5 \qquad np = 50 \qquad \sqrt{npq} = 5$$

and (4-34) yields

$$P\{40 \le \mathbf{x} \le 60\} = \mathbb{G}\left(\frac{60 - 50}{5}\right) - \mathbb{G}\left(\frac{40 - 50}{5}\right) = \mathbb{G}(2) - \mathbb{G}(-2) = 0.9545$$

Poisson. An RV $\mathbf{x}$ is *Poisson* distributed with parameter a if it takes the values $0, 1, \ldots, n \ldots$ with

$$P\{\mathbf{x} = k\} = e^{-a}\frac{a^k}{k!} \qquad k = 0, 1, \ldots \tag{4-35}$$

Thus $\mathbf{x}$ is of lattice type with density

$$f(x) = e^{-a}\sum_{k=0}^{\infty}\frac{a^k}{k!}\delta(x - k) \tag{4-36}$$

The corresponding distribution is a staircase function as in Fig. 4-10b.

With $p_k = P\{\mathbf{x} = k\}$, it follows from (4-35) that

$$\frac{p_{k-1}}{p_k} = \frac{e^{-a}a^{k-1}/(k-1)!}{e^{-a}a^k/k!} = \frac{k}{a}$$

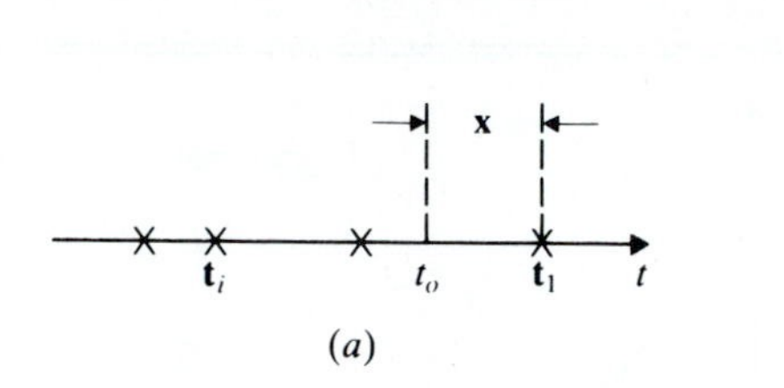

(a)

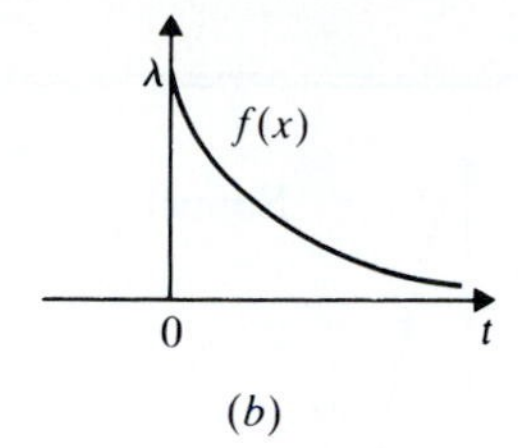

(b)

FIGURE 4-11

If the above ratio is less than 1, that is, if $k < a$, then $p_{k-1} < p_k$. This shows that, as k increases, p_k increases reaching its maximum for $k = [a]$. Hence

if $a < 1$, then p_k is maximum for $k = 0$;

if $a > 1$ but it is not an integer, then p_k increases as k increases, reaching its maximum for $k = [a]$;

if a is an integer, then p_k is maximum for $k = a - 1$ and $k = a$.

Example 4-11 Poisson points. In the Poisson points experiment, an outcome ζ is a set of points $\mathbf{t}_i$ on the t axis.

(a) Given a constant t_o, we define the RV $\mathbf{n}$ such that its value $\mathbf{n}(\zeta)$ equals the number of points $\mathbf{t}_i$ in the interval $(0, t_o)$. Clearly, $\mathbf{n} = k$ means that the number of points in the interval $(0, t_o)$ equals k. Hence [see (3-47)]

$$P\{\mathbf{n} = k\} = e^{-\lambda t_o}\frac{(\lambda t_o)^k}{k!} \tag{4-37}$$

Thus the number of Poisson points in an interval of length t_o is a Poisson distributed RV with parameter $a = \lambda t_o$ where λ is the density of the points.

(b) We denote by $\mathbf{t}_1$ the first random point to the right of the fixed point t_o and we define the RV $\mathbf{x}$ as the distance from t_o to $\mathbf{t}_1$ (Fig. 4-11a). From the definition it follows that $\mathbf{x}(\zeta) \geq 0$ for any ζ. Hence the distribution function of $\mathbf{x}$ is 0 for $x < 0$. We maintain that for $x > 0$ it is given by

$$F(x) = 1 - e^{-\lambda x}$$

Proof. As we know, $F(x)$ equals the probability that $\mathbf{x} \leq x$ where x is a specific number. But $\mathbf{x} \leq x$ means that there is at least one point between t_o and $t_o + x$. Hence $1 - F(x)$ equals the probability p_0 that there are no points in the interval $(t_o, t_o + x)$. And since the length of this interval equals x, (4-37) yields

$$p_0 = e^{-\lambda x} = 1 - F(x)$$

The corresponding density

$$f(x) = \lambda e^{-\lambda x} U(x)$$

is called *exponential* (Fig. 4-11b).

TABLE 4-1

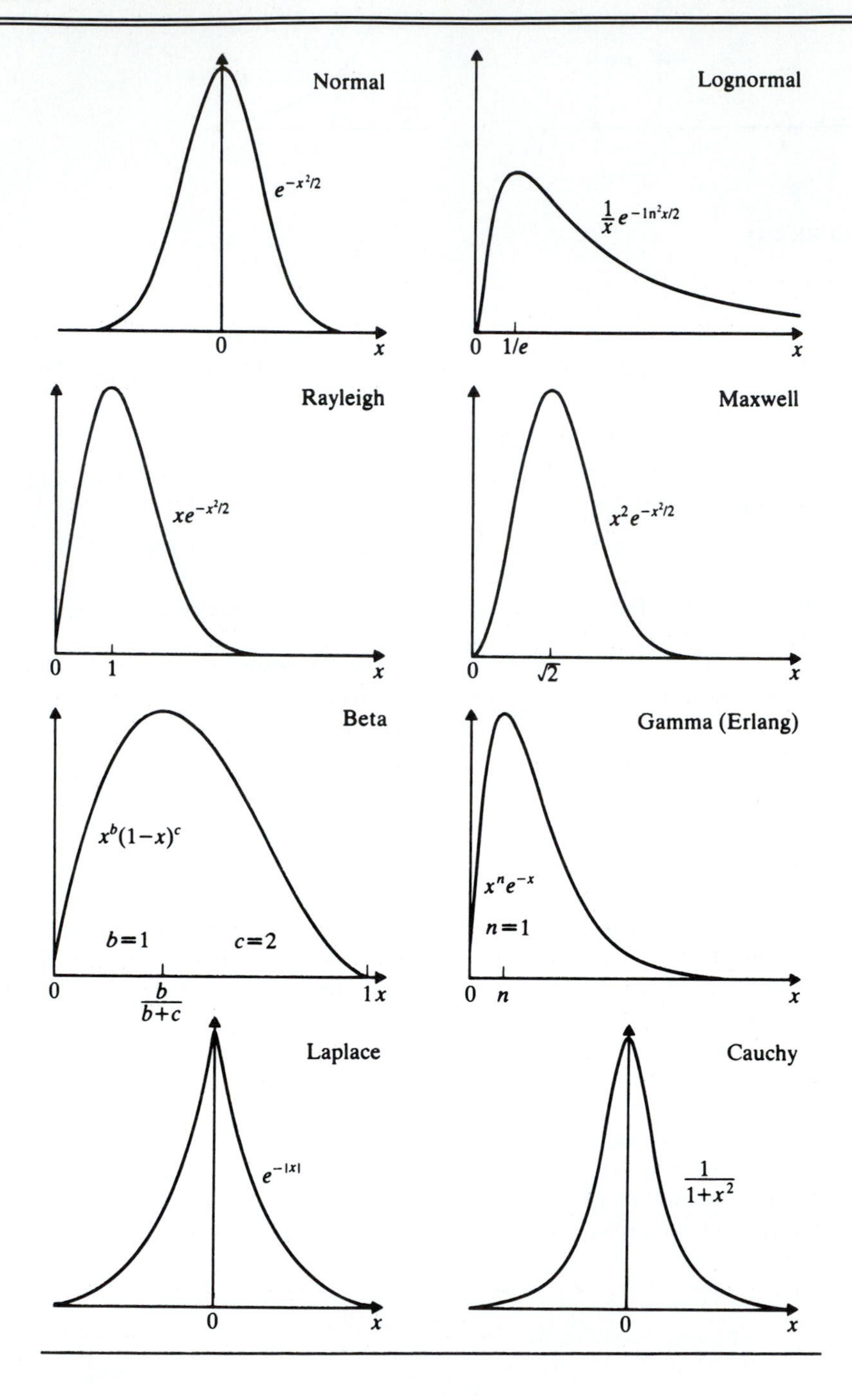
Normal
$e^{-x^2/2}$
0
x
Lognormal
$\frac{1}{x}e^{-\ln^2 x/2}$
0
1/e
x
Rayleigh
$xe^{-x^2/2}$
0
1
x
Maxwell
$x^2e^{-x^2/2}$
0
$\sqrt{2}$
x
Beta
$x^b(1-x)^c$
b=1
c=2
0
$\frac{b}{b+c}$
1
x
Gamma (Erlang)
$x^n e^{-x}$
n=1
0
n
x
Laplace
$e^{-|x|}$
0
x
Cauchy
$\frac{1}{1+x^2}$
0
x

Gamma. An RV $\mathbf{x}$ has a gamma distribution if

$$f(x) = \gamma x^{b-1} e^{-cx} U(x) \qquad \gamma = \frac{c^b}{\Gamma(b)} \tag{4-38}$$

In the above, b and c are positive numbers and

$$\Gamma(b+1) = \int_0^\infty y^b e^{-y}\, dy \qquad b > -1 \tag{4-39}$$

is the *gamma function*. This function is also called the generalized factorial because $\Gamma(b+1) = b\Gamma(b)$. If b is an integer, $\Gamma(n+1) = n\Gamma(n) = \cdots = n!$ because $\Gamma(1) = 1$. Furthermore,

$$\Gamma\left(\frac{1}{2}\right) = \int_0^\infty y^{-1/2} e^{-y}\, dy = 2\int_0^\infty e^{-z^2}\, dz = \sqrt{\pi}$$

The following densities are special cases of (4-38).

Erlang. If $b = n$ is an integer, the *Erlang* density

$$f(x) = \frac{c^n}{(n-1)!} x^{n-1} e^{-cx} U(x)$$

results. With $n = 1$, we obtain the exponential density shown in Fig. 4-11.

Chi-square. For $b = n/2$ and $c = 1/2$, (4-38) yields

$$f(x) = \frac{1}{2^{n/2}\Gamma(n/2)} x^{n/2-1} e^{-x/2} U(x) \tag{4-40}$$

This density is denoted by $\chi^2(n)$ and is called *chi-square* with n degrees of freedom. It is used extensively in statistics.

In Table 4-1, we show a number of common densities. In the formulas of the various curves, a numerical factor is omitted. The omitted factor is determined from (4-18).

4-4 CONDITIONAL DISTRIBUTIONS

We recall that the probability of an event $\mathscr{A}$ assuming $\mathscr{M}$ is given by

$$P(\mathscr{A}|\mathscr{M}) = \frac{P(\mathscr{A}\mathscr{M})}{P(\mathscr{M})} \qquad \text{where} \quad P(\mathscr{M}) \neq 0$$

The *conditional distribution* $F(x|\mathscr{M})$ of an RV $\mathbf{x}$, assuming $\mathscr{M}$ is defined as the conditional probability of the event $\{\mathbf{x} \le x\}$:

$$F(x|\mathscr{M}) = P\{\mathbf{x} \le x|\mathscr{M}\} = \frac{P\{\mathbf{x} \le x, \mathscr{M}\}}{P(\mathscr{M})} \tag{4-41}$$

In the above, $\{\mathbf{x} \le x, \mathscr{M}\}$ is the intersection of the events $\{\mathbf{x} \le x\}$ and $\mathscr{M}$, that is, the event consisting of all outcomes ζ such that $\mathbf{x}(\zeta) \le x$ and $\zeta \in \mathscr{M}$.

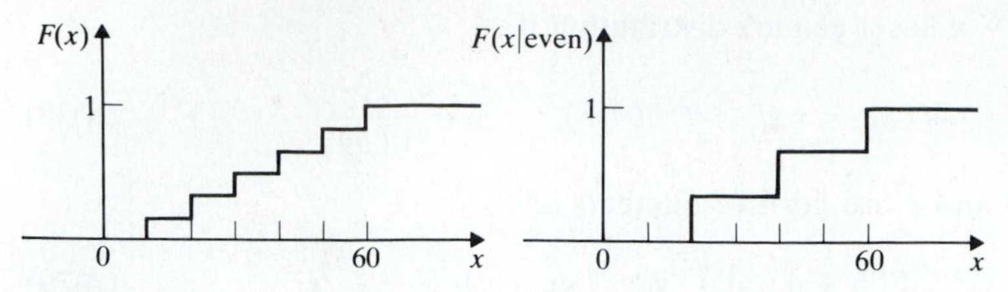

FIGURE 4-12

Thus the definition of $F(x|\mathscr{M})$ is the same as the definition (4-1) of $F(x)$, provided that all probabilities are replaced by conditional probabilities. From this it follows (see Fundamental remark, Sec. 2-3) that $F(x|\mathscr{M})$ has the same properties as $F(x)$. In particular [see (4-3) and (4-8)]

$$F(\infty|\mathscr{M}) = 1 \qquad F(-\infty|\mathscr{M}) = 0 \tag{4-42}$$

$$P\{x_1 < \mathbf{x} \le x_2|\mathscr{M}\} = F(x_2|\mathscr{M}) - F(x_1|\mathscr{M}) = \frac{P\{x_1 < \mathbf{x} \le x_2, \mathscr{M}\}}{P(\mathscr{M})} \tag{4-43}$$

The *conditional density* $f(x|\mathscr{M})$ is the derivative of $F(x|\mathscr{M})$:

$$f(x|\mathscr{M}) = \frac{dF(x|\mathscr{M})}{dx} = \lim_{\Delta x \to 0} \frac{P\{x \le \mathbf{x} \le x + \Delta x|\mathscr{M}\}}{\Delta x} \tag{4-44}$$

This function is nonnegative and its area equals 1.

Example 4-12. We shall determine the conditional $F(x|\mathscr{M})$ of the RV $\mathbf{x}(f_i) = 10i$ of the fair-die experiment (Example 4-4), where $\mathscr{M} = \{f_2, f_4, f_6\}$ is the event "even."

If $x \ge 60$, then $\{\mathbf{x} \le x\}$ is the certain event and $\{\mathbf{x} \le x, \mathscr{M}\} = \mathscr{M}$. Hence (Fig. 4-12)

$$F(x|\mathscr{M}) = \frac{P(\mathscr{M})}{P(\mathscr{M})} = 1 \qquad x \ge 60$$

If $40 \le x < 60$, then $\{\mathbf{x} \le x, \mathscr{M}\} = \{f_2, f_4\}$. Hence

$$F(x|\mathscr{M}) = \frac{P\{f_2, f_4\}}{P(\mathscr{M})} = \frac{2/6}{3/6} \qquad 40 \le x < 60$$

If $20 \le x < 40$, then $\{\mathbf{x} \le x, \mathscr{M}\} = \{f_2\}$. Hence

$$F(x|\mathscr{M}) = \frac{P\{f_2\}}{P(\mathscr{M})} = \frac{1/6}{3/6} \qquad 20 \le x < 60$$

If $x < 20$, then $\{\mathbf{x} \le x, \mathscr{M}\} = \{\emptyset\}$. Hence

$$F(x|\mathscr{M}) = 0 \qquad x < 20$$

To find $F(x|\mathscr{M})$, we must, in general, know the underlying experiment. However, if $\mathscr{M}$ is an event that can be expressed in terms of the RV $\mathbf{x}$, then, for the determination of $F(x|\mathscr{M})$, knowledge of $F(x)$ is sufficient. The following two cases are important illustrations.

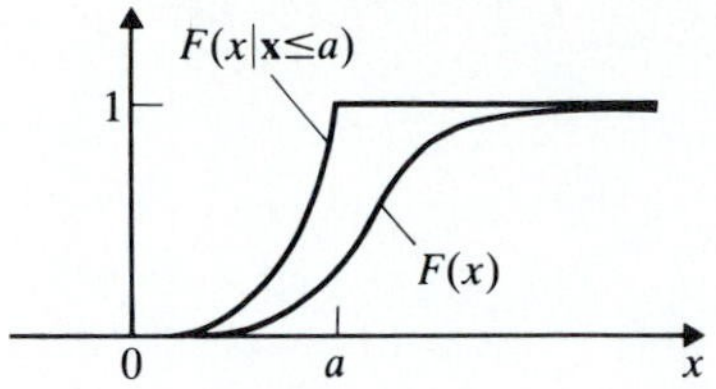

FIGURE 4-13

I. We wish to find the conditional distribution of an RV $\mathbf{x}$ assuming that $\mathbf{x} \le a$ where a is number such that $F(a) \ne 0$. This is a special case of (4-41) with

$$\mathscr{M} = \{\mathbf{x} \le a\}$$

Thus our problem is to find the function

$$F(x|\mathbf{x} \le a) = P\{\mathbf{x} \le x | \mathbf{x} \le a\} = \frac{P\{\mathbf{x} \le x, \mathbf{x} \le a\}}{P\{\mathbf{x} \le a\}}$$

If $x \ge a$, then $\{\mathbf{x} \le x, \mathbf{x} \le a\} = \{\mathbf{x} \le a\}$. Hence (Fig. 4-13)

$$F(x|\mathbf{x} \le a) = \frac{P\{\mathbf{x} \le a\}}{P\{\mathbf{x} \le a\}} = 1 \qquad x \ge a$$

If $x < a$, then $\{\mathbf{x} \le x, \mathbf{x} \le a\} = \{\mathbf{x} \le x\}$. Hence

$$F(x|\mathbf{x} \le a) = \frac{P\{\mathbf{x} \le x\}}{P\{\mathbf{x} \le a\}} = \frac{F(x)}{F(a)} \qquad x < a$$

Differentiating $F(x|\mathbf{x} \le a)$ with respect to x, we obtain the corresponding density: Since $F'(x) = f(x)$, the above yields

$$f(x|\mathbf{x} \le a) = \frac{f(x)}{F(a)} = \frac{f(x)}{\int_{-\infty}^{a} f(x)\,dx} \qquad \text{for} \quad x < a \qquad (4\text{-}45)$$

and it is 0 for $x \ge a$.

II. Suppose now that $\mathscr{M} = \{b < \mathbf{x} \le a\}$. In this case, (4-41) yields

$$F(x|b < \mathbf{x} \le a) = \frac{P\{\mathbf{x} \le x, b < \mathbf{x} \le a\}}{P\{b < \mathbf{x} \le a\}}$$

If $x \ge a$, then $\{\mathbf{x} \le x, b < \mathbf{x} \le a\} = \{b < \mathbf{x} \le a\}$. Hence

$$F(x|b < \mathbf{x} \le a) = \frac{F(a) - F(b)}{F(a) - F(b)} = 1 \qquad x \ge a$$

If $b \le x < a$, then $\{\mathbf{x} \le x, b < \mathbf{x} \le a\} = \{b < \mathbf{x} \le x\}$. Hence

$$F(x|b < \mathbf{x} \le a) = \frac{F(x) - F(b)}{F(a) - F(b)} \qquad b \le x < a$$

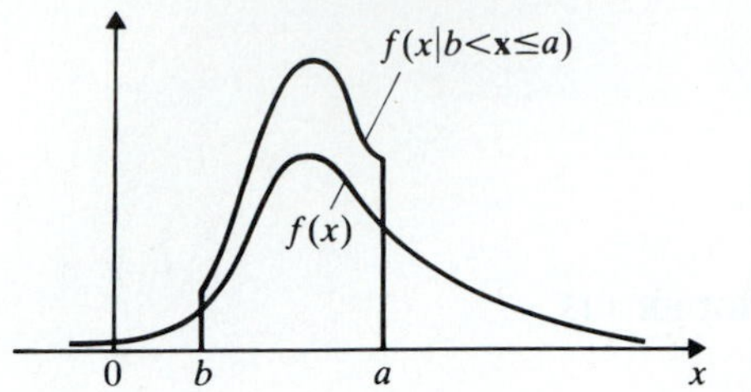

FIGURE 4-14

Finally, if $x < b$, then $\{\mathbf{x} \le x, b < \mathbf{x} \le a\} = \{\emptyset\}$. Hence

$$F(x|b < \mathbf{x} \le a) = 0 \qquad x < b$$

The corresponding density is given by

$$f(x|b < \mathbf{x} \le a) = \frac{f(x)}{F(a) - F(b)} \qquad \text{for} \quad b \le x < a \tag{4-46}$$

and it is 0 otherwise (Fig. 4-14).

Example 4-13. We shall determine the conditional density $f(x|\,|\mathbf{x} - \eta| \le k\sigma)$ of an $N(\eta; \sigma)$ RV. Since

$$P\{|\mathbf{x} - \eta| \le k\sigma\} = P\{\eta - k\sigma \le \mathbf{x} \le \eta + k\sigma\} = \mathbb{G}(k) - \mathbb{G}(-k) = 2\mathbb{G}(k) - 1$$

we conclude from (4-46) that

$$f(x|\,|\mathbf{x} - \eta| \le k\sigma) = \frac{1}{2\mathbb{G}(k) - 1} \frac{e^{-(x-\eta)^2/2\sigma^2}}{\sigma\sqrt{2\pi}}$$

for $\mathbf{x}$ between $\eta - k\sigma$ and $\eta + k\sigma$ and 0 otherwise. This density is called *truncated normal*.

Frequency interpretation In a sequence of n trials, we reject all outcomes ζ such that $\mathbf{x}(\zeta) \le b$ or $\mathbf{x}(\zeta) > a$. In the subsequence of the remaining trials, $F(x|b < \mathbf{x} \le a)$ has the same frequency interpretation as $F(x)$ [see (4-3)].

Total Probability and Bayes' Theorem

We shall now extend the results of Sec. 2-3 to random variables.

1. Setting $\mathscr{B} = \{\mathbf{x} \le x\}$ in (2-36), we obtain

$$P\{\mathbf{x} \le x\} = P\{\mathbf{x} \le x|\mathscr{A}_1\}P(\mathscr{A}_1) + \cdots + P\{\mathbf{x} \le x|\mathscr{A}_n\}P(\mathscr{A}_n)$$

Hence [see (4-41) and (4-44)]

$$F(x) = F(x|\mathscr{A}_1)P(\mathscr{A}_1) + \cdots + F(x|\mathscr{A}_n)P(\mathscr{A}_n) \tag{4-47}$$

$$f(x) = f(x|\mathscr{A}_1)P(\mathscr{A}_1) + \cdots + f(x|\mathscr{A}_n)P(\mathscr{A}_n) \tag{4-48}$$

In the above, the events $\mathscr{A}_1, \ldots, \mathscr{A}_n$ form a partition of $\mathscr{S}$.

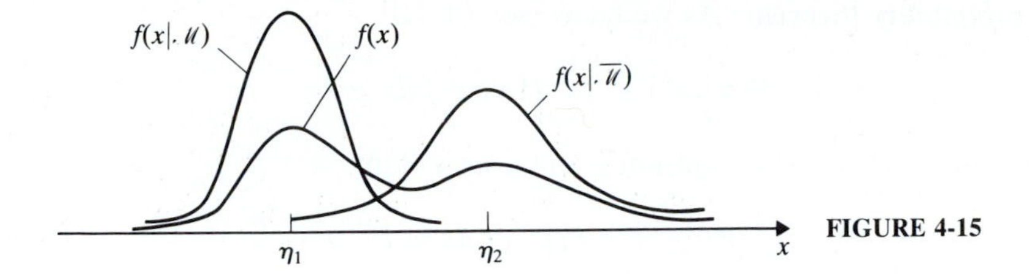

FIGURE 4-15

Example 4-14. Suppose that the RV $\mathbf{x}$ is such that $f(x|\mathscr{M})$ is $N(\eta_1;\sigma_1)$ and $f(x|\overline{\mathscr{M}})$ is $N(\eta_2;\sigma_2)$ as in Fig. 4-15. Clearly, the events $\mathscr{M}$ and $\overline{\mathscr{M}}$ form a partition of $\mathscr{S}$. Setting $\mathscr{A}_1 = \mathscr{M}$ and $\mathscr{A}_2 = \overline{\mathscr{M}}$ in (4-48), we conclude that

$$f(x) = pf(x|\mathscr{M}) + (1-p)f(x|\overline{\mathscr{M}}) = \frac{p}{\sigma_1}\mathbb{G}\left(\frac{x-\eta_1}{\sigma_1}\right) + \frac{1-p}{\sigma_2}\mathbb{G}\left(\frac{x-\eta_2}{\sigma_2}\right)$$

where $p = P(\mathscr{M})$.

2. From the identity

$$P(\mathscr{A}|\mathscr{B}) = \frac{P(\mathscr{B}|\mathscr{A})P(\mathscr{A})}{P(\mathscr{B})} \tag{4-49}$$

[see (2-38)] it follows that

$$P(\mathscr{A}|\mathbf{x} \le x) = \frac{P\{\mathbf{x} \le x|\mathscr{A}\}}{P\{\mathbf{x} \le x\}}P(\mathscr{A}) = \frac{F(x|\mathscr{A})}{F(x)}P(\mathscr{A}) \tag{4-50}$$

3. Setting $\mathscr{B} = \{x_1 < \mathbf{x} \le x_2\}$ in (4-49), we conclude with (4-43) that

$$P\{\mathscr{A}|x_1 < \mathbf{x} \le x_2\} = \frac{P\{x_1 < \mathbf{x} \le x_2|\mathscr{A}\}}{P\{x_1 < \mathbf{x} \le x_2\}}P(\mathscr{A})$$

$$= \frac{F(x_2|\mathscr{A}) - F(x_1|\mathscr{A})}{F(x_2) - F(x_1)}P(\mathscr{A}) \tag{4-51}$$

4. The conditional probability $P(\mathscr{A}|\mathbf{x} = x)$ of the event $\mathscr{A}$ assuming $\mathbf{x} = x$ cannot be defined as in (2-29) because, in general, $P\{\mathbf{x} = x\} = 0$. We shall define it as a limit

$$P\{\mathscr{A}|\mathbf{x} = x\} = \lim_{\Delta x \to 0} P\{\mathscr{A}|x < \mathbf{x} \le x + \Delta x\} \tag{4-52}$$

With $x_1 = x$, $x_2 = x + \Delta x$, we conclude from the above and (4-51) that

$$P\{\mathscr{A}|\mathbf{x} = x\} = \frac{f(x|\mathscr{A})}{f(x)}P(\mathscr{A}) \tag{4-53}$$

Total probability theorem. As we know [see (4-42)]

$$F(\infty|\mathscr{A}) = \int_{-\infty}^{\infty} f(x|\mathscr{A})\,dx = 1$$

Multiplying (4-53) by $f(x)$ and integrating, we obtain

$$\int_{-\infty}^{\infty} P(\mathscr{A}|\mathbf{x}=x) f(x)\,dx = P(\mathscr{A}) \tag{4-54}$$

This is the continuous version of the total probability theorem (2-36).

Bayes' theorem. From (4-53) and (4-54) it follows that

$$f(x|\mathscr{A}) = \frac{P(\mathscr{A}|\mathbf{x}=x)}{P(\mathscr{A})} f(x) = \frac{P(\mathscr{A}|\mathbf{x}=x) f(x)}{\int_{-\infty}^{\infty} P(\mathscr{A}|\mathbf{x}=x) f(x)\,dx} \tag{4-55}$$

This is the continuous version of Bayes' theorem (2-39).

Example 4-15. Suppose that the probability of heads in a coin-tossing experiment $\mathscr{S}$ is not a number, but an RV $\mathbf{p}$ with density $f(p)$ defined in some space $\mathscr{S}_c$. The experiment of the toss of a randomly selected coin is a cartesian product $\mathscr{S}_c \times \mathscr{S}$. In this experiment, the event $\mathscr{H} = \{\text{head}\}$ consists of all pairs of the form $\zeta_c h$ where ζ_c is any element of $\mathscr{S}_c$ and h is the element heads of the space $\mathscr{S} = \{h, t\}$. We shall show that

$$P(\mathscr{H}) = \int_0^1 p f(p)\,dp \tag{4-56}$$

Proof. The conditional probability of $\mathscr{H}$ assuming $\mathbf{p} = p$ is the probability of heads if the coin with $\mathbf{p} = p$ is tossed. In other words,

$$P\{\mathscr{H}|\mathbf{p}=p\} = p \tag{4-57}$$

Inserting into (4-54), we obtain (4-56) because $f(p) = 0$ outside the interval (0, 1).

PROBLEMS

4-1. Suppose that x_u is the u percentile of the RV $\mathbf{x}$, that is, $F(x_u) = u$. Show that if $f(-x) = f(x)$, then $x_{1-u} = -x_u$.

4-2. Show that if $f(x)$ is symmetrical about the point $x = \eta$ and $P\{\eta - a < \mathbf{x} < \eta + a\} = 1 - \alpha$, then $a = \eta - x_{\alpha/2} = x_{1-\alpha/2} - \eta$.

4-3. (*a*) Using Table 3-1 and linear interpolation, find the z_u percentile of the $N(0, 1)$ RV $\mathbf{z}$ for $u = 0.9, 0.925, 0.95, 0.975$, and 0.99. (*b*) The RV $\mathbf{x}$ is $N(\eta, \sigma)$. Express its x_u percentiles in terms of z_u.

4-4. The RV is $\mathbf{x}$ is $N(\eta, \sigma)$ and $P\{\eta - k\sigma < \mathbf{x} < \eta + k\sigma\} = p_k$. (*a*) Find p_k for $k = 1$, 2, and 3. (*b*) Find k for $p_k = 0.9, 0.99$, and 0.999. (*c*) If $P\{\eta - z_u\sigma < \mathbf{x} < \eta + z_u\sigma\} = \gamma$, express z_u in terms of γ.

4-5. Find x_u for $u = 0.1, 0.2, \ldots, 0.9$ (*a*) if $\mathbf{x}$ is uniform in the interval (0, 1); (*b*) if $f(x) = 2e^{-2x}U(x)$.

4-6. We measure for resistance $\boldsymbol{R}$ of each resistor in a production line and we accept only the units the resistance of which is between 96 and 104 ohms. Find the percentage of the accepted units (a) if $\mathbf{R}$ is uniform between 95 and 105 ohms; (b) if $\mathbf{R}$ is normal with $\eta = 100$ and $\sigma = 2$ ohms.

4-7. Show that if the RV $\mathbf{x}$ has an Erlang density with $n = 2$, then $F_x(x) = (1 - e^{-cx} - cxe^{-cx})U(x)$.

4-8. The RV $\mathbf{x}$ is $N(10; 1)$. Find $f(x|(\mathbf{x} - 10)^2 < 4)$.

4-9. Find $f(x)$ if $F(x) = (1 - e^{-\alpha x})U(x - c)$.

4-10. If $\mathbf{x}$ is $N(0, 2)$ find (a) $P\{1 \le \mathbf{x} \le 2\}$ and (b) $P\{1 \le \mathbf{x} \le 2 | \mathbf{x} \ge 1\}$.

4-11. The space $\mathscr{S}$ consists of all points t_i in the interval $(0, 1)$ and $P\{0 \le t_i \le y\} = y$ for every $y \le 1$. The function $G(x)$ is increasing from $G(-\infty) = 0$ to $G(\infty) = 1$; hence it has an inverse $G^{(-1)}(y) = H(y)$. The RV $\mathbf{x}$ is such that $\mathbf{x}(t_i) = H(t_i)$. Show that $F_x(x) = G(x)$.

4-12. If $\mathbf{x}$ is $N(1000; 20)$ find (a) $P\{\mathbf{x} < 1024\}$, (b) $P\{\mathbf{x} < 1024 | \mathbf{x} > 961\}$, and ($c$) $P\{31 < \sqrt{\mathbf{x}} \le 32\}$.

4-13. A fair coin is tossed three times and the RV $\mathbf{x}$ equals the total number of heads. Find and sketch $F_x(x)$ and $f_x(x)$.

4-14. A fair coin is tossed 900 times and the RV $\mathbf{x}$ equals the total number of heads. (a) Find $f_x(x)$: 1; exactly 2; approximately using (4-34). (b) Find $P\{435 \le \mathbf{x} \le 460\}$.

4-15. Show that, if $a \le \mathbf{x}(\zeta) \le b$ for every $\zeta \in \mathscr{S}$, then $F(x) = 1$ for $x > b$ and $F(x) = 0$ for $x < a$.

4-16. Show that if $\mathbf{x}(\zeta) \le \mathbf{y}(\zeta)$ for every $\zeta \in \mathscr{S}$, then $F_x(w) \ge F_y(w)$ for every w.

4-17. Show that if $\beta(t) = f(t|\mathbf{x} > t)$ is the conditional failure rate of the RV $\mathbf{x}$ and $\beta(t) = kt$, then $f(x)$ is a Rayleigh density (see also Sec. 7-3).

4-18. Show that $P(\mathscr{A}) = P(\mathscr{A}|\mathbf{x} \le x)F(x) + P(\mathscr{A}|\mathbf{x} > x)[1 - F(x)]$.

4-19. Show that

$$F_x(x|\mathscr{A}) = \frac{P(\mathscr{A}|\mathbf{x} \le x)F_x(x)}{P(\mathscr{A})}$$

4-20. Show that if $P(\mathscr{A}|\mathbf{x} = x) = P(\mathscr{B}|\mathbf{x} = x)$ for every $x \le x_0$, then $P(\mathscr{A}|\mathbf{x} \le x_0) = P(\mathscr{B}|\mathbf{x} \le x_0)$.

Hint: Replace in (4-54) $P(\mathscr{A})$ and $f(x)$ by $P(\mathscr{A}|\mathbf{x} \le x_0)$ and $f(x|\mathbf{x} \le x_0)$.

4-21. The probability of *heads* of a random coin is an RV $\mathbf{p}$ uniform in the interval $(0, 1)$. (a) Find $P\{0.3 \le \mathbf{p} \le 0.7\}$. ($b$) The coin is tossed 10 times and *heads* shows 6 times. Find the a posteriori probability that $\mathbf{p}$ is between 0.3 and 0.7.

4-22. The probability of *heads* of a random coin is an RV $\mathbf{p}$ uniform in the interval $(0.4, 0.6)$. (a) Find the probability that at the next tossing of the coin *heads* will show. (b) The coin is tossed 100 times and *heads* shows 60 times. Find the probability that at the next tossing *heads* will show.

CHAPTER
5

FUNCTIONS OF ONE RANDOM VARIABLE

5-1 THE RANDOM VARIABLE $g(\mathbf{x})$

Suppose that $\mathbf{x}$ is an RV and $g(x)$ is a function of the real variable x. The expression

$$\mathbf{y} = g(\mathbf{x})$$

is a new RV defined as follows: For a given ζ, $\mathbf{x}(\zeta)$ is a number and $g[\mathbf{x}(\zeta)]$ is another number specified in terms of $\mathbf{x}(\zeta)$ and $g(x)$. This number is the value $\mathbf{y}(\zeta) = g[\mathbf{x}(\zeta)]$ assigned to the RV $\mathbf{y}$. Thus a function of an RV $\mathbf{x}$ is a composite function $\mathbf{y} = g(\mathbf{x}) = g[\mathbf{x}(\zeta)]$ with domain the set $\mathscr{S}$ of experimental outcomes.

The distribution function $F_y(y)$ of the RV so formed is the probability of the event $\{\mathbf{y} \le y\}$ consisting of all outcomes ζ such that $\mathbf{y}(\zeta) = g[\mathbf{x}(\zeta)] \le y$. Thus

$$F_y(y) = P\{\mathbf{y} \le y\} = P\{g(\mathbf{x}) \le y\} \tag{5-1}$$

For a specific y, the values of x such that $g(x) \le y$ form a set on the x axis denoted by R_y. Clearly, $g[\mathbf{x}(\zeta)] \le y$ if $\mathbf{x}(\zeta)$ is a number in the set R_y. Hence

$$F_y(y) = P\{\mathbf{x} \in R_y\} \tag{5-2}$$

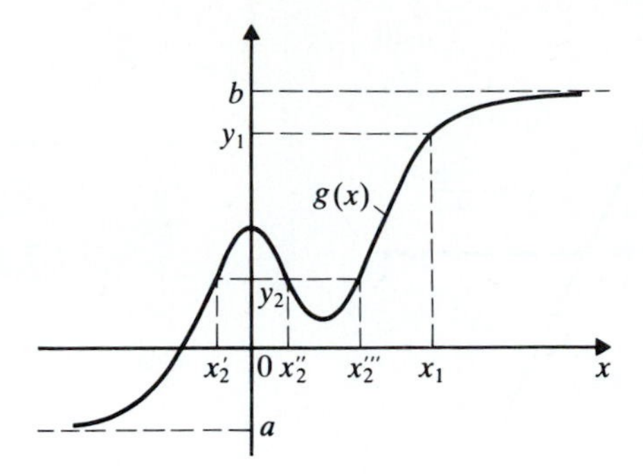

FIGURE 5-1

The above leads to the conclusion that for $g(\mathbf{x})$ to be an RV, the function $g(x)$ must have the following properties:

1. Its domain must include the range of the RV $\mathbf{x}$.
2. It must be a *Baire* function, that is, for every y, the set R_y such that $g(x) \le y$ must consist of the union and intersection of a countable number of intervals. Only then $\{\mathbf{y} \le y\}$ is an event.
3. The events $\{g(\mathbf{x}) = \pm\infty\}$ must have zero probability.

5-2 THE DISTRIBUTION OF $g(\mathbf{x})$

We shall express the distribution function $F_y(y)$ of the RV $\mathbf{y} = g(\mathbf{x})$ in terms of the distribution function $F_x(x)$ of the RV $\mathbf{x}$ and the function $g(x)$. For this purpose, we must determine the set R_y of the x axis such that $g(x) \le y$, and the probability that $\mathbf{x}$ is in this set. The method will be illustrated with several examples. Unless otherwise stated, it will be assumed that $F_x(x)$ is continuous.

1. We start with the function $g(x)$ in Fig. 5-1. As we see from the figure, $g(x)$ is between a and b for any x. This leads to the conclusion that if $y \ge b$, then $g(x) \le y$ for every x, hence $P\{\mathbf{y} \le y\} = 1$; if $y < a$, then there is no x such that $g(x) \le y$, hence $P\{\mathbf{y} \le y\} = 0$. Thus

$$F_y(y) = \begin{cases} 1 & y \ge b \\ 0 & y < a \end{cases}$$

With x_1 and $y_1 = g(x_1)$ as shown, we observe that $g(x) \le y_1$ for $x \le x_1$. Hence

$$F_y(y_1) = P\{\mathbf{x} \le x_1\} = F_x(x_1)$$

We finally note that

$$g(x) \le y_2 \qquad \text{if} \quad x \le x_2' \text{ or if } x_2'' \le x \le x_2'''$$

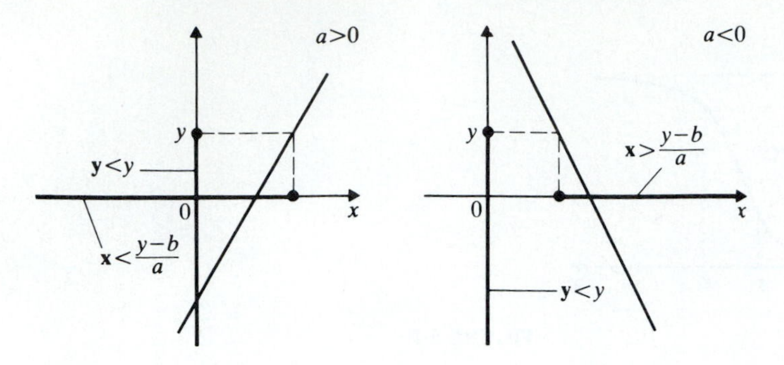

FIGURE 5-2

Hence

$$F_y(y_2) = P\{\mathbf{x} \le x_2'\} + P\{x_2'' \le \mathbf{x} \le x_2'''\} = F_x(x_2') + F_x(x_2''') - F_x(x_2'')$$

because the events $\{\mathbf{x} \le x_2'\}$ and $\{x_2'' \le \mathbf{x} \le x_2'''\}$ are mutually exclusive.

Example 5-1

$$\mathbf{y} = a\mathbf{x} + b$$

To find $F_y(y)$, we must find the values of x such that $ax + b \le y$.
(a) If $a > 0$, then $ax + b \le y$ for $x \le (y - b)/a$ (Fig. 5-2a). Hence

$$F_y(y) = P\left\{\mathbf{x} \le \frac{y-b}{a}\right\} = F_x\left(\frac{y-b}{a}\right) \qquad a > 0$$

(b) If $a < 0$, then $ax + b \le y$ for $x > (y - b)/a$ (Fig. 5-2b). Hence

$$F_y(y) = P\left\{\mathbf{x} \ge \frac{y-b}{a}\right\} = 1 - F_x\left(\frac{y-b}{a}\right) \qquad a < 0$$

Example 5-2

$$\mathbf{y} = \mathbf{x}^2$$

If $y \ge 0$, then $x^2 \le y$ for $-\sqrt{y} \le x \le \sqrt{y}$ (Fig. 5-3a). Hence

$$F_y(y) = P\{-\sqrt{y} \le \mathbf{x} \le \sqrt{y}\} = F_x(\sqrt{y}) - F_x(-\sqrt{y}) \qquad y > 0$$

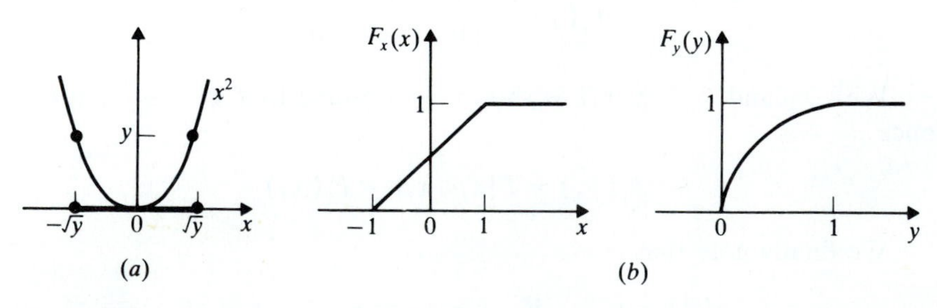

FIGURE 5-3

If $y < 0$, then there are no values of x such that $x^2 < y$. Hence

$$F_y(y) = P\{\emptyset\} = 0 \qquad y < 0$$

Special case If $\mathbf{x}$ is uniform in the interval $(-1, 1)$, then

$$F_x(x) = \frac{1}{2} + \frac{x}{2} \qquad |x| < 1$$

(Fig. 5-3*b*). Hence

$$F_y(y) = \sqrt{y} \qquad \text{for} \quad 0 \le y \le 1 \qquad \text{and} \qquad F_y(y) = \begin{cases} 1 & y > 1 \\ 0 & y < 0 \end{cases}$$

2. Suppose now that the function $g(x)$ is constant in an interval (x_0, x_1):

$$g(x) = y_1 \qquad x_0 < x \le x_1$$

In this case

$$P\{\mathbf{y} = y_1\} = P\{x_0 < \mathbf{x} \le x_1\} = F_x(x_1) - F_x(x_0) \tag{5-3}$$

Hence $F_y(y)$ is discontinuous at $\mathbf{y} = y_1$ and its discontinuity equals $F_x(x_1) - F_x(x_0)$.

Example 5-3. Consider the function (Fig. 5-4)

$$g(x) = 0 \qquad \text{for} \quad -c \le x \le c \qquad \text{and} \qquad g(x) = \begin{cases} x - c & x > c \\ x + c & x < -c \end{cases}$$

In this case, $F_y(y)$ is discontinuous for $y = 0$ and its discontinuity equals $F_x(c) - F_x(-c)$. Furthermore,

$$\text{If} \quad y \ge 0 \qquad \text{then} \quad P\{\mathbf{y} \le y\} = P\{\mathbf{x} \le y + c\} = F_x(y + c)$$

$$\text{If} \quad y < 0 \qquad \text{then} \quad P\{\mathbf{y} \le y\} = P\{\mathbf{x} \le y - c\} = F_x(y - c)$$

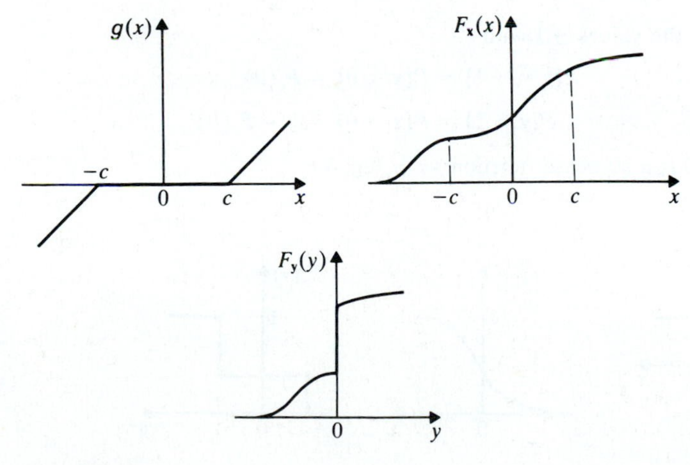

FIGURE 5-4

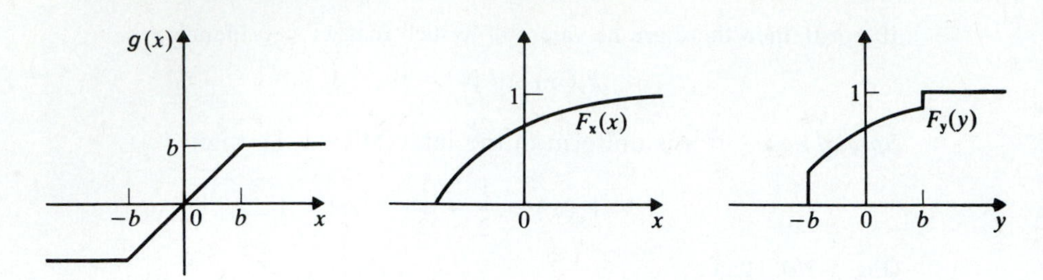

FIGURE 5-5

Example 5-4 Limiter. The curve $g(x)$ of Fig. 5-5 is constant for $x \le -b$ and $x \ge b$ and in the interval $(-b, b)$ it is a straight line. With $\mathbf{y} = g(\mathbf{x})$, it follows that $F_y(y)$ is discontinuous for $y = g(-b) = -b$ and $y = g(b) = b$ respectively. Furthermore,

If	$y \ge b$	then	$g(x) \le y$	for every x;	hence	$F_y(y) = 1$
If	$-b \le y < b$	then	$g(x) \le y$	for $x \le y$;	hence	$F_y(y) = F_x(y)$
If	$y < -b$	then	$g(x) \le y$	for no x;	hence	$F_y(y) = 0$

3. We assume next that $g(x)$ is a staircase function

$$g(x) = g(x_i) = y_i \qquad x_{i-1} < x \le x_i$$

In this case, the RV $\mathbf{y} = g(\mathbf{x})$ is of discrete type taking the values y_i with

$$P\{\mathbf{y} = y_i\} = P\{x_{i-1} < \mathbf{x} \le x_i\} = F_x(x_i) - F_x(x_{i-1})$$

Example 5-5 Hard limiter. If

$$g(x) = \begin{cases} 1 & x > 0 \\ -1 & x \le 0 \end{cases}$$

then $\mathbf{y}$ takes the values ± 1 with

$$P\{\mathbf{y} = -1\} = P\{\mathbf{x} \le 0\} = F_x(0)$$

$$P\{\mathbf{y} = 1\} = P\{\mathbf{x} > 0\} = 1 - F_x(0)$$

Hence $F_y(y)$ is a staircase function as in Fig. 5-6.

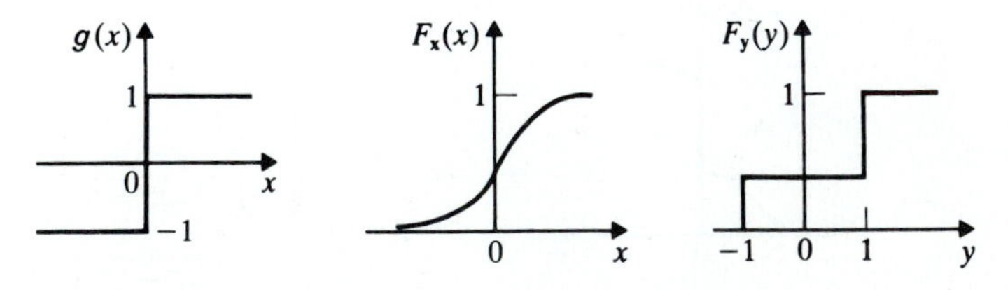

FIGURE 5-6

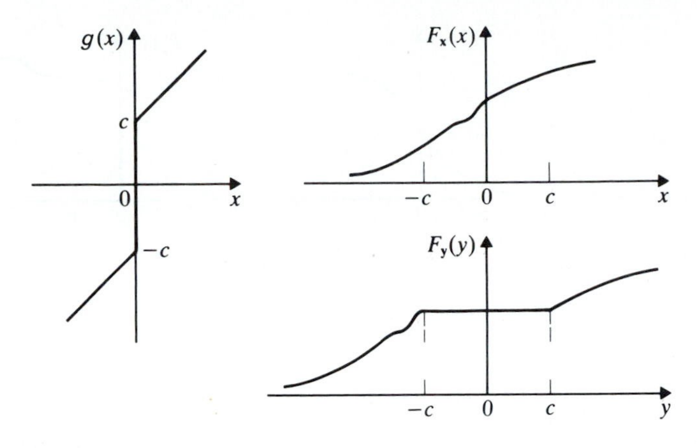

FIGURE 5-7

Example 5-6 Quantization. If

$$g(x) = ns \qquad (n-1)s < x \le ns$$

then $\mathbf{y}$ takes the values $y_n = ns$ with

$$P\{\mathbf{y} = ns\} = P\{(n-1)s < \mathbf{x} \le ns\} = F_x(ns) - F_x(ns - s)$$

4. We assume, finally, that the function $g(x)$ is discontinuous at $x = x_0$ and such that

$$g(x) < g(x_0^-) \qquad \text{for} \quad x < x_0 \qquad g(x) > g(x_0^+) \qquad \text{for} \quad x > x_0$$

In this case, if y is between $g(x_0^-)$ and $g(x_0^+)$, then $g(x) < y$ for $x \le x_0$. Hence

$$F_y(y) = P\{\mathbf{x} \le x_0\} = F_x(x_0) \qquad g(x_0^-) \le y \le g(x_0^+)$$

Example 5-7. Suppose that

$$g(x) = \begin{cases} x + c & x \ge 0 \\ x - c & x < 0 \end{cases}$$

is discontinuous (Fig. 5-7). Thus $g(x)$ is discontinuous for $x = 0$ with $g(0^-) = -c$ and $g(0^+) = c$. Hence $F_y(y) = F_x(0)$ for $|y| \le c$. Furthermore,

If	$y \ge c$	then $g(x) \le y$	for $x \le y - c$;	hence	$F_y(y) = F_x(y - c)$
If	$-c \le y \le c$	then $g(x) \le y$	for $x \le 0$;	hence	$F_y(y) = F_x(0)$
If	$y \le -c$	then $g(x) \le y$	for $x \le y + c$;	hence	$F_y(y) = F_x(y + c)$

Example 5-8. The function $g(x)$ in Fig. 5-8 equals 0 in the interval $(-c, c)$ and it is discontinuous for $x = \pm c$ with $g(c^+) = c$, $g(c^-) = 0$, $g(-c^-) = -c$, $g(-c^+) = 0$. Hence $F_y(y)$ is discontinuous for $y = 0$ and it is constant for

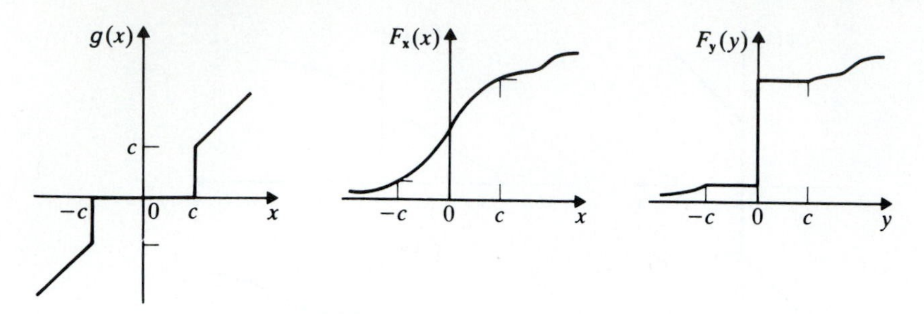

FIGURE 5-8

$0 \le y \le c$ and $-c \le y \le 0$. Thus

If	$y \ge c$	then	$g(x) \le y$	for $x \le y$;	hence	$F_y(y) = F_x(y)$
If	$0 \le y < c$	then	$g(x) \le y$	for $x < c$;	hence	$F_y(y) = F_x(c)$
If	$-c \le y < 0$	then	$g(x) \le y$	for $x \le -c$;	hence	$F_y(y) = F_x(-c)$
If	$y < -c$	then	$g(x) \le y$	for $x \le y$;	hence	$F_y(y) = F_x(y)$

5. We now assume that the RV $\mathbf{x}$ is of discrete type taking the values x_k with probability p_k. In this case, the RV $\mathbf{y} = g(\mathbf{x})$ is also of discrete type taking the values $y_k = g(x_k)$.

If $y_k = g(x)$ for only one $x = x_k$, then

$$P\{\mathbf{y} = y_k\} = P\{\mathbf{x} = x_k\} = p_k$$

If, however, $y_k = g(x)$ for $x = x_k$ and $x = x_l$, then

$$P\{\mathbf{y} = y_k\} = P\{\mathbf{x} = x_k\} + P\{\mathbf{x} = x_l\} = p_k + p_l$$

Example 5-9

$$\mathbf{y} = \mathbf{x}^2$$

(*a*) If $\mathbf{x}$ takes the values $1, 2, \ldots, 6$ with probability $1/6$, then $\mathbf{y}$ takes the values $1^2, 2^2, \ldots, 6^2$ with probability $1/6$.

(*b*) If, however, $\mathbf{x}$ takes the values $-2, -1, 0, 1, 2, 3$ with probability $1/6$, then $\mathbf{y}$ takes the values $0, 1, 4, 9$ with probabilities $1/6, 2/6, 2/6, 1/6$ respectively.

Determination of $f_y(y)$

We wish to determine the density of $\mathbf{y} = g(\mathbf{x})$ in terms of the density of $\mathbf{x}$. Suppose, first, that the set R of the y axis is not in the range of the function $g(x)$, that is, that $g(x)$ is not a point of R for any x. In this case, the probability that $g(\mathbf{x})$ is in R equals 0. Hence $f_y(y) = 0$ for $y \in R$. It suffices, therefore, to consider the values of y such that for some x, $g(x) = y$.

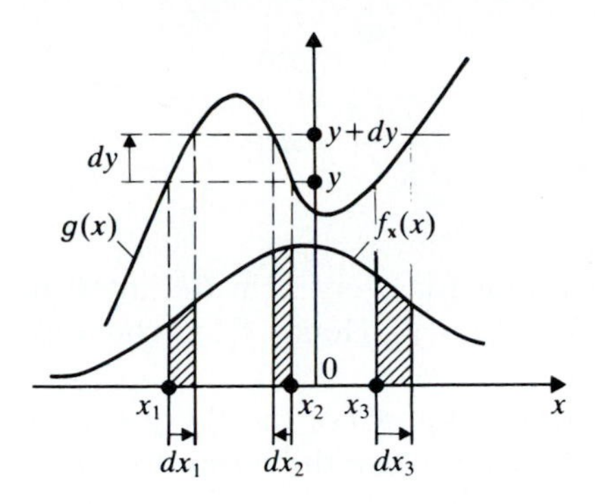

FIGURE 5-9

FUNDAMENTAL THEOREM. To find $f_y(y)$ for a specific y, we solve the equation $y = g(x)$. Denoting its real roots by x_n,

$$y = g(x_1) = \cdots = g(x_n) = \cdots \tag{5-4}$$

we shall show that

$$f_y(y) = \frac{f_x(x_1)}{|g'(x_1)|} + \cdots + \frac{f_x(x_n)}{|g'(x_n)|} + \cdots \tag{5-5}$$

where $g'(x)$ is the derivative of $g(x)$.

Proof. To avoid generalities, we assume that the equation $y = g(x)$ has three roots as in Fig. 5-9. As we know

$$f_y(y)\,dy = P\{y < \mathbf{y} \le y + dy\}$$

It suffices, therefore, the find the set of values x such that $y < g(x) \le y + dy$ and the probability that $\mathbf{x}$ is in this set. As we see from the figure, this set consists of the following three intervals

$$x_1 < x < x_1 + dx_1 \qquad x_2 + dx_2 < x < x_2 \qquad x_3 < x < x_3 + dx_3$$

where $dx_1 > 0$, $dx_3 > 0$ but $dx_2 < 0$. From the above it follows that

$$P\{y < \mathbf{y} < y + dy\} = P\{x_1 < \mathbf{x} < x_1 + dx_1\}$$
$$+ P\{x_2 + dx_2 < \mathbf{x} < x_2\} + P\{x_3 < \mathbf{x} < x_3 + dx_3\}$$

The right side equals the shaded area in Fig. 5-9. Since

$$P\{x_1 < \mathbf{x} < x_1 + dx_1\} = f_x(x_1)\,dx_1 \qquad dx_1 = dy/g'(x_1)$$
$$P\{x_2 + dx_2 < \mathbf{x} < x_2\} = f_x(x_2)|dx_2| \qquad dx_2 = dy/g'(x_2)$$
$$P\{x_3 < \mathbf{x} < x_3 + dx_3\} = f_x(x_3)\,dx_3 \qquad dx_3 = dy/g'(x_3)$$

we conclude that

$$f_y(y)\,dy = \frac{f_x(x_1)}{g'(x_1)}dy + \frac{f_x(x_2)}{|g'(x_2)|}dy + \frac{f_x(x_3)}{g'(x_3)}dy$$

and (5-5) results.

We note, finally, that if $g(x) = y_1$ = constant for every x in the interval (x_0, x_1), then [see (5-3)] $F_y(y)$ is discontinuous for $y = y_1$. Hence $f_y(y)$ contains an impulse $\delta(y - y_1)$ of area $F_x(x_1) - F_x(x_0)$.

Conditional density The conditional density $f_y(y|\mathscr{M})$ of the RV $\mathbf{y} = g(\mathbf{x})$ assuming $\mathscr{M}$ is given by (5-5) if on the right side we replace the terms $f_x(x_i)$ by $f_x(x_i|\mathscr{M})$ (see, for example, Prob. 5-17).

Illustrations

We give next several applications of (5-2) and (5-5).

1. $$\mathbf{y} = a\mathbf{x} + b \qquad g'(x) = a$$

The equation $y = ax + b$ has a single solution $x = (y - b)/a$ for every y. Hence

$$f_y(y) = \frac{1}{|a|} f_x\left(\frac{y - b}{a}\right) \tag{5-6}$$

Special case If $\mathbf{x}$ is uniform in the interval (x_1, x_2), then $\mathbf{y}$ is uniform in the interval $(ax_1 + b, ax_2 + b)$.

Example 5-10. Suppose that the voltage $\mathbf{v}$ is an RV given by

$$\mathbf{v} = i(\mathbf{r} + r_0)$$

where $i = 0.01$ A and $r_0 = 1000\ \Omega$. If the resistance $\mathbf{r}$ is an RV uniform between 900 and 1100 Ω, then $\mathbf{v}$ is uniform between 19 and 21 V.

2. $$\mathbf{y} = \frac{1}{\mathbf{x}} \qquad g'(x) = -\frac{1}{x^2}$$

The equation $y = 1/x$ has a single solution $x = 1/y$. Hence

$$f_y(y) = \frac{1}{y^2} f_x\left(\frac{1}{y}\right) \tag{5-7}$$

Special case If $\mathbf{x}$ has a *Cauchy density* with parameter α,

$$f_x(x) = \frac{\alpha/\pi}{x^2 + \alpha^2} \qquad \text{then} \qquad f_y(y) = \frac{1/\alpha\pi}{y^2 + 1/\alpha^2}$$

is also a Cauchy density with parameter $1/\alpha$.

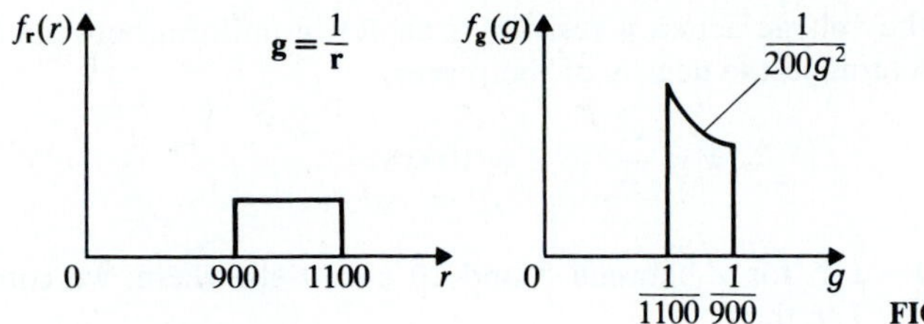

FIGURE 5-10

Example 5-11. Suppose that the resistance $\mathbf{r}$ is uniform between 900 and 1100 Ω as in Fig. 5-10. We shall determine the density of the corresponding conductance

$$\mathbf{g} = 1/\mathbf{r}$$

Since $f_r(r) = 1/200$ S for r between 900 and 1100 it follows from (5-7) that

$$f_g(g) = \frac{1}{200g^2} \qquad \text{for} \quad \frac{1}{1100} < g < \frac{1}{900}$$

and 0 elsewhere.

3.
$$\mathbf{y} = a\mathbf{x}^2 \qquad a > 0 \qquad g'(x) = 2ax$$

If $y < 0$, then the equation $y = ax^2$ has no real solutions; hence $f_y(y) = 0$. If $y > 0$, then it has two solutions

$$x_1 = \sqrt{\frac{y}{a}} \qquad x_2 = -\sqrt{\frac{y}{a}}$$

and (5-5) yields

$$f_y(y) = \frac{1}{2a\sqrt{y/a}}\left[f_x\left(\sqrt{\frac{y}{a}}\right) + f_x\left(-\sqrt{\frac{y}{a}}\right)\right] \qquad y > 0 \tag{5-8}$$

We note that $F_y(y) = 0$ for $y < 0$ and

$$F_y(y) = P\left\{-\sqrt{\frac{y}{a}} \le \mathbf{x} \le \sqrt{\frac{y}{a}}\right\} = F_x\left(\sqrt{\frac{y}{a}}\right) - F_x\left(-\sqrt{\frac{y}{a}}\right) \qquad y > 0$$

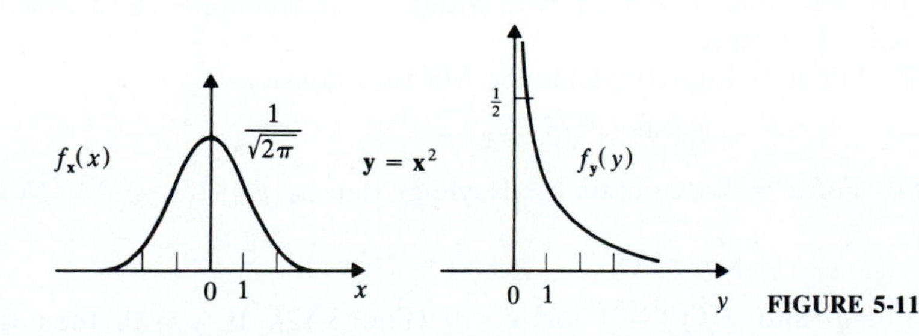

FIGURE 5-11

Example 5-12. The voltage across a resistor is an RV **e** uniform between 5 and 10 V. We shall determine the density of the power

$$\mathbf{w} = \frac{\mathbf{e}^2}{r} \qquad r = 1000\ \Omega$$

dissipated in r.

Since $f_e(e) = 1/5$ for e between 5 and 10 and 0 elsewhere, we conclude from (5-8) with $a = 1/r$ that

$$f_w(w) = \sqrt{\frac{10}{w}} \qquad \frac{1}{40} < w < \frac{1}{10}$$

and 0 elsewhere.

Special case Suppose that

$$f_x(x) = \frac{1}{\sqrt{2\pi}} e^{-x^2/2} \qquad \mathbf{y} = \mathbf{x}^2$$

With $a = 1$, it follows from (5-8) and the evenness of $f_x(x)$ that (Fig. 5-11)

$$f_y(y) = \frac{1}{\sqrt{y}} f_x(\sqrt{y}) = \frac{1}{\sqrt{2\pi y}} e^{-y/2} U(y)$$

We have thus shown that if $\mathbf{x}$ is an $N(0,1)$ RV, the RV $\mathbf{y} = \mathbf{x}^2$ has a chi-square distribution with one degree of freedom [see (4-40)].

4.

$$\mathbf{y} = \sqrt{\mathbf{x}} \qquad g'(x) = \frac{1}{2\sqrt{x}}$$

The equation $y = \sqrt{x}$ has a single solution $x = y^2$ for $y > 0$ and no solution for $y < 0$. Hence

$$f_y(y) = 2y f_x(y^2) U(y) \tag{5-9}$$

The chi density Suppose that $\mathbf{x}$ has a chi-square density as in (4-40),

$$f_x(x) = \frac{1}{2^{n/2}\Gamma(n/2)} x^{n/2-1} e^{-x/2} U(y)$$

and $y = \sqrt{x}$. In this case, (5-9) yields

$$f_y(y) = \frac{2}{2^{n/2}\Gamma(n/2)} y^{n-1} e^{-y^2/2} U(y) \tag{5-10}$$

This function is called the *chi density* with n degrees of freedom. The following cases are of special interest.

Maxwell For $n = 3$, (5-10) yields the Maxwell density

$$f_y(y) = \sqrt{2/\pi}\, y^2 e^{-y^2/2}.$$

Rayleigh For $n = 2$, we obtain the Rayleigh density $f_y(y) = y e^{-y^2/2} U(y)$.

5.

$$\mathbf{y} = \mathbf{x} U(\mathbf{x}) \qquad g'(x) = U(x)$$

Clearly, $f_y(y) = 0$ and $F_y(y) = 0$ for $y < 0$ (Fig. 5-12). If $y > 0$, then the

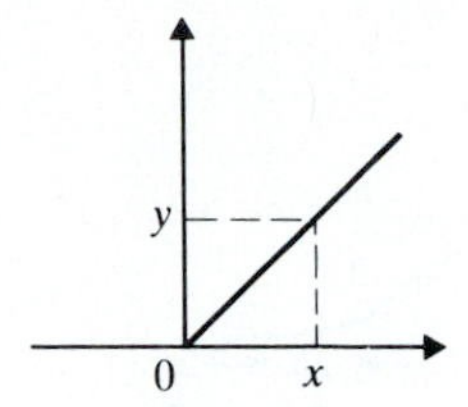

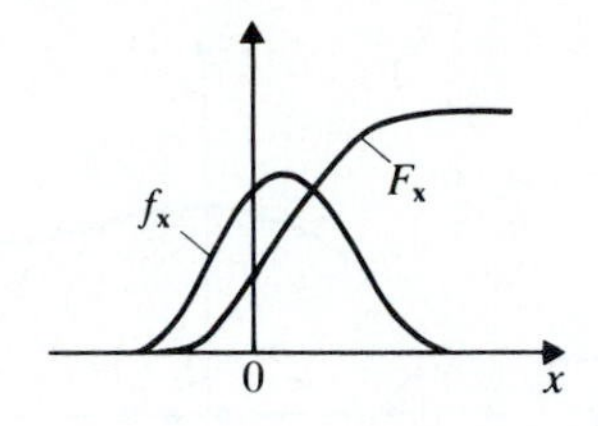

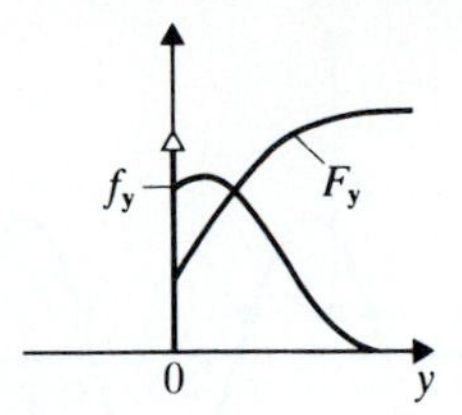

FIGURE 5-12

equation $y = xU(x)$ has a single solution $x_1 = y$. Hence

$$f_y(y) = f_x(y) \qquad F_y(y) = F_x(y) \qquad y > 0$$

Thus $F_y(y)$ is discontinuous at $y = 0$ with discontinuity $F_y(0^+) - F_y(0^-) = F_x(0)$. Hence

$$f_y(y) = f_x(y)U(y) + F_x(0)\delta(y)$$

6.
$$\mathbf{y} = e^{\mathbf{x}} \qquad g'(x) = e^x$$

If $y > 0$, then the equation $y = e^x$ has the single solution $x = \ln y$. Hence

$$f_y(y) = \frac{1}{y} f_x(\ln y) \qquad y > 0$$

If $y < 0$, then $f_y(y) = 0$.

Special case If $\mathbf{x}$ is $N(\eta; \sigma)$, then

$$f_y(y) = \frac{1}{\sigma y \sqrt{2\pi}} e^{-(\ln y - \eta)^2 / 2\sigma^2} \tag{5-11}$$

This density is called *lognormal* (see Table 4-1).

7.
$$\mathbf{y} = a \sin(\mathbf{x} + \theta) \qquad a > 0$$

If $|y| > a$, then the equation $y = a\sin(x + \theta)$ has no solutions; hence $f_y(y) = 0$. If $|y| < a$, then it has infinitely many solutions (Fig. 5-13*a*)

$$x_n = \arcsin\frac{y}{a} - \theta \qquad n = -\ldots, -1, 0, 1, \ldots$$

Since $g'(x_n) = a\cos(x_n + \theta) = \sqrt{a^2 - y^2}$, (5-5) yields

$$f_y(y) = \frac{1}{\sqrt{a^2 - y^2}} \sum_{n=-\infty}^{\infty} f_x(x_n) \qquad |y| < a \tag{5-12}$$

Special case Suppose that $\mathbf{x}$ is uniform in the interval $(-\pi, \pi)$. In this case, the equation $y = a\sin(x + \theta)$ has exactly two solutions in the interval $(-\pi, \pi)$ for any θ (Fig. 5-14). The function $f_x(x)$ equals $1/2\pi$ for these two values and it equals 0 for any x_n outside the interval $(-\pi, \pi)$. Retaining the

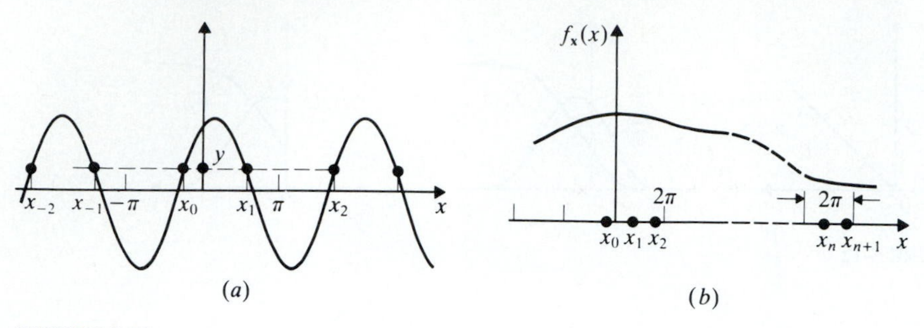

FIGURE 5-13

two nonzero terms in (5-12), we obtain

$$f_y(y) = \frac{2}{2\pi\sqrt{a^2 - y^2}} \qquad |y| < a \tag{5-13}$$

To find $F_y(y)$, we observe that $\mathbf{y} \le y$ if $\mathbf{x}$ is either between $-\pi$ and x_0 or between x_1 and π (Fig. 5-13a). Since the total length of the two intervals equals $\pi + 2x_0 + 2\theta$, we conclude, dividing by 2π, that

$$F_y(y) = \frac{1}{2} + \frac{1}{\pi}\arcsin\frac{y}{a} \qquad |y| < a \tag{5-14}$$

We note that although $f_y(\pm a) = \infty$, the probability that $\mathbf{y} = \pm a$ is 0.

Smooth phase If the density $f_x(x)$ of $\mathbf{x}$ is sufficiently smooth so that it can be approximated by a constant in any interval of length 2π (see Fig. 5-13b), then

$$\pi \sum_{n=-\infty}^{\infty} f_x(x_n) \simeq \int_{-\infty}^{\infty} f_x(x)\,dx = 1$$

because in each interval of length 2π the above sum has two terms. Inserting into (5-12), we conclude that the density of $\mathbf{x}$ is given approximately by (5-13).

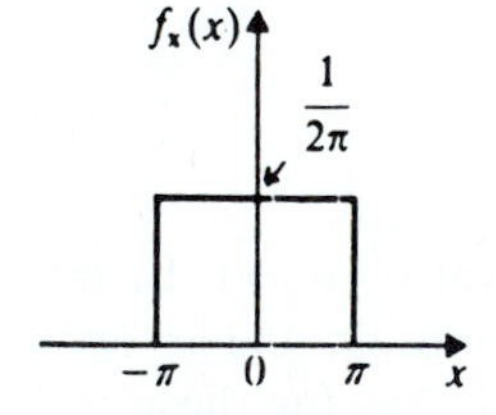

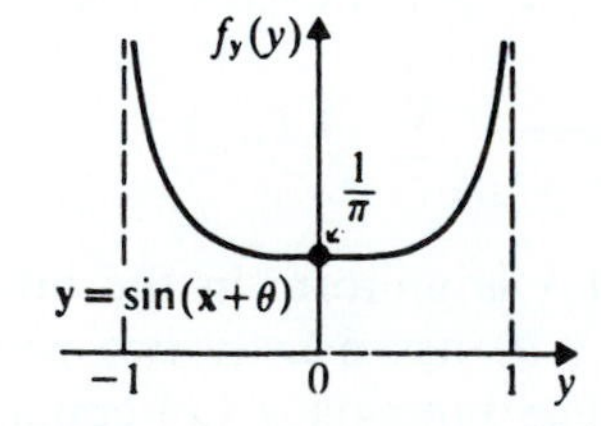

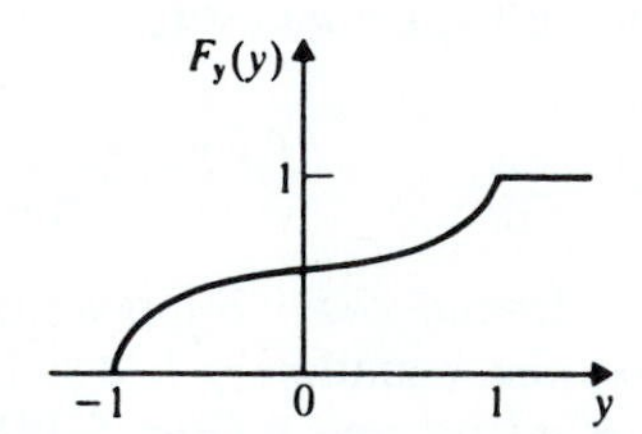

FIGURE 5-14

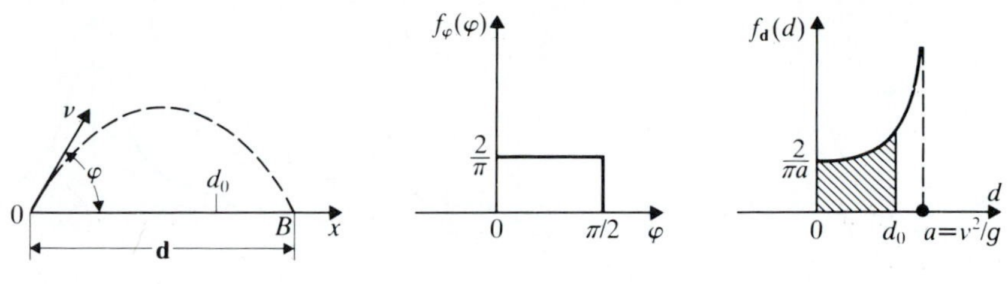

FIGURE 5-15

Example 5-13. A particle leaves the origin under the influence of the force of gravity and its initial velocity v forms an angle $\boldsymbol{\varphi}$ with the horizontal axis. The path of the particle reaches the ground at a distance

$$\mathbf{d} = \frac{v^2}{g} \sin 2\boldsymbol{\varphi}$$

from the origin (Fig. 5-15). Assuming that $\boldsymbol{\varphi}$ is an RV uniform between 0 and $\pi/2$, we shall determine: (a) the density of $\mathbf{d}$ and (b) the probability that $\mathbf{d} \le d_0$.

Solution. (a) Clearly,

$$\mathbf{d} = a \sin \mathbf{x} \qquad a = v^2/g$$

where the RV $\mathbf{x} = 2\boldsymbol{\varphi}$ is uniform between 0 and π. If $0 < d < a$, then the equation $d = a \sin x$ has exactly two solutions in the interval $(0, \pi)$. Reasoning as in (5-13), we obtain

$$f_d(d) = \frac{2}{\pi\sqrt{a^2 - d^2}} \qquad 0 < d < a$$

and 0 otherwise.

(b) The probability that $\mathbf{d} \le d_0$ equals the shaded area in Fig. 5-15:

$$P\{\mathbf{d} \le d_0\} = F_d(d_0) = \frac{2}{\pi} \arcsin \frac{d_0}{a}$$

8. $$\mathbf{y} = \tan \mathbf{x}$$

The equation $y = \tan x$ has infinitely many solutions for any y (Fig. 5-16a)

$$x_n = \arctan y \qquad n = \ldots, -1, 0, 1, \ldots$$

Since $g'(x) = 1/\cos^2 x = 1 + y^2$, (5-5) yields

$$f_y(y) = \frac{1}{1 + y^2} \sum_{n=-\infty}^{\infty} f_x(x_n) \tag{5-15}$$

Special case If $\mathbf{x}$ is uniform in the interval $(-\pi/2, \pi/2)$, then the term $f_x(x_1)$ in (5-15) equals $1/\pi$ and all others are 0 (Fig. 5-16b). Hence $\mathbf{y}$ has a

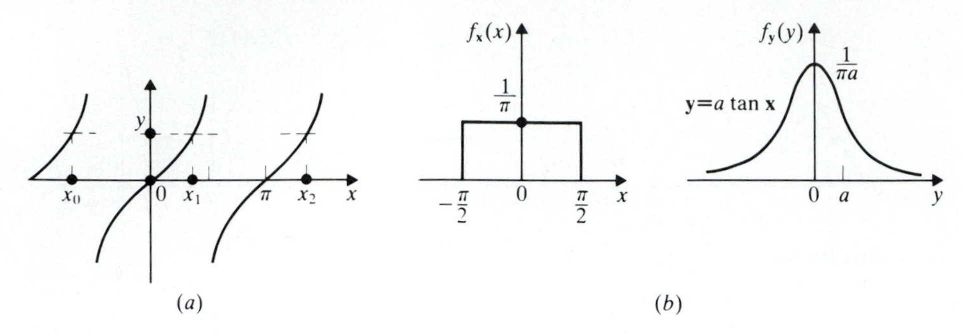

FIGURE 5-16

Cauchy density

$$f_y(y) = \frac{1/\pi}{1+y^2} \tag{5-16}$$

As we see from the figure, $\mathbf{y} \le y$ if $\mathbf{x}$ is between $-\pi/2$ and x_1. Since the length of this interval equals $x_1 + \pi/2$, we conclude, dividing by π, that

$$F_y(y) = \frac{1}{\pi}\left(x_1 + \frac{\pi}{2}\right) = \frac{1}{2} + \frac{1}{\pi}\arctan y \tag{5-17}$$

Example 5-14. A particle leaves the origin in a free motion as in Fig. 5-17 crossing the vertical line $x = d$ at

$$\mathbf{y} = d\tan\boldsymbol{\varphi}$$

Assuming that the angle $\boldsymbol{\varphi}$ is uniform in the interval $(-\theta, \theta)$, we conclude as in (5-16) that

$$f_y(y) = \frac{d/2\theta}{d^2+y^2} \qquad \text{for} \quad |y| < d\tan\theta$$

and 0 otherwise.

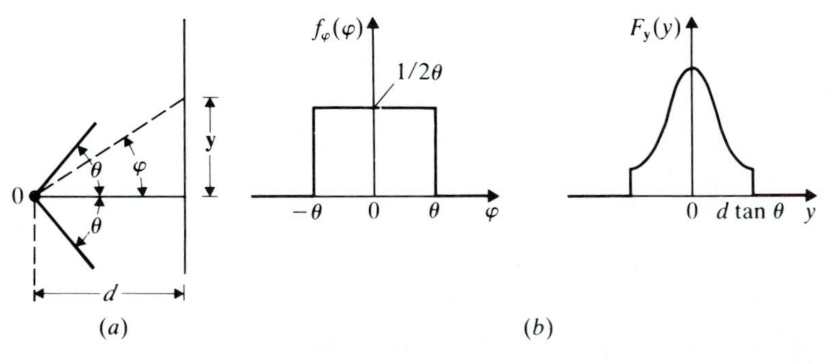

FIGURE 5-17

THE INVERSE PROBLEM. In the preceding discussion, we were given an RV $\mathbf{x}$ with known distribution $F_x(x)$ and a function $g(x)$ and we determined the distribution $F_y(y)$ of the RV $\mathbf{y} = g(\mathbf{x})$. We consider now the inverse problem: We are given the distribution of $\mathbf{x}$ and we wish to find a function $g(x)$ such that the distribution of the RV $\mathbf{y} = g(\mathbf{x})$ equals a specified function $F_y(y)$. This topic is developed further in Sec. 8-5. We start with two special cases.

From $F_x(x)$ to a uniform distribution. Given an RV $\mathbf{x}$ with distribution $F_x(x)$, we wish to find a function $g(x)$ such that the RV $\mathbf{u} = g(\mathbf{x})$ is uniformly distributed in the interval (0, 1). We maintain that $g(x) = F_x(x)$, that is, if

$$\mathbf{u} = F_x(\mathbf{x}) \qquad \text{then } F_u(u) = u \text{ for } 0 \le u \le 1 \tag{5-18}$$

Proof. Suppose that x is an arbitrary number and $u = F_x(x)$. From the monotonicity of $F_x(x)$ it follows that $\mathbf{u} \le u$ iff $\mathbf{x} \le x$. Hence

$$F_u(u) = P\{\mathbf{u} \le u\} = P\{\mathbf{x} \le x\} = F_x(x) = u$$

and (5-18) results.

The RV $\mathbf{u}$ can be considered as the output of a nonlinear memoryless system (Fig. 5-18) with input $\mathbf{x}$ and transfer characteristic $F_x(x)$. Therefore if we use $\mathbf{u}$ as the input to another system with transfer characteristic the inverse $F_x^{(-1)}(u)$ of the function $u = F_x(x)$, the resulting output will equal $\mathbf{x}$:

$$\text{If} \quad \mathbf{x} = F_x^{(-1)}(\mathbf{u}) \qquad \text{then} \quad P\{\mathbf{x} \le x\} = F_x(x)$$

From uniform to $F_y(y)$. Given an RV $\mathbf{u}$ with uniform distribution in the interval (0, 1), we wish to find a function $g(u)$ such that the distribution of the RV $\mathbf{y} = g(\mathbf{u})$ is a specified function $F_y(y)$. We maintain that $g(u)$ is the inverse of the function $u = F_y(y)$:

$$\text{If} \quad \mathbf{y} = F_y^{(-1)}(\mathbf{u}) \qquad \text{then} \quad P\{\mathbf{y} \le y\} = F_y(y) \tag{5-19}$$

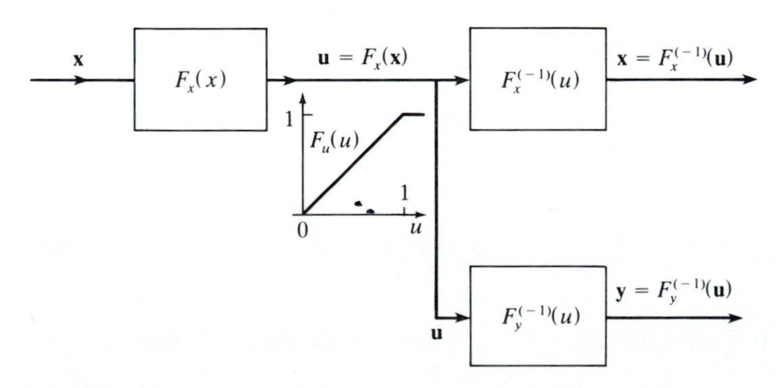

FIGURE 5-18

Proof. The RV $\mathbf{u}$ in (5-19) is uniform and the function $F_x(x)$ is arbitrary. Replacing $F_x(x)$ by $F_y(y)$, we obtain (5-19) (see also Fig. 5-18).

From $F_x(x)$ to $F_y(y)$. We consider, finally, the general case: Given $F_x(x)$ and $F_y(y)$, find $g(x)$ such that the distribution of $\mathbf{y} = g(\mathbf{x})$ equals $F_y(y)$. To solve this problem, we form the RV $\mathbf{u} = F_x(\mathbf{x})$ as in (5-18) and the RV $\mathbf{y} = F^{(-1)}(\mathbf{u})$ as in (5-19). Combining the two, we conclude:

$$\text{If} \quad \mathbf{y} = F_y^{(-1)}(F_x(\mathbf{x})) \qquad \text{then} \quad P\{\mathbf{y} \le y\} = F_y(y) \tag{5-20}$$

5-3 MEAN AND VARIANCE

The *expected value* or *mean* of an RV $\mathbf{x}$ is by definition the integral

$$E\{\mathbf{x}\} = \int_{-\infty}^{\infty} x f(x)\, dx \tag{5-21}$$

This number will also be denoted by η_x or η.

Example 5-15. If $\mathbf{x}$ is uniform in the interval (x_1, x_2), then $f(x) = 1/(x_2 - x_1)$ in this interval. Hence

$$E\{\mathbf{x}\} = \frac{1}{x_2 - x_1} \int_{x_1}^{x_2} x\, dx = \frac{x_1 + x_2}{2}$$

We note that, if the vertical line $x = a$ is an axis of symmetry of $f(x)$ then $E\{\mathbf{x}\} = a$; in particular, if $f(-x) = f(x)$, then $E\{\mathbf{x}\} = 0$. In the above example, $f(x)$ is symmetrical about the line $x = (x_1 + x_2)/2$.

Discrete type For discrete type RVs the integral in (5-21) can be written as a sum. Indeed, suppose that $\mathbf{x}$ takes the values x_i with probability p_i. In this case [see (4-15)]

$$f(x) = \sum_i p_i \delta(x - x_i)$$

Inserting into (5-21) and using the identity

$$\int_{-\infty}^{\infty} x \delta(x - x_i)\, dx = x_i$$

we obtain

$$E\{\mathbf{x}\} = \sum_i p_i x_i \qquad p_i = P\{\mathbf{x} = x_i\} \tag{5-22}$$

Example 5-16. If $\mathbf{x}$ takes the values $1, 2, \ldots, 6$ with probability $1/6$, then

$$E\{\mathbf{x}\} = \tfrac{1}{6}(1 + 2 + \cdots + 6) = 3.5$$

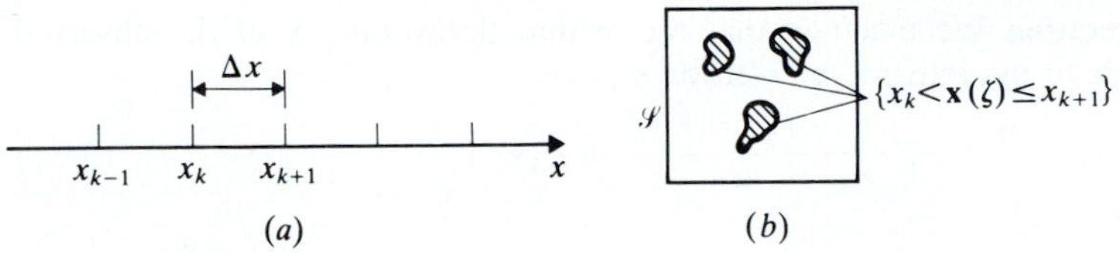

Δx x_{k-1} x_k x_{k+1} x (a) $\mathscr{S}$ $\{x_k < \mathbf{x}(\zeta) \le x_{k+1}\}$ (b)

FIGURE 5-19

Conditional mean The conditional mean of an RV $\mathbf{x}$ assuming $\mathscr{M}$ is given by the integral in (5-21) if $f(x)$ is replaced by the conditional density $f(x|\mathscr{M})$:

$$E\{\mathbf{x}|\mathscr{M}\} = \int_{-\infty}^{\infty} xf(x|\mathscr{M})\,dx \tag{5-23}$$

For discrete type RVs the above yields

$$E\{\mathbf{x}|\mathscr{M}\} = \sum_i x_i P\{\mathbf{x} = x_i|\mathscr{M}\} \tag{5-24}$$

Example 5-17. With $\mathscr{M} = \{\mathbf{x} \ge a\}$, it follows from (5-23) that

$$E\{\mathbf{x}|\mathbf{x} \ge a\} = \int_{-\infty}^{\infty} xf(x|\mathbf{x} \ge a)\,dx = \frac{\int_a^{\infty} xf(x)\,dx}{\int_a^{\infty} f(x)\,dx}$$

Lebesgue integral. The mean of an RV can be interpreted as a Lebesgue integral. This interpretation is important in mathematics but it will not be used in our development. We make, therefore, only a passing reference:

We divide the x axis into intervals (x_k, x_{k+1}) of length Δx as in Fig. 5-19a. If Δx is small, then the Riemann integral in (5-21) can be approximated by a sum

$$\int_{-\infty}^{\infty} xf(x)\,dx \simeq \sum_{k=-\infty}^{\infty} x_k f(x_k)\,\Delta x \tag{5-25}$$

And since $f(x_k)\,\Delta x \simeq P\{x_k < \mathbf{x} < x_k + \Delta x\}$, we conclude that

$$E\{\mathbf{x}\} \simeq \sum_{k=-\infty}^{\infty} x_k P\{x_k < \mathbf{x} < x_k + \Delta x\}$$

In the above, the sets $\{x_k < \mathbf{x} < x_k + \Delta x\}$ are differential events specified in terms of the RV $\mathbf{x}$, and their union is the space $\mathscr{S}$ (Fig. 5-19b). Hence, to find $E\{\mathbf{x}\}$, we multiply the probability of each differential event by the corresponding value of $\mathbf{x}$ and sum over all k. The resulting limit as $\Delta x \to 0$ is written in the form

$$E\{\mathbf{x}\} = \int_{\mathscr{S}} \mathbf{x}\,dP$$

and is called the *Lebesgue integral* of $\mathbf{x}$.

Frequency interpretation We maintain that the arithmetic average $\bar{x}$ of the observed values x_i of $\mathbf{x}$ tends to the integral in (5-21) as $n \to \infty$:

$$\bar{x} = \frac{x_1 + \cdots + x_n}{n} \to E\{\bar{\mathbf{x}}\} \tag{5-26}$$

Proof. We denote by Δn_k the number of x_i's that are between z_k and $z_k + \Delta x = z_{k+1}$. From this it follows that

$$x_1 + \cdots + x_n \simeq \sum z_k \, \Delta n_k$$

And since $f(z_k)\,\Delta x \simeq \Delta n_k / n$ [see (4-23)] we conclude that

$$\bar{x} \simeq \frac{1}{n} \sum z_k \, \Delta n_k \simeq \sum z_k f(z_k)\,\Delta x \simeq \int_{-\infty}^{\infty} x f(x)\,dx$$

and (5-26) results.

We shall use the above to express the mean of $\mathbf{x}$ in terms of its distribution. From the construction of Fig. 5-20a it follows readily that $\bar{x}$ equals the area under the empirical percentile curve of $\mathbf{x}$. Thus

$$\bar{x} = (BCD) - (OAB)$$

where (BCD) and (OAB) are the shaded areas above and below the u axis respectively. These areas equal the corresponding areas of Fig. 5-20b; hence

$$\bar{x} = \int_0^{\infty} [1 - F_n(x)]\,dx - \int_{-\infty}^{0} F_n(x)\,dx$$

where $F_n(x)$ is the empirical distribution of $\mathbf{x}$. With $n \to \infty$ this yields

$$E\{\mathbf{x}\} = \int_0^{\infty} R(x)\,dx - \int_{-\infty}^{0} F(x)\,dx \qquad R(x) = 1 - F(x) \tag{5-27}$$

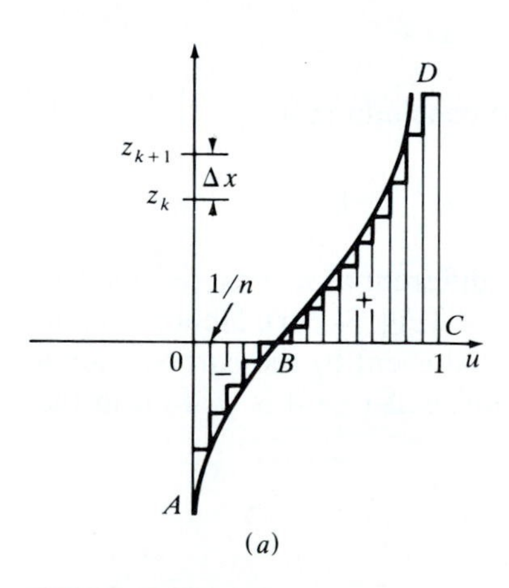

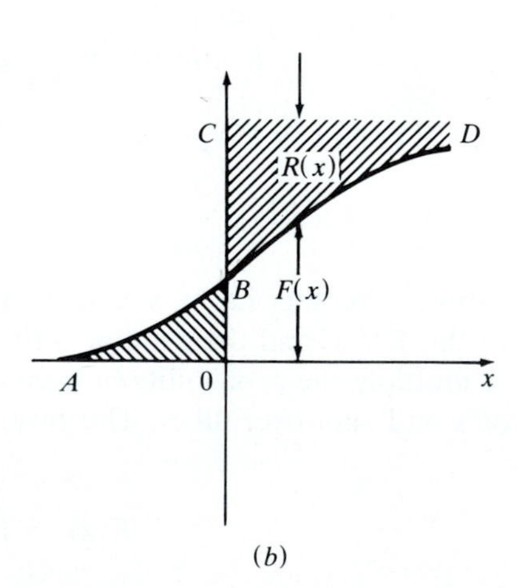

FIGURE 5-20

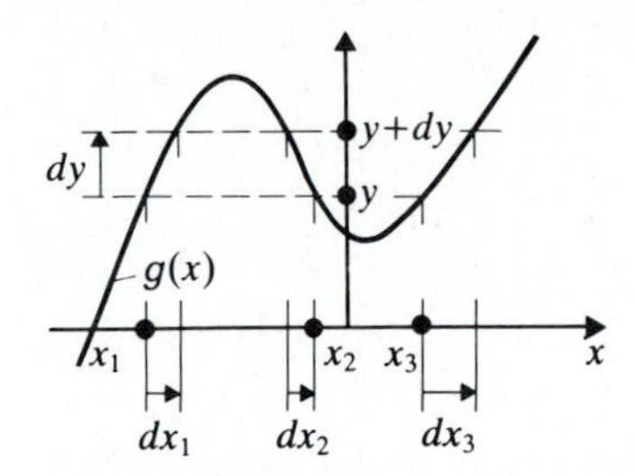

FIGURE 5-21

Mean of $g(\mathbf{x})$. Given an RV $\mathbf{x}$ and a function $g(x)$, we form the RV $\mathbf{y} = g(\mathbf{x})$. As we see from (5-21), the mean of this RV is given by

$$E\{\mathbf{y}\} = \int_{-\infty}^{\infty} y f_y(y)\, dy \tag{5-28}$$

It appears, therefore, that to determine the mean of $\mathbf{y}$, we must find its density $f_y(y)$. This, however, is not necessary. As the next basic theorem shows, $E\{\mathbf{y}\}$ can be expressed directly in terms of the function $g(x)$ and the density $f_x(x)$ of $\mathbf{x}$.

THEOREM

$$E\{g(\mathbf{x})\} = \int_{-\infty}^{\infty} g(x) f_x(x)\, dx \tag{5-29}$$

Proof. We shall sketch a proof using the curve $g(x)$ of Fig. 5-21. With $y = g(x_1) = g(x_2) = g(x_3)$ as in the figure, we see that

$$f_y(y)\, dy = f_x(x_1)\, dx_1 + f_x(x_2)\, dx_2 + f_x(x_3)\, dx_3$$

Multiplying by y, we obtain

$$y f_y(y)\, dy = g(x_1) f_x(x_1)\, dx_1 + g(x_2) f_x(x_2)\, dx_2 + g(x_3) f_x(x_3)\, dx_3$$

Thus to each differential in (5-28) there correspond one or more differentials in (5-29). As dy covers the y axis, the corresponding dx's are nonoverlapping and they cover the entire x axis. Hence the integrals in (5-28) and (5.29) are equal.

If $\mathbf{x}$ is of discrete type as in (5-22), then (5-29) yields

$$E\{g(\mathbf{x})\} = \sum_i g(x_i) P\{\mathbf{x} = x_i\} \tag{5-30}$$

Example 5-18. With x_0 an arbitrary number and $g(x)$ as in Fig. 5-22, (5-29) yields

$$E\{g(\mathbf{x})\} = \int_{-\infty}^{x_0} f_x(x)\, dx = F_x(x_0)$$

This shows that the distribution function of an RV can be expressed as expected value.

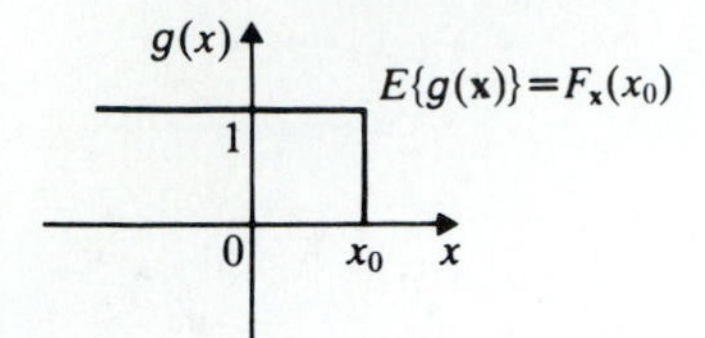

FIGURE 5-22

Example 5-19. In this example, we show that the probability of any event $\mathscr{A}$ can be expressed as expected value. For this purpose we form the zero–one RV $\mathbf{x}_{\mathscr{A}}$ associated with the event $\mathscr{A}$:

$$\mathbf{x}_{\mathscr{A}}(\zeta) = \begin{cases} 1 & \zeta \in \mathscr{A} \\ 0 & \zeta \notin \mathscr{A} \end{cases}$$

Since this RV takes the values 1 and 0 with respective probabilities $P(\mathscr{A})$ and $P(\overline{\mathscr{A}})$, (5-22) yields

$$E\{\mathbf{x}_{\mathscr{A}}\} = 1 \times P(\mathscr{A}) + 0 \times P(\overline{\mathscr{A}}) = P(\mathscr{A})$$

Linearity From (5-29) it follows that

$$E\{a_1 g_1(\mathbf{x}) + \cdots + a_n g_n(\mathbf{x})\} = a_1 E\{g_1(\mathbf{x})\} + \cdots + a_n E\{g_n(\mathbf{x})\} \tag{5-31}$$

In particular, $E\{a\mathbf{x} + b\} = aE\{\mathbf{x}\} + b$

Complex RVs If $\mathbf{z} = \mathbf{x} + j\mathbf{y}$ is a complex RV, then its expected value is by definition

$$E\{\mathbf{z}\} = E\{\mathbf{x}\} + jE\{\mathbf{y}\}$$

From this and (5-29) it follows that if

$$g(\mathbf{x}) = g_1(\mathbf{x}) + jg_2(\mathbf{x})$$

is a complex function of the real RV $\mathbf{x}$ then

$$E\{g(\mathbf{x})\} = \int_{-\infty}^{\infty} g_1(x)f(x)\,dx + j\int_{-\infty}^{\infty} g_2(x)f(x)\,dx = \int_{-\infty}^{\infty} g(x)f(x)\,dx \tag{5-32}$$

In other words, (5-29) holds even if $g(x)$ is complex.

Variance

The *variance* of an RV $\mathbf{x}$ is by definition the integral

$$\sigma^2 = \int_{-\infty}^{\infty} (x - \eta)^2 f(x)\,dx \tag{5-33}$$

where $\eta = E\{\mathbf{x}\}$. The positive constant σ, denoted also by σ_x, is called the *standard deviation* of $\mathbf{x}$.

From the definition it follows that σ^2 is the mean of the RV $(\mathbf{x} - \eta)^2$. Thus

$$\sigma^2 = E\{(\mathbf{x} - \eta)^2\} = E\{\mathbf{x}^2 - 2\mathbf{x}\eta + \eta^2\} = E\{\mathbf{x}^2\} - 2\eta E\{\mathbf{x}\} + \eta^2$$

Hence

$$\sigma^2 = E\{\mathbf{x}^2\} - E^2\{\mathbf{x}\} \tag{5-34}$$

Example 5-20. If $\mathbf{x}$ is uniform in the interval $(-c, c)$, then $\eta = 0$ and

$$\sigma^2 = E\{\mathbf{x}^2\} = \frac{1}{2c}\int_{-c}^{c} x^2\,dx = \frac{c^2}{3}$$

Example 5-21. We have written the density of a normal RV in the form

$$f(x) = \frac{1}{\sigma\sqrt{2\pi}} e^{-(x-\eta)^2/2\sigma^2}$$

where up to now η and σ^2 were two arbitrary constants. We show next that η is indeed the mean of $\mathbf{x}$ and σ^2 its variance.

Proof. Clearly, $f(x)$ is symmetrical about the line $x = \eta$; hence $E\{\mathbf{x}\} = \eta$. Furthermore,

$$\int_{-\infty}^{\infty} e^{-(x-\eta)^2/2\sigma^2}\,dx = \sigma\sqrt{2\pi}$$

because the area of $f(x)$ equals 1. Differentiating with respect to σ, we obtain

$$\int_{-\infty}^{\infty} \frac{(x-\eta)^2}{\sigma^3} e^{-(x-\eta)^2/2\sigma^2}\,dx = \sqrt{2\pi}$$

Multiplying both sides by $\sigma^2/\sqrt{2\pi}$, we conclude that $E\{\mathbf{x} - \eta)^2\} = \sigma^2$ and the proof is complete.

Discrete type. If the RV $\mathbf{x}$ is of discrete type, then

$$\sigma^2 = \sum_i p_i(x_i - \eta)^2 \qquad p_i = P\{\mathbf{x} = x_i\} \tag{5-35}$$

Example 5-22. The RV $\mathbf{x}$ takes the values 1 and 0 with probabilities p and $q = 1 - p$ respectively. In this case

$$E\{\mathbf{x}\} = 1 \times p + 0 \times q = p$$

$$E\{\mathbf{x}^2\} = 1^2 \times p + 0^2 \times q = p$$

Hence

$$\sigma^2 = E\{\mathbf{x}^2\} - E^2\{\mathbf{x}\} = p - p^2 = pq$$

Example 5-23. A Poisson distributed RV with parameter a takes the values $0, 1, \dots$ with probabilities

$$P\{\mathbf{x} = k\} = e^{-a}\frac{a^k}{k!}$$

We shall show that its mean and variance equal a:

$$E\{\mathbf{x}\} = a \qquad E\{\mathbf{x}^2\} = a^2 + a \qquad \sigma^2 = a \tag{5-36}$$

Proof. We differentiate twice the Taylor expansion of e^a:

$$e^a = \sum_{k=0}^{\infty}\frac{a^k}{k!}$$

$$e^a = \sum_{k=0}^{\infty} k\frac{a^{k-1}}{k!} = \frac{1}{a}\sum_{k=1}^{\infty} k\frac{a^k}{k!}$$

$$e^a = \sum_{k=1}^{\infty} k(k-1)\frac{a^{k-2}}{k!} = \frac{1}{a^2}\sum_{k=1}^{\infty} k^2\frac{a^k}{k!} - \frac{1}{a^2}\sum_{k=1}^{\infty} k\frac{a^k}{k!}$$

Hence

$$E\{\mathbf{x}\} = e^{-a}\sum_{k=1}^{\infty} k\frac{a^k}{k!} = a \qquad E\{\mathbf{x}^2\} = e^{-a}\sum_{k=1}^{\infty} k^2\frac{a^k}{k!} = a^2 + a$$

and (5-36) results.

Poisson points. As we have shown in (3-47), the number $\mathbf{n}$ of Poisson points in an interval of length t_0 is a Poisson distributed RV with parameter $a = \lambda t_0$. From this it follows that

$$E\{\mathbf{n}\} = \lambda t_0 \qquad \sigma_n^2 = \lambda t_0 \tag{5-37}$$

This shows that the density λ of Poisson points equals the expected number of points per unit time.

Notes 1. The variance σ^2 of an RV $\mathbf{x}$ is a measure of the concentration of $\mathbf{x}$ near its mean η. Its relative frequency interpretation (empirical estimate) is the average of $(x_i - \eta)^2$:

$$\sigma^2 \simeq \frac{1}{n}\sum(x_i - \eta)^2$$

where x_i are the observed values of $\mathbf{x}$. This average can be used as the estimate of σ^2 only if η is known. If it is unknown, we replace it by its estimate $\bar{x}$ and we change n to $n-1$. This yields the estimate

$$\sigma^2 \simeq \frac{1}{n-1}\sum(x_i - \bar{x})^2 \qquad \bar{x} = \frac{1}{n}\sum x_i$$

known as the *sample variance* of $\mathbf{x}$ [see (8–64)]. The reason for changing n to $n-1$ is explained later.

2. A simpler measure of the concentration of $\mathbf{x}$ near η is the first absolute central moment $M = E\{|\mathbf{x} - \eta|\}$. Its empirical estimate is the average of $|x_i - \eta|$:

$$M \simeq \frac{1}{n} \sum |x_i - \eta|$$

If η is unknown, it is replaced by $\bar{x}$. This estimate avoids the computation of squares.

5-4 MOMENTS

The following quantities are of interest in the study of RVs:

Moments

$$m_n = E\{\mathbf{x}^n\} = \int_{-\infty}^{\infty} x^n f(x)\, dx \tag{5-38}$$

Central moments

$$\mu_n = E\{(\mathbf{x} - \eta)^n\} = \int_{-\infty}^{\infty} (x - \eta)^n f(x)\, dx \tag{5-39}$$

Absolute moments

$$E\{|\mathbf{x}|^n\} \qquad E\{|\mathbf{x} - \eta|^n\} \tag{5-40}$$

Generalized moments

$$E\{(\mathbf{x} - a)^n\} \qquad E\{|\mathbf{x} - a|^n\} \tag{5-41}$$

We note that

$$\mu_n = E\{(\mathbf{x} - \eta)^n\} = E\left\{ \sum_{k=0}^{n} \binom{n}{k} \mathbf{x}^k (-\eta)^{n-k} \right\}$$

Hence

$$\mu_n = \sum_{k=0}^{n} \binom{n}{k} m_k (-\eta)^{n-k} \tag{5-42}$$

Similarly,

$$m_n = E\{[(\mathbf{x} - \eta) + \eta]^n\} = E\left\{ \sum_{k=0}^{n} \binom{n}{k} (\mathbf{x} - \eta)^k \eta^{n-k} \right\}$$

Hence

$$m_n = \sum_{k=0}^{n} \binom{n}{k} \mu_k \eta^{n-k} \tag{5-43}$$

In particular,

$$\mu_0 = m_0 = 1 \qquad m_1 = \eta \qquad \mu_1 = 0 \qquad \mu_2 = \sigma^2$$

and

$$\mu_3 = m_3 - 3\eta m_2 + 2\eta^3 \qquad m_3 = \mu_3 + 3\eta\sigma^2 + \eta^3$$

Notes 1. If the function $f(x)$ is interpreted as mass density on the x axis, then $E\{\mathbf{x}\}$ equals its center of gravity, $E\{\mathbf{x}^2\}$ equals the moment of inertia with respect to the origin, and σ^2 equals the central moment of inertia. The standard deviation σ is the radius of gyration.

2. The constants η and σ give only a limited characterization of $f(x)$. Knowledge of other moments provides additional information that can be used, for example, to distinguish between two densities with the same η and σ. In fact, if m_n is known for every n, then, under certain conditions, $f(x)$ is determined uniquely [see also (5-69)]. The underlying theory is known in mathematics as the *moment problem*.

3. The moments of an RV are not arbitrary numbers but must satisfy various inequalities. For example [see (5-34)]

$$\sigma^2 = m_2 - m_1^2 \geq 0$$

Similarly, since the quadratic

$$E\{(\mathbf{x}^n - a)^2\} = m_{2n} - 2am_n + a^2$$

is nonnegative for any a, its discriminant cannot be positive. Hence

$$m_{2n} \geq m_n^2$$

Normal random variables. We shall show that if

$$f(x) = \frac{1}{\sigma\sqrt{2\pi}} e^{-x^2/2\sigma^2}$$

then

$$E\{\mathbf{x}^n\} = \begin{cases} 0 & n = 2k+1 \\ 1 \cdot 3 \cdots (n-1)\sigma^n & n = 2k \end{cases} \tag{5-44}$$

$$E\{|\mathbf{x}|^n\} = \begin{cases} 1 \cdot 3 \cdots (n-1)\sigma^n & n = 2k \\ 2^k k! \sigma^{2k+1}\sqrt{2/\pi} & n = 2k+1 \end{cases} \tag{5-45}$$

The odd moments of $\mathbf{x}$ are 0 because $f(-x) = f(x)$. To prove the lower part of (5-44), we differentiate k times the identity

$$\int_{-\infty}^{\infty} e^{-\alpha x^2}\,dx = \sqrt{\frac{\pi}{\alpha}}$$

This yields

$$\int_{-\infty}^{\infty} x^{2k} e^{-\alpha x^2}\,dx = \frac{1 \cdot 3 \cdots (2k-1)}{2^k}\sqrt{\frac{\pi}{\alpha^{2k+1}}}$$

and with $\alpha = 1/2\sigma^2$, (5-44) results.

Since $f(-x) = f(x)$, we have

$$E\{|\mathbf{x}|^{2k+1}\} = 2\int_0^{\infty} x^{2k+1} f(x)\,dx = \frac{2}{\sigma\sqrt{2\pi}}\int_0^{\infty} x^{2k+1} e^{-x^2/2\sigma^2}\,dx$$

With $y = x^2/2\sigma^2$, the above yields

$$\sqrt{\frac{2}{\pi}} \frac{(2\sigma^2)^{k+1}}{2\sigma} \int_0^\infty y^k e^{-y}\, dy$$

and (5-45) results because the last integral equals $k!$

We note in particular that

$$E\{\mathbf{x}^4\} = 3\sigma^4 = 3E^2\{\mathbf{x}^2\} \tag{5-46}$$

Example 5-24. If $\mathbf{x}$ has a *Rayleigh density*

$$f(x) = \frac{x}{\alpha^2} e^{-x^2/2\alpha^2} U(x)$$

then

$$E\{\mathbf{x}^n\} = \frac{1}{\alpha^2} \int_0^\infty x^{n+1} e^{-x^2/2\alpha^2}\, dx = \frac{1}{2\alpha^2} \int_{-\infty}^\infty |x|^{n+1} e^{-x^2/2\alpha^2}\, dx$$

From this and (5-45) it follows that

$$E\{\mathbf{x}^n\} = \begin{cases} 1 \cdot 3 \cdots n\alpha^n \sqrt{\pi/2} & n = 2k+1 \\ 2^k k! \alpha^{2k} & n = 2k \end{cases} \tag{5-47}$$

In particular,

$$E\{\mathbf{x}\} = \alpha\sqrt{\pi/2} \qquad \sigma^2 = (2 - \pi/2)\alpha^2$$

Example 5-25. If $\mathbf{x}$ has a *Maxwell density*

$$f(x) = \frac{\sqrt{2}}{\alpha^3\sqrt{\pi}} x^2 e^{-x^2/2\alpha^2} U(x)$$

then

$$E\{\mathbf{x}^n\} = \frac{1}{\alpha^3\sqrt{2\pi}} \int_{-\infty}^\infty |x|^{n+2} e^{-x^2/2\alpha^2}\, dx$$

and (5-45) yields

$$E\{\mathbf{x}^n\} = \begin{cases} 1 \cdot 3 \cdots (n+1)\alpha^n & n = 2k \\ 2^k k! \alpha^{2k-1} \sqrt{2/\pi} & n = 2k-1 \end{cases} \tag{5-48}$$

In particular,

$$E\{\mathbf{x}\} = 2\alpha\sqrt{2/\pi} \qquad E\{\mathbf{x}^2\} = 3\alpha^2$$

Poisson random variables. The moments of a Poisson distributed RV are functions of the parameter a:

$$m_n(a) = E\{\mathbf{x}^n\} = e^{-a} \sum_{k=0}^{\infty} k^n \frac{a^k}{k!} \tag{5-49}$$

$$\mu_n(a) = E\{(\mathbf{x} - a)^n\} = e^{-a} \sum_{k=0}^{\infty} (k - a)^n \frac{a^k}{k!} \tag{5-50}$$

We shall show that they satisfy the recursion equations

$$m_{n+1}(a) = a[m_n(a) + m'_n(a)] \tag{5-51}$$

$$\mu_{n+1}(a) = a[n\mu_{n-1}(a) + \mu'_n(a)] \tag{5-52}$$

Proof. Differentiating (5-49) with respect to a, we obtain

$$m'_n(a) = -e^{-a} \sum_{k=0}^{\infty} k^n \frac{a^k}{k!} + e^{-a} \sum_{k=0}^{\infty} k^{n+1} \frac{a^{k-1}}{k!} = -m_n(a) + \frac{1}{a} m_{n+1}(a)$$

and (5-51) results. Similarly, from (5-50) it follows that

$$\mu'_n(a) = -e^{-a} \sum_{k=0}^{\infty} (k - a)^n \frac{a^k}{k!} - ne^{-a} \sum_{k=0}^{\infty} (k - a)^{n-1} \frac{a^k}{k!}$$

$$+ e^{-a} \sum_{k=0}^{\infty} (k - a)^n k \frac{a^{k-1}}{k!}$$

Setting $k = (k - a) + a$ in the last sum, we obtain $\mu'_n = -\mu_n - n\mu_{n-1} + (1/a)(\mu_{n+1} + a\mu_n)$ and (5-52) results.

The preceding equations lead to the recursive determination of the moments m_n and μ_n. Starting with the known moments $m_1 = a$, $\mu_1 = 0$, and $\mu_2 = a$ [see (5-36)], we obtain $m_2 = a(a + 1)$ and

$$m_3 = a(a^2 + a + 2a + 1) = a^3 + 3a^2 + a \qquad \mu_3 = a(\mu'_2 + 2\mu_1) = a$$

ESTIMATE OF THE MEAN OF $g(\mathbf{x})$. The mean of the RV $\mathbf{y} = g(\mathbf{x})$ is given by

$$E\{g(\mathbf{x})\} = \int_{-\infty}^{\infty} g(x) f(x)\, dx \tag{5-53}$$

Hence, for its determination, knowledge of $f(x)$ is required. However, if $\mathbf{x}$ is concentrated near its mean, then $E\{g(\mathbf{x})\}$ can be expressed in terms of the moments μ_n of $\mathbf{x}$.

Suppose, first, that $f(x)$ is negligible outside an interval $(\eta - \varepsilon, \eta + \varepsilon)$ and in this interval, $g(x) \simeq g(\eta)$. In this case, (5-53) yields

$$E\{g(\mathbf{x})\} \simeq g(\eta) \int_{\eta-\varepsilon}^{\eta+\varepsilon} f(x)\, dx \simeq g(\eta)$$

This estimate can be improved if $g(x)$ is approximated by a polynomial

$$g(x) \simeq g(\eta) + g'(\eta)(x-\eta) + \cdots + g^{(n)}(\eta)\frac{(x-\eta)^n}{n!}$$

Inserting into (5-53), we obtain

$$E\{g(\mathbf{x})\} \simeq g(\eta) + g''(\eta)\frac{\sigma^2}{2} + \cdots + g^{(n)}(\eta)\frac{\mu_n}{n!} \tag{5-54}$$

In particular, if $g(x)$ is approximated by a parabola, then

$$\eta_y = E\{g(\mathbf{x})\} \simeq g(\eta) + g''(\eta)\frac{\sigma^2}{2} \tag{5-55}$$

And if it is approximated by a straight line, then $\eta_y \simeq g(\eta)$. This shows that the slope of $g(x)$ has no effect on η_y; however, as we show next, it affects the variance σ_y^2 of $\mathbf{y}$.

Variance. We maintain that the first-order estimate of σ_y^2 is given by

$$\sigma_y^2 \simeq |g'(\eta)|^2\sigma^2 \tag{5-56}$$

Proof. We apply (5-55) to the function $g^2(x)$. Since its second derivative equals $2(g')^2 + 2gg''$, we conclude that

$$\sigma_y^2 + \eta_y^2 = E\{g^2(\mathbf{x})\} \simeq g^2 + \left[(g')^2 + gg''\right]\sigma^2$$

Inserting the approximation (5-55) for η_y into the above and neglecting the σ^4 term, we obtain (5-56).

Example 5-26. A voltage $E = 120$ V is connected across a resistor whose resistance is an RV $\mathbf{r}$ uniform between 900 and 1100 Ω. Using (5-55) and (5-56), we shall estimate the mean and variance of the resulting current

$$\mathbf{i} = \frac{E}{\mathbf{r}}$$

Clearly, $E\{\mathbf{r}\} = \eta = 10^3$, $\sigma^2 = 100^2/3$. With $g(r) = E/r$, we have

$$g(\eta) = 0.12 \qquad g'(\eta) = -12 \times 10^{-5} \qquad g''(\eta) = 24 \times 10^{-8}$$

Hence

$$E\{\mathbf{i}\} \simeq 0.12 + 0.0004\ A \qquad \sigma_i^2 \simeq 48 \times 10^{-6}\ A^2$$

Tchebycheff Inequality

A measure of the concentration of an RV near its mean η is its variance σ^2. In fact, as the following theorem shows, the probability that $\mathbf{x}$ is outside an arbitrary interval $(\eta - \varepsilon, \eta + \varepsilon)$ is negligible if the ratio σ/ε is sufficiently small. This result, known as the *Tchebycheff inequality*, is fundamental.

THEOREM. For any $\varepsilon > 0$,

$$P\{|\mathbf{x} - \eta| \geq \varepsilon\} \leq \frac{\sigma^2}{\varepsilon^2} \tag{5-57}$$

Proof. The proof is based on the fact that

$$P\{|\mathbf{x} - \eta| \geq \varepsilon\} = \int_{-\infty}^{-\eta-\varepsilon} f(x)\,dx + \int_{\eta+\varepsilon}^{\infty} f(x)\,dx = \int_{|x-\eta|\geq\varepsilon} f(x)\,dx$$

Indeed

$$\sigma^2 = \int_{-\infty}^{\infty} (x - \eta)^2 f(x)\,dx \geq \int_{|x-\eta|\geq\varepsilon} (x - \eta)^2 f(x)\,dx \geq \varepsilon^2 \int_{|x-\eta|\geq\varepsilon} f(x)\,dx$$

and (5-57) results because the last integral equals $P\{|\mathbf{x} - \eta| \geq \varepsilon\}$.

Notes 1. From (5-57) it follows that, if $\sigma = 0$, then the probability that $\mathbf{x}$ is outside the interval $(\eta - \varepsilon, \eta + \varepsilon)$ equals 0 for any ε; hence $\mathbf{x} = \eta$ with probability 1. Similarly, if

$$E\{\mathbf{x}^2\} = \eta^2 + \sigma^2 = 0 \qquad \text{then} \quad \eta = 0 \qquad \sigma = 0$$

hence $\mathbf{x} = 0$ with probability 1.

2. For specific densities, the bound in (5-57) is too high. Suppose, for example, that $\mathbf{x}$ is normal. In this case, $P\{|\mathbf{x} - \eta| \geq 3\sigma\} = 2 - 2\mathbb{G}(3) = 0.0027$. Inequality (5-57), however, yields $P\{|\mathbf{x} - \eta| \geq 3\sigma\} \leq 1/9$.

The significance of Tchebycheff's inequality is the fact that it holds for *any* $f(x)$ and can, therefore be used even if $f(x)$ is not known.

3. The bound in (5-57) can be reduced if various assumptions are made about $f(x)$ [see *Chernoff bound* (Prob. 5-30)].

MARKOFF INEQUALITY. If $f(x) = 0$ for $x < 0$, then, for any $\alpha > 0$,

$$P\{\mathbf{x} \geq \alpha\} \leq \frac{\eta}{\alpha} \tag{5-58}$$

Proof.

$$E\{\mathbf{x}\} = \int_0^{\infty} x f(x)\,dx \geq \int_{\alpha}^{\infty} x f(x)\,dx \geq \alpha \int_{\alpha}^{\infty} f(x)\,dx$$

and (5-58) results because the last integral equals $P\{\mathbf{x} \geq \alpha\}$.

COROLLARY. Suppose that $\mathbf{x}$ is an arbitrary RV and a and n are two arbitrary numbers. Clearly, the RV $|\mathbf{x} - a|^n$ takes only positive values. Applying (5-58), with $\alpha = \varepsilon^n$, we conclude that

$$P\{|\mathbf{x} - a|^n \geq \varepsilon^n\} \leq \frac{E\{|\mathbf{x} - a|^n\}}{\varepsilon^n}$$

Hence

$$P\{|\mathbf{x} - a| \geq \varepsilon\} \leq \frac{E\{|\mathbf{x} - a|^n\}}{\varepsilon^n} \tag{5-59}$$

This result is known as the *inequality of Bienaymé*. Tchebycheff's inequality is a special case obtained with $a = \eta$ and $n = 2$.

5-5 CHARACTERISTIC FUNCTIONS

The *characteristic function* of an RV is by definition the integral

$$\Phi(\omega) = \int_{-\infty}^{\infty} f(x) e^{j\omega x}\, dx \tag{5-60}$$

This function is maximum at the origin because $f(x) \geq 0$:

$$|\Phi(\omega)| \leq \Phi(0) = 1 \tag{5-61}$$

If $j\omega$ is changed to s, the resulting integral

$$\mathbf{\Phi}(s) = \int_{-\infty}^{\infty} f(x) e^{sx}\, dx \qquad \mathbf{\Phi}(j\omega) = \Phi(\omega) \tag{5-62}$$

is the *moment (generating) function* of $\mathbf{x}$.

The function

$$\Psi(\omega) = \ln \Phi(\omega) = \mathbf{\Psi}(j\omega) \tag{5-63}$$

is the *second characteristic function* of $\mathbf{x}$.

Clearly [see (5-32)]

$$\Phi(\omega) = E\{e^{j\omega\mathbf{x}}\} \qquad \mathbf{\Phi}(s) = E\{e^{s\mathbf{x}}\}$$

This leads to the following:

$$\text{If} \quad \mathbf{y} = a\mathbf{x} + b \qquad \text{then} \quad \Phi_y(\omega) = e^{jb\omega}\Phi_x(a\omega) \tag{5-64}$$

because

$$E\{e^{j\omega\mathbf{y}}\} = E\{e^{j\omega(a\mathbf{x}+b)}\} = e^{jb\omega}E\{e^{ja\omega\mathbf{x}}\}$$

Example 5-27. We shall show that the characteristic function of an $N(\eta, \sigma)$ RV $\mathbf{x}$ equals

$$\Phi_x(\omega) = \exp\{j\eta\omega - \tfrac{1}{2}\sigma^2\omega^2\} \tag{5-65}$$

Proof. The RV $\mathbf{z} = (\mathbf{x} - \eta)/\sigma$ is $N(0, 1)$ and its moment function equals

$$\mathbf{\Phi}_z(s) = \frac{1}{\sqrt{2\pi}} \int_{-\infty}^{\infty} e^{sz} e^{-z^2/2}\, dz$$

With

$$sz - \frac{z^2}{2} = -\frac{1}{2}(z - s)^2 + \frac{s^2}{2}$$

we conclude that

$$\mathbf{\Phi}_z(s) = e^{s^2/2}\int_{-\infty}^{\infty} \frac{1}{\sqrt{2\pi}} e^{-(z-s)^2/2}\, dz = e^{s^2/2}$$

And since $\mathbf{x} = \sigma \mathbf{z} + \eta$, (5-65) follows from (5-64).

Inversion formula As we see from (5-60), $\Phi(\omega)$ is the Fourier transform of $f(x)$. Hence the properties of characteristic functions are essentially the same as the properties of Fourier transforms. We note, in particular, that $f(x)$ can be expressed in terms of $\Phi(\omega)$

$$f(x) = \frac{1}{2\pi}\int_{-\infty}^{\infty} \Phi(\omega) e^{-j\omega x}\, d\omega \tag{5-66}$$

Moment theorem. Differentiating (5-62) n times, we obtain

$$\mathbf{\Phi}^{(n)}(s) = E\{\mathbf{x}^n e^{s\mathbf{x}}\}$$

Hence

$$\mathbf{\Phi}^{(n)}(0) = E\{\mathbf{x}^n\} = m_n \tag{5-67}$$

Thus the derivatives of $\mathbf{\Phi}(s)$ at the origin equal the moments of $\mathbf{x}$. This justifies the name "moment function" given to $\mathbf{\Phi}(s)$.

In particular,

$$\mathbf{\Phi}'(0) = m_1 = \eta \qquad \mathbf{\Phi}''(0) = m_2 = \eta^2 + \sigma^2 \tag{5-68}$$

Note Expanding $\mathbf{\Phi}(s)$ into a series near the origin and using (5-67), we obtain

$$\mathbf{\Phi}(s) = \sum_{n=0}^{\infty} \frac{m_n}{n!} s^n \tag{5-69}$$

This is valid only if all moments are finite and the series converges absolutely near $s = 0$. Since $f(x)$ can be determined in terms of $\mathbf{\Phi}(s)$, (5-69) shows that, under the stated conditions, the density of an RV is uniquely determined if all its moments are known.

Example 5-28. We shall determine the moment function and the moments of an RV $\mathbf{x}$ with *gamma distribution*:

$$f(x) = \gamma x^{b-1} e^{-cx} U(x) \qquad \gamma = \frac{c^{b+1}}{\Gamma(b+1)}$$

From (4-39) it follows that

$$\mathbf{\Phi}(s) = \gamma \int_0^{\infty} x^{b-1} e^{-(c-s)x}\, dx = \frac{\gamma \Gamma(b)}{(c-s)^b} = \frac{c^b}{(c-s)^b} \tag{5-70}$$

Differentiating with respect to s and setting $s = 0$, we obtain

$$\mathbf{\Phi}^{(n)}(0) = \frac{b(b+1)\cdots(b+n-1)}{c^n} = E\{\mathbf{x}^n\}$$

With $n = 1$ and $n = 2$, this yields

$$E\{\mathbf{x}\} = \frac{b}{c} \qquad E\{\mathbf{x}^2\} = \frac{b(b+1)}{c^2} \qquad \sigma^2 = \frac{b}{c^2}$$

The *exponential density* is a special case obtained with $b = 1$:

$$f(x) = ce^{-cx}U(x) \qquad \mathbf{\Phi}(s) = \frac{c}{c-s} \qquad E\{\mathbf{x}\} = \frac{1}{c} \qquad \sigma^2 = \frac{1}{c^2}$$

Chi square. Setting $b = m/2$ and $c = 1/2$ in (5-70), we obtain the moment function of the chi-square density $\chi^2(m)$:

$$\mathbf{\Phi}(s) = \frac{1}{\sqrt{(1-2s)^m}} \qquad E\{\mathbf{x}\} = m \qquad \sigma^2 = 2m \tag{5-71}$$

Cumulants. The cumulants λ_n of RV $\mathbf{x}$ are by definition the derivatives

$$\frac{d^n\mathbf{\Psi}(0)}{ds^n} = \lambda_n \tag{5-72}$$

of its second moment function $\mathbf{\Psi}(s)$. Clearly [see (5-63)] $\mathbf{\Psi}(0) = \lambda_0 = 0$; hence

$$\mathbf{\Psi}(s) = \lambda_1 s + \frac{1}{2}\lambda_2 s^2 + \cdots + \frac{1}{n!}\lambda_n s^n + \cdots$$

We maintain that

$$\lambda_1 = \eta \qquad \lambda_2 = \sigma^2 \tag{5-73}$$

Proof. Since $\mathbf{\Phi} = e^{\mathbf{\Psi}}$, we conclude that

$$\mathbf{\Phi}' = \mathbf{\Psi}' e^{\mathbf{\Psi}} \qquad \mathbf{\Phi}' = \left[\mathbf{\Psi}'' + (\mathbf{\Psi}')^2\right]e^{\mathbf{\Psi}}$$

With $s = 0$, this yields

$$\mathbf{\Phi}'(0) = \mathbf{\Psi}'(0) = m_1 \qquad \mathbf{\Phi}''(0) = \mathbf{\Psi}''(0) + [\mathbf{\Psi}'(0)]^2 = m_2$$

and (5-73) results.

Discrete Type

Suppose that $\mathbf{x}$ is a discrete type RV taking the values x_i with probability p_i. In this case, (5-60) yields

$$\Phi(\omega) = \sum_i p_i e^{j\omega x_i} \tag{5-74}$$

Thus $\Phi(\omega)$ is a sum of exponentials. The moment function of $\mathbf{x}$ can be defined as in (5-62). However, if $\mathbf{x}$ takes only integer values, then a definition in terms of z transforms is preferable.

LATTICE TYPE. If $\mathbf{n}$ is a lattice type RV taking integer values, then its moment function is by definition the sum

$$\Gamma(z) = E\{z^{\mathbf{n}}\} = \sum_{n=-\infty}^{\infty} p_n z^n \tag{5-75}$$

Thus $\Gamma(1/z)$ is the z transform of the sequence $p_n = P\{\mathbf{n} = n\}$. With $\Phi(\omega)$ as in (5-74), the above yields

$$\Phi(\omega) = \Gamma(e^{j\omega}) = \sum_{n=-\infty}^{\infty} p_n e^{jn\omega}$$

Thus $\Phi(\omega)$ is the *discrete Fourier transform* (DFT) of p_n and

$$\Psi(s) = \ln \Gamma(e^s) \tag{5-76}$$

Moment theorem. Differentiating (5-75) k times, we obtain

$$\Gamma^{(k)}(z) = E\{\mathbf{n}(\mathbf{n}-1)\cdots(\mathbf{n}-k+1)z^{\mathbf{n}-k}\}$$

With $z = 1$, this yields

$$\Gamma^{(k)}(1) = E\{\mathbf{n}(\mathbf{n}-1)\cdots(\mathbf{n}-k+1)\} \tag{5-77}$$

We note, in particular, that $\Gamma(1) = 1$ and

$$\Gamma'(1) = E\{\mathbf{n}\} \qquad \Gamma''(1) = E\{\mathbf{n}^2\} - E\{\mathbf{n}\} \tag{5-78}$$

Example 5-29. (*a*) If $\mathbf{n}$ takes the values 0 and 1 with $P\{\mathbf{n} = 1\} = p$ and $P\{\mathbf{n} = 0\} = q$, then

$$\Gamma(z) = pz + q$$

$$\Gamma'(1) = E\{\mathbf{n}\} = p \qquad \Gamma''(1) = E\{\mathbf{n}^2\} - E\{\mathbf{n}\} = 0$$

(*b*) If $\mathbf{n}$ has the binomial distribution

$$p_n = P\{\mathbf{n} = n\} = \binom{m}{n} p^n q^{m-n} \qquad 0 \le n \le m$$

then

$$\Gamma(z) = \sum_{n=0}^{m} \binom{m}{n} p^n q^{m-n} z^n = (pz+q)^m$$

$$\Gamma'(1) = mp \qquad \Gamma''(1) = m(m-1)p^2$$

Hence

$$E\{\mathbf{n}\} = mp \qquad \sigma^2 = mpq$$

Example 5-30. If $\mathbf{n}$ is Poisson distributed with parameter a,

$$P\{\mathbf{n} = n\} = e^{-a}\frac{a^n}{n!} \qquad n = 0, 1, \ldots$$

then

$$\Gamma(z) = e^{-a} \sum_{n=0}^{\infty} a^n \frac{z^n}{n!} = e^{a(z-1)} \tag{5-79}$$

In this case [see (5-76)]

$$\Psi(s) = a(e^s - 1) \qquad \Psi'(0) = a \qquad \Psi''(0) = a$$

and (5-73) yields $E\{\mathbf{n}\} = a$, $\sigma^2 = a$ in agreement with (5-36).

Determination of the density of $g(\mathbf{x})$. We show next that characteristic functions can be used to determine the density $f_y(y)$ of the RV $\mathbf{y} = g(\mathbf{x})$ in terms of the density $f_x(x)$ of $\mathbf{x}$.

From (5-32) it follows that the characteristic function

$$\Phi_y(\omega) = \int_{-\infty}^{\infty} e^{j\omega y} f_y(y)\, dy$$

of the RV $\mathbf{y} = g(\mathbf{x})$ equals

$$\Phi_y(\omega) = E\{e^{j\omega g(\mathbf{x})}\} = \int_{-\infty}^{\infty} e^{j\omega g(x)} f_x(x)\, dx \tag{5-80}$$

If, therefore, the above integral can be written in the form

$$\int_{-\infty}^{\infty} e^{j\omega y} h(y)\, dy$$

it will follow that (uniqueness theorem)

$$f_y(y) = h(y)$$

This method leads to simple results if the transformation $y = g(x)$ is one-to-one.

Example 5-31. Suppose that $\mathbf{x}$ is $N(0;\sigma)$ and $\mathbf{y} = a\mathbf{x}^2$. Inserting into (5-80) and using the evenness of the integrand, we obtain

$$\Phi_y(\omega) = \int_{-\infty}^{\infty} e^{j\omega a x^2} f(x)\, dx = \frac{2}{\sigma\sqrt{2\pi}} \int_0^{\infty} e^{ja\omega x^2} e^{-x^2/2\sigma^2}\, dx$$

As x increases from 0 to ∞, the transformation $y = ax^2$ is one-to-one. Since

$$dy = 2ax\, dx = 2\sqrt{ay}\, dx$$

the above yields

$$\Phi_y(\omega) = \frac{2}{\sigma\sqrt{2\pi}} \int_0^{\infty} e^{j\omega y} e^{-y/2a\sigma^2} \frac{dy}{2\sqrt{ay}}$$

Hence

$$f_y(y) = \frac{e^{-y/2a\sigma^2}}{\sigma\sqrt{2\pi ay}} U(y)$$

in agreement with (5-8).

Example 5-32. We assume finally that **x** is uniform in the interval $(-\pi/2, \pi/2)$ and $\mathbf{y} = \sin \mathbf{x}$. In this case

$$\Phi_y(\omega) = \int_{-\infty}^{\infty} e^{j\omega \sin x} f(x)\, dx = \frac{1}{\pi} \int_{-\pi/2}^{\pi/2} e^{j\omega \sin x}\, dx$$

As x increases from $-\pi/2$ to $\pi/2$, the function $y = \sin x$ increases from -1 to 1 and

$$dy = \cos x\, dx = \sqrt{1 - y^2}\, dx$$

Hence

$$\Phi_y(\omega) = \frac{1}{\pi} \int_{-1}^{1} e^{j\omega y} \frac{dy}{\sqrt{1 - y^2}}$$

This leads to the conclusion that

$$f_y(y) = \frac{1}{\pi\sqrt{1 - y^2}} \qquad \text{for} \quad |y| < 1$$

and 0 otherwise, in agreement with (5-13).

PROBLEMS

5-1. The RV **x** is $N(5, 2)$ and $\mathbf{y} = 2\mathbf{x} + 4$. Find η_y, σ_y, and $f_y(y)$.

5-2. Find $F_y(y)$ and $f_y(y)$ if $\mathbf{y} = -4\mathbf{x} + 3$ and $f_x(x) = 2e^{-2x}U(x)$.

5-3. If the RV **x** is $N(0, c)$ and $g(x)$ is the function in Fig. 5-4, find and sketch the distribution and the density of the RV $\mathbf{y} = g(\mathbf{x})$.

5-4. The RV **x** is uniform in the interval $(-2c, 2c)$. Find and sketch $f_y(y)$ and $F_y(y)$ if $\mathbf{y} = g(\mathbf{x})$ and $g(x)$ is the function in Fig. 5-3.

5-5. The RV **x** is $N(0, b)$ and $g(x)$ is the function in Fig. 5-5. Find and sketch $f_y(y)$ and $F_y(y)$.

5-6. The RV **x** is uniform in the interval $(0, 1)$. Find the density of the RV $\mathbf{y} = -\ln \mathbf{x}$.

5-7. We place at random 200 points in the interval $(0, 100)$. The distance from 0 to the first random point is an RV **z**. Find $F_z(z)$ (a) exactly and (b) using the Poisson approximation.

5-8. If $\mathbf{y} = \sqrt{2\mathbf{x}}$ and $f_x(x) = ce^{-cx}U(x)$, find $f_y(y)$.

5-9. Express the density $f_y(y)$ of the RV $\mathbf{y} = g(\mathbf{x})$ in terms of $f_x(x)$ if (a) $g(x) = |x|$; (b) $g(x) = e^{-x}U(x)$.

5-10. Find $F_y(y)$ and $f_y(y)$ if $F_x(x) = (1 - e^{-2x})U(x)$ and (a) $\mathbf{y} = (\mathbf{x} - 1)U(\mathbf{x} - 1)$; (b) $\mathbf{y} = \mathbf{x}^2$.

5-11. Show that, if the RV $\mathbf{x}$ has a Cauchy density with $\alpha = 1$ and $\mathbf{y} = \arctan \mathbf{x}$, then $\mathbf{y}$ is uniform in the interval $(-\pi/2, \pi/2)$.

5-12. The RV $\mathbf{x}$ is uniform in the interval $(-2\pi, 2\pi)$. Find $f_y(y)$ if (a) $\mathbf{y} = \mathbf{x}^3$, (b) $\mathbf{y} = \mathbf{x}^4$, and (c) $\mathbf{y} = 2\sin(3\mathbf{x} + 40°)$.

5-13. The RV $\mathbf{x}$ is uniform in the interval $(-1, 1)$. Find $g(x)$ such that if $\mathbf{y} = g(\mathbf{x})$ then $f_y(y) = 2e^{-2y}U(y)$.

5-14. Given that RV $\mathbf{x}$ of continuous type, we form the RV $\mathbf{y} = g(\mathbf{x})$. (a) Find $f_y(y)$ if $g(x) = 2F_x(x) + 4$. (b) Find $g(x)$ such that $\mathbf{y}$ is uniform in the interval $(8, 10)$.

5-15. A fair coin is tossed 10 times and $\mathbf{x}$ equals the number of heads. (a) Find $F_x(x)$. (b) Find $F_y(y)$ if $\mathbf{y} = (\mathbf{x} - 3)^2$.

5-16. If $\mathbf{t}$ is an RV of continuous type and $\mathbf{y} = a \sin \omega\mathbf{t}$, show that

$$f_y(y) \xrightarrow[\omega \to \infty]{} \begin{cases} 1/\pi\sqrt{a^2 - y^2} & |y| < a \\ 0 & |y| > a \end{cases}$$

5-17. Show that if $\mathbf{y} = \mathbf{x}^2$, then

$$f_y(y \mid \mathbf{x} \geq 0) = \frac{U(y)}{1 - F_x(0)} \frac{f_x(\sqrt{y})}{2\sqrt{y}}$$

5-18. (a) Show that if $\mathbf{y} = a\mathbf{x} + b$, then $\sigma_y = |a|\sigma_x$. (b) Find η_y and σ_y if $\mathbf{y} = (\mathbf{x} - \eta_x)/\sigma_x$.

5-19. Show that if $\mathbf{x}$ has a Rayleigh density with parameter α and $\mathbf{y} = b + c\mathbf{x}^2$, then $\sigma_y^2 = 4c^2\alpha^4$.

5-20. (a) Show that if m is the median of $\mathbf{x}$, then

$$E\{|\mathbf{x} - a|\} = E\{|\mathbf{x} - m|\} + 2\int_a^m (x - a) f(x)\, dx$$

for any a. (b) Find c such that $E\{|\mathbf{x} - c|\}$ is minimum.

5-21. Show that if the RV $\mathbf{x}$ is $N(\eta; \sigma)$, then

$$E\{|\mathbf{x}|\} = \sigma\sqrt{\frac{2}{\pi}}\, e^{-\eta^2/2\sigma^2} + 2\eta\mathbb{G}\left(\frac{\eta}{\sigma}\right) - \eta$$

5-22. If $\mathbf{x}$ is $N(0, 2)$ and $\mathbf{y} = 3\mathbf{x}^2$, find η_y, σ_y, and $f_y(y)$.

5-23. Show that if $\mathfrak{A} = [\mathscr{A}_1, \ldots, \mathscr{A}_n]$ is a partition of $\mathscr{S}$, then

$$E\{\mathbf{x}\} = E\{\mathbf{x} \mid \mathscr{A}_1\}P(\mathscr{A}_1) + \cdots + E\{\mathbf{x} \mid \mathscr{A}_n\}P(\mathscr{A}_n).$$

5-24. Show that if $\mathbf{x} \geq 0$ and $E\{\mathbf{x}\} = \eta$, then $P\{\mathbf{x} \geq \sqrt{\eta}\} \leq \sqrt{\eta}$.

5-25. Using (5-55), find $E\{\mathbf{x}^3\}$ if $\eta_x = 10$ and $\sigma_x = 2$.

5-26. If $\mathbf{x}$ is uniform in the interval $(10, 12)$ and $\mathbf{y} = \mathbf{x}^3$, (a) find $f_y(y)$; (b) find $E\{\mathbf{y}\}$: 1, exactly; 2, using (5-55).

5-27. The RV $\mathbf{x}$ is $N(100, 3)$. Find approximately the mean of the RV $\mathbf{y} = 1/\mathbf{x}$ using (5-55).

5-28. We are given an even convex function $g(x)$ and an RV $\mathbf{x}$ whose density $f(x)$ is symmetrical as in Fig. P5-28 with a single maximum at $x = \eta$. Show that the mean $E\{g(\mathbf{x} - a)\}$ of the RV $g(\mathbf{x} - a)$ is minimum if $a = \eta$.

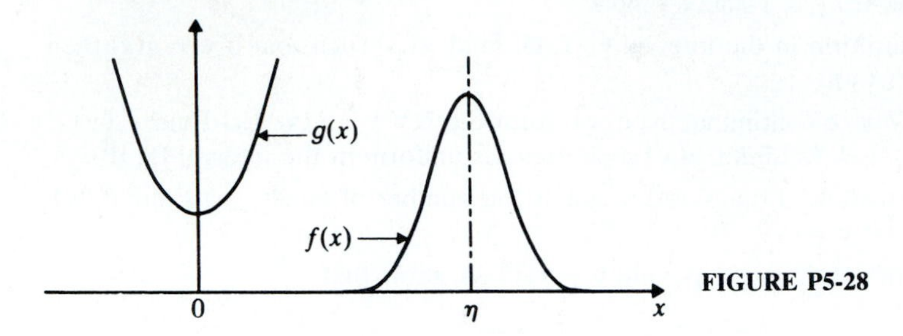

FIGURE P5-28

5-29. Show that if $\mathbf{x}$ and $\mathbf{y}$ are two RVs with densities $f_x(x)$ and $f_y(y)$ respectively, then

$$E\{\log f_x(\mathbf{x})\} \geq E\{\log f_y(\mathbf{x})\}$$

5-30. (*Chernoff bound*) (*a*) Show that for any $\alpha > 0$ and for any real s,

$$P\{e^{s\mathbf{x}} \geq \alpha\} \leq \frac{\Phi(s)}{\alpha} \qquad \text{where} \quad \Phi(s) = E\{e^{s\mathbf{x}}\} \tag{i}$$

Hint: Apply (5-58) to the RV $\mathbf{y} = e^{s\mathbf{x}}$. (*b*) For any A,

$$P\{\mathbf{x} \geq A\} \leq e^{-sA}\Phi(s) \qquad s > 0$$

$$P\{\mathbf{x} \leq A\} \leq e^{-sA}\Phi(s) \qquad s < 0$$

Hint: Set $\alpha = e^{sA}$ in (i).

5-31. Show that (*a*) if $f(x)$ is a Cauchy density, then $\Phi(\omega) = e^{-\alpha|\omega|}$; (*b*) if $f(x)$ is a Laplace density, then $\Phi(\omega) = \alpha^2/(\alpha^2 + \omega^2)$.

5-32. Show that if $E\{\mathbf{x}\} = \eta$, then

$$E\{e^{s\mathbf{x}}\} = e^{s\eta} \sum_{n=0}^{\infty} \mu_n \frac{s^n}{n!} \qquad \mu_n = E\{(\mathbf{x} - \eta)^n\}$$

5-33. Show that if $\Phi_x(\omega_1) = 1$ for some $\omega_1 \neq 0$, then the RV $\mathbf{x}$ is of lattice type taking the values $x_n = 2\pi n/\omega_1$.
Hint:

$$0 = 1 - \Phi_x(\omega_1) = \int_{-\infty}^{\infty} (1 - e^{j\omega_1 x}) f_x(x)\, dx$$

5-34. The RV $\mathbf{x}$ has zero mean, central moments μ_n, and cumulants λ_n. Show that $\lambda_3 = \mu_3$, $\lambda_4 = \mu_4 - 3\mu_2^2$; if $\mathbf{y}$ is $N(0; \sigma_y)$ and $\sigma_y = \sigma_x$, then $E\{\mathbf{x}^4\} = E\{\mathbf{y}^4\} + \lambda_4$.

5-35. An RV $\mathbf{x}$ has a *geometric* distribution if

$$P\{\mathbf{x} = k\} = pq^k \qquad k = 0, 1, \ldots \qquad p + q = 1$$

Find $\Gamma(z)$ and show that $\eta_x = q/p$, $\sigma_x^2 = q/p^2$.

5-36. An RV $\mathbf{x}$ has a *Pascal* (or *negative binomial*) distribution if

$$P\{\mathbf{x}=k\} = \binom{-n}{k} p^n(-q)^k = \binom{n+k-1}{k} p^n q^k \qquad k=0,1,\ldots$$

Find $\Gamma(z)$ and show that $\eta_x = nq/p$, $\sigma_x^2 = nq/p^2$.

5-37. The RV $\mathbf{x}$ takes the values $0,1,\ldots$ with $P\{\mathbf{x}=k\}=p_k$. Show that if

$$\mathbf{y} = (\mathbf{x}-1)U(\mathbf{x}-1) \qquad \text{then} \quad \Gamma_y(z) = p_0 + z^{-1}[\Gamma_x(z) - p_0]$$

$$\eta_y = \eta_x - 1 + p_0 \qquad E\{\mathbf{y}^2\} = E\{\mathbf{x}^2\} - 2\eta_x + 1 - p_0$$

5-38. Show that, if $\Phi(\omega) = E\{e^{j\omega\mathbf{x}}\}$, then for any a_i,

$$\sum_{i=1}^{n}\sum_{j=1}^{n} \Phi(\omega_i - \omega_j) a_i a_j^* \geq 0$$

Hint:

$$E\left\{\left|\sum_{i=1}^{n} a_i e^{j\omega_i \mathbf{x}}\right|^2\right\} \geq 0$$

5-39. The RV $\mathbf{x}$ is $N(0;\sigma)$. (*a*) Using characteristic functions, show that if $g(x)$ is a function such that $g(x)e^{-x^2/2\sigma^2} \to 0$ as $|x| \to \infty$, then (Price's theorem)

$$\frac{dE\{g(\mathbf{x})\}}{dv} = \frac{1}{2}E\left\{\frac{d^2 g(\mathbf{x})}{d\mathbf{x}^2}\right\} \qquad v = \sigma^2 \tag{i}$$

(*b*) The moments μ_n of $\mathbf{x}$ are functions of v. Using (i), show that

$$\mu_n(v) = \frac{n(n-1)}{2}\int_0^v \mu_{n-2}(\beta)\, d\beta$$

5-40. Show that, if $\mathbf{n}$ is an integer-valued RV with moment function $\Gamma(z)$ as in (5-75), then

$$P\{\mathbf{n}=k\} = \frac{1}{2\pi}\int_{-\pi}^{\pi} \Gamma(e^{j\omega}) e^{-jk\omega}\, d\omega$$

CHAPTER 6

TWO RANDOM VARIABLES

6-1 BIVARIATE DISTRIBUTIONS

We are given two RVs $\mathbf{x}$ and $\mathbf{y}$, defined as in Sec. 4-1, and we wish to determine their joint statistics, that is, the probability that the point $(\mathbf{x}, \mathbf{y})$ is in a specified region† D in the xy plane. The distribution functions $F_x(x)$ and $F_y(y)$ of the given RVs determine their separate (marginal) statistics but not their joint statistics. In particular, the probability of the event

$$\{\mathbf{x} \le x\} \cap \{\mathbf{y} \le y\} = \{\mathbf{x} \le x, \mathbf{y} \le y\}$$

cannot be expressed in terms of $F_x(x)$ and $F_y(y)$. In the following, we show that the joint statistics of the RVs $\mathbf{x}$ and $\mathbf{y}$ are completely determined if the probability of this event is known for every x and y.

Joint Distribution and Density

The joint (bivariate) distribution $F_{xy}(x, y)$ or, simply, $F(x, y)$ of two RVs $\mathbf{x}$ and $\mathbf{y}$ is the probability of the event

$$\{\mathbf{x} \le x, \mathbf{y} \le y\} = \{(\mathbf{x}, \mathbf{y}) \in D_1\}$$

where x and y are two arbitrary real numbers and D_1 is the quadrant shown in

†The region D is arbitrary subject only to the mild condition that it can be expressed as a countable union or intersection of rectangles.

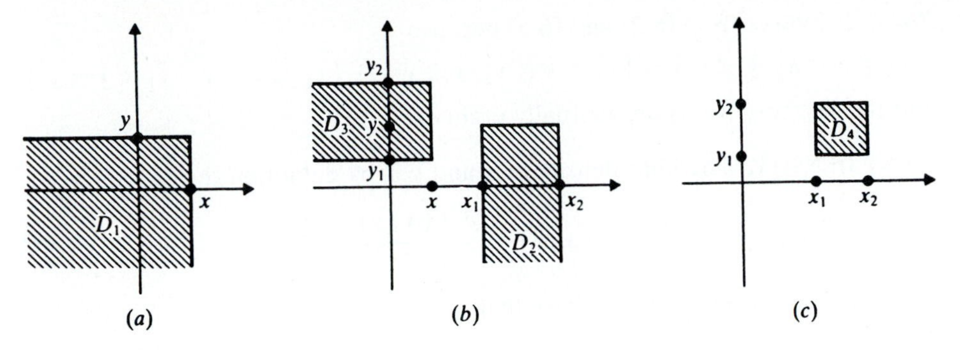

FIGURE 6-1

Fig. 6-1a:

$$F(x, y) = P\{\mathbf{x} \le x, \mathbf{y} \le y\} \tag{6-1}$$

PROPERTIES. **1**. The function $F(x, y)$ is such that

$$F(-\infty, y) = 0 \qquad F(x, -\infty) = 0 \qquad F(\infty, \infty) = 1$$

Proof. As we know, $P\{\mathbf{x} = -\infty\} = P\{\mathbf{y} = -\infty\} = 0$. And since

$$\{\mathbf{x} = -\infty, \mathbf{y} \le y\} \subset \{\mathbf{x} = -\infty\} \qquad \{\mathbf{x} \le x, \mathbf{y} = -\infty\} \subset \{\mathbf{y} = -\infty\}$$

the first two equations follow. The last is a consequence of the identities

$$\{\mathbf{x} \le \infty, \mathbf{y} \le \infty\} = \mathscr{S} \qquad P(\mathscr{S}) = 1$$

2. The event $\{x_1 < \mathbf{x} \le x_2, \mathbf{y} \le y\}$ consists of all points $(\mathbf{x}, \mathbf{y})$ in the vertical half-strip D_2 and the event $\{\mathbf{x} \le x, y_1 < \mathbf{y} \le y_2\}$ consists of all points $(\mathbf{x}, \mathbf{y})$ in the horizontal half-strip D_3 of Fig. 6-1b. We maintain that

$$P\{x_1 < \mathbf{x} \le x_2, \mathbf{y} \le y\} = F(x_2, y) - F(x_1, y) \tag{6-2}$$

$$P\{\mathbf{x} < x, y_1 < \mathbf{y} \le y_2\} = F(x, y_2) - F(x, y_1) \tag{6-3}$$

Proof. Clearly,

$$\{\mathbf{x} \le x_2, \mathbf{y} \le y\} = \{\mathbf{x} \le x_1, \mathbf{y} \le y\} + \{x_1 < \mathbf{x} \le x_2, \mathbf{y} \le y\}$$

The last two events are mutually exclusive; hence [see (2-10)]

$$P\{\mathbf{x} \le x_2, \mathbf{y} \le y\} = P\{\mathbf{x} \le x_1, \mathbf{y} \le y\} + P\{x_1 < \mathbf{x} \le x_2, \mathbf{y} \le y\}$$

and (6-2) results. The proof of (6-3) is similar.

3.
$$P\{x_1 < \mathbf{x} \le x_2, y_1 < \mathbf{y} \le y_2\} = F(x_2, y_2) - F(x_1, y_2) - F(x_2, y_1) + F(x_1, y_1) \tag{6-4}$$

This is the probability that $(\mathbf{x}, \mathbf{y})$ is in the rectangle D_4 of Fig. 6-1c.

Proof. It follows from (6-2) and (6-3) because

$$\{x_1 < \mathbf{x} \le x_2, \mathbf{y} \le y_2\} = \{x_1 < \mathbf{x} \le x_2, \mathbf{y} \le y_1\} + \{x_1 < \mathbf{x} \le x_2, y_1 < \mathbf{y} \le y_2\}$$

and the last two events are mutually exclusive.

JOINT DENSITY. The joint density of $\mathbf{x}$ and $\mathbf{y}$ is by definition the function

$$f(x, y) = \frac{\partial^2 F(x, y)}{\partial x \, \partial y} \tag{6-5}$$

From this and property 1 it follows that

$$F(x, y) = \int_{-\infty}^{x} \int_{-\infty}^{y} f(\alpha, \beta) \, d\alpha \, d\beta \tag{6-6}$$

Joint statistics. We shall now show that the probability that the point $(\mathbf{x}, \mathbf{y})$ is in a region D of the xy plane equals the integral of $f(x, y)$ in D. In other words,

$$P\{(\mathbf{x}, \mathbf{y}) \in D\} = \int_D \int f(x, y) \, dx \, dy \tag{6-7}$$

where $\{(\mathbf{x}, \mathbf{y}) \in D\}$ is the event consisting of all outcomes ζ such that the point $[\mathbf{x}(\zeta), \mathbf{y}(\zeta)]$ is in D.

Proof. As we know, the ratio

$$\frac{F(x + \Delta x, y + \Delta y) - F(x, y + \Delta y) - F(x + \Delta x, y) + F(x, y)}{\Delta x \, \Delta y}$$

tends to $\partial^2 F(x, y)/\partial x \, \partial y$ as $\Delta x \to 0$ and $\Delta y \to 0$. Hence [see (6-4) and (6-5)]

$$P\{x < \mathbf{x} \le x + \Delta x, y < \mathbf{y} \le y + \Delta y\} \simeq f(x, y) \, \Delta x \, \Delta y \tag{6-8}$$

We have thus shown that the probability that $(\mathbf{x}, \mathbf{y})$ is in a differential rectangle equals $f(x, y)$ times the area $\Delta x \, \Delta y$ of the rectangle. This proves (6-7) because the region D can be written as the limit of the union of such rectangles.

Marginal statistics. In the study of several RVs, the statistics of each are called marginal. Thus $F_x(x)$ is the *marginal distribution* and $f_x(x)$ the *marginal density* of $\mathbf{x}$. In the following, we express the marginal statistics of $\mathbf{x}$ and $\mathbf{y}$ in terms of their joint statistics $F(x, y)$ and $f(x, y)$.

We maintain that

$$F_x(x) = F(x, \infty) \qquad F_y(y) = F(\infty, y) \tag{6-9}$$

$$f_x(x) = \int_{-\infty}^{\infty} f(x, y) \, dy \qquad f_y(y) = \int_{-\infty}^{\infty} f(x, y) \, dx \tag{6-10}$$

Proof. Clearly, $\{\mathbf{x} \le \infty\} = \{\mathbf{y} \le \infty\} = \mathscr{S}$; hence

$$\{\mathbf{x} \le x\} = \{\mathbf{x} \le x, \mathbf{y} \le \infty\} \qquad \{\mathbf{y} \le y\} = \{\mathbf{x} \le \infty, \mathbf{y} \le y\}$$

The probabilities of the two sides above yield (6-9).

Differentiating (6-6), we obtain

$$\frac{\partial F(x,y)}{\partial x} = \int_{-\infty}^{y} f(x,\beta)\, d\beta \qquad \frac{\partial F(x,y)}{\partial y} = \int_{-\infty}^{x} f(\alpha,y)\, d\alpha \qquad (6\text{-}11)$$

Setting $y = \infty$ in the first and $x = \infty$ in the second equation, we obtain (6-10) because [see (6-9)]

$$f_x(x) = \frac{\partial F(x,\infty)}{\partial x} \qquad f_y(y) = \frac{\partial F(\infty,y)}{\partial y}$$

Existence theorem. From properties 1 and 3 it follows that

$$F(-\infty, y) = 0 \qquad F(x, -\infty) = 0 \qquad F(\infty,\infty) = 1 \qquad (6\text{-}12)$$

and

$$F(x_2, y_2) - F(x_1, y_2) - F(x_2, y_1) + F(x_1, y_1) \ge 0 \qquad (6\text{-}13)$$

for every $x_1 < x_2$ $y_1 < y_2$. Hence [see (6-6) and (6-8)]

$$\int_{-\infty}^{\infty}\int_{-\infty}^{\infty} f(x,y)\, dx\, dy = 1 \qquad f(x,y) \ge 0 \qquad (6\text{-}14)$$

Conversely, given $F(x, y)$ or $f(x, y)$ as above, we can find two RVs **x** and **y**, defined in some space $\mathscr{S}$, with distribution $F(x, y)$ or density $f(x, y)$. This can be done by extending the existence theorem of Sec. 4-3 to joint statistics.

Joint normality. We shall say that the RVs **x** and **y** are *jointly normal* if their joint density is given by

$$f(x,y) = A \exp\left\{ -\frac{1}{2(1-r^2)}\left[\frac{(x-\eta_1)^2}{\sigma_1^2} - 2r\frac{(x-\eta_1)(y-\eta_2)}{\sigma_1\sigma_2} + \frac{(y-\eta_2)^2}{\sigma_2^2}\right]\right\} \qquad (6\text{-}15)$$

This function is positive and its integral equals 1 if

$$A = \frac{1}{2\pi\sigma_1\sigma_2\sqrt{1-r^2}} \qquad |r| < 1 \qquad (6\text{-}16)$$

Thus $f(x, y)$ is an exponential and its exponent is a negative quadratic because $|r| < 1$. The function $f(x, y)$ will be denoted by

$$N(\eta_1, \eta_2; \sigma_1, \sigma_2; r)$$

As we shall presently see, η_1 and η_2 are the expected values of **x** and **y**, and σ_1^2 and σ_2^2 their variances. The significance of r will be given later (correlation coefficient).

We maintain that the marginal densities of **x** and **y** are given by

$$f_x(x) = \frac{1}{\sigma_1\sqrt{2\pi}} e^{-(x-\eta_1)^2/2\sigma_1^2} \qquad f_y(y) = \frac{1}{\sigma_2\sqrt{2\pi}} e^{-(y-\eta_2)^2/2\sigma_2^2} \qquad (6\text{-}17)$$

Proof. To prove the above, we must show that if (6-15) is inserted into (6-10), the result is (6-17). The bracket in (6-15) can be written in the form

$$[\cdots] = \left[\frac{x-\eta_1}{\sigma_1} - r\frac{y-\eta_2}{\sigma_2}\right]^2 + (1-r^2)\frac{(y-\eta_2)^2}{\sigma_2^2}$$

Hence

$$\int_{-\infty}^{\infty} f(x,y)\,dx = A\exp\left[-\frac{(y-\eta_2)^2}{2\sigma_2^2}\right]$$

$$\times \int_{-\infty}^{\infty} \exp\left\{-\frac{1}{2(1-r^2)}\left[\frac{x-\eta_1}{\sigma_1} - r\frac{y-\eta_2}{\sigma_2}\right]^2\right\} dx$$

The last integral is a constant B (independent of x and y). Therefore [see (6-10)]

$$f_y(y) = ABe^{-(y-\eta_2)^2/2\sigma_2^2}$$

And since $f_y(y)$ is a density, its area must equal 1. This yields $AB = 1/\sigma_2\sqrt{2\pi}$ and the second equation in (6-17) results. The proof of the first is similar.

Notes 1. From (6-17) it follows that if two RVs are jointly normal, they are also marginally normal. However, as the next example shows, *the converse is not true*.

2. Joint normality can be defined as follows: Two RVs $\mathbf{x}$ and $\mathbf{y}$ are jointly normal if the sum $a\mathbf{x} + b\mathbf{y}$ is normal for every a and b [see (8-56)].

Example 6-1. We shall construct two RVs $\mathbf{x}_1$ and $\mathbf{y}_1$ that are marginally but not jointly normal. We start with two jointly normal RVs $\mathbf{x}$ and $\mathbf{y}$ with density $f(x, y)$ as in (6-15). Adding and subtracting small masses in the region D of Fig. 6-2 consisting of four circles as shown, we obtain a new function $f_1(x, y)$ such that $f_1(x, y) = f(x, y) \pm \varepsilon$ in D and $f_1(x, y) = f(x, y)$ everywhere else. The function $f_1(x, y)$ so formed is a density; hence it defines two new RVs $\mathbf{x}_1$ and $\mathbf{y}_1$. These RVs are not jointly normal because $f_1(x, y)$ is not of the form (6-15). We maintain,

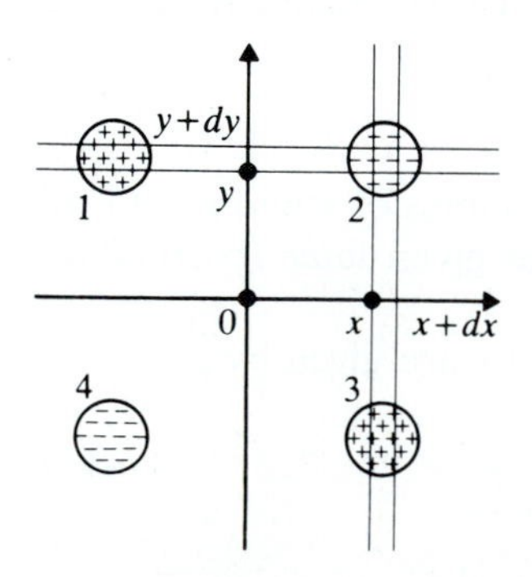

FIGURE 6-2

however, that they are marginally normal. Indeed, the densities of $\mathbf{x}_1$ and $\mathbf{y}_1$ are determined by the masses in the vertical strip $x_1 < x < x_1 + dx$ and the horizontal strip $y_1 < y < y_1 + dy$. As we see from the figure, the masses in these strips have not changed. This shows that $\mathbf{x}_1$ and $\mathbf{y}_1$ are normal because $\mathbf{x}$ and $\mathbf{y}$ are normal.

Discrete type. Suppose that the RVs $\mathbf{x}$ and $\mathbf{y}$ are of discrete type taking the values x_i and y_k with respective probabilities

$$P\{\mathbf{x} = x_i\} = p_i \qquad P\{\mathbf{y} = y_k\} = q_k \tag{6-18}$$

Their joint statistics are determined in terms of the *joint probabilities*

$$P\{\mathbf{x} = x_i, \mathbf{y} = y_k\} = p_{ik} \tag{6-19}$$

Clearly,

$$\sum_{i,k} p_{ik} = 1 \tag{6-20}$$

because, as i and k take all possible values, the events $\{\mathbf{x} = x_i, \ \mathbf{y} = y_k\}$ are mutually exclusive and their union equals the certain event.

We maintain that the *marginal probabilities* p_i and q_k can be expressed in terms of the joint probabilities p_{ik}:

$$p_i = \sum_k p_{ik} \qquad q_k = \sum_i p_{ik} \tag{6-21}$$

This is the discrete version of (6-10).

Proof. The events $\{\mathbf{y} = y_k\}$ form a partition of $\mathscr{S}$. Hence as k ranges over all possible values, the events $\{\mathbf{x} = x_i, \ \mathbf{y} = y_k\}$ are mutually exclusive and their union equals $\{\mathbf{x} = x_i\}$. This yields the first equation in (6-21) [see (2-36)]. The proof of the second is similar.

Probability Masses

The probability that the point $(\mathbf{x}, \mathbf{y})$ is in a region D of the plane can be interpreted as the probability mass in this region. Thus the mass in the entire plane equals 1. The mass in the half-plane $\mathbf{x} \le x$ to the left of the line L_x of Fig. 6-3 equals $F_x(x)$. The mass in the half-plane $\mathbf{y} \le y$ below the line L_y equals $F_y(y)$. The mass in the cross-hatched quadrant $(\mathbf{x} \le x, \ \mathbf{y} \le y)$ equals $F(x, y)$.

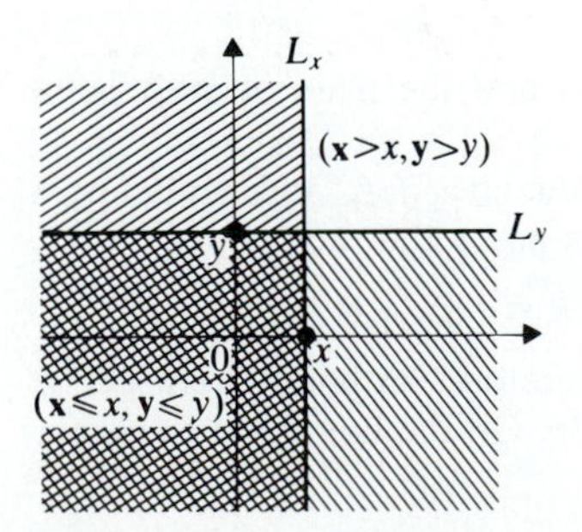

FIGURE 6-3

Finally, the mass in the clear quadrant ($\mathbf{x} > x$, $\mathbf{y} > y$) equals

$$P\{\mathbf{x} > x, \mathbf{y} > y\} = 1 - F_x(x) - F_y(y) + F(x, y) \tag{6-22}$$

The probability mass in a region D equals the integral [see (6-7)]

$$\int_D\int f(x, y)\, dx\, dy$$

If, therefore, $f(x, y)$ is a bounded function, it can be be interpreted as surface mass density.

Example 6-2. Suppose that

$$f(x, y) = \frac{1}{2\pi\sigma^2} e^{-(x^2+y^2)/2\sigma^2} \tag{6-23}$$

We shall find the mass m in the circle $x^2 + y^2 \le a^2$. Inserting (6-23) into (6-7) and using the transformation

$$x = r\cos\theta \qquad y = r\sin\theta$$

we obtain

$$m = \frac{1}{2\pi\sigma^2}\int_0^a\int_{-\pi}^{\pi} e^{-r^2/2\sigma^2} r\, dr\, d\theta = 1 - e^{-a^2/2\sigma^2} \tag{6-24}$$

POINT MASSES. If the RVs $\mathbf{x}$ and $\mathbf{y}$ are of discrete type taking the values x_i and y_k, then the probability masses are 0 everywhere except at the points (x_i, y_k). We have, thus, only point masses and the mass at each point equals p_{ik} [see (6-19)]. The probability $p_i = P\{\mathbf{x} = x_i\}$ equals the sum of all masses p_{ik} on the line $x = x_i$ in agreement with (6-21).

If $i = 1, \ldots, M$ and $k = 1, \ldots, N$, then the number of possible point masses on the plane equals MN. However, as the next example shows, some of these masses might be 0.

Example 6-3. (*a*) In the fair-die experiment, $\mathbf{x}$ equals the number of dots shown and $\mathbf{y}$ equals twice this number:

$$\mathbf{x}(f_i) = i \qquad \mathbf{y}(f_i) = 2i \qquad i = 1, \ldots, 6$$

In other words, $x_i = i$, $y_k = 2k$ and

$$p_{ik} = P\{\mathbf{x} = i, \mathbf{y} = 2k\} = \begin{cases} \frac{1}{6} & i = k \\ 0 & i \ne k \end{cases}$$

Thus there are masses only on the six points $(i, 2i)$ and the mass of each point equals $1/6$ (Fig. 6-4*a*).

(*b*) We toss the die twice obtaining the 36 outcomes $f_i f_k$ and we define $\mathbf{x}$ and $\mathbf{y}$ such that $\mathbf{x}$ equals the first number that shows and $\mathbf{y}$ the second

$$\mathbf{x}(f_i f_k) = i \qquad \mathbf{y}(f_i f_k) = k \qquad i, k = 1, \ldots, 6$$

Thus $x_i = i$, $y_k = k$, and $p_{ik} = 1/36$. We have, therefore, 36 point masses (Fig. 6-4*b*) and the mass of each of each point equals $1/36$. On the line $x = i$ there are six points with total mass $1/6$.

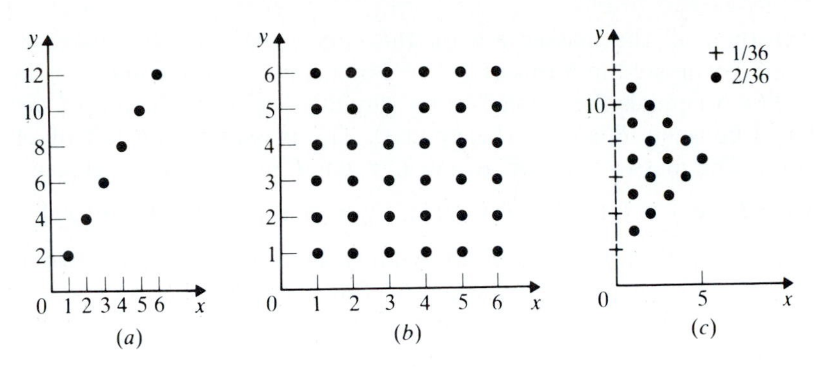

FIGURE 6-4

(c) Again the die is tossed twice but now

$$\mathbf{x}(f_i f_k) = |i - k| \qquad \mathbf{y}(f_i f_k) = i + k$$

In this case, $\mathbf{x}$ takes the values $0, 1, \ldots, 5$ and $\mathbf{y}$ the values $2, 3, \ldots, 12$. The number of possible points equals $6 \times 11 = 66$; however, only 21 have positive masses (Fig. 6-4c). Specifically, if $\mathbf{x} = 0$, then $\mathbf{y} = 2$, or $4, \ldots,$ or 12 because if $\mathbf{x} = 0$, then $i = k$ and $\mathbf{y} = 2i$. There are, therefore, six mass points on this line and the mass of each point equals $1/36$. If $\mathbf{x} = 1$, then $\mathbf{y} = 3$, or $5, \ldots,$ or 11. There are, therefore, five mass points on the line $x = 1$ and the mass of each point equals $2/36$. For example, if $\mathbf{x} = 1$ and $\mathbf{y} = 7$, then $i = 3$, $k = 4$, or $i = 4$, $k = 3$; hence $P\{\mathbf{x} = 1, \mathbf{y} = 7\} = 2/36$.

LINE MASSES. The following cases lead to line masses:

1. If $\mathbf{x}$ is of discrete type taking the values x_i and $\mathbf{y}$ is of continuous type, then all probability masses are on the vertical lines $x = x_i$ (Fig 6-5a). In particular, the mass between the points y_1 and y_2 on the line $x = x_i$ equals the probability of the event

$$\{\mathbf{x} = x_i, y_1 \leq \mathbf{y} \leq y_2\}$$

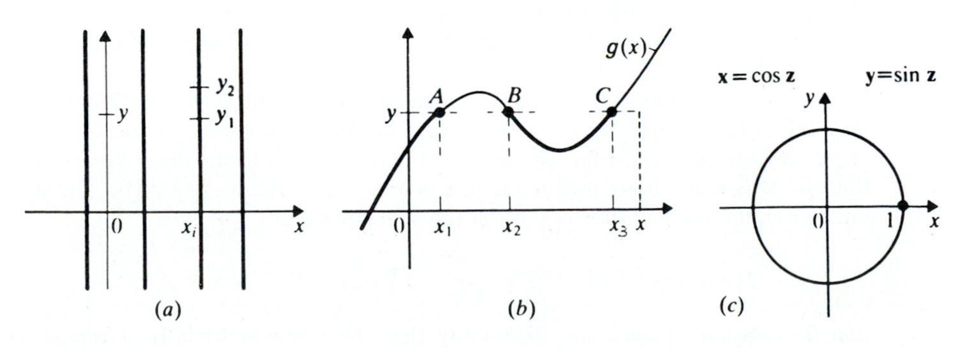

FIGURE 6-5

2. If $\mathbf{y} = g(\mathbf{x})$, then all the masses are on the curve $y = g(x)$. In this case, $F(x, y)$ can be expressed in terms of $F_x(x)$. For example, with x and y as in Fig. 6-5*b*, $F(x, y)$ equals the masses on the curve $y = g(x)$ to the left of the point A and between B and C (heavy line). The masses to the left of A equal $F_x(x_1)$. The masses between B and C equal $F_x(x_3) - F_x(x_2)$. Hence

$$F(x, y) = F_x(x_1) + F_x(x_3) - F_x(x_2) \qquad y = g(x_1) = g(x_2) = g(x_3)$$

3. If $\mathbf{x} = g(\mathbf{z})$ and $\mathbf{y} = h(\mathbf{z})$, then all probability masses are on the curve $x = g(z)$, $y = h(z)$ specified parametrically. For example, if $g(z) = \cos z$, $h(z) = \sin z$, then the curve is a circle (Fig. 6-5*c*). In this case, the joint statistics of $\mathbf{x}$ and $\mathbf{y}$ can be expressed in terms of $F_z(z)$.

Independence

Two RVs $\mathbf{x}$ and $\mathbf{y}$ are called (*statistically*) *independent* if the events $\{\mathbf{x} \in A\}$ and $\{\mathbf{y} \in B\}$ are independent [see (2-40)], that is, if

$$P\{\mathbf{x} \in A, \mathbf{y} \in B\} = P\{\mathbf{x} \in A\}P\{\mathbf{y} \in B\} \tag{6-25}$$

where A and B are two arbitrary sets on the x and y axes respectively.

Applying the above to the events $\{\mathbf{x} \le x\}$ and $\{\mathbf{y} \le y\}$, we conclude that, if the RVs $\mathbf{x}$ and $\mathbf{y}$ are independent, then

$$F(x, y) = F_x(x)F_y(y) \tag{6-26}$$

Hence

$$f(x, y) = f_x(x)f_y(y) \tag{6-27}$$

It can be shown that, if (6-26) or (6-27) is true, then (6-25) is also true; that is, the RVs $\mathbf{x}$ and $\mathbf{y}$ are independent [see (6-7)].

If the RVs $\mathbf{x}$ and $\mathbf{y}$ are of discrete type as in (6-19) and independent, then

$$p_{ik} = p_i p_k \tag{6-28}$$

This follows if we apply (6-25) to the events $\{\mathbf{x} = x_i\}$ and $\{\mathbf{y} = y_k\}$.

Example 6-4 Buffon's needle. A fine needle of length $2a$ is dropped at random on a board covered with parallel lines distance $2b$ apart where $b > a$ as in Fig. 6-6*a*. We shall show that the probability p that the needle intersects one of the lines equals $2a/\pi b$.

In terms of RVs the above experiment can be phrased as follows: We denote by $\mathbf{x}$ the distance from the center of the needle to the nearest line and by $\boldsymbol{\theta}$ the angle between the needle and the direction perpendicular to the lines. We assume that the RVs $\mathbf{x}$ and $\boldsymbol{\theta}$ are independent, $\mathbf{x}$ is uniform in the interval $(0, b)$, and $\boldsymbol{\theta}$ is uniform in the interval $(0, \pi/2)$. From this it follows that

$$f(x,\theta) = f_x(x)f_\theta(\theta) = \frac{1}{b}\frac{2}{\pi} \qquad 0 \le x \le b \qquad 0 \le \theta \le \frac{\pi}{2}$$

and 0 elsewhere. Hence the probability that the point $(\mathbf{x}, \boldsymbol{\theta})$ is in a region D included in the rectangle R of Fig. 6-6*b* equals the area of D times $2/\pi b$.

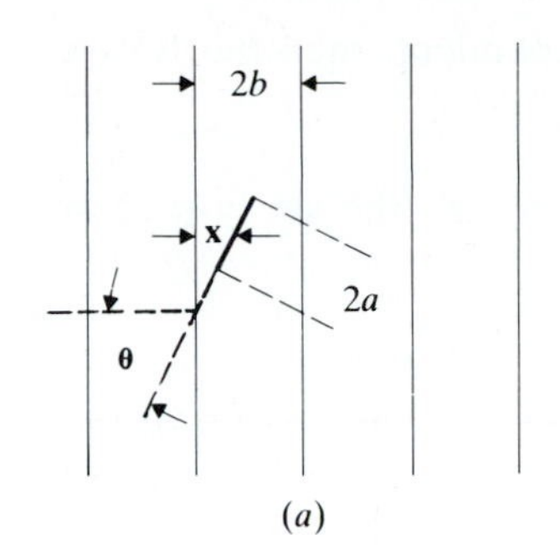

(a)

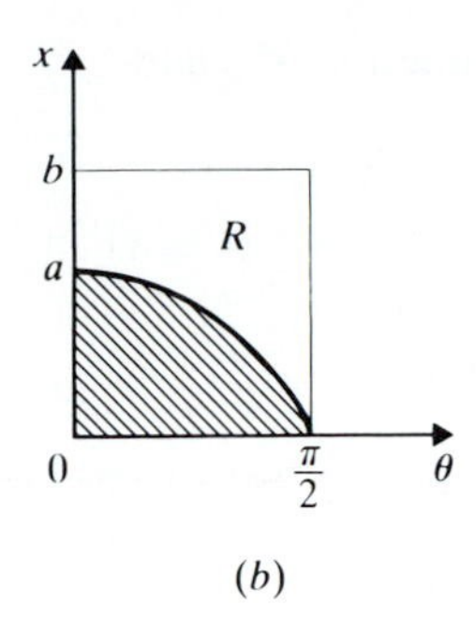

(b)

FIGURE 6-6

The needle intersects the lines if $x < a\cos\theta$. Hence p equals the shaded area of Fig. 6-6b times $2/\pi b$:

$$p = P\{\mathbf{x} < a\cos\boldsymbol{\theta}\} = \frac{2}{\pi b}\int_0^{\pi/2} a\cos\theta\,d\theta = \frac{2a}{\pi b}$$

The above can be used to determine experimentally the number π using the relative frequency interpretation of p: If the needle is dropped n times and it intersects the lines n_i times, then

$$\frac{n_i}{n} \simeq p = \frac{2a}{\pi b} \qquad \text{hence} \quad \pi \simeq \frac{2an}{bn_i}$$

THEOREM. If the RVs $\mathbf{x}$ and $\mathbf{y}$ are independent, then the RVs

$$\mathbf{z} = g(\mathbf{x}) \qquad \mathbf{w} = h(\mathbf{y})$$

are also independent.

Proof. We denote by A_z the set of points on the x axis such that $g(x) \le z$ and by B_w the set of points on the y axis such that $h(y) \le w$. Clearly,

$$\{\mathbf{z} \le z\} = \{\mathbf{x} \in A_z\} \qquad \{\mathbf{w} \le w\} = \{\mathbf{y} \in B_w\} \tag{6-29}$$

Therefore the events $\{\mathbf{z} \le z\}$ and $\{\mathbf{w} \le w\}$ are independent because the events $\{\mathbf{x} \in A_z\}$ and $\{\mathbf{y} \in B_w\}$ are independent.

INDEPENDENT EXPERIMENTS. As in the case of events (Sec. 3-1), the concept of independence is important in the study of RVs defined on product spaces. Suppose that the RV $\mathbf{x}$ is defined on a space $\mathscr{S}_1$ consisting of the outcomes ζ_1 and the RV $\mathbf{y}$ is defined on a space $\mathscr{S}_2$ consisting of the outcomes ζ_2. In the combined experiment $\mathscr{S}_1 \times \mathscr{S}_2$ the RVs $\mathbf{x}$ and $\mathbf{y}$ are such that

$$\mathbf{x}(\zeta_1\zeta_2) = \mathbf{x}(\zeta_1) \qquad \mathbf{y}(\zeta_1\zeta_2) = \mathbf{y}(\zeta_2) \tag{6-30}$$

In other words, $\mathbf{x}$ depends on the outcomes of $\mathscr{S}_1$ only, and $\mathbf{y}$ depends on the outcomes of $\mathscr{S}_2$ only.

THEOREM. If the experiments $\mathscr{S}_1$ and $\mathscr{S}_2$ are independent, then the RVs $\mathbf{x}$ and $\mathbf{y}$ are independent.

Proof. We denote by $\mathscr{A}_x$ the set $\{\mathbf{x} \le x\}$ in $\mathscr{S}_1$ and by $\mathscr{B}_y$ the set $\{\mathbf{y} \le y\}$ in $\mathscr{S}_2$. In the space $\mathscr{S}_1 \times \mathscr{S}_2$,

$$\{\mathbf{x} \le x\} = \mathscr{A}_x \times \mathscr{S}_2 \qquad \{\mathbf{y} \le y\} = \mathscr{S}_1 \times \mathscr{B}_y$$

From the independence of the two experiments, it follows that [see (3-4)] the events $\mathscr{A}_x \times \mathscr{S}_2$ and $\mathscr{S}_1 \times \mathscr{B}_y$ are independent. Hence the events $\{\mathbf{x} \le x\}$ and $\{\mathbf{y} \le y\}$ are also independent.

Example 6-5. A die with $P\{f_i\} = p_i$ is tossed twice and the RVs $\mathbf{x}$ and $\mathbf{y}$ are such that

$$\mathbf{x}(f_i f_k) = i \qquad \mathbf{y}(f_i f_k) = k$$

Thus $\mathbf{x}$ equals the first number that shows and $\mathbf{y}$ equals the second; hence the RVs $\mathbf{x}$ and $\mathbf{y}$ are independent. This leads to the conclusion that

$$p_{ik} = P\{\mathbf{x} = i, \mathbf{y} = k\} = p_i p_k$$

Circular Symmetry

We say that the joint density of two RVs $\mathbf{x}$ and $\mathbf{y}$ is circularly symmetrical if it depends only on the distance from the origin, that is, if

$$f(x, y) = g(r) \qquad r = \sqrt{x^2 + y^2} \tag{6-31}$$

THEOREM. If the RVs $\mathbf{x}$ and $\mathbf{y}$ are circularly symmetrical and independent, then they are normal with zero mean and equal variance.

Proof. From (6-31) and (6-27) it follows that

$$g\left(\sqrt{x^2 + y^2}\right) = f_x(x) f_y(y) \tag{6-32}$$

Since

$$\frac{\partial g(r)}{\partial x} = \frac{dg(r)}{dr} \frac{\partial r}{\partial x} \qquad \text{and} \qquad \frac{\partial r}{\partial x} = \frac{x}{r}$$

we conclude, differentiating (6-32) with respect to x, that

$$\frac{x}{r} g'(r) = f_x'(x) f_y(y)$$

Dividing both sides by $x g(r) = x f_x(x) f_y(y)$, we obtain

$$\frac{1}{r} \frac{g'(r)}{g(r)} = \frac{1}{x} \frac{f_x'(x)}{f_x(x)} \tag{6-33}$$

The right side above is independent of y and the left side is a function of r

$= \sqrt{x^2 + y^2}$. This shows that both sides are independent of x and y. Hence

$$\frac{1}{r}\frac{g'(r)}{g(r)} = \alpha = \text{constant}$$

From this if follows that

$$\frac{d \ln g(r)}{dr} = \alpha r \qquad g(r) = Ae^{\alpha r^2/2}$$

and (6-31) yields

$$f(x, y) = g\left(\sqrt{x^2 + y^2}\right) = Ae^{\alpha(x^2+y^2)/2} \tag{6-34}$$

Thus the RVs **x** and **y** are normal with zero mean and variance $\sigma^2 = -1/\alpha$.

6-2 ONE FUNCTION OF TWO RANDOM VARIABLES

Given two RVs **x** and **y** and a function $g(x, y)$, we form the RV

$$\mathbf{z} = g(\mathbf{x}, \mathbf{y})$$

We shall express the statistics of **z** in terms of the function $g(x, y)$ and the joint statistics of **x** and **y**.

With z a given number, we denote by D_z the region of the xy plane such that $g(x, y) \le z$. This region might not be simply connected (Fig. 6-7). Clearly,

$$\{\mathbf{z} \le z\} = \{g(\mathbf{x}, \mathbf{y}) \le z\} = \{(\mathbf{x}, \mathbf{y}) \in D_z\}$$

Hence [see (6-7)]

$$F_z(z) = P\{\mathbf{z} \le z\} = P\{(\mathbf{x}, \mathbf{y}) \in D_z\} = \int_{D_z}\!\!\int f(x, y)\, dx\, dy \tag{6-35}$$

Thus, to determine $F_z(z)$, it suffices to find the region D_z for every z and to evaluate the above integral.

The density of **z** can be determined similarly. With ΔD_z the region of the xy plane such that $z < g(x, y) \le z + dz$, we have

$$\{z < \mathbf{z} \le z + dz\} = \{(\mathbf{x}, \mathbf{y}) \in \Delta D_z\}$$

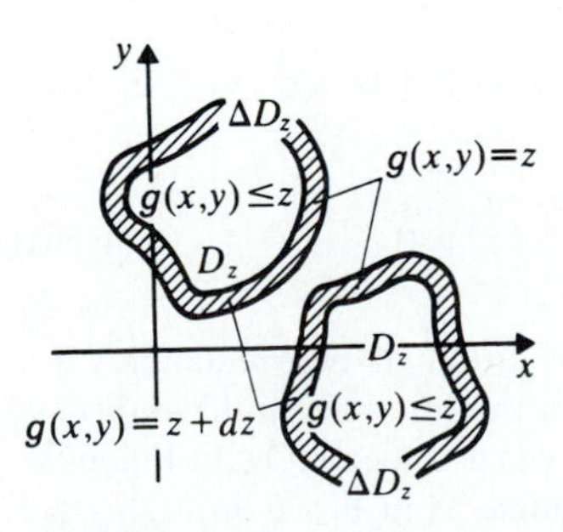

FIGURE 6-7

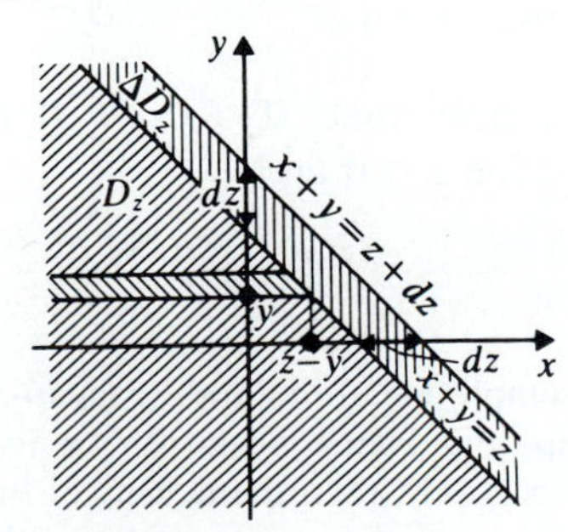

FIGURE 6-8

Hence

$$f_z(z)\,dz = P\{z < \mathbf{z} \le z + dz\} = \iint_{\Delta D_z} f(x, y)\,dx\,dy \tag{6-36}$$

Illustrations

In the following, we use (6-35) and (6-36) to find the statistics of various functions of **x** and **y**.

1. $\mathbf{z} = \mathbf{x} + \mathbf{y}$

The region D_z of the xy plane such that $x + y \le z$ is the shaded part of Fig. 6-8 to the left of the line $x + y = z$. Integrating over suitable strips, we obtain

$$F_z(z) = \int_{-\infty}^{\infty}\int_{-\infty}^{z-y} f(x, y)\,dx\,dy \tag{6-37}$$

We can find $f_z(z)$ either by differentiating $F_z(z)$ or directly from (6-36). The region ΔD_z such that $z < x + y < z + dz$ is a diagonal strip bounded by the lines $x + y = z$ and $x + y = z + dz$. The coordinates of a point of this region are $z - y$, y and the area of a differential equals $dy\,dz$. Hence

$$f_z(z)\,dz = \int_{-\infty}^{\infty} f(z - y, y)\,dy\,dz \tag{6-38}$$

INDEPENDENCE AND CONVOLUTION. If the RVs **x** and **y** are independent, then

$$f(x, y) = f_x(x) f_y(y)$$

Inserting into (6-38), we obtain

$$f_z(z) = \int_{-\infty}^{\infty} f_x(z - y) f_y(y)\,dy \tag{6-39}$$

The above integral is the convolution of the functions $f_x(x)$ and $f_y(y)$. We thus reach the following fundamental conclusion:

> If two RVs are independent, then the density of their sum equals the convolution of their densities.

We note that, if $f_x(x) = 0$ for $x < 0$ and $f_y(y) = 0$ for $y < 0$, then $f_z(0) = 0$ for $z < 0$ and

$$f_z(z) = \int_0^z f_x(z - y) f_y(y)\,dy \qquad z > 0 \tag{6-40}$$

Example 6-6. It follows from (6-39) that the convolution of two rectangles is a trapezoid. Hence, if the RVs **x** and **y** are uniform in the intervals (a, b) and (c, d) respectively, then the density of their sum $\mathbf{z} = \mathbf{x} + \mathbf{y}$ is a trapezoid as in Fig. 6-9a. If, in particular, $b - a = d - c$, then $f_z(z)$ is a triangle as in Fig. 6-9b.

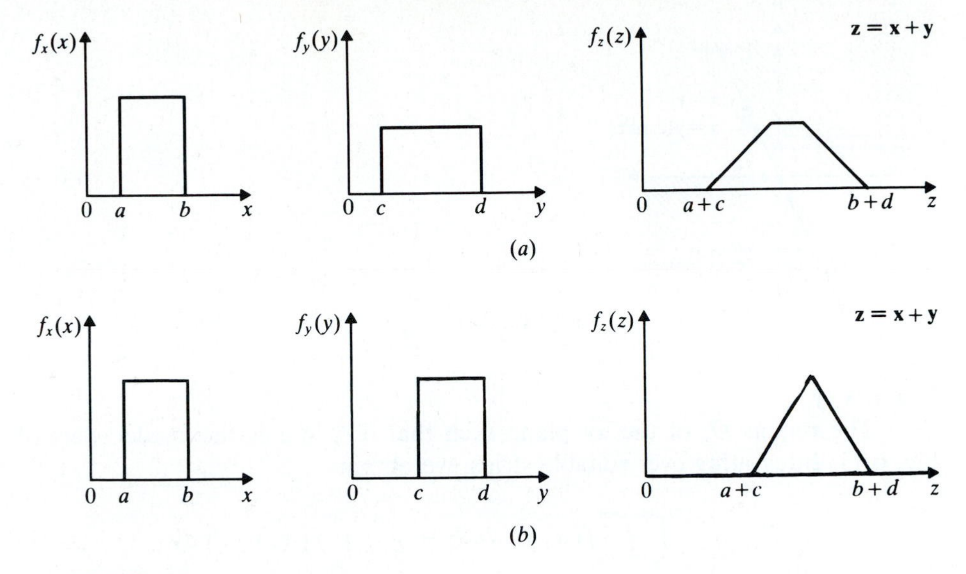

FIGURE 6-9

Suppose, for example, that resistors $\mathbf{r}_1$ and $\mathbf{r}_2$ are two independent RVs uniform between 900 and 1100 Ω. From the above it follows that, if they are connected in series, the density of the resulting resistor $\mathbf{r} = \mathbf{r}_1 + \mathbf{r}_2$ is a triangle between 1800 and 2200 Ω. In particular, the probability that $\mathbf{r}$ is between 1900 and 2100 Ω equals 0.75.

Example 6-7. If the RVs $\mathbf{x}$ and $\mathbf{y}$ are independent and

$$f_x(x) = \alpha e^{-\alpha x} U(x) \qquad f_y(y) = \beta e^{-\beta y} U(y)$$

(Fig. 6-10) then for $z > 0$,

$$f_z(z) = \alpha\beta \int_0^z e^{-\alpha(z-y)} e^{-\beta y}\, dy = \begin{cases} \dfrac{\alpha\beta}{\beta - \alpha}\left(e^{-\alpha z} - e^{-\beta z}\right) & \beta \neq \alpha \\ \alpha^2 z e^{-\alpha z} & \beta = \alpha \end{cases} \tag{6-41}$$

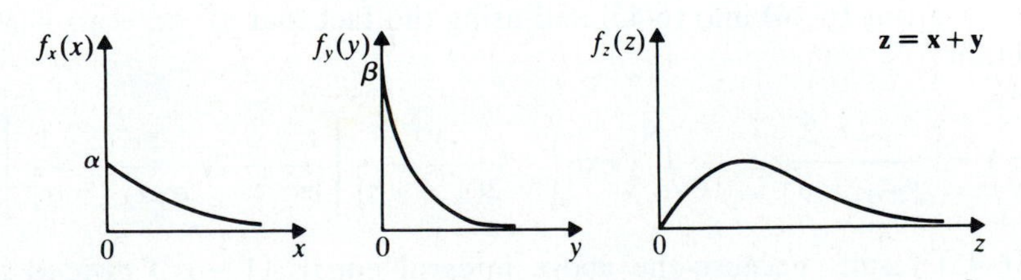

FIGURE 6-10

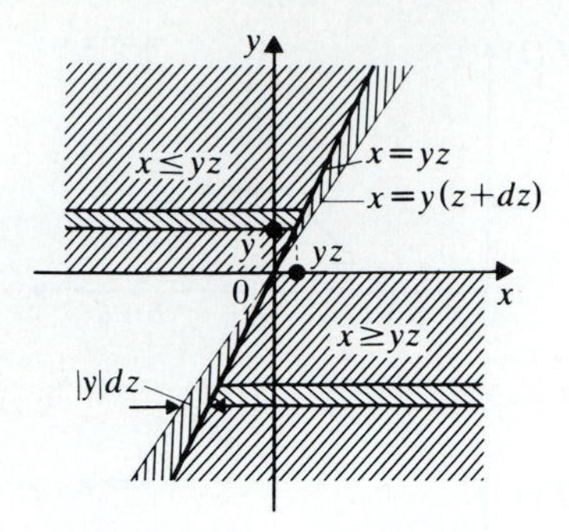

FIGURE 6-11

2. $\mathbf{z} = \mathbf{x}/\mathbf{y}$

The region D_z of the xy plane such that $x/y \le z$ is the shaded part of Fig. 6-11. Integrating over suitable strips, we obtain

$$F_z(z) = \int_0^{\infty} \int_{-\infty}^{yz} f(x, y)\, dx\, dy + \int_{-\infty}^{0} \int_{yz}^{\infty} f(x, y)\, dx\, dy \tag{6-42}$$

The region ΔD_z such that $z < x/y < z + dz$ is a triangle sector bounded by the lines $x = yz$ and $x = y(z + dz)$. The coordinates of a point in this region are zy, y and the area of a differential equals $|y|\, dy\, dz$. Inserting into (6-36) and canceling dz, we obtain

$$f_z(z) = \int_{-\infty}^{\infty} |y| f(zy, y)\, dy \tag{6-43}$$

Normal densities. We maintain that, if the RVs $\mathbf{x}$ and $\mathbf{y}$ are jointly normal with zero mean

$$f(x, y) = \frac{1}{2\pi\sigma_1\sigma_2\sqrt{1 - r^2}} \exp\left[-\frac{1}{2(1 - r^2)} \left(\frac{x^2}{\sigma_1^2} - 2r\frac{xy}{\sigma_1\sigma_2} + \frac{y^2}{\sigma_2^2} \right) \right] \tag{6-44}$$

then their ratio $\mathbf{z} = \mathbf{x}/\mathbf{y}$ has a *Cauchy* density centered at $r\sigma_1/\sigma_2$:

$$f_z(z) = \frac{\sigma_1\sigma_2\sqrt{1 - r^2}/\pi}{\sigma_2^2(z - r\sigma_1/\sigma_2)^2 + \sigma_1^2(1 - r^2)} \tag{6-45}$$

Proof. Inserting (6-44) into (6-43) and using the fact that $f(-x, -y) = f(x, y)$, we obtain

$$f_z(z) = \frac{2}{2\pi\sigma_1\sigma_2\sqrt{1 - r^2}} \int_0^{\infty} y \exp\left\{ -\frac{y^2}{2(1 - r^2)} \left[\frac{z^2}{\sigma_1^2} - 2r\frac{z}{\sigma_1\sigma_2} + \frac{1}{\sigma_1^2} \right] \right\} dy$$

and (6-45) results because the above integral equals $(1 - r^2)$ divided by the quantity in brackets.

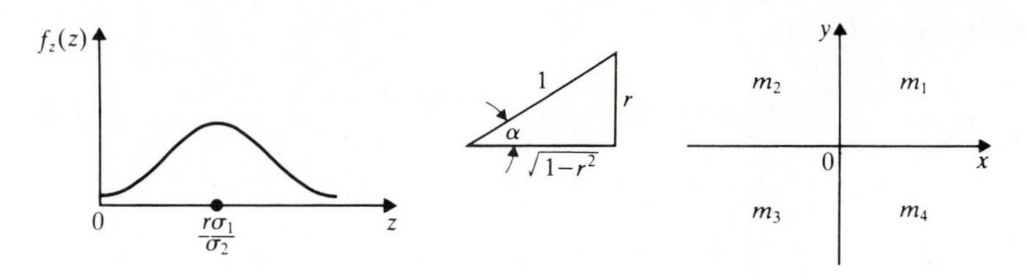

FIGURE 6-12

Integrating (6-45) from $-\infty$ to z, we obtain the corresponding distribution function

$$F_z(z) = \frac{1}{2} + \frac{1}{\pi}\arctan\frac{\sigma_2 z - r\sigma_1}{\sigma_1\sqrt{1-r^2}} \tag{6-46}$$

Quadrant masses Using (6-46), we shall show that the probability masses m_1, m_2, m_3, m_4 in the four quadrants of the xy plane are given by

$$m_1 = m_3 = \frac{1}{4} + \frac{\alpha}{2\pi} \qquad m_2 = m_4 = \frac{1}{4} - \frac{\alpha}{2\pi} \tag{6-47}$$

where (Fig. 6-12)

$$\alpha = \arcsin r = \arctan r/\sqrt{1-r^2} \qquad -\pi/2 < \alpha \le \pi/2$$

Proof. The second and fourth quadrant is the region of the plane such that $x/y \le 0$. The probability that the point $(\mathbf{x}, \mathbf{y})$ is in the region equals, therefore, the probability that the RV $\mathbf{z} = \mathbf{x}/\mathbf{y}$ is negative. Hence

$$m_2 + m_4 = P\{\mathbf{z} \le 0\} = F_z(0) = \frac{1}{2} - \frac{1}{\pi}\arctan\frac{r}{\sqrt{1-r^2}}$$

and (6-47) results because

$$m_1 = m_3 \qquad m_2 = m_4 \qquad m_1 + m_2 + m_3 + m_4 = 1$$

This useful result could have been obtained by integrating $f(x, y)$ in each quadrant; the above method is, however, simpler.

3. $\mathbf{z} = \sqrt{\mathbf{x}^2 + \mathbf{y}^2}$

The region D_z is the circle $x^2 + y^2 \le z^2$ and $F_z(z)$ equals the probability masses in this circle. If $f(x, y) = g(r)$ is circularly symmetrical, then

$$F_z(z) = 2\pi\int_0^z rg(r)\,dr \qquad z > 0.$$

Normal densities. (*a*) If

$$f(x, y) = \frac{1}{2\pi\sigma^2} e^{-(x^2+y^2)/2\sigma^2} \tag{6-48}$$

then

$$F_z(z) = \frac{1}{\sigma^2} \int_0^z r e^{-r^2/2\sigma^2} dr = 1 - e^{-z^2/2\sigma^2} \qquad z > 0 \tag{6-49}$$

Hence

$$f_z(z) = \frac{z}{\sigma^2} e^{-z^2/2\sigma^2} U(z) \tag{6-50}$$

Thus, if the RVs **x** and **y** are normal, independent with zero mean and equal variance, then the RV $\mathbf{z} = \sqrt{\mathbf{x}^2 + \mathbf{y}^2}$ has a Rayleigh density.

(*b*) Suppose now that

$$f(x, y) = \frac{1}{2\pi\sigma^2} e^{-[(x-\eta)^2+y^2]/2\sigma^2} \tag{6-51}$$

The region ΔD_z of the plane such that $z < \sqrt{x^2 + y^2} < z + dz$ is a circular ring with inner radius z and thickness dz. With

$$x = z\cos\theta \qquad y = z\sin\theta \qquad dx\,dy = z\,dz\,d\theta$$

it follows that

$$f_z(z)\,dz = \iint_{\Delta D_z} f(x, y)\,dx\,dy = \frac{1}{2\pi\sigma^2} \int_0^{2\pi} e^{-[(z\cos\theta-\eta)^2+(z\sin\theta)^2]/2\sigma^2} z\,dz\,d\theta$$

Hence

$$f_z(z) = \frac{z}{2\pi\sigma^2} e^{-(z^2+\eta^2)/2\sigma^2} \int_0^{2\pi} e^{z\eta\cos\theta/\sigma^2}\,d\theta$$

This yields

$$f_z(z) = \frac{z}{\sigma^2} I_0\left(\frac{z\eta}{\sigma^2}\right) e^{-(z^2+\eta^2)/2\sigma^2} \qquad z > 0 \tag{6-52}$$

where

$$I_0(x) = \frac{1}{2\pi} \int_0^{2\pi} e^{x\cos\theta}\,d\theta \tag{6-53}$$

is the modified Bessel function.

Example 6-8. Consider the sine wave

$$\mathbf{x}\cos\omega t + \mathbf{y}\sin\omega t = \mathbf{r}\cos(\omega t + \boldsymbol{\theta})$$

Since $\mathbf{r} = \sqrt{\mathbf{x}^2 + \mathbf{y}^2}$, if follows from the above that, if the RVs **x** and **y** are normal as in (6-48), then the density of **r** is Rayleigh as in (6-50).

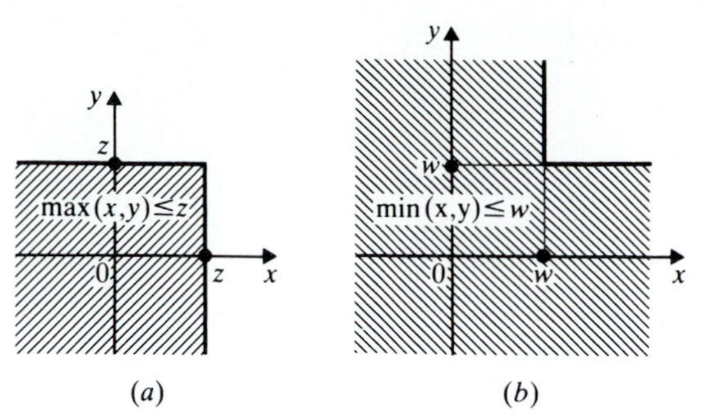

FIGURE 6-13

4. $\mathbf{z} = \max(\mathbf{x}, \mathbf{y}) \qquad \mathbf{w} = \min(\mathbf{x}, \mathbf{y})$

(*a*) The region D_z of the *xy* plane such that $\max(x, y) \le z$ is the set of points such that $x \le z$ *and* $y \le z$ (shaded in Fig. 6-13*a*). Hence

$$F_z(z) = F_{xy}(z, z) \tag{6-54}$$

If the RVs **x** and **y** are *independent*, then

$$F_z(z) = F_x(z)F_y(z) \qquad f_z(z) = f_x(z)F_y(z) + f_y(z)F_x(z) \tag{6-55}$$

(*b*) The region D_w of the *xy* plane such that $\min(x, y) \le w$ is the set of points such that $x \le w$ *or* $y \le w$ (shaded in Fig. 6-13*b*). Hence

$$F_w(w) = F_x(w) + F_y(w) - F_{xy}(w, w) \tag{6-56}$$

If the RVs **x** and **y** are *independent* then it is simpler to express the result in terms of the *reliability function*.

$$R_x(x) = P\{\mathbf{x} > x\} = 1 - F_x(x) \tag{6-57}$$

Defining $R_y(y)$ and $R_w(w)$ similarly, we conclude from (6-56) that

$$R_w(w) = R_x(w)R_y(w) \qquad f_w(w) = f_x(w)R_y(w) + f_y(w)R_x(w) \tag{6-58}$$

Discrete type. If the RVs **x** and **y** are of discrete type taking the values x_i and y_k, then the RV $\mathbf{z} = g(\mathbf{x}, \mathbf{y})$ is also of discrete type taking the values $z_r = g(x_i, y_k)$. The probability that $\mathbf{z} = z_r$ equals the sum of the point masses on the curve $g(x, y) = z_r$.

Example 6-9. A fair die is tossed twice and the RVs **x** and **y** are such that

$$\mathbf{x}(f_i f_k) = i \qquad \mathbf{y}(f_i f_k) = k$$

The *xy* plane has 36 equal point masses as in Fig. 6-14. The RV $\mathbf{z} = \mathbf{x} + \mathbf{y}$ takes the values $z_r = x_i + y_k$ with probabilities $p_r = m/36$ where m is the number of points on the line $x + y = z_r$. As we see from the figure

$z_r =$	2	3	4	5	6	7	8	9	10	11	12
$p_r =$	$\frac{1}{36}$	$\frac{2}{36}$	$\frac{3}{36}$	$\frac{4}{36}$	$\frac{5}{36}$	$\frac{6}{36}$	$\frac{5}{36}$	$\frac{4}{36}$	$\frac{3}{36}$	$\frac{2}{36}$	$\frac{1}{36}$

For example, there are four mass points on the line $x + y = 5$; hence $p_5 = 4/36$.

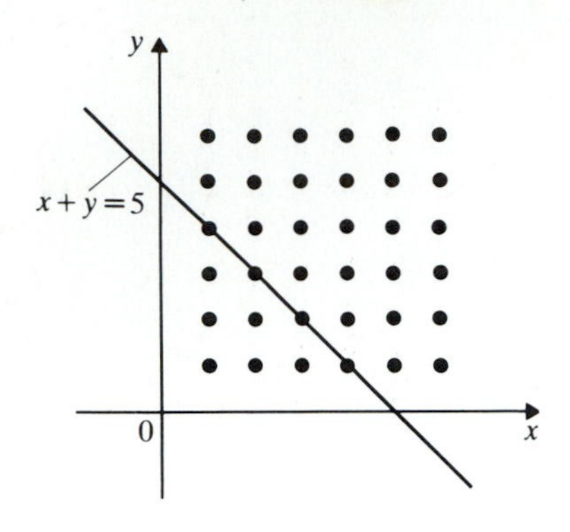

FIGURE 6-14

6-3 TWO FUNCTIONS OF TWO RANDOM VARIABLES

Given two RVs **x** and **y** and two functions $g(x, y)$ and $h(x, y)$, we form the RVs

$$\mathbf{z} = g(\mathbf{x}, \mathbf{y}) \qquad \mathbf{w} = h(\mathbf{x}, \mathbf{y}) \tag{6-59}$$

We shall express the joint statistics of **z** and **w** in terms of the functions $g(x, y)$ and $h(x, y)$ and the joint statistics of **x** and **y**.

With z and w two given numbers, we denote by D_{zw} the region of the xy plane such that $g(x, y) \le z$ and $h(x, y) \le w$. Clearly,

$$\{\mathbf{z} \le z, \mathbf{w} \le w\} = \{(\mathbf{x}, \mathbf{y}) \in D_{zw}\}$$

Hence [see (6-7)]

$$F_{zw}(z, w) = P\{(\mathbf{x}, \mathbf{y}) \in D_{zw}\} = \iint_{D_{zw}} f_{xy}(x, y)\, dx\, dy \tag{6-60}$$

Suppose, for example, that

$$\mathbf{z} = \sqrt{\mathbf{x}^2 + \mathbf{y}^2} \qquad \mathbf{w} = \mathbf{y}/\mathbf{x} \tag{6-61}$$

In this case, the set D_{zw} such that

$$\sqrt{x^2 + y^2} \le z \qquad y/x \le w$$

is the shaded region of Fig. 6-15a, and $F_{zw}(z, w)$ equals the mass in this region.

Example 6-10. If

$$f_{xy}(x, y) = \frac{1}{2\pi\sigma^2} e^{-(x^2+y^2)/2\sigma^2} \qquad \mathbf{z} = \sqrt{\mathbf{x}^2 + \mathbf{y}^2} \qquad \mathbf{w} = \mathbf{y}/\mathbf{x}$$

then [see (6-49)] the mass in the circle $x^2 + y^2 \le z^2$ equals $1 - e^{-z^2/2\sigma^2}$. Since $f_{xy}(x, y)$ has circular symmetry we conclude that for $z > 0$:

$$F_{zw}(z, w) = \frac{2\theta}{2\pi}(1 - e^{-z^2/2\sigma^2}) \qquad \theta = \frac{\pi}{2} + \arctan w$$

and $F_{zw}(z, w) = 0$ for $z < 0$. This is a product of a function of z times a function

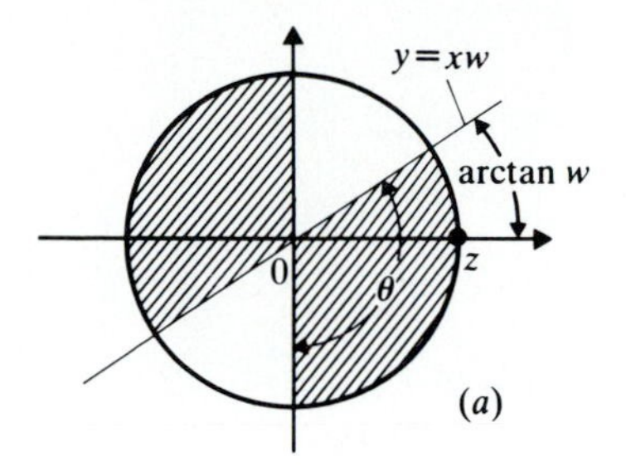

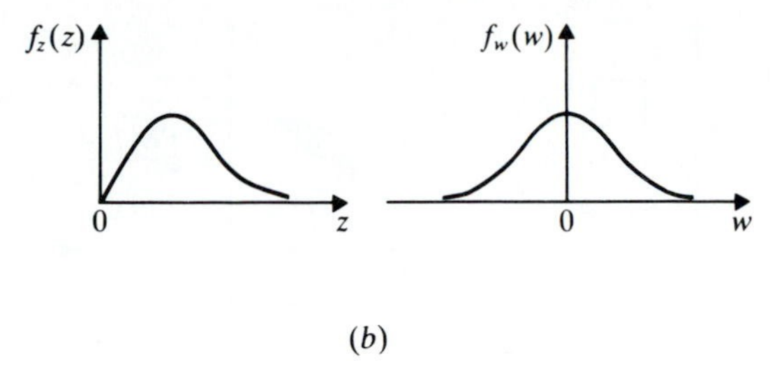

FIGURE 6-15

of w. Hence the RV **z** and **w** are *independent* with

$$F_z(z) = (1 - e^{-z^2/2\sigma})U(z) \qquad F_w(w) = \frac{1}{2} + \frac{1}{\pi}\arctan w$$

In other words, **z** has a *Rayleigh* density and **w** has a *Cauchy* density [see (5-17)] as in Fig. 6-15*b*.

Joint Density

We shall determine the joint density of the RVs

$$\mathbf{z} = g(\mathbf{x}, \mathbf{y}) \qquad \mathbf{w} = h(\mathbf{x}, \mathbf{y})$$

in terms of the joint density of **x** and **y**.

Fundamental theorem. To find $f_{zw}(z, w)$, we solve the system

$$g(x, y) = z \qquad h(x, y) = w \tag{6-62}$$

Denoting by (x_n, y_n) its real roots

$$g(x_n, y_n) = z \qquad h(x_n, y_n) = w$$

we maintain that

$$f_{zw}(z, w) = \frac{f_{xy}(x_1, y_1)}{|J(x_1, y_1)|} + \cdots + \frac{f_{xy}(x_n, y_n)}{|J(x_n, y_n)|} + \cdots \tag{6-63}$$

where

$$J(x, y) = \begin{vmatrix} \dfrac{\partial z}{\partial x} & \dfrac{\partial z}{\partial y} \\ \dfrac{\partial w}{\partial x} & \dfrac{\partial w}{\partial y} \end{vmatrix} = \begin{vmatrix} \dfrac{\partial x}{\partial z} & \dfrac{\partial x}{\partial w} \\ \dfrac{\partial y}{\partial z} & \dfrac{\partial y}{\partial w} \end{vmatrix}^{-1} \tag{6-64}$$

is the *jacobian* of the transformation (6-62).

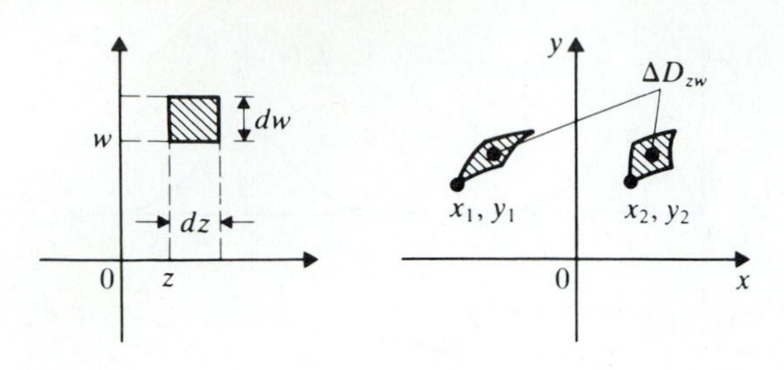

FIGURE 6-16

Proof. We denote by ΔD_{zw} the region in the xy plane such that

$$z < g(x, y) < z + dz \qquad w < h(x, y) < w + dw$$

This region consists of differential parallelograms, one for each (x_n, y_n) as in Fig. 6-16. The area of each parallelogram equals $dz\,dw/\,|J(x_n, y_n)|$ and its mass equals

$$f_{xy}(x_n, y_n)\,dz\,dw/\,|J(x_n, y_n)|$$

Since $f_{zw}(z, w)\,dz\,dw$ equals the mass in ΔD_{zw}, we conclude, summing the masses in all parallelograms, that

$$f_{zw}(z, w)\,dz\,dw = \frac{f_{xy}(x_1, y_1)\,dz\,dw}{|J(x_1, y_1)|} + \cdots + \frac{f_{xy}(x_n, y_n)\,dz\,dw}{|J(x_n, y_n)|} + \cdots$$

and (6-63) results.

If the system (6-62) has no solutions in some region of the zw plane, then $f_{zw}(z, w) = 0$ in that region.

We shall illustrate the above theorem with two special cases.

LINEAR TRANSFORMATION

$$\mathbf{z} = a\mathbf{x} + b\mathbf{y} \qquad \mathbf{w} = c\mathbf{x} + d\mathbf{y} \tag{6-65}$$

If $ad - bc \neq 0$, then the system $ax + by = z$, $cx + dy = w$ has one and only one solution

$$x = Az + Bw \qquad y = Cz + Dw$$

Since $J(x, y) = ad - bc$, (6-63) yields

$$f_{zw}(z, w) = \frac{1}{|ad - bc|} f_{xy}(Az + Bw, Cz + Dw) \tag{6-66}$$

Joint normality. From (6-66) it follows that if the RVs $\mathbf{x}$ and $\mathbf{y}$ are jointly normal and

$$\mathbf{z} = a\mathbf{x} + b\mathbf{y} \qquad \mathbf{w} = c\mathbf{x} + d\mathbf{y}$$

then $\mathbf{z}$ and $\mathbf{w}$ are also jointly normal.

Proof. Joint normality means that $f_{xy}(x, y)$ is an exponential whose exponent is a quadratic in x and y. If, in this quadratic, we replace x by $Az + Bw$ and y by $Cz + Dw$ as in (6-66), then an exponential results whose exponent is a quadratic in z and w. This shows that the RVs $\mathbf{z}$ and $\mathbf{w}$ are jointly, and therefore also marginally, normal.

From the above it follows that, if $\mathbf{x}$ and $\mathbf{y}$ are jointly normal and $\mathbf{z} = \mathbf{x} + \mathbf{y}$, then $\mathbf{z}$ is normal. We should emphasize, however, that if $\mathbf{x}$ and $\mathbf{y}$ are marginally but not jointly normal, then $\mathbf{z}$ is not, in general, normal. We give next a counter example.

Example 6-11. We shall construct two marginally normal RVs $\mathbf{x}_1$ and $\mathbf{y}_1$ such that their sum $\mathbf{z}_1 = \mathbf{x}_1 + \mathbf{y}_1$ is not normal: We start with two jointly normal RVs $\mathbf{x}$ and $\mathbf{y}$ and add and subtract masses on the four circles of Fig. 6-17. The resulting mass distribution specifies the joint density of the RVs $\mathbf{x}_1$ and $\mathbf{y}_1$. As we have shown in Example 6-1, these RVs are marginally normal. However, their sum $\mathbf{z}_1$ is not normal.

Rotation. A special case of (6-65) is the transformation

$$\mathbf{z} = \mathbf{x}\cos\varphi + \mathbf{y}\sin\varphi \qquad \mathbf{w} = -\mathbf{x}\sin\varphi + \mathbf{y}\cos\varphi \tag{6-67}$$

In this case $a = d = \cos\varphi$, $b = -c = \sin\varphi$, and $ad - bc = 1$. Hence

$$\mathbf{x} = \mathbf{z}\cos\varphi - \mathbf{w}\sin\varphi \qquad \mathbf{y} = \mathbf{z}\sin\varphi + \mathbf{w}\cos\varphi$$

and (6-66) yields

$$f_{zw}(z,w) = f_{xy}(z\cos\varphi - w\sin\varphi, z\sin\varphi + w\cos\varphi) \tag{6-68}$$

Thus, if two RVs are rotated by an angle φ, their probability masses are rotated in the opposite direction by the same angle.

Circular symmetry If $f_{xy}(x, y)$ is circularly symmetrical as in (6-31), then

$$f_{xy}(x,y) = f_{xy}(x\cos\varphi - y\sin\varphi, x\sin\varphi + y\cos\varphi) \tag{6-69}$$

because

$$(x\cos\varphi - y\sin\varphi)^2 + (x\sin\varphi + y\cos\varphi)^2 = x^2 + y^2$$

Hence [see (6-68)]

$$f_{zw}(z,w) = f_{xy}(z,w) = g\left(\sqrt{z^2 + w^2}\right) \tag{6-70}$$

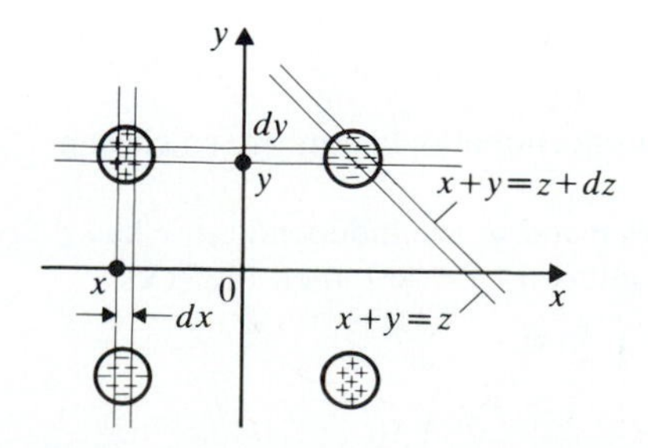

FIGURE 6-17

Conversely, if the RVs $\mathbf{x}, \mathbf{y}$ and $\mathbf{z}, \mathbf{w}$ have the same statistics for every φ, then their joint density is circularly symmetrical. From (6-34) it follows that if $\mathbf{x}$ and $\mathbf{y}$ are also independent, then they are normal.

POLAR COORDINATES. Consider the RVs

$$\mathbf{r} = \sqrt{\mathbf{x}^2 + \mathbf{y}^2} \qquad \boldsymbol{\varphi} = \arctan \mathbf{y}/\mathbf{x} \tag{6-71}$$

where we assume that $\mathbf{r} \geq 0$ and $-\pi < \boldsymbol{\varphi} \leq \pi$. With this assumption, the system $\sqrt{x^2 + y^2} = r$, $\arctan y/x = \varphi$ has a single solution

$$x = r \cos \varphi \qquad y = r \sin \varphi \qquad \text{for} \quad r > 0$$

Since [see (6-64)]

$$J(x, y) = \begin{vmatrix} \cos \varphi & -r \sin \varphi \\ \sin \varphi & r \cos \varphi \end{vmatrix}^{-1} = \frac{1}{r}$$

we conclude from (6-63) that

$$f_{r\varphi}(r, \varphi) = r f_{xy}(r \cos \varphi, r \sin \varphi) \qquad r > 0 \tag{6-72}$$

and 0 for $r < 0$.

Example 6-12. We shall show that if

$$\mathbf{x} \cos \omega t + \mathbf{y} \sin \omega t = \mathbf{r} \cos(\omega t - \boldsymbol{\varphi}) \qquad |\boldsymbol{\varphi}| < \pi$$

and the RVs $\mathbf{x}$ and $\mathbf{y}$ are $N(0, \sigma)$ and independent, then the RVs $\mathbf{r}$ and $\boldsymbol{\varphi}$ are independent, $\boldsymbol{\varphi}$ is uniform in the interval $(-\pi, \pi)$ and $\mathbf{r}$ has a Rayleigh distribution.

Proof. Since $x = r \cos \varphi$, $y = r \sin \varphi$, and

$$f_{xy}(x, y) = \frac{1}{2\pi\sigma^2} e^{-(x^2+y^2)/2\sigma^2}$$

(6-72) yields

$$f_{r\varphi}(r, \varphi) = \frac{r}{2\pi\sigma^2} e^{-r^2/2\sigma^2} \qquad r > 0 \qquad |\varphi| < \pi$$

and 0 otherwise. This is a product of a function of r times a function of φ. Hence the RVs $\mathbf{r}$ and $\boldsymbol{\varphi}$ are independent with

$$f_r(r) = \frac{r}{\sigma^2} e^{-r^2/2\sigma^2} \qquad f_\varphi(\varphi) = \frac{1}{2\pi}$$

for $r > 0$, $-\pi < \varphi \leq \pi$ and 0 otherwise. The proportionality factors are so chosen as to make the area of each term equal to 1.

From the above it follows that, if the RVs $\mathbf{r}$ and $\boldsymbol{\varphi}$ are independent, $\mathbf{r}$ has a Rayleigh distribution, and $\boldsymbol{\varphi}$ is uniform in the interval $(-\pi, \pi)$, then the RVs

$$\mathbf{x} = \mathbf{r} \cos \boldsymbol{\varphi} \qquad \mathbf{y} = \mathbf{r} \sin \boldsymbol{\varphi}$$

are $N(0, \sigma)$ and independent.

Auxiliary variables. The determination of the density of *one* function $\mathbf{z} = g(\mathbf{x}, \mathbf{y})$ of two RVs can be determined from (6-63) where $\mathbf{w}$ is a conveniently chosen auxiliary variable, for example $\mathbf{w} = \mathbf{x}$ or $\mathbf{w} = \mathbf{y}$. The density of $\mathbf{z}$ is then found by integrating the function $f_{zw}(z, w)$ so obtained.

Example 6-13. We shall find the density of the RV

$$\mathbf{z} = a\mathbf{x} + b\mathbf{y}$$

using as auxiliary variable the function $\mathbf{w} = \mathbf{y}$.

The system $z = ax + by$, $w = y$ has a single solution: $x = (z - bw)/a$, $y = w$. Since

$$J(x, y) = \begin{vmatrix} a & b \\ 0 & 1 \end{vmatrix} = a$$

it follows from (6-63) that

$$f_{zw}(z, w) = \frac{1}{|a|} f_{xy}\left(\frac{z - bw}{a}, w\right)$$

Hence

$$f_z(z) = \frac{1}{|a|} \int_{-\infty}^{\infty} f_{xy}\left(\frac{z - by}{a}, y\right) dy \tag{6-73}$$

Example 6-14. With

$$\mathbf{z} = \mathbf{xy} \qquad \mathbf{w} = \mathbf{x}$$

the system $xy = z$, $x = w$ has a single solution: $x = w$, $y = z/w$. In this case, $J = -w$ and (6-63) yields

$$f_{zw}(z, w) = \frac{1}{|w|} f_{xy}\left(w, \frac{z}{w}\right)$$

Hence the density of the RV $\mathbf{z} = \mathbf{xy}$ is given by

$$f_z(z) = \int_{-\infty}^{\infty} \frac{1}{|w|} f_{xy}\left(w, \frac{z}{w}\right) dw \tag{6-74}$$

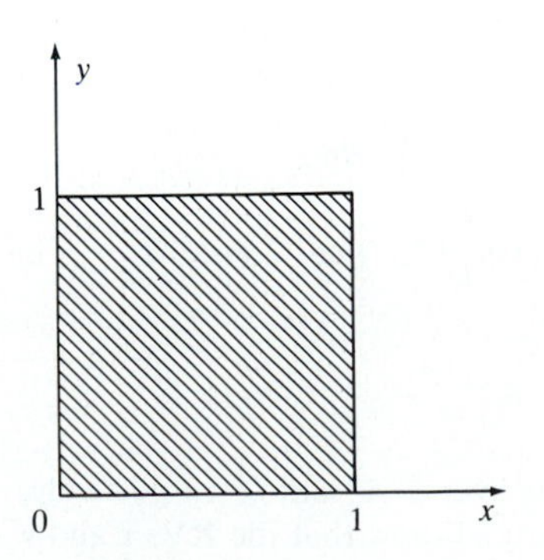

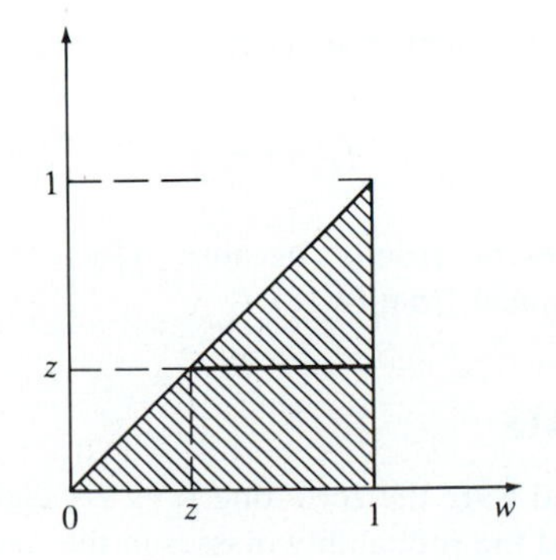

FIGURE 6-18

Special case. We now assume that the RVs **x** and **y** are independent and each is uniform in the interval (0, 1). In this case,

$$f_{xy}\left(w, \frac{z}{w}\right) = f_x(w) f_y\left(\frac{z}{w}\right) = 1$$

in the triangle $z < w < 1$, $0 < z < 1$ (shaded in Fig. 6-18) and 0 elsewhere. Inserting into (6-74), we obtain

$$f_z(z) = \int_z^1 \frac{1}{w}\, dw = \begin{cases} -\ln z & 0 < z < 1 \\ 0 & \text{elsewhere} \end{cases} \tag{6-75}$$

Example 6-15. An RV **z** has a *Student-t* distribution $t(n)$ with n degrees of freedom if

$$f_z(z) = \frac{\gamma_1}{\sqrt{(1 + z^2/n)^{n+1}}} \qquad \gamma_1 = \frac{\Gamma[(n+1)/2]}{\sqrt{\pi n}\,\Gamma(n/2)} \tag{6-76}$$

We shall show that if **x** and **y** are two independent RVs, **x** is $N(0, 1)$, and **y** is $\chi^2(n)$:

$$f_x(x) \sim e^{-x^2/2} \qquad f_y(y) \sim y^{n/2-1} e^{-y/2} U(y)$$

then the RV

$$\mathbf{z} = \frac{\mathbf{x}}{\sqrt{\mathbf{y}/n}}$$

has a $t(n)$ distribution

Proof. We introduce the RV $\mathbf{w} = \mathbf{y}$ and use (6-63) with

$$x = z\sqrt{\frac{w}{n}} \qquad y = w \qquad J(x, y) = \sqrt{\frac{n}{y}} = \sqrt{\frac{n}{w}}$$

This yields

$$f_{zw}(z, w) \sim \sqrt{y}\, f_x\left(z\sqrt{\frac{w}{n}}\right) f_y(w) \sim \frac{1}{\sqrt{w}} \exp\left\{-\frac{w}{2} z^2\right\} w^{n/2-1} e^{-w/2} U(w)$$

Integrating with respect to w, we obtain

$$f_z(z) \sim \int_0^\infty w^{(n-1)/2} \exp\left\{-\frac{w}{2}\left(1 + \frac{z^2}{n}\right)\right\} dw$$

and (6-76) results because $\int_0^\infty w^{a-1} e^{-bw}\, dw = \Gamma(a)/b^a$. The constant γ_1 is determined from (4-18).

PROBLEMS

6-1. If **x** and **y** are the zero–one RVs associated with the events $\mathscr{A}$ and $\mathscr{B}$ respectively, (a) find the probability masses in the x–y plane and (b) show that the RVs **x** and **y** are independent iff the events $\mathscr{A}$ and $\mathscr{B}$ are independent.

6-2. The RVs $\mathbf{x}$ and $\mathbf{y}$ are independent and $\mathbf{z} = \mathbf{x} + \mathbf{y}$. Find $f_y(y)$ if

$$f_x(x) = ce^{-cx}U(x) \qquad f_z(z) = c^2 z e^{-cz}U(z)$$

6-3. The RVs $\mathbf{x}$ and $\mathbf{y}$ are independent and $\mathbf{y}$ is uniform in the interval (0, 1). Show that, if $\mathbf{z} = \mathbf{x} + \mathbf{y}$, then

$$f_z(z) = F_x(z) - F_x(z-1)$$

6-4. (*a*) The function $g(x)$ is monotone increasing and $\mathbf{y} = g(\mathbf{x})$. Show that

$$F_{xy}(x, y) = \begin{cases} F_x(x) & \text{if} \quad y > g(x) \\ F_y(y) & \text{if} \quad y < g(x) \end{cases}$$

(*b*) Find $F_{xy}(x, y)$ if $g(x)$ is monotone decreasing.

6-5. Express $F_{zw}(z, w)$ in terms of $F_{xy}(x, y)$ if $\mathbf{z} = \max(\mathbf{x}, \mathbf{y})$, $\mathbf{w} = \min(\mathbf{x}, \mathbf{y})$.

6-6. The RVs $\mathbf{x}$ and $\mathbf{y}$ are $N(0, 2)$ and independent. Find $f_z(z)$ and $F_z(z)$ if (*a*) $\mathbf{z} = 2\mathbf{x} + 3\mathbf{y}$, and (*b*) $\mathbf{z} = \mathbf{x}/\mathbf{y}$.

6-7. The RVs $\mathbf{x}$ and $\mathbf{y}$ are independent with

$$f_x(x) = \frac{x}{\alpha^2} e^{-x^2/2\alpha^2} U(x) \qquad f_y(y) = \begin{cases} 1/\pi\sqrt{1-y^2} & |y| < 1 \\ 0 & |y| > 1 \end{cases}$$

Show that the RV $\mathbf{z} = \mathbf{xy}$ is $N(0, \alpha)$.

6-8. The RVs $\mathbf{x}$ and $\mathbf{y}$ are independent with Rayleigh densities

$$f_x(x) = \frac{x}{\alpha^2} e^{-x^2/2\alpha^2} U(x) \qquad f_y(y) = \frac{y}{\beta^2} e^{-y^2/2\beta^2} U(y)$$

(*a*) Show that if $\mathbf{z} = \mathbf{x}/\mathbf{y}$, then

$$f_z(z) = \frac{2\alpha^2}{\beta^2} \frac{z}{\left(z^2 + \alpha^2/\beta^2\right)^2} U(z) \tag{i}$$

(*b*) Using (i), show that for any $k > 0$,

$$P\{\mathbf{x} \le k\mathbf{y}\} = \frac{k^2}{k^2 + \alpha^2/\beta^2}$$

6-9. The RVs $\mathbf{x}$ and $\mathbf{y}$ are independent with exponential densities

$$f_x(x) = \alpha e^{-\alpha x} U(x) \qquad f_y(y) = \beta e^{-\beta y} U(y)$$

Find the densities of the following RVs:

1. $2\mathbf{x} + \mathbf{y}$ 2. $\mathbf{x} - \mathbf{y}$ 3. $\dfrac{\mathbf{x}}{\mathbf{y}}$ 4. $\max(\mathbf{x}, \mathbf{y})$ 5. $\min(\mathbf{x}, \mathbf{y})$

6-10. The RVs $\mathbf{x}$ and $\mathbf{y}$ are independent and each is uniform in the interval $(0, a)$. Find the density of the RV $\mathbf{z} = |\mathbf{x} - \mathbf{y}|$.

6-11. Show that (*a*) the convolution of two normal densities is a normal density, and (*b*) the convolution of two Cauchy densities is a Cauchy density.

6-12. The RVs $\mathbf{x}$ and $\boldsymbol{\theta}$ are independent and $\boldsymbol{\theta}$ is uniform in the interval $(-\pi, \pi)$. Show that if $\mathbf{z} = \mathbf{x}\cos(wt + \boldsymbol{\theta})$, then

$$f_z(z) = \frac{1}{\pi}\int_{-\infty}^{-|z|} \frac{f_x(y)}{\sqrt{y^2 - z^2}}\,dy + \frac{1}{\pi}\int_{|z|}^{\infty} \frac{f_x(y)}{\sqrt{y^2 - z^2}}\,dy$$

6-13. The RVs $\mathbf{x}$ and $\mathbf{y}$ are independent, $\mathbf{x}$ is $N(0, \sigma)$, and $\mathbf{y}$ is uniform in the interval $(0, \pi)$. Show that if $\mathbf{z} = \mathbf{x} + a\cos\mathbf{y}$, then

$$f_z(z) = \frac{1}{\pi\sigma\sqrt{2\pi}}\int_0^{\pi} e^{-(z - a\cos y)^2/2\sigma^2}\,dy$$

6-14. The RVs $\mathbf{x}$ and $\mathbf{y}$ are of discrete type, independent, with $P\{\mathbf{x} = n\} = a_n$, $P\{\mathbf{y} = n\} = b_n$, $n = 0, 1, \ldots$. Show that, if $\mathbf{z} = \mathbf{x} + \mathbf{y}$, then

$$P\{\mathbf{z} = n\} = \sum_{k=0}^{n} a_k b_{n-k}$$

6-15. The RV $\mathbf{x}$ is of discrete type taking the values x_n with $P\{\mathbf{x} = x_n\} = p_n$ and the RV $\mathbf{y}$ is of continuous type and independent of $\mathbf{x}$. Show that if $\mathbf{z} = \mathbf{x} + \mathbf{y}$ and $\mathbf{w} = \mathbf{xy}$, then

$$f_z(z) = \sum_n f_y(z - x_n)p_n \qquad f_w(w) = \sum_n \frac{1}{|x_n|} f_y\left(\frac{w}{x_n}\right)p_n$$

6-16. The Rvs $\mathbf{x}$ and $\mathbf{y}$ are normal, independent, with the same variance. Show that, if $\mathbf{z} = \sqrt{\mathbf{x}^2 + \mathbf{y}^2}$, then $f_z(z)$ is given by (6-52) where $\eta = \sqrt{\eta_x^2 + \eta_y^2}$.

6-17. The RVs $\mathbf{x}_1$ and $\mathbf{x}_2$ are jointly normal with zero mean. Show that their density can be written in the form

$$f(x_1, x_2) = \frac{1}{2\pi\sqrt{\Delta}}\exp\left\{-\frac{1}{2}XC^{-1}X^t\right\} \qquad C = \begin{bmatrix} \mu_{11} & \mu_{12} \\ \mu_{21} & \mu_{22} \end{bmatrix}$$

where $X: [x_1, x_2]$, $\mu_{ij} = E\{\mathbf{x}_i\mathbf{x}_j\}$, and $\Delta = \mu_{11}\mu_{22} + \mu_{12}^2$.

6-18. Show that if the RVs $\mathbf{x}$ and $\mathbf{y}$ are normal and independent, then

$$P\{\mathbf{xy} < 0\} = \mathbb{G}\left(\frac{\eta_x}{\sigma_x}\right) + \mathbb{G}\left(\frac{\eta_y}{\sigma_y}\right) - 2\mathbb{G}\left(\frac{\eta_x}{\sigma_x}\right)\mathbb{G}\left(\frac{\eta_y}{\sigma_y}\right)$$

6-19. The RVs $\mathbf{x}$ and $\mathbf{y}$ are independent with respective densities $\chi^2(m)$ and $\chi^2(n)$. Show that if

$$\mathbf{z} = \frac{\mathbf{x}/m}{\mathbf{y}/n} \qquad \text{then} \quad f_z(z) = \gamma\frac{x^{m/2-2}}{\sqrt{(1 + mx/n)^{m+n}}}U(x)$$

This distribution is denoted by $F(m, n)$ and is called the *Snedecor F* distribution. It is used in hypothesis testing (see Prob. 9-34).

CHAPTER

7

MOMENTS AND CONDITIONAL DISTRIBUTIONS

7-1 JOINT MOMENTS

Given two RVs **x** and **y** and a function $g(x, y)$, we form the RV $\mathbf{z} = g(\mathbf{x}, \mathbf{y})$. The expected value of this RV is given by

$$E\{\mathbf{z}\} = \int_{-\infty}^{\infty} z f_z(z)\, dz \tag{7-1}$$

However, as the next theorem shows, $E\{\mathbf{z}\}$ can be expressed directly in terms of the function $g(x, y)$ and the joint density $f(x, y)$ of **x** and **y**.

THEOREM

$$E\{g(\mathbf{x}, \mathbf{y})\} = \int_{-\infty}^{\infty}\int_{-\infty}^{\infty} g(x, y) f(x, y)\, dx\, dy \tag{7-2}$$

Proof. The proof is similar to the proof of (5-29). We denote by ΔD_z the region of the xy plane such that $z < g(x, y) < z + dz$. Thus to each differential in (7-1) there corresponds a region ΔD_z in the xy plane. As dz covers the z axis, the regions ΔD_z are not overlapping and they cover the entire xy plane. Hence the integrals in (7-1) and (7-2) are equal.

We note that the expected value of $g(\mathbf{x})$ can be determined either from (7-2) or from (5-29) as a single integral

$$E\{g(\mathbf{x})\} = \int_{-\infty}^{\infty}\int_{-\infty}^{\infty} g(x) f(x, y)\, dx\, dy = \int_{-\infty}^{\infty} g(x) f_x(x)\, dx$$

This is consistent with the relationship (6-10) between marginal and joint densities.

If the RVs **x** and **y** are of discrete type taking the values x_i and y_k with probability p_{ik} as in (6-19), then

$$E\{g(\mathbf{x},\mathbf{y})\} = \sum_{i,k} g(x_i, y_k) p_{ik} \tag{7-3}$$

Linearity From (7-2) it follows that

$$E\left\{\sum_{k=1}^{n} a_k g_k(\mathbf{x},\mathbf{y})\right\} = \sum_{k=1}^{n} a_k E\{g_k(\mathbf{x},\mathbf{y})\} \tag{7-4}$$

This fundamental result will be used extensively.

We note in particular that

$$E\{\mathbf{x}+\mathbf{y}\} = E\{\mathbf{x}\} + E\{\mathbf{y}\} \tag{7-5}$$

Thus the expected value of the sum of two RVs equals the sum of their expected values. We should stress, however, that in general

$$E\{\mathbf{xy}\} \neq E\{\mathbf{x}\}E\{\mathbf{y}\}$$

Frequency interpretation As in (5-26)

$$E\{\mathbf{x}+\mathbf{y}\} \simeq \frac{\mathbf{x}(\zeta_1) + \mathbf{y}(\zeta_1) + \cdots + \mathbf{x}(\zeta_n) + \mathbf{y}(\zeta_n)}{n}$$

$$= \frac{\mathbf{x}(\zeta_1) + \cdots + \mathbf{x}(\zeta_n)}{n} + \frac{\mathbf{y}(\zeta_1) + \cdots + \mathbf{y}(\zeta_n)}{n} \simeq E\{\mathbf{x}\} + E\{\mathbf{y}\}$$

However, in general,

$$E\{\mathbf{xy}\} \simeq \frac{\mathbf{x}(\zeta_1)\mathbf{y}(\zeta_1) + \cdots + \mathbf{x}(\zeta_n)\mathbf{y}(\zeta_n)}{n}$$

$$\neq \frac{\mathbf{x}(\zeta_1) + \cdots + \mathbf{x}(\zeta_n)}{n} \times \frac{\mathbf{y}(\zeta_1) + \cdots + \mathbf{y}(\zeta_n)}{n} \simeq E\{\mathbf{x}\}E\{\mathbf{y}\}$$

Covariance. The covariance C or C_{xy} of two RVs **x** and **y** is by definition the number

$$C = E\{(\mathbf{x} - \eta_x)(\mathbf{y} - \eta_y)\} \tag{7-6}$$

where $E\{\mathbf{x}\} = \eta_x$ and $E\{\mathbf{y}\} = \eta_y$. Expanding the product in (7-6) and using (7-4) we obtain

$$C = E\{\mathbf{xy}\} - E\{\mathbf{x}\}E\{\mathbf{y}\} \tag{7-7}$$

Correlation coefficient The correlation coefficient r or r_{xy} of the RVs **x** and **y** is by definition the ratio

$$r = \frac{C}{\sigma_x \sigma_y} \tag{7-8}$$

We maintain that

$$|r| \le 1 \qquad |C| \le \sigma_x \sigma_y \tag{7-9}$$

Proof. Clearly,

$$E\{[a(\mathbf{x} - \eta_x) + (\mathbf{y} - \eta_y)]^2\} = a^2\sigma_x^2 + 2aC + \sigma_y^2 \tag{7-10}$$

The above is a positive quadratic for any a; hence its discriminant is negative. In other words,

$$C^2 - \sigma_x^2\sigma_y^2 \le 0 \tag{7-11}$$

and (7-9) results.

We note that the RVs $\mathbf{x}, \mathbf{y}$ and $\mathbf{x} - \eta_x, \mathbf{y} - \eta_y$ have the same covariance and correlation coefficient.

Example 7-1. We shall show that the correlation coefficient of two jointly normal RVs is the parameter r in (6-15). It suffices to assume that $\eta_x = \eta_y = 0$ and to show that $E\{\mathbf{xy}\} = r\sigma_1\sigma_2$.

Since

$$\frac{x^2}{\sigma_1^2} - 2r\frac{xy}{\sigma_1\sigma_2} + \frac{y^2}{\sigma_2^2} = \left(\frac{x}{\sigma_1} - r\frac{y}{\sigma_2}\right)^2 + (1 - r^2)\frac{y^2}{\sigma_2^2}$$

we conclude with (6-44) that

$$E\{\mathbf{xy}\} = \frac{1}{\sigma_2\sqrt{2\pi}}\int_{-\infty}^{\infty} ye^{-y^2/2\sigma_2^2}\int_{-\infty}^{\infty}\frac{x}{\sigma_1\sqrt{2\pi(1 - r^2)}}\exp\left[-\frac{(x - ry\sigma_1/\sigma_2)^2}{2\sigma_1^2(1 - r^2)}\right]dx\,dy$$

The inner integral is a normal density with mean $ry\sigma_1/\sigma_2$ multiplied by x; hence it equals $ry\sigma_1/\sigma_2$. This yields

$$E\{\mathbf{xy}\} = \frac{r\sigma_1/\sigma_2}{\sigma_2\sqrt{2\pi}}\int_{-\infty}^{\infty} y^2 e^{-y^2/2\sigma_2^2}\,dy = r\sigma_1\sigma_2$$

Uncorrelatedness Two RVs are called uncorrelated if their covariance is 0. This can be phrased in the following equivalent forms

$$C = 0 \qquad r = 0 \qquad E\{\mathbf{xy}\} = E\{\mathbf{x}\}E\{\mathbf{y}\}$$

Orthogonality Two RVs are called orthogonal if

$$E\{\mathbf{xy}\} = 0$$

We shall use the notation

$$\mathbf{x} \perp \mathbf{y}$$

to indicate that the RVs $\mathbf{x}$ and $\mathbf{y}$ are orthogonal.

Note (a) If $\mathbf{x}$ and $\mathbf{y}$ are uncorrelated, then $\mathbf{x} - \eta_x \perp \mathbf{y} - \eta_y$. (b) If $\mathbf{x}$ and $\mathbf{y}$ are uncorrelated and $\eta_x = 0$ or $\eta_y = 0$ then $\mathbf{x} \perp \mathbf{y}$.

Vector space of random variables. We shall find it convenient to interpret RVs as vectors in an abstract space. In this space, the second moment

$$E\{\mathbf{xy}\}$$

of the RVs $\mathbf{x}$ and $\mathbf{y}$ is by definition their *inner product* and $E\{\mathbf{x}^2\}$ and $E\{\mathbf{y}^2\}$ are the squares of their lengths. The ratio

$$\frac{E\{\mathbf{xy}\}}{\sqrt{E\{\mathbf{x}^2\}E\{\mathbf{y}^2\}}}$$

is the cosine of their angle.

We maintain that

$$E^2\{\mathbf{xy}\} \le E\{\mathbf{x}^2\}E\{\mathbf{y}^2\} \tag{7-12}$$

This is the *cosine inequality* and its proof is similar to the proof of (7-11): The quadratic

$$E\{(a\mathbf{x} - \mathbf{y})^2\} = a^2E\{\mathbf{x}^2\} - 2aE\{\mathbf{xy}\} + E\{\mathbf{y}^2\}$$

is positive for every a; hence its discriminant is negative and (7-12) results. If (7-12) is an equality, then the quadratic is 0 for some $a = a_0$; hence $\mathbf{y} = a_0\mathbf{x}$. This agrees with the geometric interpretation of RVs because, if (7-12) is an equality, then the vectors $\mathbf{x}$ and $\mathbf{y}$ are on the same line.

The following illustration is an example of the correspondence between vectors and RVs: Consider two RVs $\mathbf{x}$ and $\mathbf{y}$ such that $E\{\mathbf{x}^2\} = E\{\mathbf{y}^2\}$. Geometrically, this means that the vectors $\mathbf{x}$ and $\mathbf{y}$ have the same length. If, therefore, we construct a parallelogram with sides $\mathbf{x}$ and $\mathbf{y}$, it will be a rhombus with diagonals $\mathbf{x} + \mathbf{y}$ and $\mathbf{x} - \mathbf{y}$ (Fig. 7-1). These diagonals are perpendicular because

$$E\{(\mathbf{x} + \mathbf{y})(\mathbf{x} - \mathbf{y})\} = E\{\mathbf{x}^2 - \mathbf{y}^2\} = 0$$

THEOREM. If two RVs are independent, that is, if

$$f(x, y) = f_x(x)f_y(y) \tag{7-13}$$

then they are uncorrelated.

Proof. It suffices to show that

$$E\{\mathbf{xy}\} = E\{\mathbf{x}\}E\{\mathbf{y}\} \tag{7-14}$$

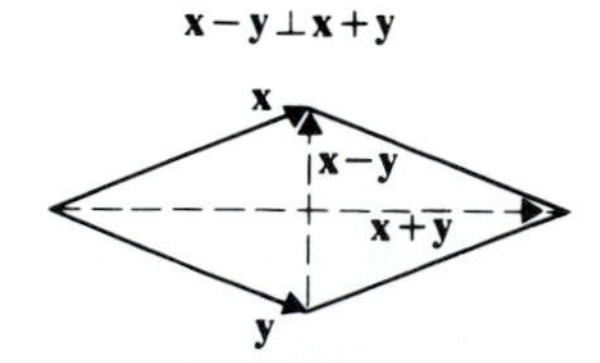

FIGURE 7-1

From (7-2) and (7-13) it follows that

$$E\{\mathbf{xy}\} = \int_{-\infty}^{\infty}\int_{-\infty}^{\infty} xy f_x(x) f_y(y)\,dx\,dy = \int_{-\infty}^{\infty} x f_x(x)\,dx \int_{-\infty}^{\infty} y f_y(y)\,dy$$

and (7-14) results.

If the RVs $\mathbf{x}$ and $\mathbf{y}$ are independent, then the RVs $g(\mathbf{x})$ and $h(\mathbf{y})$ are also independent [see (6-29)]. Hence

$$E\{g(\mathbf{x})h(\mathbf{y})\} = E\{g(\mathbf{x})\}E\{h(\mathbf{y})\} \tag{7-15}$$

This is not, in general, true if $\mathbf{x}$ and $\mathbf{y}$ are merely uncorrelated.

We note, finally, that if two RVs are uncorrelated they are not necessarily independent. However, for normal RVs uncorrelatedness is equivalent to independence. Indeed, if the RVs $\mathbf{x}$ and $\mathbf{y}$ are jointly normal and $r = 0$, then [see (6-15)] $f(x, y) = f_x(x) f_y(y)$.

Variance of the sum of two RVs If $\mathbf{z} = \mathbf{x} + \mathbf{y}$, then $\eta_z = \eta_x + \eta_y$; hence

$$\sigma_z^2 = E\{(\mathbf{z} - \eta_z)^2\} = E\{[(\mathbf{x} - \eta_x) + (\mathbf{y} - \eta_y)]^2\}$$

From this and (7-10) it follows that

$$\sigma_z^2 = \sigma_x^2 + 2r\sigma_x\sigma_y + \sigma_y^2 \tag{7-16}$$

The above leads to the conclusion that if $r = 0$ then

$$\sigma_z^2 = \sigma_x^2 + \sigma_y^2 \tag{7-17}$$

Thus, if two RVs are uncorrelated, then the variance of their sum equals the sum of their variances.

It follows from (7-14) that this is also true if $\mathbf{x}$ and $\mathbf{y}$ are independent.

Moments

The mean

$$m_{kr} = E\{\mathbf{x}^k \mathbf{y}^r\} = \int_{-\infty}^{\infty}\int_{-\infty}^{\infty} x^k y^r f(x, y)\,dx\,dy \tag{7-18}$$

of the product $\mathbf{x}^k\mathbf{y}^r$ is by definition a joint moment of the RVs $\mathbf{x}$ and $\mathbf{y}$ of order $k + r = n$.

Thus $m_{10} = \eta_x$, $m_{01} = \eta_y$ are the first-order moments and

$$m_{20} = E\{\mathbf{x}^2\} \qquad m_{11} = E\{\mathbf{xy}\} \qquad m_{02} = E\{\mathbf{y}^2\}$$

are the second-order moments.

The joint central moments of $\mathbf{x}$ and $\mathbf{y}$ are the moments of $\mathbf{x} - \eta_x$ and $\mathbf{y} - \eta_y$:

$$\mu_{kr} = E\{(\mathbf{x} - \eta_x)^k(\mathbf{y} - \eta_y)^r\} = \int_{-\infty}^{\infty}\int_{-\infty}^{\infty} (x - \eta_x)^k (y - \eta_y)^r f(x, y)\,dx\,dy \tag{7-19}$$

Clearly, $\mu_{10} = \mu_{01} = 0$ and

$$\mu_{11} = C \qquad \mu_{20} = \sigma_x^2 \qquad \mu_{02} = \sigma_y^2$$

Absolute and generalized moments are defined similarly [see (5-40) and (5-41)].

For the determination of the joint statistics of **x** and **y** knowledge of their joint density is required. However, in many applications, only the first- and second-order moments are used. These moments are determined in terms of the five parameters

$$\eta_x \quad \eta_y \quad \sigma_x \quad \sigma_y \quad r_{xy}$$

If **x** and **y** are jointly normal, then [see (6-15)] the above parameters determine uniquely $f(x, y)$.

Example 7-2. The RVs **x** and **y** are jointly normal with

$$\eta_x = 10 \qquad \eta_y = 0 \qquad \sigma_x = 2 \qquad \sigma_y = 1 \qquad r_{xy} = 0.5$$

We shall find the joint density of the RVs

$$\mathbf{z} = \mathbf{x} + \mathbf{y} \qquad \mathbf{w} = \mathbf{x} - \mathbf{y}$$

Clearly,

$$\eta_z = \eta_x + \eta_y = 10 \qquad \eta_w = \eta_x - \eta_y = 10$$

$$\sigma_z^2 = \sigma_x^2 + \sigma_y^2 + 2r_{xy}\sigma_x\sigma_y = 7 \qquad \sigma_w^2 = \sigma_x^2 + \sigma_y^2 - 2r_{xy}\sigma_x\sigma_y = 3$$

$$E\{\mathbf{zw}\} = E\{\mathbf{x}^2 - \mathbf{y}^2\} = (100 + 4) - 1 = 103$$

$$r_{zw} = \frac{E\{\mathbf{zw}\} - E\{\mathbf{z}\}E\{\mathbf{w}\}}{\sigma_z\sigma_w} = \frac{3}{\sqrt{7 \times 3}}$$

As we know [see (6-66)], the RVs **z** and **w** are jointly normal because they are linearly dependent on **x** and **y**. Hence their joint density is

$$N\left(10, 10; \sqrt{7}, \sqrt{3}; \sqrt{3/7}\right)$$

Estimate of the mean of $g(\mathbf{x}, \mathbf{y})$. If the function $g(x, y)$ is sufficiently smooth near the point (η_x, η_y), then the mean η_g and variance σ_g^2 of $g(\mathbf{x}, \mathbf{y})$ can be estimated in terms of the mean, variance, and covariance of **x** and **y**:

$$\eta_g \simeq g + \frac{1}{2}\left(\frac{\partial^2 g}{\partial x^2}\sigma_x^2 + 2\frac{\partial^2 g}{\partial x\,\partial y}r\sigma_x\sigma_y + \frac{\partial^2 g}{\partial y^2}\sigma_y^2\right) \tag{7-20}$$

$$\sigma_g^2 \simeq \left(\frac{\partial g}{\partial x}\right)^2\sigma_x^2 + 2\left(\frac{\partial g}{\partial x}\right)\left(\frac{\partial g}{\partial y}\right)r\sigma_x\sigma_y + \left(\frac{\partial g}{\partial y}\right)^2\sigma_y^2 \tag{7-21}$$

where the function $g(x, y)$ and its derivatives are evaluated at $x = \eta_x$ and $y = \eta_y$.

Proof. We expand $g(x, y)$ into a series about the point (η_x, η_y):

$$g(x, y) = g(\eta_x, \eta_y) + (x - \eta_x)\frac{\partial g}{\partial x} + (y - \eta_y)\frac{\partial g}{\partial y} + \cdots \tag{7-22}$$

Inserting the above into (7-2), we obtain the moment expansion of $E\{g(\mathbf{x}, \mathbf{y})\}$ in terms of the derivatives of $g(x, y)$ at (η_x, η_y) and the joint moments μ_{kr} of $\mathbf{x}$ and $\mathbf{y}$. Using only the first five terms in (7-22), we obtain (7-20). Equation (7-21) follows if we apply (7-20) to the function $[g(x, y) - \eta_g]^2$ and neglect moments of order higher than 2.

7-2 JOINT CHARACTERISTIC FUNCTIONS

The *joint characteristic function* of the RVs $\mathbf{x}$ and $\mathbf{y}$ is by definition the integral

$$\Phi(\omega_1, \omega_2) = \int_{-\infty}^{\infty}\int_{-\infty}^{\infty} f(x, y)e^{j(\omega_1 x + \omega_2 y)}\,dx\,dy \tag{7-23}$$

From the above and the two-dimensional *inversion formula* for Fourier transforms, it follows that

$$f(x, y) = \frac{1}{4\pi^2}\int_{-\infty}^{\infty}\int_{-\infty}^{\infty} \Phi(\omega_1, \omega_2)e^{-j(\omega_1 x + \omega_2 y)}\,d\omega_1\,d\omega_2 \tag{7-24}$$

Clearly,

$$\Phi(\omega_1, \omega_2) = E\{e^{j(\omega_1 \mathbf{x} + \omega_2 \mathbf{y})}\} \tag{7-25}$$

The logarithm

$$\Psi(\omega_1, \omega_2) = \ln \Phi(\omega_1, \omega_2) \tag{7-26}$$

of $\Phi(\omega_1, \omega_2)$ is the joint second characteristic function of $\mathbf{x}$ and $\mathbf{y}$.

The *marginal* characteristic functions

$$\Phi_x(\omega) = E\{e^{j\omega\mathbf{x}}\} \qquad \Phi_y(\omega) = E\{e^{j\omega\mathbf{y}}\} \tag{7-27}$$

of $\mathbf{x}$ and $\mathbf{y}$ can be expressed in terms of their joint characteristic function $\Phi(\omega_1, \omega_2)$. From (7-25) and (7-27) it follows that

$$\Phi_x(\omega) = \Phi(\omega, 0) \qquad \Phi_y(\omega) = \Phi(0, \omega) \tag{7-28}$$

We note that, if $\mathbf{z} = a\mathbf{x} + b\mathbf{y}$, then

$$\Phi_z(\omega) = E\{e^{j(a\mathbf{x} + b\mathbf{y})\omega}\} = \Phi(a\omega, b\omega) \tag{7-29}$$

Hence $\Phi_z(1) = \Phi(a, b)$.

Cramér–Wold theorem The above shows that if $\Phi_z(\omega)$ is known for every a and b, then $\Phi(\omega_1, \omega_2)$ is uniquely determined. In other words, if the density of $a\mathbf{x} + b\mathbf{y}$ is known for every a and b, then the joint density $f(x, y)$ of $\mathbf{x}$ and $\mathbf{y}$ is uniquely determined.

Independence and convolution. If the RVs $\mathbf{x}$ and $\mathbf{y}$ are independent, then [see (7-15)]

$$E\{e^{j(\omega_1 \mathbf{x} + \omega_2 \mathbf{y})}\} = E\{e^{j\omega_1 \mathbf{x}}\}E\{e^{j\omega_2 \mathbf{y}}\}$$

om this it follows that

$$\Phi(\omega_1, \omega_2) = \Phi_x(\omega_1)\Phi_y(\omega_2) \tag{7-30}$$

Conversely, if (7-30) is true, then the RVs **x** and **y** are independent. Indeed, inserting (7-30) into the inversion formula (7-24) and using (5-66), we conclude that $f(x, y) = f_x(x)f_y(y)$.

Convolution theorem If the RVs **x** and **y** are independent and $\mathbf{z} = \mathbf{x} + \mathbf{y}$, then

$$E\{e^{j\omega \mathbf{z}}\} = E\{e^{j\omega(\mathbf{x}+\mathbf{y})}\} = E\{e^{j\omega \mathbf{x}}\}E\{e^{j\omega \mathbf{y}}\}$$

Hence

$$\Phi_z(\omega) = \Phi_x(\omega)\Phi_y(\omega) \qquad \Psi_z(\omega) = \Psi_x(\omega) + \Psi_y(\omega) \tag{7-31}$$

As we know [see (6-39)], the density of **z** equals the convolution of $f_x(x)$ and $f_y(y)$. From this and (7-31) it follows that the characteristic function of the convolution of two densities equals the product of their characteristic functions.

Example 7-3. We shall show that if the RVs **x** and **y** are *independent* and Poisson distributed with parameters a and b respectively, then their sum $\mathbf{z} = \mathbf{x} + \mathbf{y}$ is also Poisson distributed with parameter $a + b$.

Proof. As we know (see Example 5-30)

$$\Psi_x(\omega) = a(e^{j\omega} - 1) \qquad \Psi_y(\omega) = b(e^{j\omega} - 1)$$

Hence

$$\Psi_z(\omega) = \Psi_x(\omega) + \Psi_y(\omega) = (a + b)(e^{j\omega} - 1)$$

It can be shown that the converse is also true: If the RVs **x** and **y** are *independent* and their sum is Poisson distributed, then **x** and **y** are also Poisson distributed. The proof of this difficult theorem will not be given.

Example 7-4. It was shown in Sec. 6-3 that if the RVs **x** and **y** are jointly normal, then the sum $a\mathbf{x} + b\mathbf{y}$ is also normal. In the following we reestablish a special case of the above using (7-30): If **x** and **y** are *independent* and normal, then their sum $\mathbf{z} = \mathbf{x} + \mathbf{y}$ is also normal.

Proof. In this case [see (5.65)]

$$\Psi_x(\omega) = j\eta_x\omega - \tfrac{1}{2}\sigma_x^2\omega^2 \qquad \Psi_y(\omega) = j\eta_y\omega - \tfrac{1}{2}\sigma_y^2\omega^2$$

Hence

$$\Psi_z(\omega) = j(\eta_x + \eta_y)\omega - \tfrac{1}{2}\left(\sigma_x^2 + \sigma_y^2\right)\omega^2$$

It can be shown that the converse is also true (Cramér theorem): If the RVs **x** and **y** are *independent* and their sum is normal, then they are also normal. The proof of this difficult theorem will not be given.†

---

†E. Lukacs: *Characteristic Functions*, Hafner Publishing Co., New York, 1960.

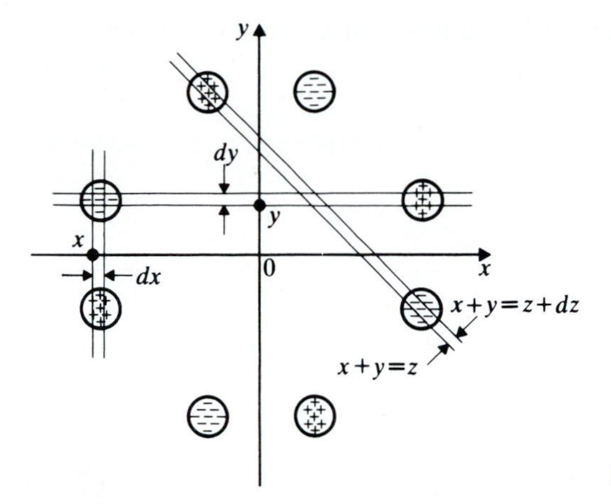

FIGURE 7-2

Normal RVs. We shall show that the joint characteristic function of two jointly normal RVs is given by

$$\Phi(\omega_1, \omega_2) = e^{j(\eta_1\omega_1 + \eta_2\omega_2)} e^{-\frac{1}{2}(\omega_1^2\sigma_1^2 + 2r\sigma_1\sigma_2\omega_1\omega_2 + \omega_2^2\sigma_2^2)} \tag{7-32}$$

Proof. This can be derived by inserting $f(x, y)$ into (7-23). The following simpler proof is based on the fact that the RV $\mathbf{z} = \omega_1\mathbf{x} + \omega_2\mathbf{y}$ is normal and

$$\Psi_z(\omega) = j\eta_z\omega - \tfrac{1}{2}\sigma_z^2\omega^2 \tag{7-33}$$

Since

$$\eta_z = \omega_1\eta_1 + \omega_2\eta_2 \qquad \sigma_z^2 = \omega_1^2\sigma_1^2 + 2r\omega_1\omega_2\sigma_1\sigma_2 + \omega_2^2\sigma_2^2$$

and $\Phi_z(\omega) = \Phi(\omega_1\omega, \omega_2\omega)$, (7-32) follows from (7-33) with $\omega = 1$.

The above proof is based on the fact that the RV $\mathbf{z} = \omega_1\mathbf{x} + \omega_2\mathbf{y}$ is normal for any ω_1 and ω_2; this leads to the following conclusion: If it is known that the sum $a\mathbf{x} + b\mathbf{y}$ is normal for every a and b, then RVs $\mathbf{x}$ and $\mathbf{y}$ are jointly normal. We should stress, however, that this is not true if $a\mathbf{x} + b\mathbf{y}$ is normal for only a finite set of values of a and b. A counterexample can be formed by a simple extension of the construction in Fig. 7-2.

Example 7-5. We shall construct two RVs $\mathbf{x}_1$ and $\mathbf{x}_2$ with the following properties: $\mathbf{x}_1$, $\mathbf{x}_2$, and $\mathbf{x}_1 + \mathbf{x}_2$ are normal but $\mathbf{x}_1$ and $\mathbf{x}_2$ are not jointly normal.

Suppose that $\mathbf{x}$ and $\mathbf{y}$ are two jointly normal RVs with mass density $f(x, y)$. Adding and subtracting small masses in the region D of Fig. 7-2 consisting of eight circles as shown, we obtain a new function $f_1(x, y)$ such that $f_1(x, y) = f(x, y) \pm \varepsilon$ in D and $f_1(x, y) = f(x, y)$ everywhere else. The function $f_1(x, y)$ is a density; hence it defines two new RVs $\mathbf{x}_1$ and $\mathbf{y}_1$. These RVs are obviously not jointly normal. However, they are marginally normal because $\mathbf{x}$ and $\mathbf{y}$ are marginally normal and the masses in any vertical or horizontal strip have not changed. Furthermore, the RV $\mathbf{z}_1 = \mathbf{x}_1 + \mathbf{y}_1$ is also normal because $\mathbf{z} = \mathbf{x} + \mathbf{y}$ is normal and the masses in any diagonal strip of the form $z \le x + y \le z + dz$ have not changed.

Moment theorem. The moment generating function of $\mathbf{x}$ and $\mathbf{y}$ is given by

$$\Phi(s_1, s_2) = E\{e^{s_1\mathbf{x}+s_2\mathbf{y}}\}$$

Expanding the exponential and using the linearity of expected values, we obtain the series

$$\Phi(s_1, s_2) = \sum_{n=0}^{\infty} \frac{1}{n!} \sum_{k=0}^{n} \binom{n}{k} E\{\mathbf{x}^k \mathbf{y}^{n-k}\} s_1^k s_2^{n-k}$$

$$= 1 + m_{10}s_1 + m_{01}s_2 + \tfrac{1}{2}\left(m_{20}s_1^2 + 2m_{11}s_1s_2 + m_{02}s_2^2\right) + \cdots \tag{7-34}$$

From this it follows that

$$\frac{\partial^k \partial^r}{\partial s_1^k \partial s_2^r}\Phi(0,0) = m_{kr} \tag{7-35}$$

The derivatives of the function $\Psi(s_1, s_2) = \ln \Phi(s_1, s_2)$ are by definition the joint cumulants λ_{kr} of $\mathbf{x}$ and $\mathbf{y}$. It can be shown that

$$\lambda_{10} = m_{10} \qquad \lambda_{01} = m_{01} \qquad \lambda_{20} = \mu_{20} \qquad \lambda_{02} = \mu_{02} \qquad \lambda_{11} = \mu_{11}$$

Hence

$$\Psi(s_1, s_2) = \eta_1 s_1 + \eta_2 s_2 + \tfrac{1}{2}\left(\sigma_1^2 s_1^2 + 2r\sigma_1\sigma_2 s_1 s_2 + \sigma_2^2 s_2^2\right) + \cdots$$

Example 7-6a. Using (7-34), we shall show that if the RVs $\mathbf{x}$ and $\mathbf{y}$ are jointly normal with zero mean, then

$$E\{\mathbf{x}^2\mathbf{y}^2\} = E\{\mathbf{x}^2\}E\{\mathbf{y}^2\} + 2E^2\{\mathbf{xy}\} \tag{7-36}$$

Proof. As we see from (7-32)

$$\Phi(s_1, s_2) = e^{-A} \qquad A = \tfrac{1}{2}\left(\sigma_1^2 s_1^2 + 2Cs_1s_2 + \sigma_2^2 s_2^2\right)$$

where $C = E\{\mathbf{xy}\} = r\sigma_1\sigma_2$. To prove (7-36), we shall equate the coefficient

$$\frac{1}{4!}\binom{4}{2}E\{\mathbf{x}^2\mathbf{y}^2\}$$

of $s_1^2 s_2^2$ in (7-34) with the corresponding coefficient of the expansion of e^{-A}. In this expansion, the factors $s_1^2 s_2^2$ appear only in the terms

$$\frac{A^2}{2} = \frac{1}{8}\left(\sigma_1^2 s_1^2 + 2Cs_1s_2 + \sigma_2^2 s_2^2\right)^2$$

Hence

$$\frac{1}{4!}\binom{4}{2}E\{\mathbf{x}^2\mathbf{y}^2\} = \frac{1}{8}\left(2\sigma_1^2\sigma_2^2 + 4C^2\right)$$

and (7-36) results.

Price's theorem.† Given two jointly normal RVs **x** and **y**, we form the mean

$$I = E\{g(\mathbf{x},\mathbf{y})\} = \int_{-\infty}^{\infty}\int_{-\infty}^{\infty} g(x,y)f(x,y)\,dx\,dy \qquad (7\text{-}37a)$$

of some function $g(\mathbf{x},\mathbf{y})$ of $(\mathbf{x},\mathbf{y})$. The above integral is a function $I(\mu)$ of the covariance μ of the RVs **x** and **y** and of four parameters specifying the joint density $f(x,y)$ of **x** and **y**. We shall show that if $g(x,y)f(x,y) \to 0$ as $(x,y) \to \infty$, then

$$\frac{\partial^n I(\mu)}{\partial \mu^n} = \int_{-\infty}^{\infty}\int_{-\infty}^{\infty} \frac{\partial^{2n} g(x,y)}{\partial x^n\,\partial y^n} f(x,y)\,dx\,dy = E\left\{\frac{\partial^{2n} g(\mathbf{x},\mathbf{y})}{\partial \mathbf{x}^n\,\partial \mathbf{y}^n}\right\} \qquad (7\text{-}37b)$$

Proof. Inserting (7-24) into (7-37a) and differentiating with respect to μ, we obtain

$$\frac{\partial^n I(\mu)}{\partial \mu^n} = \frac{(-1)^n}{4\pi^2}\int_{-\infty}^{\infty}\int_{-\infty}^{\infty} g(x,y)$$

$$\times \int_{-\infty}^{\infty}\int_{-\infty}^{\infty} \omega_1^n \omega_2^n \Phi(\omega_1,\omega_2) e^{-j(\omega_1 x + \omega_2 y)}\,d\omega_1\,d\omega_2\,dx\,dy$$

From this and the derivative theorem, it follows that

$$\frac{\partial^n I(\mu)}{\partial \mu^n} = \int_{-\infty}^{\infty}\int_{-\infty}^{\infty} g(x,y)\frac{\partial^{2n} f(x,y)}{\partial x^n\,\partial y^n}\,dx\,dy$$

Integrating by parts and using the condition at ∞, we obtain (7-37b) (see also Prob. 5-39).

Example 7-6b. Using Price's theorem, we shall rederive (7-36). Setting $g(\mathbf{x},\mathbf{y}) = \mathbf{x}^2\mathbf{y}^2$ into (7-37b), we conclude with $n = 1$ that

$$\frac{\partial I(\mu)}{\partial \mu} = E\left\{\frac{\partial^2 g(\mathbf{x},\mathbf{y})}{\partial \mathbf{x}\,\partial \mathbf{y}}\right\} = 4E\{\mathbf{xy}\} = 4\mu \qquad I(\mu) = \frac{4\mu^2}{2} + I(0)$$

If $\mu = 0$, the RVs **x** and **y** are independent; hence $I(0) = E\{\mathbf{x}^2\mathbf{y}^2\} = E\{\mathbf{x}^2\}E\{\mathbf{y}^2\}$ and (7-36) results.

†R. Price, "A Useful Theorem for Nonlinear Devices Having Gaussian Inputs," *IRE, PGIT*, Vol. IT-4, 1958. See also A. Papoulis, "On an Extension of Price's Theorem," *IEEE Transactions on Information Theory*, Vol. IT-11, 1965.

7-3 CONDITIONAL DISTRIBUTIONS

As we have noted, conditional distributions can be expressed as conditional probabilities:

$$\begin{aligned} F_z(z|\mathscr{M}) &= P\{\mathbf{z} \le z|\mathscr{M}\} = \frac{P\{\mathbf{z} \le z, \mathscr{M}\}}{P(\mathscr{M})} \\ F_{zw}(z, w|\mathscr{M}) &= P\{\mathbf{z} \le z, \mathbf{w} \le w|\mathscr{M}\} = \frac{P\{\mathbf{z} \le z, \mathbf{w} \le w, \mathscr{M}\}}{P(\mathscr{M})} \end{aligned} \tag{7-38}$$

The corresponding densities are obtained by appropriate differentiations. In this section, we evaluate these functions for various special cases.

Example 7-7. We shall first determine the conditional distribution $F_y(y|\mathbf{x} \le x)$ and density $f_y(y|\mathbf{x} \le x)$.

With $\mathscr{M} = \{\mathbf{x} \le x\}$, (7-38) yields

$$F_y(y|\mathbf{x} \le x) = \frac{P\{\mathbf{x} \le x, \mathbf{y} \le y\}}{P\{\mathbf{x} \le x\}} = \frac{F(x, y)}{F_x(x)}$$

$$f_y(y|\mathbf{x} \le x) = \frac{\partial F(x, y)/\partial y}{F_x(x)}$$

Example 7-8. We shall next determine the conditional distribution $F(x, y|\mathscr{M})$ for $\mathscr{M} = \{x_1 < \mathbf{x} \le x_2\}$. In this case, $F(x, y|\mathscr{M})$ is given by

$$F(x, y|x_1 < \mathbf{x} \le x_2) = \frac{P\{\mathbf{x} \le x, \mathbf{y} \le y, x_1 < \mathbf{x} \le x_2\}}{P(x_1 < \mathbf{x} \le x_2)}$$

$$= \begin{cases} \dfrac{F(x_2, y) - F(x_1, y)}{F_x(x_2) - F_x(x_1)} & x > x_2 \\ \dfrac{F(x, y) - F(x_1, y)}{F_x(x_2) - F_x(x_1)} & x_1 < x \le x_2 \end{cases}$$

and it equals 0 for $x < x_1$. Since $f = \partial^2 F/\partial x\, \partial y$, the above yields

$$f(x, y|x_1 < \mathbf{x} \le x_2) = \frac{f(x, y)}{F_x(x_2) - F_x(x_1)} \qquad x_1 < x \le x_2 \tag{7-39}$$

and 0 otherwise.

The determination of the conditional density of $\mathbf{y}$ assuming $\mathbf{x} = x$ is of particular interest. This density cannot be derived directly from (7-38) because, in general, the event $\{\mathbf{x} = x\}$ has zero probability. It can, however, be defined as a limit. Suppose first that

$$\mathscr{M} = \{x_1 < \mathbf{x} \le x_2\}$$

In this case, (7-38) yields

$$F_y(y|x_1 < \mathbf{x} \le x_2) = \frac{P\{x_1 < \mathbf{x} \le x_2, \mathbf{y} \le y\}}{P\{x_1 < \mathbf{x} \le x_2\}} = \frac{F(x_2, y) - F(x_1, y)}{F_x(x_2) - F_x(x_1)}$$

Differentiating with respect to y, we obtain

$$f_y(y|x_1 < \mathbf{x} \le x_2) = \frac{\int_{x_1}^{x_2} f(x, y)\, dx}{F_x(x_2) - F_x(x_1)} \tag{7-40}$$

because [see (6.6)]

$$\frac{\partial F(x, y)}{\partial y} = \int_{-\infty}^{x} f(\alpha, y)\, d\alpha$$

To determine $f_y(y|\mathbf{x} = x)$, we set $x_1 = x$ and $x_2 = x + \Delta x$ in (7-40). This yields

$$f_y(y|x < \mathbf{x} \le x + \Delta x) = \frac{\int_{x}^{x+\Delta x} f(\alpha, y)\, d\alpha}{F_x(x + \Delta x) - F_x(x)} \simeq \frac{f(x, y)\, \Delta x}{f_x(x)\, \Delta x}$$

Hence

$$f_y(y|\mathbf{x} = x) = \lim_{\Delta x \to 0} f_y(y|x < \mathbf{x} \le x + \Delta x) = \frac{f(x, y)}{f_x(x)}$$

If there is no fear of ambiguity, the function $f_y(y|\mathbf{x} = x)$ will be written in the form $f(y|x)$. Defining $f(x|y)$ similarly, we obtain

$$f(y|x) = \frac{f(x, y)}{f(x)} \qquad f(x|y) = \frac{f(x, y)}{f(y)} \tag{7-41}$$

If the RVs $\mathbf{x}$ and $\mathbf{y}$ are independent, then

$$f(x, y) = f(x)f(y) \qquad f(y|x) = f(y) \qquad f(x|y) = f(x)$$

Notes 1. For a specific x, the function $f(x, y)$ is a *profile* of $f(x, y)$; that is, it equals the intersection of the surface $f(x, y)$ by the plane $x =$ constant. The conditional density $f(y|x)$ is the equation of this curve normalized by the factor $1/f(x)$ so as to make its area 1. The function $f(x|y)$ has a similar interpretation: It is the normalized equation of the intersection of the surface $f(x, y)$ by the plane $y =$ constant.

2. As we know, the product $f(y)\, dy$ equals the probability of the event $\{y < \mathbf{y} \le y + dy\}$. Extending this to conditional probabilities, we obtain

$$f_y(y|x_1 < \mathbf{x} \le x_2)\, dy = \frac{P\{x_1 < \mathbf{x} \le x_2,\, y < \mathbf{y} \le y + dy\}}{P\{x_1 < \mathbf{x} \le x_2\}}$$

This equals the mass in the rectangle of Fig. 7-3a divided by the mass in the vertical strip $x_1 < \mathbf{x} \le x_2$. Similarly, the product $f(y|x)\, dy$ equals the ratio of the mass in the differential rectangle $dx\, dy$ of Fig. 7-3b over the mass in the vertical strip $(x, x + dx)$.

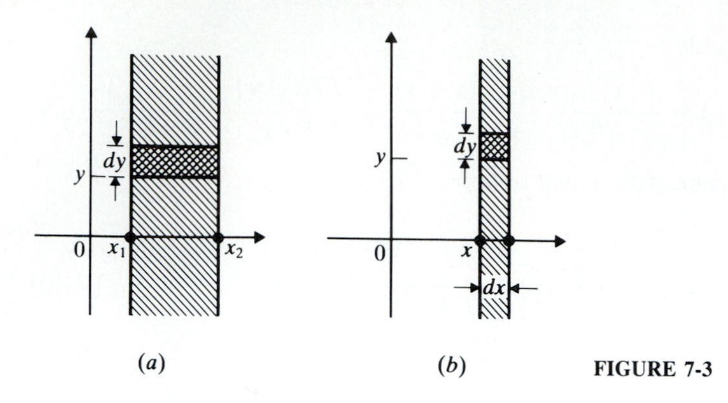

FIGURE 7-3

3. The joint statistics of **x** and **y** are determined in terms of their joint density $f(x, y)$. Since

$$f(x, y) = f(y|x)f(x)$$

we conclude that they are also determined in terms of the marginal density $f(x)$ and the conditional density $f(y|x)$.

Example 7-9. We shall show that, if the RVs **x** and **y** are jointly normal with zero mean as in (6-44), then

$$f(y|x) = \frac{1}{\sigma_2\sqrt{2\pi(1-r^2)}} \exp\left[-\frac{(y - r\sigma_2 x/\sigma_1)^2}{2\sigma_2^2(1-r^2)}\right] \tag{7-42}$$

Proof. The exponent in (6-44) equals

$$\frac{(y - r\sigma_2 x/\sigma_1)^2}{2\sigma_2^2(1-r^2)} - \frac{x^2}{2\sigma_1^2}$$

Division by $f(x)$ removes the term $-x^2/2\sigma_1^2$ and (7-42) results.

The same reasoning leads to the conclusion that if **x** and **y** are jointly normal with $E\{\mathbf{x}\} = \eta_1$ and $E\{\mathbf{y}\} = \eta_2$, then $f(y|x)$ is given by (7-42) if y and x are replaced by $y - \eta_2$ and $x - \eta_1$ respectively. In other words, for a given x, $f(y|x)$ is a normal density with mean $\eta_2 + r\sigma_2(x - \eta_1)/\sigma_1$ and variance $\sigma_2^2(1 - r^2)$.

Bayes' theorem and total probability. From (7-41) it follows that

$$f(x|y) = \frac{f(y|x)f(x)}{f(y)} \tag{7-43}$$

This is the density version of (2-38).

The denominator $f(y)$ can be expressed in terms of $f(y|x)$ and $f(x)$. Since

$$f(y) = \int_{-\infty}^{\infty} f(x, y)\, dx \qquad \text{and} \qquad f(x, y) = f(y|x)f(x)$$

we conclude that (total probability)

$$f(y) = \int_{-\infty}^{\infty} f(y|x)f(x)\,dx \tag{7-44}$$

Inserting into (7-43), we obtain Bayes' theorem for densities

$$f(x|y) = \frac{f(y|x)f(x)}{\int_{-\infty}^{\infty} f(y|x)f(x)\,dx} \tag{7-45}$$

Note As (7-44) shows, to remove the condition $\mathbf{x} = x$ from the conditional density $f(y|x)$, we multiply by the density $f(x)$ of $\mathbf{x}$ and integrate the product.

Discrete type. Suppose that the RVs $\mathbf{x}$ and $\mathbf{y}$ are of discrete type

$$P\{\mathbf{x} = x_i\} = p_i \qquad P\{\mathbf{y} = y_k\} = q_k$$

$$P\{\mathbf{x} = x_i, \mathbf{y} = y_k\} = p_{ik} \qquad i = 1, \ldots, M \qquad k = 1, \ldots, N$$

where [see (6-21)]

$$p_i = \sum_k p_{ik} \qquad q_k = \sum_i p_{ik}$$

From the above and (2-29) it follows that

$$P\{\mathbf{y} = y_k | \mathbf{x} = x_i\} = \frac{P\{\mathbf{x} = x_i, \mathbf{y} = y_k\}}{P\{\mathbf{x} = x_i\}} = \frac{p_{ik}}{p_i}$$

Markoff matrix We denote by π_{ik} the above conditional probabilities

$$P\{\mathbf{y} = y_k | \mathbf{x} = x_i\} = \pi_{ik}$$

and by Π the $M \times N$ matrix whose elements are π_{ik}. Clearly,

$$\pi_{ik} = \frac{p_{ik}}{p_i} \tag{7-46}$$

Hence

$$\pi_{ik} \geq 0 \qquad \sum_k \pi_{ik} = 1 \tag{7-47}$$

Thus the elements of the matrix Π are positive and the sum on each row equals 1. Such a matrix is called *Markoff*. The conditional probabilities

$$P\{\mathbf{x} = x_i | \mathbf{y} = y_k\} = \pi^{ki} = \frac{p_{ik}}{q_k}$$

are the elements of an $N \times M$ Markoff matrix.

If the RVs $\mathbf{x}$ and $\mathbf{y}$ are independent, then

$$p_{ik} = p_i q_k \qquad \pi_{ik} = q_k \qquad \pi^{ki} = p_i$$

We note that

$$\pi^{ki} = \pi_{ik}\frac{p_i}{q_k} \qquad q_k = \sum_i \pi_{ik} p_i \tag{7-48}$$

These equations are the discrete versions of Eqs. (7-43) and (7-44).

System Reliability

We shall use the term *system* to identify a physical device used to perform a certain function. The device might be a simple element, a light bulb, for example, or a more complicated structure. We shall call the time interval from the moment the system is put into operation until it fails the *time to failure*. This interval is, in general, random. It specifies, therefore, an RV $\mathbf{x} \geq 0$. The distribution $F(t) = P\{\mathbf{x} \leq t\}$ of this RV is the probability that the system fails prior to time t where we assume that $t = 0$ is the moment the system is put into operation. The difference

$$R(t) = 1 - F(t) = P\{\mathbf{x} > t\}$$

is the *system reliability*. It equals the probability that the system functions at time t.

The *mean time to failure* of a system is the mean of $\mathbf{x}$. Since $F(x) = 0$ for $x < 0$, we conclude from (5-27) that

$$E\{\mathbf{x}\} = \int_0^\infty x f(x)\,dx = \int_0^\infty R(t)\,dt \tag{7-49}$$

The probability that a system functioning at time t fails prior to time $x > t$ equals

$$F(x|\mathbf{x} > t) = \frac{P\{\mathbf{x} \leq x, \mathbf{x} > t\}}{P\{\mathbf{x} > t\}} = \frac{F(x) - F(t)}{1 - F(t)} \tag{7-50}$$

Differentiating with respect to x, we obtain

$$f(x|\mathbf{x} > t) = \frac{f(x)}{1 - F(t)} \qquad x > t \tag{7-51}$$

The product $f(x|\mathbf{x} > t)\,dx$ equals the probability that the system fails in the interval $(x, x + dx)$, assuming that it functions at time t.

Example 7-10. If $f(x) = ce^{-cx}$, then $F(t) = 1 - e^{-ct}$ and (7-51) yields

$$f(x|\mathbf{x} > t) = \frac{ce^{-cx}}{e^{-ct}} = f(x - t)$$

This shows that the probability that a system functioning at time t fails in the interval $(x, x + dx)$ depends only on the difference $x - t$ (Fig. 7-4). We show later that this is true only if $f(x)$ is an exponential density.

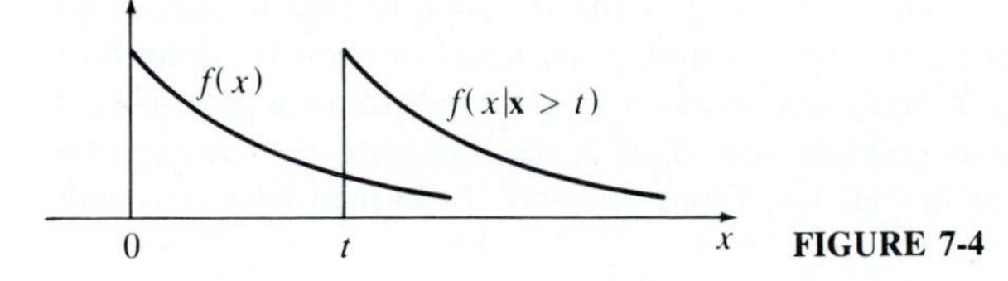

FIGURE 7-4

Conditional failure rate. The conditional density $f(x|\mathbf{x} > t)$ is a function of x and t. Its value at $x = t$ is a function only of t. This function is denoted by $\beta(t)$ and is called the *conditional failure rate* or, the *hazard rate* of the system. From (7-51) and the definition it follows that

$$\beta(t) = f(t|\mathbf{x} > t) = \frac{f(t)}{1 - F(t)} \tag{7-52}$$

The product $\beta(t)\,dt$ is the probability that a system functioning at time t fails in the interval $(t, t + dt)$. In Sec. 8-1 (Example 8-3) we interpret the function $\beta(t)$ as the expected failure rate.

Example 7-11. (*a*) If $f(x) = ce^{-cx}$, then $F(t) = 1 - e^{-ct}$ and

$$\beta(t) = \frac{ce^{-ct}}{1 - (1 - e^{-ct})} = c$$

(*b*) If $f(x) = c^2xe^{-cx}$, then $F(x) = 1 - cxe^{-cx} - e^{-cx}$ and

$$\beta(t) = \frac{c^2te^{-ct}}{cte^{-ct} + e^{-ct}} = \frac{c^2t}{1 + ct}$$

From (7-52) it follows that

$$\beta(t) = \frac{F'(t)}{1 - F(t)} = -\frac{R'(t)}{R(t)}$$

We shall use this relationship to express the distribution of $\mathbf{x}$ in terms of the function $\beta(t)$. Integrating from 0 to x and using the fact that $\ln R(0) = 0$, we obtain

$$-\int_0^x \beta(t)\,dt = \ln R(x)$$

Hence

$$R(x) = 1 - F(x) = \exp\left\{-\int_0^x \beta(t)\,dt\right\} \tag{7-53}$$

And since $f(x) = F'(x)$, this yields

$$f(x) = \beta(x)\exp\left\{-\int_0^x \beta(t)\,dt\right\} \tag{7-54}$$

Example 7-12. A system is called *memoryless* if the probability that it fails in an interval (t, x), assuming that it functions at time t, depends only on the length of this interval. In other words, if the system works a week, a month, or a year after it was put into operation, it is as good as new. This is equivalent to the assumption that $f(x|\mathbf{x} > t) = f(x - t)$ as in Fig. 7-4. From this and (7-52) it follows that with $x = t$:

$$\beta(t) = f(t|\mathbf{x} > t) = f(t - t) = f(0) = c$$

and (7-54) yields $f(x) = ce^{-cx}$. Thus a system is memoryless iff $\mathbf{x}$ has an exponential density.

Example 7-13. A special form of $\beta(t)$ of particular interest in reliability theory is the function

$$\beta(t) = ct^{b-1}$$

This is a satisfactory approximation of a variety of failure rates, at least near the origin. The corresponding $f(x)$ is obtained from (7-54):

$$f(x) = cx^{b-1}\exp\left\{-\frac{cx^b}{b}\right\} \qquad (7\text{-}55)$$

This function is called the *Weibull density*.

We conclude with the observation that the function $\beta(t)$ equals the value of the conditional density $f(x|\mathbf{x} > t)$ for $x = t$; however, $\beta(t)$ is not a density because its area is not one. In fact its area is infinite. This follows from (7-53) because $R(\infty) = 1 - F(\infty) = 0$.

Interconnection of systems. We are given two systems S_1 and S_2 with times to failure $\mathbf{x}$ and $\mathbf{y}$ respectively, and we connect them in parallel or in series or in

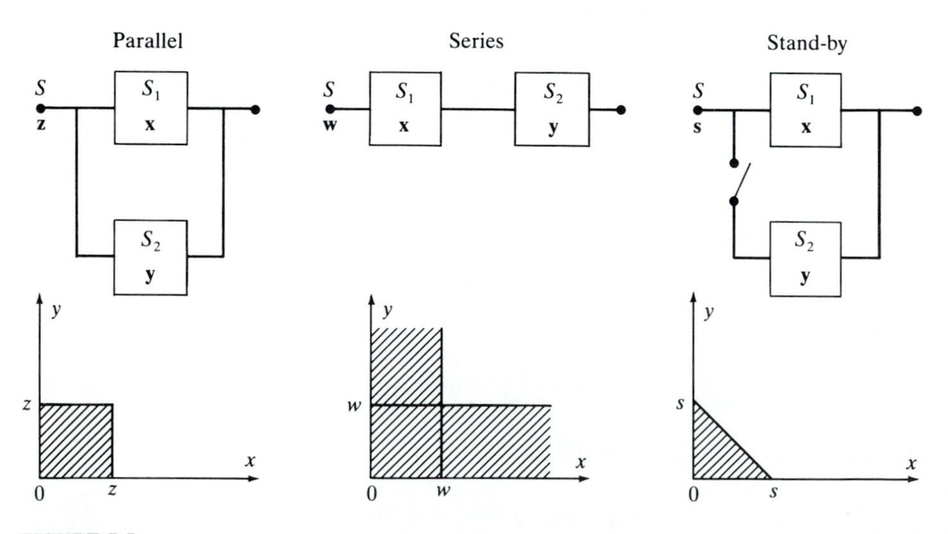

FIGURE 7-5

standby as in Fig. 7-5, forming a new system S. We shall express the properties of S in terms of the joint distribution of the RVs $\mathbf{x}$ and $\mathbf{y}$.

Parallel. We say that the two systems are connected in parallel if S fails when both systems fail. Denoting by $\mathbf{z}$ the time to failure of S, we conclude that $\mathbf{z} = t$ when the larger of the numbers $\mathbf{x}$ and $\mathbf{y}$ equals t. Hence [see (6-54)]

$$\mathbf{z} = \max(\mathbf{x}, \mathbf{y}) \qquad F_z(z) = F_{xy}(z, z)$$

If the RVs $\mathbf{x}$ and $\mathbf{y}$ are independent, $F_z(z) = F_x(z)F_y(z)$.

Series. We say that the two systems are connected in series if S fails when at least one of the two systems fails. Denoting by $\mathbf{w}$ the time to failure of S, we conclude that $\mathbf{w} = t$ when the smaller of the numbers $\mathbf{x}$ and $\mathbf{y}$ equals t. Hence [see (6-56)]

$$\mathbf{w} = \min(\mathbf{x}, \mathbf{y}) \qquad F_w(w) = F_x(w) + F_y(w) - F_{xy}(w, w)$$

If the RVs $\mathbf{x}$ and $\mathbf{y}$ are independent,

$$R_w(w) = R_x(w)R_y(w) \qquad \beta_w(t) = \beta_x(t) + \beta_y(t)$$

where $\beta_x(t)$, $\beta_y(t)$, and $\beta_w(t)$ are the conditional failure rates of systems S_1, S_2, and S respectively.

Standby. We put system S_1 into operation, keeping S_2 in reserve. When S_1 fails, we put S_2 into operation. The system S so formed fails when S_2 fails. If t_1 and t_2 are the times of operation of S_1 and S_2, $t_1 + t_2$ is the time of operation of S. Denoting by $\mathbf{s}$ the time to failure of system S, we conclude that

$$\mathbf{s} = \mathbf{x} + \mathbf{y}$$

The distribution of $\mathbf{s}$ equals the probability that the point $(\mathbf{x}, \mathbf{y})$ is in the shaded region of Fig. 7-5. If the RVs $\mathbf{x}$ and $\mathbf{y}$ are independent, the density of $\mathbf{s}$ equals

$$f_s(s) = \int_0^s f_x(t) f_y(s - t)\, dt$$

as in (6-40).

7-4 CONDITIONAL EXPECTED VALUES

Applying theorem (5-29) to conditional densities, we obtain the conditional mean of $g(\mathbf{y})$:

$$E\{g(\mathbf{y}) | \mathscr{M}\} = \int_{-\infty}^{\infty} g(y) f(y | \mathscr{M})\, dy \tag{7-56}$$

This can be used to define the conditional moments of $\mathbf{y}$.

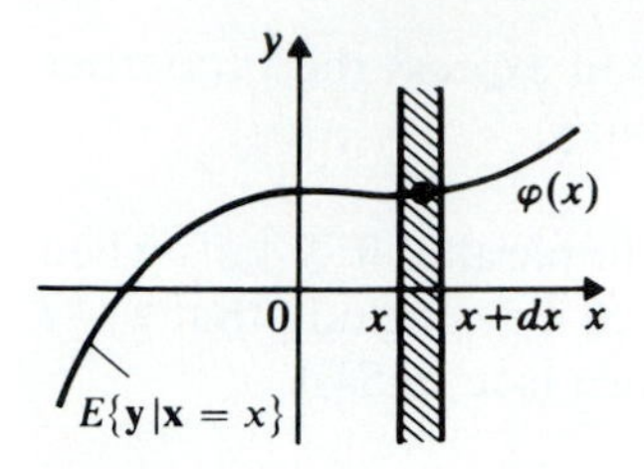

FIGURE 7-6

Using a limit argument as in (7-41), we can also define the conditional mean $E\{g(\mathbf{y})|x\}$. In particular,

$$\eta_{y|x} = E\{\mathbf{y}|x\} = \int_{-\infty}^{\infty} y f(y|x)\,dy \tag{7-57}$$

is the *conditional mean* of $\mathbf{y}$ assuming $\mathbf{x} = x$, and

$$\sigma_{y|x}^2 = E\left\{(\mathbf{y} - \eta_{y|x})^2|x\right\} = \int_{-\infty}^{\infty} (y - \eta_{y|x})^2 f(y|x)\,dy \tag{7-58}$$

is its *conditional variance*.

For a given x, the integral in (7-57) is the center of gravity of the masses in the vertical strip $(x, x + dx)$. The locus of these points, as x varies from $-\infty$ to ∞, is the function

$$\varphi(x) = \int_{-\infty}^{\infty} y f(y|x)\,dy \tag{7-59}$$

known as the *regression line* (Fig. 7-6).

Note If the RVs $\mathbf{x}$ and $\mathbf{y}$ are functionally related, that is, if $\mathbf{y} = g(\mathbf{x})$, then the probability masses on the xy plane are on the line $y = g(x)$ (see Fig. 6-5b); hence $E\{\mathbf{y}|x\} = g(x)$.

Galton's law. The term *regression* has its origin in the following observation attributed to the geneticist Sir Francis Galton (1822–1911): "Population extremes *regress* toward their mean." This observation applied to parents and their adult children means that children of tall (or short) parents are on the average shorter (or taller) than their parents. In statistical terms this can be phrased in terms of conditional expected values:

Suppose that the RVs $\mathbf{x}$ and $\mathbf{y}$ model the height of parents and their children respectively. These RVs have the same mean and variance, and they are positively correlated:

$$\eta_x = \eta_x = \eta \qquad \sigma_x = \sigma_y = \sigma \qquad r > 0$$

According to Galton's law, the conditional mean $E\{\mathbf{y}|x\}$ of the height of children whose parents height is x, is smaller (or larger) than x if $x > \eta$ (or $x < \eta$):

$$E\{\mathbf{y}|x\} = \varphi(x)\begin{cases} < x & \text{if} \quad x > \eta \\ > x & \text{if} \quad x < \eta \end{cases}$$

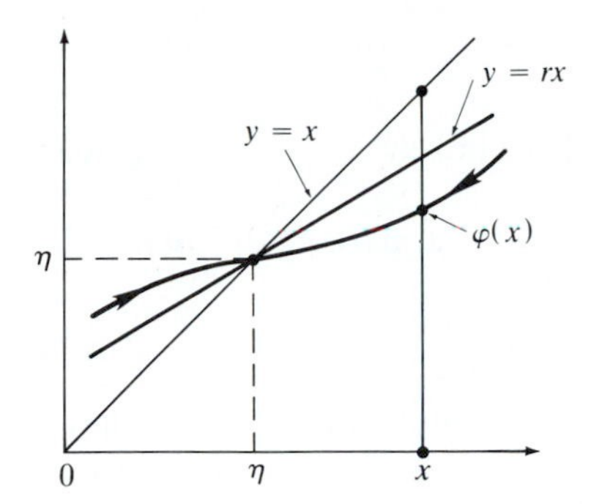

FIGURE 7-7

This shows that the regression line $\varphi(x)$ is below the line $y = x$ for $x > \eta$ and above this line if $x < \eta$ as in Fig. 7-7. If the RVs **x** and **y** are jointly normal, then [see (7-60) below] the regression line is the straight line $\varphi(x) = rx$. For arbitrary RVs, the function $\varphi(x)$ does not obey Galton's law. The term regression is used, however, to identify any conditional mean.

Example 7-14. If the RVs **x** and **y** are normal as in Example 7-9, then the function

$$E\{\mathbf{y}|x\} = \eta_2 + r\sigma_2 \frac{x - \eta_1}{\sigma_1} \tag{7-60}$$

is a straight line with slope $r\sigma_2/\sigma_1$ passing through the point (η_1, η_2). Since for normal RVs the conditional mean $E\{\mathbf{y}|x\}$ coincides with the maximum of $f(y|x)$, we conclude that the locus of the maxima of all profiles of $f(x, y)$ is the straight line (7-60).

From theorem (7-2) it follows that

$$E\{g(\mathbf{x}, \mathbf{y})|\mathscr{M}\} = \int_{-\infty}^{\infty} \int_{-\infty}^{\infty} g(x, y) f(x, y|\mathscr{M})\, dx\, dy \tag{7-61}$$

This expression can be used to determine $E\{g(\mathbf{x}, \mathbf{y})|x\}$; however, the conditional density $f(x, y|x)$ consists of line masses on the line x-constant. To avoid dealing with line masses, we shall define $E\{g(\mathbf{x}, \mathbf{y})|x\}$ as a limit:

As we have shown in Example 7-8, the conditional density $f(x, y|x < \mathbf{x} < x + \Delta x)$ is 0 outside the strip $(x, x + \Delta x)$ and in this strip it is given by (7-39) where $x_1 = x$ and $x_2 = x + \Delta x$. It follows, therefore, from (7-61) with $\mathscr{M} = \{x < \mathbf{x} \le x + \Delta x\}$ that

$$E\{g(\mathbf{x}, \mathbf{y})|x < \mathbf{x} \le x + \Delta x\} = \int_{-\infty}^{\infty} \int_{x}^{x+\Delta x} g(\alpha, y) \frac{f(\alpha, y)\, d\alpha}{F_x(x + \Delta x) - F_x(x)}\, dy$$

As $\Delta x \to 0$, the inner integral tends to $g(x, y)f(x, y)/f(x)$. Defining $E\{g(\mathbf{x}, \mathbf{y})|x\}$ as the limit of the above, we obtain

$$E\{g(\mathbf{x}, \mathbf{y})|x\} = \int_{-\infty}^{\infty} g(x, y) f(y|x)\, dy \tag{7-62}$$

We also note that

$$E\{g(x,\mathbf{y})|x\} = \int_{-\infty}^{\infty} g(x,y)f(y|x)\,dy \tag{7-63}$$

because $g(x,\mathbf{y})$ is a function of the RV $\mathbf{y}$, with x a parameter; hence its conditional expected value is given by (7-56). Thus

$$E\{g(\mathbf{x},\mathbf{y})|x\} = E\{g(x,\mathbf{y})|x\} \tag{7-64}$$

One might be tempted from the above to conclude that (7-64) follows directly from (7-56); however, this is not so. The functions $g(\mathbf{x},\mathbf{y})$ and $g(x,\mathbf{y})$ have the same expected value, assuming $\mathbf{x} = x$, but they are not equal. The first is a function $g(\mathbf{x},\mathbf{y})$ of the RVs $\mathbf{x}$ and $\mathbf{y}$, and for a specific ζ it takes the value $g[\mathbf{x}(\zeta),\mathbf{y}(\zeta)]$. The second is a function $g(x,\mathbf{y})$ of the real variable x and the RV $\mathbf{y}$, and for a specific ζ it takes the value $g[x,\mathbf{y}(\zeta)]$ where x is an arbitrary number.

Conditional Expected Values as RVs

The conditional mean of $\mathbf{y}$, assuming $\mathbf{x} = x$, is a function $\varphi(x) = E\{\mathbf{y}|x\}$ of x given by (7-59). Using this function, we can construct the RV $\varphi(\mathbf{x}) = E\{\mathbf{y}|\mathbf{x}\}$ as in Sec. 5-1. As we see from (5-29), the mean of this RV equals

$$E\{\varphi(\mathbf{x})\} = \int_{-\infty}^{\infty} \varphi(x)f(x)\,dx = \int_{-\infty}^{\infty} f(x)\int_{-\infty}^{\infty} yf(y|x)\,dy\,dx$$

Since $f(x, y) = f(x)f(y|x)$, the above yields

$$E\{E\{\mathbf{y}|\mathbf{x}\}\} = \int_{-\infty}^{\infty}\int_{-\infty}^{\infty} yf(x,y)\,dx\,dy = E\{\mathbf{y}\} \tag{7-65}$$

This basic result can be generalized: The conditional mean $E\{g(\mathbf{x},\mathbf{y})|x\}$ of $g(\mathbf{x},\mathbf{y})$, assuming $\mathbf{x} = x$, is a function of the real variable x. It defines, therefore, the function $E\{g(\mathbf{x},\mathbf{y})|\mathbf{x}\}$ of the RV $\mathbf{x}$. As we see from (7-2) and (7-61), the mean of $E\{g(\mathbf{x},\mathbf{y})|\mathbf{x}\}$ equals

$$\int_{-\infty}^{\infty} f(x)\int_{-\infty}^{\infty} g(x,y)f(y|x)\,dy\,dx = \int_{-\infty}^{\infty}\int_{-\infty}^{\infty} g(x,y)f(x,y)\,dx\,dy$$

But the last integral equals $E\{g(\mathbf{x},\mathbf{y})\}$; hence

$$E\{E\{g(\mathbf{x},\mathbf{y})|\mathbf{x}\}\} = E\{g(\mathbf{x},\mathbf{y})\} \tag{7-66}$$

We note, finally, that

$$\begin{aligned} E\{g_1(\mathbf{x})g_2(\mathbf{y})|x\} &= E\{g_1(x)g_2(\mathbf{y})|x\} = g_1(x)E\{g_2(\mathbf{y})|x\} \\ E\{g_1(\mathbf{x})g_2(\mathbf{y})\} &= E\{E\{g_1(\mathbf{x})g_2(\mathbf{y})|\mathbf{x}\}\} = E\{g_1(\mathbf{x})E\{g_2(\mathbf{y})|\mathbf{x}\}\} \end{aligned} \tag{7-67}$$

Example 7-15. Suppose that the RVs $\mathbf{x}$ and $\mathbf{y}$ are $N(0,0;\sigma_1,\sigma_2;r)$. As we know

$$E\{\mathbf{x}^2\} = \sigma_1^2 \qquad E\{\mathbf{x}^4\} = 3\sigma_1^4$$

Furthermore, $f(y|x)$ is a normal density with mean $r\sigma_2 x/\sigma_1$ and variance $\sigma_2\sqrt{1-r^2}$. Hence

$$E\{\mathbf{y}^2|x\} = \eta_{y|x}^2 + \sigma_{y|x}^2 = \left(\frac{r\sigma_2 x}{\sigma_1}\right)^2 + \sigma_2^2(1-r^2)$$

Using (7-67), we shall show that

$$E\{\mathbf{xy}\} = r\sigma_1\sigma_2 \qquad E\{\mathbf{x}^2\mathbf{y}^2\} = E\{\mathbf{x}^2\}E\{\mathbf{y}^2\} + 2E^2\{\mathbf{xy}\}$$

Proof

$$E\{\mathbf{xy}\} = E\{\mathbf{x}E\{\mathbf{y}|\mathbf{x}\}\} = E\left\{r\sigma_2\frac{\mathbf{x}^2}{\sigma_1}\right\} = r\sigma_2\frac{\sigma_1^2}{\sigma_1}$$

$$E\{\mathbf{x}^2\mathbf{y}^2\} = E\{\mathbf{x}^2E\{\mathbf{y}^2|\mathbf{x}\}\} = E\left\{\mathbf{x}^2\left[r^2\sigma_2^2\frac{\mathbf{x}^2}{\sigma_1^2} + \sigma_2^2(1-r^2)\right]\right\}$$

$$= 3\sigma_1^4 r^2\frac{\sigma_2^2}{\sigma_1^2} + \sigma_1^2\sigma_2^2(1-r^2) = \sigma_1^2\sigma_2^2 + 2r^2\sigma_1^2\sigma_2^2$$

and the proof is complete [see also (7-36)].

7-5 MEAN SQUARE ESTIMATION

The estimation problem is fundamental in the applications of probability and it will be discussed in detail later (Chap. 14). In this section, we introduce the main ideas using as illustration the estimation of an RV **y** in terms of another RV **x**. Throughout this analysis, the optimality criterion will be the minimization of the mean square value (abbreviation: MS) of the estimation error.

We start with a brief explanation of the underlying concepts in the context of repeated trials, considering first the problem of estimating the RV **y** by a constant.

Frequency interpretation As we know, the distribution function $F(y)$ of the RV **y** determines completely its statistics. This does not, of course, mean that if we know $F(y)$ we can predict the value $\mathbf{y}(\zeta)$ of **y** at some future trial. Suppose, however, that we wish to estimate the unknown $\mathbf{y}(\zeta)$ by some number c. As we shall presently see, knowledge of $F(y)$ can guide us in the selection of c.

If **y** is estimated by a constant c, then, at a particular trial, the error $\mathbf{y}(\zeta) - c$ results and our problem is to select c so as to minimize this error in some sense. A reasonable criterion for selecting c might be the condition that, in a long series of trials, the average error is close to 0:

$$\frac{\mathbf{y}(\zeta_1) - c + \cdots + \mathbf{y}(\zeta_n) - c}{n} \simeq 0$$

As we see from (5-26), this would lead to the conclusion that c should equal the *mean* of **y** (Fig. 7-8*a*).

Another criterion for selecting c might be the minimization of the average of $|\mathbf{y}(\zeta) - c|$. In this case, the optimum c is the *median* of **y** (see page 68).

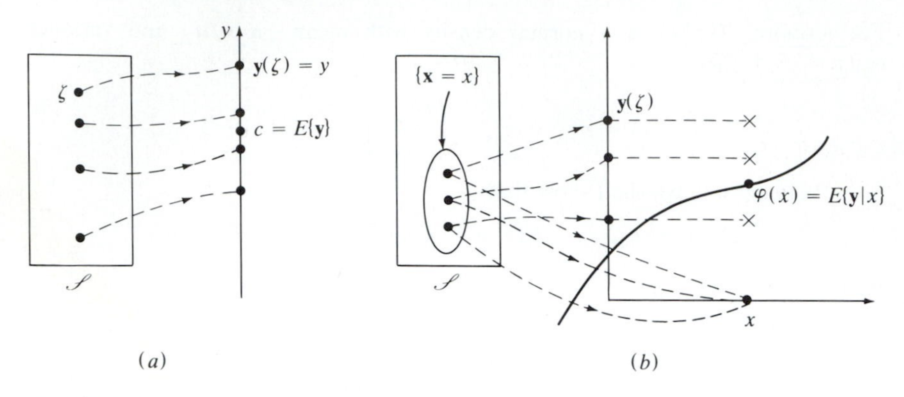

FIGURE 7-8

In our analysis, we consider only MS estimates. This means that c should be such as to minimize the average of $|\mathbf{y}(\zeta) - c|^2$. This criterion is in general useful but it is selected mainly because it leads to simple results. As we shall soon see, the best c is again the *mean* of $\mathbf{y}$.

Suppose now that at each trial we observe the value $\mathbf{x}(\zeta)$ of the RV $\mathbf{x}$. On the basis of this observation it might be best to use as the estimate of $\mathbf{y}$ not the same number c at each trial, but a number that depends on the observed $\mathbf{x}(\zeta)$. In other words, we might use as the estimate of $\mathbf{y}$ a function $c(\mathbf{x})$ of the RV $\mathbf{x}$. The resulting problem is the optimum determination of this function.

It might be argued that, if at a certain trial we observe $\mathbf{x}(\zeta)$, then we can determine the outcome ζ of this trial, and hence also the corresponding value $\mathbf{y}(\zeta)$ of $\mathbf{y}$. This, however, is not so. The same number $\mathbf{x}(\zeta) = x$ is observed for every ζ in the set $\{\mathbf{x} = x\}$ (Fig. 7-8b). If, therefore, this set has many elements and the values of $\mathbf{y}$ are different for the various elements of this set, then the observed $\mathbf{x}(\zeta)$ does not determine uniquely $\mathbf{y}(\zeta)$. However, we know now that ζ is an element of the subset $\{\mathbf{x} = x\}$. This information reduces the uncertainty about the value of $\mathbf{y}$. In the subset $\{\mathbf{x} = x\}$, the RV $\mathbf{x}$ equals x and the problem of determining $c(x)$ is reduced to the problem of determining the constant $c(x)$. As we noted, if the optimality criterion is the minimization of the MS error, then $c(x)$ must be the average of $\mathbf{y}$ in this set. In other words, $c(x)$ must equal the conditional mean of $\mathbf{y}$ assuming that $\mathbf{x} = x$.

We shall illustrate with an example. Suppose that the space $\mathcal{S}$ is the set of all children in a community and the RV $\mathbf{y}$ is the height of each child. A particular outcome ζ is a specific child and $\mathbf{y}(\zeta)$ is the height of this child. From the preceding discussion it follows that if we wish to estimate $\mathbf{y}$ by a number, this number must equal the mean of $\mathbf{y}$. We now assume that each selected child is weighed. On the basis of this observation, the estimate of the height of the child can be improved. The weight is an RV $\mathbf{x}$; hence the optimum estimate of $\mathbf{y}$ is now the conditional mean $E\{\mathbf{y}|x\}$ of $\mathbf{y}$ assuming $\mathbf{x} = x$ where x is the observed weight.

In the context of probability theory, the MS estimation of the RV $\mathbf{y}$ by a constant c can be phrased as follows: Find c such that the second moment

(MS error)

$$e = E\{(\mathbf{y} - c)^2\} = \int_{-\infty}^{\infty} (y - c)^2 f(y)\, dy \tag{7-68}$$

of the difference (error) $\mathbf{y} - c$ is minimum. Clearly, e depends on c and it is minimum if

$$\frac{de}{dc} = \int_{-\infty}^{\infty} 2(y - c) f(y)\, dy = 0$$

that is, if

$$c = \int_{-\infty}^{\infty} y f(y)\, dy$$

Thus

$$c = E\{\mathbf{y}\} = \int_{-\infty}^{\infty} y f(y)\, dy \tag{7-69}$$

This result is well known from mechanics: The moment of inertia with respect to a point c is minimum if c is the center of gravity of the masses.

NONLINEAR MS ESTIMATION. We wish to estimate $\mathbf{y}$ not by a constant but by a function $c(\mathbf{x})$ of the RV $\mathbf{x}$. Our problem now is to find the function $c(x)$ such that the MS error

$$e = E\{[\mathbf{y} - c(\mathbf{x})]^2\} = \int_{-\infty}^{\infty}\int_{-\infty}^{\infty} [y - c(x)]^2 f(x, y)\, dx\, dy \tag{7-70}$$

is minimum.

We maintain that

$$c(x) = E\{\mathbf{y}|x\} = \int_{-\infty}^{\infty} y f(y|x)\, dy \tag{7-71}$$

Proof. Since $f(x, y) = f(y|x) f(x)$, (7-70) yields

$$e = \int_{-\infty}^{\infty} f(x) \int_{-\infty}^{\infty} [y - c(x)]^2 f(y|x)\, dy\, dx$$

The integrands above are positive. Hence e is minimum if the inner integral is minimum for every x. This integral is of the form (7-68) if c is changed to $c(x)$, and $f(y)$ is changed to $f(y|x)$. Hence it is minimum if $c(x)$ equals the integral in (7-69), provided that $f(y)$ is changed to $f(y|x)$. The result is (7-71).

Thus the optimum $c(x)$ is the regression line $\varphi(x)$ of Fig. 7-6.

As we noted in the beginning of the section, if $\mathbf{y} = g(\mathbf{x})$, then $E\{\mathbf{y}|x\} = g(x)$; hence $c(x) = g(x)$ and the resulting MS error is 0. This is not surprising because, if $\mathbf{x}$ is observed and $\mathbf{y} = g(\mathbf{x})$, then $\mathbf{y}$ is determined uniquely.

If the RVs $\mathbf{x}$ and $\mathbf{y}$ are independent, then $E\{\mathbf{y}|x\} = E\{\mathbf{y}\} =$ constant. In this case, knowledge of $\mathbf{x}$ has no effect on the estimate of $\mathbf{y}$.

Linear MS Estimation

The solution of the nonlinear MS estimation problem is based on knowledge of the function $\varphi(x)$. An easier problem, using only second-order moments, is the linear MS estimation of $\mathbf{y}$ in terms of $\mathbf{x}$. The resulting estimate is not as good as the nonlinear estimate; however, it is used in many applications because of the simplicity of the solution.

The linear estimation problem is the estimation of the RV $\mathbf{y}$ in terms of a linear function $A\mathbf{x} + B$ of $\mathbf{x}$. The problem now is to find the constants A and B so as to minimize the MS error

$$e = E\{[\mathbf{y} - (A\mathbf{x} + B)]^2\} \tag{7-72}$$

We maintain that $e = e_m$ is minimum if

$$A = \frac{\mu_{11}}{\mu_{20}} = \frac{r\sigma_y}{\sigma_x} \qquad B = \eta_y - A\eta_x \tag{7-73}$$

and

$$e_m = \mu_{02} - \frac{\mu_{11}^2}{\mu_{20}} = \sigma_y^2(1 - r^2) \tag{7-74}$$

Proof. For a given A, e is the MS error of the estimation of $\mathbf{y} - A\mathbf{x}$ by the constant B. Hence e is minimum if $B = E\{\mathbf{y} - A\mathbf{x}\}$ as in (7-69). With B so determined, (7-72) yields

$$e = E\left\{\left[(\mathbf{y} - \eta_y) - A(\mathbf{x} - \eta_x)\right]^2\right\} = \sigma_y^2 - 2Ar\sigma_x\sigma_y + A^2\sigma_x^2$$

This is minimum if $A = r\sigma_y/\sigma_x$ and (7-73) results. Inserting into the above quadratic, we obtain (7-74).

Terminology. In the above, the sum $A\mathbf{x} + B$ is the *nonhomogeneous* linear estimate of $\mathbf{y}$ in terms of $\mathbf{x}$. If $\mathbf{y}$ is estimated by a straight line $a\mathbf{x}$ passing through the origin, the estimate is called *homogeneous*.

The RV $\mathbf{x}$ is the data of the estimation, the RV $\boldsymbol{\varepsilon} = \mathbf{y} - (A\mathbf{x} + B)$ is the *error* of the estimation, and the number $e = E\{\boldsymbol{\varepsilon}^2\}$ is the MS error.

Fundamental note. In general, the nonlinear estimate $\varphi(x) = E[\mathbf{y}|x\}$ of $\mathbf{y}$ in terms of $\mathbf{x}$ is not a straight line and the resulting MS error $E\{[\mathbf{y} - \varphi(\mathbf{x})]^2\}$ is smaller than the MS error e_m of the linear estimate $A\mathbf{x} + B$. However, if the RVs $\mathbf{x}$ and $\mathbf{y}$ are jointly normal, then [see (7-60)]

$$\varphi(x) = \frac{r\sigma_y x}{\sigma_x} + \eta_y - \frac{r\sigma_y\eta_x}{\sigma_x}$$

is a straight line as in (7-73). In other words:

For normal RVs, nonlinear and linear MS estimates are identical.

The Orthogonality Principle

From (7-73) it follows that

$$E\{[\mathbf{y} - (A\mathbf{x} + B)]\mathbf{x}\} = 0 \tag{7-75}$$

This result can be derived directly from (7-72). Indeed, the MS error e is a function of A and B and it is minimum if $\partial e/\partial A = 0$ and $\partial e/\partial B = 0$. The first equation yields

$$\frac{\partial e}{\partial A} = E\{2[\mathbf{y} - (A\mathbf{x} + B)](-\mathbf{x})\} = 0$$

and (7-75) results. The interchange between expected value and differentiation is equivalent to the interchange of integration and differentiation.

Equation (7-75) states that the optimum linear MS estimate $A\mathbf{x} + B$ of $\mathbf{y}$ is such that the estimation error $\mathbf{y} - (A\mathbf{x} + B)$ is orthogonal to the data $\mathbf{x}$. This is known as the *orthogonality principle*. It is fundamental in MS estimation and will be used extensively. In the following, we reestablish it for the homogeneous case.

HOMOGENEOUS LINEAR MS ESTIMATION. We wish to find a constant a such that, if $\mathbf{y}$ is estimated by $a\mathbf{x}$, the resulting MS error

$$e = E\{(\mathbf{y} - a\mathbf{x})^2\} \tag{7-76}$$

is minimum. We maintain that a must be such that

$$E\{(\mathbf{y} - a\mathbf{x})\mathbf{x}\} = 0 \tag{7-77}$$

Proof. Clearly, e is minimum if $e'(a) = 0$; this yields (7-77). We shall give a second proof: We assume that a satisfies (7-77) and we shall show that e is minimum. With $\bar{a}$ an arbitrary constant,

$$\begin{aligned} E\{(\mathbf{y} - \bar{a}\mathbf{x})^2\} &= E\{[(\mathbf{y} - a\mathbf{x}) + (a - \bar{a})\mathbf{x}]^2\} \\ &= E\{(\mathbf{y} - a\mathbf{x})^2\} + (a - \bar{a})^2 E\{\mathbf{x}^2\} + 2(a - \bar{a})E\{(\mathbf{y} - a\mathbf{x})\mathbf{x}\} \end{aligned}$$

In the above, the last term is 0 by assumption and the second term is positive. From this it follows that

$$E\{(\mathbf{y} - \bar{a}\mathbf{x})^2\} \geq E\{(\mathbf{y} - a\mathbf{x})^2\}$$

for any $\bar{a}$; hence e is minimum.

The linear MS estimate of $\mathbf{y}$ in terms of $\mathbf{x}$ will be denoted by $\hat{E}\{\mathbf{y}|\mathbf{x}\}$. Solving (7-77), we conclude that

$$\hat{E}\{\mathbf{y}|\mathbf{x}\} = a\mathbf{x} \qquad a = \frac{E\{\mathbf{x}\mathbf{y}\}}{E\{\mathbf{x}^2\}} \tag{7-78}$$

MS error Since

$$e = E\{(\mathbf{y} - a\mathbf{x})\mathbf{y}\} - E\{(\mathbf{y} - a\mathbf{x})a\mathbf{x}\} = E\{\mathbf{y}^2\} - E\{(a\mathbf{x})^2\} - 2aE\{(\mathbf{y} - a\mathbf{x})\mathbf{x}\}$$

we conclude with (7-77) that

$$e = E\{(\mathbf{y} - a\mathbf{x})\mathbf{y}\} = E\{\mathbf{y}^2\} - E\{(a\mathbf{x})^2\} \tag{7-79}$$

We note finally that (7-77) is consistent with the orthogonality principle: The error $\mathbf{y} - a\mathbf{x}$ is orthogonal to the data $\mathbf{x}$.

Geometric interpretation of the orthogonality principle. In the vector representation of RVs (see Fig. 7-9), the difference $\mathbf{y} - a\mathbf{x}$ is the vector from the point $a\mathbf{x}$ on the $\mathbf{x}$ line to the point $\mathbf{y}$, and the length of that vector equals $\sqrt{e}$. Clearly, this length is minimum if $\mathbf{y} - a\mathbf{x}$ is perpendicular to $\mathbf{x}$ in agreement with (7-77). The right side of (7-79) follows from the pythagorean theorem and the middle term states that the square of the length of $\mathbf{y} - a\mathbf{x}$ equals the inner product of $\mathbf{y}$ with the error $\mathbf{y} - a\mathbf{x}$.

Risk and loss functions. We conclude with a brief comment on other optimality criteria limiting the discussion to the estimation of an RV $\mathbf{y}$ by a constant c. We select a function $L(x)$ and we choose c so as to minimize the mean

$$R = E\{L(\mathbf{y} - c)\} = \int_{-\infty}^{\infty} L(y - c)f(y)\,dy$$

of the RV $L(\mathbf{y} - c)$. The function $L(x)$ is called the *loss function* and the constant R is called the *average risk*. The choice of $L(x)$ depends on the applications. If $L(x) = x^2$, then $R = E\{(\mathbf{y} - c)^2\}$ is the MS error and as we have shown, it is minimum if $c = E\{\mathbf{y}\}$.

If $L(x) = |x|$, then $R = E\{|\mathbf{y} - c|\}$. We maintain that in this case, c equals the *median* $y_{0.5}$ of $\mathbf{y}$ (see also Prob. 5-20).

Proof. The average risk equals

$$R = \int_{-\infty}^{\infty} |y - c| f(y)\,dy = \int_{-\infty}^{c} (c - y)f(y)\,dy + \int_{c}^{\infty} (y - c)f(y)\,dy$$

Differentiating with respect to c, we obtain

$$\frac{dR}{dc} = \int_{-\infty}^{c} f(y)\,dy - \int_{c}^{\infty} f(y)\,dy = 2F(c) - 1$$

Thus R is minimum if $F(c) = 1/2$, that is, if $c = y_{0.5}$.

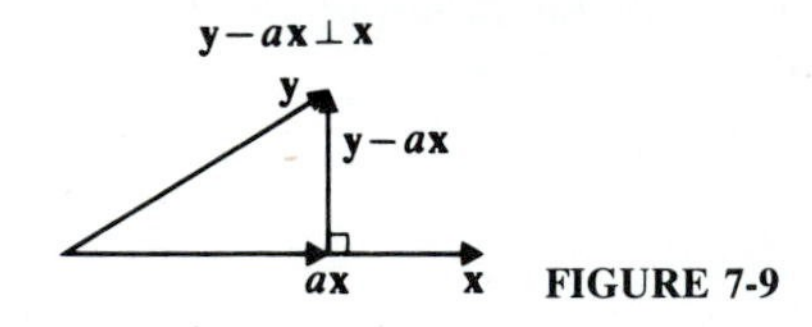

FIGURE 7-9

We note finally that in certain applications, $\mathbf{y}$ is estimated by its *mode*, that is, the value y_{max} of y for which $f(y)$ is maximum. This is based on the following: The probability that $\mathbf{y}$ is in an interval $(c, c + dy)$ of specified length dy equals $P\{c < \mathbf{y} < c + dy\} = f(c)\,dy$ This is maximum if $c = y_{max}$.

PROBLEMS

7-1. The RVs $\mathbf{x}$ and $\mathbf{y}$ are $N(0;\sigma)$ and independent. Show that if $\mathbf{z} = |\mathbf{x} - \mathbf{y}|$, then $E\{\mathbf{z}\} = 2\sigma/\sqrt{\pi}$, $E\{\mathbf{z}^2\} = 2\sigma^2$.

7-2. Show that if $\mathbf{x}$ and $\mathbf{y}$ are two independent RVs with $f_x(x) = e^{-x}U(x)$, $f_y(y) = e^{-y}U(y)$, and $\mathbf{z} = (\mathbf{x} - \mathbf{y})U(\mathbf{x} - \mathbf{y})$, then $E\{\mathbf{z}\} = 1/2$.

7-3. Show that for any $\mathbf{x}, \mathbf{y}$ real or complex
(*a*) $|E\{\mathbf{xy}\}|^2 \le E\{|\mathbf{x}|^2\}E\{|\mathbf{y}|^2\}$;
(*b*) (*triangle inequality*) $\sqrt{E\{|\mathbf{x}+\mathbf{y}|^2\}} \le \sqrt{E\{|\mathbf{x}|^2\}} + \sqrt{E\{|\mathbf{y}|^2\}}$.

7-4. Show that, if $r_{xy} = 1$, then $\mathbf{y} = a\mathbf{x} + b$.

7-5. Show that, if $E\{\mathbf{x}^2\} = E\{\mathbf{y}^2\} = E\{\mathbf{xy}\}$, then $\mathbf{x} = \mathbf{y}$.

7-6. Show that, if the RV $\mathbf{x}$ is of discrete type taking the values x_n with $P\{\mathbf{x} = x_n\} = p_n$ and $\mathbf{z} = g(\mathbf{x}, \mathbf{y})$, then

$$E\{\mathbf{z}\} = \sum_n E\{g(x_n, \mathbf{y})\}p_n \qquad f_z(z) = \sum_n f_z(z|x_n)p_n$$

7-7. The RV $\mathbf{n}$ is Poisson with parameter λ and the RV $\mathbf{x}$ is independent of $\mathbf{n}$. Show that, if $\mathbf{z} = \mathbf{nx}$ and

$$f_x(x) = \frac{\alpha}{\pi(\alpha^2 + x^2)} \qquad \text{then} \quad \Phi_z(\omega) = \exp\{\lambda e^{-\alpha|\omega|} - \lambda\}$$

7-8. Show that, if the RVs $\mathbf{x}$ and $\mathbf{y}$ are $N(0, 0; \sigma, \sigma; r)$, then

$$(a) \qquad E\{f_y(\mathbf{y}|x)\} = \frac{1}{\sigma\sqrt{2\pi(2 - r^2)}}\exp\left\{-\frac{r^2x^2}{2\sigma^2(2 - r^2)}\right\}$$

$$(b) \qquad E\{f_x(\mathbf{x})f_y(\mathbf{y})\} = \frac{1}{2\pi\sigma^2\sqrt{4 - r^2}}$$

7-9. Show that if the RVs $\mathbf{x}, \mathbf{y}$ are $N(0, 0; \sigma_1, \sigma_2; r)$ then

$$E\{|\mathbf{xy}|\} = \frac{2}{\pi}\int_0^c \arcsin\frac{\mu}{\sigma_1\sigma_2}\,d\mu + \frac{2\sigma_1\sigma_2}{\pi} = \frac{2\sigma_1\sigma_2}{\pi}(\cos\alpha + \alpha\sin\alpha)$$

where $r = \sin\alpha$ and $C = r\sigma_1\sigma_2$.
Hint: Use (7-37) with $g(x, y) = |xy|$.

7-10. The RVs $\mathbf{x}$ and $\mathbf{y}$ are uniform in the interval $(-1, 1)$ and independent. Find the conditional density $f_r(r|\mathscr{M})$ of the RV $\mathbf{r} = \sqrt{\mathbf{x}^2 + \mathbf{y}^2}$ where $\mathscr{M} = \{\mathbf{r} \le 1\}$.

7-11. We have a pile of m coins. The probability of heads of the ith coin equals p_i. We select at random one of the coins, we toss it n times and heads shows k times.

Show that the probability that we selected the rth coin equals

$$\frac{p_r^k(1-p_r)^{n-k}}{p_1^k(1-p_1)^{n-k}+\cdots+p_m^k(1-p_m)^{n-k}}$$

7-12. The RV $\mathbf{x}$ has a Student-t distribution $t(n)$. Show that $E\{\mathbf{x}^2\} = n/(n-2)$.

7-13. Show that if $\beta_x(t) = f_x(t|\mathbf{x} > t)$, $\beta_y(t|\mathbf{y} > t)$ and $\beta_x(t) = k\beta_y(t)$, then $1 - F_x(x) = [1 - F_y(x)]^k$.

7-14. Show that, for any $\mathbf{x}, \mathbf{y}$, and $\varepsilon > 0$,

$$P\{|\mathbf{x}-\mathbf{y}| > \varepsilon\} \le \frac{1}{\varepsilon^2}E\{|\mathbf{x}-\mathbf{y}|^2\}$$

7-15. Show that the RVs $\mathbf{x}$ and $\mathbf{y}$ are independent iff for any a and b:

$$E\{U(a-\mathbf{x})U(b-\mathbf{y})\} = E\{U(a-\mathbf{x})\}E\{U(b-\mathbf{y})\}$$

7-16. Show that

$$E\{\mathbf{y}|\mathbf{x} \le 0\} = \frac{1}{F_x(0)}\int_{-\infty}^{0} E\{\mathbf{y}|x\}f_x(x)\,dx$$

7-17. Show that, if the RVs $\mathbf{x}$ and $\mathbf{y}$ are independent and $\mathbf{z} = \mathbf{x} + \mathbf{y}$, then $f_z(z|x) = f_y(z-x)$.

7-18. The RVs $\mathbf{x}, \mathbf{y}$ are $N(3, 4; 1, 2; 0.5)$. Find $f(y|x)$ and $f(x|y)$.

7-19. Show that, for any $\mathbf{x}$ and $\mathbf{y}$, the RVs $\mathbf{z} = F_x(\mathbf{x})$ and $\mathbf{w} = F_y(\mathbf{y}|\mathbf{x})$ are independent and each is uniform in the interval $(0, 1)$.

7-20. The RVs $\mathbf{x}$ and $\mathbf{y}$ are $N(0, 0; 3, 5; 0.8)$. Find $g(x)$ such that $E\{[\mathbf{y} - g(\mathbf{x})]^2\}$ is minimum.

7-21. In the approximation of $\mathbf{y}$ by $\varphi(\mathbf{x})$, the "mean cost" $E\{g[\mathbf{y} - \varphi(\mathbf{x})]\}$ results, where $g(x)$ is a given function. Show that, if $g(x)$ is an even convex function as in Fig. P5-28, then the "mean cost" is minimum if $\varphi(x) = E\{\mathbf{y}|x\}$.

7-22. Show that if $\varphi(x) = E\{\mathbf{y}|x\}$ is the nonlinear MS estimate of $\mathbf{y}$ in terms of $\mathbf{x}$, then

$$E\{[\mathbf{y} - \varphi(\mathbf{x})]^2\} = E\{\mathbf{y}^2\} - E\{\varphi^2(\mathbf{x})\}$$

7-23. If $\eta_x = \eta_y = 0$, $\sigma_x = \sigma_y = 4$, and $\hat{\mathbf{y}} = 0.2\mathbf{x}$ (linear MS estimate), find $E\{(\mathbf{y} - \hat{\mathbf{y}})^2\}$.

7-24. Show that if the constants A, B, and a are such that $E\{[\mathbf{y} - (A\mathbf{x} + B)]^2\}$ and $E\{[(\mathbf{y} - \eta_y) - a(\mathbf{x} - \eta_x)]^2\}$ are minimum, then $a = A$.

7-25. Given $\eta_x = 4$, $\eta_y = 0$, $\sigma_x = 1$, $\sigma_y = 2$, $r_{xy} = 0.5$, find the parameters A, B, and a that minimize $E\{[\mathbf{y} - (A\mathbf{x} + B)]^2\}$ and $E\{(\mathbf{y} - a\mathbf{x})^2\}$.

7-26. The RVs $\mathbf{x}, \mathbf{y}$ are independent, integer-valued with $P\{\mathbf{x} = k\} = p_k$, $P\{\mathbf{y} = k\} = q_k$. Show that (a) if $\mathbf{z} = \mathbf{x} + \mathbf{y}$, then (discrete-time convolution)

$$P\{\mathbf{z} = n\} = \sum_{k=-\infty}^{\infty} p_{n-k}q_k$$

(b) if the RVs $\mathbf{x}, \mathbf{y}$ are Poisson distributed with parameters a and b respectively and $\mathbf{w} = \mathbf{x} - \mathbf{y}$, then

$$P\{\mathbf{w} = n\} = e^{-a-b}\sum_{k=m}^{\infty}\frac{a^{n+k}b^k}{(n+k)!k!} \qquad m = \begin{cases} 0 & n > 0 \\ |n| & n < 0 \end{cases}$$

7-27. If $\mathbf{y} = \mathbf{x}^3$, find the nonlinear and linear MS estimate of $\mathbf{y}$ in terms of $\mathbf{x}$ and the resulting MS errors.

7-28. The RV $\mathbf{x}$ has a Rayleigh density [see (6-50)]. Find its conditional failure rate $\beta(t)$.

7-29. Find the reliability $R(t)$ of a system if $\beta(t) = ct/(1 + ct)$.

7-30. The RV $\mathbf{x}$ is uniform in the interval $(0, T)$. Find and sketch $\beta(t)$.

7-31. Find and sketch $R(t)$ if $\beta(t) = 4U(t) + 2U(t - T)$. Find the mean time to failure of the system.

7-32. The RVs $\mathbf{x}$ and $\mathbf{y}$ are jointly normal with the zero mean, and $\sigma_x = 2$, $\sigma_y = 4$, $r_{xy} = 0.5$. (*a*) Find the regression line $E\{\mathbf{y}|x\} = \varphi(x)$. (*b*) Show that the RVs $\mathbf{x}$ and $\mathbf{y} - \varphi(\mathbf{x})$ are independent.

7-33. (*a*) Show that $E\{(\mathbf{y} - c)^2\} = \sigma_y^2 + (c - \eta_y)^2$ for any c. (*b*) Using this, show that $E\{(\mathbf{y} - c)^2\}$ is minimum if $c = \eta_y$ as in (7-69). (*c*) Reasoning similarly, show that $E\{[\mathbf{y} - g(\mathbf{x})]^2\}$ is minimum if $g(x) = E\{\mathbf{y}|x\}$ as in (7-71).

CHAPTER 8

SEQUENCES OF RANDOM VARIABLES

8-1 GENERAL CONCEPTS

A *random vector* is a vector

$$\mathbf{X} = [\mathbf{x}_1, \ldots, \mathbf{x}_n] \tag{8-1}$$

whose components $\mathbf{x}_i$ are RVs.

The probability that $\mathbf{X}$ is in a region D of the n-dimensional space equals the probability masses in D:

$$P\{\mathbf{X} \in D\} = \int_D f(X)\, dX \qquad X = [x_1, \ldots, x_n] \tag{8-2}$$

In the above

$$f(X) = f(x_1, \ldots, x_n) = \frac{\partial^n F(x_1, \ldots, x_n)}{\partial x_1, \ldots, \partial x_n} \tag{8-3}$$

is the *joint* (or, *multivariate*) *density* of the RVs $\mathbf{x}_i$ and

$$F(X) = F(x_1, \ldots, x_n) = P\{\mathbf{x} \le x_1, \ldots, \mathbf{x}_n \le x_n\} \tag{8-4}$$

is their *joint distribution*.

If we substitute in $F(x_1, \ldots, x_n)$ certain variables by ∞, we obtain the joint distribution of the remaining variables. If we integrate $f(x_1, \ldots, x_n)$ with respect to certain variables, we obtain the joint density of the remaining variables. For example

$$F(x_1, x_3) = F(x_1, \infty, x_3, \infty)$$
$$f(x_1, x_3) = \int_{-\infty}^{\infty} \int_{-\infty}^{\infty} f(x_1, x_2, x_3, x_4)\, dx_2\, dx_4 \tag{8-5}$$

Note In the above, we identify various functions in terms of their independent variables. Thus $f(x_1, x_3)$ is the joint density of the RVs $\mathbf{x}_1$ and $\mathbf{x}_3$ and it is in general *different* from the joint density $f(x_2, x_4)$ of the RVs $\mathbf{x}_2$ and $\mathbf{x}_4$. Similarly, the density $f_i(x_i)$ of the RV $\mathbf{x}_i$ will often be denoted by $f(x_i)$.

TRANSFORMATIONS. Given k functions

$$g_1(X), \ldots, g_k(X) \qquad X = [x_1, \ldots, x_n]$$

we form the RVs

$$\mathbf{y}_1 = g_1(\mathbf{X}), \ldots, \mathbf{y}_k = g_k(\mathbf{X}) \tag{8-6}$$

The statistics of these RVs can be determined in terms of the statistics of $\mathbf{X}$ as in Sec. 6-3. If $k < n$, then we could determine first the joint density of the n RVs $\mathbf{y}_1, \ldots, \mathbf{y}_k, \mathbf{x}_{k+1}, \ldots, \mathbf{x}_n$ and then use the generalization of (8-5) to eliminate the $\mathbf{x}$'s. If $k > n$, then the RVs $\mathbf{y}_{n+1}, \ldots, \mathbf{y}_k$ can be expressed in terms of $\mathbf{y}_1, \ldots, \mathbf{y}_n$. In this case, the masses in the k space are singular and can be determined in terms of the joint density of $\mathbf{y}_1, \ldots, \mathbf{y}_n$. It suffices, therefore, to assume that $k = n$.

To find the density $f_y(y_1, \ldots, y_n)$ of the random vector $\mathbf{Y} = [\mathbf{y}_1, \ldots, \mathbf{y}_n]$ for a specific set of numbers $y_1, \ldots, y_n$, we solve the system

$$g_1(X) = y_1, \ldots, g_n(X) = y_n \tag{8-7}$$

If this system has no solutions, then $f_y(y_1, \ldots, y_n) = 0$. If it has a single solution $X = [x_1, \ldots, x_n]$, then

$$f_y(y_1, \ldots, y_n) = \frac{f_x(x_1, \ldots, x_n)}{|J(x_1, \ldots, x_n)|} \tag{8-8}$$

where

$$J(x_1, \ldots, x_n) = \begin{vmatrix} \dfrac{\partial g_1}{\partial x_1} & \cdots & \dfrac{\partial g_1}{\partial x_n} \\ \cdots & \cdots & \cdots \\ \dfrac{\partial g_n}{\partial x_1} & \cdots & \dfrac{\partial g_n}{\partial x_n} \end{vmatrix} \tag{8-9}$$

is the jacobian of the transformation (8-7). If it has several solutions, then we add the corresponding terms as in (6-63).

Independence

The RVs $\mathbf{x}_1, \ldots, \mathbf{x}_n$ are called (mutually) independent if the events $\{\mathbf{x}_1 \le x_1\}, \ldots, \{\mathbf{x}_n \le x_n\}$ are independent. From this it follows that

$$\begin{aligned} F(x_1, \ldots, x_n) &= F(x_1) \cdots F(x_n) \\ f(x_1, \ldots, x_n) &= f(x_1) \cdots f(x_n) \end{aligned} \tag{8-10}$$

Example 8-1. Given n independent RVs $\mathbf{x}_i$ with respective densities $f_i(x_i)$, we form the RVs

$$\mathbf{y}_k = \mathbf{x}_1 + \cdots + \mathbf{x}_k \qquad k = 1, \ldots, n$$

We shall determine the joint density of $\mathbf{y}_k$. The system

$$x_1 = y_1, x_1 + x_2 = y_2, \ldots, x_1 + \cdots + x_n = y_n$$

has a unique solution

$$x_k = y_k - y_{k-1} \qquad 1 \le k \le n$$

and its jacobian equals 1. Hence [see (8-8) and (8-10)]

$$f_y(y_1, \ldots, y_n) = f_1(y_1) f_2(y_2 - y_1) \cdots f_n(y_n - y_{n-1}) \tag{8-11}$$

From (8-10) it follows that any subset of the set $\mathbf{x}_i$ is a set of independent RVs. Suppose, for example, that

$$f(x_1, x_2, x_3) = f(x_1) f(x_2) f(x_3)$$

Integrating with respect to x_3, we obtain $f(x_1, x_2) = f(x_1) f(x_2)$. This shows that the RVs $\mathbf{x}_1$ and $\mathbf{x}_2$ are independent. Note, however, that if the RVs $\mathbf{x}_i$ are independent in pairs, they are not necessarily independent. For example, it is possible that

$$f(x_1, x_2) = f(x_1) f(x_2) \quad f(x_1, x_3) = f(x_1) f(x_3) \quad f(x_2, x_3) = f(x_2) f(x_3)$$

but $f(x_1, x_2, x_3) \ne f(x_1) f(x_2) f(x_3)$ (see Prob. 8-2).

Reasoning as in (6-29), we can show that if the RVs $\mathbf{x}_i$ are independent, then the RVs

$$\mathbf{y}_1 = g_1(\mathbf{x}_1), \ldots, \mathbf{y}_n = g_n(\mathbf{x}_n)$$

are also independent.

INDEPENDENT EXPERIMENTS AND REPEATED TRIALS. Suppose that

$$\mathscr{S}^n = \mathscr{S}_1 \times \cdots \times \mathscr{S}_n$$

is a combined experiment and the RVs $\mathbf{x}_i$ depend only on the outcomes ζ_i of $\mathscr{S}_i$:

$$\mathbf{x}_i(\zeta_1 \cdots \zeta_i \cdots \zeta_n) = \mathbf{x}_i(\zeta_i) \qquad i = 1, \ldots, n$$

If the experiments $\mathscr{S}_i$ are independent, then the RVs $\mathbf{x}_i$ are independent [see also (6-30)]. The following special case is of particular interest.

Suppose that $\mathbf{x}$ is an RV defined on an experiment $\mathscr{S}$ and the experiment is performed n times generating the experiment $\mathscr{S}^n = \mathscr{S} \times \cdots \times \mathscr{S}$. In this experiment, we define the RVs $\mathbf{x}_i$ such that

$$\mathbf{x}_i(\zeta_1 \cdots \zeta_i \cdots \zeta_n) = \mathbf{x}(\zeta_i) \qquad i = 1, \ldots, n \tag{8-12}$$

From this it follows that the distribution $F_i(x_i)$ of $\mathbf{x}_i$ equals the distribution $F_x(x)$ of the RV $\mathbf{x}$. Thus, if an experiment is performed n times, the RVs $\mathbf{x}_i$ defined as in (8-12) are independent and they have the same distribution $F_x(x)$. These RVs are called i.i.d. (independent, identically distributed).

Example 8-2 Order statistics. The *order statistics* of the RVs $\mathbf{x}_i$ are n RVs $\mathbf{y}_k$ defined as follows: For a specific outcome ζ, the RVs $\mathbf{x}_i$ take the values $\mathbf{x}_i(\zeta)$. Ordering these numbers, we obtain the sequence

$$\mathbf{x}_{r_1}(\zeta) \le \cdots \le \mathbf{x}_{r_k}(\zeta) \le \cdots \le \mathbf{x}_{r_n}(\zeta)$$

and we define the RV $\mathbf{y}_k$ such that

$$\mathbf{y}_1(\zeta) = \mathbf{x}_{r_1}(\zeta) \le \cdots \le \mathbf{y}_k(\zeta) = \mathbf{x}_{r_k}(\zeta) \le \cdots \le \mathbf{y}_n(\zeta) = \mathbf{x}_{r_n}(\zeta) \tag{8-13}$$

We note that for a specific i, the values $\mathbf{x}_i(\zeta)$ of $\mathbf{x}_i$ occupy different locations in the above ordering as ζ changes.

We maintain that the density $f_k(y)$ of the kth statistic $\mathbf{y}_k$ is given by

$$f_k(y) = \frac{n!}{(k-1)!(n-k)!} F_x^{k-1}(y)[1 - F_x(y)]^{n-k} f_x(y) \tag{8-14}$$

where $F_x(x)$ is the distribution of the i.i.d. RVs $\mathbf{x}_i$ and $f_x(x)$ is their density.

Proof. As we know

$$f_k(y)\,dy = P\{y < \mathbf{y}_k \le y + dy\}$$

The event $\mathscr{B} = \{y < \mathbf{y}_k \le y + dy\}$ occurs iff exactly $k - 1$ of the RVs $\mathbf{x}_i$ are less than y and one is in the interval $(y, y + dy)$ (Fig. 8-1). In the original experiment $\mathscr{S}$, the events

$$\mathscr{A}_1 = \{\mathbf{x} \le y\} \qquad \mathscr{A}_2 = \{y < \mathbf{x} \le y + dy\} \qquad \mathscr{A}_3 = \{\mathbf{x} > y + dy\}$$

form a partition and

$$P(\mathscr{A}_1) = F_x(y) \qquad P(\mathscr{A}_2) = f_x(y)\,dy \qquad P(\mathscr{A}_3) = 1 - F_x(y)$$

In the experiment $\mathscr{S}^n$, the event $\mathscr{B}$ occurs iff $\mathscr{A}_1$ occurs $k - 1$ times, $\mathscr{A}_2$ occurs once, and $\mathscr{A}_3$ occurs $n - k$ times. With $k_1 = k - 1$, $k_2 = 1$, $k_3 = n - k$, it

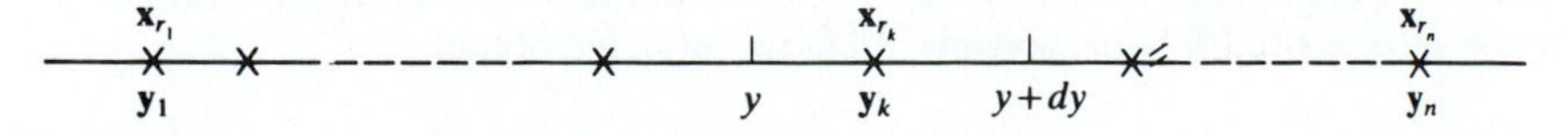

FIGURE 8-1

follows from (3-38) that

$$P(\mathscr{B}) = \frac{n!}{(k-1)!1!(n-k)!}P^{k-1}(\mathscr{A}_1)P(\mathscr{A}_2)P^{n-k}(\mathscr{A}_3)$$

and (8-14) results.

Note that

$$f_1(y) = n[1 - F_x(y)]^{n-1}f_x(y) \qquad f_n(y) = nF_x^{n-1}(y)f_x(y)$$

These are the densities of the minimum $\mathbf{y}_1$ and the maximum $\mathbf{y}_n$ of the RVs $\mathbf{x}_i$.

Special Case. If the RVs $\mathbf{x}_i$ are exponential with parameter α:

$$f_x(x) = \alpha e^{-\alpha x}U(x) \qquad F_x(x) = (1 - e^{-\alpha x})U(x)$$

then

$$f_1(y) = n\alpha e^{-\alpha n y}U(y)$$

that is, their minimum $\mathbf{y}_1$ is also exponential with parameter $n\alpha$.

Example 8-3. A system consists of m components and the time to failure of the ith component is an RV $\mathbf{x}_i$ with distribution $F_i(x)$. Thus

$$1 - F_i(t) = P\{\mathbf{x}_i > t\}$$

is the probability that the ith component is good at time t. We denote by $\mathbf{n}(t)$ the number of components that are good at time t. Clearly,

$$\mathbf{n}(t) = \mathbf{n}_1 + \cdots + \mathbf{n}_m$$

where

$$\mathbf{n}_i = \begin{cases} 1 & \mathbf{x}_i > t \\ 0 & \mathbf{x}_i < t \end{cases} \qquad E\{\mathbf{n}_i\} = 1 - F_i(t)$$

Hence the mean $E\{\mathbf{n}(t)\} = \eta(t)$ of $\mathbf{n}(t)$ is given by

$$\eta(t) = 1 - F_1(t) + \cdots + 1 - F_m(t)$$

We shall assume that the RVs $\mathbf{x}_i$ have the same distribution $F(t)$. In this case,

$$\eta(t) = m[1 - F(t)]$$

Failure rate The difference $\eta(t) - \eta(t + dt)$ is the expected number of failures in the interval $(t, t + dt)$. The derivative $-\eta'(t) = mf(t)$ of $-\eta(t)$ is the rate of failure. The ratio

$$\beta(t) = -\frac{\eta'(t)}{\eta(t)} = \frac{f(t)}{1 - F(t)} \tag{8-15}$$

is called the *relative expected failure rate*. As we see from (7-52), the function $\beta(t)$ can also be interpreted as the conditional failure rate of each component in the system. Assuming that the system is put into operation at $t = 0$, we have $\mathbf{n}(0) = m$; hence $\eta(0) = E\{\mathbf{n}(0)\} = m$. Solving (8-15) for $\eta(t)$, we obtain

$$\eta(t) = m \exp\left\{-\int_0^t \beta(\tau)\,d\tau\right\}$$

Example 8-4 Measurement errors. We measure an object of length η with n instruments of varying accuracies. The results of the measurements are n RVs

$$\mathbf{x}_i = \eta + \boldsymbol{\nu}_i \qquad E\{\boldsymbol{\nu}_i\} = 0 \qquad E\{\boldsymbol{\nu}_i^2\} = \sigma_i^2$$

where $\boldsymbol{\nu}_i$ are the measurement errors which we assume independent with zero mean. We shall determine the unbiased, minimum variance, linear estimation of η. This means the following: We wish to find n constants α_i such that the sum

$$\hat{\boldsymbol{\eta}} = \alpha_1 \mathbf{x}_1 + \cdots + \alpha_n \mathbf{x}_n$$

is an RV with mean $E\{\hat{\boldsymbol{\eta}}\} = \alpha_1 E\{\mathbf{x}_1\} + \cdots + \alpha_n E\{\mathbf{x}_n\} = \eta$ and its variance

$$V = \alpha_1^2 \sigma_1^2 + \cdots + \alpha_n^2 \sigma_n^2$$

is minimum. Thus our problem is to minimize the above sum subject to the constraint

$$\alpha_1 + \cdots + \alpha_n = 1 \tag{8-16}$$

To solve this problem, we note that

$$V = \alpha_1^2 \sigma_1^2 + \cdots + \alpha_n^2 \sigma_n^2 - \lambda(\alpha_1 + \cdots + \alpha_n - 1)$$

for any λ (Lagrange multiplier). Hence V is minimum if

$$\frac{\partial V}{\partial \alpha_i} = 2\alpha_i \sigma_i^2 - \lambda = 0 \qquad \alpha_i = \frac{\lambda}{2\sigma_i^2}$$

Inserting into (8-16) and solving for λ, we obtain

$$\frac{\lambda}{2} = V = \frac{1}{1/\sigma_1^2 + \cdots + 1/\sigma_n^2}$$

Hence

$$\hat{\eta} = \frac{x_1/\sigma_1^2 + \cdots + x_n/\sigma_n^2}{1/\sigma_1^2 + \cdots + 1/\sigma_n^2} \tag{8-17}$$

Illustration. The voltage E of a generator is measured three times. We list below the results x_i of the measurements, the standard deviations σ_i of the measurement errors, and the estimate $\hat{E}$ of E obtained from (8-17):

$$x_i = 98.6\ 98.8\ 98.9 \qquad \sigma_i = 0.20\ 0.25\ 0.28$$

$$\hat{E} = \frac{x_1/0.04 + x_2/0.0625 + x_3/0.0784}{1/0.04 + 1/0.0625 + 1/0.0784} = 98.73$$

Group independence. We say that the group G_x of the RVs $\mathbf{x}_1, \ldots, \mathbf{x}_n$ is independent of the group G_y of the RVs $\mathbf{y}_1, \ldots, \mathbf{y}_k$ if

$$f(x_1, \ldots, x_n, y_1, \ldots, y_k) = f(x_1, \ldots, x_n) f(y_1, \ldots, y_k) \tag{8-18}$$

By suitable integration as in (8-5) we conclude from (8-18) that any subgroup of G_x is independent of any subgroup of G_y. In particular, the RVs $\mathbf{x}_i$ and $\mathbf{y}_j$ are independent for any i and j.

Suppose that $\mathscr{S}$ is a combined experiment $\mathscr{S}_1 \times \mathscr{S}_2$, the RVs $\mathbf{x}_i$ depend only on the outcomes of $\mathscr{S}_1$, and the RVs $\mathbf{y}_j$ depend only on the outcomes of

$\mathscr{S}_2$. If the experiments $\mathscr{S}_1$ and $\mathscr{S}_2$ are independent, then the groups G_x and G_y are independent.

We note finally that if the RVs $\mathbf{z}_m$ depend only on the RVs $\mathbf{x}_i$ of G_x and the RVs $\mathbf{w}_r$ depend only on the RVs $\mathbf{y}_j$ of G_y, then the groups G_z and G_w are independent.

Complex random variables The statistics of the RVs

$$\mathbf{z}_1 = \mathbf{x}_1 + j\mathbf{y}_1, \ldots, \mathbf{z}_n = \mathbf{x}_n + j\mathbf{y}_n$$

are determined in terms of the joint density $f(x_1, \ldots, x_n, y_1, \ldots, y_n)$ of the $2n$ RVs $\mathbf{x}_i$ and $\mathbf{y}_j$. We say that the complex RVs $\mathbf{z}_i$ are independent if

$$f(x_1, \ldots, x_n, y_1, \ldots, y_n) = f(x_1, y_1) \cdots f(x_n, y_n) \tag{8-19}$$

Mean and Covariance

Extending (7-2) to n RVs, we conclude that the mean of $g(\mathbf{x}_1, \ldots, \mathbf{x}_n)$ equals

$$\int_{-\infty}^{\infty} \cdots \int_{-\infty}^{\infty} g(x_1, \ldots, x_n) f(x_1, \ldots, x_n)\, dx_1 \cdots dx_n \tag{8-20}$$

If the RVs $\mathbf{z}_i = \mathbf{x}_i + j\mathbf{y}_i$ are complex, then the mean of $g(\mathbf{z}_1, \ldots, \mathbf{z}_n)$ equals

$$\int_{-\infty}^{\infty} \cdots \int_{-\infty}^{\infty} g(z_1, \ldots, z_n) f(x_1, \ldots, x_n, y_1, \ldots, y_n)\, dx_1 \cdots dy_n$$

From the above it follows that (linearity)

$$E\{a_1 g_1(\mathbf{X}) + \cdots + a_m g_m(\mathbf{X})\} = a_1 E\{g_1(\mathbf{X})\} + \cdots + a_m E\{g_m(\mathbf{X})\}$$

for any random vector $\mathbf{X}$ real or complex.

CORRELATION AND COVARIANCE MATRICES. The covariance C_{ij} of two real RVs $\mathbf{x}_i$ and $\mathbf{x}_j$ is defined as in (7-6). For complex RVs

$$C_{ij} = E\{(\mathbf{x}_i - \eta_i)(\mathbf{x}_j^* - \eta_j^*)\} = E\{\mathbf{x}_i \mathbf{x}_j^*\} - E\{\mathbf{x}_i\}E\{\mathbf{x}_j^*\}$$

by definition. The variance of $\mathbf{x}_i$ is given by

$$\sigma_i^2 = C_{ii} = E\{|\mathbf{x}_i - \eta_i|^2\} = E\{|\mathbf{x}_i|^2\} - |E\{\mathbf{x}_i\}|^2$$

The RVs $\mathbf{x}_i$ are called (mutually) *uncorrelated* if $C_{ij} = 0$ for every $i \neq j$. In this case, if

$$\mathbf{x} = \mathbf{x}_1 + \cdots + \mathbf{x}_n \qquad \text{then} \quad \sigma_x^2 = \sigma_1^2 + \cdots + \sigma_n^2 \tag{8-21}$$

Example 8-5. The RVs

$$\bar{\mathbf{x}} = \frac{1}{n} \sum_{i=1}^{n} \mathbf{x}_i \qquad \bar{\mathbf{v}} = \frac{1}{n-1} \sum_{i=1}^{n} (\mathbf{x}_i - \bar{\mathbf{x}})^2$$

are by definition the *sample mean* and the *sample variance* respectively of $\mathbf{x}_i$. We shall show that, if the RVs $\mathbf{x}_i$ are uncorrelated with the same mean $E\{\mathbf{x}_i\} = \eta$ and variance $\sigma_i^2 = \sigma^2$, then

$$E\{\bar{\mathbf{x}}\} = \eta \qquad \sigma_{\bar{x}}^2 = \sigma^2/n \tag{8-22}$$

and

$$E\{\bar{\mathbf{v}}\} = \sigma^2 \tag{8-23}$$

Proof. The first equation in (8-22) follows from the linearity of expected values and the second from (8-21):

$$E\{\bar{\mathbf{x}}\} = \frac{1}{n}\sum_{i=1}^{n} E\{\mathbf{x}_i\} = \eta \qquad \sigma_{\bar{x}}^2 = \frac{1}{n^2}\sum_{i=1}^{n} \sigma_i^2 = \frac{\sigma^2}{n}$$

To prove (8-23), we observe that

$$E\{(\mathbf{x}_i - \eta)(\bar{\mathbf{x}} - \eta)\} = \frac{1}{n}E\{(\mathbf{x}_i - \eta)[(\mathbf{x}_1 - \eta) + \cdots + (\mathbf{x}_n - \eta)]\}$$

$$= \frac{1}{n}E\{(\mathbf{x}_i - \eta)(\mathbf{x}_i - \eta)\} = \frac{\sigma^2}{n}$$

because the RVs $\mathbf{x}_i$ and $\mathbf{x}_j$ are uncorrelated by assumption. Hence

$$E\{(\mathbf{x}_i - \bar{\mathbf{x}})^2\} = E\{[(\mathbf{x}_i - \eta) - (\bar{\mathbf{x}} - \eta)]^2\} = \sigma^2 + \frac{\sigma^2}{n} - 2\frac{\sigma^2}{n} = \frac{n-1}{n}\sigma^2$$

This yields

$$E\{\bar{\mathbf{v}}\} = \frac{1}{n-1}\sum_{i=1}^{n} E\{(\mathbf{x}_i - \bar{\mathbf{x}})^2\} = \frac{n}{n-1}\,\frac{n-1}{n}\sigma^2$$

and (8-23) results.

Note that if the RVs $\mathbf{x}_i$ are i.i.d. with $E\{|\mathbf{x}_i - \eta|^4\} = \mu_4$, then (see Prob. 8-21)

$$\sigma_{\bar{v}}^2 = \frac{1}{n}\left(\mu_4 - \frac{n-3}{n-1}\sigma^4\right)$$

If the RVs $\mathbf{x}_1, \ldots, \mathbf{x}_n$ are independent, they are also uncorrelated. This follows as in (7-14) for real RVs. For complex RVs the proof is similar: If the RVs $\mathbf{z}_1 = \mathbf{x}_1 + j\mathbf{y}_1$ and $\mathbf{z}_2 = \mathbf{x}_2 + j\mathbf{y}_2$ are independent, then $f(x_1, x_2, y_1, y_2) = f(x_1, y_1)f(x_2, y_2)$. Hence

$$\int_{-\infty}^{\infty} \cdots \int_{-\infty}^{\infty} z_1 z_2^* f(x_1, x_2, y_1, y_2)\, dx_1\, dy_1\, dx_2\, dy_2$$

$$= \int_{-\infty}^{\infty}\int_{-\infty}^{\infty} z_1 f(x_1, y_1)\, dx_1\, dy_1 \int_{-\infty}^{\infty}\int_{-\infty}^{\infty} z_2^* f(x_2, y_2)\, dx_2\, dy_2$$

This yields $E\{\mathbf{z}_1\mathbf{z}_2^*\} = E\{\mathbf{z}_1\}E\{\mathbf{z}_2^*\}$ therefore, $\mathbf{z}_1$ and $\mathbf{z}_2$ are uncorrelated.

Note, finally, that if the RVs $\mathbf{x}_i$ are independent, then

$$E\{g_1(\mathbf{x}_1)\cdots g_n(\mathbf{x}_n)\} = E\{g_1(\mathbf{x}_1)\}\cdots E\{g_n(\mathbf{x}_n)\} \tag{8-24}$$

Similarly, if the groups $\mathbf{x}_1,\ldots,\mathbf{x}_n$ and $\mathbf{y}_1,\ldots,\mathbf{y}_k$ are independent, then

$$E\{g(\mathbf{x}_1,\ldots,\mathbf{x}_n)h(\mathbf{y}_1,\ldots,\mathbf{y}_k)\} = E\{g(\mathbf{x}_1,\ldots,\mathbf{x}_n)\}E\{h(\mathbf{y}_1,\ldots,\mathbf{y}_k)\}$$

The correlation matrix. We introduce the matrices

$$R_n = \begin{bmatrix} R_{11} & \cdots & R_{1n} \\ \cdots & \cdots & \cdots \\ R_{n1} & \cdots & R_{nn} \end{bmatrix} \qquad C_n = \begin{bmatrix} C_{11} & \cdots & C_{1n} \\ \cdots & \cdots & \cdots \\ C_{n1} & \cdots & C_{nn} \end{bmatrix}$$

where

$$R_{ij} = E\{\mathbf{x}_i\mathbf{x}_j^*\} = R_{ji}^* \qquad C_{ij} = R_{ij} - \eta_i\eta_j^* = C_{ji}^*$$

The first is the *correlation matrix* of the random vector $\mathbf{X} = [\mathbf{x}_1,\ldots,\mathbf{x}_n]$ and the second its *covariance matrix*. Clearly,

$$R_n = E\{\mathbf{X}^t\mathbf{X}^*\}$$

where $\mathbf{X}^t$ is the transpose of $\mathbf{X}$ (column vector). We shall discuss the properties of the matrix R_n and its determinant Δ_n. The properties of C_n are similar because C_n is the correlation matrix of the "centered" RVs $\mathbf{x}_i - \eta_i$.

THEOREM. The matrix R_n is *nonnegative definite*. This means that

$$Q = \sum_{i,j} a_i a_j^* R_{ij} = AR_nA^+ \ge 0 \tag{8-25}$$

where A^+ is the conjugate transpose of the vector $A = [a_1,\ldots,a_n]$.

Proof. It follows readily from the linearity of expected values

$$E\{|a_1\mathbf{x}_1 + \cdots + a_n\mathbf{x}_n|^2\} = \sum_{i,j} a_i a_j^* E\{\mathbf{x}_i\mathbf{x}_j^*\} \tag{8-26}$$

If (8-25) is strictly positive, that is, if $Q > 0$ for any $A \ne 0$, then R_n is called *positive definite*.† The difference between $Q \ge 0$ and $Q > 0$ is related to the notion of linear dependence.

DEFINITION. The RVs $\mathbf{x}_i$ are called *linearly independent* if

$$E\{|a_1\mathbf{x}_1 + \cdots + a_n\mathbf{x}_n|^2\} > 0 \tag{8-27}$$

for any $A \ne 0$. In this case [see (8-26)], their correlation matrix R_n is positive definite.

†We shall use the abbreviation p.d. to indicate that R_n satisfies (8-25). The distinction between $Q \ge 0$ and $Q > 0$ will be understood from the context.

The RVs $\mathbf{x}_i$ are called *linearly dependent* if

$$a_1\mathbf{x}_1 + \cdots + a_n\mathbf{x}_n = 0 \tag{8-28}$$

for some $A \neq 0$. In this case, the corresponding Q equals 0 and the matrix R_n is singular [see also (8-29)].

From the definition it follows that, if the RVs $\mathbf{x}_i$ are linearly independent, then any subset is also linearly independent.

The correlation determinant. The determinant Δ_n is real because $R_{ij} = R_{ji}^*$. We shall show that it is also nonnegative

$$\Delta_n \geq 0 \tag{8-29}$$

with equality iff the RVs $\mathbf{x}_i$ are linearly dependent. The familiar inequality $\Delta_2 = R_{11}R_{22} - R_{12}^2 \geq 0$ is a special case [see (7-12)].

Suppose, first, that the RVs $\mathbf{x}_i$ are linearly independent. We maintain that, in this case, the determinant Δ_n and all its principal minors are positive

$$\Delta_k > 0 \qquad k \leq n \tag{8-30}$$

Proof. The above is true for $n = 1$ because $\Delta_1 = R_{11} > 0$. Since the RVs of any subset of the set $\{\mathbf{x}_i\}$ are linearly independent, we can assume that (8-30) is true for $k \leq n - 1$ and we shall show that $\Delta_n > 0$. For this purpose, we form the system

$$\begin{aligned} R_{11}a_1 + \cdots + R_{1n}a_n &= 1 \\ R_{21}a_1 + \cdots + R_{2n}a_n &= 0 \\ \cdots\cdots\cdots\cdots\cdots\cdots& \\ R_{n1}a_1 + \cdots + R_{nn}a_n &= 0 \end{aligned} \tag{8-31}$$

Solving for a_1, we obtain $a_1 = \Delta_{n-1}/\Delta_n$ where Δ_{n-1} is the correlation determinant of the RVs $\mathbf{x}_2, \ldots, \mathbf{x}_n$. Thus a_1 is a real number. Multiplying the jth equation by a_j^* and adding, we obtain

$$Q = \sum_{i,j} a_i a_j^* R_{ij} = a_1 = \frac{\Delta_{n-1}}{\Delta_n} \tag{8-32}$$

In the above, $Q > 0$ because the RVs $\mathbf{x}_i$ are linearly independent and the left side of (8-27) equals Q. Furthermore, $\Delta_{n-1} > 0$ by the induction hypothesis; hence $\Delta_n > 0$.

We shall now show that, if the RVs $\mathbf{x}_i$ are linearly dependent, then

$$\Delta_n = 0 \tag{8-33}$$

Proof. In this case, there exists a vector $A \neq 0$ such that $a_1\mathbf{x}_1 + \cdots + a_n\mathbf{x}_n = 0$. Multiplying by $\mathbf{x}_i^*$ and taking expected values, we obtain

$$a_1R_{i1} + \cdots + a_nR_{in} = 0 \qquad i = 1, \ldots, n$$

This is a homogeneous system satisfied by the nonzero vector A; hence $\Delta_n = 0$.

Note, finally, that [see (15-161)]

$$\Delta_n \le R_{11} R_{22} \cdots R_{nn} \tag{8-34}$$

with equality iff the RVs $\mathbf{x}_i$ are (mutually) *orthogonal*, that is, if the matrix R_n is diagonal.

8-2 CONDITIONAL DENSITIES, CHARACTERISTIC FUNCTIONS, AND NORMALITY

Conditional densities can be defined as in Sec. 7-2. We shall discuss various extensions of the equation $f(y|x) = f(x, y)/f(x)$. Reasoning as in (7-41), we conclude that the conditional density of the RVs $\mathbf{x}_n, \dots, \mathbf{x}_{k+1}$ assuming $\mathbf{x}_k, \dots, \mathbf{x}_1$ is given by

$$f(x_n, \dots, x_{k+1} | x_k, \dots, x_1) = \frac{f(x_1, \dots, x_k, \dots, x_n)}{f(x_1, \dots, x_k)} \tag{8-35}$$

The corresponding distribution function is obtained by integration:

$$F(x_n, \dots, x_{k+1} | x_k, \dots, x_1)$$

$$= \int_{-\infty}^{x_n} \cdots \int_{-\infty}^{x_{k+1}} f(\alpha_n, \dots, \alpha_{k+1} | x_k, \dots, x_1)\, d\alpha_{k+1} \cdots d\alpha_n \tag{8-36}$$

For example,

$$f(x_1 | x_2, x_3) = \frac{f(x_1, x_2, x_3)}{f(x_2, x_3)} = \frac{dF(x_1 | x_2, x_3)}{dx_1}$$

Chain rule From (8-35) it follows that

$$f(x_1, \dots, x_n) = f(x_n | x_{n-1}, \dots, x_1) \cdots f(x_2 | x_1) f(x_1) \tag{8-37}$$

Example 8-6. We have shown that [see (5-18)] if $\mathbf{x}$ is an RV with distribution $F(x)$, then the RV $\mathbf{y} = F(\mathbf{x})$ is uniform in the interval $(0, 1)$. The following is a generalization.

Given n arbitrary RVs $\mathbf{x}_i$, we form the RVs

$$\mathbf{y}_1 = F(\mathbf{x}_1) \qquad \mathbf{y}_2 = F(\mathbf{x}_2 | \mathbf{x}_1), \dots, \mathbf{y}_n = F(\mathbf{x}_n | \mathbf{x}_{n-1}, \dots, \mathbf{x}_1) \tag{8-38}$$

We shall show that these RVs are independent and each is uniform in the interval $(0, 1)$.

Proof. The RVs $\mathbf{y}_i$ are functions of the RVs $\mathbf{x}_i$ obtained with the transformation (8-38). For $0 \le y_i \le 1$, the system

$$y_1 = F(x_1) \qquad y_2 = F(x_2 | x_1), \dots, y_n = F(x_n | x_{n-1}, \dots, x_1)$$

has a unique solution $x_1, \ldots, x_n$ and its jacobian equals

$$J = \begin{vmatrix} \dfrac{\partial y_1}{\partial x_1} & 0 & 0 & & 0 \\ \dfrac{\partial y_2}{\partial x_1} & \dfrac{\partial y_2}{\partial x_2} & 0 & \cdots & 0 \\ \cdots & \cdots & \cdots & \cdots & \cdots \\ \dfrac{\partial y_n}{\partial x_1} & \cdots & \cdots & \cdots & \dfrac{\partial y_n}{\partial x_n} \end{vmatrix}$$

The above determinant is triangular; hence it equals the product of its diagonal elements

$$\frac{\partial y_k}{\partial x_k} = f(x_k | x_{k-1}, \ldots, x_1)$$

Inserting into (8-8) and using (8-37), we obtain

$$f(y_1, \ldots, y_n) = \frac{f(x_1, \ldots, x_n)}{f(x_1) f(x_2 | x_1) \cdots f(x_n | x_{n-1}, \ldots, x_1)} = 1$$

in the n-dimensional cube $0 \le y_i \le 1$, and 0 otherwise.

From (8-5) and (8-35) it follows that

$$f(x_1 | x_3) = \int_{-\infty}^{\infty} f(x_1, x_2 | x_3)\, dx_2$$

$$f(x_1 | x_4) = \int_{-\infty}^{\infty} \int_{-\infty}^{\infty} f(x_1 | x_2, x_3, x_4) f(x_2, x_3 | x_4)\, dx_2\, dx_3$$

Generalizing, we obtain the following rule for removing variables on the left or on the right of the conditional line: To remove any number of variables on the left of the conditional line, we integrate with respect to them. To remove any number of variables to the right of the line, we multiply by their conditional density with respect to the remaining variables on the right, and we integrate the product. The following special case is used extensively (Chapman–Kolmogoroff):

$$f(x_1 | x_3) = \int_{-\infty}^{\infty} f(x_1 | x_2, x_3) f(x_2 | x_3)\, dx_2 \tag{8-39}$$

Discrete type The above rule holds also for discrete type RVs provided that all densities are replaced by probabilities and all integrals by sums. We mention as an example the discrete form of (8-39): If the RVs $\mathbf{x}_1, \mathbf{x}_2, \mathbf{x}_3$ take the values a_i, b_k, c_r respectively, then

$$P\{\mathbf{x}_1 = a_i | \mathbf{x}_3 = c_r\} = \sum_k P\{\mathbf{x}_1 = a_i | b_k, c_r\} P\{\mathbf{x}_2 = b_k | c_r\} \tag{8-40}$$

CONDITIONAL EXPECTED VALUES. The conditional mean of the RVs $g(\mathbf{x}_1, \ldots, \mathbf{x}_n)$ assuming $\mathscr{M}$ is given by the integral in (8-20) provided that the density $f(x_1, \ldots, x_n)$ is replaced by the conditional density $f(x_1, \ldots, x_n | \mathscr{M})$. Note, in particular, that [see also (7-57)]

$$E\{\mathbf{x}_1 | x_2, \ldots, x_n\} = \int_{-\infty}^{\infty} x_1 f(x_1 | x_2, \ldots, x_n)\, dx_1 \tag{8-41}$$

The above is a function of $x_2, \ldots, x_n$; it defines, therefore, the RV $E\{\mathbf{x}_1 | \mathbf{x}_2, \ldots, \mathbf{x}_n\}$. Multiplying (8-41) by $f(x_2, \ldots, x_n)$ and integrating, we conclude that

$$E\{E\{\mathbf{x}_1 | \mathbf{x}_2, \ldots, \mathbf{x}_n\}\} = E\{\mathbf{x}_1\} \tag{8-42}$$

Reasoning similarly, we obtain

$$\begin{aligned} E\{\mathbf{x}_1 | x_2, x_3\} &= E\{E\{\mathbf{x}_1 | x_2, x_3, \mathbf{x}_4\}\} \\ &= \int_{-\infty}^{\infty} E\{\mathbf{x}_1 | x_2, x_3, x_4\} f(x_4 | x_2, x_3)\, dx_4 \end{aligned} \tag{8-43}$$

This leads to the following generalization: To remove any number of variables on the right of the conditional expected value line, we multiply by their conditional density with respect to the remaining variables on the right and we integrate the product. For example,

$$E\{\mathbf{x}_1 | x_3\} = \int_{-\infty}^{\infty} E\{\mathbf{x}_1 | x_2, x_3\} f(x_2 | x_3)\, dx_2 \tag{8-44}$$

and for the discrete case [see (8-40)]

$$E\{\mathbf{x}_1 | c_r\} = \sum_k E\{\mathbf{x}_1 | b_k, c_r\} P\{\mathbf{x}_2 = b_k | c_r\} \tag{8-45}$$

Example 8-7. Given a discrete type RV $\mathbf{n}$ taking the values $1, 2, \ldots$ and a sequence of RVs $\mathbf{x}_k$ independent of $\mathbf{n}$, we form the sum

$$\mathbf{s} = \sum_{k=1}^{\mathbf{n}} \mathbf{x}_k \tag{8-46}$$

This sum is an RV specified as follows: For a specific ζ, $\mathbf{n}(\zeta)$ is an integer and $\mathbf{s}(\zeta)$ equals the sum of the numbers $\mathbf{x}_k(\zeta)$ for k from 1 to $\mathbf{n}(\zeta)$. We maintain that if the RVs $\mathbf{x}_k$ have the same mean, then

$$E\{\mathbf{s}\} = \eta E\{\mathbf{n}\} \qquad \text{where} \quad E\{\mathbf{x}_k\} = \eta \tag{8-47}$$

Clearly, $E\{\mathbf{x}_k | \mathbf{n} = n\} = E\{\mathbf{x}_k\}$ because $\mathbf{x}_k$ is independent of $\mathbf{n}$. Hence

$$E\{\mathbf{s} | \mathbf{n} = n\} = E\left\{ \sum_{k=1}^{n} \mathbf{x}_k \middle| \mathbf{n} = n \right\} = \sum_{k=1}^{n} E\{\mathbf{x}_k\} = \eta n$$

From this and (7-65) it follows that

$$E\{\mathbf{s}\} = E\{E\{\mathbf{s} | \mathbf{n}\}\} = E\{\eta \mathbf{n}\}$$

and (8-47) results.

We show next that if the RVs $\mathbf{x}_k$ are uncorrelated with the same variance σ^2, then

$$E\{\mathbf{s}^2\} = \eta^2 E\{\mathbf{n}^2\} + \sigma^2 E\{\mathbf{n}\} \tag{8-48}$$

Reasoning as above, we have

$$E\{\mathbf{s}^2 | \mathbf{n} = n\} = \sum_{i=1}^{n} \sum_{k=1}^{n} E\{\mathbf{x}_i \mathbf{x}_k\} \tag{8-49}$$

where

$$E\{\mathbf{x}_i \mathbf{x}_k\} = \begin{cases} \sigma^2 + \eta^2 & i = k \\ \eta^2 & i \neq k \end{cases}$$

The double sum in (8-49) contains n terms with $i = k$ and $n^2 - n$ terms with $i \neq k$; hence it equals

$$(\sigma^2 + \eta^2)n + \eta^2(n^2 - n) = \eta^2 n^2 + \sigma^2 n$$

This yields (8-48) because

$$E\{\mathbf{s}^2\} = E\{E\{\mathbf{s}^2 | \mathbf{n}\}\} = E\{\eta^2 \mathbf{n}^2 + \sigma^2 \mathbf{n}\}$$

Special Case. The number $\mathbf{n}$ of particles emitted from a substance in t seconds is a Poisson RV with parameter λt. The energy $\mathbf{x}_k$ of the kth particle has a Maxwell distribution with mean $3kT/2$ and variance $3k^2T^2/2$ (see Prob. 8-5). The sum $\mathbf{s}$ in (8-46) is the total emitted energy in t seconds. As we know $E\{\mathbf{n}\} = \lambda t$, $E\{\mathbf{n}^2\} = \lambda^2 t^2 + \lambda t$ [see (5-37)]. Inserting into (8-47) and (8-48), we obtain

$$E\{\mathbf{s}\} = \frac{3kT\lambda t}{2} \qquad \sigma_s^2 = \frac{15k^2T^2\lambda t}{4}$$

Characteristic Functions and Normality

The characteristic function of a random vector is by definition the function

$$\Phi(\Omega) = E\{e^{j\Omega \mathbf{X}^t}\} = E\{e^{j(\omega_1 \mathbf{x}_1 + \cdots + \omega_n \mathbf{x}_n)}\} = \mathbf{\Phi}(j\Omega) \tag{8-50}$$

where

$$\mathbf{X} = [\mathbf{x}_1, \ldots, \mathbf{x}_n] \qquad \Omega = [\omega_1, \ldots, \omega_n]$$

As an application, we shall show that if the RVs $\mathbf{x}_i$ are independent with respective densities $f_i(x_i)$, then the density $f_z(z)$ of their sum $\mathbf{z} = \mathbf{x}_1 + \cdots + \mathbf{x}_n$ equals the convolution of their densities

$$f_z(z) = f_1(z) * \cdots * f_n(z) \tag{8-51}$$

Proof. Since the RVs $\mathbf{x}_i$ are independent and $e^{j\omega_i \mathbf{x}_i}$ depends only on $\mathbf{x}_i$, we conclude that from (8-24) that

$$E\{e^{j(\omega_1 \mathbf{x}_1 + \cdots + \omega_n \mathbf{x}_n)}\} = E\{e^{j\omega_1 \mathbf{x}_1}\} \cdots E\{e^{j\omega_n \mathbf{x}_n}\}$$

Hence

$$\Phi_z(\omega) = E\{e^{j\omega(\mathbf{x}_1 + \cdots + \mathbf{x}_n)}\} = \Phi_1(\omega) \cdots \Phi_n(\omega) \tag{8-52}$$

where $\Phi_i(\omega)$ is the characteristic function of $\mathbf{x}_i$. Applying the convolution theorem for Fourier transforms, we obtain (8-51).

Example 8-8. (*a*) (Bernoulli trials) Using (8-52) we shall rederive the fundamental equation (3-13). We define the RVs $\mathbf{x}_i$ as follows: $\mathbf{x}_i = 1$ if heads shows at the ith trial and $\mathbf{x}_i = 0$ otherwise. Thus

$$P\{\mathbf{x}_i = 1\} = P\{h\} = p \qquad P\{\mathbf{x}_i = 0\} = P\{t\} = q \qquad \Phi_i(\omega) = pe^{j\omega} + q \tag{8-53}$$

The RV $\mathbf{z} = \mathbf{x}_1 + \cdots + \mathbf{x}_n$ takes the values $0, 1, \dots, n$ and $\{\mathbf{z} = k\}$ is the event $\{k$ heads in n tossings$\}$. Furthermore,

$$\Phi_z(\omega) = E\{e^{j\omega\mathbf{z}}\} = \sum_{k=0}^{n} P\{\mathbf{z} = k\} e^{jk\omega} \tag{8-54}$$

The RVs $\mathbf{x}_i$ are independent because $\mathbf{x}_i$ depends only on the outcomes of the ith trial and the trials are independent. Hence [see (8-52) and (8-53)]

$$\Phi_z(\omega) = \left(pe^{j\omega} + q\right)^n = \sum_{k=0}^{n} \binom{n}{k} p^k e^{jk\omega} q^{n-k}$$

Comparing with (8-54), we conclude that

$$P\{\mathbf{z} = k\} = P\{k \text{ heads}\} = \binom{n}{k} p^k q^{n-k} \tag{8-55}$$

(*b*) (Poisson theorem) We shall show that if $p \ll 1$, then

$$P\{\mathbf{z} = k\} \simeq \frac{e^{-np}(np)^k}{k!}$$

as in (3-41). In fact, we shall establish a more general result. Suppose that the RVs $\mathbf{x}_i$ are independent and each takes the value 1 and 0 with respective probabilities p_i and $q_i = 1 - p_i$. If $p_i \ll 1$, then

$$e^{p_i(e^{j\omega} - 1)} \simeq 1 + p_i(e^{j\omega} - 1) = p_i e^{j\omega} + q_i = \Phi_i(\omega)$$

With $\mathbf{z} = \mathbf{x}_1 + \cdots + \mathbf{x}_n$, it follows from (8-52) that

$$\Phi_z(\omega) \simeq e^{p_1(e^{j\omega} - 1)} \cdots e^{p_n(e^{j\omega} - 1)} = e^{a(e^{j\omega} - 1)}$$

where $a = p_1 + \cdots + p_n$. This leads to the conclusion that [see (5-79)] the RV $\mathbf{z}$ is approximately Poisson distributed with parameter a. It can be shown that the result is exact in the limit if

$$p_i \to 0 \qquad \text{and} \qquad p_1 + \cdots + p_n \to a \qquad \text{as} \quad n \to \infty$$

NORMAL VECTORS. Joint normality of n RVs $\mathbf{x}_i$ can be defined as in (6-15): Their joint density is an exponential whose exponent is a negative quadratic. We give next an equivalent definition that expresses the normality of n RVs in terms of the normality of a single RV.

DEFINITION. The RVs $\mathbf{x}_i$ are jointly normal iff the sum

$$a_1\mathbf{x}_1 + \cdots + a_n\mathbf{x}_n = A\mathbf{X}^t \tag{8-56}$$

is a normal RV for any A.

We shall show that this definition leads to the following conclusions: If the RVs $\mathbf{x}_i$ have zero mean and covariance matrix C, then their joint characteristic function equals

$$\Phi(\Omega) = \exp\{-\tfrac{1}{2}\Omega C \Omega^t\} \tag{8-57}$$

Furthermore, their joint density equals

$$f(X) = \frac{1}{\sqrt{(2\pi)^n \Delta}} \exp\{-\tfrac{1}{2} X C^{-1} X^t\} \tag{8-58}$$

where Δ is the determinant of C.

Proof. From the definition of joint normality it follows that the RV

$$\mathbf{w} = \omega_1\mathbf{x}_1 + \cdots + \omega_n\mathbf{x}_n = \Omega\mathbf{X}^t \tag{8-59}$$

is normal. Since $E\{\mathbf{x}_i\} = 0$ by assumption, the above yields [see (8-26)]

$$E\{\mathbf{w}\} = 0 \qquad E\{\mathbf{w}^2\} = \sum_{i,j} \omega_i\omega_j C_{ij} = \sigma_w^2$$

Setting $\eta = 0$ and $\omega = 1$ in (5-65), we obtain

$$E\{e^{j\mathbf{w}}\} = \exp\left[-\frac{\sigma_w^2}{2}\right]$$

This yields

$$E\{e^{j\Omega\mathbf{X}^t}\} = \exp\left\{-\frac{1}{2}\sum_{i,j}\omega_i\omega_j C_{ij}\right\} \tag{8-60}$$

as in (8-57). The proof of (8-58) follows from (8-57) and the Fourier inversion theorem.

Note, finally, that if the RVs $\mathbf{x}_i$ are jointly normal and uncorrelated, they are independent. Indeed, in this case, their covariance matrix is diagonal and its diagonal elements equal σ_i^2. Hence C^{-1} is also diagonal with diagonal elements $1/\sigma_i^2$. Inserting into (8-58), we obtain

$$f(x_1, \ldots, x_n) = \frac{1}{\sigma_1 \cdots \sigma_n\sqrt{(2\pi)^n}} \exp\left\{-\frac{1}{2}\left(\frac{x_1^2}{\sigma_1^2} + \cdots + \frac{x_n^2}{\sigma_n^2}\right)\right\}$$

Example 8-9. Using characteristic functions, we shall show that if the RVs $\mathbf{x}_i$ are jointly normal with zero mean, and $E\{\mathbf{x}_i\mathbf{x}_j\} = C_{ij}$, then

$$E\{\mathbf{x}_1\mathbf{x}_2\mathbf{x}_3\mathbf{x}_4\} = C_{12}C_{34} + C_{13}C_{24} + C_{14}C_{23} \tag{8-61}$$

Proof. We expand the exponentials on the left and right side of (8-60) and we show explicitly only the terms containing the factor $\omega_1\omega_2\omega_3\omega_4$:

$$E\{e^{j(\omega_1\mathbf{x}_1 + \cdots + \omega_4\mathbf{x}_4)}\} = \cdots + \frac{1}{4!}E\{(\omega_1\mathbf{x}_1 + \cdots + \omega_4\mathbf{x}_4)^4\} + \cdots$$

$$= \cdots + \frac{24}{4!}E\{\mathbf{x}_1\mathbf{x}_2\mathbf{x}_3\mathbf{x}_4\}\omega_1\omega_2\omega_3\omega_4$$

$$\exp\left\{-\frac{1}{2}\sum_{i,j}\omega_i\omega_j C_{ij}\right\} = +\frac{1}{2}\left(\frac{1}{2}\sum_{i,j}\omega_i\omega_i C_{ij}\right)^2 + \cdots$$

$$= \cdots + \frac{8}{8}(C_{12}C_{34} + C_{13}C_{24} + C_{14}C_{23})\omega_1\omega_2\omega_3\omega_4$$

Equating coefficients, we obtain (8-61).

Complex normal vectors. A complex normal random vector is a vector $\mathbf{Z} = \mathbf{X} + j\mathbf{Y} = [\mathbf{z}_1, \ldots, \mathbf{z}_n]$ the components of which are n jointly normal RVs $\mathbf{z}_i = \mathbf{x}_i + j\mathbf{y}_i$. We shall assume that $E\{\mathbf{z}_i\} = 0$. The statistical properties of the vector $\mathbf{Z}$ are specified in terms of the joint density

$$f_Z(Z) \equiv f(x_1, \ldots, x_n, y_1, \ldots, y_n)$$

of the $2n$ RVs $\mathbf{x}_i$ and $\mathbf{y}_j$. This function is an exponential as in (8-58) determined in terms of the $2n$ by $2n$ matrix

$$D = \begin{bmatrix} C_{XX} & C_{XY} \\ C_{YX} & C_{YY} \end{bmatrix}$$

consisting of the $2n^2 + n$ real parameters $E\{\mathbf{x}_i\mathbf{x}_j\}$, $E\{\mathbf{y}_i\mathbf{y}_j\}$, and $E\{\mathbf{x}_i\mathbf{y}_j\}$. The corresponding characteristic function

$$\Phi_Z(\Omega) = E\{\exp(j(u_1\mathbf{x}_1 + \cdots + u_n\mathbf{x}_n + v_1\mathbf{y}_1 + \cdots + v_n\mathbf{y}_n))\}$$

is an exponential as in (8-60):

$$\Phi_Z(\Omega) = \exp\{-\tfrac{1}{2}Q\} \qquad Q = [U \quad V]\begin{bmatrix} C_{XX} & C_{XY} \\ C_{YX} & C_{YY} \end{bmatrix}\begin{bmatrix} U^t \\ V^t \end{bmatrix}$$

where $U = [u_1, \ldots, u_n]$, $V = [v_1, \ldots, v_n]$, and $\Omega = U + jV$.

The covariance matrix of the complex vector $\mathbf{Z}$ is an n by n hermitian matrix

$$C_{ZZ} = E\{\mathbf{Z}^t\mathbf{Z}^*\} = C_{XX} + C_{YY} - j(C_{XY} - C_{YX})$$

with elements $E\{\mathbf{z}_i\mathbf{z}_j^*\}$. Thus, C_{ZZ} is specified in terms of n^2 real parameters. From this it follows that, unlike the real case, the density $f_Z(Z)$ of $\mathbf{Z}$ cannot in general be determined in terms of C_{ZZ} because $f_Z(Z)$ is a normal density consisting of $2n^2 + n$ parameters. Suppose, for example, that $n = 1$. In this case, $\mathbf{Z} = \mathbf{z} = \mathbf{x} + j\mathbf{y}$ is a scalar and $C_{ZZ} = E\{|\mathbf{z}|^2\}$. Thus, C_{ZZ} is specified in terms of the single parameter $\sigma_z^2 = E\{\mathbf{x}^2 + \mathbf{y}^2\}$. However, $f_z(z) = f(x, y)$ is a bivariate normal density consisting of the three parameters σ_x, σ_y, and $E\{\mathbf{xy}\}$. In the following, we present a special class of normal vectors that are statistically determined in terms of their covariance matrix. This class is important in modulation theory (see Sec. 11-3).

Goodman's Theorem.† If the vectors **X** and **Y** are such that

$$C_{XX} = C_{YY} \qquad C_{XY} = -C_{YX}$$

and $\mathbf{Z} = \mathbf{X} + j\mathbf{Y}$, then

$$C_{ZZ} = 2(C_{XX} - jC_{XY})$$

$$f_Z(Z) = \frac{1}{\pi^n |C_{ZZ}|} \exp\{-ZC_{ZZ}^{-1}Z^+\} \tag{8-62a}$$

$$\Phi_Z(\Omega) = \exp\left\{-\frac{1}{4}\Omega C_{ZZ}\Omega^+\right\} \tag{8-62b}$$

Proof. It suffices to prove (8-62b); the proof of (8-62a) follows from (8-62b) and the Fourier inversion formula. Under the stated assumptions,

$$Q = [U \quad V]\begin{bmatrix} C_{XX} & C_{XY} \\ -C_{XY} & C_{XX} \end{bmatrix}\begin{bmatrix} U^t \\ V^t \end{bmatrix}$$

$$= UC_{XY}U^t + VC_{XY}U^t - UC_{XY}V^t + VC_{XX}V^t$$

Furthermore $C_{XX}^t = C_{XX}$ and $C_{XY}^t = -C_{XY}$. This leads to the conclusion that

$$VC_{XX}U^t = UC_{XX}V^t \qquad UC_{XY}U^t = VC_{XY}V^t = 0$$

Hence

$$\tfrac{1}{2}\Omega C_{ZZ}\Omega^+ = (U + jV)(C_{XX} - jC_{XY})(U^t - jV^t) = Q$$

and (8-62b) results.

Normal quadratic forms. Given n independent $N(0,1)$ RVs $\mathbf{z}_i$, we form the sum of their squares

$$\mathbf{x} = \mathbf{z}_1^2 + \cdots + \mathbf{z}_n^2$$

Using characteristic functions, we shall show that the RV $\mathbf{x}$ so formed has a chi-square distribution with n degrees of freedom:

$$f_x(x) = \gamma x^{n/2-1} e^{-x/2} U(x)$$

†N. R. Goodman, "Statistical Analysis Based on Certain Multivariate Complex Distribution," *Annals of Math. Statistics*, 1963, pp. 152–177.

Proof. The RVs $\mathbf{z}_i^2$ have a $\chi^2(1)$ distribution (see page 96); hence their characteristic functions are obtained from (5-71) with $m = 1$. This yields

$$\Phi_i(s) = E\{e^{s\mathbf{z}_i^2}\} = \frac{1}{\sqrt{1-2s}}$$

From (8-52) and the independence of the RVs $\mathbf{z}_i^2$, it follows therefore that

$$\Phi_x(s) = \Phi_1(s) \cdots \Phi_n(s) = \frac{1}{\sqrt{(1-2s)^n}}$$

Hence [see (5-71)] the RV $\mathbf{x}$ is $\chi^2(n)$.

Note that

$$\frac{1}{\sqrt{(1-2s)^m}} \times \frac{1}{\sqrt{(1-2s)^n}} = \frac{1}{\sqrt{(1-2s)^{m+n}}}$$

This leads to the conclusion that if the RVs $\mathbf{x}$ and $\mathbf{y}$ are independent, $\mathbf{x}$ is $\chi^2(m)$ and $\mathbf{y}$ is $\chi^2(n)$, then the RV

$$\mathbf{z} = \mathbf{x} + \mathbf{y} \quad \text{is} \quad \chi^2(m+n) \tag{8-63}$$

Conversely, if $\mathbf{z}$ is $\chi^2(m+n)$, $\mathbf{x}$ and $\mathbf{y}$ are independent, and $\mathbf{x}$ is $\chi^2(m)$, then $\mathbf{y}$ is $\chi^2(n)$. The following is an important application.

Sample variance. Given n i.i.d. $N(\eta, \sigma)$ RVs $\mathbf{x}_i$, we form their sample variance

$$\mathbf{s}^2 = \frac{1}{n-1}\sum_{i=1}^{n}(\mathbf{x}_i - \bar{\mathbf{x}})^2 \qquad \bar{\mathbf{x}} = \frac{1}{n}\sum_{i=1}^{n}\mathbf{x}_i \tag{8-64}$$

as in Example 8-4. We shall show that the RV

$$\frac{(n-1)\mathbf{s}^2}{\sigma^2} = \sum_{i=1}^{n}\left(\frac{\mathbf{x}_i - \bar{\mathbf{x}}}{\sigma}\right)^2 \quad \text{is} \quad \chi^2(n-1) \tag{8-65}$$

Proof. We sum the identity

$$(\mathbf{x}_i - \eta)^2 = (\mathbf{x}_i - \bar{\mathbf{x}} + \bar{\mathbf{x}} - \eta)^2 = (\mathbf{x}_i - \bar{\mathbf{x}})^2 + (\bar{\mathbf{x}} - \eta)^2 + 2(\mathbf{x}_i - \bar{\mathbf{x}})(\bar{\mathbf{x}} - \eta)$$

from 1 to n. Since $\Sigma(x_i - \bar{x}) = 0$, this yields

$$\sum_{i=1}^{n}\left(\frac{\mathbf{x}_i - \eta}{\sigma}\right)^2 = \sum_{i=1}^{n}\left(\frac{\mathbf{x}_i - \bar{\mathbf{x}}}{\sigma}\right)^2 + n\left(\frac{\bar{\mathbf{x}} - \eta}{\sigma}\right)^2 \tag{8-66}$$

It can be shown that the RVs $\bar{\mathbf{x}}$ and $\bar{\mathbf{s}}^2$ are independent (see Prob. 8-17). From this it follows that the two terms on the right of (8-66) are independent. Furthermore, the term

$$n\left(\frac{\bar{\mathbf{x}} - \eta}{\sigma}\right)^2 = \left(\frac{\bar{\mathbf{x}} - \eta}{\sigma/\sqrt{n}}\right)^2$$

is $\chi^2(1)$ because the RV $\bar{\mathbf{x}}$ is $N(\eta, \sigma/\sqrt{n})$. Finally, the term on the left side is $\chi^2(n)$ and the proof is complete.

From (8-65) and (5-71) it follows that the mean of the RV $(n-1)\mathbf{s}^2/\sigma^2$ equals $n-1$ and its variance equals $2(n-1)$. This leads to the conclusion that

$$E\{\mathbf{s}^2\} = (n-1)\frac{\sigma^2}{n-1} = \sigma^2 \qquad \operatorname{Var}\mathbf{s}^2 = 2(n-1)\frac{\sigma^4}{(n-1)^2} = \frac{2\sigma^4}{n-1} \tag{8-67}$$

Example 8-10. We shall verify the above for $n = 2$. In this case,

$$\bar{\mathbf{x}} = \frac{\mathbf{x}_1 + \mathbf{x}_2}{2} \qquad \mathbf{s}^2 = (\mathbf{x}_1 - \bar{\mathbf{x}})^2 + (\mathbf{x}_2 - \bar{\mathbf{x}})^2 = \frac{1}{2}(\mathbf{x}_1 - \mathbf{x}_2)^2$$

The RVs $\mathbf{x}_1 + \mathbf{x}_2$ and $\mathbf{x}_1 - \mathbf{x}_2$ are independent because they are jointly normal and $E\{\mathbf{x}_1 - \mathbf{x}_2\} = 0$, $E\{(\mathbf{x}_1 - \mathbf{x}_2)(\mathbf{x}_1 + \mathbf{x}_2)\} = 0$. From this it follows that the RVs $\bar{\mathbf{x}}$ and $\mathbf{s}^2$ are independent. But the RV $(\mathbf{x}_1 - \mathbf{x}_2)/\sigma\sqrt{2} = \mathbf{s}/\sigma$ is $N(0,1)$; hence its square $\mathbf{s}^2/\sigma^2$ is $\chi^2(1)$ in agreement with (8-65).

8-3 MEAN SQUARE ESTIMATION

In Sec. 7-5, we considered the problem of estimating an RV $\mathbf{s}$ by a linear and a nonlinear function of another RV $\mathbf{x}$. Generalizing, we consider now the problem of estimating $\mathbf{s}$ in terms of n RVs $\mathbf{x}_1, \ldots, \mathbf{x}_n$ (data). This topic is developed further in Chap. 14 in the context of infinitely many data and stochastic processes.

LINEAR ESTIMATION. The linear MS estimate of $\mathbf{s}$ in terms of the RVs $\mathbf{x}_i$ is the sum

$$\hat{\mathbf{s}} = a_1\mathbf{x}_1 + \cdots + a_n\mathbf{x}_n \tag{8-68}$$

where $a_1, \ldots, a_n$ are n constants such that the MS value

$$P = E\{(\mathbf{s} - \hat{\mathbf{s}})^2\} = E\{[\mathbf{s} - (a_1\mathbf{x}_1 + \cdots + a_n\mathbf{x}_n)]^2\} \tag{8-69}$$

of the estimation error $\mathbf{s} - \hat{\mathbf{s}}$ is minimum.

Orthogonality principle. P is minimum if the error $\mathbf{s} - \hat{\mathbf{s}}$ is orthogonal to the data $\mathbf{x}_i$:

$$E\{[\mathbf{s} - (a_1\mathbf{x}_1 + \cdots + a_n\mathbf{x}_n)]\mathbf{x}_i\} = 0 \qquad i = 1, \ldots, n \tag{8-70}$$

Proof. P is a function of the constants a_i and it is minimum if

$$\frac{\partial P}{\partial a_i} = E\{-2[\mathbf{s} - (a_1\mathbf{x}_1 + \cdots + a_n\mathbf{x}_n)]\mathbf{x}_i\} = 0$$

and (8-70) results. This important result is known also as the *projection theorem*.

Setting $i = 1, \ldots, n$ in (8-70), we obtain the system

$$\begin{aligned} R_{11}a_1 + R_{21}a_2 + \cdots + R_{n1}a_n &= R_{01} \\ R_{12}a_1 + R_{22}a_2 + \cdots + R_{n2}a_n &= R_{02} \\ \cdots\cdots\cdots\cdots\cdots\cdots\cdots\cdots\cdots \\ R_{1n}a_1 + R_{2n}a_2 + \cdots + R_{nn}a_n &= R_{0n} \end{aligned} \tag{8-71}$$

where $R_{ij} = E\{\mathbf{x}_i\mathbf{x}_j\}$ and $R_{0j} = E\{\mathbf{s}\mathbf{x}_j\}$.

To solve this system, we introduce the row vectors

$$\mathbf{X} = [\mathbf{x}_1, \ldots, \mathbf{x}_n] \qquad A = [a_1, \ldots, a_n] \qquad R_0 = [R_{01}, \ldots, R_{0n}]$$

and the data correlation matrix $R = E\{\mathbf{X}^t\mathbf{X}\}$ where $\mathbf{X}^t$ is the transpose of $\mathbf{X}$. This yields

$$AR = R_0 \qquad A = R_0 R^{-1} \tag{8-72}$$

Inserting the constants a_i so determined into (8-69), we obtain the LMS error. The resulting expression can be simplified. Since $\mathbf{s} - \hat{\mathbf{s}} \perp \mathbf{x}_i$ for every i, we conclude that $\mathbf{s} - \hat{\mathbf{s}} \perp \hat{\mathbf{s}}$; hence

$$P = E\{(\mathbf{s} - \hat{\mathbf{s}})\mathbf{s}\} = E\{\mathbf{s}^2\} - AR_0^t \tag{8-73}$$

Note that if the rank of R is $m < n$, then the data are linearly dependent. In this case, the estimate $\hat{\mathbf{s}}$ can be written as a linear sum involving a subset of m linearly independent components of the data vector $\mathbf{X}$.

Geometric interpretation. In the representation of RVs as vectors in an abstract space, the sum $\hat{\mathbf{s}} = a_1\mathbf{x}_1 + \cdots + a_n\mathbf{x}_n$ is a vector in the subspace S_n of the data $\mathbf{x}_i$ and the error $\boldsymbol{\varepsilon} = \mathbf{s} - \hat{\mathbf{s}}$ is the vector from $\mathbf{s}$ to $\hat{\mathbf{s}}$ as in Fig. 8-2a. The projection theorem states that the length of $\boldsymbol{\varepsilon}$ is minimum if $\boldsymbol{\varepsilon}$ is orthogonal to $\mathbf{x}_i$, that is, if it is perpendicular to the data subspace S_n. The estimate $\hat{\mathbf{s}}$ is thus the "projection" of $\mathbf{s}$ on S_n.

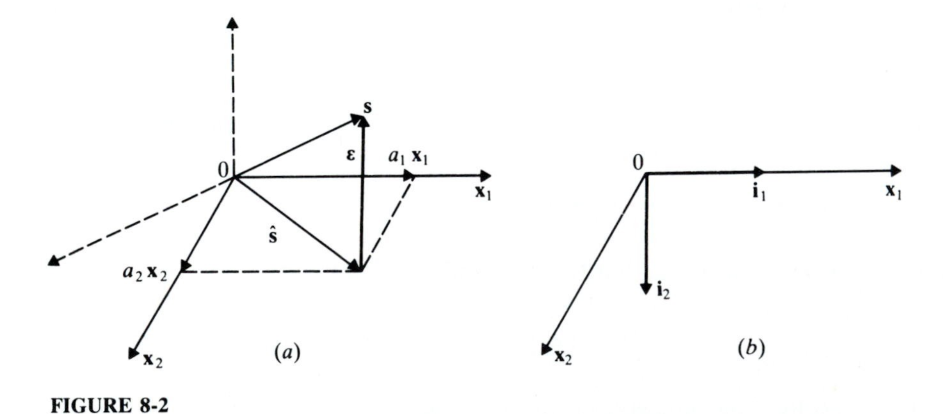

FIGURE 8-2

If $\mathbf{s}$ is a vector in S_n, then $\hat{\mathbf{s}} = \mathbf{s}$ and $P = 0$. In this case, the $n + 1$ RVs $\mathbf{s}, \mathbf{x}_1,, \ldots, \mathbf{x}_n$ are linearly dependent and the determinant Δ_{n+1} of their correlation matrix is 0. If $\mathbf{s}$ is perpendicular to S_n, then $\hat{\mathbf{s}} = 0$ and $P = E\{|\mathbf{s}|^2\}$. This is the case if $\mathbf{s}$ is orthogonal to all the data $\mathbf{x}_i$, that is, if $R_{0j} = 0$ for $j \neq 0$.

Nonhomogeneous estimation. The estimate (8-68) can be improved if a constant is added to the sum. The problem now is to determine $n + 1$ parameters α_k such that if

$$\hat{\mathbf{s}} = \alpha_0 + \alpha_1 \mathbf{x}_1 + \cdots + \alpha_n \mathbf{x}_n \tag{8-74}$$

then the resulting MS error is minimum. This problem can be reduced to the homogeneous case if we replace the term α_0 by the product $\alpha_0 \mathbf{x}_0$ where $\mathbf{x}_0 \equiv 1$. Applying (8-70) to the enlarged data set

$$\mathbf{x}_0, \mathbf{x}_1, \ldots, \mathbf{x}_n \qquad \text{where} \quad E\{\mathbf{x}_0 \mathbf{x}_i\} = \begin{cases} E\{\mathbf{x}_i\} = \eta_i & i \neq 0 \\ 1 & i = 0 \end{cases}$$

we obtain

$$\begin{aligned} \alpha_0 + \eta_1 \alpha_1 \ + \cdots + \eta_n \alpha_n &= \eta_s \\ \eta_1 \alpha_0 + R_{11} \alpha_1 + \cdots + R_{1n} \alpha_n &= R_{01} \\ \cdots\cdots\cdots\cdots\cdots\cdots & \\ \eta_n \alpha_0 + R_{n1} \alpha_1 + \cdots + R_{nn} \alpha_n &= R_{0n} \end{aligned} \tag{8.75}$$

Note that, if $\eta_s = \eta_i = 0$, then (8-75) reduces to (8-71). This yields $\alpha_0 = 0$ and $\alpha_n = a_n$.

Nonlinear estimation. The nonlinear MS estimation problem involves the determination of a function $g(\mathbf{x}_1, \ldots, \mathbf{x}_n) = g(\mathbf{X})$ of the data $\mathbf{x}_i$ such as to minimize the MS error

$$P = E\{[\mathbf{s} - g(\mathbf{X})]^2\} \tag{8-76}$$

We maintain that P is minimum if

$$g(X) = E\{\mathbf{s}|X\} = \int_{-\infty}^{\infty} s f_s(s|X)\, ds \tag{8-77}$$

The function $f_s(s|X)$ is the conditional mean (regression surface) of the RV $\mathbf{s}$ assuming $\mathbf{X} = X$.

Proof. The proof is based on the identity [see (8-42)]

$$P = E\{[\mathbf{s} - g(\mathbf{X})]^2\} = E\{E\{[\mathbf{s} - g(\mathbf{X})]^2 | \mathbf{X}\}\} \tag{8-78}$$

Since all quantities are positive, it follows that P is minimum if the conditional MS error

$$E\{[\mathbf{s} - g(\mathbf{X})]^2 | X\} = \int_{-\infty}^{\infty} [s - g(X)]^2 f_s(s|X)\, ds \tag{8-79}$$

is minimum. In the above integral, $g(X)$ is constant. Hence the integral is minimum if $g(X)$ is given by (8-77) [see also (7-71)].

The general orthogonality principle. From the projection theorem (8-70) it follows that

$$E\{[\mathbf{s} - \hat{\mathbf{s}}](c_1\mathbf{x}_1 + \cdots + c_n\mathbf{x}_n)\} = 0 \tag{8-80}$$

for any $c_1, \ldots, c_n$. This shows that if $\hat{\mathbf{s}}$ is the linear MS estimator of $\mathbf{s}$, the estimation error $\mathbf{s} - \hat{\mathbf{s}}$ is orthogonal to any linear function $\mathbf{y} = c_1\mathbf{x}_1 + \cdots + c_n\mathbf{x}_n$ of the data $\mathbf{x}_i$.

We shall now show that if $g(\mathbf{X})$ is the nonlinear MS estimator of $\mathbf{s}$, the estimation error $\mathbf{s} - g(\mathbf{X})$ is orthogonal to any function $w(\mathbf{X})$, linear or nonlinear, of the data $\mathbf{x}_i$:

$$E\{[\mathbf{s} - g(\mathbf{X})]w(\mathbf{X})\} = 0 \tag{8-81}$$

Proof. We shall use the following generalization of (7-60):

$$E\{[\mathbf{s} - g(\mathbf{X})]w(\mathbf{X})\} = E\{w(\mathbf{X})E\{\mathbf{s} - g(\mathbf{X})|\mathbf{X}\}\} \tag{8-82}$$

From the linearity of expected values and (8-77) it follows that

$$E\{\mathbf{s} - g(\mathbf{X})|X\} = E\{\mathbf{s}|X\} - E\{g(\mathbf{X})|X\} = 0$$

and (8-81) results.

Normality. Using the above, we shall show that if the RVs $\mathbf{s}, \mathbf{x}_1, \ldots, \mathbf{x}_n$ are jointly normal with zero mean, the linear and nonlinear estimators of $\mathbf{s}$ are equal:

$$\hat{\mathbf{s}} = a_1\mathbf{x}_1 + \cdots + a_n\mathbf{x}_n = g(\mathbf{X}) = E\{\mathbf{s}|\mathbf{X}\} \tag{8-83}$$

Proof. To prove (8-83), it suffices to show that $\hat{\mathbf{s}} = E\{\mathbf{s}|\mathbf{X}\}$. The RVs $\mathbf{s} - \hat{\mathbf{s}}$ and $\mathbf{x}_i$ are jointly normal with zero mean and orthogonal; hence they are independent. From this it follows that

$$E\{\mathbf{s} - \hat{\mathbf{s}}|X\} = E\{\mathbf{s} - \hat{\mathbf{s}}\} = 0 = E\{\mathbf{s}|X\} - E\{\hat{\mathbf{s}}|X\}$$

and (8-83) results because $E\{\hat{\mathbf{s}}|\mathbf{X}\} = \hat{\mathbf{s}}$.

Conditional densities of normal RVs. We shall use the preceding result to simplify the determination of conditional densities involving normal RVs. The conditional density $f_s(s|X)$ of $\mathbf{s}$ assuming $\mathbf{X}$ is the ratio of two exponentials the exponents of which are quadratics, hence it is normal. To determine it, it suffices, therefore, to find the conditional mean and variance of $\mathbf{s}$. We maintain that

$$E\{\mathbf{s}|X\} = \hat{s} \qquad E\{(\mathbf{s} - \hat{\mathbf{s}})^2|X\} = E\{(\mathbf{s} - \hat{\mathbf{s}})^2\} = P \tag{8-84}$$

The first follows from (8-83). The second follows from the fact that $\mathbf{s} - \hat{\mathbf{s}}$ is orthogonal and, therefore, independent of **X**. We thus conclude that

$$f(s|x_1, \ldots, x_n) = \frac{1}{\sqrt{2\pi P}} e^{-[s-(a_1x_1 + \cdots + a_nx_n)]^2/2P} \tag{8-85}$$

Example 8-11. The RVs $\mathbf{x}_1$ and $\mathbf{x}_2$ are jointly normal with zero mean. We shall determine their conditional density $f(x_2|x_1)$. As we know [see (7-78)]

$$E\{\mathbf{x}_2|x_1\} = ax_1 \qquad a = \frac{R_{12}}{R_{11}}$$

$$\sigma^2_{x_2|x_1} = P = E\{(\mathbf{x}_2 - a\mathbf{x}_1)\mathbf{x}_2\} = R_{22} - aR_{12}$$

Inserting into (8-85), we obtain

$$f(x_2|x_1) = \frac{1}{\sqrt{2\pi P}} e^{-(x_2 - ax_1)^2/2P}$$

Example 8-12. We now wish to find the conditional density $f(x_3|x_1, x_2)$. In this case,

$$E\{\mathbf{x}_3|\mathbf{x}_1, \mathbf{x}_2\} = a_1\mathbf{x}_1 + a_2\mathbf{x}_2$$

where the constants a_1 and a_2 are determined from the system

$$R_{11}a_1 + R_{12}a_2 = R_{13} \qquad R_{12}a_1 + R_{22}a_2 = R_{23}$$

Furthermore [see (8-84) and (8-73)]

$$\sigma^2_{x_3|x_1, x_2} = P = R_{33} - (R_{13}a_1 + R_{23}a_2)$$

and (8-85) yields

$$f(x_3|x_1, x_2) = \frac{1}{\sqrt{2\pi P}} e^{-(x_3 - a_1x_1 - a_2x_2)^2/2P}$$

Example 8-13. In this example, we shall find the two-dimensional density $f(x_2, x_3|x_1)$. This involves the evaluation of five parameters [see (6-15)]: two conditional means, two conditional variances, and the conditional covariance of the RVs $\mathbf{x}_2$ and $\mathbf{x}_3$ assuming $\mathbf{x}_1$.

The first four parameters are determined as in Example 8-11:

$$E\{\mathbf{x}_2|x_1\} = \frac{R_{12}}{R_{11}}x_1 \qquad E\{\mathbf{x}_3|x_1\} = \frac{R_{13}}{R_{11}}x_1$$

$$\sigma^2_{x_2|x_1} = R_{22} - \frac{R_{12}^2}{R_{11}} \qquad \sigma^2_{x_3|x_1} = R_{33} - \frac{R_{13}^2}{R_{11}}$$

The conditional covariance

$$C_{x_2x_3|x_1} = E\left\{\left(\mathbf{x}_2 - \frac{R_{12}}{R_{11}}\mathbf{x}_1\right)\left(\mathbf{x}_3 - \frac{R_{13}}{R_{11}}\mathbf{x}_1\right)\middle|\mathbf{x}_1 = x_1\right\} \tag{8-86}$$

is found as follows: We know that the errors $\mathbf{x}_2 - R_{12}\mathbf{x}_1/R_{11}$ and $\mathbf{x}_3 - R_{13}\mathbf{x}_1/R_{11}$

are independent of $\mathbf{x}_1$. Hence the condition $\mathbf{x}_1 = x_1$ in (8-86) can be removed. Expanding the product, we obtain

$$C_{x_2x_3|x_1} = R_{23} - \frac{R_{12}R_{13}}{R_{11}}$$

This completes the specification of $f(x_2, x_3|x_1)$.

Orthonormal Data Transformation

If the data $\mathbf{x}_i$ are orthogonal, that is, if $R_{ij} = 0$ for $i \neq j$, then R is a diagonal matrix and (8-71) yields

$$a_i = \frac{R_{0i}}{R_{ii}} = \frac{E\{\mathbf{s}\mathbf{x}_i\}}{E\{\mathbf{x}_i^2\}} \tag{8-87}$$

Thus the determination of the projection $\hat{\mathbf{s}}$ of $\mathbf{s}$ is simplified if the data $\mathbf{x}_i$ are expressed in terms of an orthonormal set of vectors. This is done as follows. We wish to find a set $\{\mathbf{i}_k\}$ of n orthonormal RVs $\mathbf{i}_k$ *linearly equivalent* to the data set $\{\mathbf{x}_k\}$. By this we mean that each $\mathbf{i}_k$ is a linear function of the elements of the set $\{\mathbf{x}_k\}$ and each $\mathbf{x}_k$ is a linear function of the elements of the set $\{\mathbf{i}_k\}$. The set $\{\mathbf{i}_k\}$ is not unique. We shall determine it using the Gram–Schmidt method (Fig. 8-2*b*). In this method, each $\mathbf{i}_k$ depends only on the first k data $\mathbf{x}_1, \ldots, \mathbf{x}_k$. Thus

$$\begin{aligned}
\mathbf{i}_1 &= \gamma_1^1 \mathbf{x}_1 \\
\mathbf{i}_2 &= \gamma_1^2 \mathbf{x}_1 + \gamma_2^2 \mathbf{x}_2 \\
&\cdots\cdots\cdots\cdots\cdots \\
\mathbf{i}_n &= \gamma_1^n \mathbf{x}_1 + \gamma_2^n \mathbf{x}_2 + \cdots + \gamma_n^n \mathbf{x}_n
\end{aligned} \tag{8-88}$$

In the notation γ_r^k, k is a superscript identifying the kth equation and r is a subscript taking the values 1 to k. The coefficient γ_1^1 is obtained from the normalization condition

$$E\{\mathbf{i}_1^2\} = (\gamma_1^1)^2 R_{11} = 1$$

To find the coefficients γ_1^2 and γ_2^2, we observe that $\mathbf{i}_2 \perp \mathbf{x}_1$ because $\mathbf{i}_2 \perp \mathbf{i}_1$ by assumption. From this it follows that

$$E\{\mathbf{i}_2\mathbf{x}_1\} = 0 = \gamma_1^2 R_{11} + \gamma_2^2 R_{21}$$

The condition $E\{\mathbf{i}_2^2\} = 1$ yields a second equation. Similarly, since $\mathbf{i}_k \perp \mathbf{i}_r$ for $r < k$, we conclude from (8-88) that $\mathbf{i}_k \perp \mathbf{x}_r$ if $r < k$. Multiplying the kth equation in (8-88) by $\mathbf{x}_r$ and using the above, we obtain

$$E\{\mathbf{i}_k\mathbf{x}_r\} = 0 = \gamma_1^k R_{1r} + \cdots + \gamma_k^k R_{kr} \qquad 1 \le r \le k-1 \tag{8-89}$$

This is a system of $k - 1$ equations for the k unknowns $\gamma_1^k, \ldots, \gamma_k^k$. The condition $E\{\mathbf{i}_k^2\} = 1$ yields one more equation.

The system (8-88) can be written in a vector form

$$\mathbf{I} = \mathbf{X}\Gamma \tag{8-90}$$

where $\mathbf{I}$ is a row vector with elements $\mathbf{i}_k$. Solving for $\mathbf{X}$, we obtain

$$\begin{aligned} \mathbf{x}_1 &= l_1^1 \mathbf{i}_1 \\ \mathbf{x}_2 &= l_1^2 \mathbf{i}_1 + l_2^2 \mathbf{i}_2 \\ &\cdots\cdots\cdots\cdots \\ \mathbf{x}_n &= l_1^n \mathbf{i}_1 + l_2^n \mathbf{i}_2 + \cdots + l_n^n \mathbf{i}_n \end{aligned} \qquad \mathbf{X} = \mathbf{I}\Gamma^{-1} = \mathbf{I}L \tag{8-91}$$

In the above, the matrix Γ and its inverse are upper triangular

$$\Gamma = \begin{bmatrix} \gamma_1^1 & \gamma_1^2 & \cdots & \gamma_1^n \\ & \gamma_2^2 & \cdots & \gamma_2^n \\ & & \cdots\cdots & \\ & 0 & & \\ & & & \gamma_n^n \end{bmatrix} \qquad L = \begin{bmatrix} l_1^1 & l_1^2 & \cdots & l_1^n \\ & l_2^2 & \cdots & l_2^n \\ & & \cdots\cdots & \\ & 0 & & \\ & & & l_n^n \end{bmatrix}$$

Since $E\{\mathbf{i}_i \mathbf{i}_j\} = \delta[i - j]$ by construction, we conclude that

$$E\{\mathbf{I}^t\mathbf{I}\} = 1_n = E\{\Gamma^t \mathbf{X}^t \mathbf{X}\Gamma\} = \Gamma^t E\{\mathbf{X}^t\mathbf{X}\}\Gamma \tag{8-92}$$

where 1_n is the identity matrix. Hence

$$\Gamma^t R \Gamma = 1_n \qquad R = L^t L \qquad R^{-1} = \Gamma\Gamma^t \tag{8-93}$$

We have thus expressed the matrix R and its inverse R^{-1} as products of an upper triangular and a lower triangular matrix [see also Cholesky factorization (14-79)].

The orthonormal base $\{\mathbf{i}_n\}$ in (8-88) is the finite version of the innovations process $\mathbf{i}[n]$ introduced in Sec. (12-1). The matrices Γ and L correspond to the whitening filter and to the innovations filter respectively and the factorization (8-93) corresponds to the spectral factorization (12-3).

From the linear equivalence of the sets $\{\mathbf{i}_k\}$ and $\{\mathbf{x}_k\}$, it follows that the estimate (8-68) of the RV $\mathbf{s}$ can be expressed in terms of the set $\{\mathbf{i}_k\}$:

$$\hat{\mathbf{s}} = b_1 \mathbf{i}_1 + \cdots + b_n \mathbf{i}_n = B\mathbf{I}^t$$

where again the coefficients b_k are such that

$$\mathbf{s} - \hat{\mathbf{s}} \perp \mathbf{i}_k \qquad 1 \le k \le n$$

This yields [see (8-92)]

$$E\{(\mathbf{s} - B\mathbf{I}^t)\mathbf{I}\} = 0 = E\{\mathbf{s}\mathbf{I}\} - B$$

from which it follows that

$$B = E\{\mathbf{s}\mathbf{I}\} = E\{\mathbf{s}\mathbf{X}\Gamma\} = R_0\Gamma \tag{8-94}$$

Returning to the estimate (8-68) of $\mathbf{s}$, we conclude that

$$\hat{\mathbf{s}} = B\mathbf{I}^t = B\Gamma^t\mathbf{X}^t = A\mathbf{X}^t \qquad A = B\Gamma^t \tag{8-95}$$

This simplifies the determination of the vector A if the matrix Γ is known.

8-4 STOCHASTIC CONVERGENCE AND LIMIT THEOREMS

A fundamental problem in the theory of probability is the determination of the asymptotic properties of random sequences. In this section, we introduce the subject, concentrating on the clarification of the underlying concepts. We start with a simple problem.

Suppose that we wish to measure the length a of an object. Due to measurement inaccuracies, the instrument reading is a sum

$$\mathbf{x} = a + \boldsymbol{\nu}$$

where $\boldsymbol{\nu}$ is the error term. If there are no systematic errors, then $\boldsymbol{\nu}$ is an RV with zero mean. In this case, if the standard deviation σ of $\boldsymbol{\nu}$ is small compared to a, then the observed value $\mathbf{x}(\zeta)$ of $\mathbf{x}$ at a single measurement is a satisfactory estimate of the unknown length a. In the context of probability, this conclusion can be phrased as follows: The mean of the RV $\mathbf{x}$ equals a and its variance equals σ^2. Applying Tchebycheff's inequality, we conclude that

$$P\{|\mathbf{x} - a| < \varepsilon\} > 1 - \frac{\sigma^2}{\varepsilon^2} \tag{8-96}$$

If, therefore, $\sigma \ll \varepsilon$, then the probability that $|\mathbf{x} - a|$ is less than that ε is close to 1. From this it follows that "almost certainly" the observed $\mathbf{x}(\zeta)$ is between $a - \varepsilon$ and $a + \varepsilon$, or equivalently, that the unknown a is between $\mathbf{x}(\zeta) - \varepsilon$ and $\mathbf{x}(\zeta) + \varepsilon$. In other words, the reading $\mathbf{x}(\zeta)$ of a single measurement is "almost certainly" a satisfactory estimate of the length a as long as $\sigma \ll a$. If σ is not small compared to a, then a single measurement does not provide an adequate estimate of a. To improve the accuracy, we perform the measurement a large number of times and we average the resulting readings. The underlying probabilistic model is now a product space

$$\mathscr{S}^n = \mathscr{S} \times \cdots \times \mathscr{S}$$

formed by repeating n times the experiment $\mathscr{S}$ of a single measurement. If the measurements are independent, then the ith reading is a sum

$$\mathbf{x}_i = a + \boldsymbol{\nu}_i$$

where the noise components $\boldsymbol{\nu}_i$ are independent RVs with zero mean and variance σ^2. This leads to the conclusion that the *sample mean*

$$\bar{\mathbf{x}} = \frac{\mathbf{x}_1 + \cdots + \mathbf{x}_n}{n} \tag{8-97}$$

of the measurements is an RV with mean a and variance σ^2/n. If, therefore, n is so large that $\sigma^2 \ll na^2$, then the value $\bar{\mathbf{x}}(\zeta)$ of the sample mean $\bar{\mathbf{x}}$ in a single performance of the experiment $\mathscr{S}^n$ (consisting of n independent measurements) is a satisfactory estimate of the unknown a.

To find a bound of the error in the estimate of a by $\bar{\mathbf{x}}$, we apply (8-96). To be concrete, we assume that n is so large that $\sigma^2/na^2 = 10^{-4}$, and we ask for the probability that $\mathbf{x}$ is between $0.9a$ and $1.1a$. The answer is given by (8-96) with $\varepsilon = 0.1a$.

$$P\{0.9a < \bar{\mathbf{x}} < 1.1a\} \geq 1 - \frac{100\sigma^2}{n} = 0.99$$

Thus, if the experiment is performed $n = 10^4\sigma^2/a^2$ times, then "almost certainly" in 99 percent of the cases, the estimate $\bar{\mathbf{x}}$ of a will be between $0.9a$ and $1.1a$. Motivated by the above, we introduce next various convergence modes involving sequences of random variables.

DEFINITION. A *random sequence* or a *discrete-time random process* is a sequence of RVs

$$\mathbf{x}_1, \ldots, \mathbf{x}_n, \ldots \tag{8-98}$$

For a specific ζ, $\mathbf{x}_n(\zeta)$ is a sequence of numbers that might or might not converge. This suggests that the notion of convergence of a random sequence might be given several interpretations:

Convergence everywhere (e) As we recall, a sequence of numbers x_n tends to a limit x if, given $\varepsilon > 0$, we can find a number n_0 such that

$$|x_n - x| < \varepsilon \qquad \text{for every} \quad n > n_0 \tag{8-99}$$

We say that a random sequence $\mathbf{x}_n$ converges everywhere if the sequence of numbers $\mathbf{x}_n(\zeta)$ converges as above for every ζ. The limit is a number that depends, in general, on ζ. In other words, the limit of the random sequence $\mathbf{x}_n$ is an RV $\mathbf{x}$:

$$\mathbf{x}_n \to \mathbf{x} \qquad \text{as} \quad n \to \infty$$

Convergence almost everywhere (a.e.) If the set of outcomes ζ such that

$$\lim \mathbf{x}_n(\zeta) = \mathbf{x}(\zeta) \qquad \text{as} \quad n \to \infty \tag{8-100}$$

exists and its probability equals 1, then we say that the sequence $\mathbf{x}_n$ converges almost everywhere (or with probability 1). This is written in the form

$$P\{\mathbf{x}_n \to \mathbf{x}\} = 1 \qquad \text{as} \quad n \to \infty \tag{8-101}$$

In the above, $\{\mathbf{x}_n \to \mathbf{x}\}$ is an event consisting of all outcomes ζ such that $\mathbf{x}_n(\zeta) \to \mathbf{x}(\zeta)$.

Convergence in the MS sense (MS) The sequence $\mathbf{x}_n$ tends to the RV $\mathbf{x}$ in the MS sense if

$$E\{|\mathbf{x}_n - \mathbf{x}|^2\} \to 0 \qquad \text{as} \quad n \to \infty \tag{8-102}$$

This is called *limit in the mean* and it is often written in the form

$$\text{l.i.m.}\, \mathbf{x}_n = \mathbf{x} \qquad n \to \infty$$

Convergence in probability (p) The probability $P\{|\mathbf{x} - \mathbf{x}_n| > \varepsilon\}$ of the event $\{|\mathbf{x} - \mathbf{x}_n| > \varepsilon\}$ is a sequence of numbers depending on ε. If this sequence tends to 0:

$$P\{|\mathbf{x} - \mathbf{x}_n| > \varepsilon\} \to 0 \qquad n \to \infty \tag{8-103}$$

for any $\varepsilon > 0$, then we say that the sequence $\mathbf{x}_n$ tends to the RV $\mathbf{x}$ in probability (or in measure). This is also called stochastic convergence.

Convergence in distribution (d) We denote by $F_n(x)$ and $F(x)$ respectively the distribution of the RVs $\mathbf{x}_n$ and $\mathbf{x}$. If

$$F_n(x) \to F(x) \qquad n \to \infty \tag{8-104}$$

for every point x of continuity of $F(x)$, then we say that the sequence $\mathbf{x}_n$ tends to the RV $\mathbf{x}$ in distribution. We note that, in this case, the sequence $\mathbf{x}_n(\zeta)$ need not converge for any ζ.

Cauchy criterion As we noted, a deterministic sequence x_n converges if it satisfies (8-99). This definition involves the limit x of x_n. The following theorem, known as the Cauchy criterion, establishes conditions for the convergence of x_n that avoid the use of x: If

$$|x_{n+m} - x_n| \to 0 \qquad \text{as} \quad n \to \infty \tag{8-105}$$

for any $m > 0$, then the sequence x_n converges.

The above theorem holds also for random sequence. In this case, the limit must be interpreted accordingly. For example, if

$$E\{|\mathbf{x}_{n+m} - \mathbf{x}_n|^2\} \to 0 \qquad \text{as} \quad n \to \infty$$

for every $m > 0$, then the random sequence $\mathbf{x}_n$ converges in the MS sense.

Comparison of convergence modes. In Fig. 8-3, we show the relationship between various convergence modes. Each point in the rectangle represents a random sequence. The letter on each curve indicates that all sequences in the interior of the curve converge in the stated mode. The shaded region consists of all sequences that do not converge in any sense. The letter d on the outer curve shows that if a sequence converges at all, then it converges also in distribution. We comment next on the less obvious comparisons:

If a sequence converges in the MS sense, then it also converges in probability. Indeed, Tchebycheff's inequality yields

$$P\{|\mathbf{x}_n - \mathbf{x}| > \varepsilon\} \le \frac{E\{|\mathbf{x}_n - \mathbf{x}|^2\}}{\varepsilon^2}$$

If $\mathbf{x}_n \to \mathbf{x}$ in the MS sense, then for a fixed $\varepsilon > 0$ the right side tends to 0; hence the left side also tends to 0 as $n \to \infty$ and (8-103) follows. The converse,

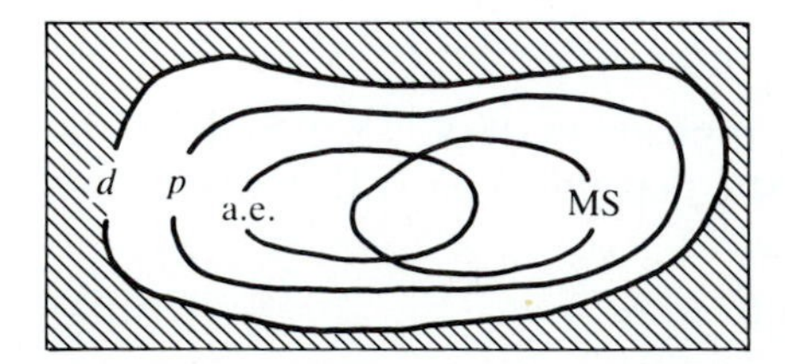

FIGURE 8-3

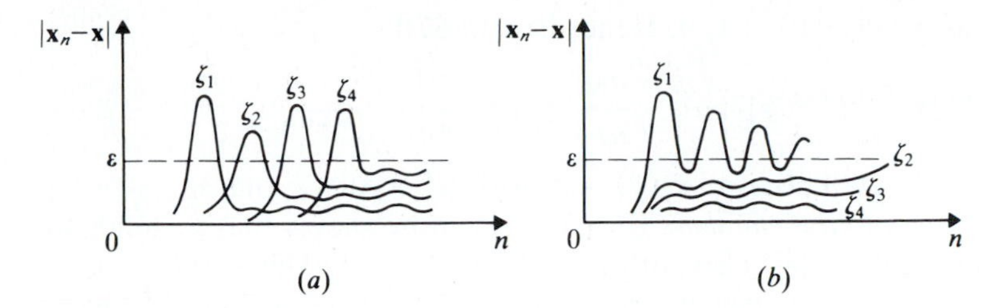

FIGURE 8-4

however, is not necessarily true. If $\mathbf{x}_n$ is not bounded, then $P\{|\mathbf{x}_n - \mathbf{x}| > \varepsilon\}$ might tend to 0 but not $E\{|\mathbf{x}_n - \mathbf{x}|^2\}$. If, however, $\mathbf{x}_n$ vanishes outside some interval $(-c, c)$ for every $n > n_0$, then p convergence and MS convergence are equivalent.

It is self-evident that a.e. convergence implies p convergence. We shall show by a heuristic argument that the converse is not true. In Fig. 8-4, we plot the difference $|\mathbf{x}_n - \mathbf{x}|$ as a function of n where, for simplicity, sequences are drawn as curves. Each curve represents, thus, a particular sequence $|\mathbf{x}_n(\zeta) - \mathbf{x}(\zeta)|$. Convergence in probability means that for a *specific* $n > n_0$, only a *small* percentage of these curves will have ordinates that exceed ε (Fig. 8-4a). It is, of course, possible that not even one of these curves will remain less than ε for *every* $n > n_0$. Convergence a.e., on the other hand, demands that most curves will be below ε for every $n > n_0$ (Fig. 8-4b).

The law of large numbers (Bernoulli). In Sec. 3-3 we showed that if the probability of an event $\mathscr{A}$ in a given experiment equals p and the number of successes of $\mathscr{A}$ in n trials equals k, then

$$P\left\{\left|\frac{k}{n} - p\right| < \varepsilon\right\} \to 1 \qquad \text{as} \quad n \to \infty \tag{8-106}$$

We shall reestablish this result as a limit of a sequence of RVs. For this purpose, we introduce the RVs

$$\mathbf{x}_i = \begin{cases} 1 & \text{if } \mathscr{A} \text{ occurs at the } i\text{th trial} \\ 0 & \text{otherwise} \end{cases}$$

We shall show that the sample mean

$$\bar{\mathbf{x}}_n = \frac{\mathbf{x}_1 + \cdots + \mathbf{x}_n}{n}$$

of these RVs tends to p in probability as $n \to \infty$.

Proof. As we know

$$E\{\mathbf{x}_i\} = E\{\bar{\mathbf{x}}_n\} = p \qquad \sigma_{x_i}^2 = pq \qquad \sigma_{\bar{x}_n}^2 = \frac{pq}{n}$$

Furthermore, $pq = p(1-p) \le 1/4$. Hence [see (5-57)]

$$P\{|\bar{\mathbf{x}}_n - p| < \varepsilon\} \ge 1 - \frac{pq}{n\varepsilon^2} \ge 1 - \frac{1}{4n\varepsilon^2} \xrightarrow[n\to\infty]{} 1$$

This reestablishes (8-106) because $\bar{\mathbf{x}}_n(\zeta) = k/n$ if $\mathscr{A}$ occurs k times.

The strong law of large numbers (Borel) It can be shown that $\bar{\mathbf{x}}_n$ tends to p not only in probability, but also with probability 1 (a.e.). This result, due to Borel, is known as the strong law of large numbers. The proof will not be given. We give below only a heuristic explanation of the difference between (8-106) and the strong law of large numbers in terms of relative frequencies.

Frequency interpretation We wish to estimate p within an error $\varepsilon = 0.1$, using as its estimate the sample mean $\bar{\mathbf{x}}_n$. If $n \ge 1000$, then

$$P\{|\bar{\mathbf{x}}_n - p| < 0.1\} \ge 1 - \frac{1}{4n\varepsilon^2} \ge \frac{39}{40}$$

Thus, if we repeat the experiment at least 1000 times, then in 39 out of 40 such runs, our error $|\bar{\mathbf{x}}_n - p|$ will be less than 0.1.

Suppose, now, that we perform the experiment 2000 times and we determine the sample mean $\bar{\mathbf{x}}_n$ not for one n but for every n between 1000 and 2000. The Bernoulli version of the law of large numbers leads to the following conclusion: If our experiment (the toss of the coin 2000 times) is repeated a large number of times, then, for a *specific* n larger than 1000, the error $|\bar{\mathbf{x}}_n - p|$ will exceed 0.1 only in one run out of 40. In other words, 97.5 percent of the runs will be "good." We cannot draw the conclusion that in the good runs the error will be less than 0.1 for *every* n between 1000 and 2000. This conclusion, however, is correct, but it can be deduced only from the strong law of large numbers.

Ergodicity. Ergodicity is a topic dealing with the relationship between statistical averages and sample averages. This topic is treated in Sec. 12-1. In the following, we discuss certain results phrased in the form of limits of random sequences.

Markoff's theorem. We are given a sequence $\mathbf{x}_i$ of RVs and we form their sample mean

$$\bar{\mathbf{x}}_n = \frac{\mathbf{x}_1 + \cdots + \mathbf{x}_n}{n}$$

Clearly, $\bar{\mathbf{x}}_n$ is an RV whose values $\bar{\mathbf{x}}_n(\zeta)$ depend on the experimental outcome ζ. We maintain that, if the RVs $\mathbf{x}_i$ are such that the mean $\bar{\eta}_n$ of $\bar{\mathbf{x}}_n$ tends to a limit η and its variance $\bar{\sigma}_n$ tends to 0 as $n \to \infty$:

$$E\{\bar{\mathbf{x}}_n\} = \bar{\eta}_n \xrightarrow[n\to\infty]{} \eta \qquad \bar{\sigma}_n^2 = E\{(\bar{\mathbf{x}}_n - \bar{\eta}_n)^2\} \xrightarrow[n\to\infty]{} 0 \tag{8-107}$$

then the RV $\bar{\mathbf{x}}_n$ tends to η in the MS sense

$$E\{(\bar{\mathbf{x}}_n - \eta)^2\} \xrightarrow[n\to\infty]{} 0 \tag{8-108}$$

Proof. The proof is based on the simple inequality

$$|\bar{\mathbf{x}}_n - \eta|^2 \le 2|\bar{\mathbf{x}}_n - \bar{\eta}_n|^2 + 2|\bar{\eta}_n - \eta|^2$$

Indeed, taking expected values of both sides, we obtain

$$E\{(\bar{\mathbf{x}}_n - \eta)^2\} \le 2E\{(\bar{\mathbf{x}}_n - \bar{\eta}_n)^2\} + 2(\bar{\eta}_n - \eta)^2$$

and (8-108) follows from (8-107).

COROLLARY (Tchebycheff's condition). If the RVs $\mathbf{x}_i$ are uncorrelated and

$$\frac{\sigma_1^2 + \cdots + \sigma_n^2}{n^2} \xrightarrow[n\to\infty]{} 0 \tag{8-109}$$

then

$$\bar{\mathbf{x}}_n \xrightarrow[n\to\infty]{} \eta = \lim_{n\to\infty} \frac{1}{n} \sum_{i=1}^{n} E\{\mathbf{x}_i\}$$

in the MS sense.

Proof. It follows from the theorem because, for uncorrelated RVs, the left side of (8-109) equals $\bar{\sigma}_n^2$.

We note that Tchebycheff's condition (8-109) is satisfied if $\sigma_i < K < \infty$ for every i. This is the case if the RVs $\mathbf{x}_i$ are i.i.d. with finite variance.

Kinchin We mention without proof that if the RVs $\mathbf{x}_i$ are i.i.d., then their sample mean $\bar{\mathbf{x}}_n$ tends to η even if nothing is known about their variance. In this case, however, $\bar{\mathbf{x}}_n$ tends to η in probability only. The following is an application:

Example 8-14. We wish to determine the distribution $F(x)$ of an RV $\mathbf{x}$ defined in a certain experiment. For this purpose we repeat the experiment n times and form the RVs $\mathbf{x}_i$ as in (8-12). As we know, these RVs are i.i.d. and their common distribution equals $F(x)$. We next form the RVs

$$\mathbf{y}_i(x) = \begin{cases} 1 & \text{if} \quad \mathbf{x}_i \le x \\ 0 & \text{if} \quad \mathbf{x}_i > x \end{cases}$$

where x is a fixed number. The RVs $\mathbf{y}_i(x)$ so formed are also i.i.d. and their mean equals

$$E\{\mathbf{y}_i(x)\} = 1 \times P\{\mathbf{y}_i = 1\} = P\{\mathbf{x}_i \le x\} = F(x)$$

Applying Kinchin's theorem to $\mathbf{y}_i(x)$, we conclude that

$$\frac{\mathbf{y}_1(x) + \cdots + \mathbf{y}_n(x)}{n} \xrightarrow[n\to\infty]{} F(x)$$

in probability. Thus, to determine $F(x)$, we repeat the original experiment n times and count the number of times the RV $\mathbf{x}$ is less than x. If this number equals k and n is sufficiently large, then $F(x) \simeq k/n$. The above is thus a restatement of the relative frequency interpretation (4-3) of $F(x)$ in the form of a limit theorem.

The Central Limit Theorem

Given n independent RVs $\mathbf{x}_i$, we form their sum

$$\mathbf{x} = \mathbf{x}_1 + \cdots + \mathbf{x}_n$$

This is an RV with mean $\eta = \eta_1 + \cdots + \eta_n$ and variance $\sigma^2 = \sigma_1^2 + \cdots + \sigma_n^2$. The central limit theorem (CLT) states that under certain general conditions, the distribution $F(x)$ of $\mathbf{x}$ approaches a normal distribution with the same mean and variance:

$$F(x) \simeq \mathbb{G}\left(\frac{x - \eta}{\sigma}\right) \tag{8-110}$$

as n increases. Furthermore, if the RVs $\mathbf{x}_i$ are of continuous type, the density $f(x)$ of $\mathbf{x}$ approaches a normal density (Fig. 8-5a):

$$f(x) \simeq \frac{1}{\sigma\sqrt{2\pi}} e^{-(x-\eta)^2/2\sigma^2} \tag{8-111}$$

This important theorem can be stated as a limit: If $\mathbf{z} = (\mathbf{x} - \eta)/\sigma$ then

$$F_z(z) \underset{n\to\infty}{\longrightarrow} \mathbb{G}(z) \qquad f_z(z) \underset{n\to\infty}{\longrightarrow} \frac{1}{\sqrt{2\pi}} e^{-z^2/2}$$

for the general and for the continuous case respectively. The proof is outlined later.

The CLT can be expressed as a property of convolutions: The convolution of a large number of positive functions is approximately a normal function [see (8-51)].

The nature of the CLT approximation and the required value of n for a specified error bound depend on the form of the densities $f_i(x)$. If the RVs $\mathbf{x}_i$ are i.i.d., the value $n = 30$ is adequate for most applications. In fact, if the

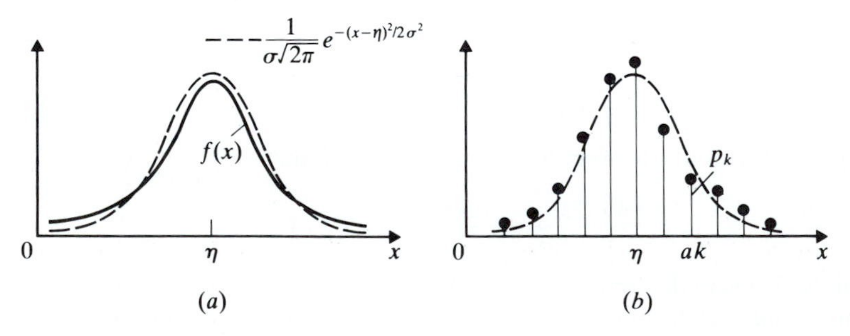

FIGURE 8-5

functions $f_i(x)$ are smooth, values of n as low as 5 can be used. The next example is an illustration.

Example 8-15. The RVs $\mathbf{x}_i$ are i.i.d. and uniformly distributed in the interval (0, 1). We shall compare the density $f_x(x)$ of their sum $\mathbf{x}$ with the normal approximation (8-111) for $n = 2$ and $n = 3$. In this problem,

$$\eta_i = \frac{T}{2} \qquad \sigma_i^2 = \frac{T^2}{12} \qquad \eta = n\frac{T}{2} \qquad \sigma^2 = n\frac{T^2}{12}$$

$n = 2$ $f(x)$ is a triangle obtained by convolving a pulse with itself (Fig. 8-6)

$$\eta = T \qquad \sigma^2 = \frac{T^2}{6} \qquad f(x) \simeq \frac{1}{T}\sqrt{\frac{3}{\pi}}\,e^{-3(x-T)^2/T^2}$$

$n = 3$ $f(x)$ consists of three parabolic pieces obtained by convolving a triangle with a pulse

$$\eta = \frac{3T}{2} \qquad \sigma^2 = \frac{T^2}{4} \qquad f(x) \simeq \frac{1}{T}\sqrt{\frac{2}{\pi}}\,e^{-2(x-1.5T)^2/T^2}$$

As we can see from the figure, the approximation error is small even for such small values of n.

For a discrete-type RV, $F(x)$ is a staircase function approaching a normal distribution. The probabilities p_k however, that $\mathbf{x}$ equals specific values x_k are, in general, unrelated to the normal density. Lattice-type RVs are an exception:

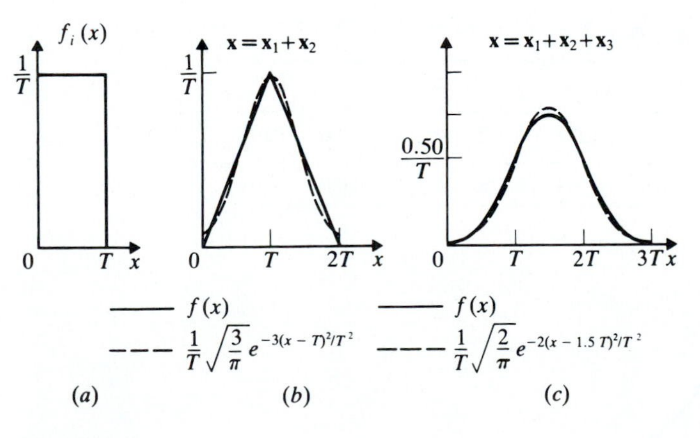

FIGURE 8-6

If the RVs $\mathbf{x}_i$ take equidistant values ak_i, then $\mathbf{x}$ takes the values ak and for large n, the discontinuities $p_k = P\{\mathbf{x} = ak\}$ of $F(x)$ at the points $x_k = ak$ equal the samples of the normal density (Fig. 8-5*b*):

$$P\{\mathbf{x} = ak\} \simeq \frac{1}{\sigma\sqrt{2\pi}} e^{-(ak-\eta)^2/2\sigma^2} \tag{8-112}$$

We give next an illustration in the context of Bernoulli trials. The RVs $\mathbf{x}_i$ of Example 8-7 are i.i.d. taking the values 1 and 0 with probabilities p and q respectively; hence their sum $\mathbf{x}$ is of lattice type taking the values $k = 0, \ldots, n$. In this case,

$$E\{\mathbf{x}\} = nE\{\mathbf{x}_i\} = np \qquad \sigma_x^2 = n\sigma_i^2 = npq$$

Inserting into (8-112), we obtain the approximation

$$P\{\mathbf{x} = k\} = \binom{n}{k} p^k q^{n-k} \simeq \frac{1}{\sqrt{2\pi npq}} e^{-(k-np)^2/2npq} \tag{8-113}$$

This shows that the *DeMoivre–Laplace theorem* (3-27) is a special case of the lattice-type form (8-112) of the central limit theorem.

Example 8-16. A fair coin is tossed six times and $\mathbf{x}_i$ is the zero–one RV associated with the event {heads at the ith toss}. The probability of k heads in six tosses equals

$$P\{\mathbf{x} = k\} = \binom{6}{k} \frac{1}{2^6} = p_k \qquad \mathbf{x} = \mathbf{x}_1 + \cdots + \mathbf{x}_6$$

In the following table we show the above probabilities and the samples of the normal curve $N(\eta, \sigma^2)$ (Fig. 8-7) where

$$\eta = np = 3 \qquad \sigma^2 = npq = 1.5$$

k	0	1	2	3	4	5	6
p_k	0.016	0.094	0.234	0.312	0.234	0.094	0.016
$N(\eta, \sigma)$	0.016	0.086	0.233	0.326	0.233	0.086	0.016

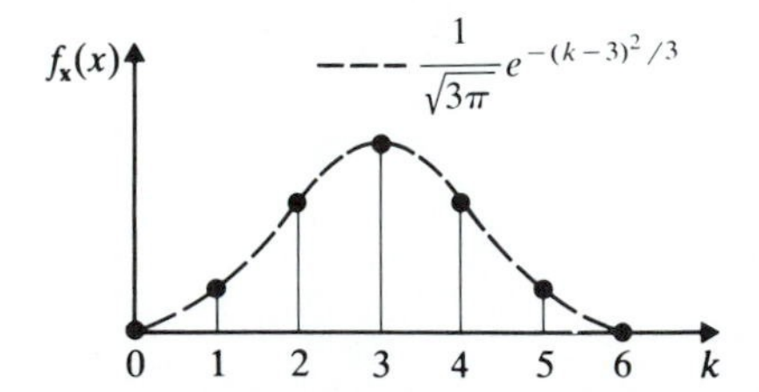

FIGURE 8-7

ERROR CORRECTION. In the approximation of $f(x)$ by the normal curve $N(\eta, \sigma^2)$, the error

$$\varepsilon(x) = f(x) - \frac{1}{\sigma\sqrt{2\pi}} e^{-x^2/2\sigma^2}$$

results where we assumed, shifting the origin, that $\eta = 0$. We shall express this error in terms of the moments

$$m_n = E\{\mathbf{x}^n\}$$

of $\mathbf{x}$ and the *Hermite polynomials*

$$H_k(x) = (-1)^k e^{x^2/2} \frac{d^k}{dx^k} e^{-x^2/2}$$
$$= x^k - \binom{k}{2} x^{k-2} + 1 \cdot 3 \binom{k}{4} x^{k-4} + \cdots \tag{8-114}$$

These polynomials form a complete orthogonal set on the real line:

$$\int_{-\infty}^{\infty} e^{-x^2/2} H_n(x) H_m(x)\, dx = \begin{cases} n!\sqrt{2\pi} & n = m \\ 0 & n \neq m \end{cases}$$

Hence $\varepsilon(x)$ can be written as a series

$$\varepsilon(x) = \frac{1}{\sigma\sqrt{2\pi}} e^{-x^2/2\sigma^2} \sum_{k=3}^{\infty} C_k H_k\left(\frac{x}{\sigma}\right) \tag{8-115}$$

The series starts with $k = 3$ because the moments of $\varepsilon(x)$ of order up to 2 are 0. The coefficients C_n can be expressed in terms of the moments m_n of $\mathbf{x}$. Equating moments of order $n = 3$ and $n = 4$, we obtain [see (5-44)]

$$3!\sigma^3 C_3 = m_3 \qquad 4!\sigma^4 C_4 = m_4 - 3\sigma^4$$

First-order correction. From (8-114) it follows that

$$H_3(x) = x^3 - 3x \qquad H_4(x) = x^4 - 6x^2 + 3$$

Retaining the first nonzero term of the sum in (8-115), we obtain

$$f(x) \simeq \frac{1}{\sigma\sqrt{2\pi}} e^{-x^2/2\sigma^2} \left[1 + \frac{m_3}{6\sigma^3}\left(\frac{x^3}{\sigma^3} - \frac{3x}{\sigma}\right)\right] \tag{8-116}$$

If $f(x)$ is even, then $m_3 = 0$ and (8-115) yields

$$f(x) \simeq \frac{1}{\sigma\sqrt{2\pi}} e^{-x^2/2\sigma^2} \left[1 + \frac{1}{24}\left(\frac{m_4}{\sigma^4} - 3\right)\left(\frac{x^4}{\sigma^4} - \frac{6x^2}{\sigma^2} + 3\right)\right] \tag{8-117}$$

Example 8-17. If the RVs $\mathbf{x}_i$ are i.i.d. with density $f_i(x)$ as in Fig. 8-8a, then $f(x)$ consists of three parabolic pieces (see also Example 8-12) and $N(0, 1/4)$ is its normal approximation. Since $f(x)$ is even and $m_4 = 13/80$ (see Prob. 8-4), (8-117)

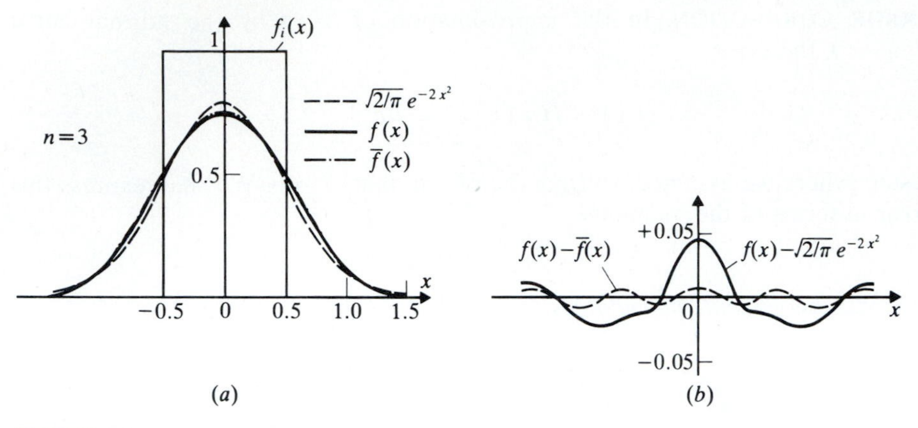

FIGURE 8-8

yields

$$f(x) \simeq \sqrt{\frac{2}{\pi}}\, e^{-2x^2}\left(1 - \frac{4x^4}{15} + \frac{2x^2}{5} - \frac{1}{20}\right) \equiv \bar{f}(x)$$

In Fig. 8-8b, we show the error $\varepsilon(x)$ of the normal approximation and the first-order correction error $f(x) - \bar{f}(x)$.

ON THE PROOF OF THE CENTRAL LIMIT THEOREM. We shall justify the approximation (8-111) using characteristic functions. We assume for simplicity that $\eta_i = 0$. Denoting by $\Phi_i(\omega)$ and $\Phi(\omega)$, respectively, the characteristic functions of the RVs $\mathbf{x}_i$ and $\mathbf{x} = \mathbf{x}_1 + \cdots + \mathbf{x}_n$, we conclude from the independence of $\mathbf{x}_i$ that

$$\Phi(\omega) = \Phi_1(\omega) \cdots \Phi_n(\omega)$$

Near the origin, the functions $\Psi_i(\omega) = \ln \Phi_i(\omega)$ can be approximated by a parabola:

$$\Psi_i(\omega) \simeq -\tfrac{1}{2}\sigma_i^2\omega^2 \qquad \Phi_i(\omega) = e^{-\sigma_i^2\omega^2/2} \quad \text{for } |\omega| < \varepsilon \tag{8-118}$$

If the RVs $\mathbf{x}_i$ are of continuous type, then [see (5-61) and Prob. 5-25]

$$\Phi_i(0) = 1 \qquad |\Phi_i(\omega)| < 1 \quad \text{for } |\omega| \neq 0 \tag{8-119}$$

Equation (8-119) suggests that for small ε and large n, the function $\Phi(\omega)$ is negligible for $|\omega| > \varepsilon$, (Fig. 8-9a). This holds also for the exponential $e^{-\sigma^2\omega^2/2}$ if $\sigma \to \infty$ as in (8-123). From the above it follows that

$$\Phi(\omega) \simeq e^{-\sigma_1^2\omega^2/2} \cdots e^{-\sigma_n^2\omega^2/2} = e^{-\sigma^2\omega^2/2} \quad \text{for all } \omega \tag{8-120}$$

in agreement with (8-111).

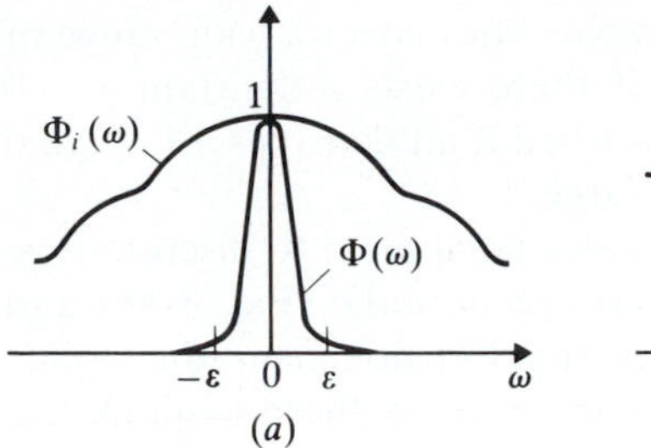

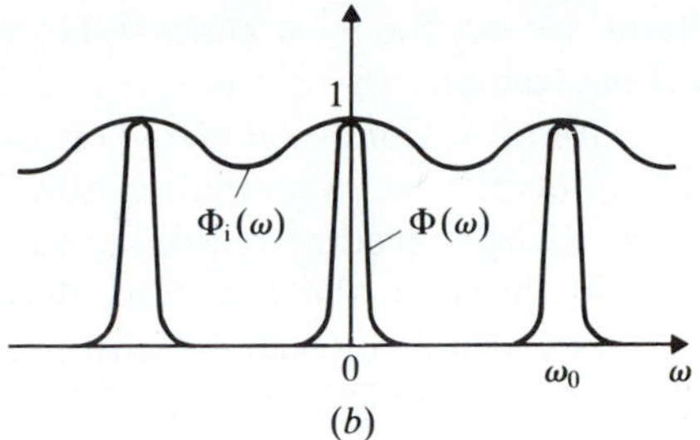

FIGURE 8-9

The exact form of the theorem states that the normalized RV

$$\mathbf{z} = \frac{\mathbf{x}_1 + \cdots + \mathbf{x}_n}{\sigma} \qquad \sigma^2 = \sigma_1^2 + \cdots + \sigma_n^2$$

tends to an $N(0, 1)$ RV as $n \to \infty$:

$$f_z(z) \underset{n\to\infty}{\longrightarrow} \frac{1}{\sqrt{2\pi}} e^{-z^2/2} \tag{8-121}$$

A general proof of the theorem is given below. In the following, we sketch a proof under the assumption that the RVs $\mathbf{x}_i$ are i.i.d. In this case

$$\Phi_1(\omega) = \cdots = \Phi_n(\omega) \qquad \sigma = \sigma_i\sqrt{n}$$

Hence,

$$\Phi_z(\omega) = \Phi_i^n\left(\frac{\omega}{\sigma_i\sqrt{n}}\right)$$

Expanding the functions $\Psi_i(\omega) = \ln \Phi_i(\omega)$ near the origin, we obtain

$$\Psi_i(\omega) = -\frac{\sigma_i^2\omega^2}{2} + O(\omega^3)$$

Hence,

$$\Psi_z(\omega) = n\Psi_i\left(\frac{\omega}{\sigma_i\sqrt{n}}\right) = -\frac{\omega^2}{2} + O\left(\frac{1}{\sqrt{n}}\right) \underset{n\to\infty}{\longrightarrow} -\frac{\omega^2}{2} \tag{8-122}$$

This shows that $\Phi_z(\omega) \to e^{-\omega^2/2}$ as $n \to \infty$ and (8-121) results.

As we noted, the theorem is not true always. The following is a set of sufficient conditions:

(a) $$\sigma_1^2 + \cdots + \sigma_n^2 \underset{n\to\infty}{\longrightarrow} \infty \tag{8-123}$$

(b) There exists a number $\alpha > 2$ and a finite constant K such that

$$\int_{-\infty}^{\infty} x^\alpha f_i(x)\, dx < K < \infty \qquad \text{for all } i \tag{8-124}$$

These conditions are not the most general. However, they cover a wide range of applications. For example, (8-123) is satisfied if there exists a constant $\varepsilon > 0$ such that $\sigma_i > \varepsilon$ for all i. Condition (8-124) is satisfied if all densities $f_i(x)$ are 0 outside a finite interval $(-c, c)$ no matter how large.

Lattice type The preceding reasoning can also be applied to discrete-type RVs. However, in this case the functions $\Phi_i(\omega)$ are periodic (Fig. 8-9b) and their product takes significant values only in a small region near the points $\omega = 2\pi n/a$. Using the approximation (8-112) in each of these regions, we obtain

$$\Phi(\omega) \simeq \sum_n e^{-\sigma^2(\omega - n\omega_0)^2/2} \qquad \omega_0 = \frac{2\pi}{a} \tag{8-125}$$

As we can see from (11A-1), the inverse of the above yields (8-112).

The Berry–Esseén theorem† This theorem states that if

$$E\{\mathbf{x}_i^3\} \le c\sigma_i^2 \quad \text{all } i \tag{8-126}$$

where c is some constant, then the distribution $\bar{F}(x)$ of the normalized sum

$$\bar{\mathbf{x}} = \frac{\mathbf{x}_1 + \cdots + \mathbf{x}_n}{\sigma} \qquad \sigma_1^2 + \cdots + \sigma_n^2 = \sigma^2$$

is close to the normal distribution $\mathbb{G}(x)$ in the following sense

$$|\bar{F}(x) - \mathbb{G}(x)| < \frac{4c}{\sigma} \tag{8-127}$$

The central limit theorem is a corollary of (8-127) because (8-127) leads to the conclusion that

$$\bar{F}(x) \to \mathbb{G}(x) \qquad \text{as} \quad \sigma \to \infty \tag{8-128}$$

This proof is based on condition (8-126). This condition, however, is not too restrictive. It holds, for example, if the RVs $\mathbf{x}_i$ are i.i.d. and their third moment is finite.

We note, finally, that whereas (8-128) establishes merely the convergence in distribution of $\bar{\mathbf{x}}$ to a normal RV, (8-127) gives also a *bound* of the deviation of $\bar{F}(x)$ from normality.

The central limit theorem for products. Given n independent positive RVs $\mathbf{x}_i$, we form their product:

$$\mathbf{y} = \mathbf{x}_1\mathbf{x}_2 \cdots \mathbf{x}_n \qquad \mathbf{x}_i > 0$$

†A. Papoulis: "Narrow-Band Systems and Gaussianity," *IEEE Transactions on Information Theory*, January 1972.

THEOREM. For large n, the density of $\mathbf{y}$ is approximately *lognormal*:

$$f_y(y) \simeq \frac{1}{y\sigma\sqrt{2\pi}} \exp\left\{-\frac{1}{2\sigma^2}(\ln y - \eta)^2\right\} U(y) \qquad (8\text{-}129)$$

where

$$\eta = \sum_{i=1}^{n} E\{\ln \mathbf{x}_i\} \qquad \sigma^2 = \sum_{i=1}^{n} \operatorname{Var}(\ln \mathbf{x}_i)$$

Proof. The RV

$$\mathbf{z} = \ln \mathbf{y} = \ln \mathbf{x}_1 + \cdots + \ln \mathbf{x}_n$$

is the sum of the RVs $\ln \mathbf{x}_i$. From the CLT it follows, therefore, that for large n, this RV is nearly normal with mean η and variance σ^2. And since $\mathbf{y} = e^{\mathbf{z}}$, we conclude from (5-11) that $\mathbf{y}$ has a lognormal density. The theorem holds if the RVs $\ln \mathbf{x}_i$ satisfy the conditions for the validity of the CLT.

Example 8-18. Suppose that the RVs $\mathbf{x}_i$ are uniform in the interval $(0, 1)$. In this case,

$$E\{\ln \mathbf{x}_i\} = \int_0^1 \ln x \, dx = -1 \qquad E\{(\ln \mathbf{x}_i)^2\} = \int_0^1 (\ln \mathbf{x})^2 \, dx = 2$$

Hence $\eta = -n$ and $\sigma^2 = n$. Inserting into (8-129), we conclude that the density of the product $\mathbf{y} = \mathbf{x}_1 \cdots \mathbf{x}_n$ equals

$$f_y(y) \simeq \frac{1}{y\sqrt{2\pi n}} \exp\left\{-\frac{1}{2n}(\ln y + n)^2\right\} U(y)$$

8-5 RANDOM NUMBERS: MEANING AND GENERATION

Random numbers (RNs) are used in a variety of applications involving computers and statistics. In this section, we explain the underlying ideas concentrating on the meaning and generation of RNs. We start with a simple illustration of the role of statistics in the numerical solution of deterministic problems.

MONTE CARLO INTEGRATION. We wish to evaluate the integral

$$I = \int_0^1 g(x) \, dx \qquad (8\text{-}130)$$

For this purpose, we introduce an RV $\mathbf{x}$ with uniform distribution in the interval $(0, 1)$ and we form the RV $\mathbf{y} = g(\mathbf{x})$. As we know,

$$E\{g(\mathbf{x})\} = \int_0^1 g(x) f_x(x) \, dx = \int_0^1 g(x) \, dx \qquad (8\text{-}131)$$

hence $\eta_y = I$. We have thus expressed the unknown I as the expected value of the RV $\mathbf{y}$. This result involves only concepts; it does not yield a numerical method for evaluating I. Suppose, however, that the RV $\mathbf{x}$ models a physical quantity in a real experiment. We can then estimate I using the relative frequency interpretation of expected values: We repeat the experiment a large number of times and observe the values x_i of $\mathbf{x}$; we compute the corresponding values $y_i = g(x_i)$ of $\mathbf{y}$ and form their average as in (5-26). This yields

$$I = E\{g(\mathbf{x})\} \simeq \frac{1}{n} \sum g(x_i) \tag{8-132}$$

The above suggests the following method for determining I:

The data x_i, no matter how they are obtained, are random numbers; that is, they are numbers having certain properties. If, therefore, we can numerically generate such numbers, we have a method for determining I. To carry out this method, we must reexamine the meaning of RNs and develop computer programs for generating them.

THE DUAL INTERPRETATION OF RNs. "What are RNs? Can they be generated by a computer? Is it possible to generate truly random number sequences?" Such questions do not have a generally accepted answer. The reason is simple. As in the case of probability (see Chap. 1), the term *random numbers* has two distinctly different meanings. The first is theoretical: RNs are mental constructs defined in terms of an abstract model. The second is empirical: RNs are sequences of real numbers generated either as physical data obtained from a random experiment or as computer output obtained from a deterministic program. The duality of interpretation of RNs is apparent in the following extensively quoted definitions†:

> A sequence of numbers is random if it has every property that is shared by all infinite sequences of independent samples of random variables from the uniform distribution (J. M. Franklin)

> A random sequence is a vague notion embodying the ideas of a sequence in which each term is unpredictable to the uninitiated and whose digits pass a certain number of tests, traditional with statisticians and depending somewhat on the uses to which the sequence is to be put. (D. H. Lehmer)

It is obvious that these definitions cannot have the same meaning. Nevertheless, both are used to define RN sequences. To avoid this confusing ambiguity, we shall give two definitions: one theoretical, the other empirical. For these definitions we shall rely solely on the uses for which RNs are intended: RNs are used to apply statistical techniques to other fields. It is natural, therefore, that they are defined in terms of the corresponding probabilistic concepts and their

†D. E. Knuth: *The Art of Computer Programming*, Addison-Wesley, Reading, MA, 1969.

properties as physically generated numbers are expressed directly in terms of the properties of real data generated by random experiments.

CONCEPTUAL DEFINITION. A sequence of numbers x_i is called random if it equals the samples $x_i = \mathbf{x}_i(\zeta)$ of a sequence $\mathbf{x}_i$ of i.i.d. RVs $\mathbf{x}_i$ defined in the space of repeated trials.

It appears that this definition is the same as Franklin's. There is, however, a subtle but important difference. Franklin says that the sequence x_i has every property shared by i.i.d. RVs; we say that x_i equals the samples of the i.i.d. RVs $\mathbf{x}_i$. In this definition, all theoretical properties of RNs are the same as the corresponding properties of RVs. There is, therefore, no need for a new theory.

EMPIRICAL DEFINITION. A sequence of numbers x_i is called random if its statistical properties are the same as the properties of random data obtained from a random experiment.

Not all experimental data lead to conclusions consistent with the theory of probability. For this to be the case, the experiments must be so designed that data obtained by repeated trials satisfy the i.i.d. condition. This condition is accepted only after the data have been subjected to a variety of tests and in any case, it can be claimed only as an approximation. The same applies to computer-generated RNs. Such uncertainties, however, cannot be avoided no matter how we define physically generated sequences. The advantage of the above definition is that it shifts the problem of establishing the randomness of a sequence of numbers to an area with which we are already familiar. We can, therefore, draw directly on our experience with random experiments and apply the well-established tests of randomness to computer-generated RNs.

Generation of RN Sequences

RNs used in Monte Carlo calculations are generated mainly by computer programs; however, they can also be generated as observations of random data obtained from real experiments: The tosses of a fair coin generate a random sequence of 0's (heads) and 1's (tails); the distance between radioactive emissions generates a random sequence of exponentially distributed samples. We accept number sequences so generated as random because of our long experience with such experiments. RN sequences experimentally generated are not, however, suitable for computer use, for obvious reasons. An efficient source of RNs is a computer program with small memory, involving simple arithmetic operations. We outline next the most commonly used programs.

Our objective is to generate RN sequences with arbitrary distributions. In the present state of the art, however, this cannot be done directly. The available algorithms only generate sequences consisting of integers z_i uniformly distributed in an interval $(0, m)$. As we show later, the generation of a sequence x_i with an arbitrary distribution is obtained indirectly by a variety of methods involving the uniform sequence z_i.

The most general algorithm for generating an RN sequence z_i is an equation of the form

$$z_n = f(z_{n-1}, \ldots, z_{n-r}) \mod m \tag{8-133}$$

where $f(z_{n-1}, \ldots, z_{n-r})$ is a function depending on the r most recent past values of z_n. In this notation, z_n is the remainder of the division of the number $f(z_{n-1}, \ldots, z_{n-r})$ by m. The above is a nonlinear recursion expressing z_n in terms of the constant m, the function f, and the initial conditions $z_1, \ldots, z_{r-1}$. The quality of the generator depends on the form of the function f. It might appear that good RN sequences result if this function is complicated. Experience has shown, however, that this is not the case. Most algorithms in use are linear recursions of order 1. We shall discuss the homogeneous case.

LEHMER'S ALGORITHM. The simplest and one of the oldest RN generators is the recursion

$$z_n = az_{n-1} \mod m \qquad z_0 = 1 \qquad n \ge 1 \tag{8-134}$$

where m is a large prime number and a is an integer. Solving, we obtain

$$z_n = a^n \mod m \tag{8-135}$$

The sequence z_n takes values between 1 and $m - 1$; hence at least two of its first m values are equal. From this it follows that z_n is a periodic sequence for $n > m$ with period $m_o \le m - 1$. A periodic sequence is not, of course, random. However, if for the applications for which it is intended the required number of sample does not exceed m_o, periodicity is irrelevant. For most applications, it is enough to choose for m a number of the order of 10^9 and to search for a constant a such that $m_o = m - 1$. A value for m suggested by Lehmer in 1951 is the prime number $2^{31} - 1$.

To complete the specification of (8-134), we must assign a value to the multiplier a. Our first condition is that the period m_o of the resulting sequence z_0 equal $m - 1$.

DEFINITION. An integer a is called the *primitive root* of m if the smallest n such that

$$a^n = 1 \mod m \text{ is } n = m - 1. \tag{8-136}$$

From the definition it follows that the sequence a^n is periodic with period $m_o = m - 1$ iff a is a primitive root of m. Most primitive roots do not generate good RN sequences. For a final selection, we subject specific choices to a variety of tests based on tests of randomness involving real experiments. Most tests are carried out not in terms of the integers z_i but in terms of the properties of the numbers

$$u_i = \frac{z_i}{m} \tag{8-137}$$

These numbers take essentially all values in the interval $(0, 1)$ and the purpose of testing is to establish whether they are the values of a sequence $\mathbf{u}_i$ of continuous-type i.i.d. RVs uniformly distributed in the interval (0, 1). The i.i.d. condition leads to the following equations:

For every u_i in the interval (0, 1) and for every n,

$$P\{\mathbf{u}_i \le u_i\} = u_i \tag{8-138a}$$

$$P\{\mathbf{u}_1 \le u_1, \ldots, \mathbf{u}_n \le u_n\} = P\{\mathbf{u}_1 \le u_1\} \cdots P\{\mathbf{u}_n \le u_n\} \tag{8-138b}$$

To establish the validity of these equations, we need an infinite number of tests. In real life, however, we can perform only a finite number of tests. Furthermore, all tests involve approximations based on the empirical interpretation of probability. We cannot, therefore, claim with certainty that a sequence of real numbers is truly random. We can claim only that a particular sequence is reasonably random for certain applications or that one sequence is more random than another. In practice, a sequence u_n is accepted as random not only because it passes the standard tests but also because it has been used with satisfactory results in many problems.

Over the years, several algorithms have been proposed for generating "good" RN sequences. Not all, however, have withstood the test of time. An example of a sequence z_n that seems to meet most requirements is obtained from (8-134) with $a = 7^5$ and $m = 2^{31} - 1$:

$$z_n = 16{,}807 z_{n-1} \mod 2{,}147{,}483{,}647 \tag{8-139}$$

This sequence meets most standard tests of randomness and has been used effectively in a variety of applications.†

We conclude with the observation that most tests of randomness are applications, direct or indirect, of well-known tests of various statistical hypotheses. For example, to establish the validity of (8-138a), we apply the Kolmogoroff–Smirnov test, page 272, or the chi-square test, page 273. These tests are used to determine whether given experimental data fit a particular distribution. To establish the validity of (8-138b), we apply the chi-square test, page 274. This test is used to determine the independence of various events.

In addition to direct testing, a variety of special methods have been proposed for testing indirectly the validity of both equations in (8-138). These methods are based on well-known properties of RVs and they are designed for particular applications. The generation of random vector sequences is an application requiring special tests.

Random vectors. We shall construct a multidimensional sequence of RNs using the following properties of subsequences. Suppose that $\mathbf{x}$ is an RV with distribution $F(x)$ and x_i is the corresponding RN sequence. It follows from

---

†S. K. Park and K. W. Miller "Random Number Generations: Good Ones Are Hard to Find," *Communications of the ACM*, vol. 31, no. 10, October 1988.

(8-138) that every subsequence of x_i is an RN sequence with distribution $F(x)$. Furthermore, if two subsequences have no common elements, they are the samples of two independent RVs. From this we conclude that the odd-subscript and even-subscript sequences

$$x_i^o = x_{2i-1} \qquad x_i^e = x_{2i} \qquad i = 1, 2, \ldots$$

are the samples of two i.i.d. RVs $\mathbf{x}^o$ and $\mathbf{x}^e$ with distribution $F(x)$. Thus, starting from a scalar RN sequence, x_i, we constructed a vector RN sequence (x_i^o, x_i^e). Proceeding similarly, we can construct RN sequences of any dimensionality. Using superscripts to identify various RVs and their samples, we conclude that the RN sequences

$$x_i^k = x_{mi-m+k} \qquad k = 1, \ldots, m \qquad i = 1, 2, \ldots \tag{8-140}$$

are the samples of m i.i.d. RVs $\mathbf{x}^1, \ldots, \mathbf{x}^m$ with distribution $F(x)$.

Note that a sequence of numbers might be sufficiently random for scalar but not for vector applications. If, therefore, an RN sequence x_i is to be used for multidimensional applications, it is desirable to subject it to special tests involving its subsequences.

RN Sequences with Arbitrary Distributions

In the following, the letter $\mathbf{u}$ will identify an RV with uniform distribution in the interval (0, 1); the corresponding RN sequence will be identified by u_i. Using the sequence u_i, we shall present a variety of methods for generating sequences with arbitrary distributions. In this analysis, we shall make frequent use of the following:

If x_i are the samples of the RV $\mathbf{x}$, then $y_i = g(x_i)$ are the samples of the RV $\mathbf{y} = g(\mathbf{x})$. For example, if x_i is an RN sequence with distribution $F_x(x)$, then $y_i = a + bx_i$ is an RN sequence with distribution $F_x[(y-a)/b]$ if $b > 0$, and $1 - F_x[(y-a)/b]$ if $b < 0$. From this it follows, for example, that $v_i = 1 - u_i$ is an RN sequence uniform in the interval (0, 1).

PERCENTILE TRANSFORMATION METHOD. Consider an RV $\mathbf{x}$ with distribution $F_x(x)$. We have shown in Sec. 5-2 that the RV $\mathbf{u} = F_x(\mathbf{x})$ is uniform in the interval (0, 1) no matter what the form of $F_x(x)$ is. Denoting by $F_x^{(-1)}(u)$ the inverse of $F_x(x)$, we conclude that $\mathbf{x} = F_x^{(-1)}(\mathbf{u})$ (see Fig. 8-10). From this it follows that

$$x_i = F_x^{(-1)}(u_i) \tag{8-141}$$

is an RN sequence with distribution $F_x(x)$, [see also (5-19)]. Thus, to find an RN sequence x_i with distribution a given function $F_x(x)$, it suffices to determine the inverse of $F_x(x)$ and to compute $F_x^{(-1)}(u_i)$. Note that the numbers x_i are the u_i percentiles of $F_x(x)$.

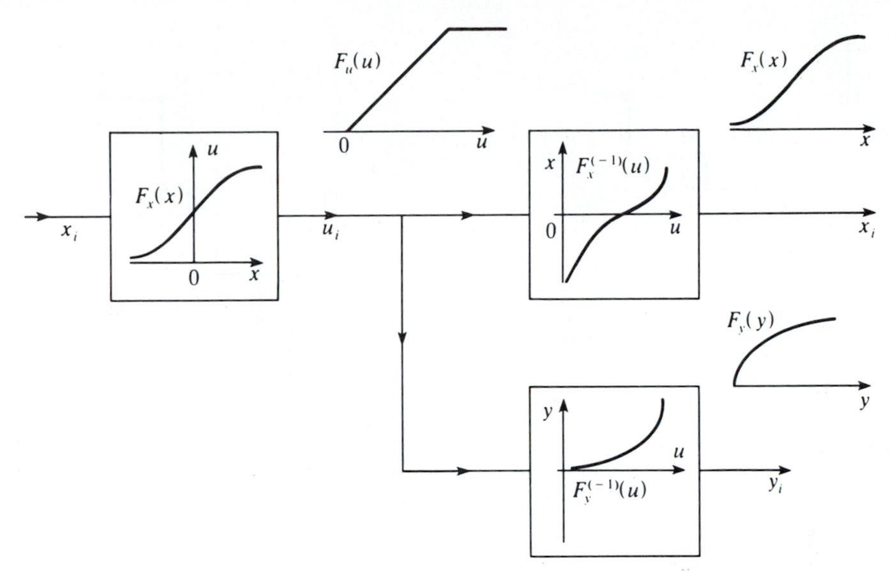

FIGURE 8-10

Example 8-19. We wish to generate an RN sequence x_i with exponential distribution. In this case,

$$F_x(x) = 1 - e^{-x/\lambda} \qquad \mathbf{x} = -\lambda \ln(1 - \mathbf{u})$$

Since $1 - \mathbf{u}$ is an RV with uniform distribution, we conclude that the sequence

$$x_i = -\lambda \ln u_i \tag{8-142}$$

has an exponential distribution.

Example 8-20. We wish to generate an RN sequence x_i with Rayleigh distribution. In this case,

$$F_x(x) = 1 - e^{-x^2/2} \qquad F_x^{(-1)}(\mathbf{u}) = \sqrt{-2\ln(1 - \mathbf{u})}$$

Replacing $1 - \mathbf{u}$ by $\mathbf{u}$, we conclude that the sequence

$$x_i = \sqrt{-2\ln u_i} \tag{8-143}$$

has a Rayleigh distribution.

Suppose now that we wish to generate the samples x_i of a discrete-type RV $\mathbf{x}$ taking the values a_k with probability

$$p_k = P\{\mathbf{x} = a_k\} \qquad k = 1, \ldots, m$$

In this case, $F_x(x)$ is a staircase function (Fig. 8-11) with discontinuities at the points a_k, and its inverse is a staircase function with discontinuities at the points $F_x(a_k) = p_1 + \cdots + p_k$. Applying (8-141), we obtain the following rule for generating the RN sequence x_i:

$$\text{Set } x_i = a_k \text{ iff } p_1 + \cdots + p_{k-1} \le u_i < p_1 + \cdots + p_k \tag{8-144}$$

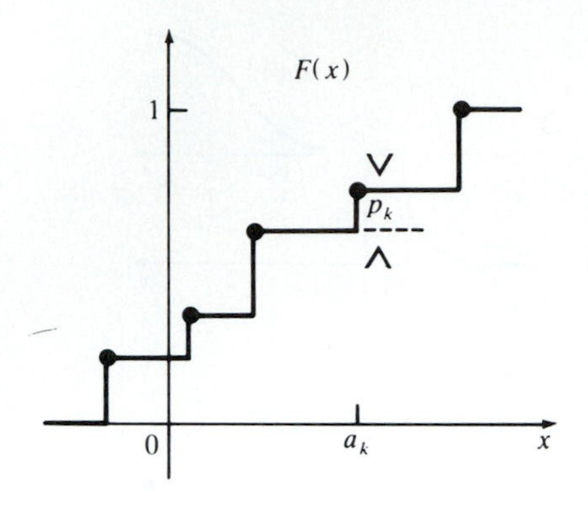

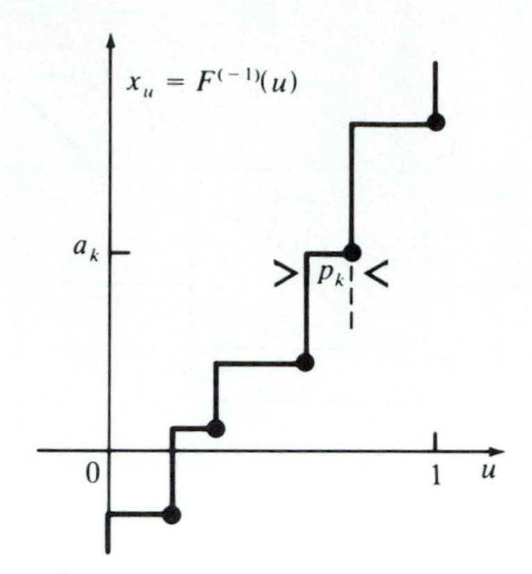

FIGURE 8-11

Example 8-21. The sequence

$$x_i = \begin{cases} 0 & \text{if } \; 0 < u_i < p \\ 1 & \text{if } \; p < u_i < 1 \end{cases}$$

takes the values 0 and 1 with probability p and $1 - p$ respectively. It specifies, therefore, a *binary* RN sequence.

The sequence

$$x_i = k \qquad \text{iff} \quad 0.1k < u_i < 0.1(k+1) \qquad k = 0, 1, \dots, 9$$

takes the values $0, 1, \dots, 9$ with equal probability. It specifies, therefore, a *decimal* RN sequence with uniform distribution.

Setting

$$a_k = k \qquad p_k = \binom{n}{k} p^k q^{n-k} \qquad k = 0, 1, \dots, m$$

into (8-15), we obtain an RN sequence with *binomial* distribution.

Setting

$$a_k = k \qquad p_k = e^{-\lambda} \frac{\lambda^k}{k!} \qquad k = 0, 1, \dots$$

into (8-15) we obtain an RN sequence with *Poisson* distribution.

Suppose now that we are given not a uniform sequence, but a sequence x_i with distribution $F_x(x)$. We wish to find a sequence y_i with distribution $F_y(y)$. As we know, $y_i = F_y^{(-1)}(u_i)$ is an RN sequence with distribution $F_y(y)$. Hence (see Fig. 8-10) the composite function

$$y_i = F_y^{(-1)}(F_x(x_i)) \tag{8-145}$$

generates an RN sequence with distribution $F_y(y)$ [see also (5-20)].

Example 8-22. We are given an RN sequence $x_i > 0$ with distribution $F_x(x) = 1 - e^{-x} - xe^{-x}$ and we wish to generate an RN sequence $y_i > 0$ with distribution $F_y(y) = 1 - e^{-y}$. In this example $F_y^{(-1)}(u) = -\ln(1-u)$; hence

$$F_y^{(-1)}(F_x(x)) = -\ln[1 - F_x(x)] = -\ln(e^{-x} + xe^{-x})$$

Inserting into (8-145), we obtain

$$y_i = -\ln(e^{-x_i} + x_i e^{-x_i})$$

REJECTION METHOD. In the percentile transformation method, we used the inverse of the function $F_x(x)$. However, inverting a function is not a simple task. To overcome this difficulty, we develop next a method that avoids inversion. The problem under consideration is the generation of an RN sequence y_i with distribution $F_y(y)$ in terms of the RN sequence x_i as in (8-145).

The proposed method is based on the relative frequency interpretation of the conditional density

$$f_x(x|\mathscr{M})\,dx = \frac{P\{x < \mathbf{x} \le x + dx, \mathscr{M}\}}{P(\mathscr{M})} \tag{8-146}$$

of an RV $\mathbf{x}$ assuming $\mathscr{M}$ (see page 80). In the following method, the event $\mathscr{M}$ is expressed in terms of the RV $\mathbf{x}$ and another RV $\mathbf{u}$, and it is so chosen that the resulting function $f_x(x|\mathscr{M})$ equals $f_y(y)$. The sequence y_i is generated by setting $y_i = x_i$ if $\mathscr{M}$ occurs, rejecting x_i otherwise. The problem has a solution only if $f_y(x) = 0$ in every interval in which $f_x(x) = 0$. We can assume, therefore, without essential loss of generality, that the ratio $f_x(x)/f_y(x)$ is bounded from below by some positive constant a:

$$\frac{f_x(x)}{f_y(x)} \ge a > 0 \qquad \text{for every} \quad x$$

Rejection theorem. If the RVs $\mathbf{x}$ and $\mathbf{u}$ are independent and

$$\mathscr{M} = \{\mathbf{u} \le r(\mathbf{x})\} \qquad \text{where} \quad r(x) = a\frac{f_y(x)}{f_x(x)} \le 1 \tag{8-147}$$

then

$$f_x(x|\mathscr{M}) = f_y(x) \tag{8-148}$$

Proof. The joint density of the RVs $\mathbf{x}$ and $\mathbf{u}$ equals $f_x(x)$ in the strip $0 < u < 1$ of the xu plane, and 0 elsewhere. The event $\mathscr{M}$ consists of all outcomes such that the point $(\mathbf{x}, \mathbf{u})$ is in the shaded area of Fig. 8-12 below the curve $u = r(x)$. Hence

$$P(\mathscr{M}) = \int_{-\infty}^{\infty} r(x) f_x(x)\,dx = a\int_{-\infty}^{\infty} f_y(x)\,dx = a$$

The event $\{x < \mathbf{x} \le x + dx, \mathscr{M}\}$ consists of all outcomes such that the point

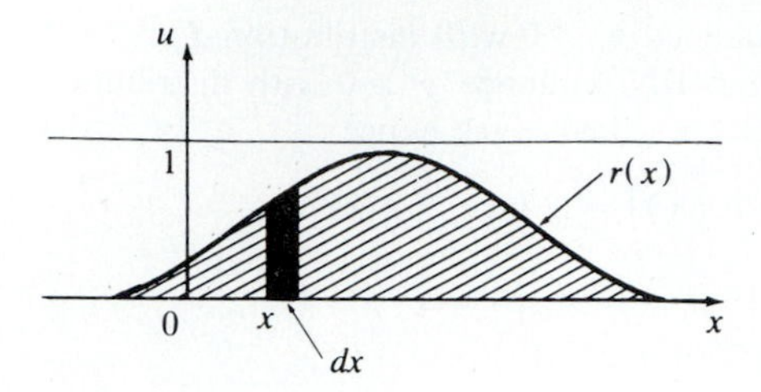

FIGURE 8-12

$(\mathbf{x}, \mathbf{u})$ is in the strip $x < \mathbf{x} \le x + dx$ below the curve $u = r(x)$. The probability masses in this strip equal $f_x(x)r(x)\,dx$. Hence

$$P\{x < \mathbf{x} \le x + dx, \mathscr{M}\} = f_x(x)r(x)\,dx$$

Inserting into (8-146), we obtain (8-148).

From the rejection theorem it follows that the subsequence of x_i such that $u_i \le r(x_i)$ forms a sequence of random numbers that are the samples of an RV $\mathbf{y}$ with density $f_x(y|\mathscr{M}) = f_y(y)$. This leads to the following rule for generating the sequence y_i: Form the two-dimensional RN sequence (x_i, u_i).

$$\text{Set} \quad y_i = x_i \qquad \text{if} \quad u_i \le a\frac{f_y(x_i)}{f_x(x_i)}; \qquad \text{reject } x_i \text{ otherwise} \tag{8-149}$$

Example 8-23. We are given an RN sequence x_i with exponential distribution and we wish to construct an RN sequence y_i with truncated normal distribution:

$$f_x(x) = e^{-x}U(x) \qquad f_y(y) = \frac{2}{\sqrt{2\pi}}e^{-y^2/2}U(y)$$

For $x > 0$,

$$\frac{f_y(x)}{f_x(x)} = \sqrt{\frac{2e}{\pi}}\,e^{-(x-1)^2/2} \le \sqrt{\frac{2e}{\pi}}$$

Setting $a = \sqrt{\pi/2e}$, we obtain the following rule for generating the sequence y_i:

$$\text{Set} \quad y_i = x_i \quad \text{if} \quad u_i < e^{-(x_i-1)^2/2}; \qquad \text{reject } x_i \text{ otherwise}$$

MIXING METHOD. We develop next a method generating an RN sequence x_i with density $f(x)$ under the following assumptions: The function $f(x)$ can be expressed as the weighted sum of m densities $f_k(m)$:

$$f(x) = p_1 f_1(x) + \cdots + p_m f_m(x) \qquad p_k > 0 \tag{8-150}$$

Each component $f_k(x)$ is the density of a known RN sequence x_i^k.

In the mixing method, we generate the sequence x_i by a mixing process involving certain subsequences of the m sequences x_i^k selected according to the following rule:

$$\text{Set} \quad x_i = x_i^k \qquad \text{if} \quad p_1 + \cdots + p_{k-1} \le u_i < p_1 + \cdots + p_k \tag{8-151}$$

Mixing theorem. If the sequences u_i and $x_i^1, \ldots, x_i^m$ are mutually independent, then the density $f_x(x)$ of the sequence x_i specified by (8-151) equals

$$f_x(x) = p_1 f_1(x) + \cdots + p_m f_m(x) \tag{8-152}$$

Proof. The sequence x_i is a mixture of m subsequences. The density of the subsequence of the kth sequence x_i^k equals $f_k(x)$. This subsequence is also a subsequence of x_i conditioned on the event

$$\mathscr{A}_k = \{p_1 + \cdots + p_{k-1} \le \mathbf{u} < p_1 + \cdots + p_k\}$$

Hence its density also equals $f_x(x|\mathscr{A}_k)$. This leads to the conclusion that

$$f_x(x|\mathscr{A}_k) = f_k(x)$$

From the total probability theorem (4-58), it follows that

$$f_x(x) = f_x(x|\mathscr{A}_1)P(\mathscr{A}_1) + \cdots + f_x(x|\mathscr{A}_m)P(\mathscr{A}_m)$$

And since $P(\mathscr{A}_k) = p_k$, (8-152) results. Comparing with (8-150), we conclude that the density $f_x(x)$ generated by (8-152) equals the given function $f(x)$.

Example 8-24. The Laplace density $0.5e^{-|x|}$ can be written as a sum

$$f(x) = 0.5e^{-x}U(x) + 0.5e^{x}U(-x)$$

This is a special case of (8-150) with

$$f_1(x) = e^{-x}U(x) \qquad f_2(x) = e^{x}U(-x) \qquad p_1 = p_2 = 0.5$$

A sequence x_i with density $f(x)$ can, therefore, be realized in terms of the samples of two RVs $\mathbf{x}^1$ and $\mathbf{x}^2$ with the above densities. As we have shown in Example 8-19, if the RV $\mathbf{v}$ is uniform in the interval (0, 1), then the density of the RV $\mathbf{x}^1 = -\ln \mathbf{v}$ equals $f_1(x)$; similarly, the density of the RV $\mathbf{x}^2 = \ln \mathbf{v}$ equals $f_2(x)$. This yields the following rule for generating an RN sequence x_i with Laplace distribution: Form two independent uniform sequences u_i and v_i:

$$\text{Set} \quad x_i = -\ln v_i \qquad \text{if} \quad 0 \le u_i < 0.5$$

$$\text{Set} \quad x_i = \ln v_i \qquad \text{if} \quad 0.5 \le u_i < 1$$

GENERAL TRANSFORMATIONS. We now give various examples for generating an RN sequence w_i with specified distribution $F_w(w)$ using the transformation

$$\mathbf{w} = g(\mathbf{x}^1, \ldots, \mathbf{x}^m)$$

where $\mathbf{x}^k$ are m RVs with known distributions. To do so, we determine g such that the distribution of $\mathbf{w}$ equals $F_w(w)$. The desired sequence is given by

$$w_i = g(x_i^1, \ldots, x_i^m)$$

Binomial RNs. If $\mathbf{x}^k$ are m i.i.d. RVs taking the values 0 and 1 with probabilities p and q respectively, their sum has a binomial distribution. From this it follows that if x_i^k are m binary sequences, their sum

$$w_i = x_i^1 + \cdots + x_i^m$$

is an RN sequence with binomial distribution. The m sequences x_i^k can be realized as subsequences of a single binary sequence x_i as in (8-140).

Erlang RNs. The sum $\mathbf{w} = \mathbf{x}^1 + \cdots + \mathbf{x}^m$ of m i.i.d. RVs $\mathbf{x}^k$ with density $e^{-cx}U(x)$ has an Erlang density [see (4-38)]:

$$f_w(w) \sim w^{m-1}e^{-w}U(w) \tag{8-153}$$

From this it follows that the sum $w_i = w_i^1 + \cdots + w_i^m$ of m exponentially distributed RN sequences w_i^k is an RN sequence with Erlang distribution.

The sequences x_i^k can be generated in terms of m subsequences of a single sequence u_i (see Example 8-19):

$$w_i = -\frac{1}{c}\left(\ln u_i^1 + \cdots + \ln u_i^m\right) \tag{8-154}$$

Chi-square RNs. We wish to generate an RN sequence w_i with density

$$f_w(w) \sim w^{n/2-1}e^{-w/2}U(w)$$

For $n = 2m$, this is a special case of (8-153) with $c = 1/2$. Hence w_i is given by (8-154).

To find w_i for $n = 2m + 1$, we observe that if $\mathbf{y}$ is $\chi^2(2m)$ and $\mathbf{z}$ is $N(0, 1)$ and independent of $\mathbf{y}$, the sum $\mathbf{w} = \mathbf{y} + \mathbf{z}^2$ is $\chi^2(2m + 1)$ [see (8-63)]; hence the sequence

$$w_i = -2\left(\ln u_i^1 + \cdots + \ln u_i^m\right) + (z_i)^2$$

has a $\chi^2(2m + 1)$ distribution.

Student-t RNs. Given two independent RVs $\mathbf{x}$ and $\mathbf{y}$ with distributions $N(0, 1)$ and $\chi^2(n)$ respectively, we form the RV $\mathbf{w} = \mathbf{x}/\sqrt{\mathbf{y}/n}$. As we know, $\mathbf{w}$ has a $t(n)$ distribution (see example 6-15). From this it follows that, if x_i and y_i are samples of $\mathbf{x}$ and $\mathbf{y}$, the sequence

$$w_i = \frac{x_i}{\sqrt{y_i/n}}$$

has a $t(n)$ distribution.

Lognormal RNs. If $\mathbf{z}$ is $N(0, 1)$ and $\mathbf{w} = e^{a+b\mathbf{z}}$, then $\mathbf{w}$ has a lognormal distribution [see (5-10)]:

$$f_w(w) = \frac{1}{bw\sqrt{2\pi}} \exp\left\{-\frac{(\ln w - a)^2}{2b^2}\right\}$$

Hence, if z_i is an $N(0, 1)$ sequence, the sequence

$$w_i = e^{a+bz_i}$$

has a lognormal distribution.

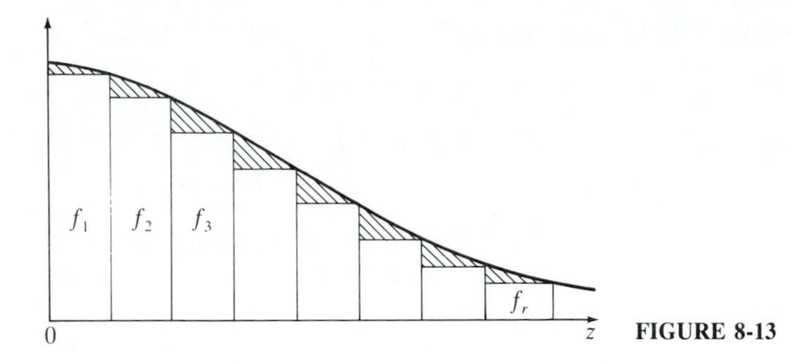

FIGURE 8-13

RN sequences with normal distributions. Several methods are available for generating normal RVs. We give next various illustrations. The percentile transformation method is not used because of the difficulty of inverting the normal distribution. The mixing method is used extensively because the normal density is a smooth curve; it can, therefore, be approximated by a sum as in (8-150). The major components (unshaded) of this sum are rectangles Fig. (8-13) that can be realized by sequences of the form $au_i + b$. The remaining components (shaded) are more complicated; however, since their areas are small, they need not be realized exactly. Other methods involve known properties of normal RVs. For example, the central limit theorem leads to the following method.

Given m independent RVs $\mathbf{u}^k$, we form the sum

$$\mathbf{z} = \mathbf{u}^1 + \cdots + \mathbf{u}^m$$

If m is large, the RV $\mathbf{z}$ is approximately normal [see (8-111)]. From this it follows that if u_1^k are m independent RN sequences their sum

$$z_i = u_i^1 + \cdots + u_i^m$$

is approximately a normal RN sequence. This method is not very efficient. The following three methods are more efficient and are used extensively.

Rejection and mixing (G. Marsaglia). In Example 8-23, we used the rejection method to generate an RV sequence y_i with a truncated normal density

$$f_y(y) = \frac{2}{\sqrt{2\pi}} e^{-y^2/2} U(y)$$

The normal density can be written as a sum

$$f_z(z) = \frac{1}{\sqrt{2\pi}} e^{-z^2/2} = \frac{1}{2} f_y(z) + \frac{1}{2} f_y(-z) \qquad (8\text{-}155)$$

The density $f_y(y)$ is realized by the sequence y_i as in Example 8-23 and the density $f_y(-y)$ by the sequence $-y_i$. Applying (8-151), we conclude that the following rule generates an $N(0, 1)$ sequence z_i:

$$\text{Set} \quad z_i = y_i \qquad \text{if} \quad 0 \le u_i < 0.5$$
$$\text{Set} \quad z_i = -y_i \qquad \text{if} \quad 0.5 \le u_i < 1 \tag{8-156}$$

Polar coordinates. We have shown that, if the RVs $\mathbf{r}$ and $\boldsymbol{\varphi}$ are independent, $\mathbf{r}$ has the Rayleigh density $f_r(r) = re^{-r^2/2}$ and $\boldsymbol{\varphi}$ is uniform in the interval $(-\pi, \pi)$, then (see Example 6-12) the RVs

$$\mathbf{z} = \mathbf{r}\cos\boldsymbol{\varphi} \qquad \mathbf{w} = \mathbf{r}\sin\boldsymbol{\varphi} \tag{8-157}$$

are $N(0, 1)$ and independent. Using this, we shall construct two independent normal RN sequences z_i and w_i as follows: Clearly, $\boldsymbol{\varphi} = \pi(2\mathbf{u} - 1)$; hence $\varphi_i = \pi(2u_i - 1)$. As we know, $\mathbf{r} = \sqrt{2\mathbf{x}} = \sqrt{-2\ln\mathbf{v}}$ where $\mathbf{x}$ is an RV with exponential distribution and $\mathbf{v}$ is uniform in the interval $(0, 1)$. Denoting by x_i and v_i the samples of the RVs $\mathbf{x}$ and $\mathbf{v}$, we conclude that $r_i = \sqrt{2x_i} = \sqrt{-2\ln v_i}$ is an RN sequence with Rayleigh distribution. From this and (8-157) it follows that if u_i and v_i are two independent RN sequences uniform in the interval $(0, 1)$, then the sequences

$$z_i = \sqrt{-2\ln v_i}\,\cos\pi(2u_i - 1) \qquad w_i = \sqrt{-2\ln v_i}\,\sin\pi(2u_i - 1) \tag{8-158}$$

are $N(0, 1)$ and independent.

The Box–Muller method. The rejection method was based on the following: If x_i is an RN sequence with distribution $F(x)$, its subsequence y_i conditioned on an event $\mathscr{M}$ is an RN sequence with distribution $F(x|\mathscr{M})$. Using this, we shall generate two independent $N(0, 1)$ sequences z_i and w_i in terms of the samples x_i, y_i of two independent RVs $\mathbf{x}, \mathbf{y}$ uniformly distributed in the interval $(-1, 1)$. We shall use for $\mathscr{M}$ the event

$$\mathscr{M} = \{\mathbf{q} \le 1\} \qquad \mathbf{q} = \sqrt{\mathbf{x}^2 + \mathbf{y}^2}$$

The joint density of $\mathbf{x}$ and $\mathbf{y}$ equals $1/4$ in the square $|x| < 1$, $|y| < 1$ of Fig. 8-14 and 0 elsewhere. Hence

$$P(\mathscr{M}) = \frac{\pi}{4} \qquad P\{\mathbf{q} \le q\} = \frac{\pi q^2}{4} \qquad \text{for} \quad q < 1$$

But $\{\mathbf{q} \le q, \mathscr{M}\} = \{\mathbf{q} \le q\}$, for $q < 1$ because $\{\mathbf{q} < q\}$ is a subset of $\mathscr{M}$. Hence

$$F_q(q|\mathscr{M}) = \frac{P\{\mathbf{q} \le q, \mathscr{M}\}}{P(\mathscr{M})} = q^2 \qquad f_q(q|\mathscr{M}) = 2q \qquad 0 \le q < 1 \tag{8-159}$$

Writing the RVs $\mathbf{x}$ and $\mathbf{y}$ in polar form:

$$\mathbf{x} = \mathbf{q}\cos\boldsymbol{\varphi} \qquad \mathbf{y} = \mathbf{q}\sin\boldsymbol{\varphi} \qquad \tan\boldsymbol{\varphi} = \mathbf{y}/\mathbf{x} \tag{8-160}$$

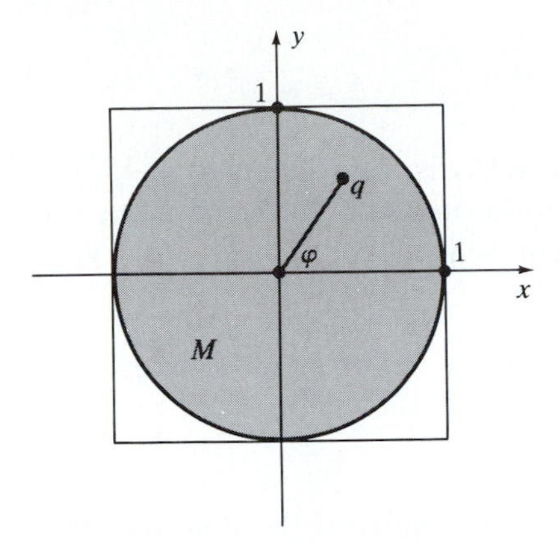

FIGURE 8-14

we conclude as in (8-159) that the joint density of the RVs $\mathbf{q}$ and $\boldsymbol{\varphi}$ is such that

$$f_{q\phi}(q, \varphi|\mathscr{M})\, dq\, d\varphi = \frac{P\{q \le \mathbf{q} < q + dq, \varphi \le \boldsymbol{\varphi} < \varphi + d\varphi\}}{P(\mathscr{M})} = \frac{q\, dq\, d\varphi/4}{\pi/4}$$

for $0 \le q < 1$ and $|\varphi| < \pi$. From this it follows that the RVs $\mathbf{q}$ and φ are conditionally independent and

$$f_q(q|\mathscr{M}) = 2q \qquad f_\varphi(\varphi) = 1/2\pi \qquad 0 \le q \le 1 \qquad -\pi < \varphi < \pi$$

THEOREM. If $\mathbf{x}$ and $\mathbf{y}$ are two independent RVs uniformly distributed in the interval $(-1, 1)$ and $\mathbf{q} = \sqrt{\mathbf{x}^2 + \mathbf{y}^2}$, then the RVs

$$\mathbf{z} = \frac{\mathbf{x}}{\mathbf{q}}\sqrt{-4\ln \mathbf{q}} \qquad \mathbf{w} = \frac{\mathbf{y}}{\mathbf{q}}\sqrt{-4\ln \mathbf{q}} \tag{8-161}$$

are conditionally $N(0, 1)$ and independent:

$$f_{zw}(z, w|\mathscr{M}) = f_z(z|\mathscr{M}) f_w(w|\mathscr{M}) = \frac{1}{2\pi} e^{-(z^2+w^2)/2}$$

Proof. From (8-160) it follows that

$$\mathbf{z} = \sqrt{-4\ln \mathbf{q}}\, \cos \boldsymbol{\varphi} \qquad \mathbf{w} = \sqrt{-4\ln \mathbf{q}}\, \sin \boldsymbol{\varphi}$$

This system is similar to the system (8-157). To prove the theorem, it suffices, therefore, to show that the conditional density of the RV $\mathbf{r} = \sqrt{-4\ln \mathbf{q}}$ assuming $\mathscr{M}$ equals $re^{-r^2/2}$. To show this, we apply (5-5). In our case,

$$q(r) = e^{-r^2/4} \qquad q'(r) = \frac{-r}{2} e^{-r^2/4} = \frac{1}{r'(q)} \qquad f_q(q|\mathscr{M}) = 2q$$

Hence

$$f_r(r|\mathscr{M}) = f_q(q|\mathscr{M})|q'(r)| = 2e^{-r^2/4}\frac{r}{2}e^{-r^2/4} = re^{-r^2/2}$$

This shows that the conditional density of the RV **r** is Rayleigh as in (8-157).

The preceding theorem leads to the following rule for generating the sequences z_i and w_i: Form two independent sequences $x_i = 2u_i - 1$, $y_i = 2v_i - 1$.

If $q_i = \sqrt{x_i^2 + y_i^2} < 1$, set $z_i = \dfrac{x_i}{q_i}\sqrt{-4\ln q_i} \qquad w_i = \dfrac{y_i}{q_i}\sqrt{-4\ln q_i}$

Reject (x_i, y_i) otherwise.

COMPUTERS AND STATISTICS. In this section, we analyzed the dual meaning of random numbers and their computer generation. We conclude with a brief outline of the general areas of interaction between computers and statistics:

1. Statistical methods are used to solve numerically a variety of deterministic problems.
   Examples include the following: evaluation of integrals, solution of differential equations; determination of various mathematical constants. The solutions are based on the availability of RN sequences. Such sequences can be obtained from random experiments; in most cases, however, they are computer generated. We shall give a simple illustration of the two approaches in the context of Buffon's needle. The objective in this problem is the statistical estimation of the number π. The method proposed in Example 6-4 involves the performance of a physical experiment. We introduce the event $\mathscr{A} = \{\mathbf{x} < a\cos\boldsymbol{\theta}\}$ where **x** (distance from the nearest line) and $\boldsymbol{\theta}$ (angle of the needle) are two independent RVs uniform in the intervals $(0, a)$ and $(0, \pi/2)$, respectively. This event occurs if the needle intersects one of the lines and its probability equals $\pi b/2a$. From this it follows that

$$P(\mathscr{A}) \simeq \frac{n_{\mathscr{A}}}{n} \qquad \pi \simeq \frac{2a}{nb}n_{\mathscr{A}} \tag{8-162}$$

   where $n_{\mathscr{A}}$ is the number of intersections in n trials. The above estimate can be obtained without experimentation. We form two independent RN sequences x_i and θ_i with distributions $F_x(x)$ and $F_\theta(\theta)$, respectively, and we denote by $n_{\mathscr{A}}$ the number of times $x_i < a\cos\theta_i$. With $n_{\mathscr{A}}$ so determined the computer generated estimate of π is obtained from (8-162).
2. Computers are used to solve a variety of deterministic problems originating in statistics.
   Examples include the following: evaluation of the mean, the variance, or other averages used in parameter estimation and hypothesis testing; classification and storage of experimental data; use of computers as instructional tools. For example, graphical demonstration of the law of large numbers or the central limit theorem. Such applications involve mostly routine computer

programs unrelated to statistics. There is, however, another class of deterministic problems the solution of which is based on statistical concepts and RN sequences. A simple illustration follows:
We are given m RVs $\mathbf{x}_1, \ldots, \mathbf{x}_n$ with known distributions and we wish to estimate the distribution of the RV $\mathbf{y} = g(\mathbf{x}_1, \ldots, \mathbf{x}_n)$. This problem can, in principle, be solved analytically; however, its solution is, in general, complex. See, for example, the problem of determining the exact distribution of the RV $\mathbf{q}$ used in the chi-square test (9-76). As we can show next, the determination of $F_y(y)$ is simplified if we use Monte Carlo techniques. Assuming for simplicity that $m = 1$, we generate an RN sequence x_i of length n with distribution the known function $F_x(x)$ and we form the RN sequence $y_i = g(x_i)$. To determine $F_y(y)$ for a specific y, we count the number n_y of samples y_i such that $y_i < y$. Inserting into (4-3), we obtain the estimate

$$F_y(y) \simeq \frac{n_y}{n} \tag{8-163}$$

A similar approach can be used to determine the u percentile x_u of $\mathbf{x}$ or to decide whether x_u is larger or smaller than a given number (see hypothesis testing, Sec. 9-2).

3. Computers are used to simulate random experiments or to verify a scientific theory.
This involves the familiar methods of simulating physical systems where now all inputs and responses are replaced by appropriate RN sequences.

PROBLEMS

8-1. Show that if $F(x, y, z)$ is a joint distribution, then for any $x_1 \le x_2$, $y_1 \le y_2$, $z_1 \le z_2$:

$$F(x_2, y_2, z_2) + F(x_1, y_1, z_1) + F(x_1, y_2, z_1) + F(x_2, y_1, z_1)$$
$$- F(x_1, y_2, z_2) - F(x_2, y_1, z_2) - F(x_2, y_2, z_1) - F(x_1, y_1, z_1) \ge 0$$

8-2. The events $\mathscr{A}, \mathscr{B}, \mathscr{C}$ are such that

$$P(\mathscr{A}) = P(\mathscr{B}) = P(\mathscr{C}) = 0.5$$
$$P(\mathscr{A}\mathscr{B}) = P(\mathscr{A}\mathscr{C}) = P(\mathscr{B}\mathscr{C}) = P(\mathscr{A}\mathscr{B}\mathscr{C}) = 0.25$$

Show that the zero–one RVs associated with these events are not independent; they are, however, independent in pairs.

8-3. Show that if the RVs $\mathbf{x}, \mathbf{y}, \mathbf{z}$ are jointly normal and independent in pairs, they are independent.

8-4. The RVs $\mathbf{x}_i$ are i.i.d. and uniform in the interval $(-0.5, 0.5)$. Show that

$$E\{(\mathbf{x}_1 + \mathbf{x}_2 + \mathbf{x}_3)^4\} = \tfrac{13}{80}$$

8-5. (*a*) Reasoning as in (6-34), show that if the RVs $\mathbf{x}, \mathbf{y}, \mathbf{z}$ are independent and their joint density has spherical symmetry:

$$f(x, y, z) = f\left(\sqrt{x^2 + y^2 + z^2}\right)$$

then they are normal with zero mean and equal variance.
(*b*) The components $\mathbf{v}_x, \mathbf{v}_y, \mathbf{v}_z$ of the velocity $\mathbf{v} = \sqrt{\mathbf{v}_x^2 + \mathbf{v}_y^2 + \mathbf{v}_z^2}$ of a particle are independent RVs with zero mean and variance kT/m. Furthermore, their joint density has spherical symmetry. Show that $\mathbf{v}$ has a Maxwell density and

$$E\{\mathbf{v}\} = 2\sqrt{\frac{2kT}{\pi m}} \qquad E\{\mathbf{v}^2\} = \frac{3kT}{m} \qquad E\{\mathbf{v}^4\} = \frac{15k^2T^2}{m^2}$$

8-6. Show that if the RVs $\mathbf{x}, \mathbf{y}, \mathbf{z}$ are such that $r_{xy} = r_{yz} = 1$, then $r_{xz} = 1$.

8-7. Show that

$$E\{\mathbf{x}_1\mathbf{x}_2|\mathbf{x}_3\} = E\{E\{\mathbf{x}_1\mathbf{x}_2|\mathbf{x}_2, \mathbf{x}_3\}|\mathbf{x}_3\} = E\{\mathbf{x}_2 E\{\mathbf{x}_1|\mathbf{x}_2, \mathbf{x}_3\}|\mathbf{x}_3\}$$

8-8. Show that $\hat{E}\{\mathbf{y}|\mathbf{x}_1\} = \hat{E}\{\hat{E}\{\mathbf{y}|\mathbf{x}_1, \mathbf{x}_2\}|\mathbf{x}_1\}$ where $\hat{E}\{\mathbf{y}|\mathbf{x}_1, \mathbf{x}_2\} = a_1\mathbf{x}_1 + a_2\mathbf{x}_2$ is the linear MS estimate of $\mathbf{y}$ terms of $\mathbf{x}_1$ and $\mathbf{x}_2$.

8-9. Show that if

$$\mathbf{x}_i \geq 0, \qquad E\{\mathbf{x}_i^2\} = M \qquad \text{and} \qquad \mathbf{s} = \sum_{i=1}^{\mathbf{n}} \mathbf{x}_i$$

then

$$E\{\mathbf{s}^2\} \leq ME\{\mathbf{n}^2\}$$

8-10. We denote by $\mathbf{x}_m$ an RV equal to the number of tosses of a coin until heads shows for the mth time. Show that if $P\{h\} = p$, then $E\{\mathbf{x}_m\} = m/p$.
Hint: $E\{\mathbf{x}_m - \mathbf{x}_{m-1}\} = E\{\mathbf{x}_1\} = p + 2pq + \cdots + npq^{n-1} + \cdots = 1/p$.

8-11. The number of daily accidents is a Poisson RV $\mathbf{n}$ with parameter a. The probability that a single accident is fatal equals p. Show that the number $\mathbf{m}$ of fatal accidents in one day is a Poisson RV with parameter ap.
Hint:

$$E\{e^{j\omega\mathbf{m}}|\mathbf{n} = n\} = \sum_{k=0}^{n} e^{j\omega k}\binom{n}{k}p^k q^{n-k} = \left(pe^{j\omega} + q\right)^n$$

8-12. The RVs $\mathbf{x}_k$ are independent with densities $f_k(x)$ and the RV $\mathbf{n}$ is independent of $\mathbf{x}_k$ with $P\{\mathbf{n} = k\} = p_k$. Show that if

$$\mathbf{s} = \sum_{k=1}^{\mathbf{n}} \mathbf{x}_k \qquad \text{then} \qquad f_s(s) = \sum_{k=1}^{\infty} p_k[f_1(s) * \cdots * f_k(s)]$$

8-13. The RVs $\mathbf{x}_i$ are i.i.d. with moment function $\Phi_x(s) = E\{e^{s\mathbf{x}_i}\}$. The RV $\mathbf{n}$ takes the values $0, 1, \ldots$ and its moment function equals $\Gamma_n(z) = E\{\mathbf{z}^{\mathbf{n}}\}$. Show that if

$$\mathbf{y} = \sum_{i=1}^{\mathbf{n}} \mathbf{x}_i \qquad \text{then} \qquad \Phi_y(s) = E\{e^{s\mathbf{y}}\} = \Gamma_n[\Phi_x(s)]$$

Hint: $E\{e^{s\mathbf{y}}|\mathbf{n} = k\} = E\{e^{s(\mathbf{x}_1 + \cdots + \mathbf{x}_k)}\} = \Phi_x^k(s)$.
Special case: If $\mathbf{n}$ is Poisson with parameter a, then $\Phi_y(s) = e^{a\Phi_x(s) - a}$.

8-14. The RVs $\mathbf{x}_i$ are i.i.d. and uniform in the interval (0, 1). Show that if $\mathbf{y} = \max \mathbf{x}_i$, then $F(y) = y^n$ for $0 \le y \le 1$.

8-15. Given an RV $\mathbf{x}$ with distribution $F_x(x)$, we form its order statistics $\mathbf{y}_k$ as in Example 8-2, and their extremes

$$\mathbf{z} = \mathbf{y}_n = \mathbf{x}_{\text{max}} \qquad \mathbf{w} = \mathbf{y}_1 = \mathbf{x}_{\text{min}}$$

Show that

$$f_{zw}(z, w) = \begin{cases} n(n-1) f_x(z) f_x(w)[F_x(z) - F_x(w)]^{n-2} & z > w \\ 0 & z < w \end{cases}$$

8-16. Given n independent $N(\eta_i, 1)$ RVs $\mathbf{z}_i$, we form the RV $\mathbf{w} = \mathbf{z}_1^2 + \cdots + \mathbf{z}_n^2$. This RV is called *noncentral chi-square* with n degrees of freedom and eccentricity $e = \eta_1^2 + \cdots + \eta_n^2$. Show that its moment generating function equals

$$\Phi_w(s) = \frac{1}{\sqrt{(1-2s)^n}} \exp\left\{\frac{es}{1-2s}\right\}$$

8-17. Show that if the RVs $\mathbf{x}_i$ are i.i.d. and normal, then their sample mean $\bar{\mathbf{x}}$ and sample variances $\mathbf{s}^2$ are two independent RVs.

8-18. Show that, if $\alpha_0 + \alpha_1 \mathbf{x}_1 + \alpha_2 \mathbf{x}_2$ is the nonhomogeneous linear MS estimate of $\mathbf{s}$ in terms of $\mathbf{x}_1$ and $\mathbf{x}_2$, then

$$\hat{E}\{\mathbf{s} - \eta_s | \mathbf{x}_1 - \eta_1, \mathbf{x}_2 - \eta_2\} = \alpha_1(\mathbf{x}_1 - \eta_1) + \alpha_2(\mathbf{x}_2 - \eta_2)$$

8-19. Shows that

$$\hat{E}\{\mathbf{y} | \mathbf{x}_1\} = \hat{E}\{\hat{E}\{\mathbf{y} | \mathbf{x}_1, \mathbf{x}_2\} | \mathbf{x}_1\}$$

8-20. We place at random n points in the interval (0, 1) and we denote by $\mathbf{x}$ and $\mathbf{y}$ the distance from the origin to the first and last point respectively. Find $F(x)$, $F(y)$, and $F(x, y)$.

8-21. Show that if the RVs $\mathbf{x}_i$ are i.i.d. with zero mean, variance σ^2, and sample variance $\bar{\mathbf{v}}$ (see Example 8-5), then

$$\sigma_{\bar{v}}^2 = \frac{1}{n}\left[E\{\mathbf{x}_i^4\} - \frac{n-3}{n-1}\sigma^4\right]$$

8-22. The RVs $\mathbf{x}_i$ are $N(0; \sigma)$ and independent. Using Prob. 7-1, show that if

$$\mathbf{z} = \frac{\sqrt{\pi}}{2n} \sum_{i=1}^{n} |\mathbf{x}_{2i} - \mathbf{x}_{2i-1}| \qquad \text{then} \quad E\{\mathbf{z}\} = \sigma \qquad \sigma_z^2 = \frac{\pi - 2}{2n}\sigma^2$$

8-23. Show that if R is the correlation matrix of the random vector $\mathbf{X}$: $[\mathbf{x}_1, \ldots, \mathbf{x}_n]$ and R^{-1} is its inverse, then

$$E\{\mathbf{X} R^{-1} \mathbf{X}'\} = n$$

8-24. Show that if the RVs $\mathbf{x}_i$ are of continuous type and independent, then, for sufficiently large n, the density of $\sin(\mathbf{x}_1 + \cdots + \mathbf{x}_n)$ is nearly equal to the density of $\sin \mathbf{x}$ where $\mathbf{x}$ is an RV uniform in the interval $(-\pi, \pi)$.

8-25. Show that if $a_n \to a$ and $E\{|\mathbf{x}_n - a_n|^2\} \to 0$, then $\mathbf{x}_n \to a$ in the MS sense as $n \to \infty$.

8-26. Using the Cauchy criterion, show that a sequence $\mathbf{x}_n$ tends to a limit in the MS sense iff the limit of $E\{\mathbf{x}_n \mathbf{x}_m\}$ as $n, m \to \infty$ exists.

8-27. An infinite sum is by definition a limit:

$$\sum_{k=1}^{\infty} \mathbf{x}_k = \lim_{n \to \infty} \mathbf{y}_n \qquad \mathbf{y}_n = \sum_{k=1}^{n} \mathbf{x}_k$$

Show that if the RVs $\mathbf{x}_k$ are independent with zero mean and variance σ_k^2, then the sum exists in the MS sense iff

$$\sum_{k=1}^{\infty} \sigma_k^2 < \infty$$

Hint:

$$E\{(\mathbf{y}_{n+m} - \mathbf{y}_n)^2\} = \sum_{k=n+1}^{n+m} \sigma_k^2$$

8-28. The RVs $\mathbf{x}_i$ are i.i.d. with density $ce^{-cx}U(x)$. Show that, if $\mathbf{x} = \mathbf{x}_1 + \cdots + \mathbf{x}_n$, then $f_x(x)$ is an Erlang density.

8-29. Using the central limit theorem, show that for large n:

$$\frac{c^n}{(n-1)!} x^{n-1} e^{-cx} \simeq \frac{c}{\sqrt{2\pi n}} e^{-(cx-n)^2/2n} \qquad x > 0$$

8-30. The resistors $\mathbf{r}_1, \mathbf{r}_2, \mathbf{r}_3, \mathbf{r}_4$ are independent RVs and each is uniform in the interval (450; 550). Using the central limit theorem, find $P\{1900 \le \mathbf{r}_1 + \mathbf{r}_2 + \mathbf{r}_3 + \mathbf{r}_4 \le 2100\}$.

8-31. Show that the central limit theorem does not hold if the RVs $\mathbf{x}_i$ have a Cauchy density.

8-32. The RVs $\mathbf{x}$ and $\mathbf{y}$ are uncorrelated with zero mean and $\sigma_x = \sigma_y = \sigma$. Show that if $\mathbf{z} = \mathbf{x} + j\mathbf{y}$, then

$$f_z(z) = f(x, y) = \frac{1}{2\pi\sigma^2} e^{-(x^2+y^2)/2\sigma^2} = \frac{1}{\pi\sigma_z^2} e^{-|z|^2/\sigma_z^2}$$

$$\Phi_z(\Omega) = \exp\left\{-\frac{1}{2}(\sigma^2 u^2 + \sigma^2 v^2)\right\} = \exp\left\{-\frac{1}{4}\sigma_z^2 |\Omega|^2\right\}$$

where $\Omega = u + jv$. This is the scalar form of (8-62).

CHAPTER 9

STATISTICS

9-1 INTRODUCTION

Probability is a mathematical discipline developed as an abstract model and its conclusions are *deductions* based on the axioms. Statistics deals with the applications of the theory to real problems and its conclusions are *inferences* based on observations. Statistics consists of two parts: analysis and design.

Analysis, or mathematical statistics, is part of probability involving mainly repeated trials and events the probability of which is close to 0 or to 1. This leads to inferences that can be accepted as near certainties (see page 12). *Design*, or applied statistics, deals with data collection and construction of experiments that can be adequately described by probabilistic models. In this chapter, we introduce the basic elements of mathematical statistics.

We start with the observation that the connection between probabilistic concepts and reality is based on the approximation

$$p \simeq \frac{n_{\mathscr{A}}}{n} \tag{9-1}$$

relating the probability $p = P(\mathscr{A})$ of an event $\mathscr{A}$ to the number $n_{\mathscr{A}}$ of successes of $\mathscr{A}$ in n trials of the underlying physical experiment. We used this empirical formula to give the relative frequency interpretation of all probabilistic concepts. For example, we showed that the mean η of an RV $\mathbf{x}$ can be approximated by the average

$$\hat{\eta} = \frac{1}{n}\sum x_i = \bar{x} \tag{9-2}$$

of the observed values x_i of $\mathbf{x}$, and its distribution $F(x)$ by the empirical

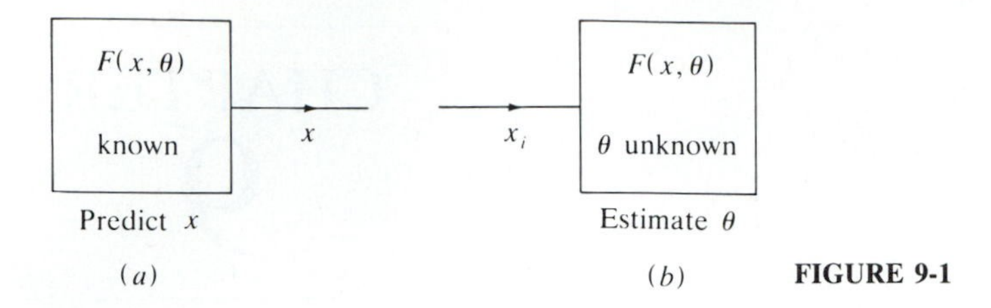

FIGURE 9-1

distribution

$$\hat{F}(x) = \frac{n_x}{n} \tag{9-3}$$

where n_x is the number of x_i's that do not exceed x. These relationships are empirical point estimates of the parameters η and $F(x)$ and a major objective of statistics is to give them an exact interpretation.

In a statistical investigation, we deal with two general classes of problems. In the first class, we assume that the probabilistic model is known and we wish to make predictions concerning future observations. For example, we know the distribution $F(x)$ of an RV $\mathbf{x}$ and we wish to predict the average $\bar{x}$ of its n future samples or we know the probability p of an event $\mathscr{A}$ and we wish to predict the number $n_{\mathscr{A}}$ of successes of $\mathscr{A}$ in n future trials. In both cases, we proceed from the model to the observations (Fig. 9-1a). In the second class, one or more parameters θ_i of the model are unknown and our objective is either to *estimate* their values (parameter estimation) or to *decide* whether θ_i is a set of known constants θ_{0i} (hypothesis testing). For example, we observe the values x_i of an RV $\mathbf{x}$ and we wish to estimate its mean η or to decide whether to accept the hypothesis that $\eta = 5.3$. We toss a coin 1000 times and heads shows 465 times. Using this information, we wish to estimate the probability p of heads or to decide whether the coin is fair. In both cases, we proceed from the observations to the model (Fig. 9-1b). In this chapter, we concentrate on parameter estimation and hypothesis testing. As a preparation, we comment briefly on the prediction problem.

Prediction. We are given an RV $\mathbf{x}$ with known distribution and we wish to predict its value x at a future trial. A *point prediction* of $\mathbf{x}$ is the determination of a constant c chosen so as to minimize in some sense the error $\mathbf{x} - c$. At a specific trial, the RV $\mathbf{x}$ can take one of many values. Hence the value that it actually takes cannot be predicted; it can only be estimated. Thus prediction of an RV $\mathbf{x}$ is the estimation of its next value x by a constant c. If we use as the criterion for selecting c the minimization of the MS error $E\{(\mathbf{x} - c)^2\}$, then $c = E\{\mathbf{x}\}$. This problem was considered in Sec. 7-3 and Sec. 8-3.

An *interval prediction* of $\mathbf{x}$ is the determination of two constants c_1 and c_2 such that

$$P\{c_1 < \mathbf{x} < c_2\} = \gamma = 1 - \delta \tag{9-4}$$

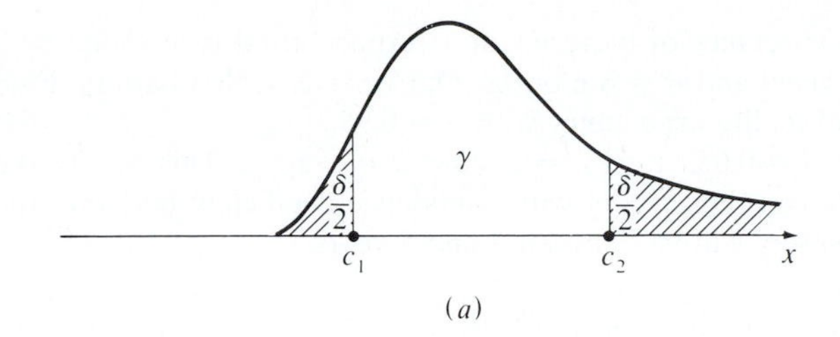

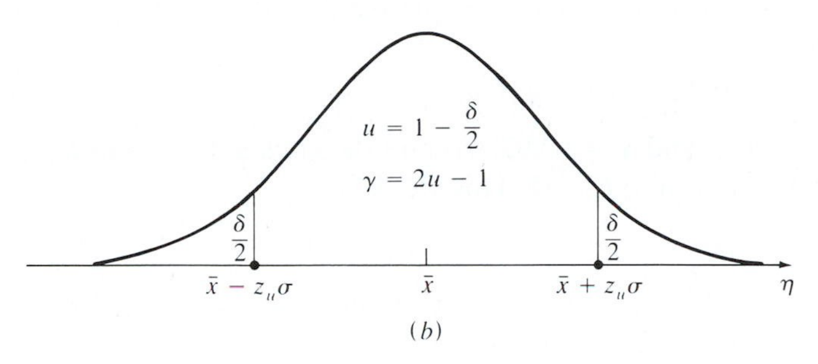

FIGURE 9-2

where γ is a given constant called the *confidence coefficient*. The above equation states that if we predict that the value x of $\mathbf{x}$ at the next trial will be in the interval (c_1, c_2), our prediction will be correct in $100\gamma\%$ of the cases. The problem in interval prediction is to find c_1 and c_2 so as to minimize the difference $c_2 - c_1$ subject to the constraint (9-4). The selection of γ is dictated by two conflicting requirements. If γ is close to 1, the prediction that x will be in the interval (c_1, c_2) is reliable but the difference $c_2 - c_1$ is large; if γ is reduced, $c_2 - c_1$ is reduced but the estimate is less reliable. Typical values of γ are 0.9, 0.95, and 0.99. For optimum prediction, we assign a value to γ and we determine c_1 and c_2 so as to minimize the difference $c_2 - c_1$ subject to the constraint (9-4). We can show that (see Prob. 9-6) if the density $f(x)$ of $\mathbf{x}$ has a single maximum, $c_2 - c_1$ is minimum if $f(c_1) = f(c_2)$. This yields c_1 and c_2 by trial and error. A simpler suboptimal solution is easily found if we determine c_1 and c_2 such that

$$P\{\mathbf{x} < c_1\} = \frac{\delta}{2} \qquad P\{\mathbf{x} > c_2\} = \frac{\delta}{2} \tag{9-5}$$

This yields $c_1 = x_{\delta/2}$ and $c_2 = x_{1-\delta/2}$ where x_u is the u percentile of $\mathbf{x}$ (Fig. 9-2a). This solution is optimum if $f(x)$ is symmetrical about its mean η because then $f(c_1) = f(c_2)$. If $\mathbf{x}$ is also normal, then $x_u = \eta + z_u\sigma$ where z_u is the standard normal percentile (Fig. 9-2b).

Example 9-1. The life expectancy of batteries of a certain brand is modeled by a normal RV with $\eta = 4$ years and $\sigma = 6$ months. Our car has such a battery. Find the prediction interval of its life expectancy with $\gamma = 0.95$.

In this example, $\delta = 0.05$, $z_{1-\delta/2} = z_{0.975} = 2 = -z_{\delta/2}$. This yields the interval $4 \pm 2 \times 0.5$. We can thus expect with confidence coefficient 0.95 that the life expectancy of our battery will be between 3 and 5 years.

As a second application, we shall estimate the number $n_{\mathscr{A}}$ of successes of an event $\mathscr{A}$ in n trials. The point estimate of $n_{\mathscr{A}}$ is the product np. The interval estimate (k_1, k_2) is determined so as to minimize the difference $k_2 - k_1$ subject to the constraint

$$P\{k_1 < n_{\mathscr{A}} < k_2\} = \gamma$$

We shall assume that n is large and $\gamma = 0.997$. To find the constants k_1 and k_2, we set $k = n_{\mathscr{A}}$ and $\varepsilon = \sqrt{pq/n}$ into (3-37). This yields

$$P\{np - 3\sqrt{npq} < n_{\mathscr{A}} < np + 3\sqrt{npq}\} = 0.997 \tag{9-6}$$

because $2\mathbb{G}(3) - 1 \simeq 0.997$. Hence we predict with confidence coefficient 0.997 that $n_{\mathscr{A}}$ will be in the interval $np \pm 3\sqrt{npq}$.

Example 9-2. We toss a fair coin 100 times and we wish to estimate the number $n_{\mathscr{A}}$ of heads with $\gamma = 0.997$. In this problem $n = 100$ and $p = 0.5$. Hence

$$k_1 = np - 3\sqrt{npq} = 35 \qquad k_2 = np - 3\sqrt{npq} = 65$$

We predict, therefore, with confidence coefficient 0.997 that the number of heads will be between 35 and 65.

The above example illustrates the role of statistics in the applications of probability to real problems: The event $\mathscr{A} = \{\text{heads}\}$ is defined in the experiment $\mathscr{S}$ of the single toss of a coin. The given information that $P(\mathscr{A}) = 0.5$ cannot be used to make a reliable prediction about the occurrence of $\mathscr{A}$ at a single performance of $\mathscr{S}$. The event

$$\mathscr{B} = \{35 < n_{\mathscr{A}} < 65\}$$

is defined in the experiment $\mathscr{S}_n$ of repeated trials and its probability equals $P(\mathscr{B}) = 0.997$. Since $P(\mathscr{B}) \simeq 1$ we can claim with near certainty that $\mathscr{B}$ will occur at a single performance of the experiment $\mathscr{S}_n$. We have thus changed the "subjective" knowledge about $\mathscr{A}$ based on the given information that $P(\mathscr{A}) = 0.5$ to the "objective" conclusion that $\mathscr{B}$ will almost certainly occur, based on the derived probability that $P(\mathscr{B}) \simeq 1$. Note, however, that both conclusions are inductive inferences; the difference between them is only quantitative.

9-2 PARAMETER ESTIMATION

Suppose that the distribution of an RV $\mathbf{x}$ is a function $F(x, \theta)$ of known form depending on a parameter θ, scalar or vector. We wish to estimate θ. To do so, we repeat the underlying physical experiment n times and we denote by x_i the observed values of **x**. Using these observations, we shall find a point estimate and an interval estimate of θ.

A *point estimate* is a function $\hat{\theta} = g(X)$ of the observation vector $X = [x_1, \ldots, x_n]$. The corresponding RV $\hat{\boldsymbol{\theta}} = g(\mathbf{X})$ is the *point estimator* of θ. Any function of the sample vector $\mathbf{X} = [\mathbf{x}_1, \ldots, \mathbf{x}_n]$ is called a *statistic*.† Thus a point estimator is a statistic.

We shall say that $\hat{\boldsymbol{\theta}}$ is an unbiased estimator of the parameter θ if $E\{\hat{\boldsymbol{\theta}}\} = \theta$. Otherwise, it is called biased with bias $b = E\{\hat{\boldsymbol{\theta}}\} - \theta$. If the function $g(X)$ is properly selected, the estimation error $\hat{\boldsymbol{\theta}} - \theta$ decreases as n increases. If it tends to 0 in probability as $n \to \infty$, then $\hat{\boldsymbol{\theta}}$ is called a *consistent* estimator. The sample mean $\bar{\mathbf{x}}$ of **x** is an unbiased estimator of its mean η. Furthermore, its variance σ^2/n tends to 0 as $n \to \infty$. From this it follows that $\bar{\mathbf{x}}$ tends to η in the MS sense, therefore, also in probability. In other words, $\bar{\mathbf{x}}$ is a consistent estimator of η. Consistency is a desirable property; however, it is a theoretical concept. In reality, the number n of trials might be large but it is finite. The objective of estimation is thus the selection of a function $g(X)$ minimizing in some sense the estimation error $g(\mathbf{X}) - \theta$. If $g(X)$ is chosen so as to minimize the MS error

$$e = E\{[g(\mathbf{X}) - \theta]^2\} = \int_R [g(X) - \theta]^2 f(X, \theta)\, dX \tag{9-7}$$

then the estimator $\hat{\boldsymbol{\theta}} = g(\mathbf{X})$ is called the *best estimator*. The determination of best estimators is not, in general, simple because the integrand in (9-7) depends not only on the function $g(X)$ but also on the unknown parameter θ. The corresponding prediction problem involves the same integral but it has a simple solution because in this case, θ is known (see Sec. 8-3).

In the following, we shall select the function $g(X)$ empirically. In this choice we are guided by the following: Suppose that θ is the mean $\theta = E\{q(\mathbf{x})\}$ of some function $q(\mathbf{x})$ of **x**. As we have noted, the sample mean

$$\hat{\boldsymbol{\theta}} = \frac{1}{n} \sum q(\mathbf{x}_i) \tag{9-8}$$

of $q(\mathbf{x})$ is a consistent estimator of θ. If, therefore, we use the sample mean $\hat{\boldsymbol{\theta}}$ of $q(\mathbf{x})$ as the point estimator of θ, our estimate will be satisfactory at least for large n. In fact, it turns out that in a number of cases it is the best estimate.

INTERVAL ESTIMATES. We measure the length θ of an object and the results are the samples $x_i = \theta + \nu_i$ of the RV $\mathbf{x} = \theta + \boldsymbol{\nu}$ where $\boldsymbol{\nu}$ is the measurement error. Can we draw with near certainty a conclusion about the true value of θ? We cannot do so if we claim that θ equals its point estimate $\hat{\theta}$ or any other

†This interpretation of the term *statistic* applies only for Chap. 9. In all other chapters, *statistics* means *statistical properties*.

constant. We can, however, conclude with near certainty that θ equals $\hat{\theta}$ within specified tolerance limits. This leads to the following concept.

An *interval estimate* of a parameter θ is an interval (θ_1, θ_2), the endpoints of which are functions $\theta_1 = g_1(X)$ and $\theta_2 = g_2(X)$ of the observation vector X. The corresponding random interval $(\boldsymbol{\theta}_1, \boldsymbol{\theta}_2)$ is the *interval estimator* of θ. We shall say that (θ_1, θ_2) is a γ *confidence interval* of θ if

$$P\{\boldsymbol{\theta}_1 < \theta < \boldsymbol{\theta}_2\} = \gamma \tag{9-9}$$

The constant γ is the *confidence coefficient* of the estimate and the difference $\delta = 1 - \gamma$ is the *confidence level*. Thus γ is a subjective measure of our confidence that the unknown θ is in the interval (θ_1, θ_2). If γ is close to 1 we can expect with near certainty that this is true. Our estimate is correct in 100γ percent of the cases. The objective of interval estimation is the determination of the functions $g_1(X)$ and $g_2(X)$ so as to minimize the length $\theta_2 - \theta_1$ of the interval (θ_1, θ_2) subject to the constraint (9-9). If $\hat{\boldsymbol{\theta}}$ is an unbiased estimator of the mean η of $\mathbf{x}$ and the density of $\mathbf{x}$ is symmetrical about η, then the optimum interval is of the form $\eta \pm a$ as in (9-10). In this section, we develop estimates of the commonly used parameters. In the selection of $\hat{\theta}$ we are guided by (9-8) and in all cases we assume that n is large. This assumption is necessary for good estimates and, as we shall see, it simplifies the analysis.

Mean

We wish to estimate the mean η of an RV $\mathbf{x}$. We use as the point estimate of η the value

$$\bar{x} = \frac{1}{n} \sum x_i$$

of the sample mean $\bar{\mathbf{x}}$ of $\mathbf{x}$. To find an interval estimate, we must determine the distribution of $\bar{\mathbf{x}}$. In general, this is a difficult problem involving multiple convolutions. To simplify it we shall assume that $\bar{\mathbf{x}}$ is normal. This is true if $\mathbf{x}$ is normal and it is approximately true for any $\mathbf{x}$ if n is large (CLT).

Known variance. Suppose first that the variance σ^2 of $\mathbf{x}$ is known. The normality assumption leads to the conclusion that the point estimator $\bar{\mathbf{x}}$ of η is $N(\eta, \sigma/\sqrt{n})$. Denoting by z_u the u percentile of the standard normal density, we conclude that

$$P\left\{\eta - z_{1-\delta/2}\frac{\sigma}{\sqrt{n}} < \bar{\mathbf{x}} < \eta + z_{1-\delta/2}\frac{\sigma}{\sqrt{n}}\right\} = \mathbb{G}(z_{1-\delta/2}) - \mathbb{G}(-z_{1-\delta/2})$$

$$= 1 - \frac{\delta}{2} - \frac{\delta}{2} \tag{9-10}$$

TABLE 9-1

$$z_{1-u} = -z_u \qquad u = \frac{1}{\sqrt{2\pi}} \int_{-\infty}^{z_u} e^{-z^2/2}\, dz$$

u	0.90	0.925	0.95	0.975	0.99	0.995	0.999	0.9995
z_u	1.282	1.440	1.645	1.967	2.326	2.576	3.090	3.291

because $z_u = -z_{1-u}$ and $\mathbb{G}(-z_{1-u}) = \mathbb{G}(z_u) = u$. This yields

$$P\left\{\bar{\mathbf{x}} - z_{1-\delta/2}\frac{\sigma}{\sqrt{n}} < \eta < \bar{\mathbf{x}} + z_{1-\delta/2}\frac{\sigma}{\sqrt{n}}\right\} = 1 - \delta = \gamma \tag{9-11}$$

We can thus state with confidence coefficient γ that η is in the interval $\bar{x} \pm z_{1-\delta/2}\sigma/\sqrt{n}$. The determination of a confidence interval for η thus proceeds as follows:

Observe the samples x_i of $\mathbf{x}$ and form their average $\bar{x}$. Select a number $\gamma = 1 - \delta$ and find the standard normal percentile z_u for $u = 1 - \delta/2$. Form the interval $\bar{x} \pm z_u\sigma/\sqrt{n}$.

This also holds for discrete-type RVs provided that n is large [see (8-110)]. The choice of the confidence coefficient γ is dictated by two conflicting requirements: If γ is close to 1, the estimate is reliable but the size $2z_u\sigma/\sqrt{n}$ of the confidence interval is large; if γ is reduced, z_u is reduced but the estimate is less reliable. The final choice is a compromise based on the applications. In Table 9-1 we list z_u for the commonly used values of u. The listed values are determined from Table 3-1 by interpolation.

Tchebycheff inequality. Suppose now that the distribution of $\bar{\mathbf{x}}$ is not known. To find the confidence interval of η, we shall use (5-57): We replace x by $\bar{x}$ and σ by $\sigma/\sqrt{n}$, and we set $\varepsilon = \sigma/n\delta$. This yields

$$P\left\{\bar{\mathbf{x}} - \frac{\sigma}{\sqrt{n\delta}} < \eta < \bar{\mathbf{x}} + \frac{\sigma}{\sqrt{n\delta}}\right\} > 1 - \delta = \gamma \tag{9-12}$$

The above shows that the exact γ confidence interval of η is contained in the interval $\bar{x} \pm \sigma/\sqrt{n\delta}$. If, therefore, we claim that η is in this interval, the probability that we are correct is larger than γ. This result holds regardless of the form of $F(x)$ and, surprisingly, it is not very different from the estimate (9-11). Indeed, suppose that $\gamma = 0.95$; in this case, $1/\sqrt{\delta} = 4.47$. Inserting into (9-12), we obtain the interval $\bar{x} \pm 4.47\sigma/\sqrt{n}$. The corresponding interval (9-11), obtained under the normality assumption, is $\bar{x} \pm 2\sigma/\sqrt{n}$ because $z_{0.975} \simeq 2$.

Unknown variance. If σ is unknown, we cannot use (9-11). To estimate η, we form the sample variance

$$s^2 = \frac{1}{n-1}\sum_{i=1}^{n}(x_i - \bar{x})^2 \tag{9-13}$$

This is an unbiased estimate of σ^2 [see (8-23)] and it tends to σ^2 as $n \to \infty$. Hence, for large n, we can use the approximation $s \simeq \sigma$ in (9-11). This yields the approximate confidence interval

$$\bar{x} - z_{1-\delta/2} \frac{s}{\sqrt{n}} < \eta < \bar{x} + z_{1-\delta/2} \frac{s}{\sqrt{n}} \tag{9-14}$$

We shall find an exact confidence interval under the assumption that $\mathbf{x}$ is normal. In this case [see (8-65)] the ratio

$$\frac{\bar{\mathbf{x}} - \eta}{\mathbf{s}/\sqrt{n}} \tag{9-15}$$

has a Student-t distribution with $n - 1$ degrees of freedom. Denoting by t_u its u percentiles, we conclude that

$$P\left\{-t_u < \frac{\bar{\mathbf{x}} - \eta}{\mathbf{s}/\sqrt{n}} < t_u\right\} = 2u - 1 = \gamma \tag{9-16}$$

This yields the interval

$$\bar{x} - t_{1-\delta/2} \frac{s}{\sqrt{n}} < \eta < \bar{x} + t_{1-\delta/2} \frac{s}{\sqrt{n}} \tag{9-17}$$

In Table 9-2 we list $t_u(n)$ for n from 1 to 20. For $n > 20$, the $t(n)$ distribution is nearly normal with zero mean and variance $n/(n-2)$ (see Prob. 7-12).

Example 9-3. The voltage V of a voltage source is measured 25 times. The results of the measurement† are the samples $x_i = V + \nu_i$ of the RV $\mathbf{x} = V + \boldsymbol{\nu}$ and their average equals $\bar{x} = 112$ V. Find the 0.95 confidence interval of V.

(a) Suppose that the standard deviation of $\mathbf{x}$ due to the error $\boldsymbol{\nu}$ is $\sigma = 0.4$ V. With $\delta = 0.05$, Table 9-1 yields $z_{0.975} \simeq 2$. Inserting into (9-11), we obtain the interval

$$\bar{x} \pm z_{0.975}\sigma/\sqrt{n} = 112 \pm 2 \times 0.4/\sqrt{25} = 112 \pm 0.16 \text{ V}$$

(b) Suppose now that σ is unknown. To estimate it, we compute the sample variance and we find $s^2 = 0.36$. Inserting into (9-14), we obtain the approximate estimate

$$\bar{x} \pm z_{0.975}s/\sqrt{n} = 112 \pm 2 \times 0.6/\sqrt{25} = 112 \pm 0.24 \text{ V}$$

Since $t_{0.975}(25) = 2.06$, the exact estimate (9-17) yields 112 ± 0.247 V.

In the following three estimates the distribution of $\mathbf{x}$ is specified in terms of a single parameter. We cannot, therefore, use (9-11) directly because the constants η and σ are related.

---

† In most examples of this chapter, we shall not list all experimental data. To avoid lengthly tables, we shall list only the relevant averages.

TABLE 9-2
Student-t Percentiles $t_u(n)$

n \ u	.9	.95	.975	.99	.995
1	3.08	6.31	12.7	31.8	63.7
2	1.89	2.92	4.30	6.97	9.93
3	1.64	2.35	3.18	4.54	5.84
4	1.53	2.13	2.78	3.75	4.60
5	1.48	2.02	2.57	3.37	4.03
6	1.44	1.94	2.45	3.14	3.71
7	1.42	1.90	2.37	3.00	3.50
8	1.40	1.86	2.31	2.90	3.36
9	1.38	1.83	2.26	2.82	3.25
10	1.37	1.81	2.23	2.76	3.17
11	1.36	1.80	2.20	2.72	3.11
12	1.36	1.78	2.18	2.68	3.06
13	1.35	1.77	2.16	2.65	3.01
14	1.35	1.76	2.15	2.62	2.98
15	1.34	1.75	2.13	2.60	2.95
16	1.34	1.75	2.12	2.58	2.92
17	1.33	1.74	2.11	2.57	2.90
18	1.33	1.73	2.10	2.55	2.88
19	1.33	1.73	2.09	2.54	2.86
20	1.33	1.73	2.09	2.53	2.85
22	1.32	1.72	2.07	2.51	2.82
24	1.32	1.71	2.06	2.49	2.80
26	1.32	1.71	2.06	2.48	2.78
28	1.31	1.70	2.05	2.47	2.76
30	1.31	1.70	2.05	2.46	2.75

For $n \geq 30$: $t_u(n) \simeq z_u \sqrt{\dfrac{n}{n-2}}$

Exponential distribution. We are given an RV $\mathbf{x}$ with density

$$f(x, \lambda) = \frac{1}{\lambda} e^{-x/\lambda} U(x)$$

and we wish to find the γ confidence interval of the parameter λ. As we know, $\eta = \lambda$ and $\sigma = \lambda$; hence, for large n, the sample mean $\bar{\mathbf{x}}$ of $\mathbf{x}$ is $N(\lambda, \lambda/\sqrt{n})$. Inserting into (9-11), we obtain

$$P\left\{\lambda - z_u \frac{\lambda}{\sqrt{n}} < \bar{\mathbf{x}} < \lambda + z_u \frac{\lambda}{\sqrt{n}}\right\} = \gamma = 2u - 1$$

This yields

$$P\left\{\frac{\bar{\mathbf{x}}}{1 + z_u/\sqrt{n}} < \lambda < \frac{\bar{\mathbf{x}}}{1 - z_u/\sqrt{n}}\right\} = \gamma \tag{9-18}$$

and the interval $\bar{x}/(1 \pm z_u/\sqrt{n})$ results.

Example 9-4. The time to failure of a light bulb is an RV **x** with exponential distribution. We wish to find the 0.95 confidence interval of λ. To do so, we observe the time to failure of 64 bulbs and we find that their average $\bar{x}$ equals 210 hours. Setting $z_u/\sqrt{n} \simeq 2/\sqrt{64} = 0.25$ into (9-18), we obtain the interval

$$168 < \lambda < 280$$

We thus expect with confidence coefficient 0.95 that the mean time to failure $E\{\mathbf{x}\} = \lambda$ of the bulb is between 168 and 280 hours.

Poisson distribution. Suppose that the RV **x** is Poisson distribution with parameter λ:

$$P\{\mathbf{x} = k\} = e^{-\lambda}\frac{\lambda^k}{k!} \qquad k = 0, 1, \ldots$$

In this case, $\eta = \lambda$ and $\sigma^2 = \lambda$; hence, for large n, the distribution of $\bar{\mathbf{x}}$ is approximately $N(\lambda, \sqrt{\lambda/n})$ [see (8-110)]. This yields

$$P\left\{|\bar{\mathbf{x}} - \lambda| < z_u\sqrt{\frac{\lambda}{n}}\right\} = \gamma$$

The points of the $\bar{x}\lambda$ plane that satisfy the inequality $|\bar{x} - \lambda| < z_u\sqrt{\lambda/n}$ are in the interior of the parabola

$$(\lambda - \bar{x})^2 = \frac{z_u^2}{n}\lambda \tag{9-19}$$

From this it follows that the γ confidence interval of λ is the vertical segment (λ_1, λ_2) of Fig. 9-3 where λ_1 and λ_2 are the roots of the quadratic (9-19).

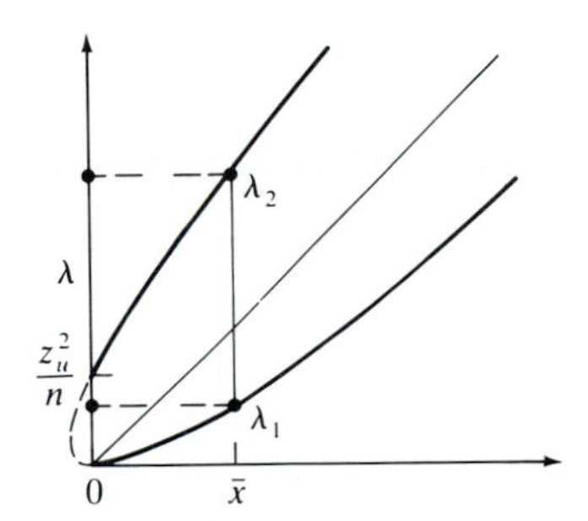

FIGURE 9-3

Example 9-5. The number of particles emitted from a radioactive substance per second is a Poisson RV $\mathbf{x}$ with parameter λ. We observe the emitted particles x_i in 64 consecutive seconds and we find that $\bar{x} = 6$. Find the 0.95 confidence interval of λ. With $z_u^2/n = 0.0625$, (9-19) yields the quadratic

$$(\lambda - 6)^2 = 0.0625\lambda$$

Solving, we obtain $\lambda_1 = 5.42$, $\lambda_2 = 6.64$. We can thus claim with confidence coefficient 0.95 that $5.42 < \lambda < 6.64$.

Probability. We wish to estimate the probability $p = P(\mathscr{A})$ of an event $\mathscr{A}$. To do so, we form the zero–one RV $\mathbf{x}$ associated with this event. As we know, $E\{\mathbf{x}\} = p$ and $\sigma_x^2 = pq$. Thus the estimation of p is equivalent to the estimation of the mean of the RV $\mathbf{x}$.

We repeat the experiment n times and we denote by k the number of successes of $\mathscr{A}$. The ratio $\bar{x} = k/n$ is the point estimate of p. To find its interval estimate, we form the sample mean $\bar{\mathbf{x}}$ of $\mathbf{x}$. For large n, the distribution of $\bar{\mathbf{x}}$ is approximately $N(p, \sqrt{pq/n})$. Hence

$$P\left\{|\bar{\mathbf{x}} - p| < z_u\sqrt{\frac{pq}{n}}\right\} = \gamma = 2u - 1$$

The points of the $\bar{x}p$ plane that satisfy the inequality $|\bar{x} - p| < z_u\sqrt{pq/n}$ are in the interior of the ellipse

$$(p - \bar{x})^2 = z_u^2\frac{p(1-p)}{n} \qquad \bar{x} = \frac{k}{n} \tag{9-20}$$

From this it follows that the γ confidence interval of p is the vertical segment (p_1, p_2) of Fig. 9-4. The endpoints p_1 and p_2 of this segment are the roots of (9-20). For $n > 100$ the following approximation can be used:

$$\begin{matrix} p_1 \\ p_2 \end{matrix} \simeq \bar{x} \pm z_u\sqrt{\frac{\bar{x}(1-\bar{x})}{n}} \qquad p_1 < p < p_2 \tag{9-21}$$

This follows from (9-20) if we replace on the right side the unknown p by its point estimate $\bar{x}$.

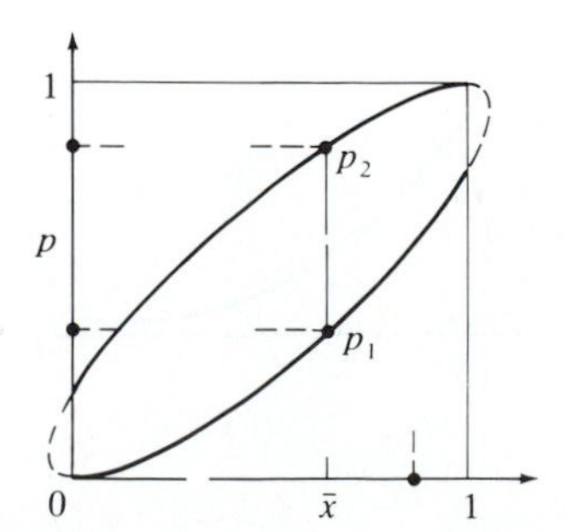

FIGURE 9-4

Example 9-6. In a preelection poll, 500 persons were questioned and 240 responded Republican. Find the 0.95 confidence interval of the probability $p = \{\text{Republican}\}$. In this example, $z_u \simeq 2$, $n = 500$, $\bar{x} = 240/500 = 0.48$, and (9-21) yields the interval 0.48 ± 0.045.

In the usual reporting of the results, the following wording is used: We estimate that 48 percent of the voters are Republican. The margin of error is ± 4.5 percent. This only specifies the point estimate and the confidence interval of the poll. The confidence coefficient (0.95 in this case) is rarely mentioned.

Variance

We wish to estimate the variance $v = \sigma^2$ of a *normal* RV $\mathbf{x}$ in terms of the n samples x_i of $\mathbf{x}$.

Known mean. We assume first that the mean η of $\mathbf{x}$ is known and we use as the point estimator of v the average

$$\hat{\mathbf{v}} = \frac{1}{n} \sum_{i=1}^{n} (\mathbf{x}_i - \eta)^2 \tag{9-22}$$

As we know,

$$E\{\hat{\mathbf{v}}\} = v \qquad \sigma_{\hat{v}}^2 = \frac{2\sigma^4}{n} \xrightarrow[n\to\infty]{} 0$$

Thus $\hat{\mathbf{v}}$ is a consistent estimator of σ^2. We shall find an interval estimate. The RV $n\hat{\mathbf{v}}/\sigma^2$ has a $\chi^2(n)$ density (see page 200). This density is not symmetrical; hence the interval estimate of σ^2 is not centered at σ^2. To determine it, we introduce two constants c_1 and c_2 such that (Fig. 9-5*a*)

$$P\left\{\frac{n\hat{\mathbf{v}}}{\sigma^2} < c_1\right\} = \frac{\delta}{2} \qquad P\left\{\frac{n\hat{\mathbf{v}}}{\sigma^2} > c_2\right\} = \frac{\delta}{2}$$

This yields $c_1 = \chi^2_{\delta/2}(n)$, $c_2 = \chi^2_{1-\delta/2}(n)$, and the interval

$$\frac{n\hat{v}}{\chi_{1-\delta/2}(n)} < \sigma^2 < \frac{n\hat{v}}{\chi_{\delta/2}(n)} \tag{9-23}$$

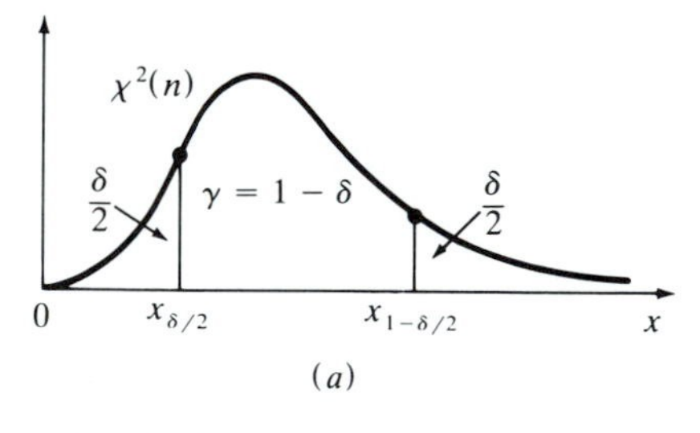

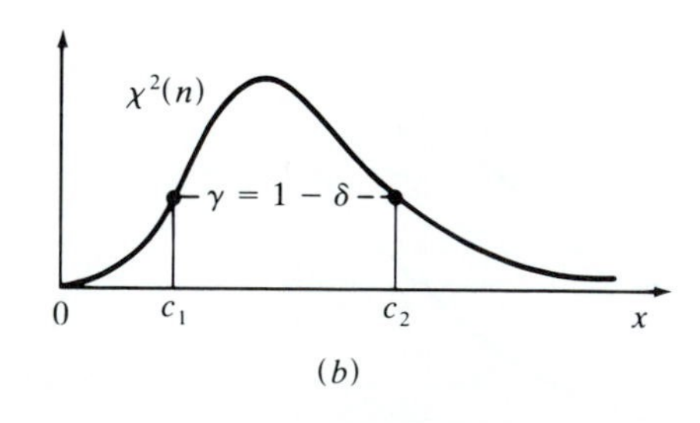

FIGURE 9-5

TABLE 9-3
Chi-square percentiles $\chi_u^2(n)$

n \ u	.005	.01	.025	.05	.1	.9	.95	.975	.99	.995
1	0.00	0.00	0.00	0.00	0.02	2.71	3.84	5.02	6.63	7.88
2	0.01	0.02	0.05	0.10	0.21	4.61	5.99	7.38	9.21	10.60
3	0.07	0.11	0.22	0.35	0.58	6.25	7.81	9.35	11.34	12.84
4	0.21	0.30	0.48	0.71	1.06	7.78	9.49	11.14	13.28	14.86
5	0.41	0.55	0.83	1.15	1.61	9.24	11.07	12.83	15.09	16.75
6	0.68	0.87	1.24	1.64	2.20	10.64	12.59	14.45	16.81	18.55
7	0.99	1.24	1.69	2.17	2.83	12.02	14.07	16.01	18.48	20.28
8	1.34	1.65	2.18	2.73	3.49	13.36	15.51	17.53	20.09	21.96
9	1.73	2.09	2.70	3.33	4.17	14.68	16.92	19.02	21.67	23.59
10	2.16	2.56	3.25	3.94	4.87	15.99	18.31	20.48	23.21	25.19
11	2.60	3.05	3.82	4.57	5.58	17.28	19.68	21.92	24.73	26.76
12	3.07	3.57	4.40	5.23	6.30	18.55	21.03	23.34	26.22	28.30
13	3.57	4.11	5.01	5.89	7.04	19.81	22.36	24.74	27.69	29.82
14	4.07	4.66	5.63	6.57	7.79	21.06	23.68	26.12	29.14	31.32
15	4.60	5.23	6.26	7.26	8.55	22.31	25.00	27.49	30.58	32.80
16	5.14	5.81	6.91	7.96	9.31	23.54	26.30	28.85	32.00	34.27
17	5.70	6.41	7.56	8.67	10.09	24.77	27.59	30.19	33.41	35.72
18	6.26	7.01	8.23	9.39	10.86	25.99	28.87	31.53	34.81	37.16
19	6.84	7.63	8.91	10.12	11.65	27.20	30.14	32.85	36.19	38.58
20	7.43	8.26	9.59	10.85	12.44	28.41	31.41	34.17	37.57	40.00
22	8.6	9.5	11.0	12.3	14.0	30.8	33.9	36.8	40.3	42.8
24	9.9	10.9	12.4	13.8	15.7	33.2	36.4	39.4	43.0	45.6
26	11.2	12.2	13.8	15.4	17.3	35.6	38.9	41.9	45.6	48.3
28	12.5	13.6	15.3	16.9	18.9	37.9	41.3	44.5	48.3	51.0
30	13.8	15.0	16.8	18.5	20.6	40.3	43.8	47.0	50.9	53.7
40	20.7	22.2	24.4	26.5	29.1	51.8	55.8	59.3	63.7	66.8
50	28.0	29.7	32.4	34.8	37.7	63.2	67.5	71.4	76.2	79.5

For $n \geq 50$: $\chi_u^2(n) \simeq \frac{1}{2}(z_u + \sqrt{2n-1})^2$

results. This interval does not have minimum length. The minimum interval is such that $f_\chi(c_1) = f_\chi(c_2)$ (Fig. 9-5*b*); however, its determination is not simple. In Table 9-3, we list the percentiles $\chi_u^2(n)$ of the $\chi^2(n)$ distribution.

Unknown mean. If η is unknown, we use as the point estimate of σ^2 the sample variance s^2 [see (9-13)]. The RV $(n-1)\mathbf{s}^2/\sigma^2$ has a $\chi^2(n-1)$ distribution. Hence

$$P\left\{\chi_{\delta/2}^2(n-1) < \frac{(n-1)\mathbf{s}^2}{\sigma^2} < \chi_{1-\delta/2}^2(n-1)\right\} = \gamma$$

This yields the interval

$$\frac{(n-1)s^2}{\chi_{1-\delta/2}^2(n-1)} < \sigma^2 < \frac{(n-1)s^2}{\chi_{\delta/2}(n-1)} \tag{9-24}$$

Example 9-7. A voltage source V is measured six times. The measurements are modeled by the RV $\mathbf{x} = V + \boldsymbol{\nu}$. We assume that the error $\boldsymbol{\nu}$ is $N(0, \sigma)$. We wish to find the 0.95 interval estimate of σ^2.

(*a*) Suppose first that the source is a known standard with $V = 110$ V. We insert the measured values $x_i = 110 + \nu_i$ of V into (9-22) and we find $\hat{v} = 0.25$. From Table 9-3 we obtain

$$\chi^2_{0.025}(6) = 1.24 \qquad \chi^2_{0.975}(6) = 14.45$$

and (9-23) yields $0.104 < \sigma^2 < 1.2$. The corresponding interval for σ is $0.332 < \sigma < 1.096$ V.

(*b*) Suppose now that V is unknown. Using the same data, we compute s^2 from (9-13) and we find $s^2 = 0.30$. From Table 9-3 we obtain

$$\chi^2_{0.025}(5) = 0.83 \qquad \chi^2_{0.975}(5) = 12.83$$

and (9-24) yields $0.117 < \sigma^2 < 1.8$. The corresponding interval for σ is $0.342 < \sigma < 1.344$ V.

PERCENTILES. The u percentile of an RV $\mathbf{x}$ is by definition a number x_u such that $F(x_u) = u$. Thus x_u is the inverse function $F^{(-1)}(u)$ of the distribution $F(x)$ of x. We shall estimate x_u in terms of the samples x_i of $\mathbf{x}$. To do so, we write the n observations x_i in ascending order and we denote by y_k the kth number so obtained. The corresponding RVs $\mathbf{y}_k$ are the order statistics of $\mathbf{x}$ [see (8-13)].

From the definition it follows that $y_k < x_u$ iff at least k of the samples x_i are less than x_u; similarly, $y_{k+r} > x_u$ iff at least $k + r$ of the samples x_i are greater than x_u. Finally, $y_k < x_u < y_{k+r}$ iff at least k and at most $k + r - 1$ of the samples x_i are less than x_u. This leads to the conclusion that the event $\{\mathbf{y}_k < x_u < \mathbf{y}_{k+r}\}$ occurs iff the number of successes of the event $\{\mathbf{x} \le x_u\}$ in n repetitions of the experiment $\mathscr{S}$ is at least k and at most $k + r - 1$. And since $P\{\mathbf{x} \le x_u\} = u$, it follows from (3-18) with $p = u$ that

$$P\{\mathbf{y}_k < x_u < \mathbf{y}_{k+r}\} = \sum_{m=k}^{k+r-1} \binom{n}{m} u^m (1-u)^{n-m} \tag{9-25}$$

Using this basic relationship, we shall find the γ confidence interval of x_u for a specific u. To do so, we must find an integer k such that the sum in (9-25) equals γ for the smallest possible r. This is a complicated task involving trial and error. A simple solution can be obtained if n is large. Using the normal approximation (3-33) with $p = nu$, we obtain

$$P\{\mathbf{y}_k < x_u < \mathbf{y}_{k+r}\} \simeq \mathbb{G}\left(\frac{k + r - 0.5 - nu}{\sqrt{nu(1-u)}}\right) - \mathbb{G}\left(\frac{k - 0.5 - nu}{\sqrt{nu(n-u)}}\right) = \gamma$$

This follows from (3-33) with $p = nu$. For a specific γ, r is minimum if nu is

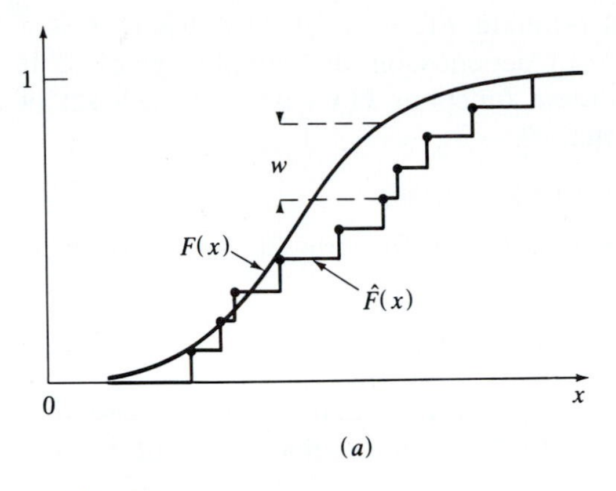

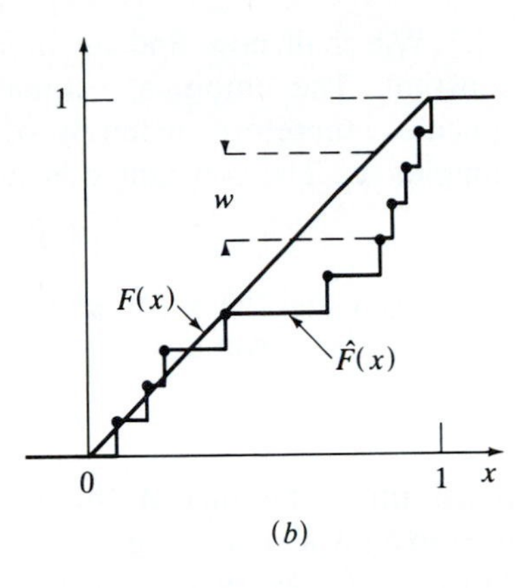

FIGURE 9-6

near the center of the interval $(k, k + r)$. This yields

$$k \simeq nu - z_{1-\delta/2}\sqrt{nu(1-u)} \qquad k + r \simeq nu + z_{1-\delta/2}\sqrt{nu(1-u)} \tag{9-26}$$

to the nearest integer.

Example 9-8. We observe 100 samples of $\mathbf{x}$ and we wish to find the 0.95 confidence interval of the median $x_{0.5}$ of $\mathbf{x}$. With $u = 0.5$, $nu = 50$, $z_{0.975} \simeq 2$, (9-26) yields $k = 40$, $k + r = 60$. Thus we can claim with confidence coefficient 0.95 that the median of $\mathbf{x}$ is between y_{40} and y_{60}.

DISTRIBUTIONS. We wish to estimate the distribution $F(x)$ of an RV $\mathbf{x}$ in terms of the samples x_i of $\mathbf{x}$. For a specific x, $F(x)$ equals the probability of the event $\{\mathbf{x} \le x\}$; hence its point estimate is the ratio n_x/n where n_x is the number of x_i's that do not exceed x. Repeating this for every x, we obtain the empirical estimate

$$\hat{F}(x) = \frac{n_x}{n}$$

of the distribution $F(x)$ [see also (4-3)]. This estimate is a staircase function (Fig. 9-6a) with discontinuities at the points x_i.

Interval estimates. For a *specific* x, the interval estimate of $F(x)$ is obtained from (9-20) with $p = F(x)$ and $\bar{x} = \hat{F}(x)$. Inserting into (9-21), we obtain the interval

$$\hat{F}(x) \pm \frac{z_u}{\sqrt{n}}\sqrt{\hat{F}(x)[1 - \hat{F}(x)]}$$

We can thus claim with confidence coefficient $\gamma = 2u - 1$ that the unknown $F(x)$ is in the above interval. Note that the length of this interval depends on x.

We shall now find an interval estimate $\hat{F}(x) \pm c$ of $F(x)$ where c is a constant. The empirical estimate $\hat{F}(x)$ depends on the samples x_i of $\mathbf{x}$. It specifies, therefore, a family of staircase functions $\hat{\mathbf{F}}(x)$, one for each set of samples x_i. The constant c is such that

$$P\{|\hat{\mathbf{F}}(x) - F(x)| \le c\} = \gamma \tag{9-27}$$

for *every* x and the γ confidence region of $F(x)$ is the strip $\hat{F}(x) \pm c$. To find c, we form the maximum

$$\mathbf{w} = \max_x |\hat{\mathbf{F}}(x) - F(x)| \tag{9-28}$$

(least upper bound) of the distance between $\hat{\mathbf{F}}(x)$ and $F(x)$. Suppose that $w = \mathbf{w}(\xi)$ is a specific value of $\mathbf{w}$. From (9-28) it follows that $w < c$ iff $\hat{F}(x) - F(x) < c$ for every x. Hence

$$\gamma = P\{\mathbf{w} \le c\} = F_w(c)$$

It suffices, therefore, to find the distribution of $\mathbf{w}$. We shall show first that the function $F_w(w)$ does not depend on $F(x)$. As we know [see (5-18)], the RV $\mathbf{y} = F(\mathbf{x})$ is uniform in the interval $(0, 1)$ for any $F(x)$. The function $y = F(x)$ transforms the points x_i to the points $y_i = F(x_i)$ and the RV $\mathbf{w}$ to itself (see Fig. 9-6b). This shows that $F_w(w)$ does not depend on the form of $F(x)$. For its determination it suffices, therefore, to assume that $\mathbf{x}$ is uniform. However, even with this simplification, it is not simple to find $F_w(w)$. We give next an approximate solution due to *Kolmogoroff*:

For large n:

$$F_w(w) \simeq 1 - 2e^{-2nw^2} \tag{9-29}$$

From this it follows that $\gamma = F_w(c) \simeq 1 - 2e^{-2nc^2}$. We can thus claim with confidence coefficient γ that the unknown $F(x)$ is between the curves $\hat{F}(x) + c$ and $\hat{F}(x) - c$ where

$$c = \sqrt{-\frac{1}{2n}\ln\frac{1-\gamma}{2}} \tag{9-30}$$

This approximation is satisfactory if $w > 1/\sqrt{n}$.

Bayesian Estimation

We return to the problem of estimating the parameter θ of a distribution $F(x, \theta)$. In our earlier approach, we viewed θ as an unknown constant and the estimate was based solely on the observed values x_i of the RV $\mathbf{x}$. This approach to estimation is called *classical*. In certain applications, θ is not totally unknown. If, for example, θ is the probability of six in the die experiment, we expect that its possible values are close to $1/6$ because most dice are reasonably fair. In *bayesian* statistics, the available prior information about θ is used in the estimation problem. In this approach, the unknown parameter θ is viewed as the value of an RV $\boldsymbol{\theta}$ and the distribution of $\mathbf{x}$ is interpreted as the conditional distribution $F_x(x|\theta)$ of $\mathbf{x}$ assuming $\boldsymbol{\theta} = \theta$. The prior information is used to assign somehow a density $f_\theta(\theta)$ to the RV $\boldsymbol{\theta}$, and the problem is to estimate the value θ of $\boldsymbol{\theta}$ in terms of the observed values x_i of $\mathbf{x}$ and the density of $\boldsymbol{\theta}$. The problem of estimating the unknown parameter θ is thus changed to the problem of estimating the value θ of the RV $\boldsymbol{\theta}$. Thus, in bayesian statistics, estimation is changed to prediction.

We shall introduce the method in the context of the following problem. We wish to estimate the inductance θ of a coil. We measure θ n times and the results are the samples $x_i = \theta + \nu_i$ of the RV $\mathbf{x} = \theta + \boldsymbol{\nu}$. If we interpret θ as an unknown number, we have a classical estimation problem. Suppose, however, that the coil is selected from a production line. In this case, its inductance θ can be interpreted as the value of an RV $\boldsymbol{\theta}$ modeling the inductances of all coils. This is a problem in bayesian estimation. To solve it, we assume first that no observations are available, that is, that the specific coil has not been measured. The available information is now the *prior* density $f_\theta(\theta)$ of $\boldsymbol{\theta}$ which we assume known and our problem is to find a constant $\hat{\theta}$ close in some sense to the unknown θ, that is, to the true value of the inductance of the particular coil. If we use the LMS criterion for selecting $\hat{\theta}$, then [see (7-62)]

$$\hat{\theta} = E\{\boldsymbol{\theta}\} = \int_{-\infty}^{\infty} \theta f_\theta(\theta)\, d\theta$$

To improve the estimate, we measure the coil n times. The problem now is to estimate θ in terms of the n samples x_i of $\mathbf{x}$. In the general case, this involves the estimation of the value θ of an RV $\boldsymbol{\theta}$ in terms of the n samples x_i of $\mathbf{x}$. Using again the MS criterion, we obtain

$$\hat{\theta} = E\{\boldsymbol{\theta}|X\} = \int_{-\infty}^{\infty} \theta f_\theta(\theta|X)\, d\theta \tag{9-31}$$

[see (8-77)] where $X = [x_1, \ldots, x_n]$ and

$$f_\theta(\theta|X) = \frac{f(X|\theta)}{f(X)} f_\theta(\theta) \tag{9-32}$$

In the above, $f(X|\theta)$ is the conditional density of the n RVs $\mathbf{x}_i$ assuming $\boldsymbol{\theta} = \theta$. If these RVs are conditionally independent, then

$$f(X|\theta) = f(x_1|\theta) \cdots f(x_n|\theta) \tag{9-33}$$

where $f(x|\theta)$ is the conditional density of the RV $\mathbf{x}$ assuming $\boldsymbol{\theta} = \theta$. These results hold in general. In the measurement problem, $f(x|\theta) = f_\nu(x - \theta)$.

We conclude with the clarification of the meaning of the various densities used in bayesian estimation, and of the underlying model, in the context of the measurement problem. The density $f_\theta(\theta)$, called *prior* (prior to the measurements), models the inductances of all coils. The density $f_\theta(\theta|X)$, called *posterior* (after the measurements), models the inductances of all coils of measured inductance x. The conditional density $f_x(x|\theta) = f_\nu(x - \theta)$ models all measurements of a particular coil of true inductance θ. This density, considered as a

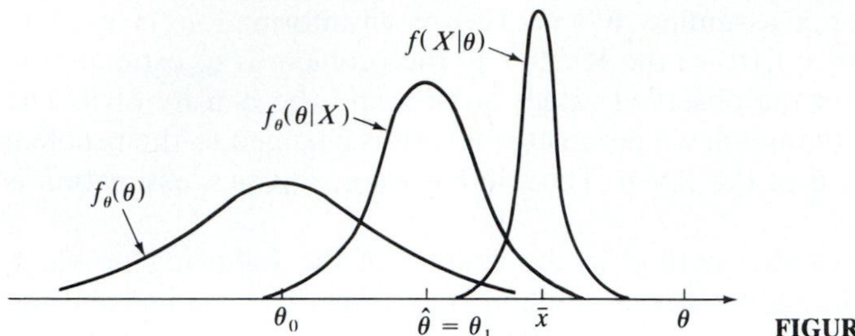

FIGURE 9-7

function of θ, is called the *likelihood* function. The unconditional density $f_x(x)$ models all measurements of all coils. Equation (9-33) is based on the reasonable assumption that the measurements of a given coil are independent.

The bayesian model is a product space $\mathscr{S} = \mathscr{S}_\theta \times \mathscr{S}_x$ where $\mathscr{S}_\theta$ is the space of the RV $\boldsymbol{\theta}$ and $\mathscr{S}_x$ is the space of the RV **x**. The space $\mathscr{S}_\theta$ is the space of all coils and $\mathscr{S}_x$ is the space of all measurements of a particular coil. Finally, $\mathscr{S}$ is the space of all measurements of all coils. The number θ has two meanings: It is the value of the RV $\boldsymbol{\theta}$ in the space $\mathscr{S}_\theta$; it is also a parameter specifying the density $f(x|\theta) = f_\nu(x - \theta)$ of the RV **x** in the space $\mathscr{S}_x$.

Example 9-9. Suppose that $\mathbf{x} = \theta + \boldsymbol{\nu}$ where $\boldsymbol{\nu}$ is an $N(0, \sigma)$ RV and θ is the value of an $N(\theta_0, \sigma_0)$ RV $\boldsymbol{\theta}$ (Fig. 9-7). Find the bayesian estimate $\hat{\theta}$ of θ.

The density $f(x|\theta)$ of **x** is $N(\theta, \sigma)$. Inserting into (9-32), we conclude that (see Prob. 9-37) the function $f_\theta(\theta|X)$ is $N(\theta_1, \sigma_1)$ where

$$\sigma_1^2 = \frac{\sigma^2}{n} \times \frac{\sigma_0^2}{\sigma_0^2 + \sigma^2/n} \qquad \theta_1 = \frac{\sigma_1^2}{\sigma_0^2}\theta_0 + \frac{n\sigma_1^2}{\sigma^2}\bar{x}$$

From the above it follows that $E\{\boldsymbol{\theta}|X\} = \theta_1$; in other words, $\hat{\theta} = \theta_1$.

Note that the classical estimate of θ is the average $\bar{x}$ of x_i. Furthermore, its prior estimate is the constant θ_0. Hence $\hat{\theta}$ is the weighted average of the prior estimate θ_0 and the classical estimate $\bar{x}$. Note further that as n tends to ∞, $\sigma_1 \to 0$ and $n\sigma_1^2/\sigma^2 \to 1$; hence $\hat{\theta}$ tends to $\bar{x}$. Thus, as the number of measurements increases, the bayesian estimate $\hat{\theta}$ approaches the classical estimate $\bar{x}$; the effect of the prior becomes negligible.

We present next the estimation of the probability $p = P(\mathscr{A})$ of an event $\mathscr{A}$. To be concrete, we assume that $\mathscr{A}$ is the event "heads" in the coin experiment. The result is based on Bayes' formula [see (4-67)]

$$f(x|\mathscr{M}) = \frac{P(\mathscr{M}|x)f(x)}{\int_{-\infty}^{\infty} P(\mathscr{M}|x)f(x)\,dx} \tag{9-34}$$

In bayesian statistics, p is the value of an RV **p** with prior density $f(p)$. In the absence of any observations, the LMS estimate $\hat{p}$ is given by

$$\hat{p} = \int_0^1 pf(p)\,dp \tag{9-35}$$

To improve the estimate, we toss the coin at hand n times and we observe that "heads" shows k times. As we know,

$$P\{\mathscr{M}|\mathbf{p} = p\} = p^k q^{n-k} \qquad \mathscr{M} = \{k \text{ heads}\}$$

Inserting into (9-34), we obtain the posterior density

$$f(p|\mathscr{M}) = \frac{p^k q^{n-k} f(p)}{\int_0^1 p^k q^{n-k} f(p)\,dp} \tag{9-36}$$

Using this function, we can estimate the probability of heads at the next toss of the coin. Replacing $f(p)$ by $f(p|\mathscr{M})$ in (9-35), we conclude that the updated estimate $\hat{p}$ of p is the conditional estimate of $\mathbf{p}$ assuming $\mathscr{M}$:

$$\int_0^1 pf(p|\mathscr{M})\,dp \tag{9-37}$$

Note that for large n, the factor $\varphi(p) = p^k(1-p)^{n-k}$ in (9-36) has a sharp maximum at $p = k/n$. Therefore, if $f(p)$ is smooth, the product $f(p)\varphi(p)$ is concentrated near k/n (Fig. 9-8a). However, if $f(p)$ has a sharp peak at $p = 0.5$ (this is the case for reasonably fair coins), then for moderate values of n, the product $f(p)\varphi(p)$ has two maxima: one near k/n and the other near 0.5 (Fig. 9-8b). As n increases, the sharpness of $\varphi(p)$ prevails and $f(p|\mathscr{M})$ is maximum near k/n (Fig. 9-8c).

Example 9-10. We toss a coin of unknown quality n times and we observe k heads. Using this information, we wish to find the bayesian estimate $\hat{p}$ of the probability p that at the next toss heads will show.

In the absence of any prior information, we assume that p is the value of an RV $\mathbf{p}$ uniformly distributed in the interval (0, 1). Setting $f(p) = 1$ in (9-36) and

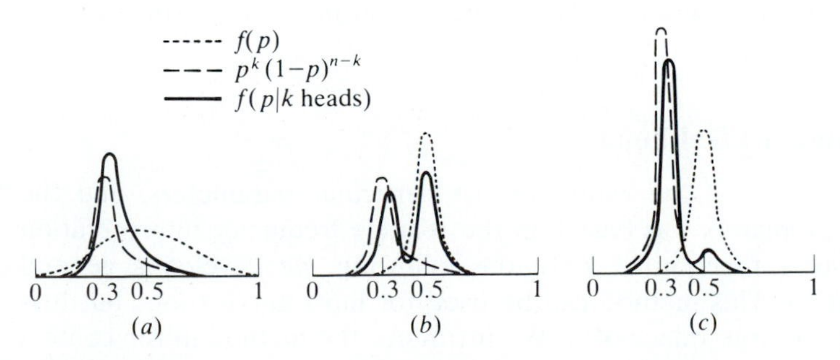

FIGURE 9-8

using the identity

$$\int_0^1 p^k(1-p)^{n-k}\,dp = \frac{k!(n-k)!}{(n+1)!}$$

we obtain

$$f(p|\mathscr{M}) = \frac{(n+1)!}{k!(n-k)!}p^k(1-p)^{n-k} \qquad 0 < p < 1$$

This function is known as the *beta density*. The updated estimate $\hat{p}$ of p is obtained from (9-37):

$$\hat{p} = \frac{(n+1)!}{k!(n-k)!}\int_0^1 p^{k+1}(1-p)^{n-k}\,dp = \frac{k+1}{n+2}$$

This result is known as the *law of succession*.

Note Bayesian estimation is a controversial subject. The controversy has its origin on the dual interpretation of the physical meaning of probability. In the first interpretation, the probability $P(\mathscr{A})$ of an event $\mathscr{A}$ is an "objective" measure of the relative frequency of the occurrence of $\mathscr{A}$ in a large number of trials. In the second interpretation, $P(\mathscr{A})$ is a "subjective" measure of our state of knowledge concerning the occurrence of $\mathscr{A}$ in a single trial. This dualism leads to two different interpretations of the meaning of parameter estimation. In the coin experiment, these interpretations take the following form:

In the classical (objective) approach, p is an unknown number. To estimate its value, we toss the coin n times and use as an estimate of p the ratio $\hat{p} = k/n$. In the bayesian (subjective) approach, p is also an unknown number, however, we interpret it as the value of an RV $\boldsymbol{\theta}$, the density of which we determine using whatever knowledge we might have about the coin. The resulting estimate of p is determined from (9-37). If we know nothing about p, we set $f(p) = 1$ and we obtain the estimate $\hat{p} = (k+1)/(n+2)$. Conceptually, the two approaches are different. However, practically, they lead in most estimates of interest to similar results if the size n of the available sample is large. In the coin problem, for example, if n is large, k is also large with high probability; hence $(k+1)/(n+2) \simeq k/n$. If n is not large, the results are different but unreliable for either method. The mathematics of bayesian estimation are also used in classical estimation problems if θ is the value of an RV the density of which can be determined objectively in terms of averages. This is the case in the problem considered in Example 9-9.

Method of Maximum Likelihood

Up to now, we considered the estimation of particular parameters, and the selection of their estimators was based on the relative frequency interpretation of the mean of some function of $\mathbf{x}$. In the following, we develop a general method of estimation. This method can be used for most applications but it is efficient primarily for large values of n. We introduce the method in the context of the following problem.

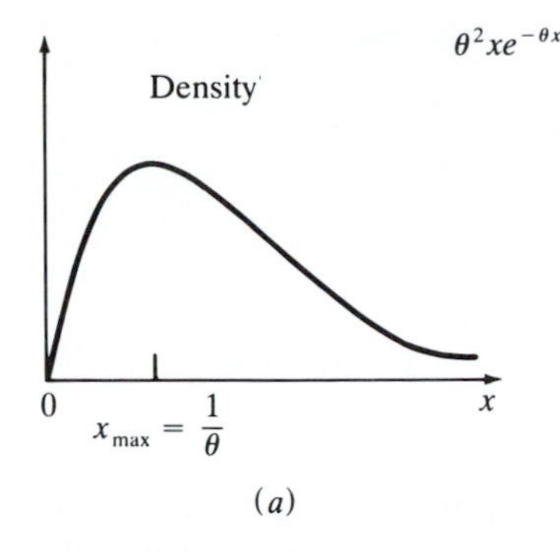

(a)

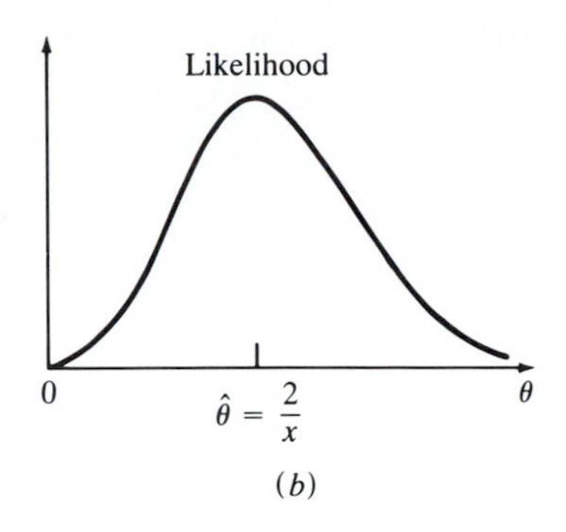

(b)

FIGURE 9-9

We have an RV **x** with density $f(x, \theta)$ and we wish to estimate θ in terms of a single observation of the RV **x**. To do so, we plot the density $f(x, \theta)$ as a function of θ, assigning to x the observed value of **x**, and we determine the value $\hat{\theta} = \theta_{max}$ of θ that maximizes $f(x, \theta)$. We shall call the curve $f(x, \theta)$ so plotted the *likelihood function* of **x** and the number $\hat{\theta}$ the *maximum likelihood* (ML) estimate of θ. This estimate is the value of θ for which the probability $f(x, \theta)\,dx$ that the RV **x** is in the interval $(x, x + dx)$ is maximum.

Example 9-11. If Fig. 9-9 we plot the Erlang density

$$f(x, \theta) = \theta^2 xe^{-\theta x} U(x)$$

as a function of x, and the corresponding likelihood function. The likelihood function is maximum for $\theta = 2/x$. Thus the ML estimate of θ in terms of the observed value x of **x** is $\hat{\theta} = 2/x$. The mode $x_{max} = 1/\theta$ of the density is the predicted value of x if θ is known (see page 179).

We shall now determine the ML estimate of θ in terms of n observations x_i of **x**. To do so, we form the joint density

$$f(X, \theta) = f(x_1, \theta) \cdots f(x_n, \theta)$$

of the n samples $\mathbf{x}_i$ of **x**. This density, considered as a function of θ is called the *likelihood function* of **X**. The value $\hat{\theta}$ of θ that maximizes $f(X, \theta)$ is the ML estimate of θ. The logarithm

$$L(X, \theta) = \ln f(X, \theta) = \sum_{i=1}^{n} \ln f(x_i, \theta) \tag{9-38}$$

is the *log-likelihood* function of **X**. From the monotonicity of the logarithm, it follows that $\hat{\theta}$ also maximizes the function $L(X, \theta)$. If $\hat{\theta}$ is in the interior of the domain Θ of θ, then $\hat{\theta}$ is a root of the equation

$$\frac{\partial L(X, \theta)}{\partial \theta} = \sum_{i=1}^{n} \frac{1}{f(x_i, \theta)} \frac{\partial f(x_i, \theta)}{\partial \theta} = 0 \tag{9-39}$$

Example 9-12. Suppose that $f(x, \theta) = \theta e^{-\theta x} U(x)$. In this case,

$$f(X, \theta) = \theta^n e^{-\theta n \bar{x}} \qquad L(X, \theta) = n \ln \theta - \theta n \bar{x}$$

Hence

$$\frac{\partial L(X, \theta)}{\partial \theta} = \frac{n}{\theta} - n\bar{x} \qquad \hat{\theta} = \frac{1}{\bar{x}}$$

Thus the ML estimator of θ equals $1/\bar{\mathbf{x}}$. This estimator is biased because $E\{1/\bar{\mathbf{x}}\} = n\theta/(n-1)$.

The ML method can be used to estimate any parameter. However, for moderate values of n, the estimate is not efficient. The method is used primarily for large values of n. This is based on the following important result.

Asymptotic properties. For large n, the distribution of the ML estimator $\hat{\boldsymbol{\theta}}$ approaches a normal curve with mean θ and variance $1/nI$ where

$$I = E\left\{\left|\frac{\partial L(\mathbf{x}, \theta)}{\partial \theta}\right|^2\right\} = \int_{-\infty}^{\infty} \left|\frac{\partial L(x, \theta)}{\partial \theta}\right|^2 f(x, \theta)\, dx \tag{9-40}$$

Thus

$$f_{\hat{\theta}}(\hat{\theta}) \simeq \sqrt{\frac{nI}{2\pi}} \exp\left\{-\frac{nI}{2}(\hat{\theta} - \theta)^2\right\} \tag{9-41}$$

The number I is called the information about θ contained in $\mathbf{x}$. Using integration by parts, we can show that (see Prob. 9-24)

$$I = -E\left\{\frac{\partial^2 L(\mathbf{x}, \theta)}{\partial \theta^2}\right\}$$

We show later [see (9-46)] that the variance of any estimator of θ cannot be smaller than $1/nI$. From this it follows that the ML estimator is asymptotically normal, unbiased, with minimum variance. In the next example, we demonstrate the validity of the above theorem. The proof will not be given.

Example 9-13. Suppose that the RV $\mathbf{x}$ is $N(\eta, \sigma)$ where η is a known constant. We wish to find the ML estimate $\hat{v}$ of its variance $v = \sigma^2$. In this problem,

$$f(X, v) = \frac{1}{(\sqrt{2\pi v})^n} \exp\left\{-\frac{1}{2v} \sum (x_i - \eta)^2\right\}$$

$$L(X, v) = -\frac{n}{2} \ln(2\pi v) - \frac{1}{2v} \sum (x_i - \eta)^2$$

Inserting into (9-39), we obtain

$$\frac{\partial L(X, v)}{\partial v} = -\frac{n}{2v} + \frac{1}{2v^2} \sum (x_i - \eta)^2 = 0$$

This yields the estimator

$$\hat{\mathbf{v}} = \frac{1}{n}\sum(\mathbf{x}_i - \eta)^2$$

As we know [see (8-67)]

$$E\{\hat{\mathbf{v}}\} = \sigma^2 \qquad \sigma_{\hat{v}}^2 = \frac{2\sigma^4}{n}$$

Furthermore, for large n the RV $\hat{\mathbf{v}}$ is nearly normal (CLT) as in (9-41). To complete the validity of (9-41), it suffices to show that $nI = 1/\sigma_{\hat{v}}^2 = n/2\sigma^4$. This follows from the identity

$$I = E\left\{-\frac{\partial^2 L(\mathbf{x}, v)}{\partial^2 v}\right\} = E\left\{-\frac{1}{2v^2} + \frac{(\mathbf{x}-\eta)^2}{v^3}\right\} = \frac{-1}{2v^2} + \frac{1}{v^2} = \frac{1}{2v^2}$$

The Rao–Cramér Bound

A basic problem in estimation is the determination of the best estimator $\hat{\boldsymbol{\theta}}$ of a parameter θ. It is easy to show that if $\hat{\boldsymbol{\theta}}$ exists, it is unique (see Prob. 9-39). However, in general, the problem of determining the best estimator of θ, or even of showing that such an estimator exists, is not simple. In the following, we determine the greatest lower bound of the variance of most estimators. This result can be used to establish whether a particular estimator is the best or that it is close to the best. We shall assume that the density $f(x, \theta)$ of $\mathbf{x}$ is differentiable with respect to θ and that the boundary of the domain of $\mathbf{x}$ does not depend on θ. Differentiating the area condition $\int f(x, \theta)\,dx = 1$ with respect to θ, we obtain the identity

$$\int_{-\infty}^{\infty} \frac{\partial f(x, \theta)}{\partial \theta}\,dx = 0 \tag{9-42}$$

A density satisfying the conditions leading to this identity will be called *regular*.

We show next that

$$E\left\{\frac{\partial L(\mathbf{X}, \theta)}{\partial \theta}\right\} = 0 \qquad E\left\{\left|\frac{\partial L(\mathbf{X}, \theta)}{\partial \theta}\right|^2\right\} = nI \tag{9-43}$$

where $L(X, \theta) = \ln f(X, \theta)$ is the log-likelihood of X and nI is the information about θ contained in X [see also (9-40)].

Proof. From the identity $L(x, \theta) = \ln f(x, \theta)$ and (9-42), it follows that

$$\int_{-\infty}^{\infty} \frac{\partial L(x, \theta)}{\partial \theta} f(x, \theta)\,dx = \int_{-\infty}^{\infty} \frac{\partial f(x, \theta)}{\partial \theta}\,dx = 0$$

This shows that the mean of the function $\partial L(\mathbf{x}, \theta)/\partial \theta$ is 0; hence its variance equals $E\{|\partial L(\mathbf{x}, \theta)/\partial \theta|^2\}$. Inserting into (9-38), we obtain (9-43) because the RVs $\ln f(\mathbf{x}_i, \theta)$ are independent.

We shall use (9-43) to determine the greatest lower bound of the variance of an arbitrary estimator $\hat{\boldsymbol{\theta}}$ of θ. Suppose first that $\hat{\boldsymbol{\theta}} = g(\mathbf{X})$ is an unbiased estimator of θ:

$$E\{\hat{\boldsymbol{\theta}}\} = \int_R g(X) f(X,\theta)\, dX = \theta$$

Differentiating with respect to θ, we obtain

$$1 = \int_R g(X) \frac{\partial f(X,\theta)}{\partial \theta}\, dX = \int_R g(X) \frac{\partial L(X,\theta)}{\partial \theta} f(X,\theta)\, dX$$

This yields

$$E\left\{ g(\mathbf{X}) \frac{\partial L(\mathbf{X},\theta)}{\partial \theta} \right\} = 1 \tag{9-44}$$

Multiplying the first equation of (9-43) by θ and subtracting from (9-44), we obtain

$$E\left\{ [g(\mathbf{X}) - \theta] \frac{\partial L(\mathbf{X},\theta)}{\partial \theta} \right\} = 1 \tag{9-45}$$

We shall use this identity to prove the following important result.

THEOREM. The variance $E\{[g(\mathbf{X}) - \theta]^2\}$ of any unbiased estimator $\hat{\boldsymbol{\theta}}$ of θ cannot be smaller than $1/nI$:

$$\sigma_{\hat{\theta}}^2 \geq \frac{1}{nI} \tag{9-46}$$

Proof. The proof is based on Schwarz's inequality

$$E^2\{\mathbf{zw}\} \leq E\{\mathbf{z}^2\} E\{\mathbf{w}^2\} \tag{9-47}$$

Squaring both sides of (9-45) and applying (9-47) to the RVs $\mathbf{z} = g(\mathbf{X}) - \theta$ and $\mathbf{w} = \partial L(\mathbf{X},\theta)/\partial\theta$ we obtain

$$1 \leq E\{[g(\mathbf{X}) - \theta]^2\} E\left\{ \left| \frac{\partial L(\mathbf{X},\theta)}{\partial \theta} \right|^2 \right\} \tag{9-48}$$

and (9-46) results.

We shall now determine the class of functions for which the estimator $\hat{\boldsymbol{\theta}}$ is best, that is, that (9-46) is an equality. As we know, (9-47) is an equality if $\mathbf{z} = c\mathbf{w}$. Hence (9-48) is an equality if $g(\mathbf{X}) - \theta = c\, \partial L(\mathbf{X},\theta)/\partial\theta$. To find c, we insert into (9-48) and use (9-43). This yields $c = 1/nI$ hence

$$\frac{\partial L(\mathbf{X},\theta)}{\partial \theta} = nI[g(\mathbf{X}) - \theta] \tag{9-49}$$

Thus the estimate $\hat{\theta} = g(\mathbf{X})$ is best if the log-likelihood function $L(X,\theta)$ satisfies (9-49).

COROLLARY. If $\hat{\boldsymbol{\theta}} = g(\mathbf{X})$ is a biased estimator of θ with mean $E\{\hat{\boldsymbol{\theta}}\} = \tau(\theta)$, then

$$\sigma_{\hat{\theta}}^2 \geq \frac{[\tau'(\theta)]^2}{nI} \tag{9-50}$$

Proof. The statistic $\hat{\boldsymbol{\theta}} = g(\mathbf{X})$ is an unbiased estimator of the parameter $\tau = \tau(\theta)$. We can, therefore, apply (9-46) provided that we replace θ by $\tau(\theta)$ and nI by the information about τ contained in X. Since

$$\frac{\partial L[X, \theta(\tau)]}{\partial \tau} = \frac{\partial L(X, \theta)}{\partial \theta} \frac{d\theta}{d\tau}$$

and $\theta'(\tau) = 1/\tau'(\theta)$ we obtain

$$E\left\{\left|\frac{\partial L[\mathbf{X}, \theta(\tau)]}{\partial \tau}\right|^2\right\} = \frac{nI}{[\tau'(\theta)]^2}$$

and (9-50) results. Reasoning as in (9-49), we conclude that (9-50) is an equality iff

$$\frac{\partial L[X, \theta(\tau)]}{\partial \tau} = \frac{nI}{[\tau'(\theta)]^2}[g(X) - \theta(\tau)] \tag{9-51}$$

Note If $f(x, \theta)$ is a density of *exponential type*, that is, if

$$f(x, \theta) = h(x)\exp\{a(\theta)q(x) + b(\theta)\} \tag{9-52}$$

then the statistic $\hat{\boldsymbol{\theta}} = (1/n)\Sigma q(\mathbf{x})$ is the best estimator of the parameter $\tau(\theta) = -b'(\theta)/a'(\theta)$. This follows readily from (9-51).

9-3 HYPOTHESIS TESTING

A statistical hypothesis is an assumption about the value of one or more parameters of a statistical model. Hypothesis testing is a process of establishing the validity of a hypothesis. This topic is fundamental in a variety of applications: Is Mendel's theory of heredity valid? Is the number of particles emitted from a radioactive substance Poisson distributed? Does the value of a parameter in a scientific investigation equal a specific constant? Are two events independent? Does the mean of an RV change if certain factors of the experiment are modified? Does smoking decrease life expectancy? Do voting patterns depend on sex? Do IQ scores depend on parental education? The list is endless.

We shall introduce the main concepts of hypothesis testing in the context of the following problem: The distribution of an RV $\mathbf{x}$ is a known function $F(x, \theta)$ depending on a parameter θ. We wish to test the assumption $\theta = \theta_0$ against the assumption $\theta \neq \theta_0$. The assumption that $\theta = \theta_0$ is denoted by H_0 and is called the *null hypothesis*. The assumption that $\theta \neq \theta_0$ is denoted by H_1 and is called the *alternative hypothesis*. The values that θ might take under the alternative hypothesis form a set Θ_1 in the parameter space. If Θ_1 consists of a single point $\theta = \theta_1$, the hypothesis H_1 is called *simple*; otherwise, it is called *composite*. The null hypothesis is in most cases simple.

The purpose of hypothesis testing is to establish whether experimental evidence supports the rejection of the null hypothesis. The decision is based on the location of the observed sample X of $\mathbf{x}$. Suppose that under hypothesis H_0 the density $f(X, \theta_0)$ of the sample vector $\mathbf{X}$ is negligible in a certain region D_c of the sample space, taking significant values only in the complement $\overline{D}_c$ of D_c. It is reasonable then to reject H_0 if X is in D_c and to accept H_0 if X is in $\overline{D}_c$. The set D_c is called the *critical region* of the test and the set $\overline{D}_c$ is called the *region of acceptance* of H_0. The test is thus specified in terms of the set D_c.

We should stress that the purpose of hypothesis testing is not to determine whether H_0 or H_1 is true. It is to establish whether the evidence supports the rejection of H_0. The terms "accept" and "reject" must, therefore, be interpreted accordingly. Suppose, for example, that we wish to establish whether the hypothesis H_0 that a coin is fair is true. To do so, we toss the coin 100 times and observe that heads show k times. If $k = 15$, we reject H_0, that is, we decide on the basis of the evidence that the fair-coin hypothesis should be rejected. If $k = 49$, we accept H_0, that is, we decide that the evidence does not support the rejection of the fair-coin hypothesis. The evidence alone, however, does not lead to the conclusion that the coin is fair. We could have as well concluded that $p = 0.49$.

In hypothesis testing two kinds of errors might occur depending on the location of X:

1. Suppose first that H_0 is true. If $X \in D_c$, we reject H_0 even though it is true. We then say that a *Type I error* is committed. The probability for such an error is denoted by α and is called the *significance level* of the test. Thus

$$\alpha = P\{\mathbf{X} \in D_c | H_0\} \tag{9-53}$$

The difference $1 - \alpha = P\{X \notin D_c | H_0\}$ equals the probability that we accept H_0 when true. In this notation, $P\{\cdots | H_0\}$ is not a conditional probability. The symbol H_0 merely indicates that H_0 is true.

2. Suppose next that H_0 is false. If $X \notin D_c$, we accept H_0 even though it is false. We then say that a *Type II error* is committed. The probability for such an error is a function $\beta(\theta)$ of θ called the *operating characteristic* (OC) of the test. Thus

$$\beta(\theta) = P\{\mathbf{X} \notin D_c | H_1\} \tag{9-54}$$

The difference $1 - \beta(\theta)$ is the probability that we reject H_0 when false. This is denoted by $P(\theta)$ and is called the *power of the test*. Thus

$$P(\theta) = 1 - \beta(\theta) = P\{\mathbf{X} \in D_c | H_1\} \tag{9-55}$$

Fundamental note Hypothesis testing is not a part of statistics. It is part of *decision theory* based on statistics. Statistical consideration alone cannot lead to a decision. They merely lead to the following probabilistic statements:

$$\text{If } H_0 \text{ is true, then } P\{\mathbf{X} \in D_c\} = \alpha \qquad \text{If } H_0 \text{ is false, then } P\{\mathbf{X} \notin D_c\} = \beta(\theta) \tag{9-56}$$

Guided by these statements, we "reject" H_0 if $\mathbf{X} \in D_c$ and we "accept" H_0 if $\mathbf{X} \notin D_c$. These decisions are not based on (9-56) alone. They take into consideration other, often subjective, factors, for example, our prior knowledge concerning the truth of H_0, or the consequences of a wrong decision.

The test of a hypothesis is specified in terms of its critical region. The region D_c is chosen so as to keep the probabilities of both types of errors small. However both probabilities cannot be arbitrarily small because a decrease in α results in an increase in β. In most applications, it is more important to control α. The selection of the region D_c proceeds thus as follows:

Assign a value to the Type I error probability α and search for a region D_c of the sample space so as to minimize the Type II error probability for a specific θ. If the resulting $\beta(\theta)$ is too large, increase α to its largest tolerable value; if $\beta(\theta)$ is still too large, increase the number n of samples.

A test is called *most powerful* if $\beta(\theta)$ is minimum. In general, the critical region of a most powerful test depends on θ. If it is the same for every $\theta \in \Theta_1$, the test is *uniformly most powerful*. Such a test does not always exist. The determination of the critical region of a most powerful test involves a search in the n-dimensional sample space. In the following, we introduce a simpler approach.

TEST STATISTIC. Prior to any experimentation, we select a function

$$\mathbf{q} = g(\mathbf{X})$$

of the sample vector $\mathbf{X}$. We then find a set R_c of the real line where under hypothesis H_0 the density of $\mathbf{q}$ is negligible, and we reject H_0 if the value $q = g(X)$ of $\mathbf{q}$ is in R_c. The set R_c is the *critical region* of the test; the RV $\mathbf{q}$ is the *test statistic*. In the selection of the function $g(X)$ we are guided by the point estimate of θ.

In a hypothesis test based on a test statistic, the two types of errors are expressed in terms of the region R_c of the real line and the density $f_q(q, \theta)$ of the test statistic $\mathbf{q}$:

$$\alpha = P\{\mathbf{q} \in R_c | H_0\} = \int_{R_c} f_q(q, \theta_0)\, dq \tag{9-57}$$

$$\beta(\theta) = P\{\mathbf{q} \notin R_c | H_1\} = \int_{\bar{R}_c} f_q(q, \theta)\, dq \tag{9-58}$$

To carry out the test, we determine first the function $f_q(q, \theta)$. We then assign a value to α and we search for a region R_c minimizing $\beta(\theta)$. The search

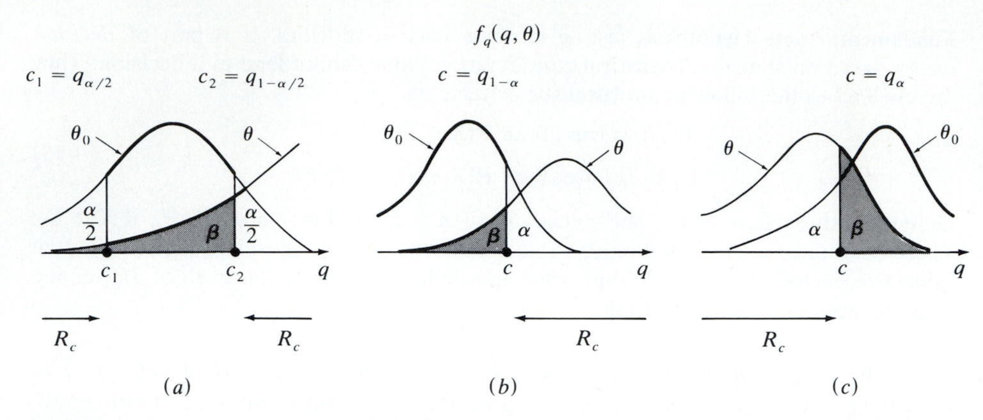

FIGURE 9-10

is now limited to the real line. We shall assume that the function $f_q(q, \theta)$ has a single maximum. This is the case for most practical tests.

Our objective is to test the hypothesis $\theta = \theta_0$ against each of the hypotheses $\theta \neq \theta_0$, $\theta > \theta_0$, and $\theta < \theta_0$. To be concrete, we shall assume that the function $f_q(q, \theta)$ is concentrated on the right of $f_q(q, \theta_0)$ for $\theta > \theta_0$ and on its left for $\theta < \theta_0$ as in Fig. 9-10.

H_1: $\theta \neq \theta_0$

Under the stated assumptions, the most likely values of $\mathbf{q}$ are on the right of $f_q(q, \theta_0)$ if $\theta > \theta_0$ and on its left if $\theta < \theta_0$. It is, therefore, desirable to reject H_0 if $\mathbf{q} < c_1$ or if $\mathbf{q} > c_2$. The resulting critical region consists of the half-lines $q < c_1$ and $q > c_2$. For convenience, we shall select the constants c_1 and c_2 such that

$$P\{\mathbf{q} < c_1 | H_0\} = \frac{\alpha}{2} \qquad P\{\mathbf{q} > c_2 | H_0\} = \frac{\alpha}{2}$$

Denoting by q_u the u percentile of $\mathbf{q}$ under hypothesis H_0, we conclude that $c_1 = q_{\alpha/2}$, $c_2 = q_{1-\alpha/2}$. This yields the following test:

$$\text{Accept } H_0 \text{ iff } q_{\alpha/2} < q < q_{1-\alpha/2} \tag{9-59a}$$

The resulting OC function equals

$$\beta(\theta) = \int_{q_{\alpha/2}}^{q_{1-\alpha/2}} f_q(q, \theta)\, dq \tag{9-60a}$$

H_1: $\theta > \theta_0$

Under hypothesis H_1, the most likely values of q are on the right of $f_q(q, \theta)$. It is, therefore, desirable to reject H_0 if $q > c$. The resulting critical region is now the half-line $q > c$ where c is such that

$$P\{\mathbf{q} > c | H_0\} = \alpha \qquad c = q_{1-\alpha}$$

and the following test results:

$$\text{Accept } H_0 \text{ iff } q < q_{1-\alpha} \tag{9-59b}$$

The resulting OC function equals

$$\beta(\theta) = \int_{-\infty}^{c} f_q(q, \theta) \, dq \tag{9-60b}$$

H_1: $\theta < \theta_0$

Proceeding similarly, we obtain the critical region $q < c$ where c is such that

$$P\{\mathbf{q} < c | H_0\} = \alpha \qquad c = q_{\alpha}$$

This yields the following test:

$$\text{Accept } H_0 \text{ iff } q > q_{\alpha} \tag{9-59c}$$

The resulting OC function equals

$$\beta(\theta) = \int_{c}^{\infty} f_q(q, \theta) \, dq \tag{9-60c}$$

The test of a hypothesis thus involves the following steps: Select a test statistic $\mathbf{q} = g(\mathbf{X})$ and determine its density. Observe the sample X and compute the function $q = g(X)$. Assign a value to α and determine the critical region R_c. Reject H_0 iff $q \in R_c$.

In the following, we give several illustrations of hypothesis testing. The results are based on (9-59) and (9-60). In certain cases, the density of $\mathbf{q}$ is known for $\theta = \theta_0$ only. This suffices to determine the critical region. The OC function $\beta(\theta)$, however, cannot be determined.

MEAN. We shall test the hypothesis H_0: $\eta = \eta_0$ that the mean η of an RV $\mathbf{x}$ equals a given constant η_0.

Known variance. We use as the test statistic the RV

$$\mathbf{q} = \frac{\bar{\mathbf{x}} - \eta_0}{\sigma / \sqrt{n}} \tag{9-61}$$

Under the familiar assumptions, $\bar{\mathbf{x}}$ is $N(\eta, \sigma/\sqrt{n})$; hence $\mathbf{q}$ is $N(\eta_q, 1)$ where

$$\eta_q = \frac{\eta - \eta_0}{\sigma / \sqrt{n}} \tag{9-62}$$

Under hypothesis H_0, $\mathbf{q}$ is $N(0, 1)$. Replacing in (9-59) and (9-60) the q_u

percentile by the standard normal percentile z_u, we obtain the following test:

H_1: $\eta \neq \eta_0$ Accept H_0 iff $z_{\alpha/2} < q < z_{1-\alpha/2}$ (9-63a)

$$\beta(\eta) = P\{|\mathbf{q}| < z_{1-\alpha/2}|H_1\} = \mathbb{G}(z_{1-\alpha/2} - \eta_q) - \mathbb{G}(z_{\alpha/2} - \eta_q) \quad (9\text{-}64a)$$

H_1: $\eta > \eta_0$ Accept H_0 iff $q < z_{1-\alpha}$ (9-63b)

$$\beta(\eta) = P\{\mathbf{q} < z_{1-\alpha}|H_1\} = \mathbb{G}(z_{1-\alpha} - \eta_q) \quad (9\text{-}64b)$$

H_1: $\eta < \eta_0$ Accept H_0 iff $q > z_\alpha$ (9-63c)

$$\beta(\eta) = P\{\mathbf{q} > z_\alpha|H_1\} = 1 - \mathbb{G}(z_\alpha - \eta_q) \quad (9\text{-}64c)$$

Unknown variance. We assume that **x** is normal and use as the test statistic the RV

$$\mathbf{q} = \frac{\bar{\mathbf{x}} - \eta_0}{\mathbf{s}/\sqrt{n}} \quad (9\text{-}65)$$

where $\mathbf{s}^2$ is the sample variance of **x**. Under hypothesis H_0, the RV **q** has a Student-t distribution with $n - 1$ degrees of freedom. We can, therefore, use (9-59) where we replace q_u by the tabulated $t_u(n - 1)$ percentile. To find $\beta(\eta)$, we must find the distribution of **q** for $\eta \neq \eta_0$.

Example 9-14. We measure the voltage V of a voltage source 25 times and we find $\bar{x} = 110.12$ V (see also Example 9-3). Test the hypothesis $V = V_0 = 110$ V against $V \neq 110$ V with $\alpha = 0.05$. Assume that the measurement error $\boldsymbol{\nu}$ is $N(0, \sigma)$.

(a) Suppose that $\sigma = 0.4$ V. In this problem, $z_{1-\alpha/2} = z_{0.975} = 2$:

$$q = \frac{110.12 - 110}{0.4/\sqrt{25}} = 1.5$$

Since 1.5 is in the interval $(-2, 2)$, we accept H_0.

(b) Suppose that σ is unknown. From the measurements we find $s = 0.6$ V. Inserting into (9-65), we obtain

$$q = \frac{110.12 - 110}{0.6/\sqrt{25}} = 1$$

Table 9-3 yields $t_{1-\alpha/2}(n - 1) = t_{0.975}(25) = 2.06 = -t_{0.025}$. Since 1 is in the interval $(-2.06, 2.06)$, we accept H_0.

PROBABILITY. We shall test the hypothesis H_0: $p = p_0 = 1 - q_0$ that the probability $p = P(\mathscr{A})$ of an event $\mathscr{A}$ equals a given constant p_0, using as data the number k of successes of $\mathscr{A}$ in n trials. The RV **k** has a binomial distribution and for large n it is $N(np, \sqrt{npq})$. We shall assume that n is large.

The test will be based on the test statistic

$$\mathbf{q} = \frac{\mathbf{k} - np_0}{\sqrt{np_0q_0}} \quad (9\text{-}66)$$

Under hypothesis H_0, **q** is $N(0, 1)$. The test thus proceeds as in (9-63).

To find the OC function $\beta(p)$, we must determine the distribution of $\mathbf{q}$ under the alternative hypothesis. Since $\mathbf{k}$ is normal, $\mathbf{q}$ is also normal with

$$\eta_q = \frac{np - np_0}{\sqrt{np_0 q_0}} \qquad \sigma_q^2 = \frac{npq}{np_0 q_0}$$

This yields the following test:

H_1: $p \neq p_0$ Accept H_0 iff $z_{\alpha/2} < q < z_{1-\alpha/2}$ (9-67a)

$$\beta(p) = P\{|\mathbf{q}| < z_{1-\alpha/2}|H_1\} = \mathbb{G}\left(\frac{z_{1-\alpha/2} - \eta_q}{\sqrt{pq/p_0 q_0}}\right) - \mathbb{G}\left(\frac{z_{\alpha/2} - \eta_q}{\sqrt{pq/p_0 q_0}}\right) \quad (9\text{-}68a)$$

H_1: $p > p_0$ Accept H_0 iff $q < z_{1-\alpha}$ (9-67b)

$$\beta(p) = P\{\mathbf{q} < z_{1-\alpha}|H_1\} = \mathbb{G}\left(\frac{z_{1-\alpha} - \eta_q}{\sqrt{pq/p_0 q_0}}\right) \quad (9\text{-}68b)$$

H_1: $p < p_0$ Accept H_0 iff $q > z_\alpha$ (9-67c)

$$\beta(p) = P\{\mathbf{q} > z_\alpha|H_1\} = 1 - \mathbb{G}\left(\frac{z_\alpha - \eta_q}{\sqrt{pq/p_0 q_0}}\right) \quad (9\text{-}68c)$$

Example 9-15. We wish to test the hypothesis that a coin is fair against the hypothesis that it is loaded in favor of "heads":

$$H_0: p = 0.5 \qquad \text{against} \qquad H_1: p > 0.5$$

We toss the coin 100 times and "heads" shows 62 times. Does the evidence support the rejection of the null hypothesis with significance level $\alpha = 0.05$?

In this example, $z_{1-\alpha} = z_{0.95} = 1.645$. Since

$$q = \frac{62 - 50}{\sqrt{25}} = 2.4 > 1.645$$

the fair-coin hypothesis is rejected.

VARIANCE. The RV $\mathbf{x}$ is $N(\eta, \sigma)$. We wish to test the hypothesis H_0: $\sigma = \sigma_0$.

Known mean. We use as test statistic the RV

$$\mathbf{q} = \sum_i \left(\frac{\mathbf{x}_i - \eta}{\sigma_0}\right)^2 \quad (9\text{-}69)$$

Under hypothesis H_0, this RV is $\chi^2(n)$. We can, therefore, use (9-59) where q_u equals the $\chi_u^2(n)$ percentile.

Unknown mean. We use as the test statistic the RV

$$\mathbf{q} = \sum_i \left(\frac{\mathbf{x}_i - \bar{\mathbf{x}}}{\sigma_0}\right)^2 \quad (9\text{-}70)$$

Under hypothesis H_0, this RV is $\chi^2(n-1)$. We can, therefore, use (9-59) with $q_u = \chi_u^2(n-1)$.

Example 9-16. Suppose that in Example 9-14, the variance σ^2 of the measurement error is unknown. Test the hypothesis H_0: $\sigma = 0.4$ against H_1: $\sigma > 0.4$ with $\alpha = 0.05$ using 20 measurements $x_i = V + \nu_i$.

(*a*) Assume that $V = 110$ V. Inserting the measurements x_i into (9-69), we find

$$q = \sum_{i=1}^{20} \left(\frac{x_i - 110}{0.4} \right)^2 = 36.2$$

Since $\chi^2_{1-\alpha}(n) = \chi^2_{0.95}(20) = 31.41 < 36.2$, we reject H_0.

(*b*) If V is unknown, we use (9-70). This yields

$$q = \sum_{i=1}^{20} \left(\frac{x_i - \bar{x}}{0.4} \right)^2 = 22.5$$

Since $\chi^2_{1-\alpha}(n-1) = \chi^2_{0.95}(19) = 30.14 > 22.5$, we accept H_0.

DISTRIBUTIONS. In this application, H_0 does not involve a parameter; it is the hypothesis that the distribution $F(x)$ of an RV **x** equals a given function $F_0(x)$. Thus

$$H_0: F(x) \equiv F_0(x) \qquad \text{against} \qquad H_1: F(x) \neq F_0(x)$$

The Kolmogoroff–Smirnov test. We form the random process $\hat{\mathbf{F}}(x)$ as in the estimation problem (see page 256) and use as the test statistic the RV

$$\mathbf{q} = \max_x |\hat{\mathbf{F}}(x) - F_0(x)| \tag{9-71}$$

This choice is based on the following observations: For a specific ζ, the function $\hat{F}(x)$ is the empirical estimate of $F(x)$ [see (4-3)]; it tends, therefore, to $F(x)$ as $n \to \infty$. From this it follows that

$$E\{\hat{\mathbf{F}}(x)\} = F(x) \qquad \hat{\mathbf{F}}(x) \underset{n\to\infty}{\longrightarrow} F(x)$$

This shows that for large n, $\mathbf{q}$ is close to 0 if H_0 is true and it is close to $\max|F(x) - F_0(x)|$ if H_1 is true. It leads, therefore, to the conclusion that we must reject H_0 if q is larger than some constant c. This constant is determined in terms of the significance level $\alpha = P\{\mathbf{q} > c|H_0\}$ and the distribution of $\mathbf{q}$. Under hypothesis H_0, the test statistic $\mathbf{q}$ equals the RV $\mathbf{w}$ in (9-28). Using the Kolmogoroff approximation (9-29), we obtain

$$\alpha = P\{\mathbf{q} > c|H_0\} \simeq 2e^{-2nc^2} \tag{9-72}$$

The test thus proceeds as follows: Form the empirical estimate $\hat{F}(x)$ of $F(x)$

and determine q from (9-71).

$$\text{Accept } H_0 \text{ iff } q < \sqrt{-\frac{1}{2n}\ln\frac{\alpha}{2}} \tag{9-73}$$

The resulting Type II error probability is reasonably small only if n is large.

Chi-Square Tests

We are given a partition $\mathfrak{A} = [\mathscr{A}_1, \ldots, \mathscr{A}_m]$ of the space $\mathscr{S}$ and we wish to test the hypothesis that the probabilities $p_i = P(\mathscr{A}_i)$ of the events $\mathscr{A}_i$ equal m given constants p_{0i}:

$$H_0\colon p_i = p_{0i}, \text{ all } i \qquad \text{against} \qquad H_1\colon p_i \neq p_{0i}, \text{ some } i \tag{9-74}$$

using as data the number of successes k_i of each of the events $\mathscr{A}_i$ in n trials. For this purpose, we introduce the sum

$$\mathbf{q} = \sum_{i=1}^{m} \frac{(\mathbf{k}_i - np_{0i})^2}{np_{0i}} \tag{9-75}$$

known as *Pearson's test statistic*. As we know, the RVs $\mathbf{k}_i$ have a binomial distribution with mean np_i and variance np_iq_i. Hence the ratio $\mathbf{k}_i/n$ tends to p_i as $n \to \infty$. From this it follows that the difference $|\mathbf{k}_i - np_{0i}|$ is small if $p_i = p_{0i}$ and it increases as $|p_i - p_{0i}|$ increases. This justifies the use of the RV $\mathbf{q}$ as a test statistic and the set $q > c$ as the critical region of the test.

To find c, we must determine the distribution of $\mathbf{q}$. We shall do so under the assumption that n is large. For moderate values of n, we use computer simulation [see (9-85)]. With this assumption, the RVs $\mathbf{k}_i$ are nearly normal with mean kp_i. Under hypothesis H_0, the RV $\mathbf{q}$ has a $\chi^2(m-1)$ distribution. This follows from the fact that the constants p_{0i} satisfy the constraint $\Sigma p_{0i} = 1$. The proof, however, is rather involved.

The above leads to the following test: Observe the numbers k_i and compute the sum q in (9-75); find $\chi^2_{1-\alpha}(m-1)$ from Table 9-3.

$$\text{Accept } H_0 \text{ iff } q < \chi^2_{1-\alpha}(m-1) \tag{9-76}$$

We note that the chi-square test is reduced to the test (9-68) involving the probability p of an event $\mathscr{A}$. In this case, the partition $\mathfrak{A}$ equals $[\mathscr{A}, \overline{\mathscr{A}}]$ and the statistic $\mathbf{q}$ in (9-75) equals $(k - np_0)^2/np_0q_0$ where $p_0 = p_{01}$, $q_0 = p_{02}$, $k = k_1$, and $n - k = k_2$ (see Prob. 9-40).

Example 9-17. We roll a die 300 times and we observe that f_i shows k_i = 55 43 44 61 40 57 times. Test the hypothesis that the die is fair with $\alpha = 0.05$. In this problem, $p_{0i} = 1/6$, $m = 6$, and $np_{0i} = 50$. Inserting into (9-75), we obtain

$$q = \sum_{i=1}^{6} \frac{(k_i - 50)^2}{50} = 7.6$$

Since $\chi^2_{0.95}(5) = 11.07 > 7.6$, we accept the fair-die hypothesis.

The chi-square test is used in *goodness-of-fit* tests involving the agreement between experimental data and theoretical models. We next give two illustrations.

TESTS OF INDEPENDENCE. We shall test the hypothesis that two events $\mathscr{B}$ and $\mathscr{C}$ are independent:

$$H_0: P(\mathscr{B} \cap \mathscr{C}) = P(\mathscr{B})P(\mathscr{C}) \qquad \text{against} \qquad H_1: P(\mathscr{B} \cap \mathscr{C}) \neq P(\mathscr{B})P(\mathscr{C}) \tag{9-77}$$

under the assumption that the probabilities $b = P(\mathscr{B})$ and $c = P(\mathscr{C})$ of these events are known. To do so, we apply the chi-square test to the partition consisting of the four events

$$\mathscr{A}_1 = \mathscr{B} \cap \mathscr{C} \qquad \mathscr{A}_2 = \mathscr{B} \cap \overline{\mathscr{C}} \qquad \mathscr{A}_3 = \overline{\mathscr{B}} \cap \mathscr{C} \qquad \mathscr{A}_4 = \overline{\mathscr{B}} \cap \overline{\mathscr{C}}$$

Under hypothesis H_0, the components of each of the events $\mathscr{A}_i$ are independent. Hence

$$p_{01} = bc \qquad p_{02} = b(1-c) \qquad p_{03} = (1-b)c \qquad p_{04} = (1-b)(1-c)$$

This yields the following test:

$$\text{Accept } H_0 \text{ iff } \sum_{k=1}^{4} \frac{(k_i - np_{0i})^2}{np_{0i}} < \chi^2_{1-\alpha}(3) \tag{9-78}$$

In the above, k_i is the number of occurrences of the event $\mathscr{A}_i$; for example, k_2 is the number of times $\mathscr{B}$ occurs but $\mathscr{C}$ does not occur.

Example 9-18. In a certain university, 60 percent of all first-year students are male and 75 percent of all entering students graduate. We select at random the records of 299 males and 101 females and we find that 168 males and 68 females graduated. Test the hypothesis that the events $\mathscr{B}$ = {male} and $\mathscr{C}$ = {graduate} are independent with $\alpha = 0.05$. In this problem, $m = 400$, $P(\mathscr{B}) = 0.6$, $P(\mathscr{C}) = 0.75$, $p_{0i} = 0.45\ 0.15\ 0.3\ 0.1$, $k_i = 168\ 68\ 131\ 33$, and (9-75) yields

$$q = \sum_{i=1}^{4} \frac{(k_i - 400p_{0i})^2}{400p_{0i}} = 4.1$$

Since $\chi^2_{0.95}(3) = 7.81 > 4.1$, we accept the independence hypothesis.

TESTS OF DISTRIBUTIONS. We introduced earlier the problem of testing the hypothesis that the distribution $F(x)$ of an RV $\mathbf{x}$ equals a given function $F_0(x)$. The resulting test is reliable only if the number of available samples x_j of $\mathbf{x}$ is very large. In the following, we test the hypothesis that $F(x) = F_0(x)$ not at every x but only at a set of $m-1$ points a_i (Fig. 9-11):

$$H_0: F(a_i) = F_0(a_i),\ 1 \le i \le m-1 \qquad \text{against} \qquad H_1: F(a_i) \neq F_0(a_i),\ \text{some } i \tag{9-79}$$

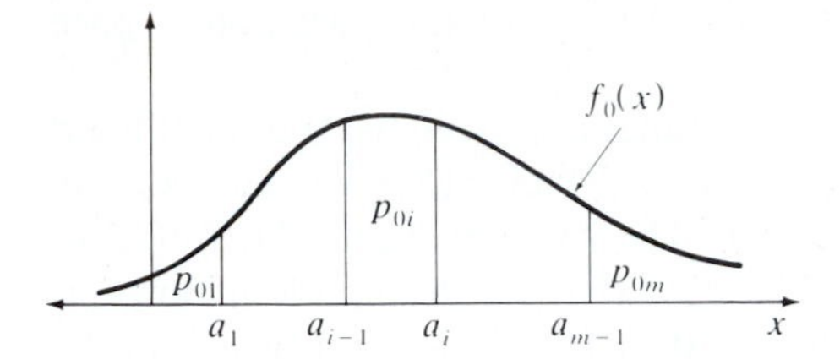

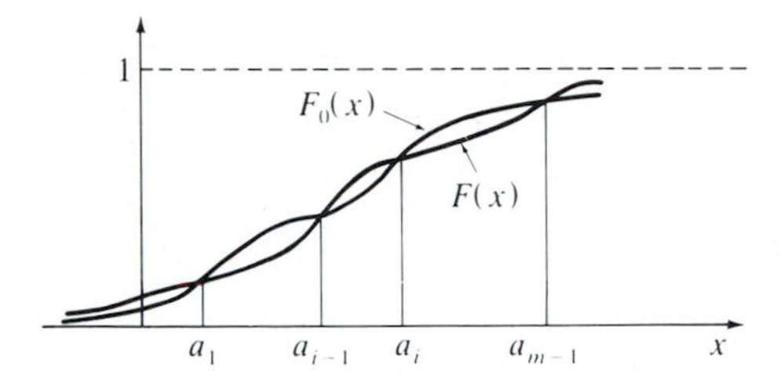

FIGURE 9-11

We introduce the m events

$$\mathscr{A}_i = \{a_{i-1} < \mathbf{x} \le a_i\} \qquad i = 1, \ldots, m$$

where $a_0 = -\infty$ and $a_m = \infty$. These events form a partition of $\mathscr{S}$. The number k_i of successes of $\mathscr{A}_i$ equals the number of samples x_j in the interval (a_{i-1}, a_i). Under hypothesis H_0,

$$P(\mathscr{A}_i) = F_0(a_i) - F_0(a_{i-1}) = p_{0i}$$

Thus, to test the hypothesis (9-79), we form the sum q in (9-75) and apply (9-76). If H_0 is rejected, then the hypothesis that $F(x) = F_0(x)$ is also rejected.

Example 9-19. We have a list of 500 computer-generated decimal numbers x_j and we wish to test the hypothesis that they are the samples of an RV $\mathbf{x}$ uniformly distributed in the interval $(0, 1)$. We divide this interval into 10 subintervals of length 0.1 and we count the number k_i of samples x_j that are in the ith subinterval. The results are

$$k_i = 43 \quad 56 \quad 42 \quad 38 \quad 59 \quad 61 \quad 41 \quad 57 \quad 46 \quad 57$$

In this problem, $m = 500$, $p_{0i} = 0.1$, and

$$q = \sum_{i=1}^{10} \frac{(k_i - 50)^2}{50} = 13.8$$

Since $\chi^2_{0.95}(9) = 16.9 > 13.8$ we accept the uniformity hypothesis.

Likelihood Ratio Test

We conclude with a general method for testing any hypothesis, simple or composite. We are given an RV $\mathbf{x}$ with density $f(x, \theta)$, where θ is an arbitrary parameter, scalar or vector, and we wish to test the hypothesis H_0: $\theta \in \Theta_0$

against H_1: $\theta \in \Theta_1$. The sets Θ_0 and Θ_1 are subsets of the parameter space $\Theta = \Theta_0 \cup \Theta_1$.

The density $f(X, \theta)$, considered as a function of θ, is the likelihood function of $\mathbf{X}$. We denote by θ_m the value of θ for which $f(\mathbf{X}, \theta)$ is maximum in the space Θ. Thus θ_m is the ML estimate of θ. The value of θ for which $f(X, \theta)$ is maximum in the set Θ_0 will be denoted by θ_{m0}. If H_0 is the simple hypothesis $\theta = \theta_0$, then $\theta_{m0} = \theta_0$. The maximum likelihood (ML) test is a test based on the statistic

$$\boldsymbol{\lambda} = \frac{f(\mathbf{X}, \theta_{m0})}{f(\mathbf{X}, \theta_m)} \tag{9-80}$$

Note that

$$0 \le \boldsymbol{\lambda} \le 1$$

because $f(X, \theta_{m0}) \le f(X, \theta_m)$. We maintain that $\boldsymbol{\lambda}$ is concentrated near 1 if H_0 is true. As we know [see (9-41)], the ML estimate θ_m of θ tends to its true value θ^* as $n \to \infty$. Furthermore, under the null hypothesis, θ^* is in the set Θ_0; hence $\lambda \to 1$ as $n \to \infty$. From this it follows that we must reject H_0 if $\lambda < c$. The constant c is determined in terms of the significance level α of the test.

Suppose, first, that H_0 is the simple hypothesis $\theta = \theta_0$. In this case,

$$\alpha = P\{\boldsymbol{\lambda} \le c \mid H_0\} = \int_0^c f_\lambda(\lambda, \theta_0)\, d\lambda \tag{9-81}$$

This leads to the following test: Using the samples x_i of $\mathbf{x}$, form the likelihood function $f(X, \theta)$. Find θ_m and θ_{m0} and form the ratio $\lambda = f(X, \theta_{m0})/f(X, \theta_m)$:

$$\text{Reject } H_0 \text{ iff } \lambda < \lambda_\alpha \tag{9-82}$$

where λ_α is the α percentile of the test statistic $\boldsymbol{\lambda}$ under hypothesis H_0.

If H_0 is a composite hypothesis, c is the smallest constant such that $P\{\boldsymbol{\lambda} \le c\} < \lambda_\alpha$ for every $\theta \in \Theta_0$.

Example 9-20. Suppose that $f(x, \theta) \sim \theta e^{-\theta x} U(x)$. We shall test the hypothesis

$$H_0: 0 < \theta \le \theta_0 \qquad \text{against} \qquad H_1: \theta > \theta_0$$

In this problem, Θ_0 is the segment $0 < \theta \le \theta_0$ of the real line and Θ is the half-line $\theta > 0$. Thus both hypotheses are composite. The likelihood function

$$f(\mathbf{X}, \theta) = \theta^n e^{-n\bar{x}\theta}$$

is shown in Fig. 9-12a for $\bar{x} > 1/\theta_0$ and $\bar{x} < 1/\theta_0$. In the half-line $\theta > 0$ this function is maximum for $\theta = 1/\bar{x}$. In the interval $0 < \theta \le \theta_0$ it is maximum for $\theta = 1/\bar{x}$ if $\bar{x} > 1/\theta_0$ and for $\theta = \theta_0$ if $\bar{x} < 1/\theta_0$. Hence

$$\theta_m = \frac{1}{\bar{x}} \qquad \theta_{m0} = \begin{cases} 1/\bar{x} & \text{for } \bar{x} > 1/\theta_0 \\ \theta_0 & \text{for } \bar{x} < 1/\theta_0 \end{cases}$$

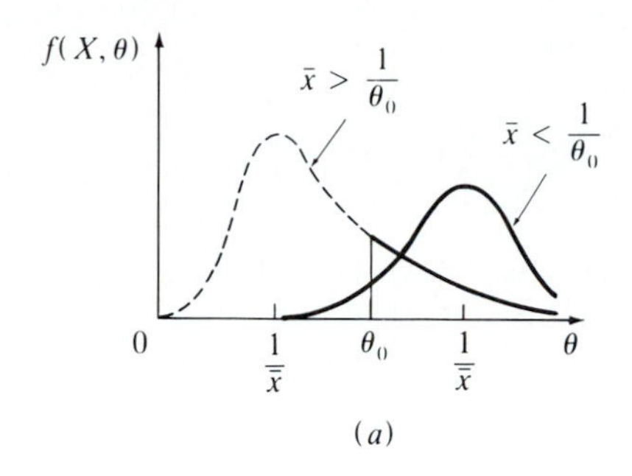

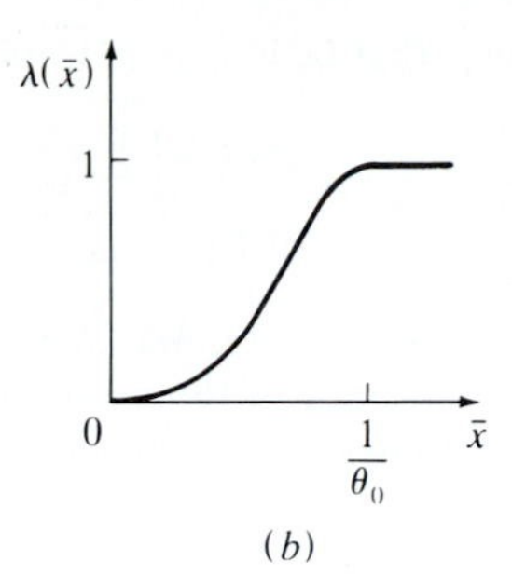

FIGURE 9-12

The likelihood ratio equals (Fig. 9-12*b*)

$$\lambda = \begin{cases} 1 & \text{for} \quad \bar{x} > 1/\theta_0 \\ (\bar{x}\theta_0)^n e^{-n\theta_0\bar{x}+n\theta_0} & \text{for} \quad \bar{x} < 1/\theta_0 \end{cases}$$

We reject H_0 if $\lambda < c$ or, equivalently, if $\bar{x} < c_1$ where c_1 equals the α percentile of the RV $\bar{\mathbf{x}}$.

To carry out a likelihood ratio test, we must determine the density of $\boldsymbol{\lambda}$. This is not always a simple task. The following theorem simplifies the problem for large n.

ASYMPTOTIC PROPERTIES. We denote by m and m_0 the number of free parameters in Θ and Θ_0 respectively, that is, the number of parameters that take noncountably many values. It can be shown that if $m > m_0$, then the distribution of the RV $\mathbf{w} = -2\ln\boldsymbol{\lambda}$ approaches a chi-square distribution with $m - m_0$ degrees of freedom as $n \to \infty$. The function $w = -2\ln\lambda$ is monotone decreasing; hence $\lambda < c$ iff $w > c_1 = -2\ln c$. From this it follows that

$$\alpha = P\{\boldsymbol{\lambda} < c\} = P\{\mathbf{w} > c_1\}$$

where $c_1 = \chi^2_{1-\alpha}(m - m_0)$, and (9-82) yields the following test

$$\text{Reject } H_0 \text{ iff } -2\ln\lambda > \chi^2_{1-\alpha}(m - m_0) \tag{9-83}$$

We give next an example illustrating the theorem.

Example 9-21. We are given an $N(\eta, 1)$ RV $\mathbf{x}$ and we wish to test the simple hypotheses $\eta = \eta_0$ against $\eta \neq \eta_0$. In this problem $\eta_{m0} = \eta_0$ and

$$f(X,\eta) = \frac{1}{\sqrt{(2\pi)^n}} \exp\left\{-\tfrac{1}{2}\sum(x_i - \eta)^2\right\}$$

This is maximum if the sum [see (8-66)]

$$\sum(x_i - \eta)^2 = \sum(x_i - \bar{x})^2 + n(\bar{x} - \eta)^2$$

is minimum, that is, if $\eta = \bar{x}$. Hence $\eta_m = \bar{x}$ and

$$\lambda = \frac{\exp\left\{-\frac{1}{2}\sum(x_i - \eta_0)^2\right\}}{\exp\left\{-\frac{1}{2}\sum(x_i - \bar{x})^2\right\}} = \exp\left\{-\frac{n}{2}(\bar{x} - \eta_0)^2\right\}$$

From the above it follows that $\lambda > c$ iff $|\bar{x} - \eta_0| < c_1$. This shows that the likelihood ratio test of the mean of a normal RV is equivalent to the test (9-63*a*).

Note that in this problem, $m = 1$ and $m_0 = 0$. Furthermore,

$$\mathbf{w} = -2\ln\boldsymbol{\lambda} = n(\bar{\mathbf{x}} - \eta_0)^2 = \left(\frac{\bar{\mathbf{x}} - \eta_0}{1/\sqrt{n}}\right)^2$$

But the right side is an RV with $\chi^2(1)$ distribution. Hence the RV $\mathbf{w}$ has a $\chi^2(m - m_0)$ distribution not only asymptotically, but for any n.

COMPUTER SIMULATION IN HYPOTHESIS TESTING. As we have seen, the test of a hypothesis H_0 involves the following steps: We determine the value X of the random vector $\mathbf{X} = [\mathbf{x}_1, \ldots, \mathbf{x}_m]$ in terms of the observations x_k of the m RVs $\mathbf{x}_k$ and compute the corresponding value $q = g(X)$ of the test statistic $\mathbf{q} = g(\mathbf{X})$. We accept H_0 if q is not in the critical region of the test, for example, if q is in the interval (q_a, q_b) where q_a and q_b are appropriately chosen values of the u percentile q_u of $\mathbf{q}$ [see (9-59)]. This involves the determination of the distribution $F(q)$ of $\mathbf{q}$ and the inverse $q_u = F^{(-1)}(u)$ of $F(q)$. The inversion problem can be avoided if we use the following approach.

The function $F(q)$ is monotone increasing. Hence,

$$q_a < q < q_b \text{ iff } a = F(q_a) < F(q) < F(q_b) = b$$

This shows that the test $q_a < q < q_b$ is equivalent to the test

$$\text{Accept } H_0 \text{ iff } a < F(q) < b \tag{9-84}$$

involving the determination of the distribution $F(q)$ of $\mathbf{q}$. As we have shown in Sec. 8-3, the function $F(q)$ can be determined by computer simulation [see (8-163)]:

To estimate numerically $F(q)$ we construct the RV vector sequence

$$X_i = [x_{1,i}, \ldots, x_{m,i}] \qquad i = 1, \ldots, n$$

where $x_{k,i}$ are the computer generated samples of the m RVs $\mathbf{x}_k$. Using the sequence X_i, we form the RN sequence $q_i = g(X_i)$ and we count the number n_q of q_i's that are smaller than the computed q. Inserting into (8-163), we obtain the estimate $F(q) \simeq n_q/n$. With $F(q)$ so determined, (9-84) yields the test

$$\text{Accept } H_0 \text{ iff } a < \frac{n_q}{n} < b \tag{9-85}$$

In the above, $q = g(X)$ is a number determined in terms of the experimental data x_k. The sequence q_i, however, is computer generated.

The above approach is used if it is difficult to determine analytically, the function $F(q)$. This is the case in the determination of Pearson's test statistic (9-75).

PROBLEMS

9-1. The diameter of cylindrical rods coming out of a production line is a normal RV $\mathbf{x}$ with $\sigma = 0.1$ mm. We measure $n = 9$ units and find that the average of the measurements is $\bar{x} = 91$ mm. (*a*) Find c such that with a 0.95 confidence coefficient, the mean η of $\mathbf{x}$ is in the interval $\bar{x} \pm c$. (*b*) We claim that η is in the interval $(90.95, 91.05)$. Find the confidence coefficient of our claim.

9-2. The length of a product is an RV $\mathbf{x}$ with $\sigma = 1$ mm and unknown mean. We measure four units and find that $\bar{x} = 203$ mm. (*a*) Assuming that $\mathbf{x}$ is a normal RV, find the 0.95 confidence interval of η. (*b*) The distribution of $\mathbf{x}$ is unknown. Using Tchebycheff's inequality, find c such that with confidence coefficient 0.95, η is in the interval $203 \pm c$.

9-3. We know from past records that the life length of type A tires is an RV $\mathbf{x}$ with $\sigma = 5000$ miles. We test 64 samples and find that their average life length is $\bar{x} = 25{,}000$ miles. Find the 0.9 confidence interval of the mean of $\mathbf{x}$.

9-4. We wish to determine the length a of an object. We use as an estimate of a the average $\bar{x}$ of n measurements. The measurement error is approximately normal with zero mean and standard deviation 0.1 mm. Find n such that with 95 percent confidence, $\bar{x}$ is within ± 0.2 mm of a.

9-5. The RV $\mathbf{x}$ is uniformly distributed in the interval $\theta - 2 < x < \theta + 2$. We observe 100 samples x_i and find that their average equals $\bar{x} = 30$. Find the 0.95 confidence interval of θ.

9-6. Consider an RV $\mathbf{x}$ with density $f(x) = xe^{-x}U(x)$. Predict with 95 percent confidence that the next value of x will be in the interval (a, b). Show that the length $b - a$ of this interval is minimum if a and b are such that

$$f(a) = f(b) \qquad P\{a < \mathbf{x} < b\} = 0.95$$

Find a and b.

9-7. (*Estimation–prediction*) The time to failure of electric bulbs of brand A is a normal RV with $\sigma = 10$ hours and unknown mean. We have used 20 such bulbs and have observed that the average $\bar{x}$ of their time to failure is 80 hours. We buy a new bulb of the same brand and wish to predict with 95 percent confidence that its time to failure will be in the interval $80 \pm c$. Find c.

9-8. Suppose that the time between arrivals of patients in a dentist's office constitutes samples of an RV $\mathbf{x}$ with density $\theta e^{-\theta x}U(x)$. The 40th patient arrived 4 hours after the first. Find the 0.95 confidence interval of the mean arrival time $\eta = 1/\theta$.

9-9. The number of particles emitted from a radioactive substance in 1 second is a Poisson distributed RV with mean λ. It was observed that in 200 seconds, 2550 particles were emitted. Find the 0.95 confidence interval of λ.

9-10. Among 4000 newborns, 2080 are male. Find the 0.99 confidence interval of the probability $p = P\{\text{male}\}$.

9-11. In an exit poll of 900 voters questioned, 360 responded that they favor a particular proposition. On this basis, it was reported that 40 percent of the voters favor the proposition. (*a*) Find the margin of error if the confidence coefficient of the results is 0.95. (*b*) Find the confidence coefficient if the margin of error is ± 2 percent.

9-12. In a market survey, it was reported that 29 percent of respondents favor product A. The poll was conducted with confidence coefficient 0.95, and the margin of error was ± 4 percent. Find the number of respondents.

9-13. We plan a poll for the purpose of estimating the probability p of Republicans in a community. We wish our estimate to be within ± 0.02 of p. How large should our sample be if the confidence coefficient of the estimate is 0.95?

9-14. A coin is tossed once, and heads shows. Assuming that the probability p of heads is the value of an RV $\mathbf{p}$ uniformly distributed in the interval $(0.4, 0.6)$, find its bayesian estimate.

9-15. The time to failure of a system is an RV $\mathbf{x}$ with density $f(x, \theta) = \theta e^{-\theta x} U(x)$. We wish to find the bayesian estimate $\hat{\theta}$ of θ in terms of the sample mean $\bar{\mathbf{x}}$ of the n samples x_i of $\mathbf{x}$. We assume that θ is the value of an RV $\boldsymbol{\theta}$ with prior density $f_\theta(\theta) = ce^{-c\theta} U(\theta)$. Show that

$$\hat{\theta} = \frac{n+1}{c + n\bar{x}} \xrightarrow[n \to \infty]{} \frac{1}{\bar{x}}$$

9-16. The RV $\mathbf{x}$ has a Poisson distribution with mean θ. We wish to find the bayesian estimate $\hat{\theta}$ of θ under the assumption that θ is the value of an RV $\boldsymbol{\theta}$ with prior density $f_\theta(\theta) \sim \theta^b e^{-c\theta} U(\theta)$. Show that

$$\hat{\theta} = \frac{n\bar{x} + b + 1}{n + c}$$

9-17. Suppose that the IQ scores of children in a certain grade are the samples of an $N(\eta, \sigma)$ RV $\mathbf{x}$. We test 10 children and obtain the following averages: $\bar{x} = 90$, $s = 5$. Find the 0.95 confidence interval of η and of σ.

9-18. The RVs $\mathbf{x}_i$ are i.i.d. and $N(0, \sigma)$. We observe that $x_1^2 + \cdots + x_{10}^2 = 4$. Find the 0.95 confidence interval of σ.

9-19. The readings of a voltmeter introduces an error $\boldsymbol{\nu}$ with mean 0. We wish to estimate its standard deviation σ. We measure a calibrated source $V = 3$ V four times and obtain the values 2.90, 3.15, 3.05, and 2.96. Assuming that $\boldsymbol{\nu}$ is normal, find the 0.95 confidence interval of σ.

9-20. The RV $\mathbf{x}$ has the Erlang density $f(x) \sim c^4 x^3 e^{-cx} U(x)$. We observe the samples $x_i = 3.1, 3.4, 3.3$. Find the ML estimate $\hat{c}$ of c.

9-21. The RV $\mathbf{x}$ has the truncated exponential density $f(x) = ce^{-c(x - x_0)} U(x - x_0)$. Find the ML estimate $\hat{c}$ of c in terms of the n samples x_i of $\mathbf{x}$.

9-22. The time to failure of a bulb is an RV $\mathbf{x}$ with density $ce^{-cx} U(x)$. We test 80 bulbs and find that 200 hours later, 62 of them are still good. Find the ML estimate of c.

9-23. The RV $\mathbf{x}$ has a Poisson distribution with mean θ. Show that the ML estimate of θ equals $\bar{x}$.

9-24. Show that if $L(x, \theta) = \ln f(x, \theta)$ is the likelihood function of an RV $\mathbf{x}$, then

$$E\left\{\left|\frac{\partial L(\mathbf{x}, \theta)}{\partial \theta}\right|^2\right\} = -E\left\{\frac{\partial^2 L(\mathbf{x}, \theta)}{\partial \theta^2}\right\}$$

9-25. We are given an RV **x** with mean η and standard deviation $\sigma = 2$, and we wish to test the hypothesis $\eta = 8$ against $\eta = 8.7$ with $\alpha = 0.01$ using as the test statistic the sample mean $\bar{x}$ of n samples. (*a*) Find the critical region R_c of the test and the resulting β if $n = 64$. (*b*) Find n and R_c if $\beta = 0.05$.

9-26. A new car is introduced with the claim that its average mileage in highway driving is at least 28 miles per gallon. Seventeen cars are tested, and the following mileage is obtained:

19 20 24 25 26 26.8 27.2 27.5
28 28.2 28.4 29 30 31 32 33.3 35

Can we conclude with significance level at most 0.05 that the claim is true?

9-27. The weights of cereal boxes are the values of an RV **x** with mean η. We measure 64 boxes and find that $\bar{x} = 7.7$ oz. and $s = 1.5$ oz. Test the hypothesis H_0: $\eta = 8$ oz. against H_1: $\eta \neq 8$ oz. with $\alpha = 0.1$ and $\alpha = 0.01$.

9-28. Brand A batteries cost more than brand B batteries. Their life lengths are two normal and independent RVs **x** and **y**. We test 16 batteries of brand A and 26 batteries of brand B and find these values, in hours:

$$\bar{x} = 4.6 \qquad s_x = 1.1 \qquad \bar{y} = 4.2 \qquad s_y = 0.9$$

Test the hypothesis $\eta_x = \eta_y$ against $\eta_x > \eta_y$ with $\alpha = 0.05$.

9-29. A coin is tossed 64 times, and heads shows 22 times. (*a*) Test the hypothesis that the coin is fair with significance level 0.05. (*b*) We toss a coin 16 times, and heads shows k times. If k is such that $k_1 \le k \le k_2$, we accept the hypothesis that the coin is fair with significance level $\alpha = 0.05$. Find k_1 and k_2 and the resulting β error.

9-30. In a production process, the number of defective units per hour is a Poisson distributed RV **x** with parameter $\lambda = 5$. A new process is introduced, and it is observed that the hourly defectives in a 22-hour period are

$$x_i = 3 \quad 0 \quad 5 \quad 4 \quad 2 \quad 6 \quad 4 \quad 1 \quad 5 \quad 3 \quad 7 \quad 4 \quad 0 \quad 8 \quad 3 \quad 2 \quad 4 \quad 3 \quad 6 \quad 5 \quad 6 \quad 9$$

Test the hypothesis $\lambda = 5$ against $\lambda < 5$ with $\alpha = 0.05$.

9-31. A die is tossed 102 times, and the ith face shows $k_i = 18, 15, 19, 17, 13$, and 20 times. Test the hypothesis that the die is fair with $\alpha = 0.05$ using the chi-square test.

9-32. A computer prints out 1000 numbers consisting of the 10 integers $j = 0, 1, \ldots, 9$. The number n_j of times j appears equals

$$n_j = 85 \quad 110 \quad 118 \quad 91 \quad 78 \quad 105 \quad 122 \quad 94 \quad 101 \quad 96$$

Test the hypothesis that the numbers j are uniformly distributed between 0 and 9, with $\alpha = 0.05$.

9-33. The number **x** of particles emitted from a radioactive substance in 1 second is a Poisson RV with mean θ. In 50 seconds, 1058 particles are emitted. Test the hypothesis $\theta_0 = 20$ against $\theta \neq 20$ with $\alpha = 0.05$ using the asymptotic approximation.

9-34. The RVs **x** and **y** are $N(\eta_x, \sigma_x)$ and $N(\eta_y, \sigma_y)$ respectively and independent. Test the hypothesis $\sigma_x = \sigma_y$ against $\sigma_x \neq \sigma_y$ using as the test statistic the ratio (see Prob. 6-19)

$$\mathbf{q} = \frac{1}{m} \sum_{i=1}^{m} (\mathbf{x}_i - \eta_x)^2 \bigg/ \frac{1}{n} \sum_{i=1}^{n} (\mathbf{y}_i - \eta_y)^2$$

9-35. Show that the variance of an RV with student-t distribution $t(n)$ equals $n/(n-2)$.

9-36. Find the probability p_5 that in a men's tennis tournament the final match will last five sets. (a) Assume that the probability p that a player wins a set equals 0.5. (b) Use bayesian statistic with uniform prior (see *law of succession*).

9-37. Show that in the measurement problem of Example 9-9, the bayesian estimate $\hat{\theta}$ of the parameter θ equals

$$\hat{\theta} = \frac{\sigma_1^2}{\sigma_0^2}\theta_0 + \frac{n\sigma_1^2}{\sigma^2}\bar{x} \quad \text{where} \quad \sigma_1^2 = \frac{\sigma^2}{n} \times \frac{\sigma_0^2}{\sigma_0^2 + \sigma^2/n}$$

9-38. Using the ML method, find the γ confidence interval of the variance $v = \sigma^2$ of an $N(\eta, \sigma)$ RV with known mean.

9-39. Show that if $\hat{\mathbf{\theta}}_1$ and $\hat{\mathbf{\theta}}_2$ are two unbiased minimum variance estimators of a parameter θ, then $\hat{\mathbf{\theta}}_1 = \hat{\mathbf{\theta}}_2$. *Hint*: Form the RV $\hat{\mathbf{\theta}} = (\hat{\mathbf{\theta}}_1 + \hat{\mathbf{\theta}}_2)/2$. Show that $\sigma_{\hat{\theta}}^2 = \sigma^2(1+r)/2 \le \sigma^2$ where σ^2 is the common variance of $\hat{\mathbf{\theta}}_1$ and $\hat{\mathbf{\theta}}_2$ and r is their correlation coefficient.

9-40. The number of successes of an event $\mathscr{A}$ in n trials equals k_1. Show that

$$\frac{(k_1 - np_1)^2}{np_1} + \frac{(k_2 - np_2)^2}{np_2} = \frac{(k_1 - np_1)^2}{np_1 p_2}$$

where $k_2 = n - k_1$ and $P(\mathscr{A}) = p_1 = 1 - p_2$.

PART II

STOCHASTIC PROCESSES

CHAPTER 10

GENERAL CONCEPTS

10-1 DEFINITIONS

As we recall, an RV $\mathbf{x}$ is a rule for assigning to every outcome ζ of an experiment $\mathscr{S}$ a *number* $\mathbf{x}(\zeta)$. A stochastic process $\mathbf{x}(t)$ is a rule for assigning to every ζ a *function* $\mathbf{x}(t, \zeta)$. Thus a stochastic process is a family of time functions depending on the parameter ζ or, equivalently, a function of t and ζ. The domain of ζ is the set of all experimental outcomes and the domain of t is a set R of real numbers.

If R is the real axis, then $\mathbf{x}(t)$ is a *continuous-time* process. If R is the set of integers, then $\mathbf{x}(t)$ is a *discrete-time* process. A discrete-time process is, thus, a sequence of random variables. Such a sequence will be denoted by $\mathbf{x}_n$ as in Sec. 8-4, or, to avoid double indices, by $\mathbf{x}[n]$.

We shall say that $\mathbf{x}(t)$ is a *discrete-state* process if its values are countable. Otherwise, it is a *continuous-state* process.

Most results in this investigation will be phrased in terms of continuous-time processes. Topics dealing with discrete-time processes will be introduced either as illustrations of the general theory, or when their discrete-time version is not self-evident.

We shall use the notation $\mathbf{x}(t)$ to represent a stochastic process omitting, as in the case of random variables, its dependence on ζ. Thus $\mathbf{x}(t)$ has the following interpretations:

1. It is a family (or an *ensemble*) of functions $\mathbf{x}(t, \zeta)$. In this interpretation, t and ζ are variables.

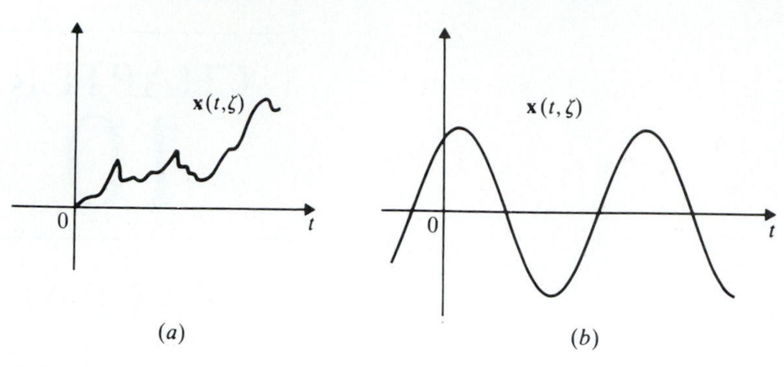

FIGURE 10-1

2. It is a single time function (or a *sample* of the given process). In this case, t is a variable and ζ is fixed.
3. If t is fixed and ζ is variable, then $\mathbf{x}(t)$ is a random variable equal to the *state* of the given process at time t.
4. If t and ζ are fixed, then $\mathbf{x}(t)$ is a *number*.

A physical example of a stochastic process is the motion of microscopic particles in collision with the molecules in a fluid (*brownian motion*). The resulting process $\mathbf{x}(t)$ consists of the motions of all particles (ensemble). A single realization $\mathbf{x}(t, \zeta_i)$ of this process (Fig. 10-1a) is the motion of a specific particle (sample). Another example is the voltage

$$\mathbf{x}(t) = \mathbf{r}\cos(\omega t + \boldsymbol{\varphi})$$

of an ac generator with random amplitude $\mathbf{r}$ and phase $\boldsymbol{\varphi}$. In this case, the process $\mathbf{x}(t)$ consists of a family of pure sine waves and a single sample is the function (Fig. 10-1b)

$$\mathbf{x}(t, \zeta_i) = \mathbf{r}(\zeta_i)\cos[\omega t + \boldsymbol{\varphi}(\zeta_i)]$$

According to our definition, both examples are stochastic processes. There is, however, a fundamental difference between them. The first example (regular) consists of a family of functions that cannot be described in terms of a finite number of parameters. Furthermore, the future of a sample $\mathbf{x}(t, \zeta)$ of $\mathbf{x}(t)$ cannot be determined in terms of its past. Finally, under certain conditions, the statistics† of a regular process $\mathbf{x}(t)$ can be determined in terms of a single sample (see Sec. 13-1). The second example (predictable) consists of a family of pure sine waves and it is completely specified in terms of the RVs $\mathbf{r}$ and $\boldsymbol{\varphi}$. Furthermore, if $\mathbf{x}(t, \zeta)$ is known for $t \le t_o$, then it is determined for $t > t_o$. Finally, a single sample $\mathbf{x}(t, \zeta)$ of $\mathbf{x}(t)$ does not specify the properties of the

†Recall that *statistics* hereafter will mean statistical properties.

entire process because it depends only on the particular values $\mathbf{r}(\zeta)$ and $\boldsymbol{\varphi}(\zeta)$ of $\mathbf{r}$ and $\boldsymbol{\varphi}$. A formal definition of regular and predictable processes is given in Sec. 12-3.

Equality. We shall say that two stochastic processes $\mathbf{x}(t)$ and $\mathbf{y}(t)$ are equal (everywhere) if their respective samples $\mathbf{x}(t, \zeta)$ and $\mathbf{y}(t, \zeta)$ are identical for every ζ. Similarly, the equality $\mathbf{z}(t) = \mathbf{x}(t) + \mathbf{y}(t)$ means that $\mathbf{z}(t, \zeta) = \mathbf{x}(t, \zeta) + \mathbf{y}(t, \zeta)$ for every ζ. Derivatives, integrals, or any other operations involving stochastic processes are defined similarly in terms of the corresponding operations for each sample.

As in the case of limits, the above definitions can be relaxed. We give below the meaning of MS equality and in App. 10A we define MS derivatives and integrals. Two processes $\mathbf{x}(t)$ and $\mathbf{y}(t)$ are equal in the MS sense iff

$$E\{|\mathbf{x}(t) - \mathbf{y}(t)|^2\} = 0 \tag{10-1}$$

for every t. Equality in the MS sense leads to the following conclusions: We denote by $\mathscr{A}_t$ the set of outcomes ζ such that $\mathbf{x}(t, \zeta) = \mathbf{y}(t, \zeta)$ for a *specific* t, and by $\mathscr{A}_\infty$ the set of outcomes ζ such that $\mathbf{x}(t, \zeta) = \mathbf{y}(t, \zeta)$ for *every* t. From (10-1) it follows that $\mathbf{x}(t, \zeta) - \mathbf{y}(t, \zeta) = 0$ with probability 1; hence $P(\mathscr{A}_t) = P(\mathscr{S}) = 1$. It does not follow, however, that $P(\mathscr{A}_\infty) = 1$. In fact, since $\mathscr{A}_\infty$ is the intersection of all sets $\mathscr{A}_t$ as t ranges over the entire axis, $P(\mathscr{A}_\infty)$ might even equal 0.

Statistics of Stochastic Processes

A stochastic process is a noncountable infinity of random variables, one for each t. For a specific t, $\mathbf{x}(t)$ is an RV with distribution

$$F(x, t) = P\{\mathbf{x}(t) \le x\} \tag{10-2}$$

This function depends on t, and it equals the probability of the event $\{\mathbf{x}(t) \le x\}$ consisting of all outcomes ζ such that, at the specific time t, the samples $\mathbf{x}(t, \zeta)$ of the given process do not exceed the number x. The function $F(x, t)$ will be called the *first-order distribution* of the process $\mathbf{x}(t)$. Its derivative with respect to x:

$$f(x, t) = \frac{\partial F(x, t)}{\partial x} \tag{10-3}$$

is the *first-order density* of $\mathbf{x}(t)$.

Frequency interpretation If the experiment is performed n times, then n functions $\mathbf{x}(t, \zeta_i)$ are observed, one for each trial (Fig. 10-2). Denoting by $n_t(x)$ the number of trials such that at time t the ordinates of the observed functions do not exceed x (solid lines), we conclude as in (4-3) that

$$F(x, t) \simeq \frac{n_t(x)}{n} \tag{10-4}$$

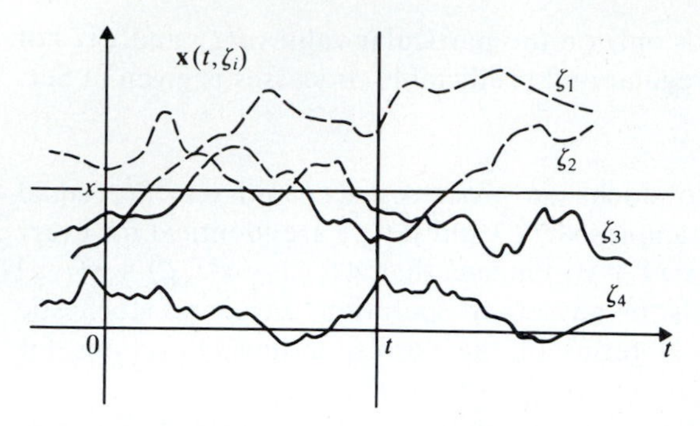

FIGURE 10-2

The *second-order distribution* of the process $\mathbf{x}(t)$ is the joint distribution

$$F(x_1, x_2; t_1, t_2) = P\{\mathbf{x}(t_1) \le x_1, \mathbf{x}(t_2) \le x_2\} \tag{10-5}$$

of the RVs $\mathbf{x}(t_1)$ and $\mathbf{x}(t_2)$. The corresponding density equals

$$f(x_1, x_2; t_1, t_2) = \frac{\partial^2 F(x_1, x_2; t_1, t_2)}{\partial x_1 \, \partial x_2} \tag{10-6}$$

We note that (consistency conditions)

$$F(x_1; t_1) = F(x_1, \infty; t_1, t_2) \qquad f(x_1, t_1) = \int_{-\infty}^{\infty} f(x_1, x_2; t_1, t_2)\, dx_2$$

as in (6-9) and (6-10).

The *nth-order distribution* of $\mathbf{x}(t)$ is the joint distribution $F(x_1, \ldots, x_n; t_1, \ldots, t_n)$ of the RVs $\mathbf{x}(t_1), \ldots, \mathbf{x}(t_n)$.

SECOND-ORDER PROPERTIES. For the determination of the statistical properties of a stochastic process, knowledge of the function $F(x_1, \ldots, x_n; t_1, \ldots, t_n)$ is required for every x_i, t_i, and n. However, for many applications, only certain averages are used, in particular, the expected value of $\mathbf{x}(t)$ and of $\mathbf{x}^2(t)$. These quantities can be expressed in terms of the second-order properties of $\mathbf{x}(t)$ defined as follows:

Mean The mean $\eta(t)$ of $\mathbf{x}(t)$ is the expected value of the RV $\mathbf{x}(t)$:

$$\eta(t) = E\{\mathbf{x}(t)\} = \int_{-\infty}^{\infty} x f(x, t)\, dx \tag{10-7}$$

Autocorrelation The autocorrelation $R(t_1, t_2)$ of $\mathbf{x}(t)$ is the expected value of the product $\mathbf{x}(t_1)\mathbf{x}(t_2)$:

$$R(t_1, t_2) = E\{\mathbf{x}(t_1)\mathbf{x}(t_2)\} = \int_{-\infty}^{\infty}\int_{-\infty}^{\infty} x_1 x_2 f(x_1, x_2; t_1, t_2)\, dx_1\, dx_2 \tag{10-8}$$

The value of $R(t_1, t_2)$ on the diagonal $t_1 = t_2 = t$ is the *average power* of $\mathbf{x}(t)$:

$$E\{\mathbf{x}^2(t)\} = R(t, t)$$

The *autocovariance* $C(t_1, t_2)$ of $\mathbf{x}(t)$ is the covariance of the RVs $\mathbf{x}(t_1)$ and $\mathbf{x}(t_2)$:

$$C(t_1, t_2) = R(t_1, t_2) - \eta(t_1)\eta(t_2) \tag{10-9}$$

and its value $C(t, t)$ on the diagonal $t_1 = t_2 = t$ equals the variance of $\mathbf{x}(t)$.

Note The following is an explanation of the reason for introducing the function $R(t_1, t_2)$ even in problems dealing only with average power: Suppose that $\mathbf{x}(t)$ is the input to a linear system and $\mathbf{y}(t)$ is the resulting output. In Sec. 10-2 we show that the mean of $\mathbf{y}(t)$ can be expressed in terms of the mean of $\mathbf{x}(t)$. However, the average power of $\mathbf{y}(t)$ cannot be found if only $E\{\mathbf{x}^2(t)\}$ is given. For the determination of $E\{\mathbf{y}^2(t)\}$, knowledge of the function $R(t_1, t_2)$ is required, not just on the diagonal $t_1 = t_2$, but for every t_1 and t_2. The following identity is a simple illustration

$$E\{[\mathbf{x}(t_1) + \mathbf{x}(t_2)]^2\} = R(t_1, t_1) + 2R(t_1, t_2) + R(t_2, t_2)$$

This follows from (10-8) if we expand the square and use the linearity of expected values.

Example 10-1. An extreme example of a stochastic process is a deterministic signal $\mathbf{x}(t) = f(t)$. In this case,

$$\eta(t) = E\{f(t)\} = f(t) \qquad R(t_1, t_2) = E\{f(t_1)f(t_2)\} = f(t_1)f(t_2)$$

Example 10-2. Suppose that $\mathbf{x}(t)$ is a process with

$$\eta(t) = 3 \qquad R(t_1, t_2) = 9 + 4e^{-0.2|t_1 - t_2|}$$

We shall determine the mean, the variance, and the covariance of the RVs $\mathbf{z} = \mathbf{x}(5)$ and $\mathbf{w} = \mathbf{x}(8)$.

Clearly, $E\{\mathbf{z}\} = \eta(5) = 3$ and $E\{\mathbf{w}\} = \eta(8) = 3$. Furthermore,

$$E\{\mathbf{z}^2\} = R(5,5) = 13 \qquad E\{\mathbf{w}^2\} = R(8,8) = 13$$

$$E\{\mathbf{zw}\} = R(5,8) = 9 + 4e^{-0.6} = 11.195$$

Thus $\mathbf{z}$ and $\mathbf{w}$ have the same variance $\sigma^2 = 4$ and their covariance equals $C(5, 8) = 4e^{-0.6} = 2.195$.

Example 10-3. The integral

$$\mathbf{s} = \int_a^b \mathbf{x}(t)\, dt$$

of a stochastic process $\mathbf{x}(t)$ is an RV $\mathbf{s}$ and its value $\mathbf{s}(\zeta)$ for a specific outcome ζ is the area under the curve $\mathbf{x}(t, \zeta)$ in the interval (a, b) (see also App. 10A). Interpreting the above as a Riemann integral, we conclude from the linearity of expected values that

$$\eta_s = E\{\mathbf{s}\} = \int_a^b E\{\mathbf{x}(t)\}\, dt = \int_a^b \eta(t)\, dt \tag{10-10}$$

Similarly, since

$$\mathbf{s}^2 = \int_a^b \int_a^b \mathbf{x}(t_1)\mathbf{x}(t_2)\, dt_1\, dt_2$$

we conclude, using again the linearity of expected values, that

$$E\{\mathbf{s}^2\} = \int_a^b \int_a^b E\{\mathbf{x}(t_1)\mathbf{x}(t_2)\}\, dt_1\, dt_2 = \int_a^b \int_a^b R(t_1, t_2)\, dt_1\, dt_2 \qquad (10\text{-}11)$$

Example 10-4. We shall determine the autocorrelation $R(t_1, t_2)$ of the process

$$\mathbf{x}(t) = \mathbf{r}\cos(\omega t + \boldsymbol{\varphi})$$

where we assume that the RVs $\mathbf{r}$ and $\boldsymbol{\varphi}$ are independent and $\boldsymbol{\varphi}$ is uniform in the interval $(-\pi, \pi)$.

Using simple trigonometric identities, we find

$$E\{\mathbf{x}(t_1)\mathbf{x}(t_2)\} = \tfrac{1}{2}E\{\mathbf{r}^2\}E\{\cos\omega(t_1 - t_2) + \cos(\omega t_1 + \omega t_2 + 2\boldsymbol{\varphi})\}$$

and since

$$E\{\cos(\omega t_1 + \omega t_2 + 2\boldsymbol{\varphi})\} = \frac{1}{2\pi}\int_{-\pi}^{\pi} \cos(\omega t_1 + \omega t_2 + 2\varphi)\, d\varphi = 0$$

we conclude that

$$R(t_1, t_2) = \tfrac{1}{2}E\{\mathbf{r}^2\}\cos\omega(t_1 - t_2) \qquad (10\text{-}12)$$

Example 10-5 Poisson process. In Sec. 3-4 we introduced the concept of Poisson points and we showed that these points are specified by the following properties:

P_1: The number $\mathbf{n}(t_1, t_2)$ of the points $\mathbf{t}_i$ in an interval (t_1, t_2) of length $t = t_2 - t_1$ is a Poisson RV with parameter λt:

$$P\{\mathbf{n}(t_1, t_2) = k\} = \frac{e^{-\lambda t}(\lambda t)^k}{k!} \qquad (10\text{-}13)$$

P_2: If the intervals (t_1, t_2) and (t_3, t_4) are nonoverlapping, then the RVs $\mathbf{n}(t_1, t_2)$ and $\mathbf{n}(t_3, t_4)$ are independent.

Using the points $\mathbf{t}_i$, we form the stochastic process

$$\mathbf{x}(t) = \mathbf{n}(0, t)$$

shown in Fig. 10-3a. This is a discrete-state process consisting of a family of increasing staircase functions with discontinuities at the points $\mathbf{t}_i$.

For a specific t, $\mathbf{x}(t)$ is a Poisson RV with parameter λt; hence

$$E\{\mathbf{x}(t)\} = \eta(t) = \lambda t$$

We shall show that its autocorrelation equals

$$R(t_1, t_2) = \begin{cases} \lambda t_2 + \lambda^2 t_1 t_2 & t_1 \ge t_2 \\ \lambda t_1 + \lambda^2 t_1 t_2 & t_1 \le t_2 \end{cases} \qquad (10\text{-}14)$$

or equivalently that

$$C(t_1, t_2) = \lambda \min(t_1, t_2) = \lambda t_1 U(t_2 - t_1) + \lambda t_2 U(t_1 - t_2)$$

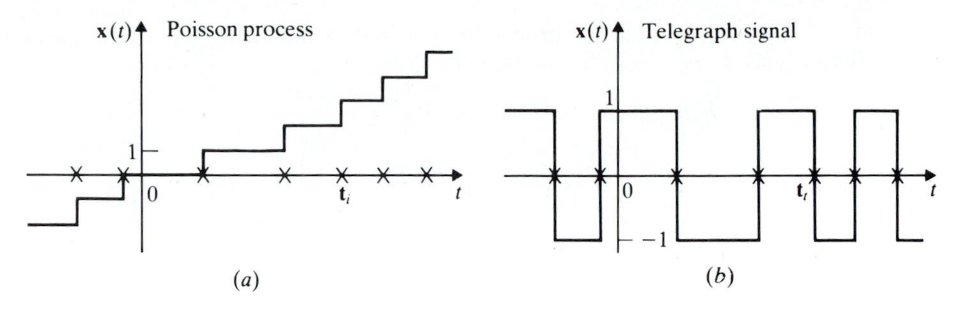

FIGURE 10-3

Proof. The above is true for $t_1 = t_2$ because [see (5-36)]

$$E\{\mathbf{x}^2(t)\} = \lambda t + \lambda^2 t^2 \tag{10-15}$$

Since $R(t_1, t_2) = R(t_2, t_1)$, it suffices to prove (10-14) for $t_1 < t_2$. The RVs $\mathbf{x}(t_1)$ and $\mathbf{x}(t_2) - \mathbf{x}(t_1)$ are independent because the intervals $(0, t_1)$ and (t_1, t_2) are nonoverlapping. Furthermore, they are Poisson distributed with parameters λt_1 and $\lambda(t_2 - t_1)$ respectively. Hence

$$E\{\mathbf{x}(t_1)[\mathbf{x}(t_2) - \mathbf{x}(t_1)]\} = E\{\mathbf{x}(t_1)\}E\{\mathbf{x}(t_2) - \mathbf{x}(t_1)\} = \lambda t_1 \lambda(t_2 - t_1)$$

Using the identity

$$\mathbf{x}(t_1)\mathbf{x}(t_2) = \mathbf{x}(t_1)[\mathbf{x}(t_1) + \mathbf{x}(t_2) - \mathbf{x}(t_1)]$$

we conclude from the above and (10-15) that

$$R(t_1, t_2) = \lambda t_1 + \lambda^2 t_1^2 + \lambda t_1 \lambda(t_2 - t_1)$$

and (10-14) results.

Nonuniform case If the points $\mathbf{t}_i$ have a nonuniform density $\lambda(t)$ as in (3-54), then the preceding results still hold provided that the product $\lambda(t_2 - t_1)$ is replaced by the integral of $\lambda(t)$ from t_1 to t_2.

Thus

$$E\{\mathbf{x}(t)\} = \int_0^t \lambda(\alpha)\, d\alpha \tag{10-16}$$

and

$$R(t_1, t_2) = \int_0^{t_1} \lambda(t)\, dt \left[1 + \int_0^{t_2} \lambda(t)\, dt\right] \qquad t_1 \le t_2 \tag{10-17}$$

Example 10-6 Telegraph signal. Using the Poisson points $\mathbf{t}_i$, we form a process $\mathbf{x}(t)$ such that $\mathbf{x}(t) = 1$ if the number of points in the interval $(0, t)$ is even, and $\mathbf{x}(t) = -1$ if this number is odd (Fig. 10-3*b*).

Denoting by $p(k)$ the probability that the number of points in the interval $(0,t)$ equals k, we conclude that [see (10-13)]

$$P\{\mathbf{x}(t) = 1\} = p(0) + p(2) + \cdots$$

$$= e^{-\lambda t}\left[1 + \frac{(\lambda t)^2}{2!} + \cdots\right] = e^{-\lambda t} \cosh \lambda t$$

$$P\{\mathbf{x}(t) = -1\} = p(1) + p(3) + \cdots$$

$$= e^{-\lambda t}\left[\lambda t + \frac{(\lambda t)^3}{3!} + \cdots\right] = e^{-\lambda t} \sinh \lambda t$$

Hence

$$E\{\mathbf{x}(t)\} = e^{-\lambda t}(\cosh \lambda t - \sinh \lambda t) = e^{-2\lambda t} \qquad (10\text{-}18)$$

To determine $R(t_1, t_2)$, we note that, if $t = t_1 - t_2 > 0$ and $\mathbf{x}(t_2) = 1$, then $\mathbf{x}(t_1) = 1$ if the number of points in the interval (t_1, t_2) is even. Hence

$$P\{\mathbf{x}(t_1) = 1 | \mathbf{x}(t_2) = 1\} = e^{-\lambda t} \cosh \lambda t \qquad t = t_1 - t_2$$

Multiplying by $P\{\mathbf{x}(t_2) = 1\}$, we obtain

$$P\{\mathbf{x}(t_1) = 1, \mathbf{x}(t_2) = 1\} = e^{-\lambda t} \cosh \lambda t e^{-\lambda t_2} \cosh \lambda t_2$$

Similarly,

$$P\{\mathbf{x}(t_1) = -1, \mathbf{x}(t_2) = -1\} = e^{-\lambda t} \cosh \lambda t e^{-\lambda t_2} \sinh \lambda t_2$$

$$P\{\mathbf{x}(t_1) = 1, \mathbf{x}(t_2) = -1\} = e^{-\lambda t} \sinh \lambda t e^{-\lambda t_2} \sinh \lambda t_2$$

$$P\{\mathbf{x}(t_1) = -1, \mathbf{x}(t_2) = 1\} = e^{-\lambda t} \sinh \lambda t e^{-\lambda t_2} \cosh \lambda t_2$$

Since the product $\mathbf{x}(t_1)\mathbf{x}(t_2)$ equals 1 or -1, we conclude omitting details that

$$R(t_1, t_2) = e^{-2\lambda|t_1 - t_2|} \qquad (10\text{-}19)$$

The above process is called *semirandom* telegraph signal because its value $\mathbf{x}(0) = 1$ at $t = 0$ is not random. To remove this certainty, we form the product

$$\mathbf{y}(t) = \mathbf{a}\mathbf{x}(t)$$

where $\mathbf{a}$ is an RV taking the values $+1$ and -1 with equal probability and is independent of $\mathbf{x}(t)$. The process $\mathbf{y}(t)$ so formed is called *random* telegraph signal. Since $E\{\mathbf{a}\} = 0$ and $E\{\mathbf{a}^2\} = 1$, the mean of $\mathbf{y}(t)$ equals $E\{\mathbf{a}\}E\{\mathbf{x}(t)\} = 0$ and its autocorrelation is given by

$$E\{\mathbf{y}(t_1)\mathbf{y}(t_2)\} = E\{\mathbf{a}^2\}E\{\mathbf{x}(t_1)\mathbf{x}(t_2)\} = e^{-2\lambda|t_1 - t_2|}$$

We note that as $t \to \infty$ the processes $\mathbf{x}(t)$ and $\mathbf{y}(t)$ have asymptotically equal statistics.

General Properties

The statistical properties of a real stochastic process $\mathbf{x}(t)$ are completely determined† in terms of its nth-order distribution

$$F(x_1,\ldots,x_n;t_1,\ldots,t_n) = P\{\mathbf{x}(t_1) \le x_1,\ldots,\mathbf{x}(t_n) \le x_n\} \qquad (10\text{-}20)$$

The joint statistics of two real processes $\mathbf{x}(t)$ and $\mathbf{y}(t)$ are determined in terms of the joint distribution of the RVs

$$\mathbf{x}(t_1),\ldots,\mathbf{x}(t_n),\mathbf{y}(t_1'),\ldots,\mathbf{y}(t_m')$$

The *complex process* $\mathbf{z}(t) = \mathbf{x}(t) + j\mathbf{y}(t)$ is specified in terms of the joint statistics of the real processes $\mathbf{x}(t)$ and $\mathbf{y}(t)$.

A *vector process* (n-dimensional process) is a family of n stochastic processes.

Correlation and covariance. The autocorrelation of a process $\mathbf{x}(t)$, real or complex, is by definition the mean of the product $\mathbf{x}(t_1)\mathbf{x}^*(t_2)$. This function, will be denoted by $R(t_1,t_2)$ or $R_x(t_1,t_2)$ or $R_{xx}(t_1,t_2)$. Thus

$$R_{xx}(t_1,t_2) = E\{\mathbf{x}(t_1)\mathbf{x}^*(t_2)\} \qquad (10\text{-}21)$$

where the conjugate term is associated with the second variable in $R_{xx}(t_1,t_2)$. From this it follows that

$$R(t_2,t_1) = E\{\mathbf{x}(t_2)\mathbf{x}^*(t_1)\} = R^*(t_1,t_2) \qquad (10\text{-}22)$$

We note, further, that

$$R(t,t) = E\{|\mathbf{x}(t)|^2\} \ge 0 \qquad (10\text{-}23)$$

The last two equations are special cases of the following: The autocorrelation $R(t_1,t_2)$ of a stochastic process $\mathbf{x}(t)$ is a *positive definite* (p.d.) function, that is, for any a_i and a_j:

$$\sum_{i,j} a_i a_j^* R(t_i,t_j) \ge 0 \qquad (10\text{-}24)$$

This is a consequence of the identity

$$0 \le E\left\{\left|\sum_i a_i\mathbf{x}(t_i)\right|^2\right\} = \sum_{i,j} a_i a_j^* E\{\mathbf{x}(t_i)\mathbf{x}^*(t_j)\}$$

We show later that the converse is also true: Given a p.d. function $R(t_1,t_2)$, we can find a process $\mathbf{x}(t)$ with autocorrelation $R(t_1,t_2)$.

---

†There are processes (nonseparable) for which this is not true. However, such processes are mainly of mathematical interest.

Example 10-7. (*a*) If $\mathbf{x}(t) = \mathbf{a}e^{j\omega t}$ then

$$R(t_1, t_2) = E\{\mathbf{a}e^{j\omega t_1}\mathbf{a}^*e^{-j\omega t_2}\} = E\{|\mathbf{a}|^2\}e^{j\omega(t_1 - t_2)}$$

(*b*) Suppose that the RVs $\mathbf{a}_i$ are uncorrelated with zero mean and variance σ_i^2. If

$$\mathbf{x}(t) = \sum_i \mathbf{a}_i e^{j\omega_i t}$$

then (10-21) yields

$$R(t_1, t_2) = \sum_i \sigma_i^2 e^{j\omega_i(t_1 - t_2)}$$

The *autocovariance* $C(t_1, t_2)$ of a process $\mathbf{x}(t)$ is the covariance of the RVs $\mathbf{x}(t_1)$ and $\mathbf{x}(t_2)$:

$$C(t_1, t_2) = R(t_1, t_2) - \eta(t_1)\eta^*(t_2) \qquad (10\text{-}25)$$

In the above, $\eta(t) = E\{\mathbf{x}(t)\}$ is the *mean* of $\mathbf{x}(t)$.

The ratio

$$r(t_1, t_2) = \frac{C(t_1, t_2)}{\sqrt{C(t_1, t_1)C(t_2, t_2)}} \qquad (10\text{-}26)$$

is the *correlation coefficient*† of the process $\mathbf{x}(t)$.

Note The autocovariance $C(t_1, t_2)$ of a process $\mathbf{x}(t)$ is the autocorrelation of the *centered process*

$$\tilde{\mathbf{x}}(t) = \mathbf{x}(t) - \eta(t)$$

Hence it is p.d.

The correlation coefficient $r(t_1, t_2)$ of $\mathbf{x}(t)$ is the autocovariance of the *normalized process* $\mathbf{x}(t)/\sqrt{C(t, t)}$; hence it is also p.d. Furthermore [see (7-9)]

$$|r(t_1, t_2)| \le 1 \qquad r(t, t) = 1 \qquad (10\text{-}27)$$

Example 10-8. If

$$\mathbf{s} = \int_a^b \mathbf{x}(t)\,dt \qquad \text{then} \qquad \mathbf{s} - \eta_s = \int_a^b \tilde{\mathbf{x}}(t)\,dt$$

where $\tilde{\mathbf{x}}(t) = \mathbf{x}(t) - \eta_x(t)$. Using (10-11), we conclude from the above note that

$$\sigma_s^2 = E\{|\mathbf{s} - \eta_s|^2\} = \int_a^b \int_a^b C_x(t_1, t_2)\,dt_1\,dt_2 \qquad (10\text{-}28)$$

The *cross-correlation* of two processes $\mathbf{x}(t)$ and $\mathbf{y}(t)$ is the function

$$R_{xy}(t_1, t_2) = E\{\mathbf{x}(t_1)\mathbf{y}^*(t_2)\} = R_{yx}^*(t_2, t_1) \qquad (10\text{-}29)$$

†In optics, $C(t_1, t_2)$ is called the *coherence function* and $r(t_1, t_2)$ is called the *complex degree of coherence* (see Papoulis, 1968).

Similarly,

$$C_{xy}(t_1, t_2) = R_{xy}(t_1, t_2) - \eta_x(t_1)\eta_y^*(t_2) \tag{10-30}$$

is their *cross-covariance*.

Two processes $\mathbf{x}(t)$ and $\mathbf{y}(t)$ are called (mutually) *orthogonal* if

$$R_{xy}(t_1, t_2) = 0 \qquad \text{for every} \quad t_1 \text{ and } t_2 \tag{10-31}$$

They are called *uncorrelated* if

$$C_{xy}(t_1, t_2) = 0 \qquad \text{for every} \quad t_1 \text{ and } t_2 \tag{10-32}$$

a-dependent processes In general, the values $\mathbf{x}(t_1)$ and $\mathbf{x}(t_2)$ of a stochastic process $\mathbf{x}(t)$ are statistically dependent for any t_1 and t_2. However, in most cases this dependence decreases as $|t_1 - t_2| \to \infty$. This leads to the following concept: A stochastic process $\mathbf{x}(t)$ is called *a-dependent* if all its values $\mathbf{x}(t)$ for $t < t_o$ and for $t > t_o + a$ are mutually *independent*. From this it follows that

$$C(t_1, t_2) = 0 \qquad \text{for} \quad |t_1 - t_2| > a \tag{10-33}$$

A process $\mathbf{x}(t)$ is called *correlation a-dependent* if its autocorrelation satisfies (10-33). Clearly, if $\mathbf{x}(t)$ is correlation a-dependent, then any linear combination of its values for $t < t_o$ is uncorrelated with any linear combination of its values for $t > t_o + a$.

White noise We shall say that a process $\boldsymbol{\nu}(t)$ is white noise if its values $\boldsymbol{\nu}(t_i)$ and $\boldsymbol{\nu}(t_j)$ are uncorrelated for every t_i and $t_j \neq t_i$:

$$C(t_i, t_j) = 0 \qquad t_i \neq t_j$$

As we explain later, the autocovariance of a nontrivial white-noise process must be of the form

$$C(t_1, t_2) = q(t_1)\delta(t_1 - t_2) \qquad q(t) \geq 0 \tag{10-34}$$

If the RVs $\boldsymbol{\nu}(t_i)$ and $\boldsymbol{\nu}(t_j)$ are not only uncorrelated but also independent, then $\boldsymbol{\nu}(t)$ will be called *strictly* white noise. Unless otherwise stated, it will be assumed that the mean of a white-noise process is identically 0.

Example 10-9. Suppose that $\boldsymbol{\nu}(t)$ is white noise and

$$\mathbf{x}(t) = \int_0^t \boldsymbol{\nu}(\alpha)\, d\alpha \tag{10-35}$$

Inserting (10-34) into (10-35), we obtain

$$E\{\mathbf{x}^2(t)\} = \int_0^t \int_0^t q(t_1)\delta(t_1 - t_2)\, dt_2\, dt_1 = \int_0^t q(t_1)\, dt_1 \tag{10-36}$$

because

$$\int_0^t \delta(t_1 - t_2)\, dt_2 = 1 \qquad \text{for} \quad 0 < t_1 < t$$

Uncorrelated and independent increments If the increments $\mathbf{x}(t_2) - \mathbf{x}(t_1)$ and $\mathbf{x}(t_4) - \mathbf{x}(t_3)$ of a process $\mathbf{x}(t)$ are uncorrelated (independent) for any

$t_1 < t_2 < t_3 < t_4$, then we say that $\mathbf{x}(t)$ is a process with uncorrelated (independent) increments. The Poisson process is a process with independent increments. The integral (10-35) of white noise is a process with uncorrelated increments.

Independent processes If two processes $\mathbf{x}(t)$ and $\mathbf{y}(t)$ are such that the RVs $\mathbf{x}(t_1), \ldots, \mathbf{x}(t_n)$ and $\mathbf{y}(t_1'), \ldots, \mathbf{y}(t_n')$ are mutually independent, then these processes are called independent.

Normal processes. A process $\mathbf{x}(t)$ is called normal, if the RVs $\mathbf{x}(t_1), \ldots, \mathbf{x}(t_n)$ are jointly normal for any n and $t_1, \ldots, t_n$.

The statistics of a normal process are completely determined in terms of its mean $\eta(t)$ and autocovariance $C(t_1, t_2)$. Indeed, since

$$E\{\mathbf{x}(t)\} = \eta(t) \qquad \sigma_x^2(t) = C(t,t)$$

we conclude that the first-order density $f(x,t)$ of $\mathbf{x}(t)$ is the normal density $N[\eta(t); \sqrt{C(t,t)}\,]$.

Similarly, since the function $r(t_1, t_2)$ in (10-26) is the correlation coefficient of the RVs $\mathbf{x}(t_1)$ and $\mathbf{x}(t_2)$, the second-order density $f(x_1, x_2; t_1, t_2)$ of $\mathbf{x}(t)$ is the jointly normal density

$$N\left[\eta(t_1), \eta(t_2); \sqrt{C(t_1,t_1)}, \sqrt{C(t_2,t_2)}; r(t_1,t_2)\right]$$

The nth-order characteristic function of the process $\mathbf{x}(t)$ is given by [see (8-60)]

$$\exp\left\{ j\sum_i \eta(t_i)\omega_i - \frac{1}{2}\sum_{i,k} C(t_i, t_k)\omega_i\omega_k \right\} \tag{10-37}$$

Its inverse $f(x_1, \ldots, x_n; t_1, \ldots, t_n)$ is the nth-order density of $\mathbf{x}(t)$.

Existence theorem. Given an arbitrary function $\eta(t)$ and a p.d. function $C(t_1, t_2)$, we can construct a normal process with mean $\eta(t)$ and autocovariance $C(t_1, t_2)$. This follows if we use in (10-37) the given functions $\eta(t)$ and $C(t_1, t_2)$. The inverse of the resulting characteristic function is a density because the function $C(t_1, t_2)$ is p.d. by assumption.

Example 10-10. Suppose that $\mathbf{x}(t)$ is a normal process with

$$\eta(t) = 3 \qquad C(t_1, t_2) = 4e^{-0.2|t_1 - t_2|}$$

(a) Find the probability that $\mathbf{x}(5) \le 2$.

Clearly, $\mathbf{x}(5)$ is a normal RV with mean $\eta(5) = 3$ and variance $C(5,5) = 4$. Hence

$$P\{\mathbf{x}(5) \le 2\} = \mathbb{G}(-1/2) = 0.309$$

(b) Find the probability that $|\mathbf{x}(8) - \mathbf{x}(5)| \le 1$.

The difference $\mathbf{s} = \mathbf{x}(8) - \mathbf{x}(5)$ is a normal RV with mean $\eta(8) - \eta(5) = 0$ and variance

$$C(8,8) + C(5,5) - 2C(8,5) = 8(1 - e^{-0.6}) = 3.608$$

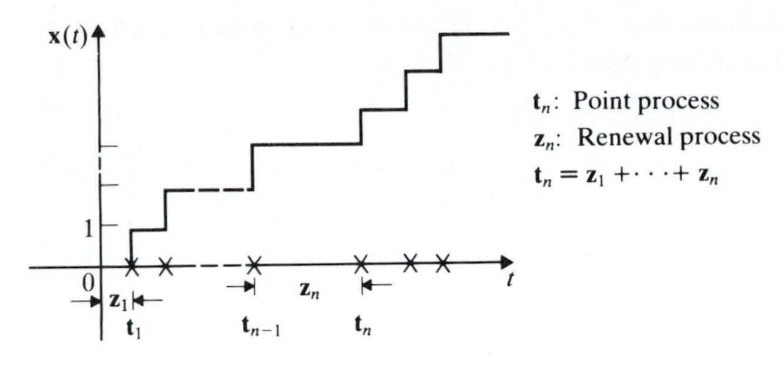

FIGURE 10-4

Hence

$$P\{|\mathbf{x}(8) - \mathbf{x}(5)| \le 1\} = 2\mathbb{G}(1/1.9) - 1 = 0.4$$

Point and renewal processes. A *point process* is a set of random points $\mathbf{t}_i$ on the time axis. To every point process we can associate a stochastic process $\mathbf{x}(t)$ equal to the number of points $\mathbf{t}_i$ in the interval $(0, t)$. An example is the Poisson process. To every point process $\mathbf{t}_i$ we can associate a sequence of RVs $\mathbf{z}_n$ such that

$$\mathbf{z}_1 = \mathbf{t}_1 \qquad \mathbf{z}_2 = \mathbf{t}_2 - \mathbf{t}_1 \cdots \mathbf{z}_n = \mathbf{t}_n - \mathbf{t}_{n-1}$$

where $\mathbf{t}_1$ is the first random point to the right of the origin. This sequence is called a *renewal process*. An example is the life history of light bulbs that are replaced as soon as they fail. In this case, $\mathbf{z}_i$ is the total time the ith bulb is in operation and $\mathbf{t}_i$ is the time of its failure.

We have thus established a correspondence between the following three concepts (Fig. 10-4): (a) a point process $\mathbf{t}_i$, (b) a discrete-state stochastic process $\mathbf{x}(t)$ increasing in unit steps at the points $\mathbf{t}_i$, (c) a renewal process consisting of the RVs $\mathbf{z}_i$ and such that $\mathbf{t}_n = \mathbf{z}_1 + \cdots + \mathbf{z}_n$. This correspondence is developed further in Sec. 16-1.

Stationary Processes

A stochastic process $\mathbf{x}(t)$ is called *strict-sense stationary* (abbreviated SSS) if its statistical properties are invariant to a shift of the origin. This means that the processes $\mathbf{x}(t)$ and $\mathbf{x}(t + c)$ have the same statistics for any c.

Two processes $\mathbf{x}(t)$ and $\mathbf{y}(t)$ are called *jointly stationary* if the joint statistics of $\mathbf{x}(t)$ and $\mathbf{y}(t)$ are the same as the joint statistics of $\mathbf{x}(t + c)$ and $\mathbf{y}(t + c)$ for any c.

A complex process $\mathbf{z}(t) = \mathbf{x}(t) + j\mathbf{y}(t)$ is stationary if the processes $\mathbf{x}(t)$ and $\mathbf{y}(t)$ are jointly stationary.

From the definition it follows that the nth-order density of an SSS process must be such that

$$f(x_1, \ldots, x_n; t_1, \ldots, t_n) = f(x_1, \ldots, x_n; t_1 + c, \ldots, t_n + c) \qquad (10\text{-}38)$$

for any c.

From the above it follows that $f(x;t) = f(x;t+c)$ for any c. Hence the first-order density of $\mathbf{x}(t)$ is independent of t:

$$f(x;t) = f(x) \tag{10-39}$$

Similarly, $f(x_1, x_2; t_1 + c, t_2 + c)$ is independent of c for any c. This leads to the conclusion that

$$f(x_1, x_2; t_1, t_2) = f(x_1, x_2; \tau) \qquad \tau = t_1 - t_2 \tag{10-40}$$

Thus the joint density of the RVs $\mathbf{x}(t+\tau)$ and $\mathbf{x}(t)$ is independent of t and it equals $f(x_1, x_2; \tau)$.

WIDE SENSE. A stochastic process $\mathbf{x}(t)$ is called *wide-sense stationary* (abbreviated WSS) if its mean is constant

$$E\{\mathbf{x}(t)\} = \eta \tag{10-41}$$

and its autocorrelation depends only on $\tau = t_1 - t_2$:

$$E\{\mathbf{x}(t+\tau)\mathbf{x}^*(t)\} = R(\tau) \tag{10-42}$$

Since τ is the distance from t to $t + \tau$, the function $R(\tau)$ can be written in the symmetrical form

$$R(\tau) = E\left\{\mathbf{x}\left(t + \frac{\tau}{2}\right)\mathbf{x}^*\left(t - \frac{\tau}{2}\right)\right\} \tag{10-43}$$

Note in particular that

$$E\{|\mathbf{x}(t)|^2\} = R(0)$$

Thus the average power of a stationary process is independent of t and it equals $R(0)$.

Example 10-11. Suppose that $\mathbf{x}(t)$ is a WSS process with autocorrelation

$$R(\tau) = Ae^{-\alpha|\tau|}$$

We shall determine the second moment of the RV $\mathbf{x}(8) - \mathbf{x}(5)$. Clearly,

$$E\{[\mathbf{x}(8) - \mathbf{x}(5)]^2\} = E\{\mathbf{x}^2(8)\} + E\{\mathbf{x}^2(5)\} - 2E\{\mathbf{x}(8)\mathbf{x}(5)\}$$
$$= R(0) + R(0) - 2R(3) = 2A - 2Ae^{-3\alpha}$$

Note As the above example suggests, the autocorrelation of a stationary process $\mathbf{x}(t)$ can be defined as average power. Assuming for simplicity that $\mathbf{x}(t)$ is real, we conclude from (10-42) that

$$E\{[\mathbf{x}(t+\tau) - \mathbf{x}(t)]^2\} = 2[R(0) - R(\tau)] \tag{10-44}$$

From (10-42) it follows that the autocovariance of a WSS process depends only on $\tau = t_1 - t_2$:

$$C(\tau) = R(\tau) - |\eta|^2 \tag{10-45}$$

and its correlation coefficient [see (10-26)] equals

$$r(\tau) = C(\tau)/C(0) \tag{10-46}$$

Thus $C(\tau)$ is the covariance, and $r(\tau)$ the correlation coefficient of the RVs $\mathbf{x}(t+\tau)$ and $\mathbf{x}(t)$.

Two processes $\mathbf{x}(t)$ and $\mathbf{y}(t)$ are called jointly WSS if each is WSS and their cross-correlation depends only on $\tau = t_1 - t_2$:

$$R_{xy}(\tau) = E\{\mathbf{x}(t+\tau)\mathbf{y}^*(t)\} \qquad C_{xy}(\tau) = R_{xy}(\tau) - \eta_x\eta_y^* \tag{10-47}$$

If $\mathbf{x}(t)$ is WSS white noise, then [see (10-34)]

$$C(\tau) = q\delta(\tau) \tag{10-48}$$

If $\mathbf{x}(t)$ is an a-dependent process, then $C(\tau) = 0$ for $|\tau| > a$. In this case, the constant a is called the *correlation time* of $\mathbf{x}(t)$. This term is also used for arbitrary processes and it is defined as the ratio

$$\tau_c = \frac{1}{C(0)}\int_0^\infty C(\tau)\,d\tau \tag{10-49}$$

In general $C(\tau) \neq 0$ for every τ. However, for most regular processes

$$C(\tau) \xrightarrow[|\tau|\to\infty]{} 0 \qquad R(\tau) \xrightarrow[|\tau|\to\infty]{} |\eta|^2$$

Example 10-12. If $\mathbf{x}(t)$ is WSS and

$$\mathbf{s} = \int_{-T}^{T} \mathbf{x}(t)\,dt$$

then [see (10-28)]

$$\sigma_s^2 = \int_{-T}^{T}\int_{-T}^{T} C(t_1 - t_2)\,dt_1\,dt_2 = \int_{-2T}^{2T}(2T - |\tau|)C(\tau)\,d\tau \tag{10-50}$$

The last equality follows with $\tau = t_1 - t_2$ (see Fig. 10-5); the details, however, are omitted [see also (10-143)].

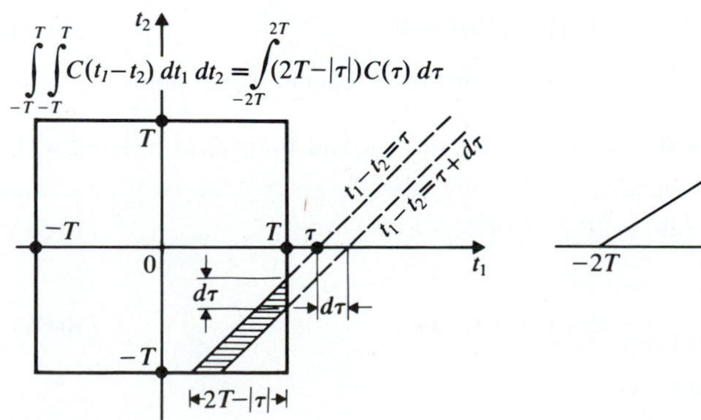

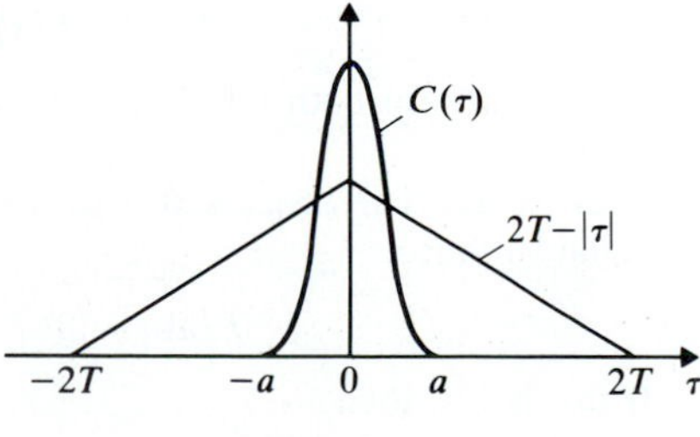

FIGURE 10-5

Special cases. (a) If $C(\tau) = q\delta(\tau)$, then

$$\sigma_s^2 = q\int_{-2T}^{2T}(2T - |\tau|)\delta(\tau)\,d\tau = 2Tq$$

(b) If the process $\mathbf{x}(t)$ is a-dependent and $a \ll T$, then (10-50) yields

$$\sigma_s^2 = \int_{-2T}^{2T}(2T - |\tau|)C(\tau)\,d\tau \simeq 2T\int_{-a}^{a} C(\tau)\,d\tau$$

This shows that, in the evaluation of the variance of $\mathbf{s}$, an a-dependent process with $a \ll T$ can be replaced by white noise as in (10-48) with

$$q = \int_{-a}^{a} C(\tau)\,d\tau$$

If a process is SSS, then it is also WSS. This follows readily from (10-39) and (10-40). The converse, however, is not in general true. As we show next, normal processes are an important exception.

Indeed, suppose that $\mathbf{x}(t)$ is a normal WSS process with mean η and autocovariance $C(\tau)$. As we see from (10-37), its nth-order characteristic function equals

$$\exp\left\{j\eta\sum_i \omega_i - \frac{1}{2}\sum_{i,k} C(t_i - t_k)\omega_i\omega_k\right\} \tag{10-51}$$

This function is invariant to a shift of the origin. And since it determines completely the statistics of $\mathbf{x}(t)$, we conclude that $\mathbf{x}(t)$ is SSS.

Example 10-13. We shall establish necessary and sufficient conditions for the stationarity of the process

$$\mathbf{x}(t) = \mathbf{a}\cos\omega t + \mathbf{b}\sin\omega t \tag{10-52}$$

The mean of this process equals

$$E\{\mathbf{x}(t)\} = E\{\mathbf{a}\}\cos\omega t + E\{\mathbf{b}\}\sin\omega t$$

This function must be independent of t. Hence the condition

$$E\{\mathbf{a}\} = E\{\mathbf{b}\} = 0 \tag{10-53}$$

is necessary for both forms of stationarity. We shall assume that it holds.

Wide sense. The process $\mathbf{x}(t)$ is WSS iff the RVs $\mathbf{a}$ and $\mathbf{b}$ are uncorrelated with equal variance:

$$E\{\mathbf{ab}\} = 0 \qquad E\{\mathbf{a}^2\} = E\{\mathbf{b}^2\} = \sigma^2 \tag{10-54}$$

If this holds, then

$$R(\tau) = \sigma^2\cos\omega\tau \tag{10-55}$$

Proof. If $\mathbf{x}(t)$ is WSS, then

$$E\{\mathbf{x}^2(0)\} = E\{\mathbf{x}^2(\pi/2\omega)\} = R(0)$$

But $\mathbf{x}(0) = \mathbf{a}$ and $\mathbf{x}(\pi/2\omega) = \mathbf{b}$; hence $E\{\mathbf{a}^2\} = E\{\mathbf{b}^2\}$. Using the above, we obtain

$$E\{\mathbf{x}(t+\tau)\mathbf{x}(t)\} = E\{[\mathbf{a}\cos\omega(t+\tau) + \mathbf{b}\sin\omega(t+\tau)][\mathbf{a}\cos\omega t + \mathbf{b}\sin\omega t]\}$$
$$= \sigma^2\cos\omega\tau + E\{\mathbf{ab}\}\sin\omega(2t+\tau) \qquad (10\text{-}56)$$

This is independent of t only if $E\{\mathbf{ab}\} = 0$ and (10-54) results.

Conversely, if (10-54) holds, then, as we see from (10-56), the autocorrelation of $\mathbf{x}(t)$ equals $\sigma^2 \cos\omega\tau$; hence $\mathbf{x}(t)$ is WSS.

Strict sense. The process $\mathbf{x}(t)$ is SSS iff the joint density $f(a,b)$ of the RVs $\mathbf{a}$ and $\mathbf{b}$ has circular symmetry, that is, if

$$f(a,b) = f\left(\sqrt{a^2+b^2}\right) \qquad (10\text{-}57)$$

Proof. If $\mathbf{x}(t)$ is SSS, then the RVs

$$\mathbf{x}(0) = \mathbf{a} \qquad \mathbf{x}(\pi/2\omega) = \mathbf{b}$$

and

$$\mathbf{x}(t) = \mathbf{a}\cos\omega t + \mathbf{b}\sin\omega t \qquad \mathbf{x}(t+\pi/2\omega) = \mathbf{b}\cos\omega t - \mathbf{a}\sin\omega t$$

have the same joint density for every t. Hence [see (6-70)], $f(a,b)$ must have circular symmetry.

We shall now show that, if $f(a,b)$ has circular symmetry, then $\mathbf{x}(t)$ is SSS. With τ a given number and

$$\mathbf{a}_1 = \mathbf{a}\cos\omega\tau + \mathbf{b}\sin\omega\tau \qquad \mathbf{b}_1 = \mathbf{b}\cos\omega\tau - \mathbf{a}\sin\omega\tau$$

we form the process

$$\mathbf{x}_1(t) = \mathbf{a}_1\cos\omega t + \mathbf{b}_1\sin\omega t = \mathbf{x}(t+\tau)$$

Clearly, the statistics of $\mathbf{x}(t)$ and $\mathbf{x}_1(t)$ are determined in terms of the joint densities $f(a,b)$ and $f(a_1,b_1)$ of the RVs $\mathbf{a}, \mathbf{b}$ and $\mathbf{a}_1, \mathbf{b}_1$. But [see (6-67)] the RVs $\mathbf{a}, \mathbf{b}$ and $\mathbf{a}_1, \mathbf{b}_1$ have the same joint density. Hence the processes $\mathbf{x}(t)$ and $\mathbf{x}(t+\tau)$ have the same statistics for every τ.

Corollary. If the process $\mathbf{x}(t)$ is SSS and the RVs $\mathbf{a}$ and $\mathbf{b}$ are independent, then they are normal.

Proof. It follows from (10-57) and (6-34).

Example 10-14. (*a*) Given an RV $\boldsymbol{\omega}$ with density $f(\omega)$ and an RV $\boldsymbol{\varphi}$ uniform in the interval $(-\pi,\pi)$ and independent of $\boldsymbol{\omega}$, we form the process

$$\mathbf{x}(t) = a\cos(\boldsymbol{\omega}t + \boldsymbol{\varphi}) \qquad (10\text{-}58)$$

We shall show that $\mathbf{x}(t)$ is WSS with zero mean and autocorrelation

$$R(\tau) = \frac{a^2}{2}E\{\cos\boldsymbol{\omega}\tau\} = \frac{a^2}{2}\operatorname{Re}\Phi_\omega(\tau) \qquad (10\text{-}59)$$

where

$$\Phi_\omega(\tau) = E\{e^{j\boldsymbol{\omega}\tau}\} = E\{\cos\boldsymbol{\omega}\tau\} + jE\{\sin\boldsymbol{\omega}\tau\} \qquad (10\text{-}60)$$

is the characteristic function of $\boldsymbol{\omega}$.

Proof. Clearly [see (7-59)]

$$E\{\cos(\boldsymbol{\omega} t + \boldsymbol{\varphi})\} = E\{E\{\cos(\boldsymbol{\omega} t + \boldsymbol{\varphi})|\boldsymbol{\omega}\}\}$$

From the independence of $\boldsymbol{\omega}$ and $\boldsymbol{\varphi}$, it follows that

$$E\{\cos(\boldsymbol{\omega} t + \boldsymbol{\varphi})|\omega\} = \cos \omega t\, E\{\cos \boldsymbol{\varphi}\} - \sin \omega t\, E\{\sin \boldsymbol{\varphi}\}$$

Hence $E\{\mathbf{x}(t)\} = 0$ because

$$E\{\cos \boldsymbol{\varphi}\} = \frac{1}{2\pi}\int_{-\pi}^{\pi} \cos \varphi \, d\varphi = 0 \qquad E\{\sin \boldsymbol{\varphi}\} = \frac{1}{2\pi}\int_{-\pi}^{\pi} \sin \varphi \, d\varphi = 0$$

Reasoning similarly, we obtain $E\{\cos(2\boldsymbol{\omega} t + \boldsymbol{\omega}\tau + 2\boldsymbol{\varphi})\} = 0$. And since

$$2\cos[\boldsymbol{\omega}(t + \tau) + \boldsymbol{\varphi}]\cos(\boldsymbol{\omega} t + \boldsymbol{\varphi}) = \cos \boldsymbol{\omega}\tau + \cos(2\boldsymbol{\omega} t + \boldsymbol{\omega}\tau + 2\boldsymbol{\varphi})$$

we conclude that

$$R(\tau) = a^2 E\{\cos[\boldsymbol{\omega}(t + \tau) + \boldsymbol{\varphi}]\cos(\boldsymbol{\omega} t + \boldsymbol{\varphi})\} = \frac{a^2}{2} E\{\cos \boldsymbol{\omega}\tau\}$$

(*b*) With $\boldsymbol{\omega}$ and $\boldsymbol{\varphi}$ as above, the process

$$\mathbf{z}(t) = a e^{j(\boldsymbol{\omega} t + \boldsymbol{\varphi})}$$

is WSS with zero mean and autocorrelation

$$E\{\mathbf{z}(t + \tau)\mathbf{z}^*(t)\} = a^2 E\{e^{j\boldsymbol{\omega}\tau}\} = a^2 \Phi_{\omega}(\tau)$$

Centering. Given a process $\mathbf{x}(t)$ with mean $\eta(t)$ and autocovariance $C_x(t_1, t_2)$, we form difference

$$\tilde{\mathbf{x}}(t) = \mathbf{x}(t) - \eta(t) \tag{10-61}$$

This difference is called the *centered process* associated with the process $\mathbf{x}(t)$. Note that

$$E\{\tilde{\mathbf{x}}(t)\} = 0 \qquad R_{\tilde{x}}(t_1, t_2) = C_x(t_1, t_2)$$

From this it follows that if the process $\mathbf{x}(t)$ is covariance stationary, that is, if $C_x(t_1, t_2) = C_x(t_1 - t_2)$, then its centered process $\tilde{\mathbf{x}}(t)$ is WSS.

Other forms of stationarity. A process $\mathbf{x}(t)$ is *asymptotically stationary* if the statistics of the RVs $\mathbf{x}(t_1 + c), \ldots, \mathbf{x}(t_n + c)$ do not depend on c if c is large. More precisely, the function

$$f(x_1, \ldots, x_n; t_1 + c, \ldots, t_n + c)$$

tends to a limit (that does not depend on c) as $c \to \infty$. The semirandom telegraph signal is an example.

A process $\mathbf{x}(t)$ is *Nth-order stationary* if (10-38) holds not for every n, but only for $n \le N$.

A process $\mathbf{x}(t)$ is *stationary in an interval* if (10-38) holds for every t_i and $t_i + c$ in this interval.

We say that $\mathbf{x}(t)$ is a process with *stationary increments* if its increments $\mathbf{y}(t) = \mathbf{x}(t + h) - \mathbf{x}(t)$ form a stationary process for every h. The Poisson process is an example.

MEAN SQUARE PERIODICITY. A process $\mathbf{x}(t)$ is called MS periodic if

$$E\{|\mathbf{x}(t+T)-\mathbf{x}(t)|^2\}=0 \tag{10-62}$$

for every t. From this it follows that, for a specific t,

$$\mathbf{x}(t+T)=\mathbf{x}(t) \tag{10-63}$$

with probability 1. It does not, however, follow that the set of outcomes ζ such that $\mathbf{x}(t+T,\zeta)=\mathbf{x}(t,\zeta)$ for all t has probability 1.

As we see from (10-63) the mean of an MS periodic process is periodic. We shall examine the properties of $R(t_1, t_2)$.

THEOREM. A process $\mathbf{x}(t)$ is MS periodic iff its autocorrelation is *doubly periodic*, that is, if

$$R(t_1+mT, t_2+nT)=R(t_1,t_2) \tag{10-64}$$

for every integer m and n.

Proof. As we know [see (7-12)]

$$E^2\{\mathbf{zw}\}\le E\{\mathbf{z}^2\}E\{\mathbf{w}^2\}$$

With $\mathbf{z}=\mathbf{x}(t_1)$ and $\mathbf{w}=\mathbf{x}(t_2+T)-\mathbf{x}(t_2)$ the above yields

$$E^2\{\mathbf{x}(t_1)[\mathbf{x}(t_2+T)-\mathbf{x}(t_2)]\}\le E\{\mathbf{x}^2(t_1)\}E\{[\mathbf{x}(t_2+T)-\mathbf{x}(t_2)]^2\}$$

If $\mathbf{x}(t)$ is MS periodic, then the last term above is 0. Equating the left side to 0, we obtain

$$R(t_1,t_2+T)-R(t_1,t_2)=0$$

Repeated application of this yields (10-64).

Conversely, if (10-64) is true, then

$$R(t+T,t+T)=R(t+T,t)=R(t,t)$$

Hence

$$E\{[\mathbf{x}(t+T)-\mathbf{x}(t)]^2\}=R(t+T,t+T)+R(t,t)-2R(t+T,t)=0$$

therefore $\mathbf{x}(t)$ is MS periodic.

10-2 SYSTEMS WITH STOCHASTIC INPUTS

Given a stochastic process $\mathbf{x}(t)$, we assign according to some rule to each of its samples $\mathbf{x}(t,\zeta_i)$ a function $\mathbf{y}(t,\zeta_i)$. We have thus created another process

$$\mathbf{y}(t)=T[\mathbf{x}(t)]$$

whose samples are the functions $\mathbf{y}(t,\zeta_i)$. The process $\mathbf{y}(t)$ so formed can be considered as the output of a *system* (transformation) with input the process $\mathbf{x}(t)$. The system is completely specified in terms of the operator T, that is, the rule of correspondence between the samples of the input $\mathbf{x}(t)$ and the output $\mathbf{y}(t)$.

The system is *deterministic* if it operates only on the variable t treating ζ as a parameter. This means that if two samples $\mathbf{x}(t, \zeta_1)$ and $\mathbf{x}(t, \zeta_2)$ of the input are identical in t, then the corresponding samples $\mathbf{y}(t, \zeta_1)$ and $\mathbf{y}(t, \zeta_2)$ of the output are also identical in t. The system is called *stochastic* if T operates on both variables t and ζ. This means that there exist two outcomes ζ_1 and ζ_2 such that $\mathbf{x}(t, \zeta_1) = \mathbf{x}(t, \zeta_2)$ identically in t but $\mathbf{y}(t, \zeta_1) \neq \mathbf{y}(t, \zeta_2)$. These classifications are based on the terminal properties of the system. If the system is specified in terms of physical elements or by an equation, then it is deterministic (stochastic) if the elements or the coefficients of the defining equations are deterministic (stochastic). Throughout this book we shall consider only deterministic systems.

In principle, the statistics of the output of a system can be expressed in terms of the statistics of the input. However, in general this is a complicated problem. We consider next two important special cases.

Memoryless Systems

A system is called memoryless if its output is given by

$$\mathbf{y}(t) = g[\mathbf{x}(t)]$$

where $g(x)$ is a function of x. Thus, at a given time $t = t_1$, the output $\mathbf{y}(t_1)$ depends only on $\mathbf{x}(t_1)$ and not on any other past or future values of $\mathbf{x}(t)$.

From the above it follows that the first-order density $f_y(y; t)$ of $\mathbf{y}(t)$ can be expressed in terms of the corresponding density $f_x(x; t)$ of $\mathbf{x}(t)$ as in Sec. 5-2. Furthermore,

$$E\{\mathbf{y}(t)\} = \int_{-\infty}^{\infty} g(x) f_x(x; t)\, dx$$

Similarly, since $\mathbf{y}(t_1) = g[\mathbf{x}(t_1)]$ and $\mathbf{y}(t_2) = g[\mathbf{x}(t_2)]$, the second-order density $f_y(y_1, y_2; t_1, t_2)$ of $\mathbf{y}(t)$ can be determined in terms of the corresponding density $f_x(x_1, x_2; t_1, t_2)$ of $\mathbf{x}(t)$ as in Sec. 6-3. Furthermore,

$$E\{\mathbf{y}(t_1)\mathbf{y}(t_2)\} = \int_{-\infty}^{\infty}\int_{-\infty}^{\infty} g(x_1) g(x_2) f_x(x_1, x_2; t_1, t_2)\, dx_1\, dx_2$$

The nth-order density $f_y(y_1, \ldots, y_n; t_1, \ldots, t_n)$ of $\mathbf{y}(t)$ can be determined from the corresponding density of $\mathbf{x}(t)$ as in (8-8) where the underlying transformation is the system

$$\mathbf{y}(t_1) = g[\mathbf{x}(t_1)], \ldots, \mathbf{y}(t_n) = g[\mathbf{x}(t_n)] \tag{10-65}$$

STATIONARITY. Suppose that the input to a memoryless system is an SSS process $\mathbf{x}(t)$. We shall show that the resulting output $\mathbf{y}(t)$ is also SSS.

Proof. To determine the nth-order density of $\mathbf{y}(t)$, we solve the system

$$g(x_1) = y_1, \ldots, g(x_n) = y_n \tag{10-66}$$

If this system has a unique solution, then [see (8-8)]

$$f_y(y_1,\ldots,y_n;t_1,\ldots,t_n) = \frac{f_x(x_1,\ldots,x_n;t_1,\ldots,t_n)}{|g'(x_1)\cdots g'(x_n)|} \tag{10-67}$$

From the stationarity of $\mathbf{x}(t)$ it follows that the numerator in (10-67) is invariant to a shift of the time origin. And since the denominator does not depend on t, we conclude that the left side does not change if t_i is replaced by $t_i + c$. Hence $\mathbf{y}(t)$ is SSS. We can similarly show that this is true even if (10-66) has more than one solution.

Notes 1. If $\mathbf{x}(t)$ is stationary of order N, then $\mathbf{y}(t)$ is stationary of order N.
2. If $\mathbf{x}(t)$ is stationary in an interval, then $\mathbf{y}(t)$ is stationary in the same interval.
3. If $\mathbf{x}(t)$ is WSS stationary, then $\mathbf{y}(t)$ might not be stationary in any sense.

Square-law detector. A square-law detector is a memoryless system whose output equals

$$\mathbf{y}(t) = \mathbf{x}^2(t)$$

We shall determine its first- and second-order densities. If $y > 0$, then the system $y = x^2$ has the two solutions $\pm\sqrt{y}$. Furthermore, $y'(x) = \pm 2\sqrt{y}$; hence

$$f_y(y;t) = \frac{1}{2\sqrt{y}}\left[f_x(\sqrt{y};t) + f_x(-\sqrt{y};t)\right]$$

If $y_1 > 0$ and $y_2 > 0$, then the system

$$y_1 = x_1^2 \qquad y_2 = x_2^2$$

has the four solutions $(\pm\sqrt{y_1}, \pm\sqrt{y_2})$. Furthermore, its jacobian equals $\pm 4\sqrt{y_1 y_2}$; hence

$$f_y(y_1, y_2; t_1, t_2) = \frac{1}{4\sqrt{y_1 y_2}} \sum f_x\left(\pm\sqrt{y_1}, \pm\sqrt{y_2}; t_1, t_2\right)$$

where the summation has four terms.

Note that, if $\mathbf{x}(t)$ is SSS, then $f_x(x;t) = f_x(x)$ is independent of t and $f_x(x_1, x_2; t_1, t_2) = f_x(x_1, x_2; \tau)$ depends only on $\tau = t_1 - t_2$. Hence $f_y(y)$ is independent of t and $f_y(y_1, y_2; \tau)$ depends only on $\tau = t_1 - t_2$.

Example 10-15. Suppose that $\mathbf{x}(t)$ is a normal stationary process with zero mean and autocorrelation $R_x(\tau)$. In this case, $f_x(x)$ is normal with variance $R_x(0)$.

If $\mathbf{y}(t) = \mathbf{x}^2(t)$ (Fig. 10-6), then $E\{\mathbf{y}(t)\} = R_x(0)$ and [see (5-8)]

$$f_y(y) = \frac{1}{\sqrt{2\pi R_x(0)y}} e^{-y/2R_x(0)} U(y)$$

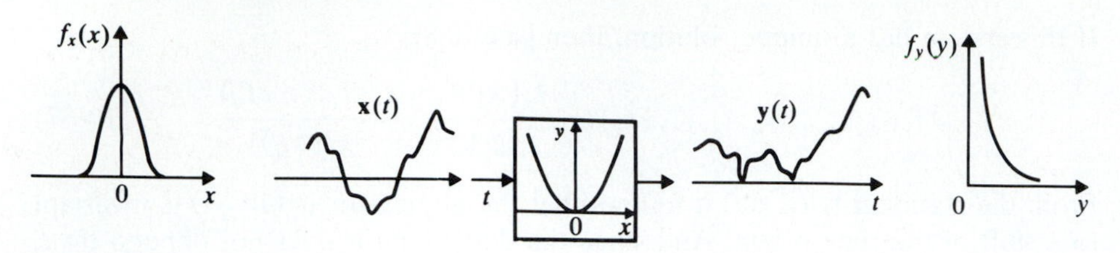

FIGURE 10-6

We shall show that

$$R_y(\tau) = R_x^2(0) + 2R_x^2(\tau) \tag{10-68}$$

Proof. The RVs $\mathbf{x}(t+\tau)$ and $\mathbf{x}(t)$ are jointly normal with zero mean. Hence [see (7-36)]

$$E\{\mathbf{x}^2(t+\tau)\mathbf{x}^2(t)\} = E\{\mathbf{x}^2(t+\tau)\}E\{\mathbf{x}^2(t)\} + 2E^2\{\mathbf{x}(t+\tau)\mathbf{x}(t)\}$$

and (10-68) results.

Note in particular that

$$E\{\mathbf{y}^2(t)\} = R_y(0) = 3R_x^2(0) \qquad \sigma_y^2 = 2R_x^2(0)$$

Hard limiter. Consider a memoryless system with

$$g(x) = \begin{cases} 1 & x > 0 \\ -1 & x < 0 \end{cases} \tag{10-69}$$

(Fig. 10-7). Its output $\mathbf{y}(t)$ takes the values ± 1 and

$$P\{\mathbf{y}(t) = 1\} = P\{\mathbf{x}(t) > 0\} = 1 - F_x(0)$$
$$P\{\mathbf{y}(t) = -1\} = P\{\mathbf{x}(t) < 0\} = F_x(0)$$

Hence

$$E\{\mathbf{y}(t)\} = 1 \times P\{\mathbf{y}(t) = 1\} - 1 \times P\{\mathbf{y}(t) = -1\} = 1 - 2F_x(0)$$

The product $\mathbf{y}(t+\tau)\mathbf{y}(t)$ equals 1 if $\mathbf{x}(t+\tau)\mathbf{x}(t) > 0$ and it equals -1 otherwise. Hence

$$R_y(\tau) = P\{\mathbf{x}(t+\tau)\mathbf{x}(t) > 0\} - P\{\mathbf{x}(t+\tau)\mathbf{x}(t) < 0\} \tag{10-70}$$

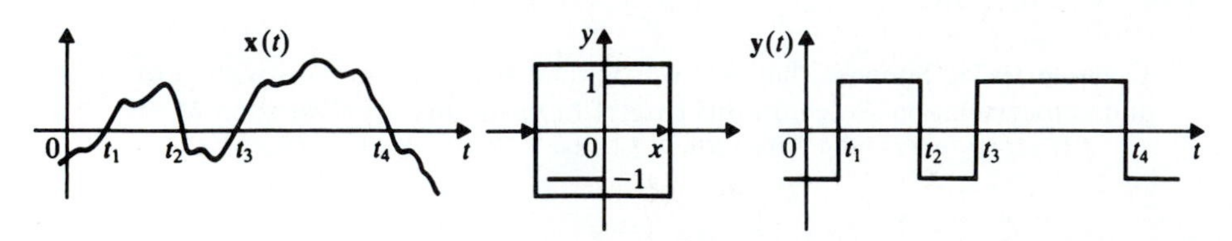

FIGURE 10-7

Thus, in the probability plane of the RVs $\mathbf{x}(t+\tau)$ and $\mathbf{x}(t)$, $R_y(\tau)$ equals the masses in the first and third quadrants minus the masses in the second and fourth quadrants.

Example 10-16. We shall show that if $\mathbf{x}(t)$ is a normal stationary process with zero mean, then the autocorrelation of the output of a hard limiter equals

$$R_y(\tau) = \frac{2}{\pi} \arcsin \frac{R_x(\tau)}{R_x(0)} \tag{10-71}$$

This result is known as the *arcsine law*.†

Proof. The RVs $\mathbf{x}(t+\tau)$ and $\mathbf{x}(t)$ are jointly normal with zero mean, variance $R_x(0)$, and correlation coefficient $R_x(\tau)/R_x(0)$. Hence [see (6-47)].

$$P\{\mathbf{x}(t+\tau)\mathbf{x}(t) > 0\} = \frac{1}{2} + \frac{\alpha}{\pi} \qquad \sin\alpha = \frac{R_x(\tau)}{R_x(0)}$$

$$P\{\mathbf{x}(t+\tau)\mathbf{x}(t) < 0\} = \frac{1}{2} - \frac{\alpha}{\pi}$$

Inserting in (10-70), we obtain

$$R_y(\tau) = \frac{1}{2} + \frac{\alpha}{\pi} - \left(\frac{1}{2} - \frac{\alpha}{\pi}\right) = \frac{2\alpha}{\pi}$$

and (10-71) follows.

Example 10-17 Bussgang's theorem. Using Price's theorem, we shall show that if the input to a memoryless system $y = g(x)$ is a zero-mean normal process $\mathbf{x}(t)$, the cross-correlation of $\mathbf{x}(t)$ with the resulting output $\mathbf{y}(t) = g[\mathbf{x}(t)]$ is proportional to $R_{xx}(\tau)$:

$$R_{xy}(\tau) = KR_{xx}(\tau) \qquad \text{where} \quad K = E\{g'[\mathbf{x}(t)]\} \tag{10-72}$$

Proof. For a specific τ, the RVs $\mathbf{x} = \mathbf{x}(t)$ and $\mathbf{z} = \mathbf{x}(t+\tau)$ are jointly normal with zero mean and covariance $\mu = E\{\mathbf{xz}\} = R_{xx}(\tau)$. With

$$I = E\{\mathbf{z}g(\mathbf{x})\} = E\{\mathbf{x}(t+\tau)\mathbf{y}(t)\} = R_{xy}(\tau)$$

it follows from (7-37) that

$$\frac{\partial I}{\partial \mu} = E\left\{\frac{\partial^2[\mathbf{z}g(\mathbf{x})]}{\partial x\,\partial z}\right\} = E\{g'[\mathbf{x}(t)]\} = K \tag{10-73}$$

If $\mu = 0$, the RVs $\mathbf{x}(t+\tau)$ and $\mathbf{x}(t)$ are independent; hence $I = 0$. Integrating (10-73) with respect to μ, we obtain $I = K\mu$ and (10-72) results.

†J. L. Lawson and G. E. Uhlenbeck: *Threshold Signals*, McGraw-Hill Book Company, New York, 1950.

Special cases.† (a) (Hard limiter) Suppose that $g(x) = \operatorname{sgn} x$ as in (10-69). In this case, $g'(x) = 2\delta(x)$; hence

$$K = E\{2\delta(\mathbf{x})\} = 2\int_{-\infty}^{\infty} \delta(x) f(x)\,dx = 2f(0)$$

where

$$f(x) = \frac{1}{\sqrt{2\pi R_{xx}(0)}} \exp\left\{-\frac{x^2}{2R_{xx}(0)}\right\}$$

is the first-order density of $\mathbf{x}(t)$. Inserting into (10-72), we obtain

$$R_{xy}(\tau) = R_{xx}(\tau)\sqrt{\frac{2}{\pi R_{xx}(0)}} \qquad \mathbf{y}(t) = \operatorname{sgn} \mathbf{x}(t) \tag{10-74}$$

(b) (Limiter) Suppose next that $\mathbf{y}(t)$ is the output of a limiter

$$g(x) = \begin{cases} x & |x| < c \\ c & |x| > c \end{cases} \qquad g'(x) = \begin{cases} 1 & |x| < c \\ 0 & |x| > c \end{cases}$$

In this case,

$$K = \int_{-c}^{c} f(x)\,dx = 2\mathbb{G}\left(\frac{c}{\sqrt{R_{xx}(0)}}\right) - 1 \tag{10-75}$$

Linear Systems

The notation

$$\mathbf{y}(t) = L[\mathbf{x}(t)] \tag{10-76}$$

will indicate that $\mathbf{y}(t)$ is the output of a *linear* system with input $\mathbf{x}(t)$. This means that

$$L[\mathbf{a}_1\mathbf{x}_1(t) + \mathbf{a}_2\mathbf{x}_2(t)] = \mathbf{a}_1 L[\mathbf{x}_1(t)] + \mathbf{a}_2 L[\mathbf{x}_2(t)] \tag{10-77}$$

for any $\mathbf{a}_1, \mathbf{a}_2, \mathbf{x}_1(t), \mathbf{x}_2(t)$.

The above is the familiar definition of linearity and it also holds if the coefficients $\mathbf{a}_1$ and $\mathbf{a}_2$ are random variables because, as we have assumed, the system is deterministic, that is, it operates only on the variable t.

Note If a system is specified by its internal structure or by a differential equation, then (10-77) holds only if $\mathbf{y}(t)$ is the *zero-state* response. The response due to the initial conditions (zero-input response) will not be considered.

A system is called *time-invariant* if its response to $\mathbf{x}(t + c)$ equals $\mathbf{y}(t + c)$. We shall assume throughout that all linear systems under consideration are time-invariant.

---

†H. E. Rowe, "Memoryless Nonlinearities with Gaussian Inputs," *BSTJ*, vol. 67, no. 7, September 1982.

It is well known that the output of a linear system is a convolution

$$\mathbf{y}(t) = \mathbf{x}(t) * h(t) = \int_{-\infty}^{\infty} \mathbf{x}(t-\alpha)h(\alpha)\,d\alpha \tag{10-78}$$

where

$$h(t) = L[\delta(t)]$$

is its impulse response. In the following, most systems will be specified by (10-78). However, we start our investigation using the operational notation (10-76) to stress the fact that various results based on the next theorem also hold for arbitrary linear operators involving one or more variables.

The following observations are immediate consequences of the linearity and time invariance of the system.

If $\mathbf{x}(t)$ is a normal process, then $\mathbf{y}(t)$ is also a normal process. This is an extension of the familiar property of linear transformations of normal RVs and can be justified if we approximate the integral in (10-78) by a sum:

$$\mathbf{y}(t_i) \simeq \sum_k \mathbf{x}(t_i - \alpha_k)h(\alpha_k)\Delta(\alpha)$$

If $\mathbf{x}(t)$ is SSS, then $\mathbf{y}(t)$ is also SSS. Indeed, since $\mathbf{y}(t+c) = L[\mathbf{x}(t+c)]$ for every c, we conclude that if the processes $\mathbf{x}(t)$ and $\mathbf{x}(t+c)$ have the same statistical properties, so do the processes $\mathbf{y}(t)$ and $\mathbf{y}(t+c)$. We show later [see (10-133)] that if $\mathbf{x}(t)$ is WSS, the processes $\mathbf{x}(t)$ and $\mathbf{y}(t)$ are jointly WSS.

Fundamental theorem. For any linear system

$$E\{L[\mathbf{x}(t)]\} = L[E\{\mathbf{x}(t)\}] \tag{10-79}$$

In other words, the mean $\eta_y(t)$ of the output $\mathbf{y}(t)$ equals the response of the system to the mean $\eta_x(t)$ of the input (Fig. 10-8a)

$$\eta_y(t) = L[\eta_x(t)] \tag{10-80}$$

The above is a simple extension of the linearity of expected values to arbitrary linear operators. In the context of (10-78) it can be deduced if we write the integral as a limit of a sum. This yields

$$E\{\mathbf{y}(t)\} = \int_{-\infty}^{\infty} E\{\mathbf{x}(t-\alpha)\}h(\alpha)\,d\alpha = \eta_x(t) * h(t) \tag{10-81}$$

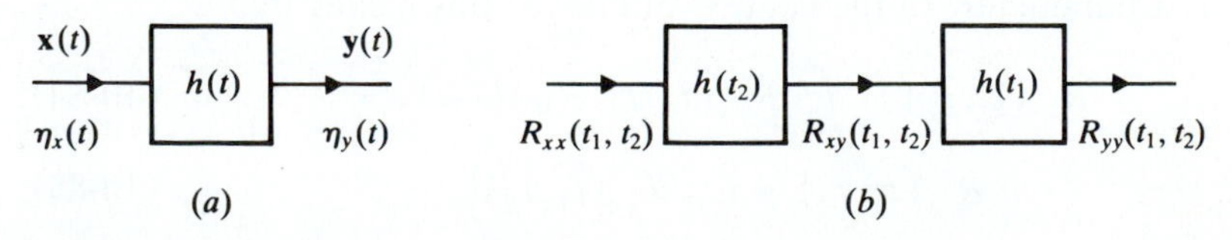

FIGURE 10-8

Frequency interpretation At the ith trial the input to our system is a function $\mathbf{x}(t, \zeta_i)$ yielding as output the function $\mathbf{y}(t, \zeta_i) = L[\mathbf{x}(t, \zeta_i)]$. For large n,

$$E\{\mathbf{y}(t)\} \simeq \frac{\mathbf{y}(t,\zeta_1) + \cdots + \mathbf{y}(t,\zeta_n)}{n} = \frac{L[\mathbf{x}(t,\zeta_1)] + \cdots + L[\mathbf{x}(t,\zeta_n)]}{n}$$

From the linearity of the system it follows that the last term above equals

$$L\left[\frac{\mathbf{x}(t,\zeta_1) + \cdots + \mathbf{x}(t,\zeta_n)}{n}\right]$$

This agrees with (10-79) because the fraction is nearly equal to $E\{\mathbf{x}(t)\}$.

Notes 1. From (10-80) it follows that if

$$\tilde{\mathbf{x}}(t) = \mathbf{x}(t) - \eta_x(t) \qquad \tilde{\mathbf{y}}(t) = \mathbf{y}(t) - \eta_y(t)$$

then

$$L[\tilde{\mathbf{x}}(t)] = L[\mathbf{x}(t)] - L[\eta_x(t)] = \tilde{\mathbf{y}}(t) \tag{10-82}$$

Thus the response of a linear system to the centered input $\tilde{\mathbf{x}}(t)$ equals the centered output $\tilde{\mathbf{y}}(t)$.

2. Suppose that

$$\mathbf{x}(t) = f(t) + \boldsymbol{\nu}(t) \qquad E\{\boldsymbol{\nu}(t)\} = 0$$

In this case, $E\{\mathbf{x}(t)\} = f(t)$; hence

$$\eta_y(t) = f(t) * h(t)$$

Thus, if $\mathbf{x}(t)$ is the sum of a deterministic signal $f(t)$ and a random component $\boldsymbol{\nu}(t)$, then for the determination of the mean of the output we can ignore $\boldsymbol{\nu}(t)$ provided that the system is linear and $E\{\boldsymbol{\nu}(t)\} = 0$.

Theorem (10-79) can be used to express the joint moments of any order of the output $\mathbf{y}(t)$ of a linear system in terms of the corresponding moments of the input. The following special cases are of fundamental importance in the study of linear systems with stochastic inputs.

OUTPUT AUTOCORRELATION. We wish to express the autocorrelation $R_{yy}(t_1, t_2)$ of the output $\mathbf{y}(t)$ of a linear system in terms of the autocorrelation $R_{xx}(t_1, t_2)$ of the input $\mathbf{x}(t)$. As we shall presently see, it is easier to find first the cross-correlation $R_{xy}(t_1, t_2)$ between $\mathbf{x}(t)$ and $\mathbf{y}(t)$.

THEOREM

(a)

$$R_{xy}(t_1, t_2) = L_2[R_{xx}(t_1, t_2)] \tag{10-83}$$

In the above notation, L_2 means that the system operates on the variable t_2, treating t_1 as a parameter. In the context of (10-78) this means that

$$R_{xy}(t_1, t_2) = \int_{-\infty}^{\infty} R_{xx}(t_1, t_2 - \alpha) h(\alpha)\, d\alpha \tag{10-84}$$

(b)

$$R_{yy}(t_1, t_2) = L_1[R_{xy}(t_1, t_2)] \tag{10-85}$$

In this case, the system operates on t_1:

$$R_{yy}(t_1, t_2) = \int_{-\infty}^{\infty} R_{xy}(t_1 - \alpha, t_2) h(\alpha) \, d\alpha \tag{10-86}$$

Proof. Multiplying (10-76) by $\mathbf{x}(t_1)$ and using (10-77), we obtain

$$\mathbf{x}(t_1)\mathbf{y}(t) = L_t[\mathbf{x}(t_1)\mathbf{x}(t)]$$

where L_t means that the system operates on t. Hence [see (10-79)]

$$E\{\mathbf{x}(t_1)\mathbf{y}(t)\} = L_t[E\{\mathbf{x}(t_1)\mathbf{x}(t)\}]$$

and (10-83) follows with $t = t_2$. The proof of (10-85) is similar: We multiply (10-76) by $\mathbf{y}(t_2)$ and use (10-79). This yields

$$E\{\mathbf{y}(t)\mathbf{y}(t_2)\} = L_t[E\{\mathbf{x}(t)\mathbf{y}(t_2)\}]$$

and (10-85) follows with $t = t_1$.

The preceding theorem is illustrated in Fig. 10-8*b*: If $R_{xx}(t_1, t_2)$ is the input to the given system and the system operates on t_2, the output equals $R_{xy}(t_1, t_2)$. If $R_{xy}(t_1, t_2)$ is the input and the system operates on t_1, the output equals $R_{yy}(t_1, t_2)$.

Inserting (10-84) into (10-86), we obtain

$$R_{yy}(t_1, t_2) = \int_{-\infty}^{\infty}\int_{-\infty}^{\infty} R_{xx}(t_1 - \alpha, t_2 - \beta) h(\alpha) h(\beta) \, d\alpha \, d\beta$$

This expresses $R_{yy}(t_1, t_2)$ directly in terms of $R_{xx}(t_1, t_2)$. However, conceptually and operationally, it is preferable to find first $R_{xy}(t_1, t_2)$.

Example 10-18. A stationary process $\boldsymbol{\nu}(t)$ with autocorrelation $R_{\nu\nu}(\tau) = q\delta(\tau)$ (white noise) is applied at $t = 0$ to a linear system with

$$h(t) = e^{-ct}U(t)$$

We shall show that the autocorrelation of the resulting output $\mathbf{y}(t)$ equals

$$R_{yy}(t_1, t_2) = \frac{q}{2c}(1 - e^{-2ct_1})e^{-c|t_2 - t_1|} \tag{10-87}$$

for $0 < t_1 < t_2$.

Proof. We can use the preceding results if we assume that the input to the system is the process

$$\mathbf{x}(t) = \boldsymbol{\nu}(t)U(t)$$

With this assumption, all correlations are 0 if $t_1 < 0$ or $t_2 < 0$. For $t_1 > 0$ and $t_2 > 0$,

$$R_{xx}(t_1, t_2) = E\{\boldsymbol{\nu}(t_1)\boldsymbol{\nu}(t_2)\} = q\delta(t_1 - t_2)$$

As we see from (10-83), $R_{xy}(t_1, t_2)$ equals the response of the system to $q\delta(t_1 - t_2)$ considered as a function of t_2. Since $\delta(t_1 - t_2) = \delta(t_2 - t_1)$ and $L[\delta(t_2 - t_1)] =$

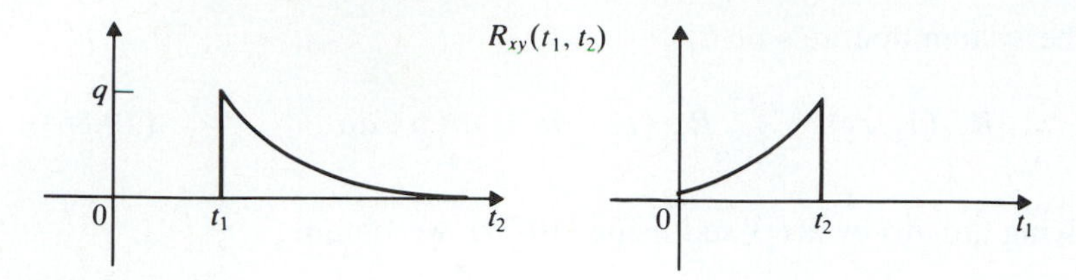

FIGURE 10-9

$h(t_2 - t_1)$ (time invariance), we conclude that

$$R_{xy}(t_1, t_2) = qh(t_2 - t_1) = qe^{-c(t_2 - t_1)}U(t_2 - t_1)U(t_1)$$

In Fig. 10-9, we show $R_{xy}(t_1, t_2)$ as a function of t_1 and t_2. Inserting into (10-86), we obtain

$$R_{yy}(t_1, t_2) = q\int_0^{t_1} e^{c(t_1 - \alpha - t_2)}e^{-c\alpha}\, d\alpha \qquad t_1 < t_2$$

and (10-87) results.

Note that

$$E\{\mathbf{y}^2(t)\} = R_{yy}(t, t) = \frac{q}{2c}(1 - e^{-2ct}) = q\int_0^t h^2(\alpha)\, d\alpha$$

COROLLARY. The autocovariance $C_{yy}(t_1, t_2)$ of $\mathbf{y}(t)$ is the autocorrelation of the process $\tilde{\mathbf{y}}(t) = \mathbf{y}(t) - \eta_y(t)$ and, as we see from (10-82), $\tilde{\mathbf{y}}(t)$ equals $L[\tilde{\mathbf{x}}(t)]$. Applying (10-84) and (10-86) to the centered processes $\tilde{\mathbf{x}}(t)$ and $\tilde{\mathbf{y}}(t)$, we obtain

$$\begin{aligned} C_{xy}(t_1, t_2) &= C_{xx}(t_1, t_2) * h(t_2) \\ C_{yy}(t_1, t_2) &= C_{xy}(t_1, t_2) * h(t_1) \end{aligned} \qquad (10\text{-}88)$$

where the convolutions are in t_1 and t_2 respectively.

Complex processes The preceding results can be readily extended to complex processes and to systems with complex-valued $h(t)$. Reasoning as in the real case, we obtain

$$\begin{aligned} R_{xy}(t_1, t_2) &= R_{xx}(t_1, t_2) * h^*(t_2) \\ R_{yy}(t_1, t_2) &= R_{xy}(t_1, t_2) * h(t_1) \end{aligned} \qquad (10\text{-}89)$$

Response to white noise. We shall determine the average power $E\{|\mathbf{y}(t)|^2\}$ of the output of a system driven by white noise. This is a special case of (10-89), however, because of its importance it is stated as a theorem.

THEOREM. If the input to a linear system is white noise with autocorrelation

$$R_{xx}(t_1, t_2) = q(t_1)\delta(t_1 - t_2)$$

then

$$E\{|\mathbf{y}(t)|^2\} = q(t) * |h(t)|^2 = \int_{-\infty}^{\infty} q(t - \alpha)|h(\alpha)|^2\, d\alpha \qquad (10\text{-}90)$$

Proof. From (10-89) it follows that

$$R_{xy}(t_1, t_2) = q(t_1)\delta(t_2 - t_1) * h^*(t_2) = q(t_1)h^*(t_2 - t_1)$$

$$R_{yy}(t_1, t_2) = \int_{-\infty}^{\infty} q(t_1 - \alpha)h^*[t_2 - (t_1 - \alpha)]h(\alpha)\, d\alpha$$

and with $t_1 = t_2 = t$, (10-90) results.

Special cases (*a*) If $\mathbf{x}(t)$ is stationary white noise, then $q(t) = q$ and (10-90) yields

$$E\{\mathbf{y}^2(t)\} = qE \qquad \text{where} \quad E = \int_{-\infty}^{\infty} |h(t)|^2\, dt$$

is the energy of $h(t)$.

(*b*) If $h(t)$ is of short duration relative to the variations of $q(t)$, then

$$E\{\mathbf{y}^2(t)\} \simeq q(t)\int_{-\infty}^{\infty} |h(\alpha)|^2\, d\alpha = Eq(t) \tag{10-91}$$

This relationship justifies the term *average intensity* used to describe the function $q(t)$.

(*c*) If $R_{\nu\nu}(\tau) = q\delta(\tau)$ and $\boldsymbol{\nu}(t)$ is applied to the system at $t = 0$, then $q(t) = qU(t)$ and (10-90) yields

$$E\{\mathbf{y}^2(t)\} = q\int_{-\infty}^{t} |h(\alpha)|^2\, d\alpha$$

Example 10-19. The integral

$$\mathbf{y} = \int_0^t \boldsymbol{\nu}(\alpha)\, d\alpha$$

can be considered as the output of a linear system with input $\mathbf{x}(t) = \boldsymbol{\nu}(t)U(t)$ and impulse response $h(t) = U(t)$. If, therefore, $\boldsymbol{\nu}(t)$ is white noise with average intensity $q(t)$, then $\mathbf{x}(t)$ is white noise with average intensity $q(t)U(t)$ and (10-90) yields

$$E\{\mathbf{y}^2(t)\} = q(t)U(t) * U(t) = \int_0^t q(\alpha)\, d\alpha$$

Differentiators. A differentiator is a linear system whose output is the derivative of the input

$$L[\mathbf{x}(t)] = \mathbf{x}'(t)$$

We can, therefore, use the preceding results to find the mean and the autocorrelation of $\mathbf{x}'(t)$.

From (10-80) it follows that

$$\eta_{x'}(t) = L[\eta_x(t)] = \eta_x'(t) \tag{10-92}$$

Similarly [see (10-83)]

$$R_{xx'}(t_1, t_2) = L_2[R_{xx}(t_1, t_2)] = \frac{\partial R_{xx}(t_1, t_2)}{\partial t_2} \tag{10-93}$$

because, in this case, L_2 means differentiation with respect to t_2. Finally,

$$R_{x'x'}(t_1, t_2) = L_1[R_{xx'}(t_1, t_2)] = \frac{\partial R_{xx'}(t_1, t_2)}{\partial t_1} \tag{10-94}$$

Combining, we obtain

$$R_{x'x'}(t_1, t_2) = \frac{\partial^2 R_{xx}(t_1, t_2)}{\partial t_1 \, \partial t_2} \tag{10-95}$$

Stationary processes If $\mathbf{x}(t)$ is WSS, then $\eta_x(t)$ is constant; hence

$$E\{\mathbf{x}'(t)\} = 0 \tag{10-96}$$

Furthermore, since $R_{xx}(t_1, t_2) = R_{xx}(\tau)$, we conclude with $\tau = t_1 - t_2$ that

$$\frac{\partial R_{xx}(t_1 - t_2)}{\partial t_2} = -\frac{dR_{xx}(\tau)}{d\tau} \qquad \frac{\partial^2 R_{xx}(t_1 - t_2)}{\partial t_1 \, \partial t_2} = -\frac{d^2 R_{xx}(\tau)}{d\tau^2}$$

Hence

$$R_{xx'}(\tau) = -R'_{xx}(\tau) \qquad R_{x'x'}(\tau) = -R''_{xx}(\tau) \tag{10-97}$$

Poisson impulses. If the input $\mathbf{x}(t)$ to a differentiator is a Poisson process, the resulting output $\mathbf{z}(t)$ is a train of impulses (Fig. 10-10)

$$\mathbf{z}(t) = \sum_i \delta(t - \mathbf{t}_i) \tag{10-98}$$

We maintain that $\mathbf{z}(t)$ is a stationary process with mean

$$\eta_z = \lambda \tag{10-99}$$

and autocorrelation

$$R_{zz}(\tau) = \lambda^2 + \lambda\delta(\tau) \tag{10-100}$$

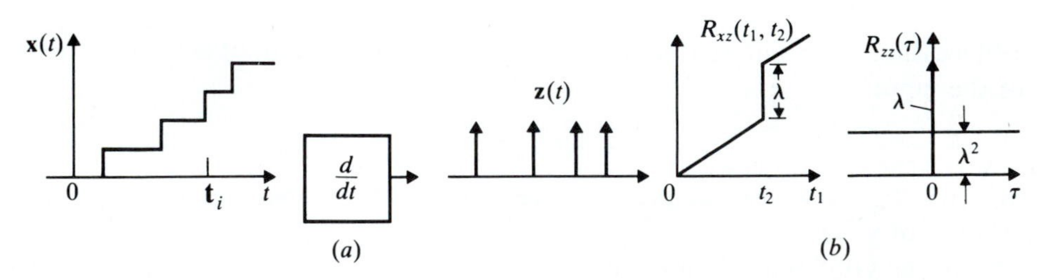

FIGURE 10-10

Proof. The first equation follows from (10-92) because $\eta_x(t) = \lambda t$. To prove the second, we observe that [see (10-14)]

$$R_{xx}(t_1, t_2) = \lambda^2 t_1 t_2 + \lambda \min(t_1, t_2) \tag{10-101}$$

And since $\mathbf{z}(t) = \mathbf{x}'(t)$, (10-93) yields

$$R_{xz}(t_1, t_2) = \frac{\partial R_{xx}(t_1, t_2)}{\partial t_2} = \lambda^2 t_1 + \lambda U(t_1 - t_2)$$

This function is plotted in Fig. 10-10b where the independent variable is t_1. As we see, it is discontinuous for $t_1 = t_2$ and its derivative with respect to t_1 contains the impulse $\lambda\delta(t_1 - t_2)$. This yields [see (10-94)]

$$R_{zz}(t_1, t_2) = \frac{\partial R_{xz}(t_1, t_2)}{\partial t_1} = \lambda^2 + \lambda\delta(t_1 - t_2)$$

DIFFERENTIAL EQUATIONS. A deterministic differential equation with random excitation is an equation of the form

$$a_n \mathbf{y}^{(m)}(t) + \cdots + a_0 \mathbf{y}(t) = \mathbf{x}(t) \tag{10-102}$$

where the coefficients a_k are given numbers and the driver $\mathbf{x}(t)$ is a stochastic process. We shall consider its solution $\mathbf{y}(t)$ under the assumption that the initial conditions are 0. With this assumption, $\mathbf{y}(t)$ is unique (zero-state response) and it satisfies the linearity condition (10-77). We can, therefore, interpret $\mathbf{y}(t)$ as the output of a linear system specified by (10-102).

In general, the determination of the complete statistics of $\mathbf{y}(t)$ is complicated. In the following, we evaluate only its second-order moments using the preceding results. The above system is an operator L specified as follows: Its output $\mathbf{y}(t)$ is a process with zero initial conditions satisfying (10-102).

Mean. As we know [see (10-80)] the mean $\eta_y(t)$ of $\mathbf{y}(t)$ is the output of L with input $\eta_x(t)$. Hence it satisfies the equation

$$a_n \eta_y^{(n)}(t) + \cdots + a_0 \eta_y(t) = \eta_x(t) \tag{10-103}$$

and the initial conditions

$$\eta_y(0) = \cdots = \eta_y^{(n-1)}(0) = 0 \tag{10-104}$$

This result can be established directly: Clearly,

$$E\{\mathbf{y}^{(k)}(t)\} = \eta_y^{(k)}(t) \tag{10-105}$$

Taking expected values of both sides of (10-102) and using the above, we obtain (10-103). Equation (10-104) follows from (10-105) because $\mathbf{y}^{(k)}(0) = 0$ by assumption.

Correlation. To determine $R_{xy}(t_1, t_2)$, we use (10-83)

$$R_{xy}(t_1, t_2) = L_2[R_{xx}(t_1, t_2)]$$

In this case, L_2 means that $R_{xy}(t_1, t_2)$ satisfies the differential equation

$$a_n \frac{\partial^n R_{xy}(t_1, t_2)}{\partial t_2^n} + \cdots + a_0 R_{xy}(t_1, t_2) = R_{xx}(t_1, t_2) \qquad (10\text{-}106)$$

with the initial conditions

$$R_{xy}(t_1, 0) = \cdots = \frac{\partial^{n-1} R_{xy}(t_1, 0)}{\partial t_2^{n-1}} = 0 \qquad (10\text{-}107)$$

Similarly, since [see (10-85)]

$$R_{yy}(t_1, t_2) = L_1[R_{xy}(t_1, t_2)]$$

we conclude as above that

$$a_n \frac{\partial^n R_{yy}(t_1, t_2)}{\partial t_1^n} + \cdots + a_0 R_{yy}(t_1, t_2) = R_{xy}(t_1, t_2) \qquad (10\text{-}108)$$

$$R_{yy}(0, t_2) = \cdots = \frac{\partial^{n-1} R_{yy}(0, t_2)}{\partial t_1^{n-1}} = 0 \qquad (10\text{-}109)$$

The preceding results can be established directly: From (10-102) it follows that

$$\mathbf{x}(t_1)\left[a_n \mathbf{y}^{(n)}(t_2) + \cdots + a_0 \mathbf{y}(t_2)\right] = \mathbf{x}(t_1)\mathbf{x}(t_2)$$

This yields (10-106) because [see (10-119)]

$$E\{\mathbf{x}(t_1)\mathbf{y}^{(k)}(t_2)\} = \partial^k R_{xy}(t_1, t_2)/\partial t_2^k$$

Similarly, (10-108) is a consequence of the identity

$$\left[a_n \mathbf{y}^{(n)}(t_1) + \cdots + a_0 \mathbf{y}(t_1)\right]\mathbf{y}(t_2) = \mathbf{x}(t_1)\mathbf{y}(t_2)$$

because

$$E\{\mathbf{y}^{(k)}(t_1)\mathbf{y}(t_2)\} = \partial^k R_{yy}(t_1, t_2)/\partial t_1^k$$

Finally, the expected values of

$$\mathbf{x}(t_1)\mathbf{y}^{(k)}(0) = 0 \qquad \mathbf{y}^{(k)}(0)\mathbf{y}(t_2) = 0$$

yield (10-107) and (10-109).

General moments. The moments of any order of the output $\mathbf{y}(t)$ of a linear system can be expressed in terms of the corresponding moments of the input $\mathbf{x}(t)$. As an illustration, we shall determine the third-order moment

$$R_{yyy}(t_1, t_2, t_3) = E\{\mathbf{y}_1(t)\mathbf{y}_2(t)\mathbf{y}_3(t)\}$$

of $\mathbf{y}(t)$ in terms of the third-order moment $R_{xxx}(t_1, t_2, t_3)$ of $\mathbf{x}(t)$. Proceeding as in (10-83), we obtain

$$E\{\mathbf{x}(t_1)\mathbf{x}(t_2)\mathbf{y}(t_3)\} = L_3[E\{\mathbf{x}(t_1)\mathbf{x}(t_2)\mathbf{x}(t_3)\}]$$

$$= \int_{-\infty}^{\infty} R_{xxx}(t_1, t_2, t_3 - \gamma)h(\gamma)\, d\gamma \quad (10\text{-}110a)$$

$$E\{\mathbf{x}(t_1)\mathbf{y}(t_2)\mathbf{y}(t_3)\} = L_2[E\{\mathbf{x}(t_1)\mathbf{x}(t_2)\mathbf{y}(t_3)\}]$$

$$= \int_{-\infty}^{\infty} R_{xxy}(t_1, t_2 - \beta, t_3)h(\beta)\, d\beta \quad (10\text{-}110b)$$

$$E\{\mathbf{y}(t_1)\mathbf{y}(t_2)\mathbf{y}(t_3)\} = L_1[E\{\mathbf{x}(t_1)\mathbf{y}(t_2)\mathbf{y}(t_3)\}]$$

$$= \int_{-\infty}^{\infty} R_{xyy}(t_1 - \alpha, t_2, t_3)h(\alpha)\, d\alpha \quad (10\text{-}110c)$$

Note that for the evaluation of $R_{yyy}(t_1, t_2, t_3)$ for specific times t_1, t_2, t_3, the function $R_{xxx}(t_1, t_2, t_3)$ must be known for every t_1, t_2, t_3.

Vector Processes and Multiterminal Systems

We consider now systems with n inputs $\mathbf{x}_i(t)$ and r outputs $\mathbf{y}_j(t)$. As a preparation, we introduce the notion of autocorrelation and cross-correlation for vector processes starting with a review of the standard matrix notation.

The expression $A = [a_{ij}]$ will mean a matrix with elements a_{ij}. The notation

$$A^t = [a_{ji}] \qquad A^* = [a_{ij}^*] \qquad A^\dagger = [a_{ji}^*]$$

will mean the transpose, the conjugate, and the conjugate transpose of A.

A column vector will be identified by $A = [a_i]$. Whether A is a vector or a general matrix will be understood from the context. If $A = [a_i]$ and $B = [b_j]$ are two vectors with m elements each, the product $A^tB = a_1b_1 + \cdots + a_mb_m$ is a number, and the product $AB^t = [a_ib_j]$ is an $m \times m$ matrix with elements a_ib_j.

A vector process $\mathbf{X}(t) = [\mathbf{x}_i(t)]$ is a vector, the components of which are stochastic processes. The mean $\eta(t) = E\{\mathbf{X}(t)\} = [\eta_i(t)]$ of $\mathbf{X}(t)$ is a vector with components $\eta_i(t) = E\{\mathbf{x}_i(t)\}$. The autocorrelation $R(t_1, t_2)$ or $R_{xx}(t_1, t_2)$ of a vector process $\mathbf{X}(t)$ is an $m \times m$ matrix

$$R(t_1, t_2) = E\{\mathbf{X}(t_1)\mathbf{X}^\dagger(t_2)\} \quad (10\text{-}111)$$

with elements $E\{\mathbf{x}_i(t_1)\mathbf{x}_j^*(t_2)\}$. We define similarly the cross-correlation matrix

$$R_{xy}(t_1, t_2) = E\{\mathbf{X}(t_1)\mathbf{Y}^\dagger(t_2)\} \quad (10\text{-}112)$$

of the vector processes

$$\mathbf{X}(t) = [\mathbf{x}_i(t)] \quad i = 1, \ldots, m \qquad \mathbf{Y}(t) = [\mathbf{y}_j(t)] \quad j = 1, \ldots, r \quad (10\text{-}113)$$

A multiterminal system with m inputs $\mathbf{x}_i(t)$ and r outputs $\mathbf{y}_j(t)$ is a rule for assigning to an m vector $\mathbf{X}(t)$ an r vector $\mathbf{Y}(t)$. If the system is linear and time-invariant, it is specified in terms of its impulse response matrix. This is an $r \times m$ matrix

$$H(t) = \left[h_{ji}(t)\right] \qquad i = 1, \ldots, m \qquad j = 1, \ldots, r \tag{10-114}$$

defined as follows: Its component $h_{ji}(t)$ is the response of the jth output when the ith input equals $\delta(t)$ and all other inputs equal 0. From this and the linearity of the system, it follows that the response $\mathbf{y}_j(t)$ of the jth output to an arbitrary input $\mathbf{X}(t) = [\mathbf{x}_i(t)]$ equals

$$\mathbf{y}_j(t) = \int_{-\infty}^{\infty} h_{j1}(\alpha)\mathbf{x}_1(t-\alpha)\,d\alpha + \cdots + \int_{-\infty}^{\infty} h_{jm}(\alpha)\mathbf{x}_m(t-\alpha)\,d\alpha$$

Hence

$$\mathbf{Y}(t) = \int_{-\infty}^{\infty} H(\alpha)\mathbf{X}(t-\alpha)\,d\alpha \tag{10-115}$$

In the above, $\mathbf{X}(t)$ and $\mathbf{Y}(t)$ are column vectors and $H(t)$ is an $r \times m$ matrix. We shall use this relationship to determine the autocorrelation $R_{yy}(t_1, t_2)$ of $\mathbf{Y}(t)$. Premultiplying the conjugate transpose of (10-115) by $\mathbf{X}(t_1)$ and setting $t = t_2$, we obtain

$$\mathbf{X}(t_1)\mathbf{Y}^{\dagger}(t_2) = \int_{-\infty}^{\infty} \mathbf{X}(t_1)\mathbf{X}^{\dagger}(t_2-\alpha)H^{\dagger}(\alpha)\,d\alpha$$

Hence

$$R_{xy}(t_1, t_2) = \int_{-\infty}^{\infty} R_{xx}(t_1, t_2-\alpha)H^{\dagger}(\alpha)\,d\alpha \tag{10-116a}$$

Postmultiplying (10-115) by $\mathbf{Y}^{\dagger}(t_2)$ and setting $t = t_1$, we obtain

$$R_{yy}(t_1, t_2) = \int_{-\infty}^{\infty} H(\alpha)R_{xy}(t_1-\alpha, t_2)\,d\alpha \tag{10-116b}$$

as in (10-89). These results can be used to express the cross-correlation of the outputs of several scalar systems in terms of the cross-correlation of their inputs. The next example is an illustration.

Example 10-20. In Fig. 10-11 we show two systems with inputs $\mathbf{x}_1(t), \mathbf{x}_2(t)$ and outputs

$$\mathbf{y}_1(t) = \int_{-\infty}^{\infty} h_1(\alpha)\mathbf{x}_1(t-\alpha)\,d\alpha \qquad \mathbf{y}_2(t) = \int_{-\infty}^{\infty} h_2(\alpha)\mathbf{x}_2(t-\alpha)\,d\alpha \tag{10-117}$$

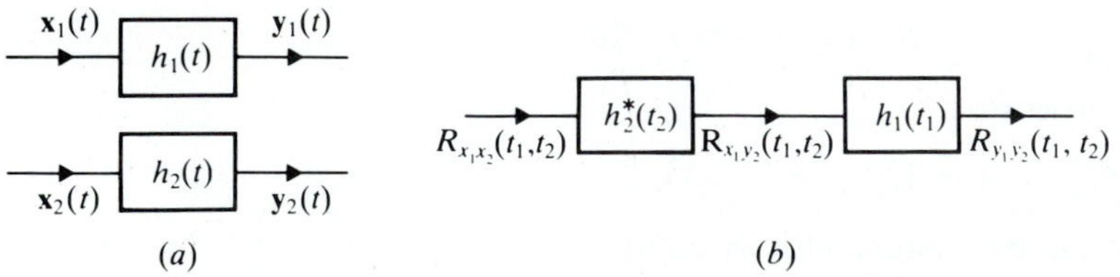

$\mathbf{x}_1(t)$ → $h_1(t)$ → $\mathbf{y}_1(t)$

$\mathbf{x}_2(t)$ → $h_2(t)$ → $\mathbf{y}_2(t)$

(*a*)

$R_{x_1x_2}(t_1,t_2)$ → $h_2^*(t_2)$ → $R_{x_1y_2}(t_1,t_2)$ → $h_1(t_1)$ → $R_{y_1y_2}(t_1,t_2)$

(*b*)

FIGURE 10-11

These signals can be considered as the components of the output vector $\mathbf{Y}'(t) = [\mathbf{y}_1(t), \mathbf{y}_2(t)]$ of a 2×2 system with input vector $\mathbf{X}'(t) = [\mathbf{x}_1(t), \mathbf{x}_2(t)]$ and impulse response matrix

$$H(t) = \begin{bmatrix} h_1(t) & 0 \\ 0 & h_2(t) \end{bmatrix}$$

Inserting into (10-116), we obtain

$$\begin{aligned} R_{x_1y_2}(t_1, t_2) &= \int_{-\infty}^{\infty} R_{x_1x_2}(t_1, t_2 - \alpha) h_2^*(\alpha)\, d\alpha \\ R_{y_1y_2}(t_1, t_2) &= \int_{-\infty}^{\infty} h_1(\alpha) R_{x_1y_2}(t_1 - \alpha, t_2)\, d\alpha \end{aligned} \tag{10-118}$$

Thus, to find $R_{x_1y_2}(t_1, t_2)$, we use $R_{x_1x_2}(t_1, t_2)$ as the input to the conjugate $h_2^*(t)$ of $h_2(t)$, operating on the variable t_2. To find $R_{y_1y_2}(t_1, t_2)$, we use $R_{x_1y_2}(t_1, t_2)$ as the input to $h_1(t)$ operating on the variable t_1 (Fig. 10-11).

Example 10-21. The derivatives $\mathbf{y}_1(t) = \mathbf{z}^{(m)}(t)$ and $\mathbf{y}_2(t) = \mathbf{w}^{(n)}(t)$ of two processes $\mathbf{z}(t)$ and $\mathbf{w}(t)$ can be considered as the responses of two differentiators with inputs $\mathbf{x}_1(t) = \mathbf{z}(t)$ and $\mathbf{x}_2(t) = \mathbf{w}(t)$. Applying (10-118) suitably interpreted, we conclude that

$$E\{\mathbf{z}^{(m)}(t_1)\mathbf{w}^{(n)}(t_2)\} = \frac{\partial^{m+n} R_{zw}(t_1, t_2)}{\partial t_1^m\, \partial t_2^n} \tag{10-119}$$

10-3 THE POWER SPECTRUM

In signal theory, spectra are associated with Fourier transforms. For deterministic signals, they are used to represent a function as a superposition of exponentials. For random signals, the notion of a spectrum has two interpretations. The first involves transforms of averages; it is thus essentially deterministic. The second leads to the representation of the process under consideration as superposition of exponentials with random coefficients. In this section, we introduce the first interpretation. The second is treated in Sec. 12-4. We shall consider only stationary processes. For nonstationary processes the notion of a spectrum is of limited interest.

DEFINITIONS. The *power spectrum* (or *spectral density*) of a WSS process $\mathbf{x}(t)$, real or complex, is the Fourier transform $S(\omega)$ of its autocorrelation $R(\tau) = E\{\mathbf{x}(t+\tau)\mathbf{x}^*(t)\}$:

$$S(\omega) = \int_{-\infty}^{\infty} R(\tau) e^{-j\omega\tau}\, d\tau \tag{10-120}$$

Since $R(-\tau) = R^*(\tau)$ it follows that $S(\omega)$ is a real function of ω.

From the Fourier inversion formula, it follows that

$$R(\tau) = \frac{1}{2\pi} \int_{-\infty}^{\infty} S(\omega) e^{j\omega\tau}\, d\omega \tag{10-121}$$

TABLE 10-1

$$R(\tau) = \frac{1}{2\pi}\int_{-\infty}^{\infty} S(\omega)e^{j\omega\tau}\,d\omega \leftrightarrow S(\omega) = \int_{-\infty}^{\infty} R(\tau)e^{-j\omega\tau}\,d\tau$$

$\delta(\tau) \leftrightarrow 1$	$1 \leftrightarrow 2\pi\delta(\omega)$
$e^{j\beta\tau} \leftrightarrow 2\pi\delta(\omega - \beta)$	$\cos\beta\tau \leftrightarrow \pi\delta(\omega - \beta) + \pi\delta(\omega + \beta)$
$e^{-\alpha\|\tau\|} \leftrightarrow \dfrac{2\alpha}{\alpha^2 + \omega^2}$	$e^{-\alpha\tau^2} \leftrightarrow \sqrt{\dfrac{\pi}{\alpha}}\, e^{-\omega^2/4\alpha}$

$$e^{-\alpha|\tau|}\cos\beta\tau \leftrightarrow \frac{\alpha}{\alpha^2 + (\omega - \beta)^2} + \frac{\alpha}{\alpha^2 + (\omega + \beta)^2}$$

$$2e^{-\alpha\tau^2}\cos\beta\tau \leftrightarrow \sqrt{\frac{\pi}{\alpha}}\,[e^{-(\omega-\beta)^2/4\alpha} + e^{-(\omega+\beta)^2/4\alpha}]$$

$$\begin{cases} 1 - \dfrac{|\tau|}{T} & |\tau| < T \\ 0 & |\tau| > T \end{cases} \leftrightarrow \frac{4\sin^2(\omega T/2)}{T\omega^2}$$

$$\frac{\sin\sigma\tau}{\pi\tau} \leftrightarrow \begin{cases} 1 & |\omega| < \sigma \\ 0 & |\omega| > \sigma \end{cases}$$

If $\mathbf{x}(t)$ is a real process, then $R(\tau)$ is real and even; hence $S(\omega)$ is also real and even. In this case,

$$S(\omega) = \int_{-\infty}^{\infty} R(\tau)\cos\omega\tau\,d\tau = 2\int_{0}^{\infty} R(\tau)\cos\omega\tau\,d\tau$$
$$R(\tau) = \frac{1}{2\pi}\int_{-\infty}^{\infty} S(\omega)\cos\omega\tau\,d\omega = \frac{1}{\pi}\int_{0}^{\infty} S(\omega)\cos\omega\tau\,d\omega \tag{10-122}$$

The *cross-power spectrum* of two processes $\mathbf{x}(t)$ and $\mathbf{y}(t)$ is the Fourier transform $S_{xy}(\omega)$ of their cross-correlation $R_{xy}(\tau) = E\{\mathbf{x}(t + \tau)\mathbf{y}^*(t)\}$:

$$S_{xy}(\omega) = \int_{-\infty}^{\infty} R_{xy}(\tau)e^{-j\omega\tau}\,d\tau \qquad R_{xy}(\tau) = \frac{1}{2\pi}\int_{-\infty}^{\infty} S_{xy}(\omega)e^{j\omega\tau}\,d\omega \tag{10-123}$$

The function $S_{xy}(\omega)$ is, in general, complex even when both processes $\mathbf{x}(t)$ and $\mathbf{y}(t)$ are real. In all cases,

$$S_{xy}(\omega) = S_{yx}^*(\omega) \tag{10-124}$$

because $R_{xy}(-\tau) = E\{\mathbf{x}(t - \tau)\mathbf{y}^*(t)\} = R_{yx}^*(\tau)$.

In Table 10-1 we list a number of frequently used autocorrelations and the corresponding spectra. Note that in all cases, $S(\omega)$ is positive. As we shall soon show, this is true for every spectrum.

Example 10-22. A random telegraph signal is a process $\mathbf{x}(t)$ taking the values $+1$ and -1 as in Example 10-6:

$$\mathbf{x}(t) = \begin{cases} 1 & \mathbf{t}_{2i} < t < \mathbf{t}_{2i+1} \\ -1 & \mathbf{t}_{2i-1} < t < \mathbf{t}_{2i} \end{cases}$$

where $\mathbf{t}_i$ is a set of Poisson points with average density λ. As we have shown in (10-19), its autocorrelation equals $e^{-2\lambda|\tau|}$. Hence

$$S(\omega) = \frac{4\lambda}{4\lambda^2 + \omega^2}$$

For most processes $R(\tau) \to \eta^2$ where $\eta = E\{\mathbf{x}(t)\}$ (see Sec. 12-4). If, therefore, $\eta \neq 0$, then $S(\omega)$ contains an impulse at $\omega = 0$. To avoid this, it is often convenient to express the spectral properties of $\mathbf{x}(t)$ in terms of the Fourier transform $S^c(\omega)$ of its autocovariance $C(\tau)$. Since $R(\tau) = C(\tau) + \eta^2$, it follows that

$$S(\omega) = S^c(\omega) + 2\pi\eta^2\delta(\omega) \tag{10-125}$$

The function $S^c(\omega)$ is called the *covariance spectrum* of $\mathbf{x}(t)$.

Example 10-23. We have shown in (10-100) that the autocorrelation of the Poisson impulses

$$\mathbf{z}(t) = \frac{d}{dt}\sum_i U(t - \mathbf{t}_i) = \sum_i \delta(t - \mathbf{t}_i)$$

equals $R_z(\tau) = \lambda^2 + \lambda\delta(\tau)$. From this it follows that

$$S_z(\omega) = \lambda + 2\pi\lambda^2\delta(\omega) \qquad S_z^c(\omega) = \lambda$$

We shall show that given an arbitrary positive function $S(\omega)$, we can find a process $\mathbf{x}(t)$ with power spectrum $S(\omega)$.

(*a*) Consider the process

$$\mathbf{x}(t) = ae^{j(\boldsymbol{\omega} t - \boldsymbol{\varphi})} \tag{10-126}$$

where a is a real constant, $\boldsymbol{\omega}$ is an RV with density $f_\omega(\omega)$, and $\boldsymbol{\varphi}$ is an RV independent of $\boldsymbol{\omega}$ and uniform in the interval $(0, 2\pi)$. As we know, this process is WSS with zero mean and autocorrelation

$$R_x(\tau) = a^2E\{e^{j\boldsymbol{\omega}\tau}\} = a^2\int_{-\infty}^{\infty} f_\omega(\omega)e^{j\omega\tau}\,d\omega$$

From this and the uniqueness property of Fourier transforms, it follows that [see (10-121)] the power spectrum of $\mathbf{x}(t)$ equals

$$S_x(\omega) = 2\pi a^2 f_\omega(\omega) \tag{10-127}$$

If, therefore,

$$f_\omega(\omega) = \frac{S(\omega)}{2\pi a^2} \qquad a^2 = \frac{1}{2\pi}\int_{-\infty}^{\infty} S(\omega)\,d\omega = R(0)$$

then $f_\omega(\omega)$ is a density and $S_x(\omega) = S(\omega)$. To complete the specification of $\mathbf{x}(t)$, it suffices to construct an RV $\boldsymbol{\omega}$ with density $S(\omega)/2\pi a^2$ and insert it into (10-126).

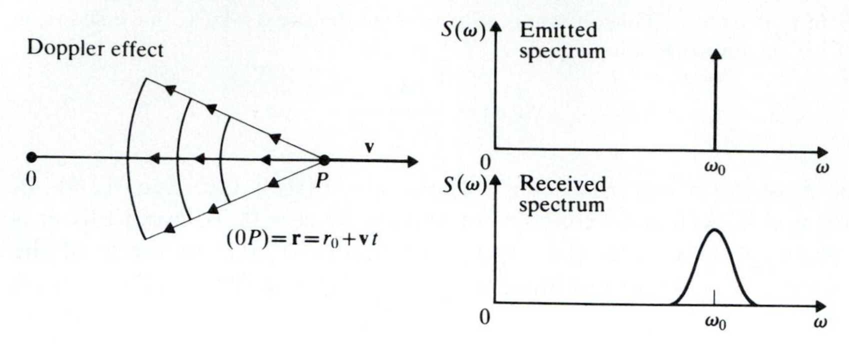

FIGURE 10-12

(*b*) We show next that if $S(-\omega) = S(\omega)$, we can find a real process with power spectrum $S(\omega)$. To do so, we form the process

$$\mathbf{y}(t) = a\cos(\boldsymbol{\omega}t + \boldsymbol{\varphi}) \qquad (10\text{-}128)$$

In this case (see Example 10-14)

$$R_y(\tau) = \frac{a^2}{2}E\{\cos\boldsymbol{\omega}\tau\} = \frac{a^2}{2}\int_{-\infty}^{\infty} f(\omega)\cos\omega\tau\, d\omega$$

From this it follows that if $f_\omega(\omega) = S(\omega)/\pi a^2$, then $S_y(\omega) = S(\omega)$.

Example 10-24 Doppler effect. A harmonic oscillator located at point P of the x axis (Fig. 10-12) moves in the x direction with velocity $\mathbf{v}$. The emitted signal equals $e^{j\omega_0 t}$ and the signal received by an observer located at point O equals

$$\mathbf{s}(t) = ae^{j\omega_0(t-\mathbf{r}/c)}$$

where c is the velocity of propagation and $\mathbf{r} = r_0 + \mathbf{v}t$. We assume that $\mathbf{v}$ is an RV with density $f_v(v)$. Clearly,

$$\mathbf{s}(t) = ae^{j(\boldsymbol{\omega}t-\varphi)} \qquad \boldsymbol{\omega} = \omega_0\left(1 - \frac{\mathbf{v}}{c}\right) \qquad \varphi = \frac{r_0\omega_0}{c}$$

hence the spectrum of the received signal is given by (10-127)

$$S(\omega) = 2\pi a^2 f_\omega(\omega) = \frac{2\pi a^2 c}{\omega_0} f_v\left[\left(1 - \frac{\omega}{\omega_0}\right)c\right] \qquad (10\text{-}129)$$

Note that if $\mathbf{v} = 0$, then

$$\mathbf{s}(t) = ae^{j(\omega_0 t-\varphi)} \qquad R(\tau) = a^2 e^{j\omega_0\tau} \qquad S(\omega) = 2\pi a^2\delta(\omega - \omega_0)$$

This is the spectrum of the emitted signal. Thus the motion causes broadening of the spectrum.

The above holds also if the motion forms an angle with the x axis provided that $\mathbf{v}$ is replaced by its projection $\mathbf{v}_x$ on OP. The following case is of special interest. Suppose that the emitter is a particle in a gas of temperature T. In this case, the x component of its velocity is a normal RV with zero mean and variance

kT/m (see Prob. 8-5). Inserting into (10-129), we conclude that

$$S(\omega) = \frac{2\pi a^2 c}{\omega_0\sqrt{2\pi kT/m}} \exp\left\{-\frac{mc^2}{2kT}\left(1 - \frac{\omega}{\omega_0}\right)^2\right\}$$

$$R(\tau) = a^2 \exp\left\{-\frac{kT\omega_0^2\tau^2}{2mc^2}\right\} e^{j\omega_0\tau}$$

Line spectra. (*a*) We have shown in Example 10-7 that the process

$$\mathbf{x}(t) = \sum_i \mathbf{c}_i e^{j\omega_i t}$$

is WSS if the RVs $\mathbf{c}_i$ are uncorrelated with zero mean. From this and Table 10-1 it follows that

$$R(\tau) = \sum_i \sigma_i^2 e^{j\omega_i\tau} \qquad S(\omega) = 2\pi\sum_i \sigma_i^2 \delta(\omega - \omega_i) \tag{10-130}$$

where $\sigma_i^2 = E\{\mathbf{c}_i^2\}$. Thus $S(\omega)$ consists of lines. In Sec. 14-2 we show that such a process is predictable, that is, its present value is uniquely determined in terms of its past.

(*b*) Similarly, the process

$$\mathbf{y}(t) = \sum_i (\mathbf{a}_i \cos\omega_i t + \mathbf{b}_i \sin\omega_i t)$$

is WSS iff the RVs $\mathbf{a}_i$ and $\mathbf{b}_i$ are uncorrelated with zero mean and $E\{\mathbf{a}_i^2\} = E\{\mathbf{b}_i^2\} = \sigma_i^2$. In this case,

$$R(\tau) = \sum_i \sigma_i^2 \cos\omega_i\tau \qquad S(\omega) = \pi\sum_i \sigma_i^2[\delta(\omega - \omega_i) + \delta(\omega + \omega_i)] \tag{10-131}$$

Linear systems. We shall express the autocorrelation $R_{yy}(\tau)$ and power spectrum $S_{yy}(\omega)$ of the response

$$\mathbf{y}(t) = \int_{-\infty}^{\infty} \mathbf{x}(t - \alpha)h(\alpha)\,d\alpha \tag{10-132}$$

of a linear system in terms of the autocorrelation $R_{xx}(\tau)$ and power spectrum $S_{xx}(\omega)$ of the input $\mathbf{x}(t)$.

THEOREM

$$R_{xy}(\tau) = R_{xx}(\tau) * h^*(-\tau) \qquad R_{yy}(\tau) = R_{xy}(\tau) * h(\tau) \tag{10-133}$$

$$S_{xy} = S_{xx}(\omega)H^*(\omega) \qquad S_{yy}(\omega) = S_{xy}(\omega)H(\omega) \tag{10-134}$$

Proof. The two equations in (10-133) are special cases of (10-184) and (10-185). However, because of their importance they will be proved directly. Multiplying the conjugate of (10-132) by $\mathbf{x}(t+\tau)$ and taking expected values, we obtain

$$E\{\mathbf{x}(t+\tau)\mathbf{y}^*(t)\} = \int_{-\infty}^{\infty} E\{\mathbf{x}(t+\tau)\mathbf{x}^*(t-\alpha)\}h^*(\alpha)\,d\alpha$$

Since $E\{\mathbf{x}(t+\tau)\mathbf{x}^*(t-\alpha)\} = R_{xx}(\tau+\alpha)$, this yields

$$R_{xy}(\tau) = \int_{-\infty}^{\infty} R_{xx}(\tau+\alpha)h^*(\alpha)\,d\alpha = \int_{-\infty}^{\infty} R_{xx}(\tau-\beta)h^*(-\beta)\,d\beta$$

Proceeding similarly, we obtain

$$E\{\mathbf{y}(t)\mathbf{y}^*(t-\tau)\} = \int_{-\infty}^{\infty} E\{\mathbf{x}(t-\alpha)\mathbf{y}^*(t-\tau)\}h(\alpha)\,d\alpha$$

$$= \int_{-\infty}^{\infty} R_{xy}(\tau-\alpha)h(\alpha)\,d\alpha$$

Equation (10-134) follows from (10-133) and the convolution theorem.

COROLLARY. Combining the two equations in (10-133) and (10-134), we obtain

$$R_{yy}(\tau) = R_{xx}(\tau) * h(\tau) * h^*(-\tau) = R_{xx}(\tau) * \rho(\tau) \qquad (10\text{-}135)$$

$$S_{yy}(\omega) = S_{xx}(\omega)H(\omega)H^*(\omega) = S_{xx}(\omega)|H(\omega)|^2 \qquad (10\text{-}136)$$

where

$$\rho(\tau) = h(\tau) * h^*(-\tau) = \int_{-\infty}^{\infty} h(t+\tau)h^*(t)\,dt \leftrightarrow |H(\omega)|^2 \qquad (10\text{-}137)$$

Note, in particular, that if $\mathbf{x}(t)$ is white noise with average power q, then

$$\begin{aligned} R_{xx}(\tau) &= q\delta(\tau) \qquad & S_{xx}(\omega) &= q \\ S_{yy}(\omega) &= q|H(\omega)|^2 \qquad & R_{yy}(\tau) &= q\rho(\tau) \end{aligned} \qquad (10\text{-}138)$$

From (10-136) and the inversion formula (10-121), it follows that

$$E\{|\mathbf{y}(t)|^2\} = R_{yy}(0) = \frac{1}{2\pi}\int_{-\infty}^{\infty} S_{xx}(\omega)|H(\omega)|^2\,d\omega \ge 0 \qquad (10\text{-}139)$$

This equation describes the filtering properties of a system when the input is a random process. It shows, for example, that if $H(\omega) = 0$ for $|\omega| > \omega_0$ and $S_{xx}(\omega) = 0$ for $|\omega| < \omega_0$, then $E\{\mathbf{y}^2(t)\} = 0$.

Note The preceding results hold if all correlations are replaced by the corresponding covariances and all spectra by the corresponding covariance spectra. This follows from the fact that the response to $\mathbf{x}(t) - \eta_x$ equals $\mathbf{y}(t) - \eta_y$. For example, (10-136) and (10-142) yield

$$S^c_{yy}(\omega) = S^c_{xx}(\omega)|H(\omega)|^2 \qquad (10\text{-}140)$$

$$\operatorname{Var}\mathbf{y}(t) = \frac{1}{2\pi}\int_{-\infty}^{\infty} S^c_{xx}(\omega)|H(\omega)|^2\,d\omega \qquad (10\text{-}141)$$

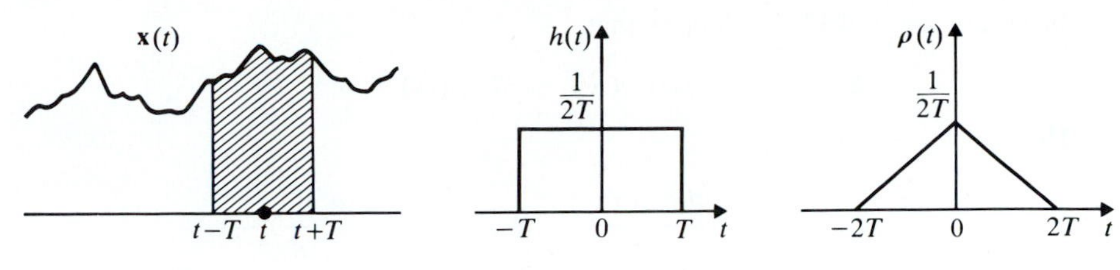

FIGURE 10-13

Example 10-25. (*a*) (*Moving average*) The integral

$$\mathbf{y}(t) = \frac{1}{2T}\int_{t-T}^{t+T} \mathbf{x}(\alpha)\, d\alpha$$

is the average of the process $\mathbf{x}(t)$ in the interval $(t - T, t + T)$. Clearly, $\mathbf{y}(t)$ is the output of a system with input $\mathbf{x}(t)$ and impulse response a rectangular pulse as in Fig. 10-13. The corresponding $\rho(\tau)$ is a triangle. In this case,

$$H(\omega) = \frac{1}{2T}\int_{-T}^{T} e^{-j\omega\tau}\, d\tau = \frac{\sin T\omega}{T\omega} \qquad S_{yy}(\omega) = S_{xx}(\omega)\frac{\sin^2 T\omega}{T^2\omega^2}$$

Thus $H(\omega)$ takes significant values only in an interval of the order of $1/T$ centered at the origin. Hence the moving average suppresses the high-frequency components of the input. It the thus a simple low-pass filter.

Since $\rho(\tau)$ is a triangle, it follows from (10-135) that

$$R_{yy}(\tau) = \frac{1}{2T}\int_{-2T}^{2T}\left(1 - \frac{|\alpha|}{2T}\right)R_{xx}(\tau - \alpha)\, d\alpha \tag{10-142}$$

We shall use this result to determine the variance of the integral

$$\boldsymbol{\eta}_T = \frac{1}{2T}\int_{-T}^{T} \mathbf{x}(t)\, dt$$

Clearly, $\boldsymbol{\eta}_T = \mathbf{y}(0)$; hence

$$\text{Var}\,\boldsymbol{\eta}_T = C_{yy}(0) = \frac{1}{2T}\int_{-2T}^{2T}\left(1 - \frac{|\alpha|}{2T}\right)C_{xx}(\alpha)\, d\alpha \tag{10-143}$$

(*b*) (*High-pass filter*) The process $\mathbf{z}(t) = \mathbf{x}(t) - \mathbf{y}(t)$ is the output of a system with input $\mathbf{x}(t)$ and system function

$$H(\omega) = 1 - \frac{\sin T\omega}{T\omega}$$

This function is nearly 0 in an interval of the order of $1/T$ centered at the origin, and it approaches 1 for large ω. It acts, therefore, as a high-pass filter suppressing the low frequencies of the input.

Example 10-26 Derivatives. The derivative $\mathbf{x}'(t)$ of a process $\mathbf{x}(t)$ can be considered as the output of a linear system with input $\mathbf{x}(t)$ and system function $j\omega$.

From this and (10-134), it follows that

$$S_{xx'}(\omega) = -j\omega S_{xx}(\omega) \qquad S_{x'x'}(\omega) = \omega^2 S_{xx}(\omega)$$

Hence

$$R_{xx'}(\tau) = -\frac{dR_{xx}(\tau)}{d\tau} \qquad R_{x'x'}(\tau) = -\frac{d^2R_{xx}(\tau)}{d\tau^2}$$

The nth derivative $\mathbf{y}(t) = \mathbf{x}^{(n)}(t)$ of $\mathbf{x}(t)$ is the output of a system with input $\mathbf{x}(t)$ and system function $(j\omega)^n$. Hence

$$S_{yy}(\omega) = |j\omega|^{2n} \qquad R_{yy}(\tau) = (-1)^n R^{(2n)}(\tau) \tag{10-144}$$

Example 10-27. (a) The differential equation

$$\mathbf{y}'(t) + c\mathbf{y}(t) = \mathbf{x}(t) \quad \text{all } t$$

specifies a linear system with input $\mathbf{x}(t)$, output $\mathbf{y}(t)$, and system function $1/(j\omega + c)$. We assume that $\mathbf{x}(t)$ is white noise with $R_{xx}(\tau) = q\delta(\tau)$. Applying (10-136), we obtain

$$S_{yy}(\omega) = \frac{S_{xx}(\omega)}{\omega^2 + c^2} = \frac{q}{\omega^2 + c^2} \qquad R_{yy}(\tau) = \frac{q}{2c}e^{-c|\tau|}$$

Note that $E\{\mathbf{y}^2(t)\} = R_{yy}(0) = q/2c$.

(b) Similarly, if

$$\mathbf{y}''(t) + b\mathbf{y}'(t) + c\mathbf{y}(t) = \mathbf{x}(t) \qquad S_{xx}(\omega) = q$$

then

$$H(\omega) = \frac{1}{-\omega^2 + jb\omega + c} \qquad S_{yy}(\omega) = \frac{q}{(c - \omega^2)^2 + b^2\omega^2}$$

To find $R_{yy}(\tau)$, we shall consider three cases:

$\underline{b^2 < 4c}$

$$R_{yy}(\tau) = \frac{q}{2bc}e^{-\alpha|\tau|}\left(\cos\beta\tau + \frac{\alpha}{\beta}\sin\beta|\tau|\right) \qquad \alpha = \frac{b}{2} \qquad \alpha^2 + \beta^2 = c$$

$\underline{b^2 = 4c}$

$$R_{yy}(\tau) = \frac{q}{2bc}e^{-\alpha|\tau|}(1 + \alpha|\tau|) \qquad \alpha = \frac{b}{2}$$

$\underline{b^2 > 4c}$

$$R_{yy}(\tau) = \frac{q}{4\gamma bc}\left[(\alpha + \gamma)e^{-(\alpha-\gamma)|\tau|} - (\alpha - \gamma)e^{-(\alpha+\gamma)|\tau|}\right]$$

$$\alpha = \frac{b}{2} \qquad \alpha^2 - \gamma^2 = c$$

In all cases, $E\{\mathbf{y}^2(t)\} = q/2bc$.

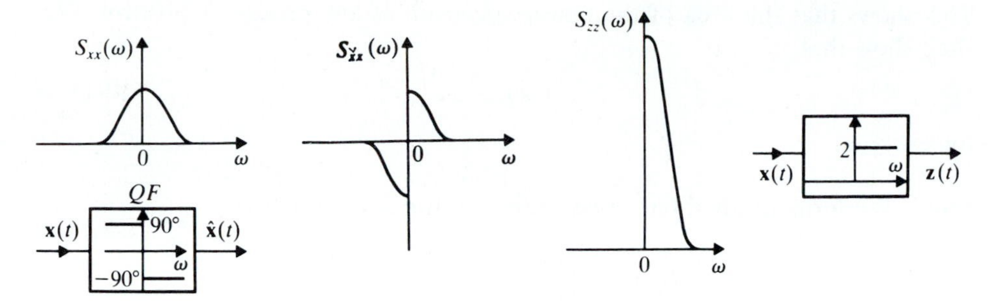

FIGURE 10-14

Example 10-28 Hilbert transforms. A system with system function (Fig. 10-14)

$$H(\omega) = -j \operatorname{sgn} \omega = \begin{cases} -j & \omega > 0 \\ j & \omega < 0 \end{cases} \tag{10-145}$$

is called a *quadrature filter*. The corresponding impulse response equals $1/\pi t$ (Papoulis, 1977). Thus $H(\omega)$ is all-pass with $-90°$ phase shift; hence its response to $\cos \omega t$ equals $\cos(\omega t - 90°) = \sin \omega t$ and its response to $\sin \omega t$ equals $\sin(\omega t - 90°) = -\cos \omega t$.

The response of a quadrature filter to a real process $\mathbf{x}(t)$ is denoted by $\check{\mathbf{x}}(t)$ and it is called the *Hilbert transform* of $\mathbf{x}(t)$. Thus

$$\check{\mathbf{x}}(t) = \mathbf{x}(t) * \frac{1}{\pi t} = \frac{1}{\pi} \int_{-\infty}^{\infty} \frac{\mathbf{x}(\alpha)}{t - \alpha} \, d\alpha \tag{10-146}$$

From (10-134) and (10-124) it follows that (Fig. 10-14)

$$\begin{aligned} S_{x\check{x}}(\omega) &= jS_{xx}(\omega) \operatorname{sgn} \omega = -S_{\check{x}x}(\omega) \\ S_{\check{x}\check{x}}(\omega) &= S_{xx}(\omega) \end{aligned} \tag{10-147}$$

The complex process

$$\mathbf{z}(t) = \mathbf{x}(t) + j\check{\mathbf{x}}(t)$$

is called the *analytic signal* associated with $\mathbf{x}(t)$. Clearly, $\mathbf{z}(t)$ is the response of the system

$$1 + j(-j \operatorname{sgn} \omega) = 2U(\omega)$$

with input $\mathbf{x}(t)$. Hence [see (10-136)]

$$S_{zz}(\omega) = 4S_{xx}(\omega)U(\omega) = 2S_{xx}(\omega) + 2jS_{\check{x}x}(\omega) \tag{10-148}$$

$$R_{zz}(\tau) = 2R_{xx}(\tau) + 2jR_{\check{x}x}(\tau) \tag{10-149}$$

THE WIENER–KHINCHIN THEOREM. From (10-121) it follows that

$$E\{\mathbf{x}^2(t)\} = R(0) = \frac{1}{2\pi} \int_{-\infty}^{\infty} S(\omega) \, d\omega \geq 0 \tag{10-150}$$

This shows that the area of the power spectrum of any process is positive. We shall show that

$$S(\omega) \geq 0 \tag{10-151}$$

for every ω.

Proof. We form an ideal bandpass system with system function

$$H(\omega) = \begin{cases} 1 & \omega_1 < \omega < \omega_2 \\ 0 & \text{otherwise} \end{cases}$$

and apply $\mathbf{x}(t)$ to its input. From (10-139) it follows that the power spectrum $S_{yy}(\omega)$ of the resulting output $\mathbf{y}(t)$ equals

$$S_{yy}(\omega) = \begin{cases} S(\omega) & \omega_1 < \omega < \omega_2 \\ 0 & \text{otherwise} \end{cases}$$

Hence

$$0 \leq E\{\mathbf{y}^2(t)\} = \frac{1}{2\pi}\int_{-\infty}^{\infty} S_{yy}(\omega)\, d\omega = \frac{1}{2\pi}\int_{\omega_1}^{\omega_2} S(\omega)\, d\omega \tag{10-152}$$

Thus the area of $S(\omega)$ in any interval is positive. This is possible only if $S(\omega) \geq 0$ everywhere.

We have shown on page 321 that if $S(\omega)$ is a positive function, then we can find a process $\mathbf{x}(t)$ such that $S_{xx}(\omega) = S(\omega)$. From this it follows that a function $S(\omega)$ is a power spectrum iff it is positive. In fact, we can find an exponential with random frequency $\boldsymbol{\omega}$ as in (10-127) with power spectrum an arbitrary positive function $S(\omega)$.

We shall use (10-152) to express the power spectrum $S(\omega)$ of a process $\mathbf{x}(t)$ as the average power of another process $\mathbf{y}(t)$ obtained by filtering $\mathbf{x}(t)$. Setting $\omega_1 = \omega_0 + \delta$ and $\omega_2 = \omega_0 - \delta$, we conclude that if δ is sufficiently small,

$$E\{\mathbf{y}^2(t)\} \simeq \frac{\delta}{\pi} S(\omega_0) \tag{10-153}$$

This shows the *localization* of the average power of $\mathbf{x}(t)$ on the frequency axis.

Integrated spectrum. In mathematics, the spectral properties of a process $\mathbf{x}(t)$ are expressed in terms of the integrated spectrum $F(\omega)$ defined as the integral of $S(\omega)$:

$$F(\omega) = \int_{-\infty}^{\omega} S(\alpha)\, d\alpha \tag{10-154}$$

From the positivity of $S(\omega)$, it follows that $F(\omega)$ is a nondecreasing function of ω. Integrating the inversion formula (10-121) by parts, we can express the autocorrelation $R(\tau)$ of $\mathbf{x}(t)$ as a Riemann–Stieltjes integral:

$$R(\tau) = \frac{1}{2\pi}\int_{-\infty}^{\infty} e^{j\omega\tau}\, dF(\omega) \qquad (10\text{-}155)$$

This approach avoids the use of singularity functions in the spectral representation of $R(\tau)$ even when $S(\omega)$ contains impulses. If $S(\omega)$ contains the terms $\beta_i\delta(\omega - \omega_i)$, then $F(\omega)$ is discontinuous at ω_i and the discontinuity jump equals β_i.

The integrated covariance spectrum $F^c(\omega)$ is the integral of the covariance spectrum. From (10-125) it follows that $F(\omega) = F^c(\omega) + 2\pi\eta^2 U(\omega)$.

Vector spectra. The vector process $\mathbf{X}(t) = [\mathbf{x}_i(t)]$ is WSS if its components $\mathbf{x}_i(t)$ are jointly WSS. In this case, its autocorrelation matrix depends only on $\tau = t_1 - t_2$. From this it follows that [see (10-116)]

$$R_{xy}(\tau) = \int_{-\infty}^{\infty} R_{xx}(\tau + \alpha) H^\dagger(\alpha)\, d\alpha \qquad R_{yy}(\tau) = \int_{-\infty}^{\infty} H(\alpha) R_{xy}(\tau - \alpha)\, d\alpha \qquad (10\text{-}156)$$

The power spectrum of a WSS vector process $\mathbf{X}(t)$ is a square matrix $S_{xx}(\omega) = [S_{ij}(\omega)]$, the elements of which are the Fourier transforms $S_{ij}(\omega)$ of the elements $R_{ij}(\tau)$ of its autocorrelation matrix $R_{xx}(\tau)$. Defining similarly the matrices $S_{xy}(\omega)$ and $S_{yy}(\omega)$, we conclude from (10-156) that

$$S_{xy}(\omega) = S_{xx}(\omega)\overline{H}^\dagger(\omega) \qquad S_{yy}(\omega) = \overline{H}(\omega) S_{xy}(\omega) \qquad (10\text{-}157)$$

where $\overline{H}(\omega) = [H_{ji}(\omega)]$ is an $m \times r$ matrix with elements the Fourier transforms $H_{ji}(\omega)$ of the elements $h_{ji}(t)$ of the impulse response matrix $H(t)$. Thus

$$S_{yy}(\omega) = \overline{H}(\omega) S_{xx}(\omega) \overline{H}^\dagger(\omega) \qquad (10\text{-}158)$$

This is the extension of (10-136) to a multiterminal system.

Example 10-29. The derivatives

$$\mathbf{y}_1(t) = \mathbf{z}^{(m)}(t) \qquad \mathbf{y}_2(t) = \mathbf{w}^{(n)}(t)$$

of two WSS processes $\mathbf{z}(t)$ and $\mathbf{w}(t)$ can be considered as the responses of two differentiators with inputs $\mathbf{z}(t)$ and $\mathbf{w}(t)$ and system functions $H_1(\omega) = (j\omega)^m$ and $H_2(\omega) = (j\omega)^n$. Proceeding as in (10-119), we conclude that the cross-power spectrum of $\mathbf{z}^{(m)}(t)$ and $\mathbf{w}^{(n)}(t)$ equals $(j\omega)^m(-j\omega)^n S_{zw}(\omega)$. Hence

$$E\{\mathbf{z}^{(m)}(t + \tau)\mathbf{z}^{(n)}(t)\} = (-1)^n \frac{d^{m+n} R_{zw}(\tau)}{d\tau^{m+n}} \qquad (10\text{-}159)$$

PROPERTIES OF CORRELATIONS. If a function $R(\tau)$ is the autocorrelation of a WSS process $\mathbf{x}(t)$, then [see (10-151)] its Fourier transform $S(\omega)$ is positive. Furthermore, if $R(\tau)$ is a function with positive Fourier transform, we can find a process $\mathbf{x}(t)$ as in (10-126) with autocorrelation $R(\tau)$. Thus a necessary and sufficient condition for a function $R(\tau)$ to be an autocorrelation is the positivity of its Fourier transform. The conditions for a function $R(\tau)$ to be an autocorrelation can be expressed directly in terms of $R(\tau)$. We have shown in (10-84) that the autocorrelation $R(\tau)$ of a process $\mathbf{x}(t)$ is p.d., that is,

$$\sum_{i,j} a_i a_j^* R(\tau_i - \tau_j) \geq 0 \tag{10-160}$$

for every a_i, a_j, τ_i, and τ_j. It can be shown that the converse is also true†: If $R(\tau)$ is a p.d. function, then its Fourier transform is positive. Thus a function $R(\tau)$ has a positive Fourier transform iff it is p.d.

A sufficient condition. To establish whether $R(\tau)$ is p.d., we must show either that it satisfies (10-160) or that its transform is positive. This is not, in general, a simple task. The following is a simple sufficient condition.

Polya's criterion. It can be shown that a function $R(\tau)$ is p.d. if it is concave for $\tau > 0$ and it tends to a finite limit as $\tau \to \infty$ (see Yaglom 1987).

Consider, for example, the function $w(\tau) = e^{-\alpha|\tau|^c}$. If $0 < c < 1$, then $w(\tau) \to 0$ as $\tau \to \infty$ and $w''(\tau) > 0$ for $\tau > 0$; hence $w(\tau)$ is p.d. because it satisfies Polya's criterion. Note, however, that it is p.d. also for $1 \leq c \leq 2$ even though it does not satisfy this criterion.

Necessary conditions. The autocorrelation $R(\tau)$ of any process $\mathbf{x}(t)$ is maximum at the origin because [see (10-121)]

$$|R(\tau)| \leq \frac{1}{2\pi} \int_{-\infty}^{\infty} S(\omega)\, d\omega = R(0) \tag{10-161}$$

We show next that if $R(\tau)$ is not periodic, it reaches its maximum only at the origin.

THEOREM. If $R(\tau_1) = R(0)$ for some $\tau_1 \neq 0$, then $R(\tau)$ is periodic with period τ_1:

$$R(\tau + \tau_1) = R(\tau) \quad \text{for all } \tau \tag{10-162}$$

Proof. From Schwarz's inequality

$$E^2\{\mathbf{zw}\} \leq E\{\mathbf{z}^2\}E\{\mathbf{w}^2\} \tag{10-163}$$

it follows that

$$E^2\{[\mathbf{x}(t+\tau+\tau_1) - \mathbf{x}(t+\tau)]\mathbf{x}(t)\}$$
$$\leq E\{[\mathbf{x}(t+\tau+\tau_1) - \mathbf{x}(t+\tau)]^2\}E\{\mathbf{x}^2(t)\}$$

Hence

$$[R(\tau+\tau_1) - R(\tau)]^2 \leq 2[R(0) - R(\tau_1)]R(0) \tag{10-164}$$

If $R(\tau_1) = R(0)$, then the right side is 0; hence the left side is also 0 for every τ. This yields (10-162).

†S. Bocher: *Lectures on Fourier Integrals*, Princeton Univ. Press, Princeton, NJ, 1959.

COROLLARY. If $R(\tau_1) = R(\tau_2) = R(0)$ and the numbers τ_1 and τ_2 are noncommensurate, that is, their ratio is irrational, then $R(\tau)$ is constant.

Proof. From the theorem it follows that $R(\tau)$ is periodic with periods τ_1 and τ_2. This is possible only if $R(\tau)$ is constant.

Continuity. If $R(\tau)$ is continuous at the origin, it is continuous for every τ.

Proof. From the continuity of $R(\tau)$ at $\tau = 0$ it follows that $R(\tau_1) \to R(0)$; hence the left side of (10-164) also tends to 0 for every τ as $\tau_1 \to 0$.

Example 10-30. Using the theorem, we shall show that the truncated parabola

$$w(\tau) = \begin{cases} a^2 - \tau^2 & |\tau| < a \\ 0 & |\tau| > a \end{cases}$$

is not an autocorrelation.

If $w(\tau)$ is the autocorrelation of some process $\mathbf{x}(t)$, then [see (10-144)] the function

$$-w''(\tau) = \begin{cases} 2 & |\tau| < a \\ 0 & |\tau| > a \end{cases}$$

is the autocorrelation of $\mathbf{x}'(t)$. This is impossible because $-w''(\tau)$ is continuous for $\tau = 0$ but not for $\tau = a$.

MS continuity and periodicity. We shall say that the process $\mathbf{x}(t)$ is MS continuous if

$$E\{[\mathbf{x}(t+\varepsilon) - \mathbf{x}(t)]^2\} \to 0 \qquad \text{as} \quad \varepsilon \to 0 \tag{10-165}$$

Since $E\{[\mathbf{x}(t+\varepsilon) - \mathbf{x}(t)]^2\} = 2[R(0) - R(\varepsilon)]$, we conclude that if $\mathbf{x}(t)$ is MS continuous, $R(0) - R(\varepsilon) \to 0$ as $\varepsilon \to 0$. Thus a WSS process $\mathbf{x}(t)$ is MS continuous iff its autocorrelation $R(\tau)$ is continuous for all τ.

We shall say that the process $\mathbf{x}(t)$ is MS periodic with period τ_1 if

$$E\{[\mathbf{x}(t+\tau_1) - \mathbf{x}(t)]^2\} = 0 \tag{10-166}$$

Since the left side equals $2[R(0) - R(\tau_1)]$, we conclude that $R(\tau_1) = R(0)$; hence [see (10-162)] $R(\tau)$ is periodic. This leads to the conclusion that a WSS process $\mathbf{x}(t)$ is MS periodic iff its autocorrelation is periodic.

Cross-correlation. Using (10-163), we shall show that the cross-correlation $R_{xy}(\tau)$ of two WSS processes $\mathbf{x}(t)$ and $\mathbf{y}(t)$ satisfies the inequality

$$R_{xy}^2(\tau) \le R_{xx}(0) R_{yy}(0) \tag{10-167}$$

Proof. From (10-163) it follows that

$$E^2\{\mathbf{x}(t+\tau)\mathbf{y}^*(t)\} \le E\{|\mathbf{x}(t+\tau)|^2\}E\{|\mathbf{y}(t)|^2\} = R_{xx}(0) R_{yy}(0)$$

and (10-167) results.

COROLLARY. For any a and b,

$$\left|\int_a^b S_{xy}(\omega)\,d\omega\right|^2 \le \int_a^b S_{xx}(\omega)\,d\omega \int_a^b S_{yy}(\omega)\,d\omega \qquad (10\text{-}168)$$

Proof. Suppose that $\mathbf{x}(t)$ and $\mathbf{y}(t)$ are the inputs to the ideal filters

$$H_1(\omega) = H_2(\omega) = \begin{cases} 1 & a < \omega < b \\ 0 & \text{otherwise} \end{cases}$$

Denoting by $\mathbf{z}(t)$ and $\mathbf{w}(t)$ respectively the resulting outputs, we conclude that

$$R_{zz}(0) = \frac{1}{2\pi}\int_a^b S_{xx}(\omega)\,d\omega \qquad R_{ww}(0) = \frac{1}{2\pi}\int_a^b S_{yy}(\omega)\,d\omega$$

$$R_{zw}(0) = \frac{1}{2\pi}\int_a^b S_{zw}(\omega)\,d\omega$$

and (10-168) follows because $R_{zw}^2(0) \le R_{zz}(0)R_{ww}(0)$.

10-4 DIGITAL PROCESSES

A digital (or discrete-time) process is a sequence $\mathbf{x}_n$ of RVs. To avoid double subscripts, we shall use also the notation $\mathbf{x}[n]$ where the brackets will indicate that n is an integer. Most results involving analog (or continuous-time) processes can be readily extended to digital processes. We outline the main concepts.

The autocorrelation and autocovariance of $\mathbf{x}[n]$ are given by

$$R[n_1, n_2] = E\{\mathbf{x}[n_1]\mathbf{x}^*[n_2]\} \qquad C[n_1, n_2] = R[n_1, n_2] - \eta[n_1]\eta^*[n_2] \qquad (10\text{-}169)$$

respectively where $\eta[n] = E\{\mathbf{x}[n]\}$ is the mean of $\mathbf{x}[n]$.

A process $\mathbf{x}[n]$ is SSS if its statistical properties are invariant to a shift of the origin. It is WSS if $\eta[n] = \eta$ = constant and

$$R[n+m, n] = E\{\mathbf{x}[n+m]\mathbf{x}^*[n]\} = R[m] \qquad (10\text{-}170)$$

A process $\mathbf{x}[n]$ is strictly white noise if the RVs $\mathbf{x}[n_i]$ are independent. It is white noise if the RVs $\mathbf{x}[n_i]$ are uncorrelated. The autocorrelation of a white-noise process with zero mean is thus given by

$$R[n_1, n_2] = q[n_1]\delta[n_1 - n_2] \qquad \text{where} \quad \delta[n] = \begin{cases} 1 & n = 0 \\ 0 & n \ne 0 \end{cases} \qquad (10\text{-}171)$$

and $q[n] = E\{\mathbf{x}^2[n]\}$. If $\mathbf{x}[n]$ is also stationary, then $R[m] = q\delta[m]$. Thus a WSS white noise is a sequence of i.i.d. RVs with variance q.

The delta response $h[n]$ of a linear system is its response to the delta sequence $\delta[n]$. Its system function is the z transform of $h[n]$:

$$\mathbf{H}(z) = \sum_{n=-\infty}^{\infty} h[n]z^{-n} \qquad (10\text{-}172)$$

If $\mathbf{x}[n]$ is the input to a digital system, the resulting output is the digital convolution of $\mathbf{x}[n]$ with $h[n]$:

$$\mathbf{y}[n] = \sum_{k=-\infty}^{\infty} \mathbf{x}[n-k]h[k] = \mathbf{x}[n] * h[n] \tag{10-173}$$

From this it follows that $\eta_y[n] = \eta_x[n] * h[n]$. Furthermore,

$$R_{xy}[n_1, n_2] = \sum_{k=-\infty}^{\infty} R_{xx}[n_1, n_2 - k]h^*[k] \tag{10-174}$$

$$R_{yy}[n_1, n_2] = \sum_{r=-\infty}^{\infty} R_{xy}[n_1 - r, n_2]h[r] \tag{10-175}$$

If $\mathbf{x}[n]$ is white noise with average intensity $q[n]$ as in (10-171), then, [see (10-90)],

$$E\{\mathbf{y}^2[n]\} = q[n] * |h[n]|^2 \tag{10-176}$$

If $\mathbf{x}[n]$ is WSS, then $\mathbf{y}[n]$ is also WSS with $\eta_y = \eta_x \mathbf{H}(1)$. Furthermore,

$$R_{xy}[m] = R_{xx}[m] * h^*[-m] \qquad R_{yy}[m] = R_{xy}[m] * h[m]$$

$$R_{yy}[m] = R_{xx}[m] * \rho[m] \qquad \rho[m] = \sum_{k=-\infty}^{\infty} h[m+k]h^*[k] \tag{10-177}$$

as in (10-133) and (10-135).

THE POWER SPECTRUM. Given a WSS process $\mathbf{x}[n]$, we form the z transform $\mathbf{S}(z)$ of its autocorrelation $R[m]$:

$$\mathbf{S}(z) = \sum_{m=-\infty}^{\infty} R[m]z^{-m} \tag{10-178}$$

The power spectrum of $\mathbf{x}[n]$ is the function

$$S(\omega) = \mathbf{S}(e^{j\omega}) = \sum_{m=-\infty}^{\infty} R[m]e^{-jm\omega} \tag{10-179}$$

Thus $\mathbf{S}(e^{j\omega})$ is the DFT of $R[m]$. The function $\mathbf{S}(e^{j\omega})$ is periodic with period 2π and Fourier series coefficients $R[m]$. Hence

$$R[m] = \frac{1}{2\pi}\int_{-\pi}^{\pi} \mathbf{S}(e^{j\omega})e^{jm\omega}\,d\omega \tag{10-180}$$

It suffices, therefore, to specify $\mathbf{S}(e^{j\omega})$ for $|\omega| < \pi$ only (see Fig. 10-15).

If $\mathbf{x}[n]$ is a real process, then $R[-m] = R[m]$ and (10-179) yields

$$\mathbf{S}(e^{j\omega}) = R[0] + 2\sum_{m=0}^{\infty} R[m]\cos m\omega \tag{10-181}$$

This shows that the power spectrum of a real process is a function of $\cos\omega$ because $\cos m\omega$ is a function of $\cos\omega$.

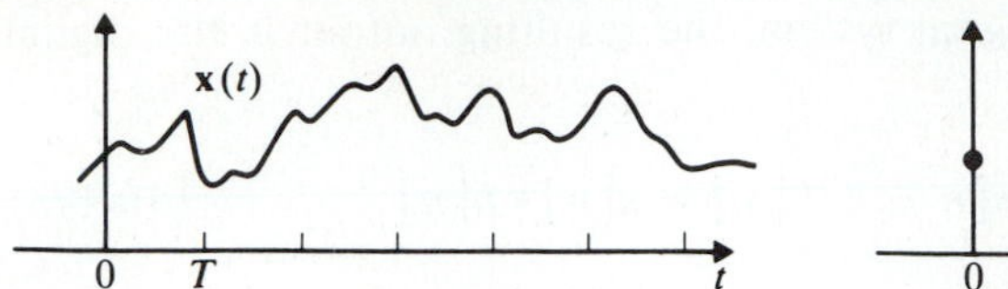

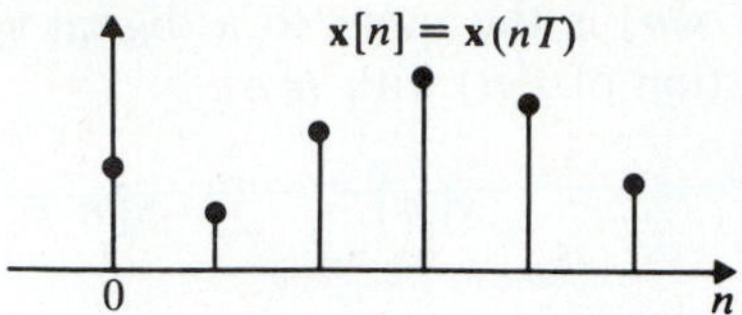

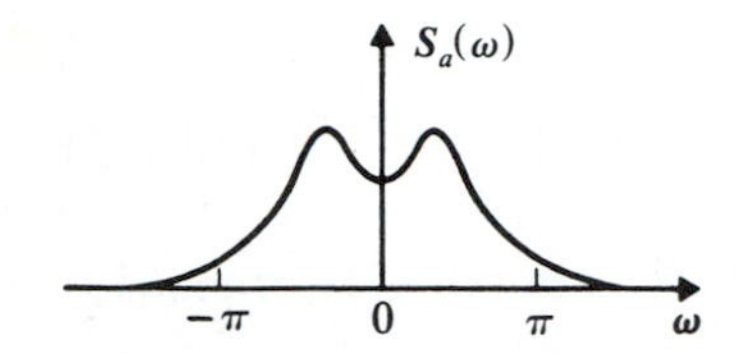

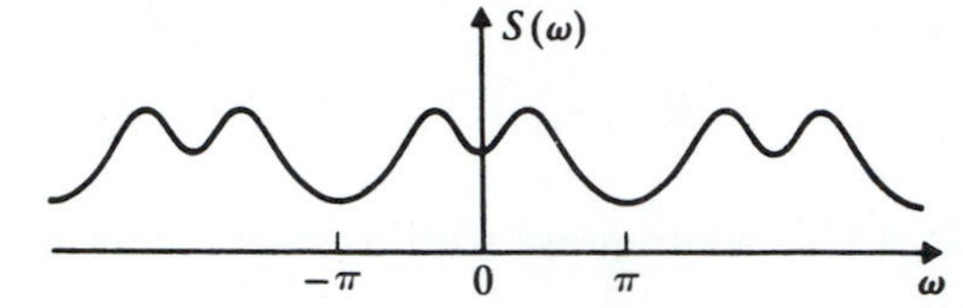

FIGURE 10-15

Example 10-31. If $R[m] = a^{|m|}$, then

$$\mathbf{S}(z) = \sum_{m=-\infty}^{-1} a^{-m}z^{-m} + \sum_{m=0}^{\infty} a^{m}z^{-m} = \frac{az}{1-az} + \frac{z}{z-a}$$

$$= \frac{a^{-1}-a}{(a^{-1}+a)-(z^{-1}+z)}$$

Hence

$$\mathbf{S}(e^{j\omega}) = \frac{a^{-1}-a}{a^{-1}+a-2\cos\omega}$$

Example 10-32. Proceeding as in the analog case, we can show that the process

$$\mathbf{x}[n] = \sum_i \mathbf{c}_i e^{j\omega_i n}$$

is WSS iff the coefficients $\mathbf{c}_i$ are uncorrelated with zero mean. In this case,

$$R[m] = \sum_i \sigma_i^2 e^{j\beta_i|m|} \qquad S(\omega) = 2\pi\sum_i \sigma_i^2\delta(\omega-\beta_i) \qquad |\omega| < \pi \quad (10\text{-}182)$$

where $\sigma_i^2 = E\{\mathbf{c}_i^2\}$, $\omega_i = 2\pi k_i + \beta_i$, and $|\beta_i| < \pi$.

From (10-177) and the convolution theorem, it follows that if $\mathbf{y}[n]$ is the output of a linear system with input $\mathbf{x}[n]$, then

$$\begin{aligned} \mathbf{S}_{xy}(e^{j\omega}) &= \mathbf{S}_{xx}(e^{j\omega})\mathbf{H}^*(e^{j\omega}) \\ \mathbf{S}_{yy}(e^{j\omega}) &= \mathbf{S}_{xy}(e^{j\omega})\mathbf{H}(e^{j\omega}) \\ \mathbf{S}_{yy}(e^{j\omega}) &= \mathbf{S}_{xx}(e^{j\omega})|\mathbf{H}(e^{j\omega})|^2 \end{aligned} \qquad (10\text{-}183)$$

If $h[n]$ is real, $\mathbf{H}^*(e^{j\omega}) = \mathbf{H}(e^{-j\omega})$. In this case

$$\mathbf{S}_{yy}(z) = \mathbf{S}_{xx}(z)\mathbf{H}(z)\mathbf{H}(1/z) \tag{10-184}$$

Example 10-33. The first difference

$$\mathbf{y}[n] = \mathbf{x}[n] - \mathbf{x}[n-1]$$

of a process $\mathbf{x}[n]$ can be considered as the output of a linear system with input $\mathbf{x}[n]$ and system function $\mathbf{H}(z) = 1 - z^{-1}$. Applying (10-184), we obtain

$$\mathbf{S}_{yy}(z) = \mathbf{S}_{xx}(z)(1 - z^{-1})(1 - z) = \mathbf{S}_{xx}(z)(2 - z - z^{-1})$$

$$R_{yy}[m] = -R_{xx}[m+1] + 2R_{xx}[m] - R_{xx}[m-1]$$

If $\mathbf{x}[n]$ is white noise with $\mathbf{S}_{xx}(z) = q$, then

$$\mathbf{S}_{yy}(e^{j\omega}) = q(2 - e^{j\omega} - e^{-j\omega}) = 2q(1 - \cos\omega)$$

Example 10-34. The recursion equation

$$\mathbf{y}[n] - a\mathbf{y}[n-1] = \mathbf{x}[n]$$

specifies a linear system with input $\mathbf{x}[n]$ and system function $\mathbf{H}(z) = 1/(1 - az^{-1})$. If $\mathbf{S}_{xx}(z) = q$, then (see Example 10-31)

$$\mathbf{S}_{yy}(z) = \frac{q}{(1 - az^{-1})(1 - az)} \qquad R_{yy}[m] = \frac{q}{a^{-1} - a}a^{|m|}$$

From (10-183) it follows that

$$E\{|\mathbf{y}[n]|^2\} = R_{yy}[0] = \frac{1}{2\pi}\int_{-\pi}^{\pi} \mathbf{S}_{xx}(e^{j\omega})|\mathbf{H}(e^{j\omega})|^2\,d\omega \tag{10-185}$$

Using this identity, we shall show that the power spectrum of a process $\mathbf{x}[n]$ real or complex is a positive function:

$$\mathbf{S}_{xx}(e^{j\omega}) \geq 0 \tag{10-186}$$

Proof. We form an ideal bandpass filter with center frequency ω_0 and bandwidth 2Δ and apply (10-185). For small Δ,

$$E\{|\mathbf{y}[n]|^2\} = \frac{1}{2\pi}\int_{\omega_0-\Delta}^{\omega_0+\Delta} \mathbf{S}_{xx}(e^{j\omega})\,d\omega \simeq \frac{\Delta}{\pi}\mathbf{S}_{xx}(e^{j\omega_0})$$

and (10-186) results because $E\{\mathbf{y}^2[n]\} \geq 0$ and ω_0 is arbitrary.

SAMPLING. In many applications, the digital processes under consideration are obtained by sampling various analog processes. We relate next the corresponding correlations and spectra.

Given an analog process $\mathbf{x}(t)$, we form the digital process

$$\mathbf{x}[n] = \mathbf{x}(nT)$$

where T is a given constant. From this it follows that

$$\eta[n] = \eta_a(nT) \qquad R[n_1, n_2] = R_a(n_1T, n_2T) \tag{10-187}$$

where $\eta_a(t)$ is the mean and $R_a(t_1, t_2)$ the autocorrelation of $\mathbf{x}(t)$. If $\mathbf{x}(t)$ is a stationary process, then $\mathbf{x}[n]$ is also stationary with mean $\eta = \eta_a$ and autocorrelation

$$R[m] = R_a(mT)$$

From this it follows that the power spectrum of $\mathbf{x}[n]$ equals (Fig. 10-15)

$$\mathbf{S}(e^{j\omega}) = \sum_{m=-\infty}^{\infty} R_a(mT)e^{-jm\omega} = \frac{1}{T}\sum_{n=-\infty}^{\infty} S_a\left(\frac{\omega + 2\pi n}{T}\right) \tag{10-188}$$

where $S_a(\omega)$ is the power spectrum of $\mathbf{x}(t)$. The above is a consequence of Poisson's sum formula [see (11A-1)].

Example 10-35. Suppose that $\mathbf{x}(t)$ is a WSS process consisting of M exponentials as in (10-130):

$$\mathbf{x}(t) = \sum_{i=1}^{M} \mathbf{c}_i e^{j\omega_i t} \qquad S_a(\omega) = 2\pi \sum_{i=1}^{M} \sigma_i^2 \delta(\omega - \omega_i)$$

where $\sigma_i^2 = E\{\mathbf{c}_i^2\}$. We shall determine the power spectrum $\mathbf{S}(e^{j\omega})$ of the process $\mathbf{x}[n] = \mathbf{x}(nT)$. Since $\delta(\omega/T) = T\delta(\omega)$, it follows from (10-188) that

$$\mathbf{S}(e^{j\omega}) = 2\pi \sum_{n=-\infty}^{\infty} \sum_{i=1}^{M} \sigma_i^2 \delta(\omega - T\omega_i + 2\pi n)$$

In the interval $(-\pi, \pi)$, this consists of M lines:

$$\mathbf{S}(e^{j\omega}) = 2\pi \sum_{i=1}^{M} \sigma_i^2 \delta(\omega - \beta_i) \qquad |\omega| < \pi$$

where $\beta_i = T\omega_i - 2\pi n_i$ and such that $|\beta_i| < \pi$.

APPENDIX 10A
CONTINUITY, DIFFERENTIATION, INTEGRATION

In the earlier discussion, we routinely used various limiting operations involving stochastic processes, with the tacit assumption that these operations hold for every sample involved. This assumption is, in many cases, unnecessarily restrictive. To give some idea of the notion of limits in a more general case, we discuss next conditions for the existence of MS limits and we show that these conditions can be phrased in terms of second-order moments (see also Sec. 8-4).

STOCHASTIC CONTINUITY. A process $\mathbf{x}(t)$ is called MS continuous if

$$E\{[\mathbf{x}(t+\varepsilon) - \mathbf{x}(t)]^2\} \underset{\varepsilon \to 0}{\longrightarrow} 0 \tag{10A-1}$$

THEOREM. We maintain that $\mathbf{x}(t)$ is MS continuous if its autocorrelation is continuous.

Proof. Clearly,

$$E\{[\mathbf{x}(t+\varepsilon) - \mathbf{x}(t)]^2\} = R(t+\varepsilon, t+\varepsilon) - 2R(t+\varepsilon, t) + R(t,t)$$

If, therefore, $R(t_1, t_2)$ is continuous, then the right side tends to 0 as $\varepsilon \to 0$ and (10A-1) results.

Note Suppose that (10A-1) holds for *every* t in an interval I. From this it follows that [see (10-1)] almost all samples of $\mathbf{x}(t)$ will be continuous at a *particular* point of I. It does not follow, however, that these samples will be continuous for *every* point in I. We mention as illustrations the Poisson process and the Wiener process. As we see from (10-14) and (11-5), both processes are MS continuous. However, the samples of the Poisson process are discontinuous at the points $\mathbf{t}_i$, whereas almost all samples of the Wiener process are continuous.

COROLLARY. If $\mathbf{x}(t)$ is MS continuous, then its mean is continuous

$$\eta(t+\varepsilon) \to \eta(t) \qquad \varepsilon \to 0 \tag{10A-2}$$

Proof. As we know

$$E\{[\mathbf{x}(t+\varepsilon) - \mathbf{x}(t)]^2\} \geq E^2\{[\mathbf{x}(t+\varepsilon) - \mathbf{x}(t)]\}$$

Hence (10A-2) follows that (10A-1).

The above shows that

$$\lim_{\varepsilon \to 0} E\{\mathbf{x}(t+\varepsilon)\} = E\left\{\lim_{\varepsilon \to 0} \mathbf{x}(t+\varepsilon)\right\} \tag{10A-3}$$

STOCHASTIC DIFFERENTIATION. A process $\mathbf{x}(t)$ is MS differentiable if

$$\frac{\mathbf{x}(t+\varepsilon) - \mathbf{x}(t)}{\varepsilon} \underset{\varepsilon \to 0}{\longrightarrow} \mathbf{x}'(t) \tag{10A-4}$$

in the MS sense, that is, if

$$E\left\{\left[\frac{\mathbf{x}(t+\varepsilon) - \mathbf{x}(t)}{\varepsilon} - \mathbf{x}'(t)\right]^2\right\} \underset{\varepsilon \to 0}{\longrightarrow} 0 \tag{10A-5}$$

THEOREM. The process $\mathbf{x}(t)$ is MS differentiable if $\partial^2 R(t_1, t_2)/\partial t_1 \, \partial t_2$ exists.

Proof. It suffices to show that (Cauchy criterion)

$$E\left\{\left[\frac{\mathbf{x}(t+\varepsilon_1) - \mathbf{x}(t)}{\varepsilon_1} - \frac{\mathbf{x}(t+\varepsilon_2) - \mathbf{x}(t)}{\varepsilon_2}\right]^2\right\} \underset{\varepsilon_1, \varepsilon_2 \to 0}{\longrightarrow} 0 \tag{10A-6}$$

We use this criterion because, unlike (10A-5), it does not involve the unknown

$\mathbf{x}'(t)$. Clearly,

$$E\{[\mathbf{x}(t+\varepsilon_1)-\mathbf{x}(t)][\mathbf{x}(t+\varepsilon_2)-\mathbf{x}(t)]\}$$
$$= R(t+\varepsilon_1, t+\varepsilon_2) - R(t+\varepsilon_1, t) - R(t, t+\varepsilon_2) + R(t,t)$$

The right side divided by $\varepsilon_1\varepsilon_2$ tends to $\partial^2 R(t,t)/\partial t\,\partial t$ which, by assumption, exists. Expanding the square in (10A-6), we conclude that its left side tends to

$$\frac{\partial^2 R(t,t)}{\partial t\,\partial t} - 2\frac{\partial^2 R(t,t)}{\partial t\,\partial t} + \frac{\partial^2 R(t,t)}{\partial t\,\partial t} = 0$$

COROLLARY. The above yields

$$E\{\mathbf{x}'(t)\} = E\left\{\lim_{\varepsilon\to 0}\frac{\mathbf{x}(t+\varepsilon)-\mathbf{x}(t)}{\varepsilon}\right\} = \lim_{\varepsilon\to 0} E\left\{\frac{\mathbf{x}(t+\varepsilon)-\mathbf{x}(t)}{\varepsilon}\right\}$$

Note The autocorrelation of a Poisson process $\mathbf{x}(t)$ is discontinuous at the points $\mathbf{t}_i$; hence $\mathbf{x}'(t)$ does not exist at these points. However, as in the case of deterministic signals, it is convenient to introduce random impulses and to interpret $\mathbf{x}'(t)$ as in (10-98).

STOCHASTIC INTEGRALS. A process $\mathbf{x}(t)$ is MS integrable if the limit

$$\int_a^b \mathbf{x}(t)\,dt = \lim_{\Delta t_i\to 0}\sum_i \mathbf{x}(t_i)\,\Delta t_i \tag{10A-7}$$

exists in the MS sense.

THEOREM. The process $\mathbf{x}(t)$ is MS integrable if

$$\int_a^b\int_a^b |R(t_1,t_2)|\,dt_1\,dt_2 < \infty \tag{10A-8}$$

Proof. Using again the Cauchy criterion, we must show that

$$E\left\{\left|\sum_i \mathbf{x}(t_i)\,\Delta t_i - \sum_k \mathbf{x}(t_k)\,\Delta t_k\right|^2\right\} \xrightarrow[\Delta t_i,\,\Delta t_k\to 0]{} 0$$

This follows if we expand the square and use the identity

$$E\left\{\sum_i \mathbf{x}(t_i)\,\Delta t_i \sum_k \mathbf{x}(t_k)\,\Delta t_k\right\} = \sum_{i,k} R(t_i,t_k)\,\Delta t_i\Delta t_k$$

because the right side tends to the integral of $R(t_1, t_2)$ as Δt_i and Δt_k tend to 0.

COROLLARY. From the above it follows that

$$E\left\{\left|\int_a^b \mathbf{x}(t)\,dt\right|^2\right\} = \int_a^b\int_a^b R(t_1,t_2)\,dt_1\,dt_2 \tag{10A-9}$$

as in (10-11).

APPENDIX 10B SHIFT OPERATORS AND STATIONARY PROCESSES

An SSS process can be generated by a succession of shifts $T\mathbf{x}$ of a single RV $\mathbf{x}$ where T is a one-to-one measure preserving transformation (mapping) of the probability space $\mathscr{S}$ into itself. This difficult topic is of fundamental importance in mathematics. In the following, we give a brief explanation of the underlying concept, limiting the discussion to the discrete-time case.

A *transformation* T of $\mathscr{S}$ into itself is a rule for assigning to each element ζ_i of $\mathscr{S}$ another element of $\mathscr{S}$:

$$\tilde{\zeta}_i = T\zeta_i \tag{10B-1}$$

called the *image* of ζ_i. The images $\tilde{\zeta}_i$ of all elements ζ_i of a subset $\mathscr{A}$ of $\mathscr{S}$ form another subset

$$\tilde{\mathscr{A}} = T\mathscr{A}$$

of $\mathscr{S}$ called the image of $\mathscr{A}$.

We shall assume that the transformation T has the following properties.

P_1: It is one-to-one. This means that

$$\text{if} \quad \zeta_i \neq \zeta_j \qquad \text{then} \quad \tilde{\zeta}_i \neq \tilde{\zeta}_j$$

P_2: It is measure preserving. This means that if $\mathscr{A}$ is an event, then its image $\tilde{\mathscr{A}}$ is also an event and

$$P(\tilde{\mathscr{A}}) = P(\mathscr{A}) \tag{10B-2}$$

Suppose that $\mathbf{x}$ is an RV and that T is a transformation as above. The expression $T\mathbf{x}$ will mean another RV

$$\mathbf{y} = T\mathbf{x} \qquad \text{such that} \quad \mathbf{y}(\tilde{\zeta}_i) = \mathbf{x}(\zeta_i) \tag{10B-3}$$

where ζ_i is the unique inverse of $\tilde{\zeta}_i$. This specifies $\mathbf{y}$ for every element of $\mathscr{S}$ because (see P_1) the set of elements $\tilde{\zeta}_i$ equals $\mathscr{S}$.

The expression $\mathbf{z} = T^{-1}\mathbf{x}$ will mean that $\mathbf{x} = T\mathbf{z}$. Thus

$$\mathbf{z} = T^{-1}\mathbf{x} \qquad \text{iff} \quad \mathbf{z}(\zeta_i) = \mathbf{x}(\tilde{\zeta}_i)$$

We can define similarly $T^2\mathbf{x} = T(T\mathbf{x}) = T\mathbf{y}$ and

$$T^n\mathbf{x} = T(T^{n-1}\mathbf{x}) = T^{-1}(T^{n+1}\mathbf{x})$$

for any n positive or negative.

From (10B-3) it follows that if, for some ζ_i, $\mathbf{x}(\zeta_i) \le w$, then $\mathbf{y}(\tilde{\zeta}_i) = \mathbf{x}(\zeta_i) \le w$. Hence the event $\{\mathbf{y} \le w\}$ is the image of the event $\{\mathbf{x} \le w\}$. This yields [see (10B-2)]

$$P\{\mathbf{x} \le w\} = P\{\mathbf{y} \le w\} \qquad \mathbf{y} = T\mathbf{x} \tag{10B-4}$$

for any w. We thus conclude that the RVs $\mathbf{x}$ and $T\mathbf{x}$ have the same distribution $F_x(x)$.

Given an RV $\mathbf{x}$ and a transformation T as above, we form the random process

$$\mathbf{x}_0 = \mathbf{x} \qquad \mathbf{x}_n = T^n\mathbf{x} \qquad n = -\infty, \ldots, \infty \tag{10B-5}$$

It follows from (10B-4) that the random variables $\mathbf{x}_n$ so formed have the same distribution. We can similarly show that their joint distributions of any order are invariant to a shift of the origin. Hence the process $\mathbf{x}_n$ so formed is SSS.

It can be shown that the converse is also true: Given an SSS process $\mathbf{x}_n$, we can find an RV $\mathbf{x}$ and a one-to-one measuring preserving transformation of the space $\mathscr{S}$ into itself such that for all essential purposes, $\mathbf{x}_n = T^n\mathbf{x}$. The proof of this difficult result will not be given.

PROBLEMS

10-1. In the fair-coin experiment, we define the process $\mathbf{x}(t)$ as follows: $\mathbf{x}(t) = \sin \pi t$ if heads shows, $\mathbf{x}(t) = 2t$ if tails shows. (*a*) Find $E\{\mathbf{x}(t)\}$. (*b*) Find $F(x, t)$ for $t = 0.25$, $t = 0.5$, and $t = 1$.

10-2. The process $\mathbf{x}(t) = e^{\mathbf{a}t}$ is a family of exponentials depending on the RV $\mathbf{a}$. Express the mean $\eta(t)$, the autocorrelation $R(t_1, t_2)$, and the first-order density $f(x, t)$ of $\mathbf{x}(t)$ in terms of the density $f_a(a)$ of $\mathbf{a}$.

10-3. Suppose that $\mathbf{x}(t)$ is a Poisson process as in Fig. 10-3 such that $E\{\mathbf{x}(9)\} = 6$. (*a*) Find the mean and the variance of $\mathbf{x}(8)$. (*b*) Find $P\{(\mathbf{x}(2) \le 3\}$. (*c*) Find $P\{\mathbf{x}(4) \le 5 | \mathbf{x}(2) \le 3\}$.

10-4. The RV $\mathbf{c}$ is uniform in the interval $(0, T)$. Find $R_x(t_1, t_2)$ if (*a*) $\mathbf{x}(t) = U(t - \mathbf{c})$, (*b*) $\mathbf{x}(t) = \delta(t - \mathbf{c})$.

10-5. The RVs $\mathbf{a}$ and $\mathbf{b}$ are independent $N(0; \sigma)$ and p is the probability that the process $\mathbf{x}(t) = \mathbf{a} - \mathbf{b}t$ crosses the t axis in the interval $(0, T)$. Show that $\pi p = \arctan T$.

Hint: $p = P\{0 \le \mathbf{a}/\mathbf{b} \le T\}$.

10-6. Show that if

$$R_v(t_1, t_2) = q(t_1)\delta(t_1 - t_2)$$

$\mathbf{w}''(t) = \mathbf{v}(t)U(t)$ and $\mathbf{w}(0) = \mathbf{w}'(0) = 0$, then

$$E\{\mathbf{w}^2(t)\} = \int_0^t (t - \tau)q(\tau)\, d\tau$$

10-7. The process $\mathbf{x}(t)$ is real with autocorrelation $R(\tau)$. (*a*) Show that

$$P\{|\mathbf{x}(t + \tau) - \mathbf{x}(t)| \ge a\} \le 2[R(0) - R(\tau)]/a^2$$

(*b*) Express $P\{|\mathbf{x}(t + \tau) - \mathbf{x}(t)| \ge a\}$ in terms of the second-order density $f(x_1, x_2; \tau)$ of $\mathbf{x}(t)$.

10-8. The process $\mathbf{x}(t)$ is WSS and normal with $E\{\mathbf{x}(t)\} = 0$ and $R(\tau) = 4e^{-2|\tau|}$. (*a*) Find $P\{\mathbf{x}(t) \le 3\}$. (*b*) Find $E\{[\mathbf{x}(t + 1) - \mathbf{x}(t - 1)]^2\}$.

10-9. Show that the process $\mathbf{x}(t) = \mathbf{c}w(t)$ is WSS iff $E\{\mathbf{c}\} = 0$ and $w(t) = e^{j(\omega t + \theta)}$.

10-10. The process $\mathbf{x}(t)$ is normal WSS and $E\{\mathbf{x}(t)\} = 0$. Show that if $\mathbf{z}(t) = \mathbf{x}^2(t)$, then $C_{zz}(\tau) = 2C_{xx}^2(\tau)$.

10-11. Find $E\{\mathbf{y}(t)\}$, $E\{\mathbf{y}^2(t)\}$, and $R_{yy}(\tau)$ if

$$\mathbf{y}''(t) + 4\mathbf{y}'(t) + 13\mathbf{y}(t) = 26 + \boldsymbol{\nu}(t) \qquad R_{\nu\nu}(\tau) = 10\delta(\tau)$$

Find $P\{\mathbf{y}(t) \le 3\}$ if $\boldsymbol{\nu}(t)$ is normal.

10-12. Show that: If $\mathbf{x}(t)$ is a process with zero mean and autocorrelation $f(t_1)f(t_2)w(t_1 - t_2)$, then the process $\mathbf{y}(t) = \mathbf{x}(t)/f(t)$ is WSS with autocorrelation $w(\tau)$. If $\mathbf{x}(t)$ is white noise with autocorrelation $q(t_1)\delta(t_1 - t_2)$, then the process $\mathbf{z}(t) = \mathbf{x}(t)/\sqrt{q(t)}$ is WSS white noise with autocorrelation $\delta(\tau)$.

10-13. Show that $|R_{xy}(\tau)| \le \frac{1}{2}[R_{xx}(0) + R_{yy}(0)]$.

10-14. Show that if the processes $\mathbf{x}(t), \mathbf{y}(t)$ are WSS and $E\{|\mathbf{x}(0) - \mathbf{y}(0)|^2\} = 0$, then $R_{xx}(\tau) \equiv R_{xy}(\tau) \equiv R_{yy}(\tau)$.

Hint: Set $\mathbf{z} = \mathbf{x}(t + \tau)$, $\mathbf{w} = \mathbf{x}^*(t)) - \mathbf{y}^*(t)$ in (10-163).

10-15. Show that if $\mathbf{x}(t)$ is a complex WSS process, then

$$E\{|\mathbf{x}(t + \tau) - \mathbf{x}(t)|^2\} = 2\,\mathrm{Re}[R(0) - R(\tau)]$$

10-16. Show that if $\boldsymbol{\varphi}$ is an RV with $\Phi(\lambda) = E\{e^{j\lambda\boldsymbol{\varphi}}\}$ and $\Phi(1) = \Phi(2) = 0$, then the process $\mathbf{x}(t) = \cos(\omega t + \boldsymbol{\varphi})$ is WSS. Find $E\{\mathbf{x}(t)\}$ and $R_x(\tau)$ if $\boldsymbol{\varphi}$ is uniform in the interval $(-\pi, \pi)$.

10-17. Given a process $\mathbf{x}(t)$ with orthogonal increments and such that $\mathbf{x}(0) = 0$, show that (*a*) $R(t_1, t_2) = R(t_1, t_1)$ for $t_1 \le t_2$, and (*b*) if $E\{[\mathbf{x}(t_1) - \mathbf{x}(t_2)]^2\} = q|t_1 - t_2|$ then the process $\mathbf{y}(t) = [\mathbf{x}(t + \varepsilon) - \mathbf{x}(t)]/\varepsilon$ is WSS and its autocorrelation is a triangle with area q and base 2ε.

10-18. Show that if $R_{xx}(t_1, t_2) = q(t_1)\delta(t_1 - t_2)$ and $\mathbf{y}(t) = \mathbf{x}(t) * h(t)$ then

$$E\{\mathbf{x}(t)\mathbf{y}(t)\} = h(0)q(t)$$

10-19. The process $\mathbf{x}(t)$ is normal with $\eta_x = 0$ and $R_x(\tau) = 4e^{-3|\tau|}$. Find a memoryless system $g(x)$ such that the first-order density $f_y(y)$ of the resulting output $\mathbf{y}(t) = g[\mathbf{x}(t)]$ is uniform in the interval $(6, 9)$.

Answer: $g(x) = 3\mathbb{G}(x/2) + 6$.

10-20. Show that if $\mathbf{x}(t)$ is an SSS process and $\boldsymbol{\varepsilon}$ is an RV independent of $\mathbf{x}(t)$, then the process $\mathbf{y}(t) = \mathbf{x}(t - \boldsymbol{\varepsilon})$ is SSS.

10-21. Show that if $\mathbf{x}(t)$ is a stationary process with derivative $\mathbf{x}'(t)$, then for a given t the RVs $\mathbf{x}(t)$ and $\mathbf{x}'(t)$ are orthogonal and uncorrelated.

10-22. Given a normal process $\mathbf{x}(t)$ with $\eta_x = 0$ and $R_x(\tau) = 4e^{-2|\tau|}$, we form the RVs $\mathbf{z} = \mathbf{x}(t + 1)$, $\mathbf{w} = \mathbf{x}(t - 1)$, (*a*) find $E\{\mathbf{zw}\}$ and $E\{(\mathbf{z} + \mathbf{w})^2\}$, (*b*) find

$$f_z(z) \qquad P\{\mathbf{z} < 1\} \qquad f_{zw}(z, w)$$

10-23. Show that if $\mathbf{x}(t)$ is normal with autocorrelation $R(\tau)$, then

$$P\{\mathbf{x}'(t) \le a\} = \mathbb{G}\left[\frac{a}{\sqrt{-R''(0)}}\right]$$

10-24. Show that if $\mathbf{x}(t)$ is a normal process with zero mean and $\mathbf{y}(t) = \operatorname{sgn}\mathbf{x}(t)$, then

$$R_y(\tau) = \frac{2}{\pi}\sum_{n=1}^{\infty}\frac{1}{n}[J_0(n\pi) - (-1)^n]\sin\left[n\pi\frac{R_x(\tau)}{R_x(0)}\right]$$

where $J_0(x)$ is the Bessel function.

Hint: Expand the arcsine in (10-71) into a Fourier series.

10-25. Show that if $\mathbf{x}(t)$ is a normal process with zero mean and $\mathbf{y}(t) = Ie^{a\mathbf{x}(t)}$, then

$$\eta_y == I\exp\left\{\frac{a^2}{2}R_x(0)\right\} \qquad R_y(\tau) = I^2\exp\{a^2[R_x(0) + R_x(\tau)]\}$$

10-26. Show that (a) if

$$\mathbf{y}(t) = a\mathbf{x}(ct) \qquad \text{then} \quad R_y(\tau) = a^2R_x(c\tau)$$

(b) if $R_x(\tau) \to 0$ as $\tau \to \infty$ and

$$\mathbf{z}(t) = \lim_{\varepsilon\to\infty}\sqrt{\varepsilon}\,\mathbf{x}(\varepsilon t) \qquad \text{then} \quad R_z(\tau) = q\delta(\tau) \qquad q = \int_{-\infty}^{\infty} R_x(\tau)\,d\tau$$

10-27. Show that if $\mathbf{x}(t)$ is white noise, $h(t) = 0$ outside the interval $(0, T)$, and $\mathbf{y}(t) = \mathbf{x}(t) * h(t)$ then $R_{yy}(t_1, t_2) = 0$ for $|t_1 - t_2| > T$.

10-28. Show that if

$$R_{xx}(t_1, t_2) = q(t_1)\delta(t_1 - t_2) \qquad E\{\mathbf{y}^2(t)\} = I(t)$$

and

$$(a) \qquad \mathbf{y}(t) = \int_0^t h(t,\alpha)\mathbf{x}(\alpha)\,d\alpha \qquad \text{then} \quad I(t) = \int_0^t h^2(t,\alpha)q(\alpha)\,d\alpha$$

$$(b) \qquad \mathbf{y}'(t) + c(t)\mathbf{y}(t) = \mathbf{x}(t) \qquad \text{then} \quad I'(t) + 2c(t)I(t) = q(t)$$

10-29. Find $E\{\mathbf{y}^2(t)\}$ (a) if $R_{xx}(\tau) = 5\delta(\tau)$ and

$$\mathbf{y}'(t) + 2\mathbf{y}(t) = \mathbf{x}(t) \qquad \text{all} \quad t \tag{i}$$

(b) if (i) holds for $t > 0$ only and $\mathbf{y}(t) = 0$ for $t \le 0$.

Hint: Use (10-90).

10-30. The input to a linear system with $h(t) = Ae^{-\alpha t}U(t)$ is a process $\mathbf{x}(t)$ with $R_x(\tau) = N\delta(\tau)$ applied at $t = 0$ and disconnected at $t = T$. Find and sketch $E\{\mathbf{y}^2(t)\}$.

Hint: Use (10-90) with $q(t) = N$ for $0 < t < T$ and 0 otherwise.

10-31. Show that if

$$\mathbf{s} = \int_0^{10}\mathbf{x}(t)\,dt \qquad \text{then} \quad E\{\mathbf{s}^2\} = \int_{-10}^{10}(10 - |\tau|)R_x(\tau)\,d\tau$$

Find the mean and variance of $\mathbf{s}$ if $E\{\mathbf{x}(t)\} = 8$, $R_x(\tau) = 64 + 10e^{-2|\tau|}$.

10-32. The process $\mathbf{x}(t)$ is WSS with $R_{xx}(\tau) = 5\delta(\tau)$ and

$$\mathbf{y}'(t) + 2\mathbf{y}(t) = \mathbf{x}(t) \tag{i}$$

Find $E\{\mathbf{y}^2(t)\}$, $R_{xy}(t_1, t_2)$, $R_{yy}(t_1, t_2)$ (a) if (i) holds for all t, (b) if $\mathbf{y}(0) = 0$ and (i) holds for $t \ge 0$.

10-33. Find $S(\omega)$ if (a) $R(\tau) = e^{-\alpha\tau^2}$, (b) $R(\tau) = e^{-\alpha\tau^2}\cos\omega_0\tau$.

10-34. Show that the power spectrum of an SSS process $\mathbf{x}(t)$ equals

$$S(\omega) = \int_{-\infty}^{\infty}\int_{-\infty}^{\infty} x_1x_2G(x_1, x_2; \omega)\,dx_1\,dx_2$$

where $G(x_1, x_2; \omega)$ is the Fourier transform in the variable τ of the second-order density $f(x_1, x_2; \tau)$ of $\mathbf{x}(t)$.

10-35. Show that if $\mathbf{y}(t) = \mathbf{x}(t+a) - \mathbf{x}(t-a)$, then

$$R_y(\tau) = 2R_x(\tau) - R_x(\tau + 2a) - R_x(\tau - 2a) \qquad S_y(\omega) = 4S_x(\omega)\sin^2 a\omega$$

10-36. Using (10-122), show that

$$R(0) - R(\tau) \geq \frac{1}{4^n}[R(0) - R(2^n\tau)]$$

Hint:

$$1 - \cos\theta = 2\sin^2\frac{\theta}{2} \geq 2\sin^2\frac{\theta}{2}\cos^2\frac{\theta}{2} = \frac{1}{4}(1 - \cos 2\theta)$$

10-37. The process $\mathbf{x}(t)$ is normal with zero mean and $R_x(\tau) = Ie^{-\alpha|\tau|}\cos\beta\tau$. Show that if $\mathbf{y}(t) = \mathbf{x}^2(t)$, then $C_y(\tau) = I^2e^{-2\alpha|\tau|}(1 + \cos 2\beta\tau)$. Find $S_y(\omega)$.

10-38. Show that if $R(\tau)$ is the inverse Fourier transform of a function $S(\omega)$ and $S(\omega) \geq 0$, then, for any a_i,

$$\sum_{i,k} a_i a_k^* R(\tau_i - \tau_k) \geq 0$$

Hint:

$$\int_{-\infty}^{\infty} S(\omega)\left|\sum_i a_i e^{j\omega\tau_i}\right|^2 d\omega \geq 0$$

10-39. Find $R(\tau)$ if (*a*) $S(\omega) = 1/(1 + \omega^4)$, (*b*) $S(\omega) = 1/(4 + \omega^2)^2$.

10-40. Show that, for complex systems, (10-136) and (10-181) yield

$$\mathbf{S}_{yy}(s) = \mathbf{S}_{xx}(s)\mathbf{H}(s)\mathbf{H}^*(-s^*) \qquad \mathbf{S}_{yy}(z) = \mathbf{S}_{xx}(z)\mathbf{H}(z)\mathbf{H}^*(1/z^*)$$

10-41. The process $\mathbf{x}(t)$ is normal with zero mean. Show that if $\mathbf{y}(t) = \mathbf{x}^2(t)$, then

$$S_y(\omega) = 2\pi R_x^2(0)\delta(\omega) + 2S_x(\omega) * S_x(\omega)$$

Plot $S_y(\omega)$ if $S_x(\omega)$ is (*a*) ideal LP, (*b*) ideal BP.

10-42. The process $\mathbf{x}(t)$ is WSS with $E\{\mathbf{x}(t)\} = 5$ and $R_{xx}(\tau) = 25 + 4e^{-2|\tau|}$. If $\mathbf{y}(t) = 2\mathbf{x}(t) + 3\mathbf{x}'(t)$, find η_y, $R_{yy}(\tau)$, and $S_{yy}(\omega)$.

10-43. The process $\mathbf{x}(t)$ is WSS and $R_{xx}(\tau) = 5\delta(\tau)$. (*a*) Find $E\{\mathbf{y}^2(t)\}$ and $S_{yy}(\omega)$ if $\mathbf{y}'(t) + 3\mathbf{y}(t) = \mathbf{x}(t)$. (*b*) Find $E\{\mathbf{y}^2(t)\}$ and $R_{xy}(t_1, t_2)$ if $\mathbf{y}'(t) + 3\mathbf{y}(t) = \mathbf{x}(t)U(t)$. Sketch the functions $R_{xy}(2, t_2)$ and $R_{xy}(t_1, 3)$.

10-44. Given a complex process $\mathbf{x}(t)$ with autocorrelation $R(\tau)$, show that if $|R(\tau_1)| = |R(0)|$, then

$$R(\tau) = e^{j\omega_0\tau}w(\tau) \qquad \mathbf{x}(t) = e^{j\omega_0 t}\mathbf{y}(t)$$

where $w(\tau)$ is a periodic function with period τ_1 and $\mathbf{y}(t)$ is an MS periodic process with the same period.

10-45. Show that (*a*) $E\{\mathbf{x}(t)\check{\mathbf{x}}(t)\} = 0$, (*b*) $\check{\check{\mathbf{x}}}(t) = -\mathbf{x}(t)$.

10-46. (*Stochastic resonance*) The input to the system

$$\mathbf{H}(s) = \frac{1}{s^2 + 2s + 5}$$

is a WSS process $\mathbf{x}(t)$ with $E\{\mathbf{x}^2(t)\} = 10$. Find $S_x(\omega)$ such that the average power $E\{\mathbf{y}^2(t)\}$ of the resulting output $\mathbf{y}(t)$ is maximum.

Hint: $|\mathbf{H}(j\omega)|$ is maximum for $\omega = \sqrt{3}$.

10-47. Show that if $R_x(\tau) = Ae^{j\omega_0\tau}$, then $R_{xy}(\tau) = Be^{j\omega_0\tau}$ for any $\mathbf{y}(t)$.

Hint: Use (10-167).

10-48. Given a system $H(\omega)$ with input $\mathbf{x}(t)$ and output $\mathbf{y}(t)$, show that (a) if $\mathbf{x}(t)$ is WSS and $R_{xx}(\tau) = e^{j\alpha\tau}$, then

$$R_{yx}(\tau) = e^{j\alpha\tau}H(\alpha) \qquad R_{yy}(\tau) = e^{j\alpha\tau}|H(\alpha)|^2$$

(b) if $R_{xx}(t_1, t_2) = e^{j(\alpha t_1 - \beta t_2)}$, then

$$R_{yx}(t_1, t_2) = e^{j(\alpha t_1 - \beta t_2)}H(\alpha) \qquad R_{yy}(t_1, t_2) = e^{j(\alpha t_1 - \beta t_2)}H(\alpha)H^*(\beta)$$

10-49. Show that if $S_{xx}(\omega)S_{yy}(\omega) \equiv 0$, then $S_{xy}(\omega) \equiv 0$.

10-50. Show that if $\mathbf{x}[n]$ is WSS and $R_x[1] = R_x[0]$, then $R_x[m] = R_x[0]$ for every m.

10-51. Show that if $R[m] = E\{\mathbf{x}[n+m]\mathbf{x}[n]\}$, then

$$R[0]R[2] > 2R^2[1] - R^2[0]$$

10-52. Given an RV $\boldsymbol{\omega}$ with density $f(\omega)$ such that $f(\omega) = 0$ for $|\omega| > \pi$, we form the process $\mathbf{x}[n] = Ae^{jn\boldsymbol{\omega}}\pi$. Show that $S_x(\omega) = 2\pi A^2 f(\omega)$ for $|\omega| < \pi$.

10-53. (a) Find $E\{\mathbf{y}^2(t)\}$ if $\mathbf{y}(0) = \mathbf{y}'(0) = 0$ and

$$\mathbf{y}''(t) + 7\mathbf{y}'(t) + 10\mathbf{y}(t) = \mathbf{x}(t) \qquad R_x(\tau) = 5\delta(\tau)$$

(b) Find $E\{\mathbf{y}^2[n]\}$ if $\mathbf{y}[-1] = \mathbf{y}[-2] = 0$ and

$$8\mathbf{y}[n] - 6\mathbf{y}[n-1] + \mathbf{y}[n-2] = \mathbf{x}[n] \qquad R_x[m] = 5\delta[m]$$

10-54. The process $\mathbf{x}[n]$ is WSS with $R_{xx}[m] = 5\delta[m]$ and

$$\mathbf{y}[n] - 0.5\mathbf{y}[n-1] = \mathbf{x}[n] \qquad \text{(i)}$$

Find $E\{\mathbf{y}^2[n]\}$, $R_{xy}[m_1, m_2]$, $R_{yy}[m_1, m_2]$ (a) if (i) holds for all n, (b) if $\mathbf{y}[-1] = 0$ and (i) holds for $n \geq 0$.

10-55. Show that (a) if $R_x[m_1, m_2] = q[m_1]\delta[m_1 - m_2]$ and

$$\mathbf{s} = \sum_{n=0}^{N} a_n \mathbf{x}[n] \qquad \text{then} \qquad E\{\mathbf{s}^2\} = \sum_{n=0}^{N} a_n^2 q[n]$$

(b) If $R_{xx}(t_1, t_2) = q(t_1)\delta(t_1 - t_2)$ and

$$\mathbf{s} = \int_0^T a(t)\mathbf{x}(t)\,dt \qquad \text{then} \qquad E\{\mathbf{s}^2\} = \int_0^T a^2(t)q(t)\,dt$$

CHAPTER
11

BASIC APPLICATIONS

11-1 RANDOM WALK, BROWNIAN MOTION, AND THERMAL NOISE

We toss a fair coin every T seconds and after each toss we take instantly a step of length s, to the right if heads shows, to the left if tails shows. The process starts at $t = 0$ and our location at time t is a staircase function with discontinuities at the points $t = nT$ (Fig. 11-1a). We have thus created a discrete-state stochastic process $\mathbf{x}(t)$ whose samples $\mathbf{x}(t, \zeta)$ depend on the particular sequence of heads and tails. This process is called the *random walk*.

Suppose that at the first n tosses we observe k heads and $n - k$ tails. In this case, our walk consists of k steps to the right and $n - k$ steps to the left. Hence our position at time $t = nT$ is

$$\mathbf{x}(nT) = ks - (n - k)s = ms \qquad m = 2k - n$$

Thus $\mathbf{x}(nT)$ is an RV taking the values ms, where m equals n, or $n - 2, \ldots,$ or $-n$. Furthermore,

$$P\{\mathbf{x}(nT) = ms\} = \binom{n}{k}\frac{1}{2^n} \qquad k = \frac{m + n}{2} \tag{11-1}$$

This is the probability of k heads in n tosses.

We note that $\mathbf{x}(nT)$ can be written as a sum

$$\mathbf{x}(nT) = \mathbf{x}_1 + \cdots + \mathbf{x}_n$$

where $\mathbf{x}_i$ equals the size of the ith step. Thus the RVs $\mathbf{x}_i$ are independent

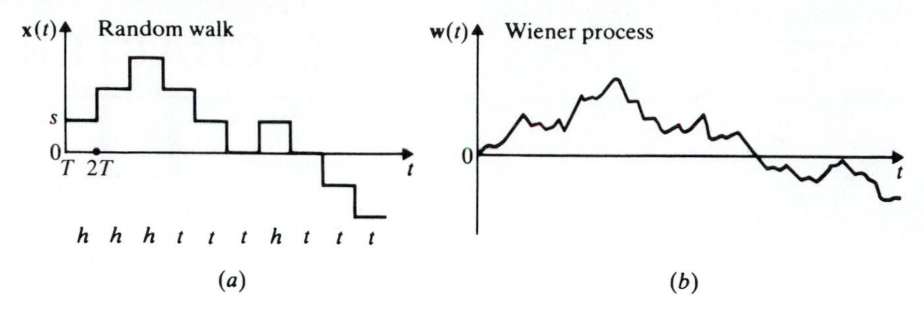

FIGURE 11-1

taking the values $\pm s$ and $E\{\mathbf{x}_i\} = 0$, $E\{\mathbf{x}_i^2\} = s^2$. From this it follows that

$$E\{\mathbf{x}(nT)\} = 0 \qquad E\{\mathbf{x}^2(nT)\} = ns^2 \tag{11-2}$$

Large t. As we know, if n is large and k is in the $\sqrt{npq}$ vicinity of np, then [see (3-27)]

$$\binom{n}{k} p^k q^{n-k} \simeq \frac{1}{\sqrt{2\pi npq}} e^{-(k-np)^2/2npq}$$

From this and (11-1) it follows with $p = q = 0.5$ and $m = 2k - n$ that

$$P\{\mathbf{x}(nT) = ms\} \simeq \frac{1}{\sqrt{n\pi/2}} e^{-m^2/2n}$$

for m of the order of $\sqrt{n}$. Hence

$$P\{\mathbf{x}(t) \le ms\} \simeq \mathbb{G}(m/\sqrt{n}) \qquad nT - T < t \le nT \tag{11-3}$$

where $\mathbb{G}(x)$ is the $N(0, 1)$ distribution [see (3-34)].

Note that if $n_1 < n_2 \le n_3 < n_4$ then the increments $\mathbf{x}(n_4 T) - \mathbf{x}(n_3 T)$ and $\mathbf{x}(n_2 T) - \mathbf{x}(n_1 T)$ of $\mathbf{x}(t)$ are independent.

The Wiener process. We shall now examine the limiting form of the random walk as $n \to \infty$ or, equivalently, as $T \to 0$. As we have shown

$$E\{\mathbf{x}^2(t)\} = ns^2 = \frac{ts^2}{T} \qquad t = nT$$

Hence, to obtain meaningful results, we shall assume that s tends to 0 as $\sqrt{T}$:

$$s^2 = \alpha T$$

The limit of $\mathbf{x}(t)$ as $T \to 0$ is then a continuous-state process (Fig. 11-1b)

$$\mathbf{w}(t) = \lim \mathbf{x}(t) \qquad T \to 0$$

known as the *Wiener process*.

We shall show that the first-order density $f(w, t)$ of $\mathbf{w}(t)$ is normal with zero mean and variance αt:

$$f(w, t) = \frac{1}{\sqrt{2\pi\alpha t}} e^{-w^2/2\alpha t} \tag{11-4}$$

Proof. If $w = ms$ and $t = nT$, then

$$\frac{m}{\sqrt{n}} = \frac{w/s}{\sqrt{t/T}} = \frac{w}{\sqrt{\alpha t}}$$

Inserting into (11-3), we conclude that

$$P\{\mathbf{w}(t) \le w\} = \mathbb{G}\left(\frac{w}{\sqrt{\alpha t}}\right)$$

and (11-4) results.

We show next that the autocorrelation of $\mathbf{w}(t)$ equals

$$R(t_1, t_2) = \alpha \min(t_1, t_2) \tag{11-5}$$

Indeed, if $t_1 < t_2$, then the difference $\mathbf{w}(t_2) - \mathbf{w}(t_1)$ is independent of $\mathbf{w}(t_1)$. Hence

$$E\{[\mathbf{w}(t_2) - \mathbf{w}(t_1)]\mathbf{w}(t_1)\} = E\{[\mathbf{w}(t_2) - \mathbf{w}(t_1)]\}E\{\mathbf{w}(t_1)\} = 0$$

This yields

$$E\{\mathbf{w}(t_1)\mathbf{w}(t_2)\} = E\{\mathbf{w}^2(t_1)\} = \frac{t_1 s^2}{T} = \alpha t_1$$

as in (11-5). The proof is similar if $t_1 > t_2$.

Note finally that if $t_1 < t_2 < t_3 < t_4$ then the increments $\mathbf{w}(t_4) - \mathbf{w}(t_3)$ and $\mathbf{w}(t_2) - \mathbf{w}(t_1)$ of $\mathbf{w}(t)$ are independent.

Generalized random walk. The random walk can be written as a sum

$$\mathbf{x}(t) = \sum_{k=1}^{n} \mathbf{c}_k U(t - kT) \qquad (n-1)T < t \le nT \tag{11-6}$$

where $\mathbf{c}_k$ is a sequence of i.i.d. RVs taking the values s and $-s$ with equal probability. In the generalized random walk, the RVs $\mathbf{c}_k$ take the values s and $-s$ with probability p and q respectively. In this case,

$$E\{\mathbf{c}_k\} = (p - q)s \qquad E\{\mathbf{c}_k^2\} = s^2 \qquad \sigma_{c_k}^2 = 4pqs^2$$

From this it follows that

$$E\{\mathbf{x}(t)\} = n(p - q)s \qquad \operatorname{Var} \mathbf{x}(t) = 4npqs^2 \tag{11-7}$$

For large n, the process $\mathbf{x}(t)$ is nearly normal with

$$E\{\mathbf{x}(t)\} \simeq \frac{t}{T}(p - q)s \qquad \operatorname{Var} \mathbf{x}(t) \simeq \frac{4t}{T}pqs^2 \tag{11-8}$$

Brownian Motion

The term *brownian motion* is used to describe the movement of a particle in a liquid, subjected to collisions and other forces. Macroscopically, the position $\mathbf{x}(t)$ of the particle can be modeled as a stochastic process satisfying a second-order differential equation:

$$m\mathbf{x}''(t) + f\mathbf{x}'(t) + c\mathbf{x}(t) = \mathbf{F}(t) \tag{11-9}$$

where $\mathbf{F}(t)$ is the collision force, m is the mass of the particle, f is the coefficient of friction, and $c\mathbf{x}(t)$ is an external force which we assume proportional to $\mathbf{x}(t)$. On a macroscopic scale, the process $\mathbf{F}(t)$ can be viewed as *normal* white noise with zero mean and power spectrum

$$S_F(\omega) = 2kTf \tag{11-10}$$

where T is the absolute temperature of the medium and $k = 1.37 \times 10^{-23}$ Joule-degrees is the Boltzmann constant. We shall determine the statistical properties of $\mathbf{x}(t)$ for various cases.

Bound motion. We assume first that the restoring force $c\mathbf{x}(t)$ is different from 0. For sufficiently large t, the position $\mathbf{x}(t)$ of the particle approaches a stationary state with zero mean and power spectrum (see Example 10-27)

$$S_x(\omega) = \frac{2kTf}{(c - m\omega^2)^2 + f^2\omega^2} \tag{11-11}$$

To determine the statistical properties of $\mathbf{x}(t)$, it suffices to find its autocorrelation. We shall do so under the assumption that the roots of the equation $ms^2 + fs + c = 0$ are complex

$$s_{1,2} = -\alpha \pm j\beta \qquad \alpha = \frac{f}{2m} \qquad \alpha^2 + \beta^2 = \frac{c}{m}$$

Replacing b, c, and q in Example 10-27b by f/m, c/m, and $2kTf/m^2$ respectively, we obtain

$$R_x(\tau) = \frac{kT}{c} e^{-\alpha|\tau|}\left(\cos\beta\tau + \frac{\alpha}{\beta}\sin\beta|\tau|\right) \tag{11-12}$$

Thus, for a specific t, $\mathbf{x}(t)$ is a normal RV with mean 0 and variance $R_x(0) = kT/c$. Hence its density equals

$$f_x(x) = \sqrt{\frac{c}{2\pi kT}}\, e^{-cx^2/2kT} \tag{11-13}$$

The conditional density of $\mathbf{x}(t)$ assuming $\mathbf{x}(t_0) = x_0$ is a normal curve with mean ax_0 and variance P where (see Example 8-11)

$$a = \frac{R_x(\tau)}{R_x(0)} \qquad P = R_x(0)(1 - a^2) \qquad \tau = t - t_0$$

FREE MOTION. We say that a particle is in free motion if the restoring force is 0. In this case, (11-9) yields

$$m\mathbf{x}''(t) + f\mathbf{x}'(t) = \mathbf{F}(t) \tag{11-14}$$

The solution of this equation is not a stationary process. We shall express its properties in terms of the properties of the velocity $\mathbf{v}(t)$ of the particle. Since $\mathbf{v}(t) = \mathbf{x}'(t)$, (11-17) yields

$$m\mathbf{v}'(t) + f\mathbf{v}(t) = \mathbf{F}(t) \tag{11-15}$$

The steady state solution of this equation is a stationary process with

$$S_v(\omega) = \frac{2kTf}{m^2\omega^2 + f^2} \qquad R_v(\tau) = \frac{kT}{m}e^{-f|\tau|/m} \tag{11-16}$$

From the preceding, it follows that $\mathbf{v}(t)$ is a normal process with zero mean, variance kT/m, and density

$$f_v(v) = \sqrt{\frac{m}{2\pi kT}}\, e^{-mv^2/2kT} \tag{11-17}$$

The conditional density of $\mathbf{v}(t)$ assuming $\mathbf{v}(0) = v_0$ is normal with mean av_0 and variance P (see Example 8-11) where

$$a = \frac{R_v(t)}{R_v(0)} = e^{-ft/m} \qquad P = \frac{kT}{m}(1 - a^2) = \frac{kT}{m}(1 - e^{-2ft/m})$$

In physics, (11-15) is called the *Langevin equation*, its solution the *Ornstein–Uhlenbeck* process, and its spectrum *lorenzian*.

The position $\mathbf{x}(t)$ of the particle is the integral of its velocity:

$$\mathbf{x}(t) = \int_0^t \mathbf{v}(\alpha)\, d\alpha \tag{11-18}$$

From this and (10-11) it follows that

$$E\{\mathbf{x}^2(t)\} = \int_0^t\int_0^t R_v(\alpha - \beta)\, d\alpha\, d\beta = \frac{kT}{m}\int_0^t\int_0^t e^{-f|\alpha-\beta|/m}\, d\alpha\, d\beta$$

Hence

$$E\{\mathbf{x}^2(t)\} = \frac{2kT}{f}\left(t - \frac{m}{f} + \frac{m}{f}e^{-ft/m}\right) \tag{11-19}$$

Thus, the position of a particle in free motion is a nonstationary normal process with zero mean and variance the right side of (11-19).

For $t \gg m/f$, (11-19) yields

$$E\{\mathbf{x}^2(t)\} \simeq \frac{2kT}{f}t = 2D^2t \qquad D^2 \equiv \frac{kT}{f} \tag{11-20}$$

The parameter D is the *diffusion constant*. This result will be presently rederived.

THE WIENER PROCESS. We now assume that the acceleration term $m\mathbf{x}''(t)$ of a particle in free motion is small compared to the friction term $f\mathbf{x}'(t)$; this is the case if $f \gg m/t$. Neglecting the term $m\mathbf{x}''(t)$ in (11-14), we conclude that

$$f\mathbf{x}'(t) = \mathbf{F}(t) \qquad \mathbf{x}(t) = \frac{1}{f}\int_0^t \mathbf{F}(\alpha)\,d\alpha$$

Because $\mathbf{F}(t)$ is white noise with spectrum $2kTf$, it follows from (10-36) with $\mathbf{v}(t) = \mathbf{F}(t)/f$ and $q(t) = 2kT/f$ that

$$E\{\mathbf{x}^2(t)\} = \frac{2kT}{f}t = \alpha t \qquad \alpha \equiv \frac{2kT}{f} = 2D^2$$

Thus, $\mathbf{x}(t)$ is a nonstationary normal process with density

$$f_{x(t)}(x) = \frac{1}{\sqrt{2\pi\alpha t}}e^{-x^2/2\alpha t}$$

We maintain that it is also a process with independent increments. Because it is normal, it suffices to show that it is a process with orthogonal increments, that is

$$E\{[\mathbf{x}(t_2) - \mathbf{x}(t_1)][\mathbf{x}(t_4) - \mathbf{x}(t_3)]\} = 0 \tag{11-21}$$

for $t_1 < t_2 < t_3 < t_4$. This follows from the fact that $\mathbf{x}(t_i) - \mathbf{x}(t_j)$ depends only on the values of $\mathbf{F}(t)$ in the interval (t_i, t_j) and $\mathbf{F}(t)$ is white noise. Using this, we shall show that

$$R_x(t_1, t_2) = \alpha \min(t_1, t_2) \tag{11-22}$$

To do so, we observe from (11-21) that if $t_1 < t_2$, then

$$E\{\mathbf{x}(t_1)\mathbf{x}(t_2)\} = E\{\mathbf{x}(t_1)[\mathbf{x}(t_2) - \mathbf{x}(t_1) + \mathbf{x}(t_1)]\} = E\{\mathbf{x}^2(t_1)\} = \alpha t_1$$

and (11-22) results. Thus the position of a particle in free motion with negligible acceleration has the following properties:

It is normal with zero mean, variance αt and autocorrelation $\alpha \min(t_1, t_2)$. It is a process with independent increments.

A process with these properties is called the *Wiener process*. As we have seen, it is the limiting form of the position of a particle in free motion as $t \to \infty$; it is also the limiting form of the random walk process as $n \to \infty$.

We note finally that the conditional density of $\mathbf{x}(t)$ assuming $\mathbf{x}(t_0) = x_0$ is normal with mean ax_0 and variance P where (see Example 8-11)

$$a = \frac{R_x(t, t_0)}{R_x(t_0, t_0)} = 1 \qquad P = R(t,t) - aR(t, t_0) = \alpha t - \alpha t_0$$

Hence

$$f_{x(t)}(x|\mathbf{x}(t_0) = x_0) = \frac{1}{\sqrt{2\pi\alpha(t - t_0)}}e^{-(x-x_0)^2/2\alpha(t-t_0)} \tag{11-23}$$

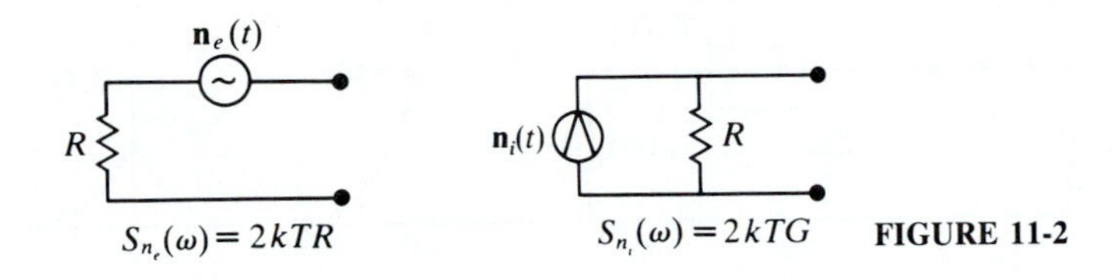

FIGURE 11-2

Diffusion equations. The right side of (11-23) is a function depending on the four parameters x, x_0, t, and t_0. Denoting this function by $\pi(x, x_0; t, t_0)$ we conclude by repeated differentiation that

$$\frac{\partial \pi}{\partial t} = D^2 \frac{\partial^2 \pi}{\partial x^2} \qquad \frac{\partial \pi}{\partial t_0} = -D^2 \frac{\partial^2 \pi}{\partial x_0^2} \tag{11-24}$$

where $D^2 = \alpha/2$. These equations are called *diffusion equations*. They are reestablished in Sec. 16-4 in the context of Markoff processes.

Thermal noise

Thermal noise is the distribution of voltages and currents in a network due to the thermal electron agitation. In the following, we discuss the statistical properties of thermal noise ignoring the underlying physics. The analysis is based on a model consisting of noiseless reactive elements and noisy resistors.

A noisy resistor is modeled by a noiseless resistor R in series with a voltage source $\mathbf{n}_e(t)$ or in parallel with a current source $\mathbf{n}_i(t) = \mathbf{n}_e(t)/R$ as in Fig. 11-2. It is assumed that $\mathbf{n}_e(t)$ is a normal process with zero mean and flat spectrum

$$S_{n_e}(\omega) = 2kTR \qquad S_{n_i}(\omega) = \frac{S_{n_e}(\omega)}{R^2} = 2kTG \tag{11-25}$$

where k is the Boltzmann constant, T is the absolute temperature of the resistor, and $G = 1/R$ is its conductance. Furthermore, the noise sources of the various network resistors are mutually independent processes. Note the similarity between the spectrum (11-25) of thermal noise and the spectrum (11-10) of the collision forces in brownian motion.

Using the above and the properties of linear systems, we shall derive the spectral properties of general network responses starting with an example.

Example 11-1. The circuit of Fig. 11-3 consists of a resistor R and a capacitor C. We shall determine the spectrum of the voltage $\mathbf{v}(t)$ across the capacitor due to thermal noise.

The voltage $\mathbf{v}(t)$ can be considered as the output of a system with input the noise voltage $\mathbf{n}_e(t)$ and system function

$$\mathbf{H}(s) = \frac{1}{1 + RCs}$$

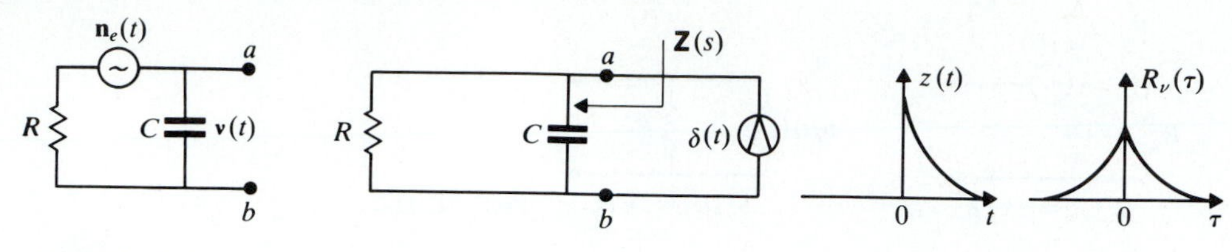

FIGURE 11-3

Applying (10-136), we obtain

$$S_v(\omega) = S_{n_e}(\omega)|H(\omega)|^2 = \frac{2kTR}{1+\omega^2R^2C^2}$$
$$R_v(\tau) = \frac{kT}{C}e^{-|\tau|/RC} \tag{11-26}$$

The following consequences are illustrations of Nyquist's theorem to be discussed presently: We denote by $\mathbf{Z}(s)$ the impedance across the terminals a, b and by $z(t)$ its inverse transform

$$\mathbf{Z}(s) = \frac{R}{1+RCs} \qquad z(t) = \frac{1}{C}e^{-t/RC}U(t)$$

The function $z(t)$ is the voltage across C due to an impulse current $\delta(t)$ (Fig. 11-3). Comparing with (11-26), we obtain

$$S_v(\omega) = 2kT\,\mathrm{Re}\,\mathbf{Z}(j\omega) \qquad \mathrm{Re}\,\mathbf{Z}(j\omega) = \frac{R}{1+\omega^2R^2C^2}$$
$$R_v(\tau) = kTz(\tau) \qquad \tau > 0 \qquad R_v(0) = kTz(0^+)$$
$$E\{\mathbf{v}^2(t)\} = R_v(0) = \frac{kT}{C} \qquad \frac{1}{C} = \lim_{\omega\to\infty} j\omega\mathbf{Z}(j\omega)$$

Given a passive, reciprocal network, we denote by $\mathbf{v}(t)$ the voltage across two arbitrary terminals a, b and by $\mathbf{Z}(s)$ the impedance from a to b (Fig. 11-4).

NYQUIST THEOREM. The power spectrum of $\mathbf{v}(t)$ equals

$$S_v(\omega) = 2kT\,\mathrm{Re}\,\mathbf{Z}(j\omega) \tag{11-27}$$

Proof. We shall assume that there is only one resistor in the network. The general case can be established similarly if we use the independence of the noise sources. The resistor is represented by a noiseless resistor in parallel with

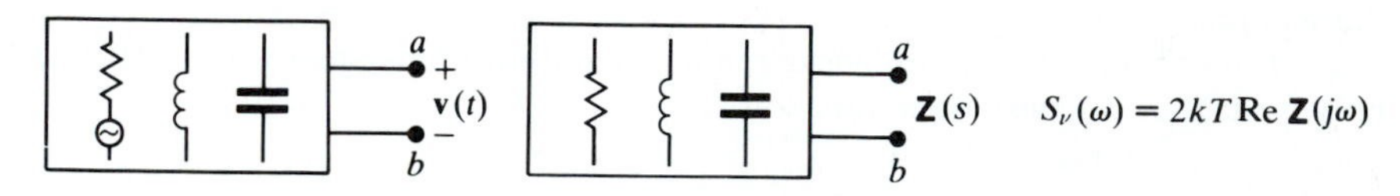

FIGURE 11-4

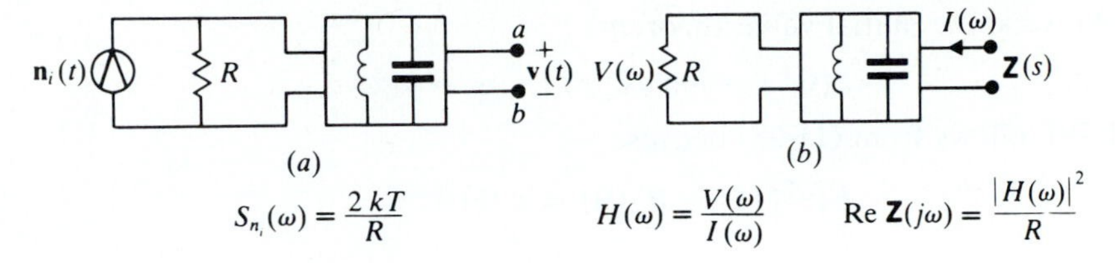

FIGURE 11-5

a current source $\mathbf{n}_i(t)$ and the remaining network contains only reactive elements (Fig. 11-5*a*). Thus $\mathbf{v}(t)$ is the output of a system with input $\mathbf{n}_i(t)$ and system function $H(\omega)$. From the reciprocity theorem it follows that $H(\omega) = V(\omega)/I(\omega)$ where $I(\omega)$ is the amplitude of a sine wave from a to b (Fig. 11-5*b*) and $V(\omega)$ is the amplitude of the voltage across R. The input power equals $|I(\omega)|^2 \,\text{Re}\,\mathbf{Z}(j\omega)$ and the power delivered to the resistance equals $|V(\omega)|^2/R$. Since the connecting network is lossless by assumption, we conclude that

$$|I(\omega)|^2 \,\text{Re}\,\mathbf{Z}(j\omega) = \frac{|V(\omega)|^2}{R}$$

Hence

$$|H(\omega)|^2 = \frac{|V(\omega)|^2}{|I(\omega)|^2} = R\,\text{Re}\,\mathbf{Z}(j\omega)$$

and (11-27) results because

$$S_v(\omega) = S_{n_i}(\omega)|H(\omega)|^2 \qquad S_{n_i}(\omega) = \frac{2kT}{R}$$

COROLLARY 1. The autocorrelation of $\mathbf{v}(t)$ equals

$$R_v(\tau) = kTz(\tau) \qquad \tau > 0 \tag{11-28}$$

where $z(t)$ is the inverse transform of $\mathbf{Z}(s)$.

Proof. Since $\mathbf{Z}(-j\omega) = \mathbf{Z}^*(j\omega)$, it follows from (11-27) that

$$S_v(\omega) = kT[\mathbf{Z}(j\omega) + \mathbf{Z}(-j\omega)]$$

and (11-28) results because the inverse of $\mathbf{Z}(-j\omega)$ equals $z(-t)$ and $z(-t) = 0$ for $t > 0$.

COROLLARY 2. The average power of $\mathbf{v}(t)$ equals

$$E\{\mathbf{v}^2(t)\} = \frac{kT}{C} \qquad \text{where} \quad \frac{1}{C} = \lim_{\omega \to \infty} j\omega \mathbf{Z}(j\omega) \tag{11-29}$$

where C is the input capacity.

Proof. As we know (initial value theorem)

$$z(0^+) = \lim s\mathbf{Z}(s) \qquad s \to \infty$$

and (11-29) follows from (11-28) because

$$E\{\mathbf{v}^2(t)\} = R_v(0) = kTz(0^+)$$

Currents. From Thévenin's theorem it follows that, terminally, a noisy network is equivalent to a noiseless network with impedance $\mathbf{Z}(s)$ in series with a voltage source $\mathbf{v}(t)$. The power spectrum $S_v(\omega)$ of $\mathbf{v}(t)$ is the right side of (11-27). This leads to the following version of Nyquist's theorem:

The power spectrum of the short-circuit current $\mathbf{i}(t)$ from a to b due to thermal noise equals

$$S_i(\omega) = 2kT \operatorname{Re} \mathbf{Y}(j\omega) \qquad \mathbf{Y}(s) = \frac{1}{\mathbf{Z}(s)} \tag{11-30}$$

Proof. From Thévenin's theorem it follows that

$$S_i(\omega) = S_v(\omega) |\mathbf{Y}(j\omega)|^2 = \frac{2kT \operatorname{Re} \mathbf{Z}(j\omega)}{|\mathbf{Z}(j\omega)|^2}$$

and (11-30) results.

The current version of the corollaries is left as an exercise.

11-2 POISSON POINTS AND SHOT NOISE

Given a set of Poisson points $\mathbf{t}_i$ and a fixed point t_0, we form the RV $\mathbf{z} = \mathbf{t}_1 - t_0$ where $\mathbf{t}_1$ is the first random point to the right of t_0 (Fig. 11-6). We shall show that $\mathbf{z}$ has an exponential distribution:

$$f_z(z) = \lambda e^{-\lambda z} \qquad F_z(z) = 1 - e^{-\lambda z} \qquad z > 0 \tag{11-31}$$

Proof. For a given $z > 0$, the function $F_z(z)$ equals the probability of the event $\{\mathbf{z} \le z\}$. This event occurs if $\mathbf{t}_1 < t_0 + z$, that is, if there is at least one random point in the interval $(t_0, t_0 + z)$. Hence

$$F_z(z) = P\{\mathbf{z} \le z\} = P\{\mathbf{n}(t_0, t_0 + z) > 0\} = 1 - P\{\mathbf{n}(t_0, t_0 + z) = 0\}$$

and (11-31) results because the probability that there are no points in the interval $(t_0, t_0 + z)$ equals $e^{-\lambda z}$.

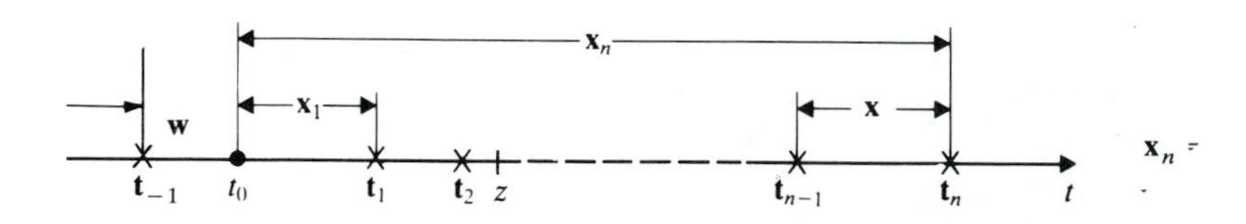

We can show similarly that if $\mathbf{w} = t_0 - \mathbf{t}_{-1}$ is the distance from t_0 to the first point $\mathbf{t}_{-1}$ to the left of t_0 then

$$f_w(w) = \lambda e^{-\lambda w} \qquad F_w(w) = 1 - e^{-\lambda w} \qquad w > 0 \tag{11-32}$$

We shall now show that the distance $\mathbf{x}_n = \mathbf{t}_n - t_0$ from t_0 the nth random point $\mathbf{t}_n$ to the right of t_0 (Fig. 11-6) has an Erlang distribution:

$$f_n(x) = \frac{\lambda^n}{(n-1)!} x^{n-1} e^{-\lambda x} \tag{11-33}$$

Proof. The event $\{\mathbf{x}_n \le x\}$ occurs if there are at least n points in the interval $(t_0, t_0 + x)$. Hence

$$F_n(x) = P\{\mathbf{x}_n \le x\} = 1 - P\{\mathbf{n}(t_0, t_0 + x) < n\} = 1 - \sum_{k=0}^{n-1} \frac{(\lambda x)^k}{k!} e^{-\lambda x}$$

Differentiating, we obtain (11-33).

Distance between random points. We show next that the distance

$$\mathbf{x} = \mathbf{x}_n - \mathbf{x}_{n-1} = \mathbf{t}_n - \mathbf{t}_{n-1}$$

between two consecutive points $\mathbf{t}_{n-1}$ and $\mathbf{t}_n$ has an exponential distribution:

$$f_x(x) = \lambda e^{-\lambda x} \tag{11-34}$$

Proof. From (11-33) and (5-70) it follows that the moment function of $\mathbf{x}_n$ equals

$$\mathbf{\Phi}_n(s) = \frac{\lambda^n}{(\lambda - s)^n} \tag{11-35}$$

Furthermore, the RVs $\mathbf{x}$ and $\mathbf{x}_{n-1}$ are independent and $\mathbf{x}_n = \mathbf{x} + \mathbf{x}_{n-1}$. Hence, if $\mathbf{\Phi}_x(s)$ is the moment function of $\mathbf{x}$, then

$$\mathbf{\Phi}_n(s) = \mathbf{\Phi}_x(s)\mathbf{\Phi}_{n-1}(s)$$

Comparing with (11-35), we obtain $\mathbf{\Phi}_x(s) = \lambda/(\lambda - s)$ and (11-34) results.

An apparent paradox. We should stress that the notion of the "distance $\mathbf{x}$ between two consecutive points of a point process" is ambiguous. In Fig. 11-6, we interpreted $\mathbf{x}$ as the distance between $\mathbf{t}_{n-1}$ and $\mathbf{t}_n$ where $\mathbf{t}_n$ was the nth random point to the right of some *fixed* point t_0. This interpretation led to the conclusion that the density of $\mathbf{x}$ is an exponential as in (11-34). The same density is obtained if we interpret $\mathbf{x}$ as the distance between consecutive points to the left of t_0. Suppose, however, that $\mathbf{x}$ is interpreted as follows:

Given a fixed point t_a, we denote by $\mathbf{t}_l$ and $\mathbf{t}_r$ the random points nearest to t_a on its left and right respectively (Fig. 11-7a). We maintain that the density of the distance $\mathbf{x} = \mathbf{t}_r - \mathbf{t}_l$ between these two points equals

$$f(x) = \lambda^2 x e^{-\lambda x} \tag{11-36}$$

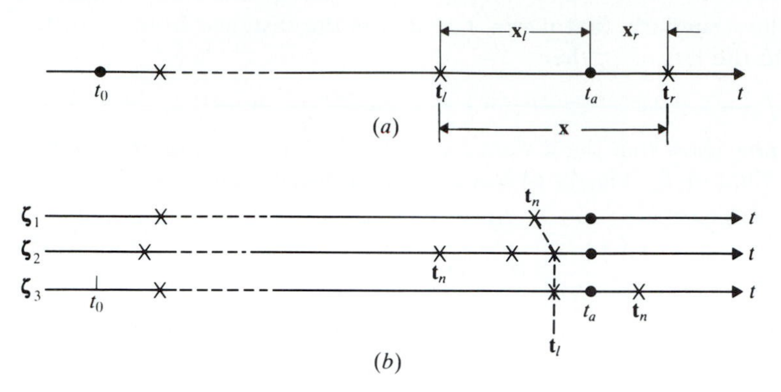

FIGURE 11-7

Indeed the RVs

$$\mathbf{x}_l = t_a - \mathbf{t}_l \qquad \text{and} \qquad \mathbf{x}_r = \mathbf{t}_r - t_a$$

are independent with exponential density as in (11-31); furthermore, $\mathbf{x} = \mathbf{x}_r + \mathbf{x}_l$. This yields (11-36) because the convolution of two exponentials is the density in (11-36).

Thus, although $\mathbf{x}$ is again the "distance between two consecutive points," its density is not an exponential. This apparent paradox is a consequence of the ambiguity in the specification of the identity of random points. Suppose, for example, that we identify the points $\mathbf{t}_i$ by their order i, where the count starts from some fixed point t_0, and we observe that in one particular realization of the point process, the point $\mathbf{t}_l$, defined as above, equals $\mathbf{t}_n$. In other realizations of the process, the RVs $\mathbf{t}_l$ might equal some other point in this identification (Fig. 11-7*b*). The same argument shows that the point $\mathbf{t}_r$ does not coincide with the ordered point $\mathbf{t}_{n+1}$ for all realizations. Hence we should not expect that the RV $\mathbf{x} = \mathbf{t}_r - \mathbf{t}_l$ has the same density as the RV $\mathbf{t}_{n+1} - \mathbf{t}_n$.

CONSTRUCTIVE DEFINITION. Given a sequence $\mathbf{w}_n$ of positive i.i.d. (independent, identically distributed) RVs with density

$$f(w) = \lambda e^{-\lambda w} \tag{11-37}$$

we form a set of points $\mathbf{t}_n$ as in Fig. 11-8*a* where $t = 0$ is an arbitrary origin and

$$\mathbf{t}_n = \mathbf{w}_1 + \mathbf{w}_2 + \cdots + \mathbf{w}_n \tag{11-38}$$

We maintain that the points so formed are Poisson distributed with parameter λ.

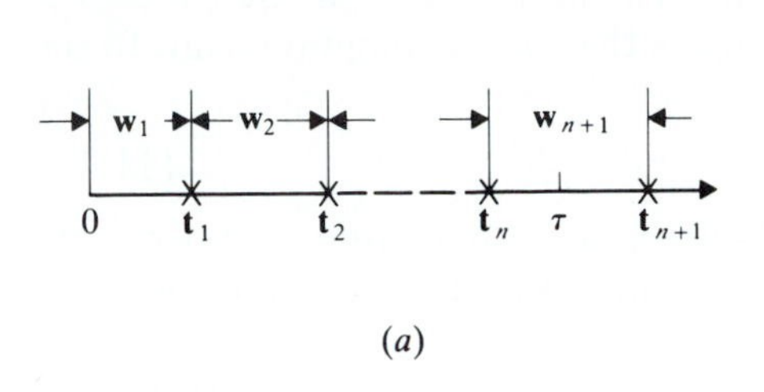

(a)

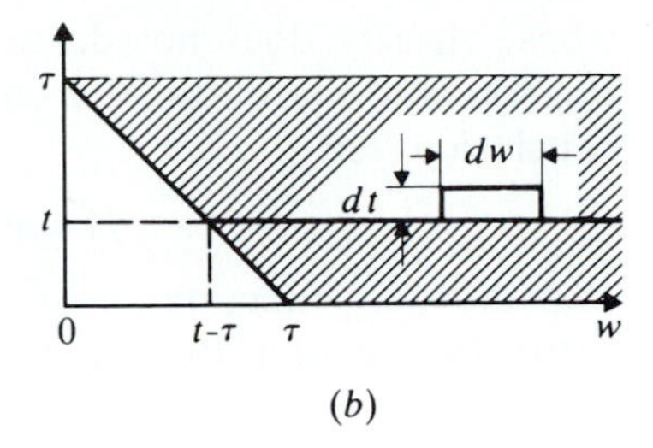

(b)

FIGURE 11-8

Proof. From the independence of the RVs $\mathbf{w}_n$, it follows that the RVs $\mathbf{t}_n$ and $\mathbf{w}_{n+1}$ are independent, the density $f_n(t)$ of $\mathbf{t}_n$ is given by (11-33)

$$f_n(t) = \frac{\lambda^n}{(n-1)!} t^{n-1} e^{-\lambda t} \tag{11-39}$$

and the joint density of $\mathbf{t}_n$ and $\mathbf{w}_{n+1}$ equals the product $f_n(t)f(w)$. If $\mathbf{t}_n < \tau$ and $\mathbf{t}_{n+1} = \mathbf{t}_n + \mathbf{w}_{n+1} > \tau$ then there are exactly n points in the interval $(0, \tau)$. As we see from Fig. 11-8b, the probability of this event equals

$$\int_0^\tau \int_{\tau - t}^\infty \lambda e^{-\lambda w} \frac{\lambda^n}{(n-1)!} t^{n-1} e^{-\lambda t}\, dw\, dt$$

$$= \int_0^\tau e^{-\lambda(t-\tau)} \frac{\lambda^n}{(n-1)!} t^{n-1} e^{-\lambda t}\, dt = e^{-\lambda t} \frac{(\lambda\tau)^n}{n!}$$

Thus the points $\mathbf{t}_n$ so constructed have property P_1. We can show similarly that they have also property P_2.

POISSON POINTS REDEFINED. Poisson points are realistic models for a large class of point processes: photon count, electron emission, telephone calls, data communication, visits to a doctor, arrivals at a park. The reason is that in these and other applications, the properties of the underlying points can be derived from certain general conditions that lead to Poisson distributions. As we show next, these conditions can be stated in a variety of forms that are equivalent to the two conditions used in Sec. 3-4 to specify random Poisson points (see page 59).

I. If we place at random N points in an interval of length T where $N \gg 1$, then the resulting point process is nearly Poisson with parameter N/T. This is exact in the limit as N and T tend to ∞ [see (3-47)].

II. If the distances $\mathbf{w}_n$ between two consecutive points $\mathbf{t}_{n-1}$ and $\mathbf{t}_n$ of a point process are independent and exponentially distributed, as in (11-37), then this process is Poisson.

The above can be phrased in an equivalent form: If the distance $\mathbf{w}$ from an arbitrary point t_0 to the next point of a point process is an RV whose density does not depend on the choice of t_0, then the process is Poisson. The reason for this equivalence is that this assumption leads to the conclusion that

$$f(w|\mathbf{w} \geq t_0) = f(w - t_0) \tag{11-40}$$

and the only function satisfying (11-40) is an exponential (see Example 7-10). In queueing theory, the above is called the *Markoff* or *memoryless property*.

III. If the number of points $\mathbf{n}(t, t + dt)$ in an interval $(t, t + dt)$ is such that:
   (*a*) $P\{\mathbf{n}(t, t + dt) = 1\}$ is of the order of dt;
   (*b*) $P\{\mathbf{n}(t, t + dt) > 1\}$ is of order higher than dt;
   (*c*) the above probabilities do not depend on the state of the point process outside the interval $(t, t + dt)$;
   then the process is Poisson (see Sec. 16-4).

IV. Suppose, finally, that:
   (*a*) $P\{\mathbf{n}(a, b) = k\}$ depends only on k and on the length of the interval (a, b);
   (*b*) if the intervals (a_i, b_i) are nonoverlapping, then the RVs $\mathbf{n}(a_i, b_i)$ are independent;
   (*c*) $P\{\mathbf{n}(a, b) = \infty\} = 0$.
   These conditions lead again to the conclusion that the probability $p_k(\tau)$ of having k points in any interval of length τ equals

$$p_k(\tau) = e^{-\lambda\tau}(\lambda\tau)^k/k! \tag{11-41}$$

The proof is omitted.

Linear interpolation. The process

$$\mathbf{x}(t) = t - \mathbf{t}_n \qquad \mathbf{t}_n \leq t < \mathbf{t}_{n+1} \tag{11-42}$$

of Fig. 11-9 consists of straight line segments of slope 1 between two consecutive random points $\mathbf{t}_n$ and $\mathbf{t}_{n+1}$. For a specific t, $\mathbf{x}(t)$ equals the distance $\mathbf{w} = t - \mathbf{t}_n$ from t to the nearest point $\mathbf{t}_n$ to the left of t; hence the first-order distribution of $\mathbf{x}(t)$ is exponential as in (11-32). From this it follows that

$$E\{\mathbf{x}(t)\} = \frac{1}{\lambda} \qquad E\{\mathbf{x}^2(t)\} = \frac{2}{\lambda^2} \tag{11-43}$$

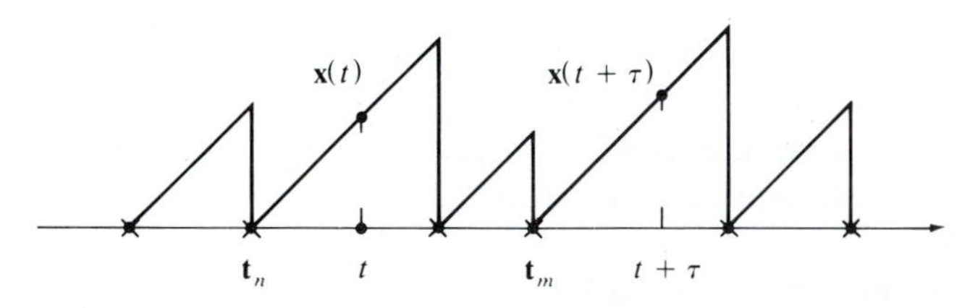

FIGURE 11-9

THEOREM. The autocovariance of $\mathbf{x}(t)$ equals

$$C(\tau) = \frac{1}{\lambda^2}(1 + \lambda|\tau|)e^{-\lambda|\tau|} \tag{11-44}$$

Proof. We denote by $\mathbf{t}_m$ and $\mathbf{t}_n$ the random points to the left of the points $t + \tau$ and t respectively. Suppose first, that $\mathbf{t}_m = \mathbf{t}_n$; in this case $\mathbf{x}(t + \tau) = t + \tau - \mathbf{t}_n$ and $\mathbf{x}(t) = t - \mathbf{t}_n$. Hence [see (11-42)]

$$C(\tau) = E\{(t + \tau - \mathbf{t}_n)(t - \mathbf{t}_n)\} = E\{(t - \mathbf{t}_n)^2\} + \tau E\{t - \mathbf{t}_n\} = \frac{2}{\lambda^2} + \frac{\tau}{\lambda}$$

Suppose, next, that $\mathbf{t}_m \neq \mathbf{t}_n$; in this case

$$C(\tau) = E\{(t + \tau - \mathbf{t}_m)(t - \mathbf{t}_n)\} = E\{t + \tau - \mathbf{t}_m\}E\{t - \mathbf{t}_n\} = \frac{1}{\lambda^2}$$

Clearly, $\mathbf{t}_m = \mathbf{t}_n$ if there are no random points in the interval $(t + \tau, t)$; hence $P\{\mathbf{t}_m = \mathbf{t}_n\} = e^{-\lambda\tau}$. Similarly, $\mathbf{t}_m \neq \mathbf{t}_n$ if there is at least one random point in the interval $(t + \tau, t)$; hence $P\{\mathbf{t}_n \neq \mathbf{t}_m\} = 1 - e^{-\lambda\tau}$. And since [see (4-48)]

$$\begin{aligned} R(\tau) &= E\{\mathbf{x}(t + \tau)\mathbf{x}(t) | \mathbf{t}_m = \mathbf{t}_n\}P\{\mathbf{t}_m = \mathbf{t}_n\} \\ &\quad + E\{\mathbf{x}(t + \tau)\mathbf{x}(t) | \mathbf{t}_n \neq \mathbf{t}_m\}P\{\mathbf{t}_n \neq \mathbf{t}_m\} \end{aligned}$$

we conclude that

$$R(\tau) = \left(\frac{2}{\lambda^2} + \frac{\tau}{\lambda}\right)e^{-\lambda\tau} + \frac{1}{\lambda^2}(1 - e^{-\lambda\tau})$$

for $\tau > 0$. Subtracting $1/\lambda^2$, we obtain (11-44).

Shot Noise

Given a set of Poisson points $\mathbf{t}_i$ with average density λ and a real function $h(t)$, we form the sum

$$\mathbf{s}(t) = \sum_i h(t - \mathbf{t}_i) \tag{11-45}$$

This sum is an SSS process known as *shot noise*. In the following, we discuss its second order properties. The general statistics are developed in Sec. 16-3.

From the definition it follows that $\mathbf{s}(t)$ can be represented as the output of a linear system (Fig. 11-10) with impulse response $h(t)$ and input the Poisson impulses

$$\mathbf{z}(t) = \sum_i \delta(t - \mathbf{t}_i) \tag{11-46}$$

This representation agrees with the generation of shot noise in physical problems: The process $\mathbf{s}(t)$ is the output of a dynamic system activated by a sequence of impulses (particle emissions, for example) occurring at the random times $\mathbf{t}_i$.

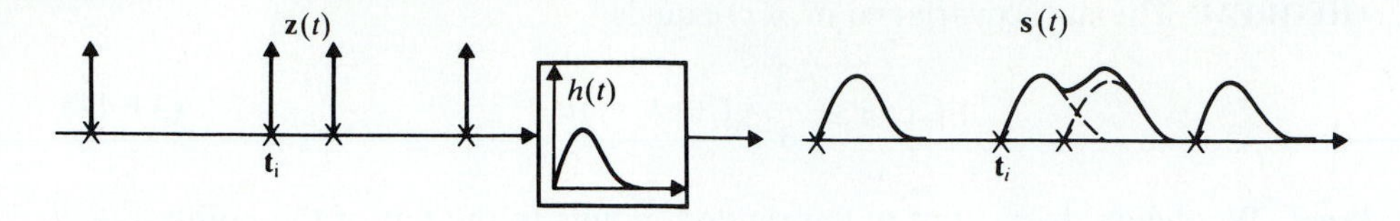

FIGURE 11-10

As we know, $\eta_z = \lambda$; hence

$$E\{\mathbf{s}(t)\} = \lambda \int_{-\infty}^{\infty} h(t)\, dt = \lambda H(0) \tag{11-47}$$

Furthermore, since (see Example 10-22)

$$S_{zz}(\omega) = 2\pi\lambda^2\delta(\omega) + \lambda \tag{11-48}$$

it follows from (10-136) that

$$S_{ss}(\omega) = 2\pi\lambda^2 H^2(0)\delta(\omega) + \lambda|H(\omega)|^2 \tag{11-49}$$

because $|H(\omega)|^2\delta(\omega) = H^2(0)\delta(\omega)$. The inverse of the above yields

$$R_{ss}(\tau) = \lambda^2 H^2(0) + \lambda\rho(\tau) \qquad C_{ss}(\tau) = \lambda\rho(\tau) \tag{11-50}$$

Campbell's theorem. The mean η_s and variance σ_s^2 of the shot-noise process $\mathbf{s}(t)$ equal

$$\eta_s = \lambda \int_{-\infty}^{\infty} h(t)\, dt \qquad \sigma_s^2 = \lambda\rho(0) = \lambda \int_{-\infty}^{\infty} h^2(t)\, dt \tag{11-51}$$

Proof. It follows from (11-50) because $\sigma_s^2 = C_{ss}(0)$.

Example 11-2. If

$$h(t) = e^{-\alpha t}U(t) \qquad H(\omega) = \frac{1}{\alpha + j\omega}$$

then

$$\eta_s = \frac{\lambda}{\alpha} \qquad \sigma_s^2 = \frac{\lambda}{2\alpha}$$

$$S_{ss}(\omega) = \frac{2\pi\lambda^2}{\alpha^2}\delta(\omega) + \frac{\lambda}{\alpha^2 + \omega^2} \qquad C_{ss}(\tau) = \frac{\lambda}{2\alpha}e^{-\alpha|\tau|}$$

Example 11-3 Electron transit. Suppose that $h(t)$ is a triangle as in Fig. 11-11a. Since

$$\int_0^T kt\, dt = \frac{kT^2}{2} \qquad \int_0^T k^2t^2\, dt = \frac{k^2T^3}{3}$$

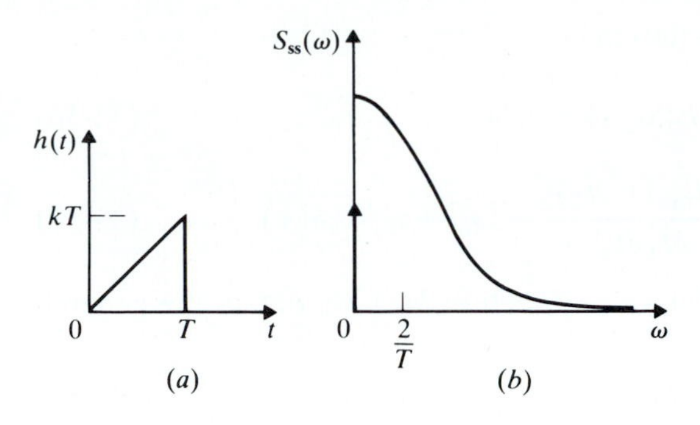

FIGURE 11-11

it follows from (11-51) that

$$\eta_s = \frac{\lambda k T^2}{2} \qquad \sigma_s^2 = \frac{\lambda k^2 T^3}{3}$$

In this case

$$H(\omega) = \int_0^T kte^{-j\omega t}\, dt = e^{-j\omega T/2}\frac{2k \sin \omega T/2}{j\omega^2} - e^{-j\omega T}\frac{kT}{j\omega}$$

Inserting into (11-49), we obtain (Fig. 11-11*b*).

$$S_{ss}(\omega) = 2\pi\eta_s^2\delta(\omega) + \frac{\lambda k^2}{\omega^4}(2 - 2\cos \omega T + \omega^2 T^2 - 2\omega T \sin \omega T)$$

Generalized Poisson process and shot noise. Given a set of Poisson points $\mathbf{t}_i$ with average density λ, we form the process

$$\mathbf{x}(t) = \sum_i \mathbf{c}_i U(t - \mathbf{t}_i) = \sum_{i=1}^{\mathbf{n}(t)} \mathbf{c}_i \tag{11-52}$$

where $\mathbf{c}_i$ is a sequence of i.i.d. RVs independent of the points $\mathbf{t}_i$ with mean η_c and variance σ_c^2. Thus $\mathbf{x}(t)$ is a staircase function as in Fig. 10-3 with jumps at the points $\mathbf{t}_i$ equal to $\mathbf{c}_i$. The process $\mathbf{n}(t)$ is the number of Poisson points in the interval $(0, t)$; hence $E\{\mathbf{n}(t)\} = \lambda t$ and $E\{\mathbf{n}^2(t)\} = \lambda^2 + \lambda t$. For a specific t, $\mathbf{x}(t)$ is a sum as in (8-46). From this it follows that

$$\begin{aligned} E\{\mathbf{x}(t)\} &= \eta_c E\{\mathbf{n}(t)\} = \eta_c \lambda t \\ E\{\mathbf{x}^2(t)\} &= \eta_c^2 E\{\mathbf{n}^2(t)\} + \sigma_c^2 E\{\mathbf{n}(t)\} = \eta_c^2(\lambda t + \lambda^2 t^2) + \sigma_c^2 \lambda t \end{aligned} \tag{11-53}$$

Proceeding as in Example 10-5, we obtain

$$C_{xx}(t_1, t_2) = (\eta_c^2 + \sigma_c^2)\lambda \min(t_1, t_2) \tag{11-54}$$

We next form the impulse train

$$\mathbf{z}(t) = \mathbf{x}'(t) = \sum_i \mathbf{c}_i \delta(t - \mathbf{t}_i) \tag{11-55}$$

From (11-54) it follows as in (10-100) that

$$E\{\mathbf{z}(t)\} = \frac{d}{dt}E\{\mathbf{x}(t)\} = \eta_c\lambda \tag{11-56}$$

$$C_{zz}(t_1, t_2) = \frac{\partial^2 C_{xx}(t_1, t_2)}{\partial t_1 \partial t_2} = (\eta_c^2 + \sigma_c^2)\lambda\delta(\tau) \tag{11-57}$$

where $\tau = t_2 - t_1$. Convolving $\mathbf{z}(t)$ with a function $h(t)$, we obtain the generalized shot noise

$$\mathbf{s}(t) = \sum_i \mathbf{c}_i h(t - \mathbf{t}_i) = \mathbf{z}(t) * h(t) \tag{11-58}$$

This yields

$$E\{\mathbf{s}(t)\} = E\{\mathbf{z}(t)\} * h(t) = \eta_c\lambda\int_{-\infty}^{\infty} h(t)\,dt \tag{11-59}$$

$$C_{ss}(\tau) = C_{zz}(\tau) * h(\tau) * h(-\tau) = (\eta_c^2 + \sigma^2)\lambda\rho(\tau) \tag{11-60}$$

$$\operatorname{Var}\mathbf{s}(t) = C_{ss}(0) = (\eta_c^2 + \sigma_c^2)\lambda\int_{-\infty}^{\infty} h^2(t)\,dt \tag{11-61}$$

The above is the extension of Campbell's theorem to a shot-noise process with random coefficients.

11-3 MODULATION†

Given two real jointly WSS processes $\mathbf{a}(t)$ and $\mathbf{b}(t)$ with zero mean and a constant ω_0, we form the process

$$\begin{aligned} \mathbf{x}(t) &= \mathbf{a}(t)\cos\omega_0 t - \mathbf{b}(t)\sin\omega_0 t \\ &= \mathbf{r}(t)\cos[\omega_0 t + \boldsymbol{\varphi}(t)] \end{aligned} \tag{11-62}$$

where

$$\mathbf{r}(t) = \sqrt{\mathbf{a}^2(t) + \mathbf{b}^2(t)} \qquad \tan\boldsymbol{\varphi}(t) = \frac{\mathbf{b}(t)}{\mathbf{a}(t)}$$

This process is called modulated with *amplitude modulation* $\mathbf{r}(t)$ and *phase modulation* $\boldsymbol{\varphi}(t)$.

We shall show that $\mathbf{x}(t)$ is WSS iff the processes $\mathbf{a}(t)$ and $\mathbf{b}(t)$ are such that

$$R_{aa}(\tau) = R_{bb}(\tau) \qquad R_{ab}(\tau) = -R_{ba}(\tau) \tag{11-63}$$

Proof. Clearly,

$$E\{\mathbf{x}(t)\} = E\{\mathbf{a}(t)\}\cos\omega_0 t - E\{\mathbf{b}(t)\}\sin\omega_0 t = 0$$

---

†A. Papoulis: "Random Modulation: A Review," *IEEE Transactions on Acoustics, Speech, and Signal Processing*, vol. ASSP-31, 1983.

Furthermore,

$$\mathbf{x}(t+\tau)\mathbf{x}(t) = \left[\mathbf{a}(t+\tau)\cos\omega_0(t+\tau) - \mathbf{b}(t+\tau)\sin\omega_0(t+\tau)\right] \times \left[\mathbf{a}(t)\cos\omega_0 t - \mathbf{b}(t)\sin\omega_0 t\right]$$

Multiplying, taking expected values, and using appropriate trigonometric identities, we obtain

$$\begin{aligned} 2E\{\mathbf{x}(t+\tau)\mathbf{x}(t)\} = {} & [R_{aa}(\tau) + R_{bb}(\tau)]\cos\omega_0\tau + [R_{ab}(\tau) - R_{ba}(\tau)]\sin\omega_0\tau \\ & + [R_{aa}(\tau) - R_{bb}(\tau)]\cos\omega_0(2t+\tau) \\ & - [R_{ab}(\tau) + R_{ba}(\tau)]\sin\omega_0(2t+\tau) \end{aligned} \tag{11-64}$$

If (11-63) is true, then the above yields

$$R_{xx}(\tau) = R_{aa}(\tau)\cos\omega_0\tau + R_{ab}(\tau)\sin\omega_0\tau \tag{11-65}$$

Conversely, if $\mathbf{x}(t)$ is WSS, then the second and third lines in (11-64) must be independent of t. This is possible only if (11-63) is true.

We introduce the "dual" process

$$\mathbf{y}(t) = \mathbf{b}(t)\cos\omega_0 t + \mathbf{a}(t)\sin\omega_0 t \tag{11-66}$$

This process is also WSS and

$$R_{yy}(\tau) = R_{xx}(\tau) \qquad R_{xy}(\tau) = -R_{yx}(\tau) \tag{11-67}$$

$$R_{xy}(\tau) = R_{ab}(\tau)\cos\omega_0\tau - R_{aa}(\tau)\sin\omega_0\tau \tag{11-68}$$

The above follows from (11-64) if we change one or both factors of the product $\mathbf{x}(t+\tau)\mathbf{x}(t)$ with $\mathbf{y}(t+\tau)$ or $\mathbf{y}(t)$.

Complex representation. We introduce the processes

$$\begin{aligned} \mathbf{w}(t) &= \mathbf{a}(t) + j\mathbf{b}(t) = \mathbf{r}(t)e^{j\boldsymbol{\varphi}(t)} \\ \mathbf{z}(t) &= \mathbf{x}(t) + j\mathbf{y}(t) = \mathbf{w}(t)e^{j\omega_0 t} \end{aligned} \tag{11-69}$$

Thus

$$\mathbf{x}(t) = \operatorname{Re}\mathbf{z}(t) = \operatorname{Re}\left[\mathbf{w}(t)e^{j\omega_0 t}\right] \tag{11-70}$$

and

$$\mathbf{a}(t) + j\mathbf{b}(t) = \mathbf{w}(t) = \mathbf{z}(t)e^{-j\omega_0 t}$$

This yields

$$\begin{aligned} \mathbf{a}(t) &= \mathbf{x}(t)\cos\omega_0 t + \mathbf{y}(t)\sin\omega_0 t \\ \mathbf{b}(t) &= \mathbf{y}(t)\cos\omega_0 t - \mathbf{x}(t)\sin\omega_0 t \end{aligned} \tag{11-71}$$

Correlations and spectra. The autocorrelation of the complex process $\mathbf{w}(t)$ equals

$$R_{ww}(\tau) = E\{[\mathbf{a}(t+\tau) + j\mathbf{b}(t+\tau)][\mathbf{a}(t) - j\mathbf{b}(t)]\}$$

Expanding and using (11-63), we obtain

$$R_{ww}(\tau) = 2R_{aa}(\tau) - 2jR_{ab}(\tau) \tag{11-72}$$

Similarly,

$$R_{zz}(\tau) = 2R_{xx}(\tau) - 2jR_{xy}(\tau) \tag{11-73}$$

We note, further, that

$$R_{zz}(\tau) = e^{j\omega_0\tau}R_{ww}(\tau) \tag{11-74}$$

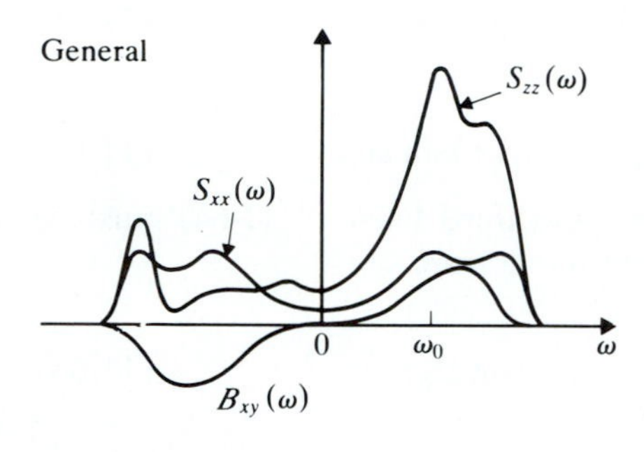

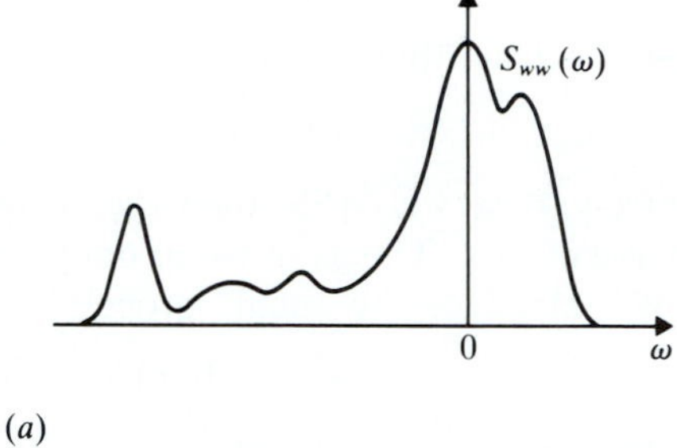

(*a*)

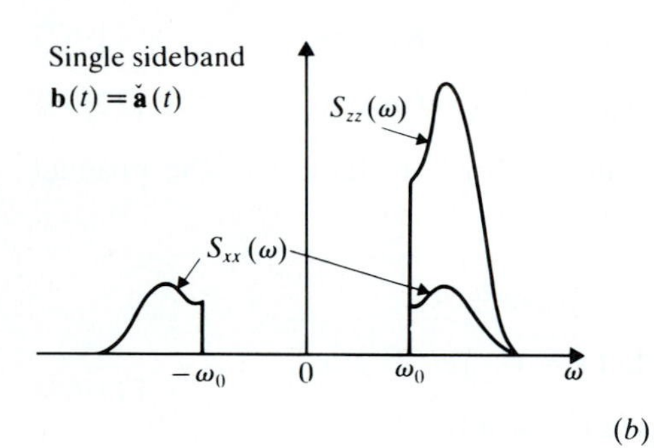

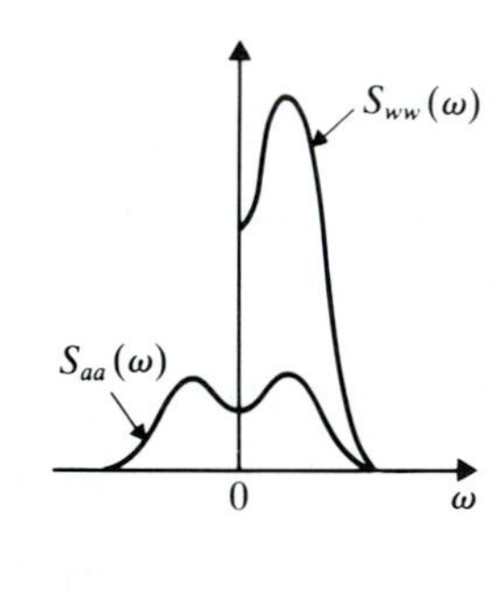

(*b*)

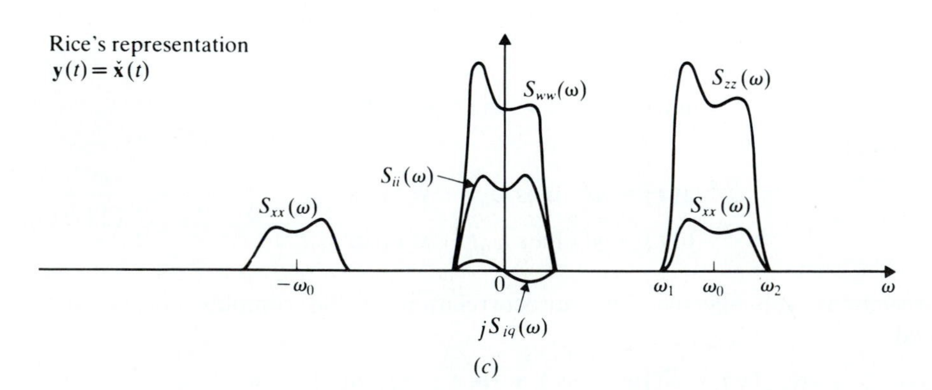

(*c*)

FIGURE 11-12

From the above it follows that

$$S_{ww}(\omega) = 2S_{aa}(\omega) - 2jS_{ab}(\omega)$$
$$S_{zz}(\omega) = 2S_{xx}(\omega) - 2jS_{xy}(\omega) \qquad (11\text{-}75)$$

$$S_{zz}(\omega) = S_{ww}(\omega - \omega_0) \qquad (11\text{-}76)$$

The functions $S_{xx}(\omega)$ and $S_{zz}(\omega)$ are real and positive. Furthermore [see (11-67)]

$$R_{xy}(-\tau) = -R_{yx}(-\tau) = -R_{xy}(\tau)$$

This leads to the conclusion that the function $-jS_{xy}(\omega) = B_{xy}(\omega)$ is real and (Fig. 11-12*a*)

$$|B_{xy}(\omega)| \leq S_{xx}(\omega) \qquad B_{xy}(-\omega) = -B_{xy}(\omega) \qquad (11\text{-}77)$$

And since $S_{xx}(-\omega) = S_{xx}(\omega)$, we conclude from the second equation in (11-75) that

$$4S_{xx}(\omega) = S_{zz}(\omega) + S_{zz}(-\omega)$$
$$4jS_{xy}(\omega) = S_{zz}(-\omega) - S_{zz}(\omega) \qquad (11\text{-}78)$$

Single sideband If $\mathbf{b}(t) = \check{\mathbf{a}}(t)$ is the Hilbert transform of $\mathbf{a}(t)$, then [see (10-147)] the constraint (11-63) is satisfied and the first equation in (11-75) yields

$$S_{ww}(\omega) = 4S_{aa}(\omega)U(\omega)$$

(Fig. 11-12*b*) because

$$S_{a\hat{a}}(\omega) = jS_{aa}(\omega)\operatorname{sgn}\omega$$

The resulting spectra are shown in Fig. 11-12*b*. We note, in particular, that $S_{xx}(\omega) = 0$ for $|\omega| < \omega_0$.

RICE'S REPRESENTATION. In (11-62) we assumed that the carrier frequency ω_0 and the processes $\mathbf{a}(t)$ and $\mathbf{b}(t)$ were given. We now consider the converse problem: Given a WSS process $\mathbf{x}(t)$ with zero mean, find a constant ω_0 and two processes $\mathbf{a}(t)$ and $\mathbf{b}(t)$ such that $\mathbf{x}(t)$ can be written in the form (11-62). To do so, it suffices to find the constant ω_0 and the dual process $\mathbf{y}(t)$ [see (11-71)]. This shows that the representation of $\mathbf{x}(t)$ in the form (11-62) is not unique because, not only ω_0 is arbitrary, but also the process $\mathbf{y}(t)$ can be chosen arbitrarily subject only to the constraint (11-67). The question then arises whether, among all possible representations of $\mathbf{x}(t)$, there is one that is optimum. The answer depends, of course, on the optimality criterion. As we shall presently explain, if $\mathbf{y}(t)$ equals the Hilbert transform $\check{\mathbf{x}}(t)$ of $\mathbf{x}(t)$, then (11-62) is optimum in the sense of minimizing the average rate of variation of the envelope of $\mathbf{x}(t)$.

Hilbert transforms. As we know [see (10-147)]

$$R_{\check{x}\check{x}}(\tau) = R_{xx}(\tau) \qquad R_{x\check{x}}(\tau) = -R_{x\check{x}}(\tau) \qquad (11\text{-}79)$$

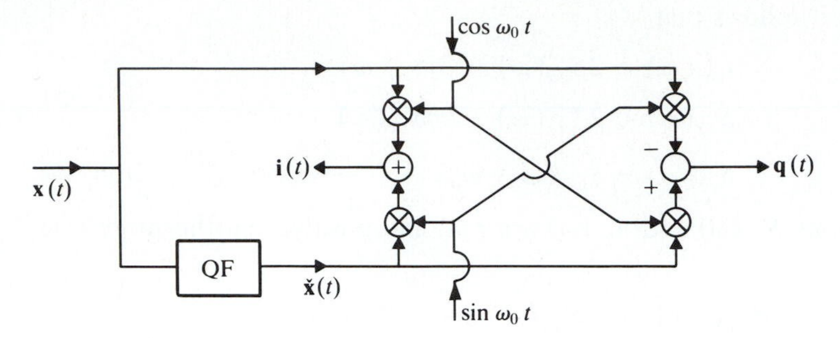

FIGURE 11-13

We can, therefore, use $\check{\mathbf{x}}(t)$ to form the processes

$$\mathbf{z}(t) = \mathbf{x}(t) + j\check{\mathbf{x}}(t) = \mathbf{w}(t)e^{j\omega_0 t}$$
$$\mathbf{w}(t) = \mathbf{i}(t) + j\mathbf{q}(t) = \mathbf{z}(t)e^{-j\omega_0 t} \qquad (11\text{-}80)$$

as in (11-69) where now (Fig. 11-12c)

$$\mathbf{y}(t) = \check{\mathbf{x}}(t) \qquad \mathbf{a}(t) = \mathbf{i}(t) \qquad \mathbf{b}(t) = \mathbf{q}(t)$$

Inserting into (11-62), we obtain

$$\mathbf{x}(t) = \mathbf{i}(t)\cos\omega_0 t - \mathbf{q}(t)\sin\omega_0 t \qquad (11\text{-}81)$$

This is known as *Rice's representation*. The process $\mathbf{i}(t)$ is called the *inphase* component and the process $\mathbf{q}(t)$ the *quadrature* component of $\mathbf{x}(t)$. Their realization is shown in Fig. 11-13 [see (11-71)]. These processes depend, not only on $\mathbf{x}(t)$, but also on the choice of the carrier frequency ω_0.

From (10-136) and (11-75) it follows that

$$S_{zz}(\omega) = 4S_{xx}(\omega)U(\omega) \qquad (11\text{-}82)$$

Bandpass processes. A process $\mathbf{x}(t)$ is called bandpass (Fig. 11-12c) if its spectrum $S_{xx}(\omega)$ is 0 outside an interval (ω_1, ω_2). It is called narrowband or *quasimonochromatic* if its bandwidth $\omega_2 - \omega_1$ is small compared with the center frequency. It is called *monochromatic* if $S_{xx}(\omega)$ is an impulse function. The process $\mathbf{a}\cos\omega_0 t + \mathbf{b}\sin\omega_0 t$ is monochromatic.

The representations (11-62) or (11-81) hold for an arbitrary $\mathbf{x}(t)$. However, they are useful mainly if $\mathbf{x}(t)$ is bandpass. In this case, the complex envelope $\mathbf{w}(t)$ and the processes $\mathbf{i}(t)$ and $\mathbf{q}(t)$ are low-pass because

$$S_{ww}(\omega) = S_{zz}(\omega + \omega_0)$$
$$S_{ii}(\omega) = S_{qq}(\omega) = \tfrac{1}{4}[S_{ww}(\omega) + S_{ww}(-\omega)] \qquad (11\text{-}83)$$

We shall show that if the process $\mathbf{x}(t)$ is bandpass and $\omega_1 + \omega_c \le 2\omega_0$, then the inphase component $\mathbf{i}(t)$ and the quadrature component $\mathbf{q}(t)$ can be obtained as responses of the system of Fig. 11-14a where the LP filters are ideal with cutoff

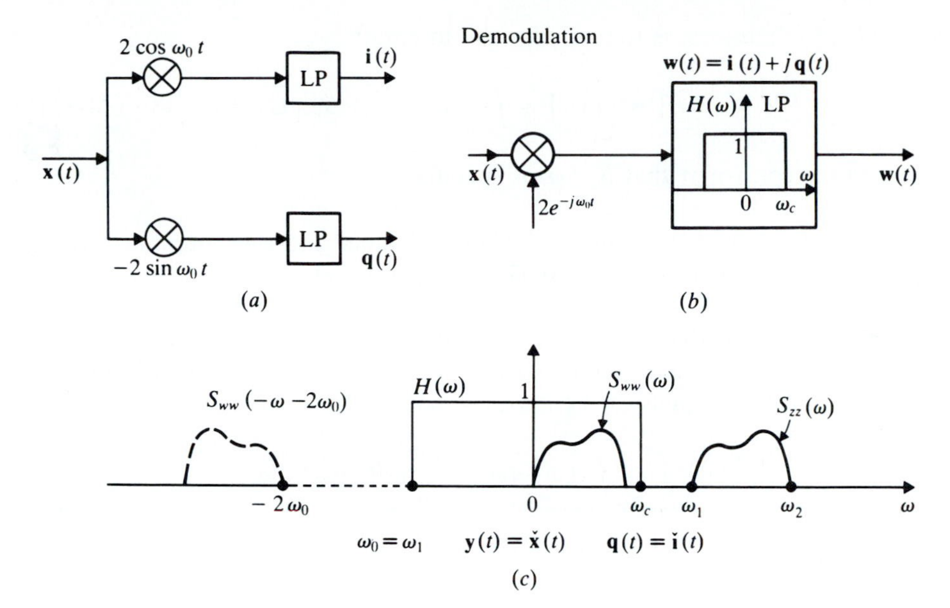

FIGURE 11-14

frequency ω_c such that

$$\omega_2 - \omega_0 < \omega_c \qquad \omega_1 - \omega_0 > -\omega_c \tag{11-84}$$

Proof. It suffices to show that (linearity) the response of the system of Fig. 11-14*b* equals $\mathbf{w}(t)$. Clearly,

$$2\mathbf{x}(t) = \mathbf{z}(t) + \mathbf{z}^*(t) \qquad \mathbf{w}^*(t) = \mathbf{z}^*(t)e^{j\omega_0 t}$$

Hence

$$2\mathbf{x}(t)e^{-j\omega_0 t} = \mathbf{w}(t) + \mathbf{w}^*(t)e^{-j2\omega_0 t}$$

The spectra of the processes $\mathbf{w}(t)$ and $\mathbf{w}^*(t)e^{-j2\omega_0 t}$ equal $S_{ww}(\omega)$ and $S_{ww}(-\omega - 2\omega_0)$ respectively. Under the stated assumptions, the first is in the band of the LP filter $H(\omega)$ and the second outside the band. Therefore, the response of the filter equals $\mathbf{w}(t)$.

We note, finally, that if $\omega_0 \le \omega_1$, then $S_{ww}(\omega) = 0$ for $\omega < 0$. In this case, $\mathbf{q}(t)$ is the Hilbert transform of $\mathbf{i}(t)$. Since $\omega_2 - \omega_1 \le 2\omega_0$, this is possible only if $\omega_2 \le 3\omega_1$. In Fig. 11-14*c*, we show the corresponding spectra for $\omega_0 = \omega_1$.

Optimum envelope. We are given an arbitrary process $\mathbf{x}(t)$ and we wish to determine a constant ω_0 and a process $\mathbf{y}(t)$ so that, in the resulting representation (11-62), the complex envelope $\mathbf{w}(t)$ of $\mathbf{x}(t)$ is smooth in the sense of minimizing $E\{|\mathbf{w}'(t)|^2\}$. As we know, the power spectrum of $\mathbf{w}'(t)$ equals

$$\omega^2 S_{ww}(\omega) = \omega^2 S_{zz}(\omega + \omega_0)$$

Our problem, therefore, is to minimize the integral†

$$M = 2\pi E\{|\mathbf{w}'(t)|^2\} = \int_{-\infty}^{\infty} (\omega - \omega_0)^2 S_{zz}(\omega)\, d\omega \tag{11-85}$$

subject to the constraint that $S_{xx}(\omega)$ is specified.

THEOREM. Rice's representation (11-81) is optimum and the optimum carrier frequency ω_0 is the center of gravity $\overline{\omega}_0$ of $S_{xx}(\omega)U(\omega)$.

Proof. Suppose, first, that $S_{xx}(\omega)$ is specified. In this case, M depends only on ω_0. Differentiating the right side of (11-85) with respect to ω_0, we conclude that M is minimum if ω_0 equals the center of gravity

$$\overline{\omega}_0 = \frac{\int_{-\infty}^{\infty} \omega S_{zz}(\omega)\, d\omega}{\int_{-\infty}^{\infty} S_{zz}(\omega)\, d\omega} = \frac{\int_0^{\infty} \omega B_{xy}(\omega)\, d\omega}{\int_0^{\infty} S_{xx}(\omega)\, d\omega} \tag{11-86}$$

of $S_{zz}(\omega)$. The second equality above follows from (11-75) and (11-77). Inserting (11-86) into (11-85), we obtain

$$M = \int_{-\infty}^{\infty} (\omega^2 - \overline{\omega}_0^2) S_{zz}(\omega)\, d\omega = 2\int_{-\infty}^{\infty} (\omega^2 - \overline{\omega}_0^2) S_{xx}(\omega)\, d\omega \tag{11-87}$$

We wish now to choose $S_{zz}(\omega)$ so as to minimize M. Since $S_{xx}(\omega)$ is given, M is minimum if $\overline{\omega}_0$ is maximum. As we see from (11-86), this is the case if $|B_{xy}(\omega)| = S_{xx}(\omega)$ because $|B_{xy}(\omega)| \le S_{xx}(\omega)$. We thus conclude that $-jS_{xy}(\omega) = S_{xx}(\omega)\operatorname{sgn}\omega$ and (11-75) yields

$$S_{zz}(\omega) = 4S_{xx}(\omega)U(\omega)$$

Instantaneous frequency. With $\boldsymbol{\varphi}(t)$ as in (11-62), the process

$$\boldsymbol{\omega}_i(t) = \omega_0 + \boldsymbol{\varphi}'(t) \tag{11-88}$$

is called the instantaneous frequency of $\mathbf{x}(t)$. Since

$$\mathbf{z} = \mathbf{r}e^{j(\omega_0 t + \boldsymbol{\varphi})} = \mathbf{x} + j\mathbf{y}$$

we have

$$\mathbf{z}'\mathbf{z}^* = \mathbf{r}\mathbf{r}' + j\mathbf{r}^2\boldsymbol{\omega}_i = (\mathbf{x}' + j\mathbf{y}')(\mathbf{x} - j\mathbf{y}) \tag{11-89}$$

---

†L. Mandel: "Complex Representation of Optical Fields in Coherence Theory," *Journal of the Optical Society of America*, vol. 57, 1967. See also N. M. Blachman: *Noise and Its Effect on Communication*, Krieger Publishing Company, Malabar, FL, 1982.

This yields $E\{\mathbf{r}\mathbf{r}'\} = 0$ and

$$E\{\mathbf{r}^2\boldsymbol{\omega}_i\} = \frac{1}{2\pi}\int_{-\infty}^{\infty} \omega S_{zz}(\omega)\,d\omega \tag{11-90}$$

because the cross-power spectrum of $\mathbf{z}'$ and $\mathbf{z}$ equals $j\omega S_{zz}(\omega)$.

The instantaneous frequency of a process $\mathbf{x}(t)$ is not a uniquely defined process because the dual process $\mathbf{y}(t)$ is not unique. In Rice's representation $\mathbf{y} = \check{\mathbf{x}}$, hence

$$\boldsymbol{\omega}_i = \frac{\mathbf{x}\check{\mathbf{x}}' - \mathbf{x}'\check{\mathbf{x}}}{\mathbf{r}^2} \qquad \mathbf{r}^2 = \mathbf{x}^2 + \check{\mathbf{x}}^2 \tag{11-91}$$

In this case [see (11-82) and (11-86)] the optimum carrier frequency $\overline{\omega}_0$ equals the weighted average of $\boldsymbol{\omega}_i$:

$$\overline{\omega}_0 = \frac{E\{\mathbf{r}^2\boldsymbol{\omega}_i\}}{E\{\mathbf{r}^2\}}$$

Frequency Modulation

The process

$$\mathbf{x}(t) = \cos[\omega_0 t + \lambda\boldsymbol{\varphi}(t) + \boldsymbol{\varphi}_0] \qquad \boldsymbol{\varphi}(t) = \int_0^t \mathbf{c}(\alpha)\,d\alpha \tag{11-92}$$

is FM with instantaneous frequency $\omega_0 + \lambda\mathbf{c}(t)$ and modulation index λ. The corresponding complex processes equal

$$\mathbf{w}(t) = e^{j\lambda\boldsymbol{\varphi}(t)} \qquad \mathbf{z}(t) = \mathbf{w}(t)e^{j(\omega_0 t + \boldsymbol{\varphi}_0)} \tag{11-93}$$

We shall study their spectral properties.

THEOREM. If the process $\mathbf{c}(t)$ is SSS and the RV $\boldsymbol{\varphi}_0$ is independent of $\mathbf{c}(t)$ and such that

$$E\{e^{j\boldsymbol{\varphi}_0}\} = E\{e^{j2\boldsymbol{\varphi}_0}\} = 0 \tag{11-94}$$

then the process $\mathbf{x}(t)$ is WSS with zero mean. Furthermore,

$$\begin{gathered} R_{xx}(\tau) = \tfrac{1}{2}\operatorname{Re} R_{zz}(\tau) \\ R_{zz}(\tau) = R_{ww}(\tau)e^{j\omega_0\tau} \qquad R_{ww}(\tau) = E\{\mathbf{w}(\tau)\} \end{gathered} \tag{11-95}$$

Proof. From (11-94) it follows that $E\{\mathbf{x}(t)\} = 0$ because

$$E\{\mathbf{z}(t)\} = E\{e^{j[\omega_0 t + \lambda\boldsymbol{\varphi}(t)]}\}E\{e^{j\boldsymbol{\varphi}_0}\} = 0$$

Furthermore,

$$E\{\mathbf{z}(t+\tau)\mathbf{z}(t)\} = E\{e^{j[\omega_0(2t+\tau) + \lambda\boldsymbol{\varphi}(t+\tau) + \lambda\boldsymbol{\varphi}(t)]}\}E\{e^{j2\boldsymbol{\varphi}_0}\} = 0$$

$$E\{\mathbf{z}(t+\tau)\mathbf{z}^*(t)\} = e^{j\omega_0\tau}E\left\{\exp\left[j\lambda\int_t^{t+\tau}\mathbf{c}(\alpha)\,d\alpha\right]\right\} = e^{j\omega_0\tau}E\{\mathbf{w}(\tau)\}$$

The last equality is a consequence of the stationarity of the process $\mathbf{c}(t)$. Since $2\mathbf{x}(t) = \mathbf{z}(t) + \mathbf{z}^*(t)$, we conclude from the above that

$$4E\{\mathbf{x}(t+\tau)\mathbf{x}(t)\} = R_{zz}(\tau) + R_{zz}(-\tau)$$

and (11-95) results because $R_{zz}(-\tau) = R_{zz}^*(\tau)$.

Definitions A process $\mathbf{x}(t)$ is *phase modulated* if the statistics of $\boldsymbol{\varphi}(t)$ are known. In this case, its autocorrelation can simply be found because

$$E\{\mathbf{w}(t)\} = E\{e^{j\lambda\boldsymbol{\varphi}(t)}\} = \Phi_\varphi(\lambda, t) \qquad (11\text{-}96)$$

where $\Phi_\varphi(\lambda, t)$ is the characteristic function of $\boldsymbol{\varphi}(t)$.

A process $\mathbf{x}(t)$ is *frequency modulated* if the statistics of $\mathbf{c}(t)$ are known. To determine $\Phi_\varphi(\lambda, t)$, we must now find the statistics of the integral of $\mathbf{c}(t)$. However, in general this is not simple. The normal case is an exception because then $\Phi_\varphi(\lambda, t)$ can be expressed in terms of the mean and variance of $\boldsymbol{\varphi}(t)$ and, as we know [see (10-143)]

$$E\{\boldsymbol{\varphi}(t)\} = \int_0^t E\{\mathbf{c}(\alpha)\}\,d\alpha = \eta_c t$$
$$E\{\boldsymbol{\varphi}^2(t)\} = 2\int_0^t R_c(\alpha)(t-\alpha)\,d\alpha \qquad (11\text{-}97)$$

For the determination of the power spectrum $S_{xx}(\omega)$ of $\mathbf{x}(t)$, we must find the function $\Phi_\varphi(\lambda, t)$ and its Fourier transform. In general, this is difficult. However, as the next theorem shows, if λ is large, then $S_{xx}(\omega)$ can be expressed directly in terms of the density $f_c(c)$ of $\mathbf{c}(t)$.

WOODWARD'S THEOREM.† If the process $\mathbf{c}(t)$ is continuous and its density $f_c(c)$ is bounded, then for large λ:

$$S_{xx}(\omega) \simeq \frac{\pi}{2\lambda}\left[f_c\left(\frac{\omega-\omega_0}{\lambda}\right) + f_c\left(\frac{-\omega-\omega_0}{\lambda}\right)\right] \qquad (11\text{-}98)$$

Proof. If τ_0 is sufficiently small, then $\mathbf{c}(t) \simeq \mathbf{c}(0)$, and

$$\boldsymbol{\varphi}(t) = \int_0^t \mathbf{c}(\alpha)\,d\alpha \simeq \mathbf{c}(0)t \qquad |t| < \tau_0 \qquad (11\text{-}99)$$

Inserting into (11-96), we obtain

$$E\{\mathbf{w}(\tau)\} \simeq E\{e^{j\lambda\tau\mathbf{c}(0)}\} = \Phi_c(\lambda\tau) \qquad |\tau| < \tau_0 \qquad (11\text{-}100)$$

†P. M. Woodward: "The Spectrum of Random Frequency Modulation," *Telecommunications Research*, Great Malvern, Worcs., England, Memo 666, 1952.

where

$$\Phi_c(\mu) = E\{e^{j\mu\mathbf{c}(t)}\}$$

is the characteristic function of $\mathbf{c}(t)$. From this and (11-95) it follows that

$$R_{zz}(\tau) \simeq \Phi_c(\lambda\tau)e^{j\omega_0 t} \qquad |\tau| < \tau_0 \tag{11-101}$$

If λ is sufficiently large, then $\Phi_c(\lambda\tau) \simeq 0$ for $|\tau| > \tau_0$ because $\Phi_c(\mu) \to 0$ as $\mu \to \infty$. Hence (11-101) is a satisfactory approximation for every τ in the region where $\Phi_c(\lambda\tau)$ takes significant values. Transforming both sides of (11-101) and using the inversion formula

$$f_c(c) = \frac{1}{2\pi}\int_{-\infty}^{\infty}\Phi_c(\mu)e^{-j\mu c}\,d\mu$$

we obtain

$$S_{zz}(\omega) = \int_{-\infty}^{\infty}\Phi_c(\lambda\tau)e^{j\omega_0\tau}e^{-j\omega\tau}\,d\tau = \frac{2\pi}{\lambda}f_c\left(\frac{\omega-\omega_0}{\lambda}\right)$$

and (11-98) follows from (11-78).

NORMAL PROCESSES. Suppose now that $\mathbf{c}(t)$ is normal with zero mean. In this case $\boldsymbol{\varphi}(t)$ is also normal with zero mean. Hence [see (11-97)]

$$\begin{gathered}\Phi_\varphi(\lambda,\tau) = \exp\left\{-\tfrac{1}{2}\lambda^2\sigma_\varphi^2(\tau)\right\} \\ \sigma_\varphi^2(\tau) = 2\int_0^\tau R_c(\alpha)(\tau-\alpha)\,d\alpha\end{gathered} \tag{11-102}$$

In general, the Fourier transform of $\Phi_\varphi(\lambda,\tau)$ is found only numerically. However, as we show next, explicit formulas can be obtained if λ is large or small. We introduce the "correlation time" τ_c of $\mathbf{c}(t)$:

$$\tau_c = \frac{1}{\rho}\int_0^{\infty}R_c(\alpha)\,d\alpha \qquad \rho = R_c(0) \tag{11-103}$$

and we select two constants τ_0 and τ_1 such that

$$R_c(\tau) \simeq \begin{cases} 0 & |\tau| > \tau_1 \\ \rho & |\tau| < \tau_0 \end{cases}$$

Inserting into (11-102), we obtain (Fig. 11-15)

$$\sigma_\varphi^2(\tau) \simeq \left\{\begin{matrix} \rho\tau^2 & |\tau| < \tau_0 & e^{-\rho\lambda^2\tau^2/2} \\ 2\rho\tau\tau_c & \tau > \tau_1 & e^{-\rho\lambda^2\tau\tau_c} \end{matrix}\right\} \simeq R_{ww}(\tau) \tag{11-104}$$

It is known from the asymptotic properties of Fourier transforms that the behavior of $R_{ww}(\tau)$ for small (large) τ determines the behaviors of $S_{ww}(\omega)$ for

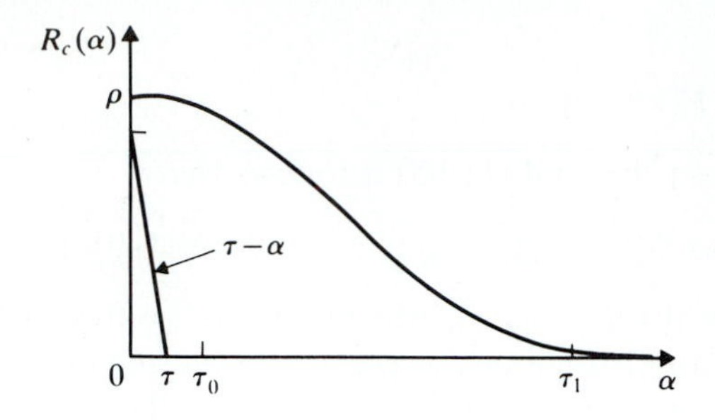

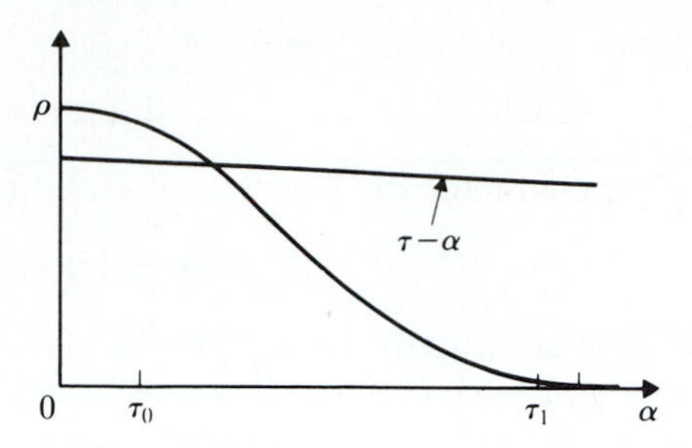

FIGURE 11-15

large (small) ω. Since

$$e^{-\rho\lambda^2\tau^2/2} \leftrightarrow \frac{1}{\lambda}\sqrt{\frac{2\pi}{\rho}}\,e^{-\omega^2/2\rho\lambda^2}$$

$$e^{-\rho\lambda^2\tau_c|\tau|} \leftrightarrow \frac{2\rho\tau_c\lambda^2}{\omega^2+\rho^2\tau_c^2\lambda^4} \tag{11-105}$$

we conclude that $S_{ww}(\omega)$ is *lorenzian* near the origin and it is asymptotically *normal* as $\omega \to \infty$. As we show next, these limiting cases give an adequate description of $S_{ww}(\omega)$ for large or small λ.

Wideband FM. If λ is such that

$$\rho\lambda^2\tau_0^2 \gg 1$$

then $R_{ww}(\tau) \simeq 0$ for $|\tau| > \tau_0$. This shows that we can use the upper approximation in (11-104) for every significant value of τ. The resulting spectrum equals

$$S_{ww}(\omega) \simeq \frac{1}{\lambda}\sqrt{\frac{2\pi}{\rho}}\,e^{-\omega^2/2\rho\lambda^2} = \frac{2\pi}{\lambda}f_c\!\left(\frac{\omega}{\lambda}\right) \tag{11-106}$$

in agreement with Woodward's theorem. The last equality in (11-106) follows because $\mathbf{c}(t)$ is normal with variance $E\{\mathbf{c}^2(t)\} = \rho$.

Narrowband FM. If λ is such that

$$\rho\lambda^2\tau_1\tau_c \ll 1$$

then $R_{ww}(\tau) \simeq 1$ for $|\tau| < \tau_1$. This shows that we can use the lower approximation in (11-104) for every significant value of τ. Hence

$$S_{ww}(\omega) \simeq \frac{2\rho\tau_c\lambda^2}{\omega^2+\rho^2\tau_c^2\lambda^4} \tag{11-107}$$

11-4 CYCLOSTATIONARY PROCESSES†

A process $\mathbf{x}(t)$ is called strict-sense cyclostationary (SSCS) with period T if its statistical properties are invariant to a shift of the origin by integer multiples of T, or, equivalently, if

$$F(x_1,\ldots,x_n;t_1+mT,\ldots,t_n+mT)=F(x_1,\ldots,x_n;t_1,\ldots,t_n) \quad (11\text{-}108)$$

for every integer m.

A process $\mathbf{x}(t)$ is called wide-sense cyclostationary (WSCS) if

$$\eta(t+mT)=\eta(t) \qquad R(t_1+mT,t_2+mT)=R(t_1,t_2) \quad (11\text{-}109)$$

for every integer m.

It follows from the definition that if $\mathbf{x}(t)$ is SSCS, it is also WSCS. The following theorems show the close connection between stationary and cyclostationary processes.

THEOREM 1. If $\mathbf{x}(t)$ is an SSCS process and $\boldsymbol{\theta}$ is an RV uniform in the interval $(0,T)$ and independent of $\mathbf{x}(t)$, then the process

$$\bar{\mathbf{x}}(t)=\mathbf{x}(t-\boldsymbol{\theta}) \quad (11\text{-}110)$$

obtained by a random shift of the origin is SSS and its nth-order distribution equals

$$\bar{F}(x_1,\ldots,x_n;t_1,\ldots,t_n)=\frac{1}{T}\int_0^T F(x_1,\ldots,x_n;t_1-\alpha,\ldots,t_n-\alpha)\,d\alpha \quad (11\text{-}111)$$

Proof. To prove the theorem, it suffices to show that the probability of the event

$$\mathscr{A}=\{\bar{\mathbf{x}}(t_1+c)\le x_1,\ldots,\bar{\mathbf{x}}(t_n+c)\le x_n\}$$

is independent of c and it equals the right side of (11-111). As we know [see 4-54)]

$$P(\mathscr{A})=\frac{1}{T}\int_0^T P(\mathscr{A}|\boldsymbol{\theta}=\theta)\,d\theta \quad (11\text{-}112)$$

Furthermore,

$$P(\mathscr{A}|\boldsymbol{\theta}=\theta)=P\{\mathbf{x}(t_1+c-\theta)\le x_1,\ldots,\mathbf{x}(t_n+c-\theta)\le x_n|\theta\}$$

And since $\boldsymbol{\theta}$ is independent of $\mathbf{x}(t)$, we conclude that

$$P\{\mathscr{A}|\boldsymbol{\theta}=\theta\}=F(x_1,\ldots,x_n;t_1+c-\theta,\ldots,t_n+c-\theta)$$

Inserting into (11-112) and using (11-108), we obtain (11-111).

†N. A. Gardner and L. E. Franks: Characteristics of Cyclostationary Random Signal Processes, *IEEE Transactions in Information Theory*, vol. IT-21, 1975.

THEOREM 2. If $\mathbf{x}(t)$ is a WSCS process, then the shifted process $\bar{\mathbf{x}}(t)$ is WSS with mean

$$\bar{\eta} = \frac{1}{T}\int_0^T \eta(t)\,dt \tag{11-113}$$

and autocorrelation

$$\bar{R}(\tau) = \frac{1}{T}\int_0^T R(t+\tau, t)\,dt \tag{11-114}$$

Proof. From (7-66) and the independence of $\boldsymbol{\theta}$ from $\mathbf{x}(t)$, it follows that

$$E\{\mathbf{x}(t-\boldsymbol{\theta})\} = E\{\eta(t-\boldsymbol{\theta})\} = \frac{1}{T}\int_0^T \eta(t-\theta)\,d\theta$$

and (11-113) results because $\eta(t)$ is periodic. Similarly,

$$\begin{aligned} E\{\mathbf{x}(t+\tau-\boldsymbol{\theta})\mathbf{x}(t-\boldsymbol{\theta})\} &= E\{R(t+\tau-\boldsymbol{\theta}, t-\boldsymbol{\theta})\} \\ &= \frac{1}{T}\int_0^T R(t+\tau-\theta, t-\theta)\,d\theta \end{aligned}$$

This yields (11-114) because $R(t+\tau, t)$ is a periodic function of t.

Pulse-Amplitude Modulation (PAM)

An important example of a cyclostationary process is the random signal

$$\mathbf{x}(t) = \sum_{n=-\infty}^{\infty} \mathbf{c}_n h(t-nT) \tag{11-115}$$

where $h(t)$ is a given function with Fourier transform $H(\omega)$ and $\mathbf{c}_n$ is a stationary sequence of RVs with autocorrelation $R_c[m] = E\{\mathbf{c}_{n+m}\mathbf{c}_n\}$ and power spectrum

$$\mathsf{S}_c(e^{j\omega}) = \sum_{m=-\infty}^{\infty} R_c[m]e^{-jm\omega} \tag{11-116}$$

THEOREM. The power spectrum $\bar{S}_x(\omega)$ of the shifted process $\bar{\mathbf{x}}(t)$ equals

$$\bar{S}_x(\omega) = \frac{1}{T}\mathsf{S}_c(e^{j\omega})|H(\omega)|^2 \tag{11-117}$$

Proof. We form the impulse train

$$\mathbf{z}(t) = \sum_{n=-\infty}^{\infty} \mathbf{c}_n \delta(t-nT) \tag{11-118}$$

Clearly, $\mathbf{z}(t)$ is the derivative of the process $\mathbf{w}(t)$ of Fig. 11-16:

$$\mathbf{w}(t) = \sum_{n=-\infty}^{\infty} \mathbf{c}_n U(t-nT) \qquad \mathbf{z}(t) = \mathbf{w}'(t) \tag{11-119}$$

Pulse-amplitude modulation

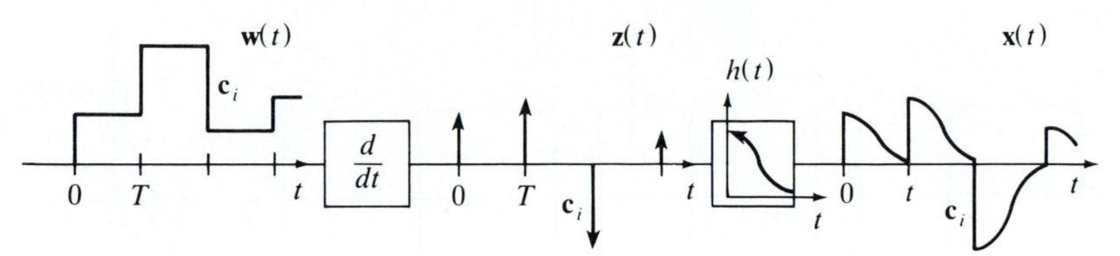

FIGURE 11-16

The process $\mathbf{w}(t)$ is cyclostationary with autocorrelation

$$R_w(t_1, t_2) = \sum_n \sum_r R_c(n - r)U(t_1 - nT)U(t_2 - rT)$$

From (10-94) it follows that

$$R_z(t_1, t_2) = \frac{\partial^2 R_w(t_1, t_2)}{\partial t_1 \partial t_2} = \sum_n \sum_r R_c[n - r]\delta(t_1 - nT)\delta(t_2 - rT)$$

This yields

$$R_z(t + \tau, t) = \sum_{m=-\infty}^{\infty} R_c[m] \sum_{r=-\infty}^{\infty} \delta[t + \tau - (m + r)T)\delta(t - rT] \quad (11\text{-}120)$$

We shall find first the autocorrelation $\bar{R}_z(\tau)$ and the power spectrum $\bar{S}_z(\omega)$ of the shifted process $\bar{\mathbf{z}}(t) = \mathbf{z}(t - \boldsymbol{\theta})$. Inserting (11-120) into (11-114) and using the identity

$$\int_0^T \delta[t + \tau - (m + r)T]\delta(t - rT)\, dt = \delta(\tau - mT)$$

we obtain

$$\bar{R}_z(\tau) = \frac{1}{T} \sum_{m=-\infty}^{\infty} R_c[m]\delta(\tau - mT) \quad (11\text{-}121)$$

From this it follows that

$$\bar{S}_z(\omega) = \frac{1}{T} \sum_{n=-\infty}^{\infty} R_c[m]e^{-jmT\omega} = \frac{1}{T} S_c(\omega) \quad (11\text{-}122)$$

The process $\mathbf{x}(t)$ is the output of a linear system with input $\mathbf{z}(t)$. Thus

$$\mathbf{x}(t) = \mathbf{z}(t) * h(t) \qquad \bar{\mathbf{x}}(t) = \bar{\mathbf{z}}(t) * h(t)$$

Hence [see (11-122) and (10-136)] the power spectrum of the shifted PAM process $\bar{\mathbf{x}}(t)$ is given by (11-117).

COROLLARY. If the process $\mathbf{c}_n$ is white noise with $S_c(\omega) = q$, then

$$\bar{S}_x(\omega) = \frac{q}{T}|H(\omega)|^2 \qquad \bar{R}_x(\tau) = \frac{q}{T} h(t) * h(-t) \quad (11\text{-}123)$$

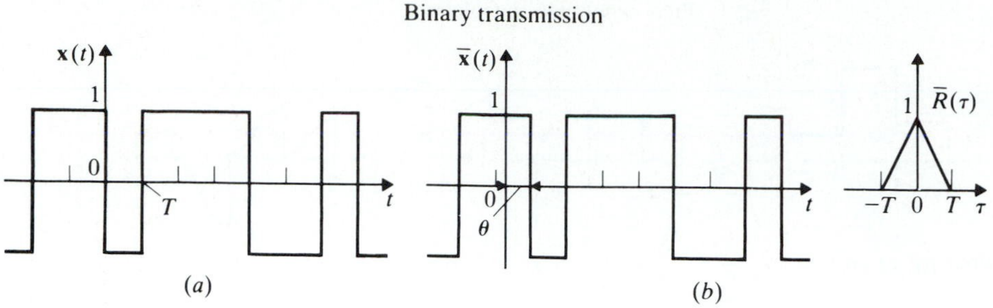

Binary transmission

(a) (b)

FIGURE 11-17

Example 11-4. Suppose that $h(t)$ is a pulse and $\mathbf{c}_n$ is a white-noise process taking the values ± 1 with equal probability:

$$h(t) = \begin{cases} 1 & 0 \le t < T \\ 0 & \text{otherwise} \end{cases} \qquad \mathbf{c}_n = \mathbf{x}(nT) \qquad R_c[m] = \delta[m]$$

The resulting process $\mathbf{x}(t)$ is called *binary transmission*. It is SSCS taking the values ± 1 in every interval $(nT - T, nT)$, the shifted process $\bar{\mathbf{x}}(t) = \mathbf{x}(t - \boldsymbol{\theta})$ is stationary. From (11-117) it follows that

$$\bar{S}_x(\omega) = \frac{4\sin^2(\omega T/2)}{T\omega^2}$$

because $\mathsf{S}_c(z) = 1$. Hence $\bar{R}_x(\tau)$ is a triangle as in Fig. 11-17.

11-5 BANDLIMITED PROCESSES AND SAMPLING THEORY

A process $\mathbf{x}(t)$ is called bandlimited (abbreviated BL) if it has finite power and its spectrum vanishes for $|\omega| > \sigma$:

$$R(0) < \infty \qquad S(\omega) = 0 \qquad |\omega| > \sigma \tag{11-124}$$

In this section we establish various identities involving linear functionals of BL processes. To do so, we express the two sides of each identity as responses of linear systems. The underlying reasoning is based on the following:

THEOREM. Suppose that $\mathbf{w}_1(t)$ and $\mathbf{w}_2(t)$ are the responses of the systems $T_1(\omega)$ and $T_2(\omega)$ to a BL process $\mathbf{x}(t)$ (Fig. 11-18). We shall show that if

$$T_1(\omega) = T_2(\omega) \qquad \text{for} \quad |\omega| \le \sigma \tag{11-125}$$

then

$$\mathbf{w}_1(t) = \mathbf{w}_2(t) \tag{11-126}$$

Proof. The difference $\mathbf{w}_1(t) - \mathbf{w}_2(t)$ is the response of the system $T_1(\omega) - T_2(\omega)$ to the input $\mathbf{x}(t)$. Since $S(\omega) = 0$ for $|\omega| > \sigma$, we conclude from (10-139) and

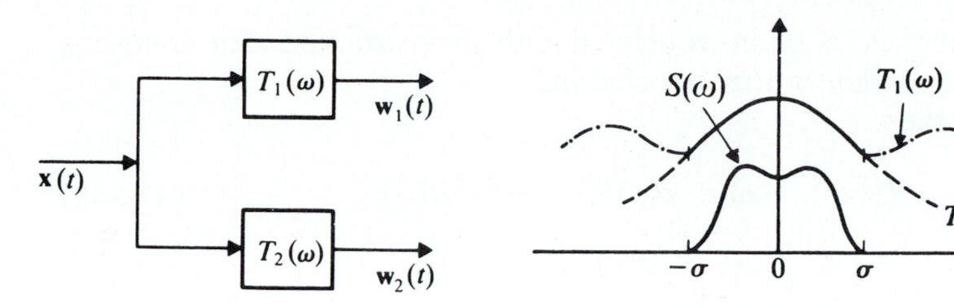

FIGURE 11-18

(11-125) that

$$E\{|\mathbf{w}_1(t) - \mathbf{w}_2(t)|^2\} = \frac{1}{2\pi}\int_{-\sigma}^{\sigma} S(\omega)|T_1(\omega) - T_2(\omega)|^2 d\omega = 0$$

Hence† $\mathbf{w}_1(t) = \mathbf{w}_2(t)$.

Taylor series. If $\mathbf{x}(t)$ is BL, then [see (10-121)]

$$R(\tau) = \frac{1}{2\pi}\int_{-\sigma}^{\sigma} S(\omega)e^{j\omega\tau}\,d\omega \tag{11-127}$$

In the above, the limits of integration are finite and the area $2\pi R(0)$ of $S(\omega)$ is also finite. We can therefore differentiate under the integral sign

$$R^{(n)}(\tau) = \frac{1}{2\pi}\int_{-\sigma}^{\sigma} (j\omega)^n S(\omega)e^{j\omega\tau}\,d\omega \tag{11-128}$$

This shows that the autocorrelation of a BL process is an entire function; that is, it has derivatives of any order for every τ. From this it follows that $\mathbf{x}^{(n)}(t)$ exists for any n (see App. 10A).

We maintain that

$$\mathbf{x}(t+\tau) = \sum_{n=0}^{\infty} \mathbf{x}^{(n)}(t)\frac{\tau^n}{n!} \tag{11-129}$$

Proof. We shall prove (11-129) using (11-126). As we know

$$e^{j\omega\tau} = \sum_{n=0}^{\infty} (j\omega)^n \frac{\tau^n}{n!} \qquad \text{all} \quad \omega \tag{11-130}$$

The processes $\mathbf{x}(t+\tau)$ and $\mathbf{x}^{(n)}(t)$ are the responses of the systems $e^{j\omega\tau}$ and $(j\omega)^n$ respectively to the input $\mathbf{x}(t)$. If, therefore, we use as systems $T_1(\omega)$ and $T_2(\omega)$ in (11-125) the two sides of (11-130), the resulting responses will equal the two sides of (11-129). And since (11-130) is true for all ω, (11-129) follows from (11-126).

†All identities in this section are interpreted in the MS sense.

Bounds. Bandlimitedness is often associated with slow variation. The following is an analytical formulation of this association.

If $\mathbf{x}(t)$ is BL, then

$$E\{[\mathbf{x}(t+\tau)-\mathbf{x}(t)]^2\} \le \sigma^2\tau^2 R(0) \tag{11-131}$$

or, equivalently,

$$2[R(0)-R(\tau)] \le \sigma^2\tau^2 R(0) \tag{11-132}$$

Proof. The familiar inequality $|\sin\varphi| \le |\varphi|$ yields

$$1-\cos\omega\tau = 2\sin^2\frac{\omega\tau}{2} \le \frac{\omega^2\tau^2}{2}$$

Since $S(\omega) \ge 0$, it follows from the above and (10-122) that

$$R(0)-R(\tau) = \frac{1}{2\pi}\int_{-\sigma}^{\sigma} S(\omega)(1-\cos\omega\tau)\,d\omega$$

$$\le \frac{1}{2\pi}\int_{-\sigma}^{\sigma} S(\omega)\frac{\omega^2\tau^2}{2}d\omega \le \frac{\sigma^2\tau^2}{4\pi}\int_{-\sigma}^{\sigma} S(\omega)\,d\omega = \frac{\sigma^2\tau^2}{2}R(0)$$

as in (11-132).

Sampling Expansions

The sampling theorem for deterministic signals states that if $f(t) \leftrightarrow F(\omega)$ and $F(\omega)=0$ for $|\omega|>\sigma$, then the function $f(t)$ can be expressed in terms of its samples $f(nT)$ where $T=\pi/\sigma$ is the *Nyquist interval*. The resulting expansion applied to the autocorrelation $R(\tau)$ of a BL process $\mathbf{x}(t)$ takes the following form:

$$R(\tau) = \sum_{n=-\infty}^{\infty} R(nT)\frac{\sin\sigma(\tau-nT)}{\sigma(\tau-nT)} \tag{11-133}$$

We shall establish a similar expansion for the process $\mathbf{x}(t)$.

THEOREM. If $\mathbf{x}(t)$ is a BL process, then

$$\mathbf{x}(t+\tau) = \sum_{n=-\infty}^{\infty} \mathbf{x}(t+nT)\frac{\sin\sigma(\tau-nT)}{\sigma(\tau-nT)} \qquad T=\frac{\pi}{\sigma} \tag{11-134}$$

for every t and τ. This is a slight extension of (11-133). This extension will permit us to base the proof of the theorem on (11-126).

Proof. We consider the exponential $e^{j\omega\tau}$ as a function of ω, viewing τ as a parameter, and we expand it into a Fourier series in the interval $(-\sigma \le \omega \le \sigma)$.

The coefficients of this expansion equal

$$a_n = \frac{1}{2\sigma}\int_{-\sigma}^{\sigma} e^{j\omega\tau}e^{-jnT\omega}\,d\omega = \frac{\sin\sigma(\tau - nT)}{\sigma(\tau - nT)}$$

Hence

$$e^{j\omega\tau} = \sum_{n=-\infty}^{\infty} e^{jnT\omega}\frac{\sin\sigma(\tau - nT)}{\sigma(\tau - nT)} \qquad |\omega| \le \sigma \qquad (11\text{-}135)$$

We denote by $T_1(\omega)$ and $T_2(\omega)$ the left and right side respectively of (11-134). Clearly, $T_1(\omega)$ is a delay line and its response $\mathbf{w}_1(t)$ to $\mathbf{x}(t)$ equals $\mathbf{x}(t+\tau)$. Similarly, the response $\mathbf{w}_2(t)$ of $T_2(\omega)$ to $\mathbf{x}(t)$ equals the right side of (11-134). Since $T_1(\omega) = T_2(\omega)$ for $|\omega| < \sigma$, (11-134) follows from (11-126).

Past samples. A deterministic BL signal is determined only if all its samples, past and future, are known. This is not necessary for random signals. We show next that a BL process $\mathbf{x}(t)$ can be approximated arbitrarily closely by a sum involving only its past samples $\mathbf{x}(nT_0)$ provided that $T_0 < T$. We illustrate first with an example.†

Example 11-5. Consider the process

$$\hat{\mathbf{x}}(t) = n\mathbf{x}(t - T_0) - \binom{n}{2}\mathbf{x}(t - 2T_0) + \cdots - (-1)^n\mathbf{x}(t - nT_0) \qquad (11\text{-}136)$$

The difference

$$\mathbf{y}(t) = \mathbf{x}(t) - \hat{\mathbf{x}}(t) = \sum_{k=0}^{n}(-1)^k\binom{n}{k}\mathbf{x}(t - kT_0)$$

is the response of the system

$$H(\omega) = \sum_{k=0}^{n}(-1)^k\binom{n}{k}e^{-jkT_0\omega} = (1 - e^{-j\omega T_0})^n$$

with input $\mathbf{x}(t)$. Since $|H(\omega)| = |2\sin(\omega T_0/2)|^n$, we conclude from (10-36) that

$$E\{\mathbf{y}^2(t)\} = \frac{1}{2\pi}\int_{-\sigma}^{\sigma} S(\omega)\left(2\sin\frac{\omega T_0}{2}\right)^{2n} d\omega \qquad (11\text{-}137)$$

If $T_0 < \pi/3\sigma$, then $2\sin|\omega T_0/2| < 2\sin(\pi/6) = 1$ for $|\omega| < \sigma$. From this it follows that the integrand in (11-137) tends to 0 as $n \to \infty$. Therefore, $E\{\mathbf{y}^2(t)\} \to 0$ and

$$\hat{\mathbf{x}}(t) \to \mathbf{x}(t) \qquad \text{as} \quad n \to \infty$$

Note that this holds only if $T_0 < T/3$; furthermore, the coefficients $\binom{n}{k}$ of $\hat{\mathbf{x}}(t)$ tend to ∞ as $n \to \infty$.

†L. A. Wainstein and V. Zubakov: *Extraction of Signals in Noise*, Prentice-Hall, Englewood Cliffs, NJ, 1962.

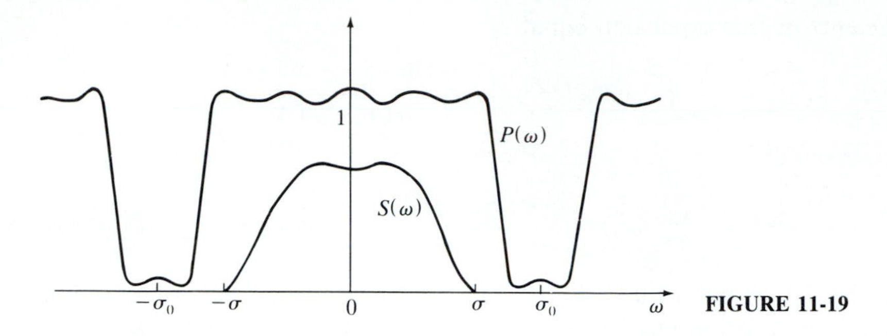

FIGURE 11-19

We show next that $\mathbf{x}(t)$ can be approximated arbitrarily closely by a sum involving only its past samples $\mathbf{x}(t - kT_0)$ where T_0 is a number smaller than T but otherwise arbitrary.

THEOREM. Given a number $T_0 < T$ and a constant $\varepsilon > 0$, we can find a set of coefficients a_k such that

$$E\{|\mathbf{x}(t) - \hat{\mathbf{x}}(t)|^2\} < \varepsilon \qquad \hat{\mathbf{x}}(t) = \sum_{k=1}^{n} a_k \mathbf{x}(t - kT_0) \tag{11-138}$$

where n is a sufficiently large constant.

Proof. The process $\hat{\mathbf{x}}(t)$ is the response of the system

$$P(\omega) = \sum_{k=1}^{n} a_k e^{-jkT_0\omega} \tag{11-139}$$

with input $\mathbf{x}(t)$. Hence

$$E\{|\mathbf{x}(t) - \hat{\mathbf{x}}(t)|^2\} = \frac{1}{2\pi}\int_{-\sigma}^{\sigma} S(\omega)|1 - P(\omega)|^2 d\omega$$

It suffices, therefore, to find a sum of exponentials with positive exponents only, approximating 1 arbitrarily closely. This cannot be done for every $|\omega| < \sigma_0 = \pi/T_0$ because $P(\omega)$ is periodic with period $2\sigma_0$. We can show, however, that if $\sigma_0 > \sigma$, we can find $P(\omega)$ such that the differences $|1 - P(\omega)|$ can be made arbitrarily small for $|\omega| < \sigma$ as in Fig. 11-19. The proof follows from the Weierstrass approximation theorem and the Fejer–Riesz factorization theorem; the details, however, are not simple.†

Note that, as in Example 11-5, the coefficients a_k tend to ∞ as $\varepsilon \to 0$. This is based on the fact that we cannot find a sum $P(\omega)$ of exponentials as in

†A. Papoulis: "A Note on the Predictability of Band-Limited Processes," *Proceedings of the IEEE*, vol. 13, no. 8, 1985.

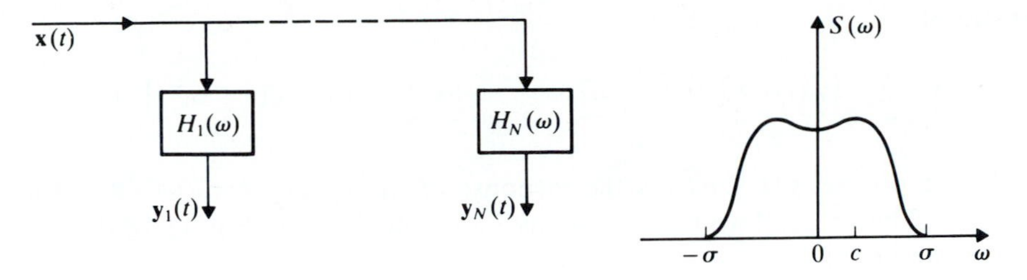

FIGURE 11-20

(11-139) such that $|1 - P(\omega)| = 0$ for every ω in an interval. This would violate the Paley–Wiener condition (12-9).

THE PAPOULIS SAMPLING EXPANSION.† The sampling expansion holds only if $T \le \pi/\sigma$. The following theorem states that if we have access to the samples of the outputs $\mathbf{y}_1(t), \ldots, \mathbf{y}_N(t)$ of N linear systems $H_1(\omega), \ldots, H_N(\omega)$ driven by $\mathbf{x}(t)$ (Fig. 11-20), then we can increase the sampling interval from π/σ to $N\pi/\sigma$.

We introduce the constants

$$c = \frac{2\sigma}{N} = \frac{2\pi}{\bar{T}} \qquad \bar{T} = NT \tag{11-140}$$

and the N functions

$$P_1(\omega, t), \ldots, P_N(\omega, t)$$

defined as the solutions of the system

$$\begin{aligned}
H_1(\omega)P_1(\omega,\tau) + \cdots + H_N(\omega)P_N(\omega,\tau) &= 1 \\
H_1(\omega + c)P_1(\omega,\tau) + \cdots + H_N(\omega + c)P_N(\omega,\tau) &= e^{jc\tau} \\
\cdots\cdots\cdots\cdots\cdots\cdots\cdots\cdots\cdots\cdots\cdots\cdots & \\
H_1(\omega + Nc - c)P_1(\omega,\tau) + \cdots + H_N(\omega + Nc - c)P_N(\omega,\tau) &= e^{j(N-1)c\tau}
\end{aligned} \tag{11-141}$$

In the above, ω takes all values in the interval $(-\sigma, -\sigma + c)$ and τ is arbitrary.

We next form the N functions

$$p_k(\tau) = \frac{1}{c}\int_{-\sigma}^{-\sigma+c} P_k(\omega,\tau)e^{j\omega\tau}\,d\omega \qquad 1 \le k \le N \tag{11-142}$$

---

†A. Papoulis: "New Results in Sampling Theory," *Hawaii Intern Conf. System Sciences*, January, 1968. (See also Papoulis, 1968, pp. 132–137).

THEOREM

$$\mathbf{x}(t+\tau) = \sum_{n=-\infty}^{\infty} \left[\mathbf{y}_1(t+n\overline{T})p_1(\tau - n\overline{T}) + \cdots + \mathbf{y}_N(t+n\overline{T})p_N(\tau - n\overline{T})\right] \quad (11\text{-}143)$$

Proof. The process $\mathbf{y}_i(t + n\overline{T})$ is the response of the system $H_i(\omega)e^{jn\overline{T}\omega}$ to the input $\mathbf{x}(t)$. Therefore, if we use as systems $T_1(\omega)$ and $T_2(\omega)$ in Fig. 11-18 the two sides of the identity

$$e^{j\omega\tau} = H_1(\omega) \sum_{n=-\infty}^{\infty} p_1(\tau - n\overline{T})e^{jn\omega\overline{T}} + \cdots + H_N(\omega) \sum_{n=-\infty}^{\infty} p_N(\tau - n\overline{T})e^{jn\omega\overline{T}} \quad (11\text{-}144)$$

the resulting responses will equal the two sides of (11-143). To prove (11-143), it suffices, therefore, to show that (11-144) is true for every $|\omega| \le \sigma$.

The coefficients $H_k(\omega + kc)$ of the system (11-141) are independent of τ and the right side consists of periodic functions of τ with period $\overline{T} = 2\pi/c$ because $e^{jkc\overline{T}} = 1$. Hence the solutions $P_k(\omega, \tau)$ are periodic

$$P_k(\omega, \tau - n\overline{T}) = P_k(\omega, \tau)$$

From the above and (11-142) it follows that

$$p_k(\tau - n\overline{T}) = \frac{1}{c}\int_{-\sigma}^{-\sigma+c} P_k(\omega, \tau)e^{j\omega(\tau - n\overline{T})}\,d\omega$$

This shows that if we expand the function $P_k(\omega, \tau)e^{j\omega\tau}$ into a Fourier series in the interval $(-\sigma, -\sigma + c)$, the coefficient of the expansion will equal $p_k(\tau - n\overline{T})$. Hence

$$P_k(\omega, \tau)e^{j\omega\tau} = \sum_{n=-\infty}^{\infty} p_k(\tau - n\overline{T})e^{jn\omega\overline{T}} \qquad -\sigma < \omega < -\sigma + c \quad (11\text{-}145)$$

Multiplying each of the equations in (11-141) by $e^{j\omega\tau}$ and using (11-145) and the identity

$$e^{jn(\omega+kc)\overline{T}} = e^{jn\omega\overline{T}}$$

we conclude that (11-144) is true for every ω in the interval $(-\sigma, \sigma)$.

Random Sampling

We wish to estimate the Fourier transform $F(\omega)$ of a deterministic signal $f(t)$ in terms of a sum involving the samples of $f(t)$. If we approximate the integral of $f(t)e^{-j\omega t}$ by its Riemann sum, we obtain the estimate

$$F(\omega) \simeq F_*(\omega) \equiv \sum_{n=-\infty}^{\infty} Tf(nT)e^{-jn\omega T} \quad (11\text{-}146)$$

From the Poisson sum formula (11A-1), it follows that $F_*(\omega)$ equals the sum of

$F(\omega)$ and its displacements

$$F_*(\omega) = \sum_{n=-\infty}^{\infty} F(\omega + 2n\sigma) \qquad \sigma = \frac{\pi}{T}$$

Hence $F_*(\omega)$ can be used as the estimate of $F(\omega)$ in the interval $(-\sigma, \sigma)$ only if $F(\omega)$ is negligible outside this interval. The difference $F(\omega) - F_*(\omega)$ is called *aliasing error*. In the following, we replace in (11-146) the equidistant samples $f(nT)$ of $f(t)$ by its samples $f(\mathbf{t}_i)$ at a random set of points $\mathbf{t}_i$ and we examine the nature of the resulting error.†

We maintain that if $\mathbf{t}_i$ is a Poisson point process with average density λ, then the sum

$$\mathbf{P}(\omega) = \frac{1}{\lambda} \sum_i f(\mathbf{t}_i) e^{-j\omega \mathbf{t}_i} \tag{11-147}$$

is an unbiased estimate of $F(\omega)$. Furthermore, if the energy

$$E = \int_{-\infty}^{\infty} f^2(t)\,dt$$

of $f(t)$ is finite, then $\mathbf{P}(\omega) \to F(\omega)$ as $\lambda \to \infty$. To prove the above, it suffices to show that

$$E\{\mathbf{P}(\omega)\} = F(\omega) \qquad \sigma^2_{P(\omega)} = \frac{E}{\lambda} \tag{11-148}$$

Proof. Clearly,

$$\int_{-\infty}^{\infty} f(t) e^{-j\omega t} \sum_i \delta(t - \mathbf{t}_i)\,dt = \sum_i f(\mathbf{t}_i) e^{-j\omega \mathbf{t}_i} \tag{11-149}$$

Comparing with (11-147), we obtain

$$\mathbf{P}(\omega) = \frac{1}{\lambda} \int_{-\infty}^{\infty} f(t)\mathbf{z}(t) e^{-j\omega t}\,dt \qquad \text{where} \quad \mathbf{z}(t) = \sum_i \delta(t - \mathbf{t}_i) \tag{11-150}$$

is a Poisson impulse train as in (10-98) with

$$E\{\mathbf{z}(t)\} = \lambda \qquad C_z(t_1, t_2) = \lambda\delta(t_1 - t_2) \tag{11-151}$$

†E. Masry: "Poisson Sampling and Spectral Estimation of Continuous-Time Processes," *IEEE Transactions on Information Theory*, vol. IT-24, 1978. See also F. J. Beutler: "Alias Free Randomly Timed Sampling of Stochastic Processes," *IEEE Transactions on Information Theory*, vol. IT-16, 1970.

Hence

$$E\{\mathbf{P}(\omega)\} = \frac{1}{\lambda}\int_{-\infty}^{\infty} f(t)E\{\mathbf{z}(t)\}e^{-j\omega t}\,dt = F(\omega)$$

$$\sigma_{P(\omega)}^2 = \frac{1}{\lambda^2}\int_{-\infty}^{\infty}\int_{-\infty}^{\infty} f(t_1)f(t_2)\lambda\delta(t_1 - t_2)\,dt_1\,dt_2 = \frac{1}{\lambda}\int_{-\infty}^{\infty} f^2(t_2)\,dt_2$$

and (11-148) results.

From (11-148) it follows that, for a satisfactory estimate of $F(\omega)$, λ must be such that

$$|F(\omega)| \gg \sqrt{\frac{E}{\lambda}} \tag{11-152}$$

Example 11-6. Suppose that $f(t)$ is a sum of sine waves in the interval $(-a, a)$:

$$f(t) = \sum_k c_k e^{j\omega_k t} \qquad |t| < a$$

and it equals 0 for $|t| > a$. In this case,

$$F(\omega) = \sum_k 2c_k \frac{\sin a(\omega - \omega_k)}{\omega - \omega_k} \qquad E \simeq 2a\sum_k |c_k|^2 \tag{11-153}$$

where we neglected cross-products in the evaluation of E. If a is sufficiently large, then

$$F(\omega_k) \simeq 2ac_k$$

This shows that if

$$\sum_i |c_i|^2 \ll 2a\lambda|c_k|^2 \qquad \text{then} \quad \mathbf{P}(\omega_k) \simeq F(\omega_k)$$

Thus with random sampling we can detect line spectra of any frequency even if the average rate λ is small, provided that the observation interval $2a$ is large.

11-6 DETERMINISTIC SIGNALS IN NOISE

A central problem in the applications of stochastic processes is the estimation of a signal in the presence of noise. This problem has many aspects (see Chap. 14). In the following, we discuss two cases that lead to simple solutions. In both cases the signal is a deterministic function $f(t)$ and the noise is a random process $\boldsymbol{\nu}(t)$ with zero mean.

The Matched Filter Principle

The following problem is typical in radar: A signal of known form is reflected from a distant target. The received signal is a sum

$$\mathbf{x}(t) = f(t) + \boldsymbol{\nu}(t) \qquad E\{\boldsymbol{\nu}(t)\} = 0$$

where $f(t)$ is a shifted and scaled version of the transmitted signal and $\boldsymbol{\nu}(t)$ is a

WSS process with known power spectrum $S(\omega)$. We assume that $f(t)$ is known and we wish to establish its presence and location. To do so, we apply the process $\mathbf{x}(t)$ to a linear filter with impulse response $h(t)$ and system function $H(\omega)$. The resulting output $\mathbf{y}(t) = \mathbf{x}(t) * h(t)$ is a sum

$$\mathbf{y}(t) = \int_{-\infty}^{\infty} \mathbf{x}(t-\alpha)h(\alpha)d(\alpha) = y_f(t) + \mathbf{y}_\nu(t) \qquad (11\text{-}154)$$

where

$$y_f(t) = \int_{-\infty}^{\infty} f(t-\alpha)h(\alpha)\,d\alpha = \frac{1}{2\pi}\int_{-\infty}^{\infty} F(\omega)H(\omega)e^{j\omega t}\,d\omega \qquad (11\text{-}155)$$

is the response due to the signal $f(t)$, and $\mathbf{y}_\nu(t)$ is a random component with average power

$$E\{\mathbf{y}_\nu^2(t)\} = \frac{1}{2\pi}\int_{-\infty}^{\infty} S(\omega)|H(\omega)|^2\,d\omega \qquad (11\text{-}156)$$

Since $\mathbf{y}_\nu(t)$ is due to $\boldsymbol{\nu}(t)$ and $E\{\boldsymbol{\nu}(t)\} = 0$, we conclude that $E\{\mathbf{y}_\nu(t)\} = 0$ and $E\{\mathbf{y}(t)\} = y_f(t)$. Our objective is to find $H(\omega)$ so as to maximize the *signal-to-noise* ratio

$$r = \frac{|y_f(t_0)|}{\sqrt{E\{\mathbf{y}_\nu^2(t_0)\}}} \qquad (11\text{-}157)$$

at a specific time t_0.

White noise. Suppose, first, that $S(\omega) = S_0$. Applying Schwarz's inequality (11B-1) to the second integral in (11-155), we conclude that

$$r^2 \le \frac{\int |F(\omega)e^{j\omega t_0}|^2\,d\omega \int |H(\omega)|^2\,d\omega}{2\pi S_0 \int |H(\omega)|^2\,d\omega} = \frac{E_f}{S_0} \qquad (11\text{-}158)$$

where $E_f = (1/2\pi)\int |F(\omega)|^2\,d\omega$ is the energy of $f(t)$. The above is an equality if [see (11B-2)]

$$H(\omega) = kF^*(\omega)e^{-j\omega t_0} \qquad h(t) = kf(t_0 - t) \qquad (11\text{-}159)$$

This determines the optimum $H(\omega)$ within a constant factor k. The system so obtained is called the *matched filter*. The resulting signal-to-noise ratio is maximum and it equals $\sqrt{E_f/S_0}$.

Colored noise. The solution is not so simple if $S(\omega)$ is not a constant. In this case, we use a trick. We first multiply and divide the integrand of (11-155) by

$\sqrt{S(\omega)}$ and then apply Schwarz's inequality. This yields

$$|2\pi y_f(t_0)|^2 = \left| \int \frac{F(\omega)}{\sqrt{S(\omega)}} \sqrt{S(\omega)}\, H(\omega) e^{j\omega t_0}\, d\omega \right|^2$$

$$\leq \int \frac{|F(\omega)|^2}{S(\omega)} d\omega \int S(\omega)|H(\omega)|^2\, d\omega$$

Inserting into (11-157), we obtain

$$r^2 \leq \frac{\displaystyle\int \frac{|F(\omega)|^2}{S(\omega)} d\omega \int S(\omega)|H(\omega)|^2\, d\omega}{2\pi \displaystyle\int S(\omega)|H(\omega)|^2\, d\omega} = \frac{1}{2\pi} \int \frac{|F(\omega)|^2}{S(\omega)} d\omega$$

Equality holds if

$$\sqrt{S(\omega)}\, H(\omega) = k \frac{F^*(\omega) e^{-j\omega t_0}}{\sqrt{S(\omega)}}$$

Thus the signal-to-noise ratio is maximum if

$$H(\omega) = k \frac{F^*(\omega)}{S(\omega)} e^{-j\omega t_0} \tag{11-160}$$

Tapped delay line. The matched filter is in general noncausal and difficult to realize. A suboptimal but simpler solution results if $H(\omega)$ is a tapped delay line:

$$H(\omega) = a_0 + a_1 e^{-j\omega T} + \cdots + a_m e^{-jm\omega T} \tag{11-161}$$

In this case,

$$y_f(t_0) = \sum_{i=0}^{m} a_i f(t_0 - iT) \qquad \mathbf{y}_\nu(t) = \sum_{i=0}^{m} a_i \boldsymbol{\nu}(t - iT) \tag{11-162}$$

and our problem is to find the $m + 1$ constants a_i so as to maximize the resulting signal-to-noise ratio. It can be shown that (see Prob. 11-26) the unknown constants are the solutions of the system

$$\sum_{i=0}^{m} a_i R(nT - iT) = kf(t_0 - nT) \qquad n = 0, \ldots, m \tag{11-163}$$

where $R(\tau)$ is the autocorrelation of $\boldsymbol{\nu}(t)$ and k is an arbitrary constant.

Smoothing

We wish to estimate an unknown signal $f(t)$ in terms of the observed value of the sum $\mathbf{x}(t) = f(t) + \boldsymbol{\nu}(t)$. We assume that the noise $\boldsymbol{\nu}(t)$ is white with known autocorrelation $R(t_1, t_2) = q(t_1)\delta(t_1 - t_2)$. Our estimator is again the response

$\mathbf{y}(t)$ of the filter $h(t)$:

$$\mathbf{y}(t) = \int_{-\infty}^{\infty} \mathbf{x}(t - \tau)h(\tau)\,d\tau \tag{11-164}$$

The estimator is biased with bias

$$b = y_f(t) - f(t) = \int_{-\infty}^{\infty} f(t - \tau)h(\tau)\,d\tau - f(t) \tag{11-165}$$

and variance [see (10-90)]

$$\sigma^2 = E\{\mathbf{y}_\nu^2(t)\} = \int_{-\infty}^{\infty} q(t - \tau)h^2(\tau)\,d\tau \tag{11-166}$$

Our objective is to find $h(t)$ so as to minimize the MS error

$$e = E\{[\mathbf{y}(t) - f(t)]^2\} = b^2 + \sigma^2$$

We shall assume that $h(t)$ is an even positive function of unit area and finite duration:

$$h(-t) = h(t) \qquad \int_{-T}^{T} h(t)\,dt = 1 \qquad h(t) > 0 \tag{11-167}$$

where T is a constant to be determined. If T is small, $y_f(t) \simeq f(t)$, hence the bias is small; however, the variance is large. As T increases, the variance decreases but the bias increases. The determination of the optimum shape and duration of $h(t)$ is in general complicated. We shall develop a simple solution under the assumption that the functions $f(t)$ and $q(t)$ are smooth in the sense that $f(t)$ can be approximated by a parabola and $q(t)$ by a constant in any interval of length $2T$. From this assumption it follows that (Taylor expansion)

$$f(t - \tau) \simeq f(t) - \tau f'(t) + \frac{\tau^2}{2} f''(t) \qquad q(t - \tau) \simeq q(t) \tag{11-168}$$

for $|\tau| < T$. And since the interval of integration in (11-165) and (11-166) is $(-T, T)$, we conclude that

$$b \simeq \frac{f''(t)}{2} \int_{-T}^{T} \tau^2 h(\tau)\,d\tau \qquad \sigma^2 \simeq q(t) \int_{-T}^{T} h^2(\tau)\,d\tau \tag{11-169}$$

because the function $h(t)$ is even and its area equals 1. The resulting MS error equals

$$e \simeq \tfrac{1}{4} M^2 [f''(t)]^2 + Eq(t) \tag{11-170}$$

where $M = \int_{-T}^{T} t^2 h(t)\,dt$ and $E = \int_{-T}^{T} h^2(t)\,d\tau$.

To separate the effects of the shape and the size of $h(t)$ on the MS error, we introduce the normalized filter

$$w(t) = Th(Tt) \tag{11-171}$$

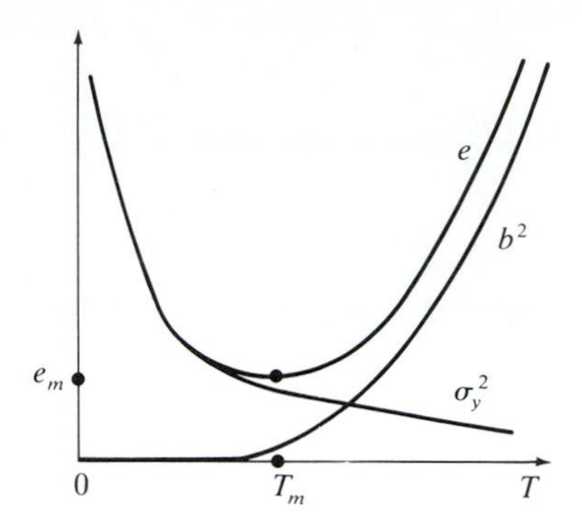

FIGURE 11-21

The function $w(t)$ is of unit area and $w(t) = 0$ for $|t| > 1$. With

$$M_w = \int_{-1}^{1} t^2 w(t)\, dt = \frac{M}{T^2} \qquad E_w = \int_{-1}^{1} w^2(t)\, dt = TE$$

it follows from (11-167) and (11-170) that

$$b \simeq \frac{T^2}{2} M_w f''(t) \qquad \sigma^2 = \frac{E_w}{T} q(t) \tag{11-172}$$

$$e = \frac{1}{4} T^2 M_w^2 [f''(t)]^2 + \frac{E_w}{T} q(t) \tag{11-173}$$

Thus e depends on the shape of $w(t)$ and on the constant T.

The two-to-one rule.† We assume first that $w(t)$ is specified. In Fig. 11-21 we plot the bias b, the variance σ^2, and the MS error e as functions of T. As T increases, b increases, and σ^2 decreases. Their sum e is minimum for

$$T = T_m = \left(\frac{E_w q(t)}{M_w^2 [f''(t)]^2} \right)^{1/5} \tag{11-174}$$

Inserting into (11-172), we conclude, omitting the simple algebra, that

$$\sigma = 2b \tag{11-175}$$

Thus if $w(t)$ is of specified shape and T is chosen so as to minimize the MS error e, then the standard deviation of the estimation error equals twice its bias.

Moving average. A simple estimator of $f(t)$ is the moving average

$$\mathbf{y}(t) = \frac{1}{2T} \int_{t-T}^{t+T} \mathbf{x}(\tau)\, d\tau$$

†A. Papoulis, Two-to-One Rule in Data Smoothing, *IEEE Trans. Inf. Theory*, September, 1977.

of $\mathbf{x}(t)$. This is a special case of (11-164) where the normalized filter $w(t)$ equals a pulse of width 2. In this case

$$M_w = \frac{1}{2}\int_{-1}^{1} t^2\,dt = \frac{1}{3} \qquad E_w = \frac{1}{4}\int_{-1}^{1} dt = \frac{1}{2}$$

Inserting into (11-174), we obtain

$$T_m = \sqrt[5]{\frac{9q(t)}{2[f''(t)]^2}} \qquad e = 5b^2 = \frac{5q(t)}{8T_m} \tag{11-176}$$

The parabolic window. We wish now to determine the shape of $w(t)$ so as to minimize the sum in (11-173). Since $h(t)$ needs to be determined within a scale factor, it suffices to assume that E_w has a constant value. Thus our problem is to find a positive even function $w(t)$ vanishing for $|t| > 1$ and such that its second moment M_w is minimum. It can be shown that (see page 388n)

$$w(t) = \begin{cases} 0.75(1 - t^2) & |t| < 1 \\ 0 & |t| > 1 \end{cases} \quad E_w = \tfrac{3}{5} \qquad M_w = \tfrac{1}{5} \tag{11-177}$$

Thus the optimum $w(t)$ is a truncated parabola. With $w(t)$ so determined, the optimum filter is

$$h(t) = \frac{1}{T_m} w\left(\frac{t}{T_m}\right)$$

where T_m is the constant in (11-174). This filter is, of course, time varying because the scaling factor T_m depends on t.

11-7 BISPECTRA AND SYSTEM IDENTIFICATION†

Correlations and spectra are the most extensively used concepts in the applications of stochastic processes. These concepts involve only second-order moments. In certain applications, moments of higher order are also used. In the following, we introduce the transform of the third-order moment

$$R_{xxx}(t_1, t_2, t_3) = E\{\mathbf{x}(t_1)\mathbf{x}(t_2)\mathbf{x}(t_3)\} \tag{11-178}$$

of a process $\mathbf{x}(t)$ and we apply it to the phase problem in system identification. We assume that $\mathbf{x}(t)$ is a real SSS process with zero mean. From the stationarity of $\mathbf{x}(t)$ it follows that the function $R_{xxx}(t_1, t_2, t_3)$ depends only on the differences

$$t_1 - t_3 = \mu \qquad t_2 - t_3 = \nu$$

†D. R. Brillinger: "An Introduction to Polyspectra," *Annals of Math Statistics*, vol. 36. Also C. L. Nikias and M. R. Raghuveer (1987): "Bispectrum Estimation; Digital Processing Framework," *IEEE Proceedings*, vol. 75, 1965.

Setting $t_3 = t$ in (11-178) and omitting subscripts, we obtain

$$R(t_1, t_2, t_3) = R(\mu, \nu) = E\{\mathbf{x}(t+\mu)\mathbf{x}(t+\nu)\mathbf{x}(t)\} \tag{11-179}$$

DEFINITION. The bispectrum $S(u, v)$ of the process $\mathbf{x}(t)$ is the two-dimensional Fourier transform of its third-order moment $R(\mu, \nu)$:

$$S(u, v) = \iint_{-\infty}^{\infty} R(\mu, \nu) e^{-j(u\mu + v\nu)}\, d\mu\, d\nu \tag{11-180}$$

The function $R(\mu, \nu)$ is real; hence

$$S(-u, -v) = S^*(u, v) \tag{11-181}$$

If $\mathbf{x}(t)$ is white noise then

$$R(\mu, \nu) = Q\delta(\mu)\delta(\nu) \qquad S(u, v) = Q \tag{11-182}$$

Notes 1. The third-order moment of a normal process with zero mean is identically zero. This is a consequence of the fact that the joint density of three jointly normal RVs with zero mean is symmetrical with respect to the origin.

2. The autocorrelation of a white noise process with third-order moment as in (11-182) is an impulse $q\delta(\tau)$; in general, however, $q \neq Q$. For example if $\mathbf{x}(t)$ is normal white noise, then $Q = 0$ but $q \neq 0$. Furthermore, whereas $q > 0$ for all nontrivial processes, Q might be negative.

Symmetries. The function $R(t_1, t_2, t_3)$ is invariant to the six permutations of the numbers t_1, t_2, and t_3. For stationary processes,

$$t_1 - t_3 = \mu \qquad t_2 - t_3 = \nu \qquad t_1 - t_2 = \mu - \nu$$

1	t_1, t_2, t_3	μ, ν	u, v	4	t_3, t_2, t_1	$-\mu, -\mu+\nu$	$-u-v, v$
2	t_2, t_1, t_3	ν, μ	v, u	5	t_2, t_3, t_1	$-\mu+\nu, -\mu$	$v, -u-v$
3	t_3, t_1, t_2	$-\nu, \mu-\nu$	$-u-v, u$	6	t_1, t_3, t_2	$\mu-\nu, -\nu$	$u, -u-v$

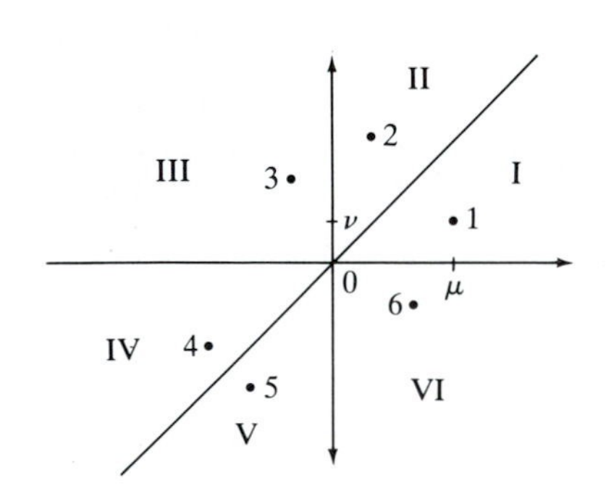

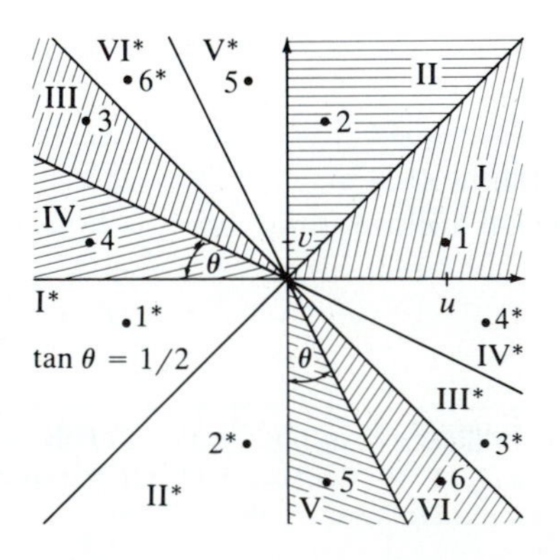

FIGURE 11-22

This yields the identities

$$R(\mu,\nu) = R(\nu,\mu) = R(-\nu,\mu-\nu) = R(-\mu,-\mu+\nu)$$
$$= R(-\mu+\nu,-\mu) = R(\mu-\nu,-\nu) \qquad (11\text{-}183)$$

Hence if we know the function $R(\mu,\nu)$ in any one of the six regions of Fig. 11-22, we can determine it everywhere.

From (11-180) and (11-183) it follows that

$$S(u,v) = S(v,u) = S(-u-v,u) = S(-u-v,v)$$
$$= S(v,-u-v) = S(u,-u-v) \qquad (11\text{-}184)$$

Combining with (11-181), we conclude that if we know $S(u,v)$ in any one of the 12 regions of Fig. 11-22, we can determine it everywhere.

Linear Systems

We have shown in (10-110) that if $\mathbf{x}(t)$ is the input to a linear system, the third-order moment of the resulting output $\mathbf{y}(t)$ equals

$$R_{yyy}(t_1,t_2,t_3)$$
$$= \iiint_{-\infty}^{\infty} R_{xxx}(t_1-\alpha,t_2-\beta,t_3-\gamma)h(\alpha)h(\beta)h(\gamma)\,d\alpha\,d\beta\,d\gamma \qquad (11\text{-}185)$$

For stationary processes, $R_{xxx}(t_1-\alpha,t_2-\beta,t_3-\gamma) = R_{xxx}(\mu+\gamma-\alpha,\nu+\gamma-\beta)$; hence

$$R_{yyy}(\mu,\nu) = \iiint_{-\infty}^{\infty} R_{xxx}(\mu+\gamma-\alpha,\nu+\gamma-\beta)h(\alpha)h(\beta)h(\gamma)\,d\alpha\,d\beta\,d\gamma \qquad (11\text{-}186)$$

Using this relationship, we shall express the bispectrum $S_{yyy}(u,v)$ of $\mathbf{y}(t)$ in terms of the bispectrum $S_{xxx}(u,v)$ of $\mathbf{x}(t)$.

THEOREM

$$S_{yyy}(u,v) = S_{xxx}(u,v)H(u)H(v)H^*(u+v) \qquad (11\text{-}187)$$

Proof. Taking transformations of both sides of (11-185) and using the identity

$$\iint_{-\infty}^{\infty} R_{xxx}(\mu+\gamma-\alpha,\nu+\gamma-\beta)e^{-j(u\mu+v\nu)}\,d\mu\,d\nu$$
$$= S_{xxx}(u,v)e^{j[u(\gamma-\alpha)+v(\gamma-\beta)]}$$

we obtain

$$S_{yyy}(u,v) = S_{xxx}(u,v)\iiint_{-\infty}^{\infty} e^{j[u(\gamma-\alpha)+v(\gamma-\beta)]}h(\alpha)h(\beta)h(\gamma)\,d\alpha\,d\beta\,d\gamma$$

Expressing the above integral as a product of three one-dimensional integrals, we obtain (11-187).

Example 11-7. Using (11-185), we shall determine the bispectrum of the shot noise

$$\mathbf{s}(t) = \sum_i h(t-\mathbf{t}_i) = \mathbf{z}(t)*h(t) \qquad \mathbf{z}(t) = \sum_i \delta(t-\mathbf{t}_i)$$

where $\mathbf{t}_i$ is a Poisson point process with average density λ.

To do so, we form the centered impulse train $\tilde{\mathbf{z}}(t) = \mathbf{z}(t) - \lambda$ and the centered shot noise $\tilde{\mathbf{s}}(t) = \tilde{\mathbf{z}}(t) * h(t)$. As we know (see Prob. 11-28)

$$R_{\tilde{z}\tilde{z}\tilde{z}}(\mu, \nu) = \lambda\delta(\mu)\delta(\nu) \quad \text{hence} \quad S_{\tilde{z}\tilde{z}\tilde{z}}(u, v) = \lambda$$

From this it follows that

$$S_{\tilde{s}\tilde{s}\tilde{s}}(u, v) = \lambda H(u)H(v)H^*(u+v)$$

and since $S_{\tilde{s}\tilde{s}}(\omega) = \lambda|H(\omega)|^2$, we conclude from Prob. 11-27 with $c = E\{\mathbf{s}(t)\} = \lambda H(0)$ that

$$\begin{aligned} S_{sss}(u, v) = {} & \lambda H(u)H(v)H^*(u+v) \\ & + 2\pi\lambda^2 H(0)\left[|H(u)|^2\delta(v) + |H(v)|^2\delta(u) + |H(u)|^2\delta(u+v)\right] \\ & + 4\pi^2\lambda^4 H^3(0)\delta(u)\delta(v) \end{aligned}$$

System Identification

A linear system is specified terminally in terms of its system function

$$H(\omega) = A(\omega)e^{j\varphi(\omega)}$$

System identification is the problem of determining $H(\omega)$. This problem is central in system theory and it has been investigated extensively. In the following, we apply the notion of spectra and polyspectra in the determination of $A(\omega)$ and $\varphi(\omega)$.

Spectra. Suppose that the input to the system $H(\omega)$ is a WSS process $\mathbf{x}(t)$ with power spectrum $S_{xx}(\omega)$. As we know,

$$S_{xy}(\omega) = S_{xx}(\omega)H^*(\omega) \tag{11-188}$$

This relationship expresses $H(\omega)$ in terms of the spectra $S_{xx}(\omega)$ and $S_{xy}(\omega)$ or, equivalently, in terms of the second-order moments $R_{xx}(\tau)$ and $R_{xy}(\tau)$. The problem of estimating these functions is considered in Chap. 13. In a number of applications, we cannot estimate $R_{xy}(\tau)$ either because we do not have access to the input $\mathbf{x}(t)$ of the system or because we cannot form the product $\mathbf{x}(t+\tau)\mathbf{y}(t)$ in real time. In such cases, an alternative method is used based on the assumption that $\mathbf{x}(t)$ is white noise. With this assumption (10-136) yields

$$S_{yy}(\omega) = S_{xx}(\omega)|H(\omega)|^2 = qA^2(\omega) \tag{11-189}$$

This relationship determines the amplitude $A(\omega)$ of $H(\omega)$ in terms of $S_{yy}(\omega)$ within a constant factor. It involves, however, only the estimation of the power spectrum $S_{yy}(\omega)$ of the output of the system. If the system is minimum phase (see page 401), then $H(\omega)$ is completely determined from (11-189) because, then, $\varphi(\omega)$ can be expressed in terms of $A(\omega)$. In general, however, this is not the case. The phase of an arbitrary system cannot be determined in terms of second-order moment of its output. It can, however, be determined if the third-order moment of $\mathbf{y}(t)$ is known.

Phase determination. We assume that $\mathbf{x}(t)$ is an SSS white-noise process with $S_{xxx}(u, v) = Q$. Inserting into (11-187), we obtain

$$S_{yyy}(u, v) = QH(u)H(v)H^*(u+v) \tag{11-190}$$

The function $S_{yyy}(u,v)$ is, in general, complex:

$$S_{yyy}(u,v) = B(u,v)e^{j\theta(u,v)} \tag{11-191}$$

Inserting (11-191) into (11-190) and equating amplitudes and phases, we obtain

$$B(u,v) = QA(u)A(v)A(u+v) \tag{11-192}$$

$$\theta(u,v) = \varphi(u) + \varphi(v) - \varphi(u+v) \tag{11-193}$$

We shall use these equations to express $A(\omega)$ in terms of $B(u,v)$ and $\varphi(\omega)$ in terms of $\theta(u,v)$. Setting $v = 0$ in (11-192), we obtain

$$QA^2(\omega) = \frac{1}{A(0)}B(\omega,0) \qquad QA^3(0) = B(0,0) \tag{11-194}$$

Since Q is in general unknown, $A(\omega)$ can be determined only within a constant factor. The phase $\varphi(\omega)$ can be determined only within a linear term because if it satisfies (11-193), so does the sum $\varphi(\omega) + c\omega$ for any c. We can assume therefore that $\varphi'(0) = 0$. To find $\varphi(\omega)$, we differentiate (11-193) with respect to v and we set $v = 0$. This yields

$$\theta_v(u,0) = -\varphi'(u) \qquad \varphi(\omega) = -\int_0^{\omega}\theta_v(u,0)\,du \tag{11-195}$$

where $\theta_v(u,v) = \partial\theta(u,v)/\partial v$. The above is the solution of (11-193).

In a numerical evaluation of $\varphi(\omega)$, we proceed as follows: Clearly, $\theta(u,0) = \varphi(u) + \varphi(0) - \varphi(u) = \varphi(0) = 0$ for every u. From this it follows that

$$\theta_v(u,0) = \lim \frac{1}{\Delta}\theta(u,\Delta) \quad \text{as } \Delta \to 0$$

Hence $\theta_v(u,0) \simeq \theta(u,\Delta)/\Delta$ for sufficiently small Δ. Inserting into (11-195), we obtain the approximations

$$\varphi(\omega) \simeq -\frac{1}{\Delta}\int_0^{\omega}\theta(u,\Delta)\,du \qquad \varphi(n\Delta) \simeq -\sum_{k=1}^{n}\theta(k\Delta,\Delta) \tag{11-196}$$

This is the solution of the digital version

$$\theta(k\Delta, r\Delta) = \varphi(k\Delta) + \varphi(r\Delta) - \varphi(k\Delta + r\Delta) \tag{11-197}$$

of (11-193) where $(k\Delta, r\Delta)$ are points in the sector I of Fig. 11-22. As we see from (11-196) $\varphi(n\Delta)$ is determined in terms of the values of $\theta(k\Delta,\Delta)$ of $\theta(u,\Delta)$ on the horizontal line $v = \Delta$. Hence the system (11-197) is overdetermined. This is used to improve the estimate of $\varphi(\omega)$ if $\theta(u,v)$ is not known exactly but it is estimated in terms of a single sample of $\mathbf{y}(t)$.† The corresponding problem of spectral estimation is considered in Chap. 13.

†T. Matsuoka and T. J. Ulrych: "Phase Estimation Using the Bispectrum," *IEEE Proceedings*, vol. 72, 1984.

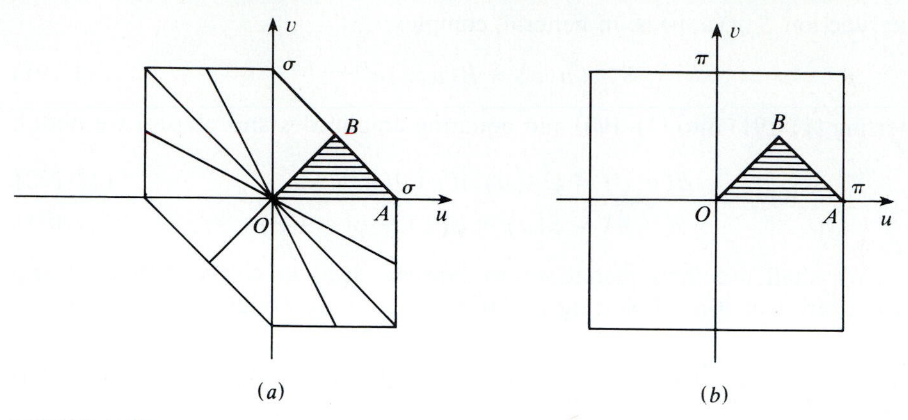

FIGURE 11-23

Note If the bispectrum $S(u, v)$ of a process $\mathbf{x}(t)$ equals the right side of (11-190) and $H(\omega) = 0$ for $|\omega| > \sigma$, then

$$S(u, v) = 0 \quad \text{for } |u| > \sigma \quad \text{or} \quad |v| > \sigma \quad \text{or} \quad |u + v| > \sigma$$

Thus, $S(u, v) = 0$ outside the hexagon of Fig. 11-23a. From this and the symmetries of Fig. 11-22 it follows that $S(u, v)$ is uniquely determined in terms of its values in the triangle OAB of Fig. 11-23a.

Digital processes. The preceding concepts can be readily extended to digital processes. We cite only the definition of bispectra.

Given an SSS digital process $\mathbf{x}[n]$, we form its third-order moment

$$R[k, r] = E\{\mathbf{x}[n + k]\mathbf{x}[n + r]\mathbf{x}[n]\} \tag{11-198}$$

The bispectrum of $\mathbf{x}[n]$ is the two-dimensional DFT of $R[k, r]$:

$$S(u, v) = \sum_{k=-\infty}^{\infty} \sum_{r=-\infty}^{\infty} R[k, r]e^{-j(uk+vr)} \tag{11-199}$$

This function is doubly periodic with period 2π:

$$S(u + 2\pi m, v + 2\pi n) = S(u, v) \tag{11-200}$$

It is therefore determined in terms of its values in the square $|u| \le \pi, |v| \le \pi$ of Fig. 11-23b. Furthermore, it has the 12 symmetries of Fig. 11-22.

Suppose now that $\mathbf{x}[n]$ is the output of a system $H(z)$ with input a white noise process. Proceeding as in (11-187) and (11-190) we conclude that its bispectrum equals

$$S(u, v) = QH(e^{ju})H(e^{jv})H(e^{-j(u+v)})$$

From the above it follows that, in this case, $S(u, v)$ is determined in terms of its values in the triangle OAB of Fig. 11-23b (see Prob. 11-29).

APPENDIX 10A THE POISSON SUM FORMULA

If

$$F(u) = \int_{-\infty}^{\infty} f(x) e^{-jux}\, dx$$

is the Fourier transform of $f(x)$ then for any c

$$\sum_{n=-\infty}^{\infty} f(x + nc) = \frac{1}{c} \sum_{n=-\infty}^{\infty} F(nu_0) e^{jnu_0 x} \qquad u_0 = \frac{2\pi}{c} \tag{11A-1}$$

Proof. Clearly

$$\sum_{n=-\infty}^{\infty} \delta(x + nc) = \frac{1}{c} \sum_{n=-\infty}^{\infty} e^{jnu_0 x} \tag{11A-2}$$

because the left side is periodic and its Fourier series coefficients equal

$$\frac{1}{c} \int_{-c/2}^{c/2} \delta(x) e^{-jnu_0 x}\, dx = \frac{1}{c}$$

Furthermore, $\delta(x + nc) * f(x) = f(x + nc)$ and

$$e^{jnu_0 x} * f(x) = \int_{-\infty}^{\infty} e^{jnu_0(x-\alpha)} f(\alpha)\, d\alpha = e^{jnu_0 x} F(nu_0)$$

Convolving both sides of (11A-2) with $f(x)$ and using the above, we obtain (11A-1).

APPENDIX 10B THE SCHWARZ INEQUALITY

We shall show that

$$\left| \int_a^b f(x) g(x)\, dx \right|^2 \le \int_a^b |f(x)|^2\, dx \int_a^b |g(x)|^2\, dx \tag{11B-1}$$

with equality iff

$$f(x) = kg^*(x) \tag{11B-2}$$

Proof. Clearly

$$\left|\int_a^b f(x)g(x)\,dx\right| \le \int_a^b |f(x)||g(x)|\,dx$$

Equality holds only if the product $f(x)g(x)$ is real. This is the case if the angles of $f(x)$ and $g(x)$ are opposite as in (11B-2). It suffices, therefore, to assume that the functions $f(x)$ and $g(x)$ are real. The quadratic

$$I(z) = \int_a^b [f(x) - zg(x)]^2\,dx$$
$$= z^2\int_a^b g^2(x)\,dx - 2z\int_a^b f(x)g(x)\,dx + \int_a^b f^2(x)\,dx$$

is nonnegative for every real z. Hence, its discriminant cannot be positive. This yields (11B-1). If the discriminant of $I(z)$ is zero, then $I(z)$ has a real (double) root $z = k$. This shows that $I(k) = 0$ and (11B-2) follows.

PROBLEMS

11-1. Find the first-order characteristic function (a) of a Poisson process, and (b) of a Wiener process.

Answer: (a) $e^{\lambda t(e^{j\omega}-1)}$; ($b$) $e^{-\alpha t\omega^2/2}$.

11-2. (*Two-dimensional random walk*). The coordinates $\mathbf{x}(t)$ and $\mathbf{y}(t)$ of a moving object are two independent random-walk processes with the same s and T as in Fig. 11-1a. Show that if $\mathbf{z}(t) = \sqrt{\mathbf{x}^2(t) + \mathbf{y}^2(t)}$ is the distance of the object from the origin and $t \gg T$, then for z of the order of $\sqrt{\alpha t}$:

$$f_z(z,t) \simeq \frac{z}{\alpha t}e^{-z^2/2\alpha t}U(z) \qquad \alpha = \frac{s^2}{T}$$

11-3. In the circuit of Fig. P11-3, $\mathbf{n}_e(t)$ is the voltage due to thermal noise. Show that

$$S_v(\omega) = \frac{2kTR}{(1-\omega^2LC)^2 + \omega^2R^2C^2} \qquad S_i(\omega) = \frac{2kTR}{R^2 + \omega^2L^2}$$

and verify Nyquist's theorems (11-27) and (11-30).

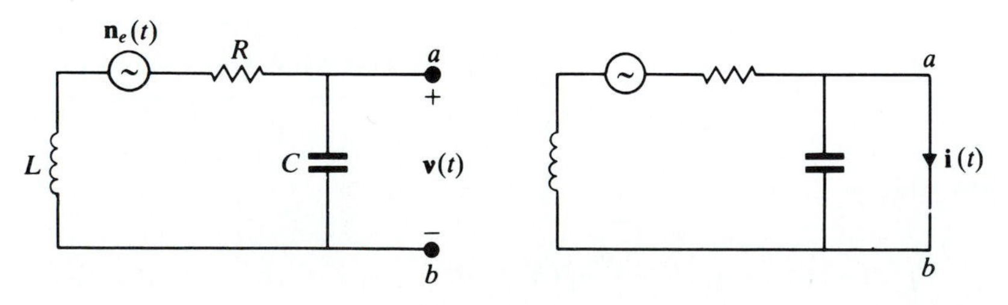

FIGURE P11-3

11-4. A particle in free motion satisfies the equation

$$m\mathbf{x}''(t) + f\mathbf{x}'(t) = \mathbf{F}(t) \qquad S_F(\omega) = 2kTf$$

Show that if $\mathbf{x}(0) = \mathbf{x}'(0) = 0$, then

$$E\{\mathbf{x}^2(t)\} = 2D^2\left(t - \frac{3}{4\alpha} + \frac{1}{\alpha}e^{-2\alpha t} - \frac{1}{4\alpha}e^{-4\alpha t}\right)$$

where $D^2 = kT/f$ and $\alpha = f/2m$.

Hint: Use (10-90) with

$$h(t) = \frac{1}{f}(1 - e^{-2\alpha t})U(t) \qquad q(t) = 2kTfU(t)$$

11-5. The position of a particle in underdamped harmonic motion is a normal process with autocorrelation as in (11-12). Show that its conditional density assuming $\mathbf{x}(0) = x_0$ and $\mathbf{x}'(0) = \mathbf{v}(0) = v_0$ equals

$$f_{x(t)}(x|x_0, v_0) = \frac{1}{\sqrt{2\pi P}}e^{-(x - ax_0 - bv_0)^2/2P}$$

Find the constants a, b, and P.

11-6. Given a Wiener process $\mathbf{w}(t)$ with parameter α, we form the processes

$$\mathbf{x}(t) = \mathbf{w}(t^2) \qquad \mathbf{y}(t) = \mathbf{w}^2(t) \qquad \mathbf{z}(t) = |\mathbf{w}(t)|$$

Show that $\mathbf{x}(t)$ is normal with zero mean. Furthermore, if $t_1 < t_2$, then

$$R_x(t_1, t_2) = \alpha t_1^2 \qquad R_y(t_1, t_2) = \alpha^2 t_1(2t_1 + t_2)$$

$$R_z(t_1, t_2) = \frac{2\alpha}{\pi}\sqrt{t_1 t_2}(\cos\theta + \theta\sin\theta) \qquad \sin\theta = \sqrt{\frac{t_1}{t_2}}$$

11-7. The process $\mathbf{s}(t)$ is shot noise with $\lambda = 3$ as in (11-45) where $h(t) = 2$ for $0 \le t \le 10$ and $h(t) = 0$ otherwise. Find $E\{\mathbf{s}(t)\}$, $E\{\mathbf{s}^2(t)\}$, and $P\{\mathbf{s}(7) = 0\}$.

11-8. The input to a real system $H(\omega)$ is a WSS process $\mathbf{x}(t)$ and the output equals $\mathbf{y}(t)$. Show that if

$$R_{xx}(\tau) = R_{yy}(\tau) \qquad R_{xy}(-\tau) = -R_{xy}(\tau)$$

as in (11-67), then $H(\omega) = jB(\omega)$ where $B(\omega)$ is a function taking only the values $+1$ and -1.

Special case: If $\mathbf{y}(t) = \check{\mathbf{x}}(t)$, then $B(\omega) = -\operatorname{sgn}\omega$.

11-9. Show that if $\check{\mathbf{x}}(t)$ is the Hilbert transform of $\mathbf{x}(t)$ and

$$\mathbf{i}(t) = \mathbf{x}(t)\cos\omega_0 t + \check{\mathbf{x}}(t)\sin\omega_0 t \qquad \mathbf{q}(t) = \check{\mathbf{x}}(t)\cos\omega_0 t - \mathbf{x}(t)\sin\omega_0 t$$

then (Fig. P11-9)

$$S_i(\omega) = S_q(\omega) = \frac{S_w(\omega) + S_w(-\omega)}{4} \qquad S_{qi}(\omega) = \frac{S_w(\omega) - S_w(-\omega)}{4j}$$

where $S_w(\omega) = 4S_x(\omega + \omega_0)U(\omega + \omega_0)$.

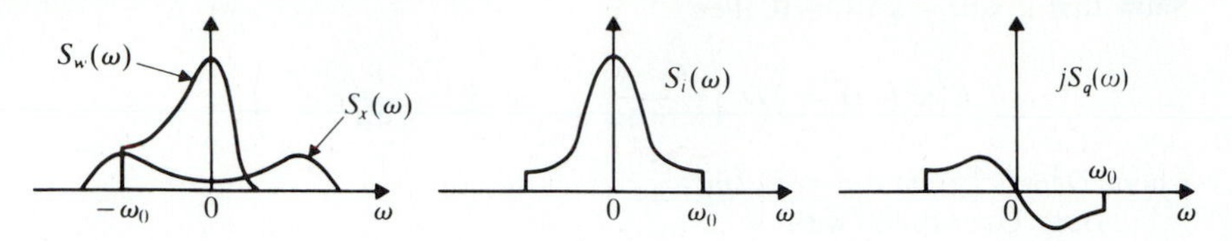

FIGURE P11-9

11-10. Show that if $\mathbf{w}(t)$ and $\mathbf{w}_\tau(t)$ are the complex envelopes of the processes $\mathbf{x}(t)$ and $\mathbf{x}(t-\tau)$ respectively, then $\mathbf{w}_\tau(t) = \mathbf{w}(t-\tau)e^{-j\omega_0\tau}$.

11-11. Show that if $\mathbf{w}(t)$ is the optimum complex envelope of $\mathbf{x}(t)$ [see (11-85)], then

$$E\{|\mathbf{w}'(t)|^2\} = -2\left[R_x''(0) + \omega_0^2 R_x(0)\right]$$

11-12. Show that if the process $\mathbf{x}(t)\cos\omega t + \mathbf{y}(t)\sin\omega t$ is normal and WSS, then its statistical properties are determined in terms of the variance of the process $\mathbf{z}(t) = \mathbf{x}(t) + j\mathbf{y}(t)$.

11-13. Show that if $\boldsymbol{\theta}$ is an RV uniform in the interval $(0, T)$ and $f(t)$ is a periodic function with period T, then the process $\mathbf{x}(t) = f(t - \boldsymbol{\theta})$ is stationary and

$$S_x(\omega) = \frac{1}{T}\left|\int_0^T f(t)e^{-j\omega t}\,dt\right|^2 \sum_{m=-\infty}^{\infty} \delta\left(\omega - \frac{2\pi}{T}m\right)$$

11-14. Show that if

$$\boldsymbol{\varepsilon}_N(t) = \mathbf{x}(t) - \sum_{n=-N}^{N} \mathbf{x}(nT)\frac{\sin\sigma(t-nT)}{\sigma(t-nT)} \qquad \sigma = \frac{\pi}{T}$$

then

$$E\{\boldsymbol{\varepsilon}_N^2(t)\} = \frac{1}{2\pi}\int_{-\infty}^{\infty} S(\omega)\left|e^{j\omega t} - \sum_{n=-N}^{N}\frac{\sin\sigma(t-nT)}{\sigma(t-nT)}e^{jn\omega T}\right|^2 d\omega$$

and if $S(\omega) = 0$ for $|\omega| > \sigma$, then $E\{\boldsymbol{\varepsilon}_N^2(t)\} \to 0$ as $N \to \infty$.

11-15. Show that if $\mathbf{x}(t)$ is BL as in (11-124), then† for $|\tau| < \pi/\sigma$:

$$\frac{2\tau^2}{\pi^2}|R''(0)| \le R(0) - R(\tau) \le \frac{\tau^2}{2}|R''(0)|$$

$$E\{[\mathbf{x}(t+\tau) - \mathbf{x}(t)]^2\} \ge \frac{4\tau^2}{\pi^2}E\{[\mathbf{x}'(t)]^2\}$$

Hint: If $0 < \varphi < \pi/2$ then $2\varphi/\pi < \sin\varphi < \varphi$.

11-16. A WSS process $\mathbf{x}(t)$ is BL as in (11-124) and its samples $\mathbf{x}(n\pi/\sigma)$ are uncorrelated. Find $S_x(\omega)$ if $E\{\mathbf{x}(t)\} = \eta$ and $E\{\mathbf{x}^2(t)\} = I$.

11-17. Find the power spectrum $S(\omega)$ of a process $\mathbf{x}(t)$ if $S(\omega) = 0$ for $|\omega| > \pi$ and

$$E\{\mathbf{x}(n+m)\mathbf{x}(n)\} = N\delta[m]$$

---

†A. Papoulis: "An Estimation of the Variation of a Bandlimited process," *IEEE, PGIT*, 1984.

11-18. Show that if $S(\omega) = 0$ for $|\omega| > \sigma$, then

$$R(\tau) \geq R(0)\cos \sigma\tau \qquad \text{for} \quad |\tau| < \pi/2\sigma$$

11-19. Show that if $\mathbf{x}(t)$ is BL as in (11-124) and $\Delta = 2\pi/\sigma$, then

$$\mathbf{x}(t) = 4\sin^2\frac{\sigma t}{2} \sum_{n=-\infty}^{\infty} \left[\frac{\mathbf{x}(n\Delta)}{(\sigma t - 2n\pi)^2} + \frac{\mathbf{x}'(n\Delta)}{\sigma(\sigma t - 2n\pi)} \right]$$

Hint: Use (11-143) with $N = 2$, $H_1(\omega) = 1$, $H_2(\omega) = j\omega$.

11-20. Find the mean and the variance of $\mathbf{P}(\omega_0)$ it $\mathbf{t}_i$ is a Poisson point process and

$$\mathbf{P}(\omega) = \frac{1}{\lambda} \sum_i \cos \omega_0 \mathbf{t}_i \cos \omega \mathbf{t}_i \qquad |\mathbf{t}_i| < a$$

11-21. Given a WSS process $\mathbf{x}(t)$ and a set of Poisson points $\mathbf{t}_i$ independent of $\mathbf{x}(t)$ and with average density λ, we form the sum

$$\mathbf{X}_c(\omega) = \sum_{|\mathbf{t}_i| < c} \mathbf{x}(\mathbf{t}_i) e^{-j\omega \mathbf{t}_i}$$

Show that if $E\{\mathbf{x}(t)\} = 0$ and $\int_{-\infty}^{\infty} |R_x(\tau)|\, d\tau < \infty$, then for large c,

$$E\{|\mathbf{X}_c(\omega)|^2\} = 2cS_x(\omega) + \frac{2c}{\lambda} R_x(0)$$

11-22. We are given the data $\mathbf{x}(t) = f(t) + \mathbf{n}(t)$ where $R_n(\tau) = N\delta(\tau)$ and $E\{\mathbf{n}(t) = 0\}$. We wish to estimate the integral

$$g(t) = \int_0^t f(\alpha)\, d\alpha$$

knowing that $g(T) = 0$. Show that if we use as the estimate of $g(t)$ the process $\mathbf{w}(t) = \mathbf{z}(t) - \mathbf{z}(T)t/T$ where

$$\mathbf{z}(t) = \int_0^t \mathbf{x}(\alpha)\, d\alpha \qquad \text{then} \quad E\{\mathbf{w}(t)\} = g(t) \qquad \sigma_w^2 = Nt\left(1 - \frac{t}{T}\right)$$

11-23. (*Cauchy inequality*) Show that

$$\left|\sum_i a_i b_i\right|^2 \leq \sum_i |a_i|^2 \sum_i |b_i|^2 \tag{i}$$

with equality iff $a_i = kb_i^*$.

11-24. The input to a system $\mathbf{H}(z)$ is the sum $\mathbf{x}[n] = f[n] + \mathbf{v}[n]$ where $f[n]$ is a known sequence with z transform $\mathbf{F}(z)$. We wish to find $\mathbf{H}(z)$ such that the ratio $y_f^2[0]/E\{\mathbf{y}_v^2[n]\}$ of the output $\mathbf{y}[n] = y_f[n] + \mathbf{y}_v[n]$ is maximum. Show that (a) if $\mathbf{v}[n]$ is white noise, then $\mathbf{H}(z) = k\mathbf{F}(z^{-1})$, and ($b$) if $\mathbf{H}(z)$ is an FIR filter that is, if $\mathbf{H}(z) = a_0 + a_1 z^{-1} + \cdots + a_N z^{-N}$, then its weights a_m are the solutions of the system

$$\sum_{m=0}^{N} R_v[n-m] a_m = kf[-n] \qquad n = 0, \ldots, N$$

11-25. If $R_n(\tau) = N\delta(\tau)$ and

$$\mathbf{x}(t) = A\cos\omega_0 t + \mathbf{n}(t) \qquad H(\omega) = \frac{1}{\alpha + j\omega}$$

$$\mathbf{y}(t) = B\cos(\omega_0 + t + \varphi) + \mathbf{y}_n(t)$$

where $\mathbf{y}_n(t)$ is the component of the output $\mathbf{y}(t)$ due to $\mathbf{n}(t)$, find the value of α that maximizes the signal-to-noise ratio

$$\frac{|B|^2}{E\{\mathbf{y}_n^2(t)\}}$$

Answer: $\alpha = \omega_0$.

11-26. In the detection problem of page 386, we apply the process $\mathbf{x}(t) = f(t) + \mathbf{v}(t)$ to the tapped delay line (11-161). Show that: (*a*) The S/N ratio r is maximum if the coefficients a_i satisfy (11-162); (*b*) the maximum r equals $\sqrt{y_f(t_0)/k}$.

11-27. Given an SSS process $\mathbf{x}(t)$ with zero mean, power spectrum $S(\omega)$, and bispectrum $S(u,v)$, we form the process $\mathbf{y}(t) = \mathbf{x}(t) + c$. Show that

$$S_{yyy}(u,v) = S(u,v) + 2\pi c[S(u)\delta(v) + S(v)\delta(u) + S(u)\delta(u+v)] + 4\pi^2 c^3\delta(u)\delta(v)$$

11-28. Given a Poisson process $\mathbf{x}(t)$, we form its centered process $\tilde{\mathbf{x}}(t) = \mathbf{x}(t) - \lambda t$ and the centered Poisson impulses

$$\tilde{\mathbf{z}}(t) = \frac{d\tilde{\mathbf{x}}(t)}{dt} = \sum_i \delta(t - \mathbf{t}_i) - \lambda$$

Show that

$$E\{\tilde{\mathbf{x}}(t_1)\tilde{\mathbf{x}}(t_2)\tilde{\mathbf{x}}(t_3)\} = \lambda \min(t_1, t_2, t_3)$$

$$E\{\tilde{\mathbf{z}}(t_1)\tilde{\mathbf{z}}(t_2)\tilde{\mathbf{z}}(t_3)\} = \lambda\delta(t_1 - t_2)\delta(t_1 - t_3)$$

Hint Use (10-94) and the identity

$$\min(t_1, t_2, t_3) = t_1 U(t_2 - t_1)U(t_3 - t_1) + t_2 U(t_1 - t_2)U(t_3 - t_2) + t_3 U(t_1 - t_3)U(t_2 - t_3)$$

11-29. Show that the function

$$S(u,v) = H(e^{ju})H(e^{jv})H(e^{-j(u+v)})$$

is determined in terms of its values in the triangle of Fig. 11-23*b*.

Outline: Form the function

$$S_a(u,v) = H_a(u)H_a(v)H_a(-ju - jv) \quad \text{where } H_a(\omega) = \begin{cases} H(e^{j\omega}) & |\omega| \le \pi \\ 0 & |\omega| > 0 \end{cases}$$

Clearly, $S_a(u,v) = S(u,v)$ for $|u|$, $|v|$, $|u+v| < \pi$ and 0 otherwise. The function $S_a(u,v)$ is a bispectrum of a bandlimited process, $\mathbf{x}(t)$ with $\sigma = \pi$; hence (see note page 394) it is determined from its values in the triangle of Fig. 11-23*a*. Inserting into (11-194) and (11-195) we obtain $H_a(\omega)$. This yields $H(e^{j\omega})$ and $S(u,v)$.

CHAPTER 12

SPECTRAL REPRESENTATION

12-1 FACTORIZATION AND INNOVATIONS

In this section, we consider the problem of representing a real WSS process $\mathbf{x}(t)$ as the response of a minimum-phase system $\mathbf{L}(s)$ with input a white-noise process $\mathbf{i}(t)$. The term *minimum-phase* has the following meaning: The system $\mathbf{L}(s)$ is causal and its impulse response $l(t)$ has finite energy; the system $\boldsymbol{\Gamma}(s) = 1/\mathbf{L}(s)$ is causal and its impulse response $\gamma(t)$ has finite energy. Thus a system $\mathbf{L}(s)$ is minimum-phase if the functions $\mathbf{L}(s)$ and $1/\mathbf{L}(s)$ are analytic in the right-hand plane Re $s > 0$. A process $\mathbf{x}(t)$ that can be so represented will be called *regular*. From the definition it follows that $\mathbf{x}(t)$ is a regular process if it is linearly equivalent with a white-noise process $\mathbf{i}(t)$ in the sense that (see Fig. 12-1)

$$\mathbf{i}(t) = \int_0^\infty \gamma(\alpha)\mathbf{x}(t-\alpha)\,d\alpha \qquad R_{ii}(\tau) = \delta(\tau) \tag{12-1}$$

$$\mathbf{x}(t) = \int_0^\infty l(\alpha)\mathbf{i}(t-\alpha)\,d\alpha \qquad E\{\mathbf{x}^2(t)\} = \int_0^\infty l^2(t)\,dt < \infty \tag{12-2}$$

The last equality follows from (10-91). The above shows that the power spectrum $\mathbf{S}(s)$ of a regular process can be written as a product

$$\mathbf{S}(s) = \mathbf{L}(s)\mathbf{L}(-s) \qquad S(\omega) = |\mathbf{L}(j\omega)|^2 \tag{12-3}$$

where $\mathbf{L}(s)$ is a minimum-phase function uniquely determined in terms of $S(\omega)$. The function $\mathbf{L}(s)$ will be called the *innovations filter* of $\mathbf{x}(t)$ and its inverse $\boldsymbol{\Gamma}(s)$

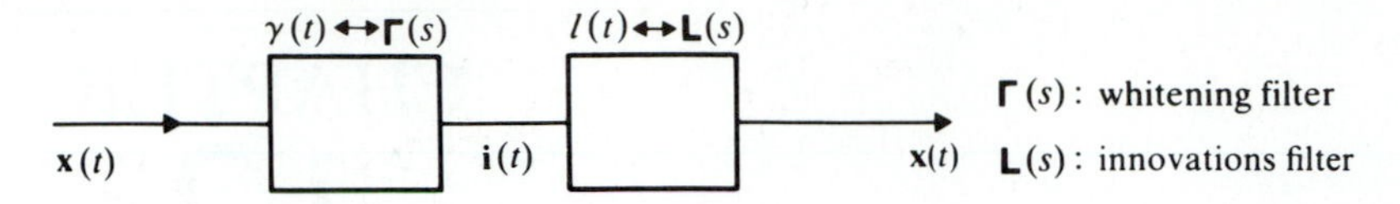

FIGURE 12-1

the *whitening filter* of $\mathbf{x}(t)$. The process $\mathbf{i}(t)$ will be called the *innovations* of $\mathbf{x}(t)$. It is the output of the filter $\mathbf{L}(s)$ with input $\mathbf{x}(t)$.

The problem of determining the function $\mathbf{L}(s)$ can be phrased as follows: Given a positive even function $S(\omega)$ of finite area, find a minimum-phase function $\mathbf{L}(s)$ such that $|\mathbf{L}(j\omega)|^2 = S(\omega)$. It can be shown that this problem has a solution if $S(\omega)$ satisfies the *Paley–Wiener condition*†

$$\int_{-\infty}^{\infty} \frac{|\ln S(\omega)|}{1+\omega^2}\, d\omega < \infty \tag{12-4}$$

This condition is not satisfied if $S(\omega)$ consists of lines, or, more generally, if it is bandlimited. As we show later, processes with such spectra are predictable. In general, the problem of factoring $S(\omega)$ as in (12-3) is not simple. In the following, we discuss an important special case.

Rational spectra. A rational spectrum is the ratio of two polynomials in ω^2 because $S(-\omega) = S(\omega)$:

$$S(\omega) = \frac{A(\omega^2)}{B(\omega^2)} \qquad \mathbf{S}(s) = \frac{A(-s^2)}{B(-s^2)} \tag{12-5}$$

This shows that if s_i is a root (zero or pole) of $\mathbf{S}(s)$, $-s_i$ is also a root. Furthermore, all roots are either real or complex conjugate. From this it follows that the roots of $\mathbf{S}(s)$ are symmetrical with respect to the $j\omega$ axis (Fig. 12-2a). Hence they can be separated into two groups: The "left" group consists of all roots s_i with $\operatorname{Re} s_i < 0$, and the "right" group consists of all roots with $\operatorname{Re} s_i > 0$. The minimum-phase factor $\mathbf{L}(s)$ of $\mathbf{S}(s)$ is a ratio of two polynomials formed with the left roots of $\mathbf{S}(s)$:

$$\mathbf{S}(s) = \frac{N(s)N(-s)}{D(s)D(-s)} \qquad \mathbf{L}(s) = \frac{N(s)}{D(s)} \qquad \mathbf{L}^2(0) = \mathbf{S}(0)$$

Example 12-1. If $S(\omega) = N/(\alpha^2 + \omega^2)$ then

$$\mathbf{S}(s) = \frac{N}{\alpha^2 - s^2} = \frac{N}{(\alpha+s)(\alpha-s)} \qquad \mathbf{L}(s) = \frac{\sqrt{N}}{\alpha + s}$$

†N. Wiener, R. E. A. C. Paley: Fourier Transforms in the Complex Domain, *American Mathematical Society College*, 1934 (see also Papoulis, 1962).

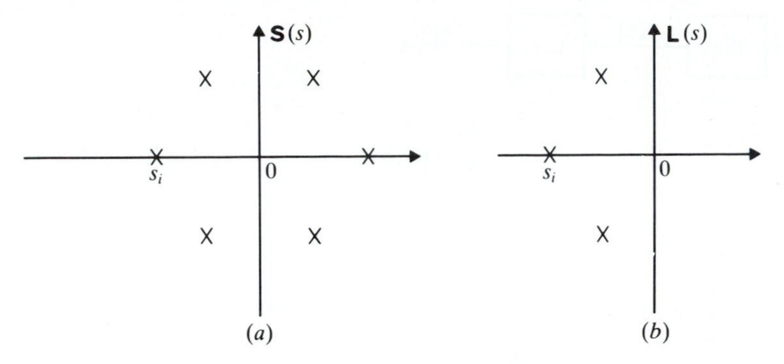

FIGURE 12-2

Example 12-2. If $S(\omega) = (49 + 25\omega^2)/(\omega^4 + 10\omega^2 + 9)$ then

$$\mathbf{S}(s) = \frac{49 - 25s^2}{(1 - s^2)(9 - s^2)} \qquad \mathbf{L}(s) = \frac{7 + 5s}{(1 + s)(3 + s)}$$

Example 12-3. If $S(\omega) = 25/(\omega^4 + 1)$ then

$$\mathbf{S}(s) = \frac{25}{s^4 + 1} = \frac{25}{(s^2 + \sqrt{2}s + 1)(s^2 - \sqrt{2}s + 1)} \qquad \mathbf{L}(s) = \frac{5}{s^2 + \sqrt{2}s + 1}$$

Digital Processes

A digital system is minimum-phase if its system function $\mathbf{L}(z)$ and its inverse $\boldsymbol{\Gamma}(z) = 1/\mathbf{L}(z)$ are analytic in the exterior $|z| > 1$ of the unit circle. A real WSS digital process $\mathbf{x}[n]$ is regular if its spectrum $\mathbf{S}(z)$ can be written as a product

$$\mathbf{S}(z) = \mathbf{L}(z)\mathbf{L}(1/z) \qquad \mathbf{S}(e^{j\omega}) = |\mathbf{L}(e^{j\omega})|^2 \tag{12-6}$$

Denoting by $l[n]$ and $\gamma[n]$ respectively the delta responses of $\mathbf{L}(z)$ and $\boldsymbol{\Gamma}(z)$, we conclude that a regular process $\mathbf{x}[n]$ is linearly equivalent with a white-noise process $\mathbf{i}[n]$ (see Fig. 12-3):

$$\mathbf{i}[n] = \sum_{k=0}^{\infty} \gamma[k]\mathbf{x}[n-k] \qquad R_{ii}[m] = \delta[m] \tag{12-7}$$

$$\mathbf{x}[n] = \sum_{k=0}^{\infty} l[k]\mathbf{i}[n-k] \qquad E\{\mathbf{x}^2[n]\} = \sum_{k=0}^{\infty} l^2[k] < \infty \tag{12-8}$$

The process $\mathbf{i}[n]$ is the innovations of $\mathbf{x}[n]$ and the function $\mathbf{L}(z)$ its innovations filter. The whitening filter of $\mathbf{x}[n]$ is the function $\boldsymbol{\Gamma}(z) = 1/\mathbf{L}(z)$.

It can be shown that the power spectrum $\mathbf{S}(e^{j\omega})$ of a process $\mathbf{x}[n]$ can be factored as in (12-6) if it satisfies the Paley–Wiener condition

$$\int_{-\pi}^{\pi} |\ln S(\omega)\, d\omega| < \infty \tag{12-9}$$

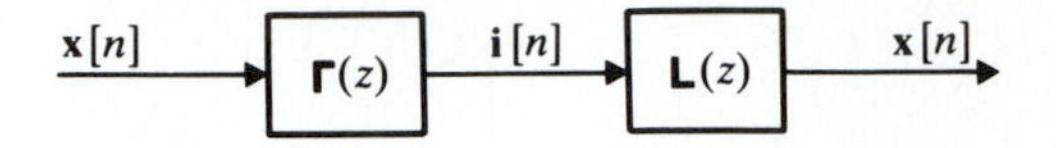

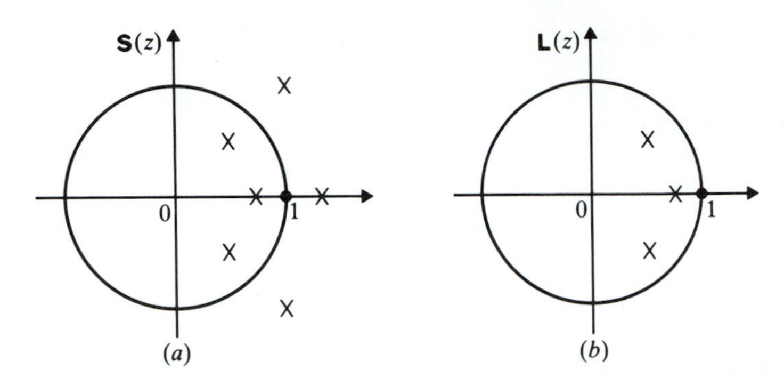

FIGURE 12-3

Rational spectra. The power spectrum $\mathbf{S}(e^{j\omega})$ of a real process is a function of $\cos \omega = (e^{j\omega} + e^{-j\omega})/2$ [see (10-180)]. From this it follows that $\mathbf{S}(z)$ is a function of $z + 1/z$. If therefore, z_i is a root of $\mathbf{S}(z)$, $1/z_i$ is also a root. We thus conclude that the roots of $\mathbf{S}(z)$ are symmetrical with respect to the unit circle (Fig. 12-3); hence they can be separated into two groups: The "inside" group consists of all roots z_i such that $|z_i| < 1$ and the "outside" group consists of all roots such that $|z_i| > 1$. The minimum-phase factor $\mathbf{L}(z)$ of $\mathbf{S}(z)$ is a ratio of two polynomials consisting of the inside roots of $\mathbf{S}(z)$:

$$\mathbf{S}(z) = \frac{N(z)N(1/z)}{D(z)D(1/z)} \qquad \mathbf{L}(z) = \frac{N(z)}{D(z)} \qquad \mathbf{L}^2(1) = \mathbf{S}(1)$$

Example 12-4. If $S(\omega) = (5 - 4\cos\omega)/(10 - 6\cos\omega)$ then

$$\mathbf{S}(z) = \frac{5 - 2(z + z^{-1})}{10 - 3(z + z^{-1})} = \frac{2(z - 1/2)(z - 2)}{3(z - 1/3)(z - 3)} \qquad \mathbf{L}(z) = \frac{2z - 1}{3z - 1}$$

12-2 FINITE-ORDER SYSTEMS AND STATE VARIABLES

In this section, we consider systems specified in terms of differential equations or recursion equations. As a preparation, we review briefly the meaning of finite-order systems and state variables starting with the analog case. The systems under consideration are multiterminal with m inputs $\mathbf{x}_i(t)$ and r outputs $\mathbf{y}_j(t)$ forming the column vectors $\mathbf{X}(t) = [\mathbf{x}_i(t)]$ and $\mathbf{Y}(t) = [\mathbf{y}_j(t)]$ as in (10-113).

At a particular time $t = t_1$, the output $\mathbf{Y}(t)$ of a system is in general specified only if the input $\mathbf{X}(t)$ is known for every t. Thus, to determine $\mathbf{Y}(t)$ for

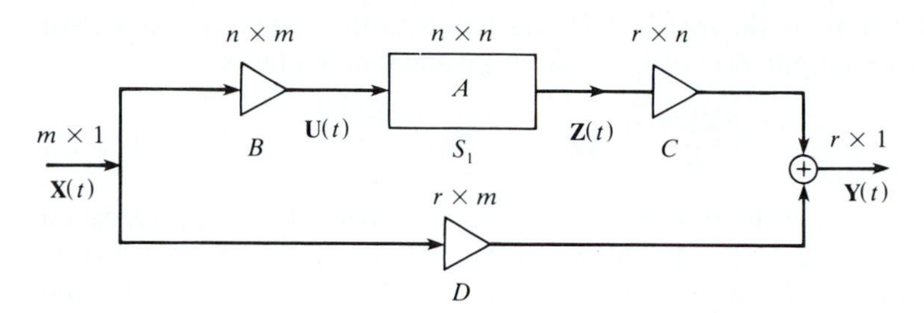

FIGURE 12-4

$t > t_0$, we must know $\mathbf{X}(t)$ for $t > t_0$ and for $t \le t_0$. For a certain class of systems, this is not necessary. The values of $\mathbf{Y}(t)$ for $t > t_0$ are completely specified if we know $\mathbf{X}(t)$ for $t > t_0$ and, in addition, the values of a finite number of parameters. These parameters specify the "state" of the system at time $t = t_0$ in the sense that their values determine the effect of the past $t < t_0$ of $\mathbf{X}(t)$ on the future $t > t_0$ of $\mathbf{Y}(t)$. The values of these parameters depend on t_0; they are, therefore, functions $\mathbf{z}_i(t)$ of t. These functions are called *state variables*. The number n of state variables is called the *order* of the system. The vector

$$\mathbf{Z}(t) = [\mathbf{z}_i(t)] \qquad i = 1, \ldots, n$$

is called the *state vector*; this vector is not unique. We shall say that the system is in *zero state* at $t = t_0$ if $\mathbf{Z}(t_0) = 0$.

We shall consider here only linear, time-invariant, real, causal systems. Such systems are specified in terms of the following equations:

$$\frac{d\mathbf{Z}(t)}{dt} = A\mathbf{Z}(t) + B\mathbf{X}(t) \tag{12-10a}$$

$$\mathbf{Y}(t) = C\mathbf{Z}(t) + D\mathbf{X}(t) \tag{12-10b}$$

In the above, A, B, C, and D are matrices with real constant elements, of order $n \times n$, $n \times m$, $r \times n$, and $r \times m$ respectively. In Fig. 12-4 we show a block diagram of the system S specified terminally in terms of these equations. It consists of a dynamic system S_1 with input $\mathbf{U}(t) = B\mathbf{X}(t)$ and output $\mathbf{Z}(t)$, and of three memoryless systems (multipliers). If the input $\mathbf{X}(t)$ of the system S is specified for every t, or, if $\mathbf{X}(t) = 0$ for $t < 0$ and the system is in zero state at $t = 0$, then the response $\mathbf{Y}(t)$ of S for $t > 0$ equals

$$\mathbf{Y}(t) = \int_0^\infty H(\alpha)\mathbf{X}(t - \alpha)\, d\alpha \tag{12-11}$$

where $H(t)$ is the impulse response matrix of S. This follows from (10-78) and the fact that $H(t) = 0$ for $t < 0$ (causality assumption).

We shall determine the matrix $H(t)$ starting with the system S_1. As we see from (12-10a), the output $\mathbf{Z}(t)$ of this system satisfies the equation

$$\frac{d\mathbf{Z}(t)}{dt} - A\mathbf{Z}(t) = \mathbf{U}(t) \qquad (12\text{-}12)$$

The impulse response of the system S_1 is an $n \times n$ matrix $\Phi(t) = [\varphi_{ji}(t)]$ called the *transition matrix* of S. The function $\varphi_{ji}(t)$ equals the value of the jth state variable $\mathbf{z}_j(t)$ when the ith element $\mathbf{u}_i(t)$ of the input $\mathbf{U}(t)$ of S_1 equals $\delta(t)$ and all other elements are 0. From this it follows that [see (10-115)]

$$\mathbf{Z}(t) = \int_0^\infty \Phi(\alpha)\mathbf{U}(t-\alpha)\,d\alpha = \int_0^\infty \Phi(\alpha)B\mathbf{X}(t-\alpha)\,d\alpha \qquad (12\text{-}13)$$

Inserting into (12-10b), we obtain

$$\mathbf{Y}(t) = \int_0^\infty C\Phi(\alpha)B\mathbf{X}(t-\alpha)\,d\alpha + D\mathbf{X}(t)$$

$$= \int_0^\infty [C\Phi(\alpha)B\mathbf{X}(t-\alpha) + \delta(\alpha)D\mathbf{X}(t-\alpha)]\,d\alpha \qquad (12\text{-}14)$$

where $\delta(t)$ is the (scalar) impulse function. Comparing with (12-11), we conclude that the impulse response matrix of the system S equals

$$H(t) = C\Phi(t)B + \delta(t)D \qquad (12\text{-}15)$$

From the definition of $\Phi(t)$ it follows that

$$\frac{d\Phi(t)}{dt} - A\Phi(t) = \delta(t)1_n \qquad (12\text{-}16)$$

where 1_n is the identity matrix of order n. The Laplace transform $\mathbf{\Phi}(s)$ of $\Phi(t)$ is the system function of the system S_1. Taking transforms of both sides of (12-16), we obtain

$$s\mathbf{\Phi}(s) - A\mathbf{\Phi}(s) = 1_n \qquad \mathbf{\Phi}(s) = (s1_n - A)^{-1} \qquad (12\text{-}17)$$

Hence

$$\Phi(t) = e^{At} \qquad t > 0 \qquad (12\text{-}18)$$

This is a direct generalization of the scalar case; however, the determination of the elements $\varphi_{ji}(t)$ of $\Phi(t)$ is not trivial. Each element is a sum of exponentials of the form

$$\varphi_{ji}(t) = \sum_k p_{ji,k}(t)e^{s_k t} \qquad t > 0$$

where s_k are the eigenvalues of the matrix A and $p_{ji,k}(t)$ are polynomials in t of degree equal to the multiplicity of s_k. There are several methods for determining these polynomials. For small n, it is simplest to replace (12-16) by n systems of n scalar equations.

Inserting $\Phi(t)$ into (12-15), we obtain

$$H(t) = Ce^{At}B + \delta(t)D$$
$$\mathbf{H}(s) = C(s1_n - A)^{-1}B + D \qquad (12\text{-}19)$$

Suppose now that the input to the system S is a WSS process $\mathbf{X}(t)$. We shall comment briefly on the spectral properties of the resulting output, limiting the discussion to the state vector $\mathbf{Z}(t)$. The system S_1 is a special case of S obtained with $B = C = 1_n$ and $D = 0$. In this case, $\mathbf{Z}(t) = \mathbf{Y}(t)$ and

$$\frac{d\mathbf{Y}(t)}{dt} - A\mathbf{Y}(t) = \mathbf{X}(t) \qquad \mathbf{H}(s) = (s1_n - A)^{-1} \qquad (12\text{-}20)$$

Inserting into (10-157), we conclude that

$$\mathbf{S}_{xy}(s) = \mathbf{S}_{xx}(s)(-s1_n - A)^{-1}$$
$$\mathbf{S}_{yy}(s) = (s1_n - A^t)^{-1}\mathbf{S}_{xy}(s) \qquad (12\text{-}21)$$
$$\mathbf{S}_{yy}(s) = (s1_n - A^t)^{-1}\mathbf{S}_{xx}(s)(-s1_n - A)^{-1}$$

Differential equations. The equation

$$\mathbf{y}^{(n)}(t) + a_1\mathbf{y}^{(n-1)}(t) + \cdots + a_n\mathbf{y}(t) = \mathbf{x}(t) \qquad (12\text{-}22)$$

specifies a system S with input $\mathbf{x}(t)$ and output $\mathbf{y}(t)$. This system is of finite order because $\mathbf{y}(t)$ is determined for $t > 0$ in terms of the values of $\mathbf{x}(t)$ for $t \geq 0$ and the initial conditions

$$\mathbf{y}(0), \mathbf{y}'(0), \ldots, \mathbf{y}^{(n-1)}(0)$$

It is, in fact, a special case of the system of Fig. 12-4 if we set $m = r = 1$,

$$\mathbf{z}_1(t) = \mathbf{y}(t) \qquad \mathbf{z}_2(t) = \mathbf{y}'(t) \cdots \mathbf{z}_n(t) = \mathbf{y}^{(n-1)}(t)$$

$$A = \begin{bmatrix} 0 & 1 & \cdots & 0 \\ 0 & 0 & \cdots & 0 \\ \cdot & \cdot & \cdots & \cdot \\ -a_n & -a_{n-1} & \cdots & -a_1 \end{bmatrix} \qquad B = \begin{bmatrix} 0 \\ 0 \\ \cdots \\ 1 \end{bmatrix} \qquad C^t = \begin{bmatrix} 1 \\ 0 \\ \cdots \\ 0 \end{bmatrix}$$

and $D = 0$. Inserting the above into (12-19), we conclude after some effort that

$$\mathbf{H}(s) = \frac{1}{s^n + a_1 s^{n-1} + \cdots + a_n}$$

This result can be derived simply from (12-22).

Multiplying both sides of (12-22) by $\mathbf{x}(t - \tau)$ and $\mathbf{y}(t + \tau)$, we obtain

$$R_{yx}^{(n)}(\tau) + a_1 R_{yx}^{(n-1)}(\tau) + \cdots + a_n R_{yx}(\tau) = R_{xx}(\tau) \qquad (12\text{-}23)$$

$$R_{yy}^{(n)}(\tau) + a_1 R_{yy}^{(n-1)}(\tau) + \cdots + a_n R_{yy}(\tau) = R_{xy}(\tau) \qquad (12\text{-}24)$$

for all τ. This is a special case of (10-133).

Finite-order processes. We shall say that a process $\mathbf{x}(t)$ is of finite order if its innovations filter $\mathbf{L}(s)$ is a rational function of s:

$$\mathbf{S}(s) = \mathbf{L}(s)\mathbf{L}(-s) \qquad \mathbf{L}(s) = \frac{b_0 s^m + b_1 s^{m-1} + \cdots + b_m}{s^n + a_1 s^{n-1} + \cdots + a_n} = \frac{N(s)}{D(s)} \tag{12-25}$$

where $N(s)$ and $D(s)$ are two Hurwitz polynomials. The process $\mathbf{x}(t)$ is the response of the filter $\mathbf{L}(s)$ with input the white-noise process $\mathbf{i}(t)$:

$$\mathbf{x}^{(n)}(t) + a_1\mathbf{x}^{(n-1)}(t) + \cdots + a_n\mathbf{x}(t) = b_0\mathbf{i}^{(m)}(t) + \cdots + b_m\mathbf{i}(t) \tag{12-26}$$

The past $\mathbf{x}(t-\tau)$ of $\mathbf{x}(t)$ depends only on the past of $\mathbf{i}(t)$; hence it is orthogonal to the right side of (12-26) for every $\tau > 0$. From this it follows as in (12-24) that

$$R^{(n)}(\tau) + a_1 R^{(n-1)}(\tau) + \cdots + a_n R(\tau) = 0 \qquad \tau > 0 \tag{12-27}$$

Assuming that the roots s_i of $D(s)$ are simple, we conclude from (12-27) that

$$R(\tau) = \sum_{i=1}^{n} \alpha_i e^{s_i \tau} \qquad \tau > 0$$

The coefficients α_i can be determined from the initial value theorem. Alternatively, to find $R(\tau)$, we expand $\mathbf{S}(s)$ into partial fractions:

$$\mathbf{S}(s) = \sum_{i=1}^{n} \frac{\alpha_i}{s - s_i} + \sum_{i=1}^{n} \frac{\alpha_i}{-s - s_i} = \mathbf{S}^+(s) + \mathbf{S}^-(s) \tag{12-28}$$

The first sum is the transform of the causal part $R^+(\tau) = R(\tau)U(\tau)$ of $R(\tau)$ and the second of its anticausal part $R^-(\tau) = R(\tau)U(-\tau)$. Since $R(-\tau) = R(\tau)$, this yields

$$R(\tau) = R^+(|\tau|) = \sum_{i=1}^{n} \alpha_i e^{s_i|\tau|} \tag{12-29}$$

Example 12-5. If $\mathbf{L}(s) = 1/(s+\alpha)$, then

$$\mathbf{S}(s) = \frac{1}{(s+\alpha)(-s+\alpha)} = \frac{1/2\alpha}{s+\alpha} + \frac{1/2\alpha}{-s+\alpha}$$

Hence $R(\tau) = (1/2\alpha)e^{-\alpha|\tau|}$.

Example 12-6. The differential equation

$$\mathbf{x}''(t) + 3\mathbf{x}'(t) + 2\mathbf{x}(t) = \mathbf{i}(t) \qquad R_{ii}(\tau) = \delta(\tau)$$

specifies a process $\mathbf{x}(t)$ with autocorrelation $R(\tau)$. From (12-27) it follows that

$$R''(\tau) + 3R'(\tau) + 2R(\tau) = 0 \qquad \text{hence} \quad R(\tau) = c_1 e^{-\tau} + c_2 e^{-2\tau}$$

for $\tau > 0$. To find the constants c_1 and c_2, we shall determine $R(0)$ and $R'(0)$. Clearly,

$$\mathbf{S}(s) = \frac{1}{(s^2+3s+2)(s^2-3s+2)} = \frac{s/12 + 1/4}{s^2+3s+2} + \frac{-s/12 + 1/4}{s^2-3s+2}$$

The first fraction on the right is the transform of $R^+(\tau)$; hence

$$R^+(0^+) = \lim_{s\to\infty} s\mathbf{S}^+(s) = \tfrac{1}{12} = c_1 + c_2 = R(0)$$

Similarly,

$$R'(0^+) = \lim_{s\to\infty} s\left(s\mathbf{S}^+(s) - \tfrac{1}{12}\right) = 0 = -c_1 - 2c_2$$

This yields $R(\tau) = \frac{1}{6}e^{-|\tau|} - \frac{1}{12}e^{-2|\tau|}$.

Note finally that $R(\tau)$ can be expressed in terms of the impulse response $l(t)$ of the innovations filter $\mathbf{L}(s)$:

$$R(\tau) = l(\tau) * l(-\tau) = \int_0^\infty l(|\tau| + \alpha)l(\alpha)\,d\alpha \qquad (12\text{-}30)$$

Digital Systems

The digital version of the system of Fig. 11-4 is a finite-order system S specified by the equations:

$$\mathbf{Z}[k+1] = A\mathbf{Z}[k] + B\mathbf{X}[k] \qquad (12\text{-}31a)$$

$$\mathbf{Y}[k] = C\mathbf{Z}[k] + D\mathbf{X}[k] \qquad (12\text{-}31b)$$

where k is the discrete time, $\mathbf{X}[k]$ the input vector, $\mathbf{Y}[k]$ the output vector, and $\mathbf{Z}[k]$ the state vector. The system is stable if the eigenvalues z_i of the $n \times n$ matrix A are such that $|z_i| < 1$. The preceding results can be readily extended to digital systems. Note, in particular, that the system function of S is the z transform

$$\mathbf{H}(z) = C(z1_n - A)^{-1}B + D \qquad (12\text{-}32)$$

of the delta response matrix

$$H[k] = C\Phi[k]B + \delta[k]D \qquad k \ge 0 \qquad (12\text{-}33)$$

We shall discuss in some detail scalar systems driven by white noise. This material is used in Sec. 13-3.

Finite-order processes. Consider a real digital process $\mathbf{x}[n]$ with innovations filter $\mathbf{L}(z)$ and power spectrum $\mathbf{S}(z)$:

$$\mathbf{S}(z) = \mathbf{L}(z)\mathbf{L}(1/z) \qquad \mathbf{L}(z) = \sum_{n=0}^{\infty} l[n]z^{-n} \qquad (12\text{-}34)$$

where n is now the discrete time. If we know $\mathbf{L}(z)$, we can find the autocorrelation $R[m]$ of $\mathbf{x}[n]$ either from the inversion formula (10-179) or from the convolution theorem

$$R[m] = l[m] * l[-m] = \sum_{k=0}^{\infty} l[|m| + k]l[k] \qquad (12\text{-}35)$$

We shall discuss the properties of $R[m]$ for the class of finite-order processes.

The power spectrum $S(\omega)$ of a finite-order process $\mathbf{x}[n]$ is a rational function of $\cos\omega$; hence its innovations filter is a rational function of z:

$$\mathbf{L}(z) = \frac{N(z)}{D(z)} = \frac{b_0 + b_1 z^{-1} + \cdots + b_M z^{-M}}{1 + a_1 z^{-1} + \cdots + a_N z^{-N}} \tag{12-36}$$

To find its autocorrelation, we determine $l[n]$ and insert the result into (12-35). Assuming that the roots z_i of $D(z)$ are simple and $M \le N$, we obtain

$$\mathbf{L}(z) = \sum_i \frac{\gamma_i}{1 - z_i z^{-1}} \qquad l[n] = \sum_i \gamma_i z_i^n U[n]$$

Alternatively, we expand $\mathbf{S}(z)$:

$$\mathbf{S}(z) = \sum_i \frac{\alpha_i}{1 - z_i z^{-1}} + \sum_i \frac{\alpha_i}{1 - z_i z} \qquad R[m] = \sum_i \alpha_i z_i^{|m|} \tag{12-37}$$

Note that $\alpha_i = \gamma_i L(1/z_i)$.

The process $\mathbf{x}[n]$ satisfies the recursion equation

$$\mathbf{x}[n] + a_1\mathbf{x}[n-1] + \cdots + a_N\mathbf{x}[n-N] = b_0\mathbf{i}[n] + \cdots + b_M\mathbf{i}[n-M] \tag{12-38}$$

where $\mathbf{i}[n]$ is its innovations. We shall use this equation to relate the coefficients of $\mathbf{L}(z)$ to the sequence $R[m]$ starting with two special cases.

Autoregressive processes. The process $\mathbf{x}[n]$ is called autoregressive (AR) if

$$\mathbf{L}(z) = \frac{b_0}{1 + a_1 z^{-1} + \cdots + a_N z^{-N}} \tag{12-39}$$

In this case, (12-38) yields

$$\mathbf{x}[n] + a_1\mathbf{x}[n-1] + \cdots + a_N\mathbf{x}[n-N] = b_0\mathbf{i}[n] \tag{12-40}$$

The past $\mathbf{x}[n-m]$ of $\mathbf{x}[n]$ depends only on the past of $\mathbf{i}[n]$; furthermore, $E\{\mathbf{i}^2[n]\} = 1$. From this it follows that $E\{\mathbf{x}[n]\mathbf{i}[n]\} = b_0$ and $E\{\mathbf{x}[n-m]\mathbf{i}[n]\} = 0$ for $m > 0$. Multiplying (12-40) by $\mathbf{x}[n-m]$ and setting $m = 0, 1, \ldots,$ we obtain the equations

$$\begin{aligned} R[0] + a_1 R[1] + \cdots + a_N R[N] &= b_0^2 \\ R[1] + a_1 R[0] + \cdots + a_N R[N-1] &= 0 \\ \cdots\cdots\cdots\cdots\cdots\cdots\cdots\cdots\cdots \\ R[N] + a_1 R[N-1] + \cdots + a_N R[0] &= 0 \end{aligned} \tag{12-41a}$$

and

$$R[m] + a_1 R[m-1] + \cdots + a_N R[m-N] = 0 \tag{12-41b}$$

for $m > N$. The first $N+1$ of these are called the *Yule–Walker* equations. They are used in Sec. 13-3 to express the $N+1$ parameters a_k and b_0 in terms of the first $N+1$ values of $R[m]$. Conversely, if $\mathbf{L}(z)$ is known, we find $R[m]$

for $|m| \le N$ solving the system (12-41a) and we determine $R[m]$ recursively from (12-41b) for $m > N$.

Example 12-7. Suppose that

$$\mathbf{x}[n] - a\mathbf{x}[n-1] = \boldsymbol{\nu}[n] \qquad R_{\nu\nu}[m] = b\delta[m]$$

This is a special case of (12-40) with $D(z) = 1 - az^{-1}$ and $z_1 = a$. Hence

$$R[0] - aR[1] = b \qquad R[m] = \alpha a^{|m|} \qquad \alpha = \frac{b}{1-a^2}$$

Line spectra. Suppose that $\mathbf{x}[n]$ satisfies the homogeneous equation

$$\mathbf{x}[n] + a_1\mathbf{x}[n-1] + \cdots + a_N\mathbf{x}[n-N] = 0 \tag{12-42}$$

This is a special case of (12-40) if we set $b_0 = 0$. Solving for $\mathbf{x}[n]$, we obtain

$$\mathbf{x}[n] = \mathbf{c}_1 z_1^n + \cdots + \mathbf{c}_N z_N^n \qquad D(z_i) = 0 \tag{12-43}$$

If $\mathbf{x}[n]$ is a stationary process, only the terms with $z_i = e^{j\omega_i}$ can appear. Furthermore, their coefficients $\mathbf{c}_k$ must be uncorrelated with zero mean. From this it follows that if $\mathbf{x}[n]$ is a WSS process satisfying (12-42), its autocorrelation must be a sum of exponentials as in Example 10-31:

$$R[m] = \sum \alpha_i e^{j\omega_i|m|} \qquad S(\omega) = 2\pi \sum \alpha_i \delta(\omega - \beta_i) \qquad |\omega| < \pi \tag{12-44}$$

where $\alpha_i = E\{\mathbf{c}_i^2\}$ and $\beta_i = \omega_i - 2\pi k_i$ as in (10-182).

Moving average processes. A process $\mathbf{x}[n]$ is a moving average (MA) if

$$\mathbf{x}[n] = b_0\mathbf{i}[n] + \cdots + b_M\mathbf{i}[n-M] \tag{12-45}$$

In this case, $\mathbf{L}(z)$ is a polynomial and its inverse $l[n]$ has a finite length (FIR filter):

$$\mathbf{L}(z) = b_0 + b_1 z^{-1} + \cdots + b_M z^{-M} \qquad l[n] = b_0\delta[n] + \cdots + b_M\delta[n-M] \tag{12-46}$$

Since $l[n] = 0$ for $n > m$, (12-35) yields

$$R[m] = \sum_{k=0}^{M-m} l[m+k]l[k] = \sum_{k=0}^{M-m} b_{k+m}b_k \tag{12-47}$$

for $0 \le m \le M$ and 0 for $m > M$. Explicitly,

$$\begin{aligned}
R[0] &= b_0^2 + b_1^2 + \cdots + b_M^2 \\
R[1] &= b_0 b_1 + b_1 b_2 + \cdots + b_{M-1}b_M \\
&\cdots\cdots\cdots\cdots\cdots\cdots \\
R[M] &= b_0 b_M
\end{aligned}$$

Example 12-8. Suppose that $\mathbf{x}[n]$ is the arithmetic average of the M values of $\mathbf{i}[n]$:

$$\mathbf{x}[n] = \frac{1}{M}(\mathbf{i}[n] + \mathbf{i}[n-1] + \cdots + \mathbf{i}[n-M+1])$$

In this case,

$$\mathbf{L}(z) = \frac{1}{M}(1 + z^{-1} + \cdots + z^{-M+1}) = \frac{1 - z^{-M}}{M(1 - z^{-1})}$$

$$R[m] = \frac{1}{M^2}\sum_{k=0}^{M-1-|m|} 1 = \frac{M - |m|}{M^2} = \frac{1}{M}\left(1 - \frac{|m|}{M}\right) \qquad |m| \le M$$

$$\mathbf{S}(z) = \mathbf{L}(z)\mathbf{L}(1/z) = \frac{2 - z^{-M} - z^{M}}{M^2(2 - z^{-1} - z)} \qquad \mathbf{S}(e^{j\omega}) = \frac{\sin^2 \dfrac{M\omega}{2}}{M^2 \sin^2 \dfrac{\omega}{2}}$$

Autoregressive moving average. We shall say that $\mathbf{x}[n]$ is an ARMA process if it satisfies the equation

$$\mathbf{x}[n] + a_1\mathbf{x}[n-1] + \cdots + a_N\mathbf{x}[n-N] = b_0\mathbf{i}[n] + \cdots + b_M\mathbf{i}[n-M] \tag{12-48}$$

Its innovations filter $\mathbf{L}(z)$ is the fraction in (12-36). Again, $\mathbf{i}[n]$ is white noise; hence

$$E\{\mathbf{x}[n-m]\mathbf{i}[n-r]\} = 0 \qquad \text{for} \quad m < r$$

Multiplying (12-48) by $\mathbf{x}[n-m]$ and using the above, we conclude that

$$R[m] + a_1 R[m-1] + \cdots + a_N R[m-N] = 0 \qquad m > M \tag{12-49}$$

Note that, unlike the AR case, this is true only for $m > M$.

12-3 FOURIER SERIES AND KARHUNEN–LOÈVE EXPANSIONS

A process $\mathbf{x}(t)$ is MS periodic with period T if $E\{|\mathbf{x}(t+T) - \mathbf{x}(t)|^2\} = 0$ for all t. A WSS process is MS periodic if its autocorrelation $R(\tau)$ is periodic with period $T = 2\pi/\omega_0$ [see (10-165)]. Expanding $R(\tau)$ into Fourier series, we obtain

$$R(\tau) = \sum_{n=-\infty}^{\infty} \gamma_n e^{jn\omega_0\tau} \qquad \gamma_n = \frac{1}{T}\int_0^T R(\tau)e^{-jn\omega_0\tau}\,d\tau \tag{12-50}$$

Given a WSS periodic process $\mathbf{x}(t)$ with period T, we form the sum

$$\hat{\mathbf{x}}(t) = \sum_{n=-\infty}^{\infty} \mathbf{c}_n e^{jn\omega_0 t} \qquad \mathbf{c}_n = \frac{1}{T}\int_0^T \mathbf{x}(t)e^{-jn\omega_0 t}\,dt \tag{12-51}$$

THEOREM. The above sum equals $\mathbf{x}(t)$ in the MS sense:

$$E\{|\mathbf{x}(t) - \hat{\mathbf{x}}(t)|^2\} = 0 \tag{12-52}$$

Furthermore, the RVs $\mathbf{c}_n$ are uncorrelated with zero mean for $n \neq 0$, and their variance equals γ_n:

$$E\{\mathbf{c}_n\} = \begin{cases} \eta_x & n = 0 \\ 0 & n \neq 0 \end{cases} \qquad E\{\mathbf{c}_n\mathbf{c}_m^*\} = \begin{cases} \gamma_n & n = m \\ 0 & n \neq m \end{cases} \tag{12-53}$$

Proof. We form the products

$$\mathbf{c}_n\mathbf{x}^*(\alpha) = \frac{1}{T}\int_0^T \mathbf{x}(t)\mathbf{x}^*(\alpha)e^{-jn\omega_0 t}\,dt$$

$$\mathbf{c}_n\mathbf{c}_m^* = \frac{1}{T}\int_0^T \mathbf{c}_n\mathbf{x}^*(t)e^{jm\omega_0 t}\,dt$$

and we take expected values. This yields

$$E\{\mathbf{c}_n\mathbf{x}^*(\alpha)\} = \frac{1}{T}\int_0^T R(t-\alpha)e^{-jn\omega_0 t}\,dt = \gamma_n e^{-jn\omega_0 \alpha}$$

$$E\{\mathbf{c}_n\mathbf{c}_m^*\} = \frac{1}{T}\int_0^T \gamma_n e^{-jn\omega_0 t}e^{jm\omega_0 t}\,dt = \begin{cases} \gamma_n & n = m \\ 0 & n \neq m \end{cases}$$

and (12-53) results.

To prove (12-52), we observe, using the above, that

$$E\{|\hat{\mathbf{x}}(t)|^2\} = \sum E\{|\mathbf{c}_n|^2\} = \sum \gamma_n = R(0) = E\{|\mathbf{x}(t)|^2\}$$

$$E\{\hat{\mathbf{x}}(t)\mathbf{x}^*(t)\} = \sum E\{\mathbf{c}_n\mathbf{x}^*(t)\}e^{jn\omega_0 t} = \sum \gamma_n = E\{\hat{\mathbf{x}}^*(t)\mathbf{x}(t)\}$$

and (12-51) follows readily.

Suppose now that the WSS process $\mathbf{x}(t)$ is not periodic. Selecting an arbitrary constant T, we form again the sum $\hat{\mathbf{x}}(t)$ as in (12-51). It can be shown that (see Prob. 12-12) $\hat{\mathbf{x}}(t)$ equals $\mathbf{x}(t)$ not for all t, but only in the interval $(0, T)$:

$$E\{|\hat{\mathbf{x}}(t) - \mathbf{x}(t)|^2\} = 0 \qquad 0 < t < T \tag{12-54}$$

Unlike the periodic case, however, the coefficients $\mathbf{c}_n$ of this expansion are not orthogonal (they are nearly orthogonal for large n). In the following, we show that an arbitrary process $\mathbf{x}(t)$, stationary or not, can be expanded into a series with orthogonal coefficients.

The Karhunen–Loève Expansion

The Fourier series is a special case of the expansion of a process $\mathbf{x}(t)$ into a series of the form

$$\hat{\mathbf{x}}(t) = \sum_{n=1}^{\infty} \mathbf{c}_n\varphi_n(t) \qquad 0 < t < T \tag{12-55}$$

where $\varphi_n(t)$ is a set of orthonormal functions in the interval $(0, T)$:

$$\int_0^T \varphi_n(t)\varphi_m^*(t)\,dt = \delta[n-m] \qquad (12\text{-}56)$$

and the coefficients $\mathbf{c}_n$ are RVs given by

$$\mathbf{c}_n = \int_0^T \mathbf{x}(t)\varphi_n^*(t)\,dt \qquad (12\text{-}57)$$

In the following, we consider the problem of determining a set of orthonormal functions $\varphi_n(t)$ such that: (a) the sum in (12-55) equals $\mathbf{x}(t)$; (b) the coefficients $\mathbf{c}_n$ are orthogonal.

To solve this problem, we form the *integral equation*

$$\int_0^T R(t_1, t_2)\varphi(t_2)\,dt_2 = \lambda\varphi(t_1) \qquad 0 < t_1 < T \qquad (12\text{-}58)$$

where $R(t_1, t_2)$ is the autocorrelation of the process $\mathbf{x}(t)$. It is well known from the theory of integral equations that the eigenfunctions $\varphi_n(t)$ of (12-58) are orthonormal as in (12-56) and they satisfy the identity

$$R(t,t) = \sum_{n=1}^{\infty} \lambda_n |\varphi_n(t)|^2 \qquad (12\text{-}59)$$

where λ_n are the corresponding eigenvalues. This is a consequence of the p.d. character of $R(t_1, t_2)$.

Using the above, we shall show that if $\varphi_n(t)$ are the eigenfunctions of (12-58) then

$$E\{|\mathbf{x}(t) - \hat{\mathbf{x}}(t)|^2\} = 0 \qquad 0 < t < T \qquad (12\text{-}60)$$

and

$$E\{\mathbf{c}_n \mathbf{c}_m^*\} = \lambda_n \delta[n-m] \qquad (12\text{-}61)$$

Proof. From (12-57) and (12-58) it follows that

$$E\{\mathbf{c}_n \mathbf{x}^*(\alpha)\} = \int_0^T R^*(\alpha, t)\varphi_n^*(t)\,dt = \lambda_n \varphi_n^*(\alpha)$$

$$E\{\mathbf{c}_n \mathbf{c}_m^*\} = \lambda_m \int_0^T \varphi_n^*(t)\varphi_m(t)\,dt = \lambda_n \delta[n-m] \qquad (12\text{-}62)$$

Hence

$$E\{\mathbf{c}_n \hat{\mathbf{x}}^*(t)\} = \sum_{m=1}^{\infty} E\{\mathbf{c}_n \mathbf{c}_m^*\}\varphi_m^*(t) = \lambda_n \varphi_n^*(t)$$

$$E\{\hat{\mathbf{x}}(t)\mathbf{x}^*(t)\} = \sum_{n=1}^{\infty} \lambda_n \varphi_n(t)\varphi_n^*(t) = R(t,t)$$

$$= E\{\hat{\mathbf{x}}^*(t)\mathbf{x}(t)\} = E\{|\mathbf{x}(t)|^2\} = E\{|\hat{\mathbf{x}}(t)|^2\}$$

and (12-60) results.

It is of interest to note that the converse of the above is also true: If $\varphi_n(t)$ is an orthonormal set of functions and

$$\mathbf{x}(t) = \sum_{n=1}^{\infty} \mathbf{c}_n \varphi_n(t) \qquad E\{\mathbf{c}_n \mathbf{c}_m^*\} = \begin{cases} \sigma_n^2 & n = m \\ 0 & n \neq m \end{cases}$$

then the functions $\varphi_n(t)$ must satisfy (12-58) with $\lambda = \sigma_n^2$.

Proof. From the assumptions it follows that $\mathbf{c}_n$ is given by (12-57). Furthermore,

$$E\{\mathbf{x}(t)\mathbf{c}_m^*\} = \sum_{n=1}^{\infty} E\{\mathbf{c}_n \mathbf{c}_m^*\} \varphi_n(t) = \sigma_m^2 \varphi_m(t)$$

$$E\{\mathbf{x}(t)\mathbf{c}_m^*\} = \int_0^T E\{\mathbf{x}(t)\mathbf{x}^*(\alpha)\} \varphi_m(\alpha)\, d\alpha = \int_0^T R(t, \alpha) \varphi_m(\alpha)\, d\alpha$$

This completes the proof.

The sum in (12-55) is called the Karhunen–Loève (K–L) expansion of the process $\mathbf{x}(t)$. In this expansion, $\mathbf{x}(t)$ need not be stationary. If it is stationary, then the origin can be chosen arbitrarily. We shall illustrate with two examples.

Example 12-9. Suppose that the process $\mathbf{x}(t)$ is ideal low-pass with autocorrelation

$$R(\tau) = \frac{\sin a\tau}{\pi\tau}$$

We shall find its K–L expansion. Shifting the origin appropriately, we conclude from (12-58) that the functions $\varphi_n(t)$ must satisfy the integral equation

$$\int_{-T/2}^{T/2} \frac{\sin a(t - \tau)}{\pi(t - \tau)} \varphi_n(\tau)\, d\tau = \lambda_n \varphi_n(t) \tag{12-63}$$

The solutions of this equation are known as *prolate-spheroidal* functions.†

Example 12-10. We shall determine the K–L expansion (12-55) of the Wiener process $\mathbf{w}(t)$ introduced in Sec. 11-1. In this case [see (11-5)]

$$R(t_1, t_2) = \alpha \min(t_1, t_2) = \begin{cases} \alpha t_2 & t_2 < t_1 \\ \alpha t_1 & t_2 > t_1 \end{cases}$$

Inserting into (12-58), we obtain

$$\alpha \int_0^{t_1} t_2 \varphi(t_2)\, dt_2 + \alpha t_1 \int_{t_1}^{T} \varphi(t_2)\, dt_2 = \lambda \varphi(t_1) \tag{12-64}$$

To solve the above integral equation, we evaluate the appropriate endpoint

†D. Slepian, H. J. Landau, and H. O. Pollack: "Prolate Spheroidal Wave Functions," *Bell System Technical Journal*, vol. 40, 1961.

conditions and differentiate twice. This yields

$$\varphi(0) = 0 \qquad \alpha \int_{t_1}^{T} \varphi(t_2)\, dt_2 = \lambda \varphi'(t_1)$$

$$\varphi'(T) = 0 \qquad \lambda \varphi''(t) + \alpha \varphi(t) = 0$$

Solving the last equation, we obtain

$$\varphi_n(t) = \sqrt{\frac{2}{T}} \sin \omega_n t \qquad \omega_n = \sqrt{\frac{\alpha}{\lambda_n}} = \frac{(2n+1)\pi}{2T}$$

Thus, in the interval $(0, T)$, the Wiener process can be written as a sum of sine waves

$$\mathbf{w}(t) = \sqrt{\frac{2}{T}} \sum_{n=1}^{\infty} \mathbf{c}_n \sin \omega_n t \qquad \mathbf{c}_n = \sqrt{\frac{2}{T}} \int_0^T \mathbf{w}(t) \sin \omega_n t\, dt$$

where the coefficients $\mathbf{c}_n$ are uncorrelated with variance $E\{\mathbf{c}_n^2\} = \lambda_n$.

12-4 SPECTRAL REPRESENTATION OF RANDOM PROCESSES

The Fourier transform of a stochastic process $\mathbf{x}(t)$ is a stochastic process $\mathbf{X}(\omega)$ given by

$$\mathbf{X}(\omega) = \int_{-\infty}^{\infty} \mathbf{x}(t) e^{-j\omega t}\, dt \tag{12-65}$$

The integral is interpreted as an MS limit. Reasoning as in (12-52), we can show that (inversion formula)

$$\mathbf{x}(t) = \frac{1}{2\pi} \int_{-\infty}^{\infty} \mathbf{X}(\omega) e^{j\omega t}\, d\omega \tag{12-66}$$

in the MS sense. The properties of Fourier transforms also hold for random signals. For example, if $\mathbf{y}(t)$ is the output of a linear system with input $\mathbf{x}(t)$ and system function $H(\omega)$, then $\mathbf{Y}(\omega) = \mathbf{X}(\omega)H(\omega)$.

The mean of $\mathbf{X}(\omega)$ equals the Fourier transform of the mean of $\mathbf{x}(t)$. We shall express the autocorrelation of $\mathbf{X}(\omega)$ in terms of the two-dimensional Fourier transform:

$$\Gamma(u, v) = \int_{-\infty}^{\infty} \int_{-\infty}^{\infty} R(t_1, t_2) e^{-j(ut_1 + vt_2)}\, dt_1\, dt_2 \tag{12-67}$$

of the autocorrelation $R(t_1, t_2)$ of $\mathbf{x}(t)$. Multiplying (12-65) by its conjugate and taking expected values, we obtain

$$E\{\mathbf{X}(u)\mathbf{X}^*(v)\} = \int_{-\infty}^{\infty} \int_{-\infty}^{\infty} E\{\mathbf{x}(t_1)\mathbf{x}^*(t_2)\} e^{-j(ut_1 - vt_2)}\, dt_1\, dt_2$$

Hence

$$E\{\mathbf{X}(u)\mathbf{X}^*(v)\} = \Gamma(u, -v) \tag{12-68}$$

Using (12-68), we shall show that, if $\mathbf{x}(t)$ is nonstationary white noise with average power $q(t)$, then $\mathbf{X}(\omega)$ is a stationary process and its autocorrelation equals the Fourier transform $Q(\omega)$ of $q(t)$:

THEOREM. If $R(t_1, t_2) = q(t_1)\delta(t_1 - t_2)$, then

$$E\{\mathbf{X}(\omega + \alpha)\mathbf{X}^*(\alpha)\} = Q(\omega) = \int_{-\infty}^{\infty} q(t)e^{-j\omega t}\,dt \tag{12-69}$$

Proof. From the identity

$$\int_{-\infty}^{\infty}\int_{-\infty}^{\infty} q(t_1)\delta(t_1 - t_2)e^{-j(ut_1 + vt_2)}\,dt_1\,dt_2 = \int_{-\infty}^{\infty} q(t_2)e^{-j(u+v)t_2}\,dt_2$$

it follows that $\Gamma(u, v) = Q(u + v)$. Hence [see (12-68)]

$$E\{\mathbf{X}(\omega + \alpha)\mathbf{X}^*(\alpha)\} = \Gamma(\omega + \alpha, -\alpha) = Q(\omega)$$

Note that if the process $\mathbf{x}(t)$ is *real*, then

$$E\{\mathbf{X}(u)\mathbf{X}(v)\} = \Gamma(u, v) \tag{12-70}$$

Furthermore,

$$\mathbf{X}(-\omega) = \mathbf{X}^*(\omega) \qquad \Gamma(-u, -v) = \Gamma^*(u, v) \tag{12-71}$$

Covariance of energy spectrum. To find the autocovariance of $|\mathbf{X}(\omega)|^2$, we must know the fourth-order moments of $\mathbf{X}(\omega)$. However, if the process $\mathbf{x}(t)$ is normal, the results can be expressed in terms of the function $\Gamma(u, v)$. We shall assume that the process $\mathbf{x}(t)$ is *real* with

$$\mathbf{X}(\omega) = \mathbf{A}(\omega) + j\mathbf{B}(\omega) \qquad \Gamma(u, v) = \Gamma_r(u, v) + j\Gamma_i(u, v) \tag{12-72}$$

From (12-68) and (12-70) it follows that

$$\begin{aligned}
2E\{\mathbf{A}(u)\mathbf{A}(v)\} &= \Gamma_r(u, v) + \Gamma_r(u, -v) \\
2E\{\mathbf{A}(v)\mathbf{B}(u)\} &= \Gamma_i(u, v) + \Gamma_i(u, -v) \\
2E\{\mathbf{B}(u)\mathbf{B}(v)\} &= \Gamma_r(u, v) - \Gamma_r(u, -v) \\
2E\{\mathbf{A}(u)\mathbf{B}(v)\} &= \Gamma_i(u, v) - \Gamma_i(u, -v)
\end{aligned} \tag{12-73}$$

THEOREM. If $\mathbf{x}(t)$ is a real normal process with zero mean, then

$$\text{Cov}\{|\mathbf{X}(u)|^2, |\mathbf{X}(v)|^2\} = \Gamma^2(u, -v) + \Gamma^2(u, v) \tag{12-74}$$

Proof. From the normality of $\mathbf{x}(t)$ it follows that the processes $\mathbf{A}(\omega)$ and $\mathbf{B}(\omega)$ are jointly normal with zero mean. Hence [see (7-36)]

$$\begin{aligned} E\{|\mathbf{X}(u)|^2|\mathbf{X}(v)|^2\} &- E\{|\mathbf{X}(u)|^2\}E\{|\mathbf{X}(v)|^2\} \\ &= E\{[\mathbf{A}^2(u) + \mathbf{B}^2(u)][\mathbf{A}^2(v) + \mathbf{B}^2(v)]\} \\ &\quad - E\{\mathbf{A}^2(u) + \mathbf{B}^2(u)\}E\{\mathbf{A}^2(v) + \mathbf{B}^2(v)\} \\ &= 2E^2\{\mathbf{A}(u)\mathbf{A}(v)\} + 2E^2\{\mathbf{B}(u)\mathbf{B}(v)\} \\ &\quad + 2E^2\{\mathbf{A}(u)\mathbf{B}(v)\} + 2E^2\{\mathbf{A}(v)\mathbf{B}(u)\} \end{aligned}$$

Inserting (12-73) into the above, we obtain (12-74).

STATIONARY PROCESSES. Suppose that $\mathbf{x}(t)$ is a stationary process with autocorrelation $R(t_1, t_2) = R(t_1 - t_2)$ and power spectrum $S(\omega)$. We shall show that

$$\Gamma(u,v) = 2\pi S(u)\delta(u+v) \tag{12-75}$$

Proof. With $t_1 = t_2 + \tau$, it follows from (12-67) that the two-dimensional transform of $R(t_1 - t_2)$ equals

$$\int_{-\infty}^{\infty}\int_{-\infty}^{\infty} R(t_1 - t_2)e^{-j(ut_1+vt_2)}\,dt_1\,dt_2 = \int_{-\infty}^{\infty} e^{-j(u+v)t_2}\int_{-\infty}^{\infty} R(\tau)e^{-ju\tau}\,d\tau\,dt_2$$

Hence

$$\Gamma(u,v) = S(u)\int_{-\infty}^{\infty} e^{-j(u+v)t_2}\,dt_2$$

This yields (12-75) because $\int e^{-j\omega t}\,dt = 2\pi\delta(\omega)$.

From (12-75) and (12-68) it follows that

$$E\{\mathbf{X}(u)\mathbf{X}^*(v)\} = 2\pi S(u)\delta(u-v) \tag{12-76}$$

This shows that the Fourier transform of a stationary process is nonstationary white noise with average power $2\pi S(u)$. It can be shown that the converse is also true (see Prob. 12-12): The process $\mathbf{x}(t)$ in (12-66) is WSS iff $E\{\mathbf{X}(\omega)\} = 0$ for $\omega \neq 0$, and

$$E\{\mathbf{X}(u)\mathbf{X}^*(v)\} = Q(u)\delta(u-v) \tag{12-77}$$

Real processes. If $\mathbf{x}(t)$ is real, then $\mathbf{A}(-\omega) = \mathbf{A}(\omega)$, $\mathbf{B}(-\omega) = \mathbf{B}(\omega)$, and

$$\mathbf{x}(t) = \frac{1}{\pi}\int_0^{\infty}\mathbf{A}(\omega)\cos\omega t\,d\omega - \frac{1}{\pi}\int_0^{\infty}\mathbf{B}(\omega)\sin\omega t\,d\omega \tag{12-78}$$

It suffices, therefore, to specify $\mathbf{A}(\omega)$ and $\mathbf{B}(\omega)$ for $\omega \geq 0$ only. From (12-68) and (12-70) it follows that

$$E\{[\mathbf{A}(u) + j\mathbf{B}(u)][\mathbf{A}(v) \pm j\mathbf{B}(v)]\} = 0 \qquad u \neq \pm v$$

Equating real and imaginary parts, we obtain

$$E\{\mathbf{A}(u)\mathbf{A}(v)\} = E\{\mathbf{A}(u)\mathbf{B}(v)\} = E\{\mathbf{B}(u)\mathbf{B}(v)\} = 0 \qquad \text{for} \quad u \neq v \tag{12-79a}$$

With $u = \omega$ and $v = -\omega$, (12-9) yields $E\{\mathbf{X}(\omega)\mathbf{X}(\omega)\} = 0$ for $\omega \neq 0$; hence

$$E\{\mathbf{A}^2(\omega)\} = E\{\mathbf{B}^2(\omega)\} \qquad E\{\mathbf{A}(\omega)\mathbf{B}(\omega)\} = 0 \tag{12-79b}$$

It can be shown that the converse is also true (see Prob. 12-13). Thus a real process $\mathbf{x}(t)$ is WSS if the coefficients $\mathbf{A}(\omega)$ and $\mathbf{B}(\omega)$ of its expansion (12-78) satisfy (12-79) and $E\{\mathbf{A}(\omega)\} = E\{\mathbf{B}(\omega)\} = 0$ for $\omega \neq 0$.

Windows. Given a WSS process $\mathbf{x}(t)$ and a function $w(t)$ with Fourier transform $W(\omega)$, we form the process $\mathbf{y}(t) = w(t)\mathbf{x}(t)$. This process is nonstationary with autocorrelation

$$R_{yy}(t_1, t_2) = w(t_1)w^*(t_2)R(t_1 - t_2)$$

The Fourier transform of $R_{yy}(t_1, t_2)$ equals

$$\Gamma_{yy}(u, v) = \int_{-\infty}^{\infty}\int_{-\infty}^{\infty} w(t_1)w^*(t_2)R(t_1 - t_2)e^{-j(ut_1 + vt_2)}\, dt_1\, dt_2$$

Proceeding as in the proof of (12-75), we obtain

$$\Gamma_{yy}(u, v) = \frac{1}{2\pi}\int_{-\infty}^{\infty} W(u - \beta)W^*(-v - \beta)S(\beta)\, d\beta \tag{12-80}$$

From (12-68) and the above it follows that the autocorrelation of the Fourier transform

$$\mathbf{Y}(\omega) = \int_{-\infty}^{\infty} w(t)\mathbf{x}(t)e^{-j\omega t}\, dt \tag{12-81}$$

of $\mathbf{y}(t)$ equals

$$E\{\mathbf{Y}(u)\mathbf{Y}^*(v)\} = \Gamma_{yy}(u, -v) = \frac{1}{2\pi}\int_{-\infty}^{\infty} W(u - \beta)W^*(v - \beta)S(\beta)\, d\beta$$

Hence

$$E\{|\mathbf{Y}(\omega)|^2\} = \frac{1}{2\pi}\int_{-\infty}^{\infty} |W(\omega - \beta)|^2 S(\beta)\, d\beta \tag{12-82}$$

Example 12-11. The integral

$$\mathbf{X}_T(\omega) = \int_{-T}^{T} \mathbf{x}(t)e^{-j\omega t}\, dt$$

is the transform of the segment $\mathbf{x}(t)p_T(t)$ of the process $\mathbf{x}(t)$. This is a special case of (12-81) with $w(t) = p_T(t)$ and $W(\omega) = 2\sin T\omega/\omega$. If, therefore, $\mathbf{x}(t)$ is a stationary process, then [see (12-82)]

$$E\{|\mathbf{X}_T(\omega)|^2\} = S(\omega) * \frac{2\sin^2 T\omega}{\pi\omega^2} \tag{12-83}$$

Fourier-Stieltjes Representation of WSS Processes†

We shall express the spectral representation of a WSS process $\mathbf{x}(t)$ in terms of the integral

$$\mathbf{Z}(\omega) = \int_0^{\omega} \mathbf{X}(\alpha)\, d\alpha \tag{12-84}$$

We have shown that the Fourier transform $\mathbf{X}(\omega)$ of $\mathbf{x}(t)$ is nonstationary white noise with average power $2\pi S(\omega)$. From (12-76) it follows that, $\mathbf{Z}(\omega)$ is a process with orthogonal increments:

For any $\omega_1 < \omega_2 < \omega_3 < \omega_4$:

$$E\{[\mathbf{Z}(\omega_2) - \mathbf{Z}(\omega_1)][\mathbf{Z}^*(\omega_4) - \mathbf{Z}^*(\omega_3)]\} = 0 \tag{12-85a}$$

$$E\{|\mathbf{Z}(\omega_2) - \mathbf{Z}(\omega_1)|^2\} = 2\pi\int_{\omega_1}^{\omega_2} S(\omega)\, d\omega \tag{12-85b}$$

Clearly,

$$d\mathbf{Z}(\omega) = \mathbf{X}(\omega)\, d\omega \tag{12-86}$$

hence the inversion formula (12-66) can be written as a Fourier-Stieltjes integral:

$$\mathbf{x}(t) = \frac{1}{2\pi}\int_{-\infty}^{\infty} e^{j\omega t}\, d\mathbf{Z}(\omega) \tag{12-87}$$

With $\omega_1 = u$, $\omega_2 = u + du$ and $\omega_3 = v$, $\omega_4 = v + dv$, (12-85) yields

$$\begin{aligned} E\{d\mathbf{Z}(u)\, d\mathbf{Z}^*(v)\} &= 0 \qquad u \neq v \\ E\{|d\mathbf{Z}(u)|^2\} &= 2\pi S(u)\, du \end{aligned} \tag{12-88}$$

The last equation can be used to define the spectrum $S(\omega)$ of WSS process $\mathbf{x}(t)$ in terms of the process $\mathbf{Z}(\omega)$.

WOLD'S DECOMPOSITION. Using (12-85), we shall show that an arbitrary WSS process $\mathbf{x}(t)$ can be written as a sum:

$$\mathbf{x}(t) = \mathbf{x}_r(t) + \mathbf{x}_p(t) \tag{12-89}$$

where $\mathbf{x}_r(t)$ is a *regular* process and $\mathbf{x}_p(t)$ is a *predictable* process consisting of

---

†H. Cramer: *Mathematical Methods of Statistics*. Princeton Univ. Press, Princeton, N.J., 1946.

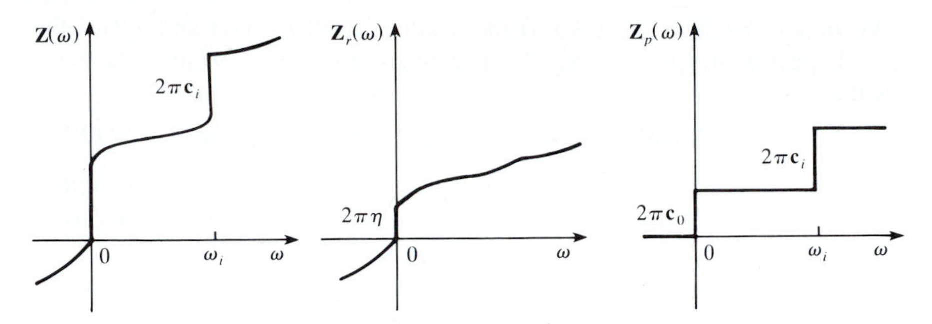

FIGURE 12-5

exponentials:

$$\mathbf{x}_p(t) = \mathbf{c}_0 + \sum_i \mathbf{c}_i e^{j\omega_i t} \qquad E\{\mathbf{c}_i\} = 0 \tag{12-90}$$

Furthermore, the two processes are orthogonal:

$$E\{\mathbf{x}_r(t+\tau)\mathbf{x}_p^*(t)\} = 0 \tag{12-91}$$

This expansion is called *Wold's decomposition*. In Sec. 14-2, we determine the processes $\mathbf{x}_r(t)$ and $\mathbf{x}_p(t)$ as the responses of two linear systems with input $\mathbf{x}(t)$. We also show that $\mathbf{x}_p(t)$ is predictable in the sense that it is determined in terms of its past; the process $\mathbf{x}_r(t)$ is not predictable.

We shall prove (12-89) using the properties of the integrated transform $\mathbf{Z}(\omega)$ of $\mathbf{x}(t)$. The process $\mathbf{Z}(\omega)$ is a family of functions. In general, these functions are discontinuous at a set of points ω_i for almost every outcome. We expand $\mathbf{Z}(\omega)$ as a sum (Fig. 12-5)

$$\mathbf{Z}(\omega) = \mathbf{Z}_r(\omega) + \mathbf{Z}_p(\omega) \tag{12-92}$$

where $\mathbf{Z}_r(\omega)$ is a continuous process for $\omega \neq 0$ and $\mathbf{Z}_p(\omega)$ is a staircase function with discontinuities at ω_i. We denote by $2\pi\mathbf{c}_i$ the discontinuity jumps at $\omega_i \neq 0$. These jumps equal the jumps of $\mathbf{Z}_p(\omega)$. We write the jump at $\omega = 0$ as a sum $2\pi(\eta + \mathbf{c}_0)$ where $\eta = E\{\mathbf{x}(t)\}$, and we associate the term $2\pi\eta$ with $\mathbf{Z}_r(\omega)$. Thus at $\omega = 0$ the process $\mathbf{Z}_r(\omega)$ is discontinuous with jump equal to $2\pi\eta$. The jump of $\mathbf{Z}_p(\omega)$ at $\omega = 0$ equals $2\pi\mathbf{c}_0$. Inserting (12-92) into (12-87), we obtain the decomposition (12-89) of $\mathbf{x}(t)$ where $\mathbf{x}_r(t)$ and $\mathbf{x}_p(t)$ are the components due to $\mathbf{Z}_r(\omega)$ and $\mathbf{Z}_p(\omega)$ respectively.

From (12-85) it follows that $\mathbf{Z}_r(\omega)$ and $\mathbf{Z}_p(\omega)$ are two processes with orthogonal increments and such that

$$E\{\mathbf{Z}_r(u)\mathbf{Z}_p^*(v)\} = 0 \qquad E\{\mathbf{c}_i\mathbf{c}_j^*\} = \begin{cases} k_i & i = j \\ 0 & i \neq j \end{cases} \tag{12-93}$$

The first equation shows that the processes $\mathbf{x}_r(t)$ and $\mathbf{x}_p(t)$ are orthogonal as in (12-89); the second shows that the coefficients $\mathbf{c}_i$ of $\mathbf{x}_p(t)$ are orthogonal. This also follows from the stationarity of $\mathbf{x}_p(t)$.

We denote by $S_r(\omega)$ and $S_p(\omega)$ the spectra and by $F_r(\omega)$ and $F_p(\omega)$ the integrated spectra of $\mathbf{x}_r(t)$ and $\mathbf{x}_p(t)$ respectively. From (12-89) and (12-91) it follows that

$$S(\omega) = S_r(\omega) + S_p(\omega) \qquad F(\omega) = F_r(\omega) + F_p(\omega) \tag{12-94}$$

The term $F_r(\omega)$ is continuous for $\omega \neq 0$; for $\omega = 0$ it is discontinuous with a jump equal to $2\pi\eta^2$. The term $F_p(\omega)$ is a staircase function, discontinuous at the points ω_i with jumps equal to $2\pi k_i$. Hence

$$S_p(\omega) = 2\pi k_0 \delta(\omega) + 2\pi \sum_i k_i \delta(\omega - \omega_i) \tag{12-95}$$

The impulse at the origin of $S(\omega)$ equals $2\pi(k_0 + \eta^2)\delta(\omega)$.

Example 12-12. Consider the process

$$\mathbf{y}(t) = \mathbf{a}\mathbf{x}(t) \qquad E\{\mathbf{a}\} = 0$$

where $\mathbf{x}(t)$ is a regular process independent of $\mathbf{a}$. We shall determine its Wold decomposition.

From the assumptions it follows that

$$E\{\mathbf{y}(t)\} = 0 \qquad R_{yy}(\tau) = E\{\mathbf{a}^2\mathbf{x}(t+\tau)\mathbf{x}(t)\} = \sigma_a^2 R_{xx}(\tau)$$

The spectrum of $\mathbf{x}(t)$ equals $S_{xx}^c(\omega) + 2\pi\eta_x^2\delta(\omega)$. Hence

$$S_{yy}(\omega) = \sigma_a^2 S_{xx}^c(\omega) + 2\pi\sigma_a^2\eta_x^2\delta(\omega)$$

From the regularity of $\mathbf{x}(t)$ it follows that its covariance spectrum $S_{xx}^c(\omega)$ has no impulses. Since $\eta_y = 0$, we conclude from (12-95) that $S_p(\omega) = 2\pi k_0 \delta(\omega)$ where $k_0 = \sigma_a^2\eta_x^2$. This yields

$$\mathbf{y}_p(t) = \eta_x \mathbf{a} \qquad \mathbf{y}_r(t) = \mathbf{a}[\mathbf{x}(t) - \eta_x]$$

DISCRETE-TIME PROCESSES. Given a discrete-time process $\mathbf{x}[n]$, we form its discrete Fourier transform (DFT)

$$\mathbf{X}(\omega) = \sum_{n=-\infty}^{\infty} \mathbf{x}[n]e^{-jn\omega} \tag{12-96}$$

This yields

$$\mathbf{x}[n] = \frac{1}{2\pi}\int_{-\pi}^{\pi} \mathbf{X}(\omega)e^{jn\omega}\,d\omega \tag{12-97}$$

From the definition it follows that the process $\mathbf{X}(\omega)$ is periodic with period 2π. It suffices, therefore, to study its properties for $|\omega| < \pi$ only. The preceding results properly modified also hold for discrete-time processes. We shall discuss only the digital version of (12-76):

If $\mathbf{x}[n]$ is a WSS process with power spectrum $S(\omega)$, then its DFT $\mathbf{X}(\omega)$ is nonstationary white noise with autocovariance

$$E\{\mathbf{X}(u)\mathbf{X}^*(v)\} = 2\pi S(u)\delta(u - v) \qquad -\pi < u, v < \pi \tag{12-98}$$

Proof. The proof is based on the identity

$$\sum_{n=-\infty}^{\infty} e^{-jn\omega} = 2\pi\delta(\omega) \qquad |\omega| < \pi$$

Clearly,

$$E\{\mathbf{X}(u)\mathbf{X}^*(v)\} = \sum_{n=-\infty}^{\infty}\sum_{m=-\infty}^{\infty} E\{\mathbf{x}[n+m]\mathbf{x}^*[m]\}\exp\{-j[(m+n)u - nv]\}$$

$$= \sum_{n=-\infty}^{\infty} e^{-jn(u-v)} \sum_{m=-\infty}^{\infty} R[m]e^{-jmu}$$

and (12-98) results.

BISPECTRA AND THIRD ORDER MOMENTS. Consider a real SSS process $\mathbf{x}(t)$ with Fourier transform $\mathbf{X}(\omega)$ and third-order moment $R(\mu, \nu)$ [see (11-179)]. Generalizing (12-76), we shall express the third-order moment of $\mathbf{X}(\omega)$ in terms of the bispectrum $S(u, v)$ of $\mathbf{x}(t)$.

THEOREM.

$$E\{\mathbf{X}(u)\mathbf{X}(v)\mathbf{X}^*(w)\} = 2\pi S(u,v)\delta(u+v-w) \tag{12-99}$$

Proof. From (12-65) it follows that the left side of (12-99) equals

$$\int_{-\infty}^{\infty}\int_{-\infty}^{\infty}\int_{-\infty}^{\infty} E\{\mathbf{x}(t_1)\mathbf{x}(t_2)\mathbf{x}(t_3)\}e^{-j(ut_1+vt_2-wt_3)}\,dt_1\,dt_2\,dt_3$$

With $t_1 = t_3 + \mu$ and $t_2 = t_3 + \nu$, the above yields

$$\int_{-\infty}^{\infty}\int_{-\infty}^{\infty} R(\mu,\nu)e^{-j(u\mu+v\nu)}\,d\mu\,d\nu \int_{-\infty}^{\infty} e^{-j(u+v-w)t_3}\,dt_3$$

and (12-99) results because the last integral equals $2\pi\delta(u+v-w)$.

We have thus shown that the third-order moment of $\mathbf{X}(\omega)$ is 0 everywhere in the uvw space except on the plane $w = u + v$ where it equals a surface singularity with density $2\pi S(u, v)$. Using this result, we shall determine the third-order moment of the increments

$$\mathbf{Z}(\omega_i) - \mathbf{Z}(\omega_k) = \int_{\omega_i}^{\omega_k} \mathbf{X}(\omega)\,d\omega \tag{12-100}$$

of the integrated transforms $\mathbf{Z}(\omega)$ of $\mathbf{x}(t)$.

THEOREM.

$$E\{[\mathbf{Z}(\omega_2) - \mathbf{Z}(\omega_1)]\{\mathbf{Z}(\omega_4) - \mathbf{Z}(\omega_3)][\mathbf{Z}^*(\omega_6) - \mathbf{Z}^*(\omega_5)]\}$$

$$= 2\pi\int_R\int S(u,v)\,du\,dv \tag{12-101}$$

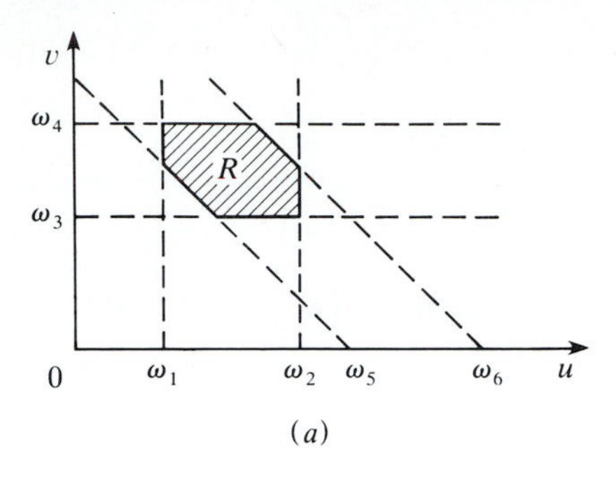

(a)

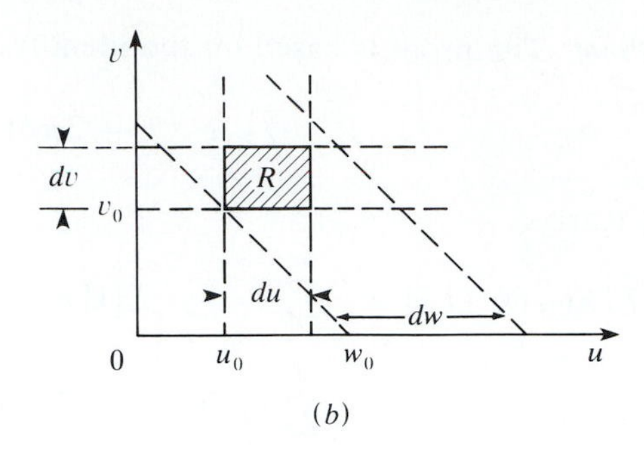

(b)

FIGURE 12-6

where R is the set of points common to the three regions

$$\omega_1 < u < \omega_2 \qquad \omega_3 < v < \omega_4 \qquad \omega_5 < w < \omega_6$$

(shaded in Fig. 12-6a) of the uv plane.

Proof. From (12-99) and (12-100) it follows that the left side of (12-101) equals

$$\int_{\omega_1}^{\omega_2}\int_{\omega_3}^{\omega_4}\int_{\omega_5}^{\omega_6} 2\pi S(u,v)\delta(u+v-w)\,du\,dv\,dw$$
$$= 2\pi\int_{\omega_1}^{\omega_2}\int_{\omega_3}^{\omega_4} S(u,v)\,du\,dv\int_{\omega_5}^{\omega_6}\delta(u+v-w)\,dw$$

The last integral equals one for $\omega_5 < u+v < \omega_6$ and 0 otherwise. Hence the right side equals the integral of $2\pi S(u,v)$ in the set R as in (12-101).

COROLLARY. Consider the differentials

$$d\mathbf{Z}(u_0) = \mathbf{X}(u_0)\,du \qquad d\mathbf{Z}(v_0) = \mathbf{X}(v_0)\,dv \qquad d\mathbf{Z}(w_0) = \mathbf{X}(w_0)\,dw$$

We maintain that

$$E\{d\mathbf{Z}(u_0)\,d\mathbf{Z}(v_0)\,d\mathbf{Z}^*(w_0)\} = 2\pi S(u_0,v_0)\,du\,dv \qquad (12\text{-}102)$$

if $w_0 = u_0 + v_0$ and $dw \ge du + dv$; it is zero if $w_0 \ne u_0 + v_0$.

Proof. Setting

$$\omega_1 = u_0 \qquad \omega_3 = v_0 \qquad \omega_5 = w_0 = u_0 + v_0$$
$$\omega_2 = u_0 + du \qquad \omega_4 = v_0 + dv \qquad \omega_6 \ge w_0 + du + dv$$

into (12-101), we obtain (12-102) because the set R is the shaded rectangle of Fig. 12-6b.

We conclude with the observation that equation (12-102) can be used to define the bispectrum of a SSS process $\mathbf{x}(t)$ in terms of $\mathbf{Z}(\omega)$.

PROBLEMS

12-1. Find $R_x[m]$ and the whitening filter of $\mathbf{x}[n]$ if

$$\mathbf{S}_x(\omega) = \frac{\cos 2\omega + 1}{12\cos 2\omega - 70\cos\omega + 62}$$

12-2. Find the innovations filter of the process $\mathbf{x}(t)$ if

$$S_x(\omega) = \frac{\omega^4 + 64}{\omega^4 + 10\omega^2 + 9}$$

12-3. Show that if $l_s[n]$ is the delta response of the innovations filter of $\mathbf{s}[n]$, then

$$R_s[0] = \sum_{n=0}^{\infty} l_s^2[n]$$

12-4. The process $\mathbf{x}(t)$ is WSS and

$$\mathbf{y}''(t) + 3\mathbf{y}'(t) + 2\mathbf{y}(t) = \mathbf{x}(t)$$

Show that (a)

$$\left.\begin{aligned} R''_{yx}(\tau) + 3R'_{yx}(\tau) + 2R_{yx}(\tau) &= R_{xx}(\tau) \\ R''_{yy}(\tau) + 3R'_{yy}(\tau) + 2R_{yy}(\tau) &= R_{xy}(\tau), \end{aligned}\right\} \quad \text{all } \tau$$

(b) If $R_{xx}(\tau) = q\delta(\tau)$, then $R_{yx}(\tau) = 0$ for $\tau < 0$ and for $\tau > 0$:

$$R''_{yx}(\tau) + 3R'_{yx}(\tau) + 2R_{yx}(\tau) = 0 \qquad R_{yx}(0) = 0 \qquad R'_{yx}(0^+) = q$$

$$R''_{yy}(\tau) + 3R'_{yy}(\tau) + 2R_{yy}(\tau) = 0 \qquad R_{yy}(0) = \frac{q}{12} \qquad R'_{yy}(0) = 0$$

12-5. Show that if $\mathbf{s}[n]$ is AR and $\mathbf{v}[n]$ is white noise orthogonal to $\mathbf{s}[n]$, then the process $\mathbf{x}[n] = \mathbf{s}[n] + \mathbf{v}[n]$ is ARMA. Find $\mathbf{S}_x(z)$ if $R_s[m] = 2^{-|m|}$ and $\mathbf{S}_v(z) = 5$.

12-6. Show that if $\mathbf{x}(t)$ is a WSS process and

$$\mathbf{s} = \frac{1}{n}\sum_{k=1}^{n} \mathbf{x}(kT) \qquad \text{then} \quad E\{\mathbf{s}^2\} = \frac{1}{2\pi n^2}\int_{-\infty}^{\infty} S_x(\omega)\frac{\sin^2 n\omega T/2}{\sin^2 \omega T/2}\,d\omega$$

12-7. Show that if $R_x(\tau) = e^{-c|\tau|}$, then the Karhunen–Loève expansion of $\mathbf{x}(t)$ in the interval $(-a, a)$ is the sum

$$\hat{\mathbf{x}}(t) = \sum_{n=1}^{\infty} (\beta_n \mathbf{b}_n \cos\omega_n t + \beta'_n \mathbf{b}'_n \sin\omega'_n t)$$

where

$$\tan a\omega_n = \frac{c}{\omega_n} \qquad \cot a\omega'_n = \frac{-c}{\omega'_n} \qquad \beta_n = (a + c\lambda_n)^{-1/2} \qquad \beta'_n = (a - c\lambda'_n)^{-1/2}$$

$$E\{\mathbf{b}_n^2\} = \lambda_n = \frac{2c}{c^2 + \omega_n^2} \qquad E\{\mathbf{b}_n'^2\} = \lambda'_n = \frac{2c}{c^2 + \omega_n'^2}$$

12-8. Show that if $\mathbf{x}(t)$ is WSS and

$$\mathbf{X}_T(\omega) = \int_{-T/2}^{T/2} \mathbf{x}(t)e^{-j\omega t}\,dt \qquad \text{then} \quad E\left\{\frac{\partial}{\partial T}|\mathbf{X}_T(\omega)|^2\right\} = \int_{-T}^{T} R_x(\tau)e^{-j\omega\tau}\,d\tau$$

12-9. Find the mean and the variance of the integral

$$\mathbf{X}(\omega) = \int_{-a}^{a} [5\cos 3t + \mathbf{v}(t)]e^{-j\omega t}\,dt$$

if $E\{\mathbf{v}(t)\} = 0$ and $R_v(\tau) = 2\delta(\tau)$.

12-10. Show that if

$$E\{\mathbf{x}_n\mathbf{x}_k\} = \sigma_n^2\delta[n-k] \qquad \mathbf{X}(\omega) = \sum_{n=-\infty}^{\infty} \mathbf{x}_n e^{-jn\omega T}$$

and $E\{\mathbf{x}_n\} = 0$, then $E\{\mathbf{X}(\omega)\} = 0$ and

$$E\{\mathbf{X}(u)\mathbf{X}^*(v)\} = \sum_{n=-\infty}^{\infty} \sigma_n^2 e^{-jn(u-v)T}$$

12-11. Given a nonperiodic WSS process $\mathbf{x}(t)$, we form the sum $\hat{\mathbf{x}}(t) = \sum \mathbf{c}_n e^{jn\omega_0 t}$ as in (12-51). Show that (*a*) $E\{|\mathbf{x}(t) - \hat{\mathbf{x}}(t)|^2\} = 0$ for $0 < t < T$. (*b*) $E\{\mathbf{c}_n\mathbf{c}_m^*\} = (1/T)\int_0^T \beta_n(\alpha)e^{jn\omega_0\alpha}\,d\alpha$ where $\beta_n(\alpha) = (1/T)\int_0^T R(\tau-\alpha)e^{-jn\omega_0\tau}\,d\tau$ are the coefficients of the Fourier expansion of $R(\tau-\alpha)$ in the interval $(0, T)$. (*c*) For large T, $E\{\mathbf{c}_n\mathbf{c}_m^*\} \simeq S(n\omega_0)\delta(n-m)/T$.

12-12. Show that, if the process $\mathbf{X}(\omega)$ is white noise with zero mean and autocovariance $Q(u)\delta(u-v)$, then its inverse Fourier transform $\mathbf{x}(t)$ is WSS with power spectrum $Q(\omega)/2\pi$.

12-13. Given a real process $\mathbf{x}(t)$ with Fourier transform $\mathbf{X}(\omega) = \mathbf{A}(\omega) + j\mathbf{B}(\omega)$, show that if the processes $\mathbf{A}(\omega)$ and $\mathbf{B}(\omega)$ satisfy (12-79) and $E\{\mathbf{A}(\omega)\} = E\{\mathbf{B}(\omega)\} = 0$, then $\mathbf{x}(t)$ is WSS.

12-14. We use as an estimate of the Fourier transform $F(\omega)$ of a signal $f(t)$ the integral

$$\mathbf{X}_T(\omega) = \int_{-T}^{T} [f(t) + \boldsymbol{\nu}(t)]e^{-j\omega t}\,dt$$

where $\boldsymbol{\nu}(t)$ is the measurement noise. Show that if $S_{\nu\nu}(\omega) = q$ and $E\{\boldsymbol{\nu}(t)\} = 0$, then

$$E\{\mathbf{X}_T(\omega)\} = \int_{-\infty}^{\infty} F(y)\frac{\sin T(\omega - y)}{\pi(\omega - y)}\,dy \qquad \operatorname{Var}\mathbf{X}_T(\omega) = 2qT$$

CHAPTER 13

SPECTRAL ESTIMATION

13-1 ERGODICITY

A central problem in the applications of stochastic processes is the estimation of various statistical parameters in terms of real data. Most parameters can be expressed as expected values of some functional of a process $\mathbf{x}(t)$. The problem of estimating the mean of a given process $\mathbf{x}(t)$ is, therefore, central in this investigation. We start with this problem.

For a specific t, $\mathbf{x}(t)$ is an RV; its mean $\eta(t) = E\{\mathbf{x}(t)\}$ can, therefore, be estimated as in Sec. 9-2: We observe n samples $\mathbf{x}(t, \zeta_i)$ of $\mathbf{x}(t)$ and use as the point estimate of $E\{\mathbf{x}(t)\}$ the average

$$\hat{\eta}(t) = \frac{1}{n} \sum_i \mathbf{x}(t, \zeta_i)$$

As we know, $\hat{\eta}(t)$ is a consistent estimate of $\eta(t)$; however, it can be used only if a large number of realizations $\mathbf{x}(t, \zeta_i)$ of $\mathbf{x}(t)$ are available. In many applications, we know only a *single sample* of $\mathbf{x}(t)$. Can we then estimate $\eta(t)$ in terms of the time average of the given sample? This is not possible if $E\{\mathbf{x}(t)\}$ depends on t. However, if $\mathbf{x}(t)$ is a regular stationary process, its time average tends to $E\{\mathbf{x}(t)\}$ as the length of the available sample tends to ∞. Ergodicity is a topic dealing with the underlying theory.

Mean-Ergodic Processes

We are given a real stationary process $\mathbf{x}(t)$ and we wish to estimate its mean $\eta = E\{\mathbf{x}(t)\}$. For this purpose, we form the *time average*

$$\boldsymbol{\eta}_T = \frac{1}{2T}\int_{-T}^{T} \mathbf{x}(t)\,dt \qquad (13\text{-}1)$$

Clearly, $\boldsymbol{\eta}_T$ is an RV with mean

$$E\{\boldsymbol{\eta}_T\} = \frac{1}{2T}\int_{-T}^{T} E\{\mathbf{x}(t)\}\,dt = \eta$$

Thus $\boldsymbol{\eta}_T$ is an unbiased estimator of η. If its variance $\sigma_T^2 \to 0$ as $T \to \infty$, then $\boldsymbol{\eta}_T \to \eta$ in the MS sense. In this case, the time average $\boldsymbol{\eta}_T(\zeta)$ computed from a single realization of $\mathbf{x}(t)$ is close to η with probability close to 1. If this is true, we shall say that the process $\mathbf{x}(t)$ is *mean-ergodic*. Thus a process $\mathbf{x}(t)$ is mean-ergodic if its time average $\boldsymbol{\eta}_T$ tends to the ensemble average η as $T \to \infty$.

To establish the ergodicity of a process, it suffices to find σ_T and to examine the conditions under which $\sigma_T \to 0$ as $T \to \infty$. As the following examples show, not all processes are mean-ergodic.

Example 13-1. Suppose that $\mathbf{c}$ is an RV with mean η_c and

$$\mathbf{x}(t) = \mathbf{c} \qquad \eta = E\{\mathbf{x}(t)\} = E\{\mathbf{c}\} = \eta_c$$

In this case, $\mathbf{x}(t)$ is a family of straight lines and $\boldsymbol{\eta}_T = \mathbf{c}$. For a specific sample, $\boldsymbol{\eta}_T(\zeta) = \mathbf{c}(\zeta)$ is a constant different from η if $\mathbf{c}(\zeta) \neq \eta$. Hence $\mathbf{x}(t)$ is not mean-ergodic.

Example 13-2. Given two mean-ergodic processes $\mathbf{x}_1(t)$ and $\mathbf{x}_2(t)$ with means η_1 and η_2, we form the sum

$$\mathbf{x}(t) = \mathbf{x}_1(t) + \mathbf{c}\mathbf{x}_2(t)$$

where $\mathbf{c}$ is an RV independent of $\mathbf{x}_2(t)$ taking the values 0 and 1 with probability 0.5. Clearly,

$$E\{\mathbf{x}(t)\} = E\{\mathbf{x}_1(t)\} + E\{\mathbf{c}\}E\{\mathbf{x}_2(t)\} = \eta_1 + 0.5\eta_2$$

If $\mathbf{c}(\zeta) = 0$ for a particular ζ, then $\mathbf{x}(t) = \mathbf{x}_1(t)$ and $\boldsymbol{\eta}_T \to \eta_1$ as $T \to \infty$. If $\mathbf{c}(\zeta) = 1$ for another ζ, then $\mathbf{x}(t) = \mathbf{x}_1(t) + \mathbf{x}_2(t)$ and $\boldsymbol{\eta}_T \to \eta_1 + \eta_2$ as $T \to \infty$. Hence $\mathbf{x}(t)$ is not mean-ergodic.

VARIANCE. To determine the variance σ_T^2 of the time average $\boldsymbol{\eta}_T$ of $\mathbf{x}(t)$, we start with the observation that

$$\boldsymbol{\eta}_T = \mathbf{w}(0) \qquad \text{where} \quad \mathbf{w}(t) = \frac{1}{2T}\int_{t-T}^{t+T} \mathbf{x}(\alpha)\,d\alpha \qquad (13\text{-}2)$$

is the moving average of $\mathbf{x}(t)$. As we know, $\mathbf{w}(t)$ is the output of a linear system with input $\mathbf{x}(t)$ and with impulse response a pulse centered at $t = 0$. Hence $\mathbf{w}(t)$

is stationary and its autocovariance equals

$$C_{ww}(\tau) = \frac{1}{2T}\int_{-2T}^{2T} C(\tau - \alpha)\left(1 - \frac{|\alpha|}{2T}\right) d\alpha \tag{13-3}$$

where $C(\tau)$ is the autocovariance of $\mathbf{x}(t)$ [see (10-142)]. Since $\sigma_T^2 = \operatorname{Var} \mathbf{w}(0) = C_{ww}(0)$ and $C(-\alpha) = C(\alpha)$, this yields

$$\sigma_T^2 = \frac{1}{2T}\int_{-2T}^{2T} C(\alpha)\left(1 - \frac{|\alpha|}{2T}\right) d\alpha = \frac{1}{T}\int_{0}^{2T} C(\alpha)\left(1 - \frac{\alpha}{2T}\right) d\alpha \tag{13-4}$$

This fundamental result leads to the following conclusion: A process $\mathbf{x}(t)$ with autocovariance $C(\tau)$ is mean-ergodic iff

$$\frac{1}{T}\int_{0}^{2T} C(\alpha)\left(1 - \frac{\alpha}{2T}\right) d\alpha \xrightarrow[T\to\infty]{} 0 \tag{13-5}$$

The determination of the variance of $\boldsymbol{\eta}_T$ is useful not only in establishing the ergodicity of $\mathbf{x}(t)$ but also in determining a confidence interval for the estimate η_T of η. Indeed, from Tchebycheff's inequality it follows that the probability that the unknown η is in the interval $\boldsymbol{\eta}_T \pm 10\sigma_T$ is larger than 0.99 [see (5-57)]. Hence η_T is a satisfactory estimate of η if T is such that $\sigma_T \ll \eta$.

Example 13-3. Suppose that $C(\tau) = qe^{-c|\tau|}$ as in (11-15). In this case,

$$\sigma_T^2 = \frac{q}{T}\int_{0}^{2T} e^{-c\tau}\left(1 - \frac{\tau}{2T}\right) d\tau = \frac{q}{cT}\left(1 - \frac{1 - e^{-2cT}}{2cT}\right)$$

Clearly, $\sigma_T \to 0$ as $T \to \infty$; hence $\mathbf{x}(t)$ is mean-ergodic. If $T \gg 1/c$, then $\sigma_T^2 \simeq q/cT$.

Example 13-4. Suppose that $\mathbf{x}(t) = \eta + \boldsymbol{\nu}(t)$ where $\boldsymbol{\nu}(t)$ is white noise with $R_{\nu\nu}(\tau) = q\delta(\tau)$. In this case, $C(\tau) = R_{\nu\nu}(\tau)$ and (13-4) yields

$$\sigma_T^2 = \frac{1}{2T}\int_{-2T}^{2T} q\delta(\tau)\left(1 - \frac{|\tau|}{2T}\right) d\tau = \frac{q}{2T}$$

Hence $\mathbf{x}(t)$ is mean-ergodic.

It is clear from (13-5) that the ergodicity of a process depends on the behavior of $C(\tau)$ for large τ. If $C(\tau) = 0$ for $\tau > a$, that is, if $\mathbf{x}(t)$ is a-dependent and $T \gg a$, then

$$\sigma_T^2 = \frac{1}{T}\int_{0}^{a} C(\tau)\left(1 - \frac{\tau}{2T}\right) d\tau \simeq \frac{1}{T}\int_{0}^{a} C(\tau)\, d\tau < \frac{a}{T}C(0) \xrightarrow[T\to\infty]{} 0$$

because $|C(\tau)| < C(0)$; hence $\mathbf{x}(t)$ is mean-ergodic.

In many applications, the RVs $\mathbf{x}(t + \tau)$ and $\mathbf{x}(t)$ are nearly uncorrelated for large τ, that is, $C(\tau) \to 0$ as $\tau \to \infty$. The above suggests that if this is the case, then $\mathbf{x}(t)$ is mean-ergodic and for large T the variance of $\boldsymbol{\eta}_T$ can be approximated by

$$\sigma_T^2 \simeq \frac{1}{T}\int_0^{2T} C(\tau)\,d\tau \simeq \frac{1}{T}\int_0^{\infty} C(\tau)\,d\tau = \frac{\tau_c}{T}C(0) \tag{13-6}$$

where τ_c is the correlation time of $\mathbf{x}(t)$ defined in (10-49). This result will be justified presently.

SLUTSKY'S THEOREM. A process $\mathbf{x}(t)$ is mean-ergodic iff

$$\frac{1}{T}\int_0^{T} C(\tau)\,d\tau \xrightarrow[T\to\infty]{} 0 \tag{13-7}$$

Proof. (*a*) We show first that if $\sigma_T \to 0$ as $T \to \infty$, then (13-7) is true. The covariance of the RVs $\boldsymbol{\eta}_T$ and $\mathbf{x}(0)$ equals

$$\text{Cov}[\boldsymbol{\eta}_T, \mathbf{x}(0)] = E\left\{\frac{1}{2T}\int_{-T}^{T}[\mathbf{x}(t) - \eta][\mathbf{x}(0) - \eta]\,dt\right\} = \frac{1}{2T}\int_{-T}^{T} C(t)\,dt$$

But [see (7-9)]

$$\text{Cov}^2[\boldsymbol{\eta}_T, \mathbf{x}(0)] \le \text{Var}\,\boldsymbol{\eta}_T\,\text{Var}\,\mathbf{x}(0) = \sigma_T^2 C(0)$$

Hence (13-7) holds if $\sigma_T \to 0$.

(*b*) We show next that if (13-7) is true, then $\sigma_T \to 0$ as $T \to \infty$. From (13-7) it follows that given $\varepsilon > 0$, we can find a constant c_0 such that

$$\frac{1}{t}\int_c^{t} C(\tau)\,d\tau < \varepsilon \qquad \text{for every} \quad c > c_0 \tag{13-8}$$

The variance of $\boldsymbol{\eta}_T$ equals [see (13-4)]

$$\sigma_T^2 = \frac{1}{T}\int_0^{2T_0} + \frac{1}{T}\int_{2T_0}^{2T} C(\tau)\left(1 - \frac{\tau}{2T}\right)d\tau$$

The integral from 0 to $2T_0$ is less than $2T_0C(0)/T$ because $|C(\tau)| \le C(0)$. Hence

$$\sigma_T^2 < \frac{2T_0}{T}C(0) + \frac{1}{T}\int_{2T_0}^{2T} C(\tau)\left(1 - \frac{\tau}{2T}\right)d\tau$$

But (see Fig. 13-1)

$$\int_{2T_0}^{2T} C(\tau)(2T - \tau)\,d\tau = \int_{2T_0}^{2T} C(\tau)\int_{\tau}^{2T} dt\,d\tau = \int_{2T_0}^{2T}\int_{2T_0}^{t} C(\tau)\,d\tau\,dt$$

From (13-8) it follows that the inner integral on the right is less than εt; hence

$$\sigma_T^2 < \frac{2T_0}{T}C(0) + \frac{\varepsilon}{T^2}\int_{2T_0}^{2T} t\,dt \xrightarrow[T\to\infty]{} 2\varepsilon$$

and since ε is arbitrary, we conclude that $\sigma_T \to 0$ as $T \to \infty$.

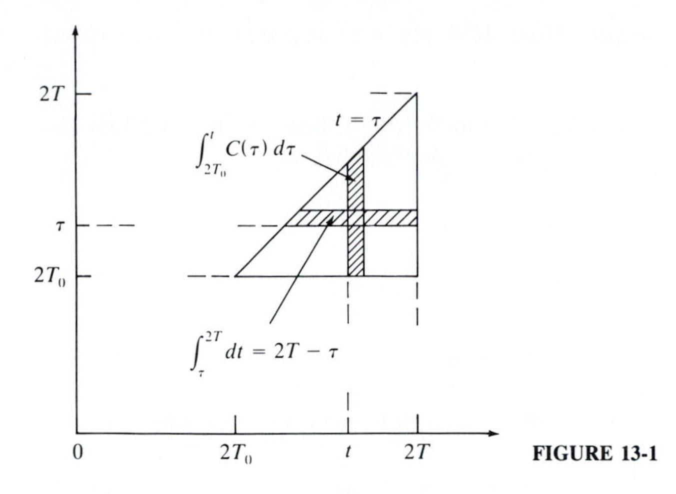

FIGURE 13-1

Example 13-5. Consider the process

$$\mathbf{x}(t) = \mathbf{a} \cos \omega t + \mathbf{b} \sin \omega t + c$$

where **a** and **b** are two uncorrelated RVs with zero mean and equal variance. As we know [see (10-55)], the process $\mathbf{x}(t)$ is WSS with mean c and autocovariance $\sigma^2 \cos \omega\tau$. We shall show that it is mean-ergodic. This follows from (13-7) and the fact that

$$\frac{1}{T}\int_0^T C(\tau)\, d\tau = \frac{\sigma^2}{T}\int_0^T \cos \omega\tau \, d\tau = \frac{\sigma^2}{\omega T} \sin \omega T \xrightarrow[T\to\infty]{} 0$$

Sufficient conditions. (a) If

$$\int_0^\infty C(\tau)\, d\tau < \infty \tag{13-9}$$

then (13-7) holds; hence the process $\mathbf{x}(t)$ is mean-ergodic.

(b) If $R(\tau) \to \eta^2$ or, equivalently, if

$$C(\tau) \to 0 \qquad \text{as} \quad \tau \to \infty \tag{13-10}$$

then $\mathbf{x}(t)$ is mean-ergodic.

Proof. If (13-10) is true, then given $\varepsilon > 0$, we can find a constant T_0 such that $|C(\tau)| < \varepsilon$ for $\tau > T_0$; hence

$$\frac{1}{T}\int_0^T C(\tau)\, d\tau = \frac{1}{T}\int_0^{T_0} C(\tau)\, d\tau + \frac{1}{T}\int_{T_0}^T C(\tau)\, d\tau$$

$$< \frac{T_0}{T} C(0) + \varepsilon \frac{T - T_0}{T} \xrightarrow[T\to\infty]{} \varepsilon$$

and since ε is arbitrary, we conclude that (13-7) is true.

Condition (13-10) is satisfied if the RVs $\mathbf{x}(t + \tau)$ and $\mathbf{x}(t)$ are *uncorrelated* for large τ.

Note The time average $\boldsymbol{\eta}_T$ is an unbiased estimator of η; however, it is not best. An estimator with smaller variance results if we use the weighted average

$$\boldsymbol{\eta}_w = \int_{-T}^{T} w(t)\mathbf{x}(t)\, dt$$

and select the function $w(t)$ appropriately (see also Example 8-4).

DISCRETE-TIME PROCESSES. We outline next, without elaboration, the discrete-time version of the preceding results. We are given a real stationary process $\mathbf{x}[n]$ with autocovariance $C[m]$ and we form the time average

$$\boldsymbol{\eta}_M = \frac{1}{N} \sum_{n=-M}^{M} \mathbf{x}[n] \qquad N = 2M + 1 \tag{13-11}$$

This is an unbiased estimator of the mean of $\mathbf{x}[n]$ and its variance equals

$$\sigma_M^2 = \frac{1}{N} \sum_{m=-2M}^{2M} C[m]\left(1 - \frac{|m|}{N}\right) \tag{13-12}$$

The process $\mathbf{x}[n]$ is mean-ergodic if the right side of (13-12) tends to 0 as $M \to \infty$.

SLUTSKY'S THEOREM. The process $\mathbf{x}[n]$ is mean-ergodic iff

$$\frac{1}{M} \sum_{m=0}^{M} C[m] \xrightarrow[m \to \infty]{} 0 \tag{13-13}$$

We can show as in (13-10) that if $C[m] \to 0$ as $m \to \infty$, then $\mathbf{x}[n]$ is mean-ergodic.

For large M,

$$\sigma_M^2 \simeq \frac{1}{M} \sum_{m=0}^{M} C[m] \tag{13-14}$$

Example 13-6. (a) Suppose that the centered process $\tilde{\mathbf{x}}[n] = \mathbf{x}[n] - \eta$ is white noise with autocovariance $P\delta[m]$. In this case,

$$C[m] = P\delta[m] \qquad \sigma_M^2 \simeq \frac{1}{N} \sum_{m=-M}^{M} P\delta[m] = \frac{P}{N}$$

Thus $\mathbf{x}[n]$ is mean-ergodic and the variance of $\boldsymbol{\eta}_M$ equals P/N. This agrees with (8-22): The RVs $\mathbf{x}[n]$ are i.i.d. with variance $C[0] = P$, and the time average $\boldsymbol{\eta}_M$ is their sample mean.

(*b*) Suppose now that $C[m] = Pa^{|m|}$ as in Example 10-31. In this case, (13-14) yields

$$\sigma_M^2 \simeq \frac{1}{N} \sum_{m=-\infty}^{\infty} Pa^{|m|} = \frac{P(1+a)}{N(1-a)}$$

Note that if we replace $\mathbf{x}[n]$ by white noise as in (*a*) with the same P and use as estimate of η the time average of N_1 terms, the variance P/N_1 of the resulting estimator will equal σ_M^2 if

$$N_1 = N\frac{1-a}{1+a}$$

Sampling. In a numerical estimate of the mean of a continuous-time process $\mathbf{x}(t)$, the time-average $\boldsymbol{\eta}_T$ is replaced by the average

$$\boldsymbol{\eta}_N = \frac{1}{N} \sum \mathbf{x}(t_n)$$

of the N samples $\mathbf{x}(t_n)$ of $\mathbf{x}(t)$. This is an unbiased estimate of η and its variance equals

$$\sigma_N^2 = \frac{1}{N^2} \sum_n \sum_k C(t_n - t_k)$$

where $C(\tau)$ is the autocovariance of $\mathbf{x}(t)$. If the samples are equidistant, then the RVs $\mathbf{x}(t_n) = \mathbf{x}(nT_0)$ form a discrete-time process with autovariance $C(mT_0)$. In this case, the variance σ_N^2 of $\boldsymbol{\eta}_N$ is given by (13-12) if we replace $C[m]$ by $C(mT_0)$.

SPECTRAL INTERPRETATION OF ERGODICITY. We shall express the ergodicity conditions in terms of the properties of the covariance spectrum

$$S^c(\omega) = S(\omega) - 2\pi\eta^2\delta(\omega)$$

of the process $\mathbf{x}(t)$. The variance σ_T^2 of $\boldsymbol{\eta}_T$ equals the variance of the moving average $\mathbf{w}(t)$ of $\mathbf{x}(t)$ [see (13-2)]. As we know,

$$S_{ww}^c(\omega) = S^c(\omega)\frac{\sin^2 T\omega}{T^2\omega^2} \tag{13-15}$$

hence

$$\sigma_T^2 = \frac{1}{2\pi}\int_{-\infty}^{\infty} S^c(\omega)\frac{\sin^2 T\omega}{T^2\omega^2}\,d\omega \tag{13-16}$$

The fraction in (13-16) takes significant values only in an interval of the order of $1/T$ centered at the origin. The ergodicity conditions of $\mathbf{x}(t)$ depend, therefore, only on the behavior of $S^c(\omega)$ near the origin.

Suppose first that the process $\mathbf{x}(t)$ is regular. In this case, $S^c(\omega)$ does not have an impulse at $\omega = 0$. If, therefore, T is sufficiently large, we can use the approximation $S^c(\omega) \simeq S^c(0)$ in (13-16). This yields

$$\sigma_T^2 \simeq \frac{S^c(0)}{2\pi}\int_{-\infty}^{\infty}\frac{\sin^2 T\omega}{T^2\omega^2}\,d\omega = \frac{S^c(0)}{2T} \xrightarrow[T\to\infty]{} 0 \tag{13-17}$$

Hence $\mathbf{x}(t)$ is mean-ergodic.

Suppose now that

$$S^c(\omega) = S_1^c(\omega) + 2\pi k_0\delta(\omega) \qquad S_1^c(0) < \infty \tag{13-18}$$

Inserting into (13-16), we conclude as in (13-17) that

$$\sigma_T^2 \simeq \frac{1}{2T}S_1(0) + k_0 \xrightarrow[T\to\infty]{} k_0$$

Hence $\mathbf{x}(t)$ is not mean-ergodic. This case arises if in Wold's decomposition (12-89) the constant term $\mathbf{c}_0$ is different from 0, or, equivalently, if the Fourier transform $\mathbf{X}(\omega)$ of $\mathbf{x}(t)$ contains the impulse $2\pi\mathbf{c}_0\delta(\omega)$.

Example 13-7. Consider the process

$$\mathbf{y}(t) = \mathbf{a}\mathbf{x}(t) \qquad E\{\mathbf{a}\} = 0$$

where $\mathbf{x}(t)$ is a mean-ergodic process independent of the RV $\mathbf{a}$. Clearly, $E\{\mathbf{y}(t)\} = 0$ and

$$S_{yy}^c(\omega) = \sigma_a^2 S_{xx}^2(\omega) + 2\pi\sigma_a^2\eta_x^2\delta(\omega)$$

as in Example 12-12. This shows that the process $\mathbf{y}(t)$ is not mean-ergodic.

The preceding discussion leads to the following equivalent conditions for mean ergodicity:

1. σ_T must tend to 0 as $T \to \infty$.
2. In Wold's decomposition (12-89) the constant random term $\mathbf{c}_0$ must be 0.
3. The integrated power spectrum $F^c(\omega)$ must be continuous at the origin.
4. The integrated Fourier transform $\mathbf{Z}(\omega)$ must be continuous at the origin.

Analog estimators. The mean η of a process $\mathbf{x}(t)$ can be estimated by the response of a physical system with input $\mathbf{x}(t)$. A simple example is a normalized integrator of finite integration time. This is a linear device with impulse response the rectangular pulse $p(t)$ of Fig. 13-2. For $t > T_0$ the output of the integrator equals

$$\mathbf{y}(t) = \frac{1}{T_0}\int_{t-T_0}^{t}\mathbf{x}(\alpha)\,d\alpha$$

If T_0 is large compared to the correlation time τ_c of $\mathbf{x}(t)$, then the variance of $\mathbf{y}(t)$ equals $2\tau_c C(0)/T_0$. This follows from (13-6) with $T_0 = 2T$.

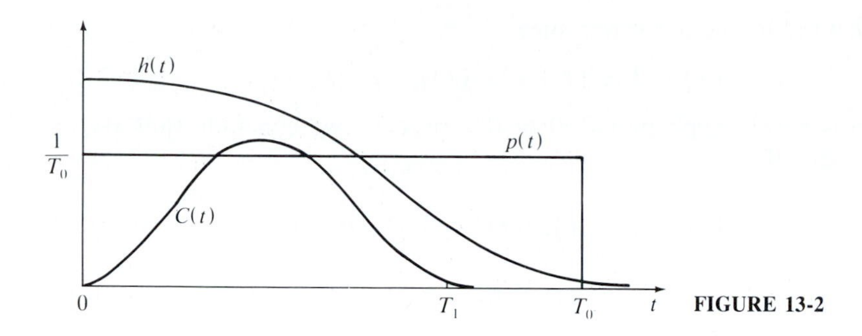

FIGURE 13-2

Suppose now that $\mathbf{x}(t)$ is the input to a system with impulse response $h(t)$ of unit area and energy E:

$$\mathbf{w}(t) = \int_0^t \mathbf{x}(\alpha)h(t-\alpha)\,d\alpha \qquad E = \int_0^\infty h^2(t)\,dt$$

We assume that $C(\tau) \simeq 0$ for $\tau > T_1$ and $h(t) \simeq 0$ for $t > T_0 > T_1$ as in Fig. 13-2. From these assumptions it follows that $E\{\mathbf{w}(t)\} = \eta$ and $\sigma_w^2 \simeq EC(0)\tau_c$ for $t > T_0$. If, therefore, $EC(0)\tau_c \ll \eta^2$ then $\mathbf{w}(t) \simeq \eta$ for $t > T_0$. The above conditions are satisfied if the system is low-pass, that is, if $H(\omega) \simeq 0$ for $|\omega| < \omega_c$ and $\omega_c \ll \eta^2/C(0)\tau_c$.

Covariance-Ergodic Processes

We shall now determine the conditions that an SSS process $\mathbf{x}(t)$ must satisfy such that its autocovariance $C(\lambda)$ can be estimated as a time average. The results are essentially the same for the estimates of the autocorrelation $R(\lambda)$ of $\mathbf{x}(t)$.

VARIANCE. We start with the estimate of the variance

$$V = C(0) = E\{|\mathbf{x}(t) - \eta|^2\} = E\{\mathbf{x}^2(t)\} - \eta^2 \tag{13-19}$$

of $\mathbf{x}(t)$.

Known mean. Suppose, first, that η is known. We can then assume, replacing the process $\mathbf{x}(t)$ by its centered process $\mathbf{x}(t) - \eta$, that

$$E\{\mathbf{x}(t)\} = 0 \qquad V = E\{\mathbf{x}^2(t)\}$$

Our problem is thus to estimate the mean V of the process $\mathbf{x}^2(t)$. Proceeding as in (13-1), we use as the estimate of V the time average

$$\mathbf{V}_T = \frac{1}{2T}\int_{-T}^{T} \mathbf{x}^2(t)\,dt \tag{13-20}$$

This estimate is unbiased and its variance is given by (13-4) where we replace the function $C(\tau)$ by the autocovariance

$$C_{x^2x^2}(\tau) = E\{\mathbf{x}^2(t+\tau)\mathbf{x}^2(t)\} - E^2\{\mathbf{x}^2(t)\} \tag{13-21}$$

of the process $\mathbf{x}^2(t)$. Applying (13-7) to this process, we conclude that $\mathbf{x}(t)$ is variance-ergodic iff

$$\frac{1}{T}\int_0^T E\{\mathbf{x}^2(t+\tau)\mathbf{x}^2(t)\}\,dt \xrightarrow[T\to\infty]{} C^2(0) \tag{13-22}$$

To test the validity of (13-22), we need the fourth-order moments of $\mathbf{x}(t)$. If, however, $\mathbf{x}(t)$ is a normal process, then [see (10-68)]

$$C_{x^2x^2}(\tau) = 2C^2(\tau) \tag{13-23}$$

From this and (13-22) it follows that a normal process is variance-ergodic iff

$$\frac{1}{T}\int_0^T C^2(\tau)\,d\tau \xrightarrow[T\to\infty]{} 0 \tag{13-24}$$

Using the simple inequality (see Prob. 13-10)

$$\left|\frac{1}{T}\int_0^T C(\tau)\,d\tau\right|^2 \le \frac{1}{T}\int_0^T C^2(\tau)\,d\tau$$

we conclude with (13-7) and (13-24) that if a normal process is variance-ergodic, it is also mean-ergodic. The converse, however, is not true. This theorem has the following spectral interpretation: The process $\mathbf{x}(t)$ is mean-ergodic iff $S^c(\omega)$ has no impulses at the origin; it is variance-ergodic iff $S^c(\omega)$ has no impulses anywhere.

Example 13-8. Suppose that the process

$$\mathbf{x}(t) = \mathbf{a}\cos\omega t + \mathbf{b}\sin\omega t + \eta$$

is normal and stationary. Clearly, $\mathbf{x}(t)$ is mean-ergodic because it does not contain a random constant. However, it is not variance-ergodic because the square

$$|\mathbf{x}(t) - \eta|^2 = \tfrac{1}{2}(\mathbf{a}^2 + \mathbf{b}^2) + \tfrac{1}{2}(\mathbf{a}^2\cos 2\omega t - \mathbf{b}^2\cos 2\omega t) + \mathbf{ab}\sin 2\omega t$$

of $\mathbf{x}(t) - \eta$ contains the random constant $(\mathbf{a}^2 + \mathbf{b}^2)/2$.

Unknown mean. If η is unknown, we evaluate its estimator $\boldsymbol{\eta}_T$ from (13-1) and form the average

$$\hat{\mathbf{V}}_T = \frac{1}{2T}\int_{-T}^{T}[\mathbf{x}(t) - \boldsymbol{\eta}_T]^2\,dt = \frac{1}{2T}\int_{-T}^{T}\mathbf{x}^2(t)\,dt - \boldsymbol{\eta}_T^2$$

The determination of the statistical properties of $\hat{\mathbf{V}}_T$ is difficult. The following observations, however, simplify the problem. In general, $\hat{\mathbf{V}}_T$ is a biased estimator of the variance V of $\mathbf{x}(t)$. However, if T is large, the bias can be neglected in the determination of the estimation error; furthermore, the variance of $\hat{\mathbf{V}}_T$ can be approximated by the variance of the known-mean estimator $\mathbf{V}_T$. In many cases, the MS error $E\{(\hat{\mathbf{V}}_T - V)^2\}$ is smaller than $E\{(\mathbf{V}_T - V)^2\}$ for moderate values of T. It might thus be preferable to use $\hat{\mathbf{V}}_T$ as the estimator of V even when η is known.

AUTOCOVARIANCE. We shall establish the ergodicity conditions for the autocovariance $C(\lambda)$ of the process $\mathbf{x}(t)$ under the assumption that $E\{\mathbf{x}(t)\} = 0$. We can do so, replacing $\mathbf{x}(t)$ by $\mathbf{x}(t) - \eta$ if η is known. If it is unknown, we replace $\mathbf{x}(t)$ by $\mathbf{x}(t) - \boldsymbol{\eta}_T$. In this case, the results are approximately correct if T is large.

For a specific λ, the product $\mathbf{x}(t + \lambda)\mathbf{x}(t)$ is an SSS process with mean $C(\lambda)$. We can, therefore, use as the estimate of $C(\lambda)$ the time average

$$\mathbf{C}_T(\lambda) = \frac{1}{2T}\int_{-T}^{T} \mathbf{z}(t)\, dt \qquad \mathbf{z}(t) = \mathbf{x}(t + \lambda)\mathbf{x}(t) \tag{13-25}$$

This is an unbiased estimator of $C(\lambda)$ and its variance is given by (13-4) if we replace the autocovariance of $\mathbf{x}(t)$ by the autocovariance

$$C_{zz}(\tau) = E\{\mathbf{x}(t + \lambda + \tau)\mathbf{x}(t + \tau)\mathbf{x}(t + \lambda)\mathbf{x}(t)\} - C^2(\lambda)$$

of the process $\mathbf{z}(t)$. Applying Slutsky's theorem, we conclude that the process $\mathbf{x}(t)$ is covariance-ergodic iff

$$\frac{1}{T}\int_0^T C_{zz}(\tau)\, d\tau \xrightarrow[T\to\infty]{} 0 \tag{13-26}$$

If $\mathbf{x}(t)$ is a normal process,

$$C_{zz}(\tau) = C(\lambda + \tau)C(\lambda - \tau) + C^2(\tau) \tag{13-27}$$

In this case, (13-6) yields

$$\operatorname{Var} \mathbf{C}_T(\lambda) \simeq \frac{1}{T}\int_0^{2T} \left[C(\lambda + \tau)C(\lambda - \tau) + C^2(\tau)\right] d\tau \tag{13-28}$$

From (13-27) it follows that if $C(\tau) \to 0$, then $C_{zz}(\tau) \to 0$ as $\tau \to \infty$; hence $\mathbf{x}(t)$ is covariance-ergodic.

Cross-covariance. We comment briefly on the estimate of the cross-covariance $C_{xy}(\tau)$ of two zero-mean processes $\mathbf{x}(t)$ and $\mathbf{y}(t)$. As in (13-25), the time average

$$\hat{\mathbf{C}}_{xy}(\tau) = \frac{1}{2T}\int_{-T}^{T} \mathbf{x}(t + \tau)\mathbf{y}(t)\, dt \tag{13-29}$$

is an unbiased estimate of $C_{xy}(\tau)$ and its variance is given by (13-4) if we replace $C(\tau)$ by $C_{xy}(\tau)$. If the functions $C_{xx}(\tau), C_{yy}(\tau)$, and $C_{xy}(\tau)$ tend to 0 as $\tau \to \infty$ then the processes $\mathbf{x}(t)$ and $\mathbf{y}(t)$ are cross-covariance-ergodic (see Prob. 13-9).

NONLINEAR ESTIMATORS. The numerical evaluation of the estimate $C_T(\lambda)$ of $C(\lambda)$ involves the evaluation of the integral of the product $\mathbf{x}(t + \lambda)\mathbf{x}(t)$ for various values of λ. We show next that the computations can in certain cases be simplified if we replace one or both factors of this product by some function† of $\mathbf{x}(t)$. We shall assume that the process $\mathbf{x}(t)$ is normal with zero mean.

The arcsine law. We have shown in (10-71) that if $\mathbf{y}(t)$ is the output of a hard limiter with input $\mathbf{x}(t)$:

$$\mathbf{y}(t) = \operatorname{sgn} \mathbf{x}(t) = \begin{cases} 1 & \mathbf{x}(t) > 0 \\ -1 & \mathbf{x}(t) < 0 \end{cases}$$

then

$$C_{yy}(\tau) = \frac{2}{\pi} \arcsin \frac{C_{xx}(\tau)}{C_{xx}(0)} \tag{13-30}$$

The estimate of $C_{yy}(\tau)$ is given by

$$\hat{\mathbf{C}}_{yy}(\tau) = \frac{1}{2T} \int_{-T}^{T} \operatorname{sgn} \mathbf{x}(t+\tau) \operatorname{sgn} \mathbf{x}(t)\, dt \tag{13-31}$$

This integral is simple to determine because the integrand equals ± 1. Thus

$$\hat{\mathbf{C}}_{yy}(\tau) = \left(\frac{\mathbf{T}_{\tau}^{+}}{T} - 1 \right)$$

where T_{τ}^{+} is the total time that $\mathbf{x}(t+\tau)\mathbf{x}(t) > 0$. This yields the estimate

$$\hat{\mathbf{C}}_{xx}(\tau) = \hat{\mathbf{C}}_{xx}(0) \sin\left[\frac{\pi}{2} \hat{\mathbf{C}}_{yy}(\tau) \right]$$

of $C_{xx}(\tau)$ within a factor.

Bussgang's theorem. We have shown in (10-72) that the cross-covariance of the processes $\mathbf{x}(t)$ and $\mathbf{y}(t) = \operatorname{sgn} \mathbf{x}(t)$ is proportional to $C_{xx}(\tau)$:

$$C_{xy}(\tau) = K C_{xx}(\tau) \qquad K = \sqrt{\frac{2}{\pi C_{xx}(0)}} \tag{13-32}$$

To estimate $C_{xx}(\tau)$, it suffices, therefore, to estimate $C_{xy}(\tau)$. Using (13-29), we obtain

$$\hat{\mathbf{C}}_{xx}(\tau) = \frac{1}{K} \hat{\mathbf{C}}_{xy}(\tau) = \frac{1}{2KT} \int_{-T}^{T} \mathbf{x}(t+\tau) \operatorname{sgn} \mathbf{x}(t)\, dt \tag{13-33}$$

CORRELOMETERS AND SPECTROMETERS. A correlometer is a physical device measuring the autocorrelation $R(\lambda)$ of a process $\mathbf{x}(t)$. In Fig. 13-3 we show two correlometers. The first consists of a delay element, a multiplier, and a low-pass

†S. Cambanis and E. Masry: "On the Reconstruction of the Covariance of Stationary Gaussian Processes Through Zero-Memory Nonlinearities," *IEEE Transactions on Information Theory*, Vol. IT-24, 1978.

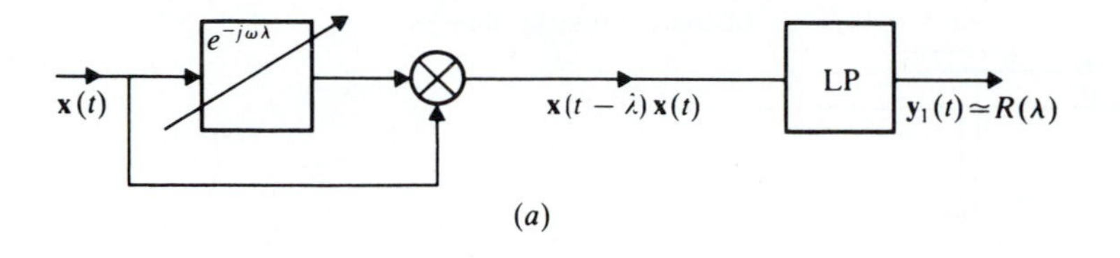

(a)

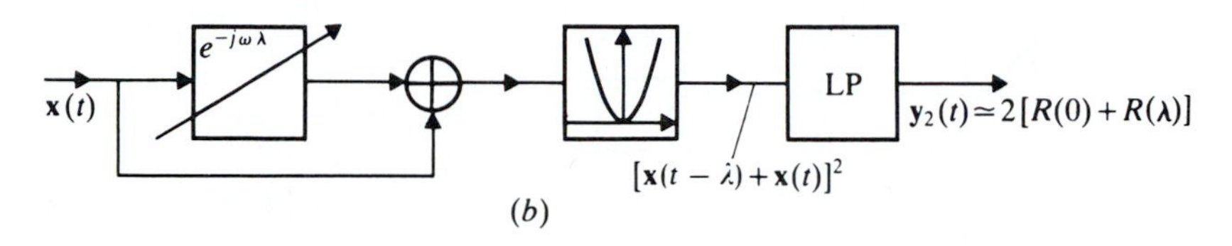

(b)

FIGURE 13-3

(LP) filter. The input to the LP filter is the process $\mathbf{x}(t - \lambda)\mathbf{x}(t)$; the output $\mathbf{y}_1(t)$ is the estimate of the mean $R(\lambda)$ of the input. The second consists of a delay element, an adder, a square-law detector, and an LP filter. The input to the LP filter is the process $[\mathbf{x}(t - \lambda) + \mathbf{x}(t)]^2$; the output $\mathbf{y}_2(t)$ is the estimate or the mean $2[R(0) + R(\lambda)]$ of the input.

A spectrometer is a physical device measuring the Fourier transform $S(\omega)$ of $R(\lambda)$. This device consists of a bandpass filter $B(\omega)$ with input $\mathbf{x}(t)$ and output $\mathbf{y}(t)$, in series with a square-law detector and an LP filter (Fig. 13-4). The input to the LP filter is the process $\mathbf{y}^2(t)$; its output $\mathbf{z}(t)$ is the estimate of the mean $E\{\mathbf{y}^2(t)\}$ of the input. Suppose that $B(\omega)$ is a narrow-band filter of unit energy with center frequency ω_0 and bandwidth $2c$. If the function $S(\omega)$ is continuous at ω_0 and c is sufficiently small, then $S(\omega) \simeq S(\omega_0)$ for $|\omega - \omega_0| < c$; hence [see (10-139)]

$$E\{\mathbf{y}^2(t)\} = \frac{1}{2\pi}\int_{-\infty}^{\infty} S(\omega)B^2(\omega)\,d\omega \simeq \frac{S(\omega_0)}{2\pi}\int_{\omega_0-c}^{\omega_0+c} B^2(\omega)\,d\omega = S(\omega_0)$$

as in (10-153). This yields

$$\mathbf{z}(t) \simeq E\{\mathbf{y}^2(t)\} \simeq S(\omega_0)$$

We give next the optical realization of the correlometer of Fig. 13-3*b* and the spectrometer of Fig. 13-4.

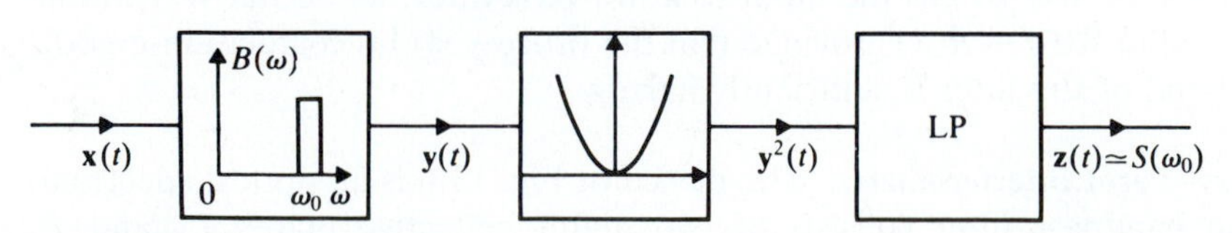

FIGURE 13-4

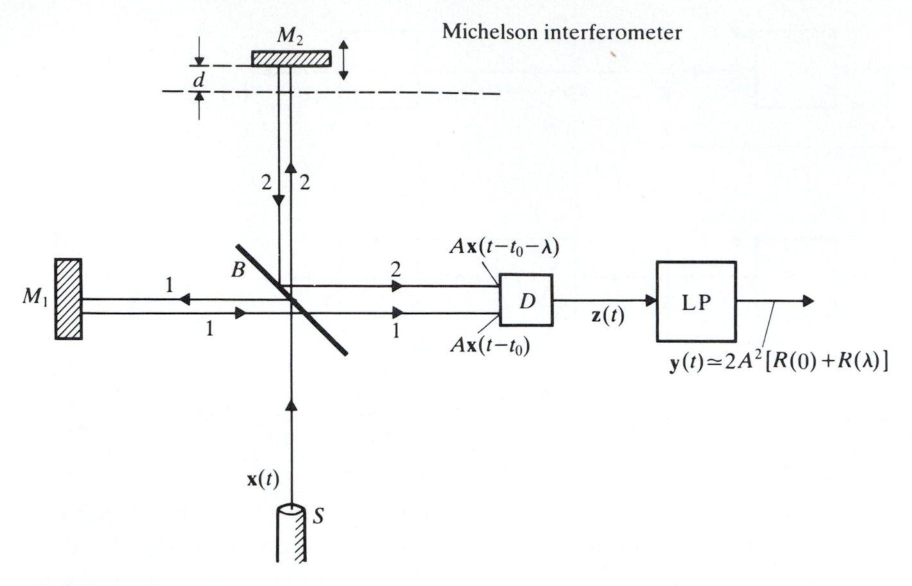

FIGURE 13-5

The Michelson interferometer. The device of Fig. 13-5 is an optical correlometer. It consists of a light source S, a beam-splitting surface B, and two mirrors. Mirror M_1 is in a fixed position and mirror M_2 is movable. The light from the source S is a random signal $\mathbf{x}(t)$ traveling with velocity c and it reaches a square-law detector D along paths 1 and 2 as shown. The lengths of these paths equal l and $l + 2d$ respectively, where d is the displacement of mirror M_2 from its equilibrium position.

The signal reaching the detector is thus the sum

$$A\mathbf{x}(t - t_0) + A\mathbf{x}(t - t_0 - \lambda)$$

where A is the attenuation in each path, $t_0 = l/c$ is the delay along path 1, and $\lambda = 2d/c$ is the additional delay due to the displacement of mirror M_2. The detector output is the signal

$$\mathbf{z}(t) = A^2[\mathbf{x}(t - t_0 - \lambda) + \mathbf{x}(t - t_0)]^2$$

Clearly,

$$E\{\mathbf{z}(t)\} = 2A^2[R(0) + R(\lambda)]$$

If, therefore, we use $\mathbf{z}(t)$ as the input to a low-pass filter, its output $\mathbf{y}(t)$ will be proportional to $R(0) + R(\lambda)$ provided that the process $\mathbf{x}(t)$ is correlation-ergodic and the band of the filter is sufficiently narrow.

The Fabry–Perot interferometer. The device of Fig. 13-6 is an optical spectrometer. The bandpass filter consists of two highly reflective plates P_1 and P_2 distance d apart and the input is a light beam $\mathbf{x}(t)$ with power spectrum $S(\omega)$.

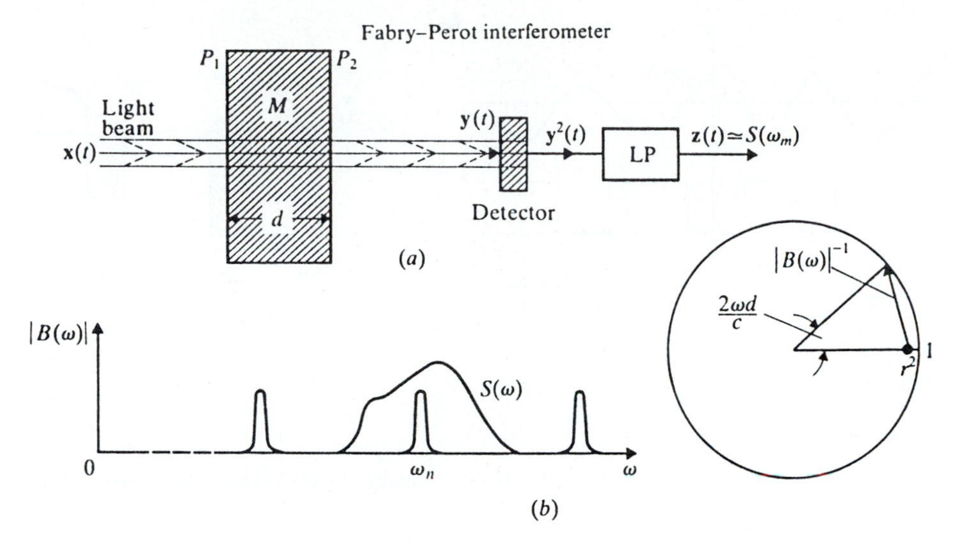

FIGURE 13-6

The frequency response of the filter is proportional to

$$B(\omega) = \frac{1}{1 - r^2 e^{-j2\omega d/c}} \qquad r \simeq 1$$

where r is the reflection coefficient of each plate and c is the velocity of light in the medium M between the plates. The function $B(\omega)$ is shown in Fig. 10-10b. It consists of a sequence of bands centered at

$$\omega_n = \frac{\pi n d}{c}$$

whose bandwidth tends to 0 as $r \to 1$. If only the mth band of $B(\omega)$ overlaps with $S(\omega)$ and $r \simeq 1$, then the output $\mathbf{z}(t)$ of the LP filter is proportional to $S(\omega_m)$. To vary ω_m, we can either vary the distance d between the plates or the dielectric constant of the medium M.

Distribution-Ergodic Processes

Any parameter of a probabilistic model that can be expressed as the mean of some function of an SSS process $\mathbf{x}(t)$ can be estimated by a time average. For a specific x, the distribution of $\mathbf{x}(t)$ is the mean of the process $\mathbf{y}(t) = U[x - \mathbf{x}(t)]$:

$$\mathbf{y}(t) = \begin{cases} 1 & \mathbf{x}(t) \le x \\ 0 & \mathbf{x}(t) > x \end{cases} \qquad E\{\mathbf{y}(t)\} = P\{\mathbf{x}(t) \le x\} = F(x)$$

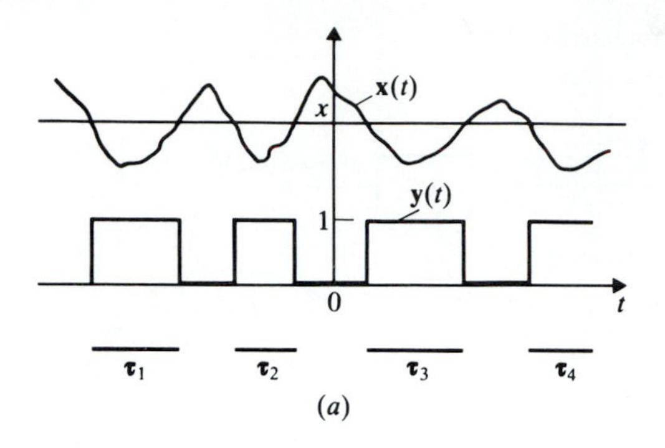

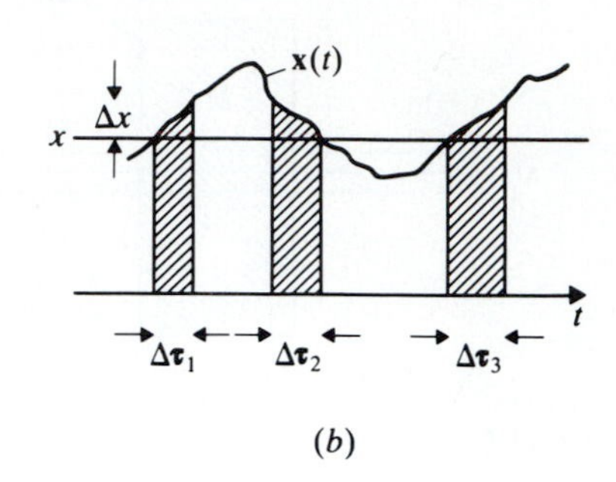

FIGURE 13-7

Hence $F(x)$ can be estimated by the time average of $\mathbf{y}(t)$. Inserting into (13-1), we obtain the estimator

$$\mathbf{F}_T(x) = \frac{1}{2T}\int_{-T}^{T} \mathbf{y}(t)\,dt = \frac{\boldsymbol{\tau}_1 + \cdots + \boldsymbol{\tau}_n}{2T} \tag{13-34}$$

where $\boldsymbol{\tau}_i$ are the lengths of the time intervals during which $\mathbf{x}(t)$ is less than x (Fig. 13-7a).

To find the variance of $\mathbf{F}_T(x)$, we must first find the autocovariance of $\mathbf{y}(t)$. The product $\mathbf{y}(t+\tau)\mathbf{y}(t)$ equals 1 if $\mathbf{x}(t+\tau) \le x$ and $\mathbf{x}(t) \le x$; otherwise, it equals 0. Hence

$$R_y(\tau) = P\{\mathbf{x}(t+\tau) \le x, \mathbf{x}(t) \le x\} = F(x, x; \tau)$$

where $F(x, x; \tau)$ is the second-order distribution of $\mathbf{x}(t)$. The variance of $\mathbf{F}_T(x)$ is obtained from (13-4) if we replace $C(\tau)$ by the autocovariance $F(x, x; \tau) - F^2(x)$ of $\mathbf{y}(t)$. From (13-7) it follows that a process $\mathbf{x}(t)$ is distribution-ergodic iff

$$\frac{1}{T}\int_0^T F(x, x; \tau)\,d\tau \xrightarrow[T\to\infty]{} F^2(x) \tag{13-35}$$

A sufficient condition is obtained from (13-10): A process $\mathbf{x}(t)$ is distribution-ergodic if $F(x, x; \tau) \to F^2(x)$ as $\tau \to \infty$. This is the case if the RVs $\mathbf{x}(t)$ and $\mathbf{x}(t+\tau)$ are *independent* for large τ.

Density. To estimate the density of $\mathbf{x}(t)$, we form the time intervals $\Delta\tau_i$ during which $\mathbf{x}(t)$ is between x and $x + \Delta x$ (Fig. 13-7b). From (13-34) it follows that

$$f(x)\,\Delta x \simeq F(x + \Delta x) - F(x) \simeq \frac{1}{2T}\sum_i \Delta\tau_i$$

Thus $f(x)\,\Delta x$ equals the percentage of time that a single sample of $\mathbf{x}(t)$ is between x and $x + \Delta x$. This can be used to design an analog estimator of $f(x)$.

13-2 SPECTRAL ESTIMATION

We wish to estimate the power spectrum $S(\omega)$ of a real process $\mathbf{x}(t)$ in terms of a single realization of a finite segment

$$\mathbf{x}_T(t) = \mathbf{x}(t)p_T(t) \qquad p_T(t) = \begin{cases} 1 & |t| < T \\ 0 & |t| > T \end{cases} \tag{13-36}$$

of $\mathbf{x}(t)$. The spectrum $S(\omega)$ is not the mean of some function of $\mathbf{x}(t)$. It cannot, therefore, be estimated directly as a time average. It is, however, the Fourier transform of the autocorrelation

$$R(\tau) = E\left\{\mathbf{x}\left(t + \frac{\tau}{2}\right)\mathbf{x}\left(t - \frac{\tau}{2}\right)\right\}$$

It will be determined in terms of the estimate of $R(\tau)$. This estimate cannot be computed from (13-25) because the product $\mathbf{x}(t + \tau/2)\mathbf{x}(t - \tau/2)$ is available only for t in the interval $(-T + |\tau|/2, T - |\tau|/2)$ (Fig. 13-8). Changing $2T$ to $2T - |\tau|$, we obtain the estimate

$$\mathbf{R}^T(\tau) = \frac{1}{2T - |\tau|}\int_{-T+|\tau|/2}^{T-|\tau|/2} \mathbf{x}\left(t + \frac{\tau}{2}\right)\mathbf{x}\left(t - \frac{\tau}{2}\right) dt \tag{13-37}$$

This integral specifies $\mathbf{R}^T(\tau)$ for $|\tau| < 2T$; for $|\tau| > 2T$ we set $\mathbf{R}^T(\tau) = 0$. The above estimate is unbiased; however, its variance increases as $|\tau|$ increases because the length $2T - |\tau|$ of the integration interval decreases. Instead of $\mathbf{R}^T(\tau)$, we shall use the product

$$\mathbf{R}_T(\tau) = \left(1 - \frac{|\tau|}{2T}\right)\mathbf{R}^T(\tau) \tag{13-38}$$

This estimator is biased; however, its variance is smaller than the variance of

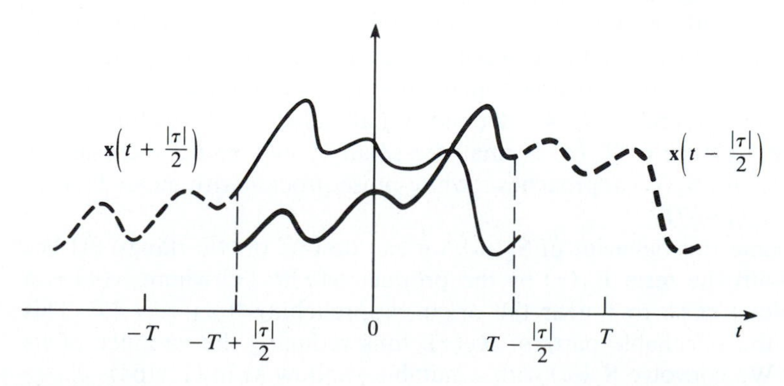

FIGURE 13-8

$\mathbf{R}^T(\tau)$. The main reason we use it is that its transform is proportional to the energy spectrum of the segment $\mathbf{x}_T(t)$ of $\mathbf{x}(t)$ [see (13-39)].

The periodogram

The periodogram of a process $\mathbf{x}(t)$ is by definition the process

$$\mathbf{S}_T(\omega) = \frac{1}{2T}\left|\int_{-T}^{T} \mathbf{x}(t)e^{-j\omega t}\,dt\right|^2 \tag{13-39}$$

The above integral is the Fourier transform of the known segment $\mathbf{x}_T(t)$ of $\mathbf{x}(t)$:

$$\mathbf{S}_T(\omega) = \frac{1}{2T}|\mathbf{X}_T(\omega)|^2 \qquad \mathbf{X}_T(\omega) = \int_{-T}^{T} \mathbf{x}(t)e^{-j\omega t}\,dt$$

We shall express $\mathbf{S}_T(\omega)$ in terms of the estimator $\mathbf{R}_T(\tau)$ of $R(\tau)$.

THEOREM

$$\mathbf{S}_T(\omega) = \int_{-2T}^{2T} \mathbf{R}_T(\tau)e^{-j\omega\tau}\,d\tau \tag{13-40}$$

Proof. The integral in (13-37) is the convolution of $\mathbf{x}_T(t)$ with $\mathbf{x}_T(-t)$ because $\mathbf{x}_T(t) = 0$ for $|t| > T$. Hence

$$\mathbf{R}_T(\tau) = \frac{1}{2T}\mathbf{x}_T(\tau) * \mathbf{x}_T(-\tau) \tag{13-41}$$

Since $\mathbf{x}_T(t)$ is real, the transform of $\mathbf{x}_T(-t)$ equals $\mathbf{X}_T^*(\omega)$. This shows that (convolution theorem) the transform of $\mathbf{R}_T(\tau)$ equals the right side of (13-39).

In the early years of signal analysis, the spectral properties of random processes were expressed in terms of their periodogram. This approach yielded reliable results so long as the integrations were based on analog techniques of limited accuracy. With the introduction of digital processing, the accuracy was improved and, paradoxically, the computed spectra exhibited noisy behavior. This apparent paradox can be readily explained in terms of the properties of the periodogram: The integral in (13-40) depends on all values of $\mathbf{R}_T(\tau)$ for τ large and small. The variance of $\mathbf{R}_T(\tau)$ is small for small τ only, and it increases as $\tau \to 2T$. As a result, $\mathbf{S}_T(\omega)$ approaches a white-noise process with mean $S(\omega)$ as T increases [see (13-57)].

To overcome this behavior of $\mathbf{S}_T(\omega)$, we can do one of two things: (1) We replace in (13-40) the term $\mathbf{R}_T(\tau)$ by the product $w(\tau)\mathbf{R}_T(\tau)$ where $w(\tau)$ is a function (window) close to 1 near the origin, approaching 0 as $\tau \to 2T$. This deemphasizes the unreliable parts of $\mathbf{R}_T(\tau)$, thus reducing the variance of its transform. (2) We convolve $\mathbf{S}_T(\omega)$ with a suitable window as in (11-164).

We continue with the determination of the bias and the variance of $\mathbf{S}_T(\omega)$.

Bias. From (13-38) and (13-40) it follows that

$$E\{\mathbf{S}_T(\omega)\} = \int_{-2T}^{2T}\left(1 - \frac{|\tau|}{2T}\right)R(\tau)e^{-j\omega\tau}\,d\tau$$

Since

$$\left(1 - \frac{|\tau|}{2T}\right)p_T(\tau) \leftrightarrow \frac{2\sin^2 T\omega}{T\omega^2}$$

we conclude that [see also (12-83)]

$$E\{\mathbf{S}_T(\omega)\} = \int_{-\infty}^{\infty}\frac{\sin^2 T(\omega - y)}{\pi T(\omega - y)^2}S(y)\,dy \tag{13-42}$$

The above shows that the mean of the periodogram is a smoothed version of $S(\omega)$; however, the smoothing kernel $\sin^2 T(\omega - y)/\pi T(\omega - y)^2$ takes significant values only in an interval of the order of $1/T$ centered at $y = \omega$. If, therefore, T is sufficiently large, we can set $S(y) \simeq S(\omega)$ in (13-42) for every point of continuity of $S(\omega)$. Hence for large T,

$$E\{\mathbf{S}_T(\omega)\} \simeq S(\omega)\int_{-\infty}^{\infty}\frac{\sin^2 T(\omega - y)}{\pi T(\omega - y)^2}\,dy = S(\omega) \tag{13-43}$$

From this it follows that $\mathbf{S}_T(\omega)$ is asymptotically an unbiased estimator of $S(\omega)$.

Data window. If $S(\omega)$ is not nearly constant in an interval of the order of $1/T$, the periodogram is a biased estimate of $S(\omega)$. To reduce the bias, we replace in (13-39) the process $\mathbf{x}(t)$ by the product $c(t)\mathbf{x}(t)$. This yields the *modified periodogram*

$$\mathbf{S}_c(\omega) = \frac{1}{2T}\left|\int_{-T}^{T}c(t)\mathbf{x}(t)e^{-j\omega t}\,dt\right|^2 \tag{13-44}$$

The factor $c(t)$ is called the *data window*. Denoting by $C(\omega)$ its Fourier transform, we conclude that [see (12-82)]

$$E\{\mathbf{S}_c(\omega)\} = \frac{1}{4\pi T}S(\omega) * C^2(\omega) \tag{13-45}$$

VARIANCE. For the determination of the variance of $\mathbf{S}_T(\omega)$, knowledge of the fourth-order moments of $\mathbf{x}(t)$ is required. For normal processes, all moments can be expressed in terms of $R(\tau)$. Furthermore, as $T \to \infty$, the fourth-order moments of most processes approach the corresponding moments of a normal process with the same autocorrelation (see Papoulis 1977). We can assume, therefore, without essential loss of generality, that $\mathbf{x}(t)$ is normal with zero mean.

THEOREM. For large T:

$$\operatorname{Var} \mathbf{S}_T(\omega) \simeq \begin{cases} 2S^2(0) & \omega = 0 \\ S^2(\omega) & |\omega| \gg 1/T \end{cases} \tag{13-46}$$

at every point of continuity of $S(\omega)$.

Proof. The Fourier transform of the autocorrelation $R(t_1 - t_2)p_T(t_1)p_T(t_2)$ of the process $\mathbf{x}_T(t)$ equals

$$\Gamma(u, v) = \int_{-\infty}^{\infty} \frac{2 \sin T\alpha \sin T(u + v - \alpha)}{\pi\alpha(u + v - \alpha)} S(u - \alpha)\, d\alpha \tag{13-47}$$

This follows from (12-80) with $W(\omega) = 2 \sin T\omega/\omega$. The fraction in (13-47) takes significant values only if the terms αT and $(u + v - \alpha)T$ are of the order of 1; hence, the entire fraction is negligible if $|u + v| \gg 1/T$. Setting $u = v = \omega$, we conclude that $\Gamma(\omega, \omega) \simeq 0$ and

$$\Gamma(\omega, -\omega) = \int_{-\infty}^{\infty} \frac{2 \sin^2 T\alpha}{\pi\alpha^2} S(\omega - \alpha)\, d\alpha$$

$$\simeq S(\omega) \int_{-\infty}^{\infty} \frac{2 \sin^2 T\alpha}{\pi\alpha^2}\, d\alpha = 2TS(\omega) \tag{13-48}$$

for $|\omega| \gg 1/T$ and since [see (12-74)]

$$\operatorname{Var} \mathbf{S}_T(\omega) = \frac{1}{4T^2}\left[\Gamma^2(\omega, -\omega) + \Gamma^2(\omega, \omega)\right]$$

and $\Gamma(0, 0) = S(0)$, (13-46) follows.

Note For a specific τ, no matter how large, the estimate $\mathbf{R}_T(\tau) \to R(\tau)$ as $T \to \infty$. Its transform $\mathbf{S}_T(\omega)$, however, does not tend to $S(\omega)$ as $T \to \infty$. The reason is that the convergence of $\mathbf{R}_T(\tau)$ to $R(\tau)$ is not uniform in τ, that is, given $\varepsilon > 0$, we cannot find a constant T_0 independent of τ such that $|\mathbf{R}_T(\tau) - R(\tau)| < \varepsilon$ for every τ, and every $T > T_0$.

Proceeding similarly, we can show that the variance of the spectrum $\mathbf{S}_c(\omega)$ obtained with the data window $c(t)$ is essentially equal to the variance of $\mathbf{S}_T(\omega)$. This shows that use of data windows does not reduce the variance of the estimate. To improve the estimation, we must replace in (13-40) the sample autocorrelation $\mathbf{R}_T(\tau)$ by the product $w(\tau)\mathbf{R}_T(\tau)$, or, equivalently, we must smooth the periodogram $\mathbf{S}_T(\omega)$.

Note Data windows might be useful if we smooth $\mathbf{S}_T(\omega)$ by an ensemble average: Suppose that we have access to N independent samples $\mathbf{x}(t, \zeta_i)$ of $\mathbf{x}(t)$, or, we divide a single long sample into N essentially independent pieces, each of duration $2T$. We form the periodograms $\mathbf{S}_T(\omega, \zeta_i)$ of each sample and their average

$$\overline{\mathbf{S}}_T(\omega) = \frac{1}{N}\sum \mathbf{S}_T(\omega, \zeta_i) \tag{13-49}$$

As we know,

$$E\{\overline{\mathbf{S}}_T(\omega)\} = S(\omega) * \frac{\sin^2 \omega T}{\pi T \omega^2} \qquad \operatorname{Var} \overline{\mathbf{S}}_T(\omega) \simeq \frac{1}{N} S^2(\omega) \qquad \omega \neq 0 \tag{13-50}$$

If N is large, the variance of $\overline{\mathbf{S}}_T(\omega)$ is small. However, its bias might be significant. Use of data windows is in this case desirable.

Smoothed Spectrum

We shall assume as before that T is large and $\mathbf{x}(t)$ is normal. To improve the estimate, we form the smoothed spectrum

$$\mathbf{S}_w(\omega) = \frac{1}{2\pi}\int_{-\infty}^{\infty} \mathbf{S}_T(\omega - y) W(y)\, dy = \int_{-2T}^{2T} w(\tau)\mathbf{R}_T(\tau) e^{-j\omega\tau}\, d\tau \tag{13-51}$$

where

$$w(t) = \frac{1}{2\pi}\int_{-\infty}^{\infty} W(\omega) e^{j\omega t}\, d\omega$$

The function $w(t)$ is called the *lag window* and its transform $W(\omega)$ the *spectral window*. We shall assume that $W(-\omega) = W(\omega)$ and

$$w(0) = 1 = \frac{1}{2\pi}\int_{-\infty}^{\infty} W(\omega)\, d\omega \qquad W(\omega) \geq 0 \tag{13-52}$$

Bias. From (13-42) it follows that

$$E\{\mathbf{S}_w(\omega)\} = \frac{1}{2\pi} E\{\mathbf{S}_T(\omega)\} * W(\omega) = \frac{1}{2\pi} S(\omega) * \frac{\sin^2 T\omega}{\pi T \omega^2} * W(\omega)$$

Assuming that $W(\omega)$ is nearly constant in any interval of length $1/T$, we obtain the large T approximation

$$E\{\mathbf{S}_w(\omega)\} \simeq \frac{1}{2\pi} S(\omega) * W(\omega) \tag{13-53}$$

Variance. We shall determine the variance of $\mathbf{S}_w(\omega)$ using the identity [see (12-74)]

$$\operatorname{Cov}[\mathbf{S}_T(u), \mathbf{S}_T(v)] = \frac{1}{4T^2}[\Gamma^2(u, -v) + \Gamma^2(u, v)] \tag{13-54}$$

This problem is in general complicated. We shall outline an approximate solution based on the following assumptions: The constant T is large in the sense that the functions $S(\omega)$ and $W(\omega)$ are nearly constant in any interval of length $1/T$. The width of $W(\omega)$, that is, the constant σ such that $W(\omega) \simeq 0$ for

$|\omega| > \sigma$, is small in the sense that $S(\omega)$ is nearly constant in any interval of length 2σ.

Reasoning as in the proof of (13-48), we conclude from (13-47) that $\Gamma(u, v) \simeq 0$ for $u + v \gg 1/T$ and

$$\Gamma(u, -v) \simeq S(u) \int_{-\infty}^{\infty} \frac{2 \sin T(u - v - \alpha) \sin T\alpha}{\pi(u - v - \alpha)\alpha} \, d\alpha = S(u) \frac{2 \sin T(u - v)}{u - v}$$

This is the generalization of (13-48). Inserting into (13-54), we obtain

$$\operatorname{Cov}[\mathbf{S}_T(u), \mathbf{S}_T(v)] \simeq \frac{\sin^2 T(u - v)}{T^2(u - v)^2} S^2(u) \tag{13-55}$$

Equation (13-46) is a special case obtained with $u = v = \omega$.

THEOREM. For $|\omega| \gg 1/T$

$$\operatorname{Var} \mathbf{S}_w(\omega) \simeq \frac{E_w}{2T} S^2(\omega) \tag{13-56}$$

where

$$E_w = \frac{1}{2\pi} \int_{-\infty}^{\infty} W^2(\omega) \, d\omega$$

Proof. The smoothed spectrum $\mathbf{S}_w(\omega)$ equals the convolution of $\mathbf{S}_T(\omega)$ with the spectral window $W(\omega)/2\pi$. From this and (10-87) it follows mutatis mutandis that the variance of $\mathbf{S}_w(\omega)$ is a double convolution involving the covariance of $\mathbf{S}_T(\omega)$ and the window $W(\omega)$. The fraction in (13-55) is negligible for $|u - v| \gg 1/T$. In any interval of length $1/T$, the function $W(\omega)$ is nearly constant by assumption. This leads to the conclusion that in the evaluation of the variance of $\mathbf{S}_w(\omega)$, the covariance of $\mathbf{S}_T(\omega)$ can be approximated by an impulse of area equal to the area

$$S^2(u) \int_{-\infty}^{\infty} \frac{\sin^2 T(u - v)}{T^2(u - v)^2} \, dv = \frac{\pi}{T} S^2(u)$$

of the right side of (13-55). This yields

$$\operatorname{Cov}[\mathbf{S}_T(u), \mathbf{S}_T(v)] = q(u)\delta(u - v) \qquad q(u) = \frac{\pi}{T} S^2(u) \tag{13-57}$$

From the above and (10-91) it follows that

$$\operatorname{Var} \mathbf{S}_w(\omega) \simeq \frac{\pi}{T} \int_{-\infty}^{\infty} S^2(\omega - y) \frac{W^2(y)}{4\pi^2} \, dy = \frac{S^2(\omega)}{2T} \int_{-\infty}^{\infty} \frac{W^2(y)}{2\pi} \, dy$$

and (13-56) results.

WINDOW SELECTION. The selection of the window pair $w(t) \leftrightarrow W(\omega)$ depends on two conflicting requirements: For the variance of $\mathbf{S}_w(\omega)$ to be small, the energy E_w of the lag window $w(t)$ must be small compared to T. From this it follows that $w(t)$ must approach 0 as $t \to 2T$. We can assume, therefore, without essential loss of generality that $w(t) = 0$ for $|t| > M$ where M is a fraction of $2T$. Thus

$$\mathbf{S}_w(\omega) = \int_{-M}^{M} w(t)\mathbf{R}_T(t)e^{-j\omega t}\,dt \qquad M < 2T$$

The mean of $\mathbf{S}_w(\omega)$ is a smoothed version of $S(\omega)$. To reduce the effect of the resulting bias, we must use a spectral window $W(\omega)$ of short duration. This is in conflict with the requirement that M be small (uncertainty principle). The final choice of M is a compromise between bias and variance. The quality of the estimate depends on M and on the shape of $w(t)$. To separate the shape factor from the size factor, we express $w(t)$ as a scaled version of a normalized window $w_0(t)$ of size 2:

$$w(t) = w_0\left(\frac{t}{M}\right) \leftrightarrow W(\omega) = MW_0(M\omega) \tag{13-58}$$

where

$$w_0(t) = 0 \qquad \text{for} \quad |t| > 1$$

The critical parameter in the selection of a window is the scaling factor M. In the absence of any prior information, we have no way of determining the optimum size of M. The following considerations, however, are useful: A reasonable measure of the reliability of the estimation is the ratio

$$\frac{\operatorname{Var}\mathbf{S}_w(\omega)}{S^2(\omega)} \simeq \frac{E_w}{2T} = \alpha \tag{13-59}$$

For most windows in use, E_w is between $0.5M$ and $0.8M$ (see Table 13-1). If we set $\alpha = 0.2$ as the largest acceptable α, we must set $M \le T/2$. If nothing is known about $S(\omega)$, we estimate it several times using windows of decreasing size. We start with $M = T/2$ and observe the form of the resulting estimate $\mathbf{S}_w(\omega)$. This estimate might not be very reliable; however, it gives us some idea of the form of $S(\omega)$. If we see that the estimate is nearly constant in any interval of the order of $1/M$, we conclude that the initial choice $M = T/2$ is too large. A reduction of M will not appreciably affect the bias but it will yield a smaller variance. We repeat this process until we obtain a balance between bias and variance. As we show later, for optimum balance, the standard deviation of the estimate must equal twice its bias. The quality of the estimate depends, of course, on the size of the available sample. If, for the given T, the resulting $\mathbf{S}_w(\omega)$ is not smooth for $M = T/2$, we conclude that T is not large enough for a satisfactory estimate.

To complete the specification of the window, we must select the form of $w_0(t)$. In this selection, we are guided by the following considerations:

1. The window $W(\omega)$ must be positive and its area must equal 2π as in (13-52). This ensures the positivity and consistency of the estimation.

TABLE 13-1

$w(t)$	$W(\omega)$
1. Bartlett	
$1 - \|t\|$	$\dfrac{4\sin^2 \omega/2}{\omega^2}$
$m_2 = \infty \qquad E_w = \frac{2}{3} \qquad n = 2$	
2. Tukey	
$\frac{1}{2}(1 + \cos \pi t)$	$\dfrac{\pi^2 \sin \omega}{\omega(\pi^2 - \omega^2)}$
$m_2 = \dfrac{\pi^2}{2} \qquad E_w = \frac{3}{4} \qquad n = 3$	
3. Parzen	
$[3(1 - 2\|t\|)p_1(t)] * [3(1 - 2\|t\|)p_1(t)]$	$\frac{3}{4}\left(\dfrac{\sin \omega/4}{\omega/4}\right)^4$
$m_2 = 12 \qquad E_w = 0.539 \qquad n = 4$	
4. Papoulis†	
$\dfrac{1}{\pi}\|\sin \pi t\| + (1 - \|t\|)\cos \pi t$	$8\pi^2 \dfrac{\cos^2(\omega/2)}{(\pi^2 - \omega^2)^2}$
$m_2 = \pi^2 \qquad E_w = 0.587 \qquad n = 4$	

†A. Papoulis: "Minimum Bias Windows for High Resolution Spectral Estimates," *IEEE Transactions on Information Theory*, vol. IT-19, 1973.

2. For small bias, the "duration" of $W(\omega)$ must be small. A measure of duration is the second moment

$$m_2 = \frac{1}{2\pi}\int_{-\infty}^{\infty} \omega^2 W(\omega)\, d\omega \tag{13-60}$$

3. The function $W(\omega)$ must go to 0 rapidly as ω increases (small sidelobes). This reduces the effect of distant peaks in the estimate of $S(\omega)$. As we know, the asymptotic properties of $W(\omega)$ depend on the continuity properties of its inverse $w(t)$. Since $w(t) = 0$ for $|t| > M$, the condition that $W(\omega) \to 0$ as A/ω^n as $n \to \infty$ leads to the requirement that the derivatives of $w(t)$ of order up to $n - 1$ be zero at the end-points $\pm M$ of the lag window $w(t)$:

$$w(\pm M) = w'(\pm M) = \cdots = w^{(n-1)}(\pm M) = 0 \tag{13-61}$$

4. The energy E_w of $w(t)$ must be small. This reduces the variance of the estimate.

Over the years, a variety of windows have been proposed. They meet more or less the stated requirements but most of them are selected empirically. Optimality criteria leading to windows that do not depend on the form of the

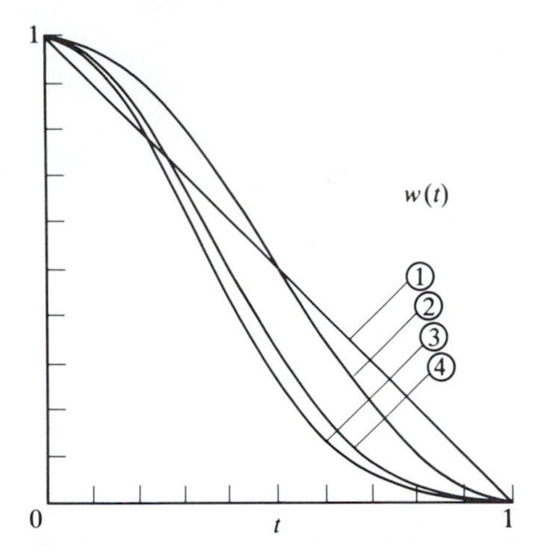

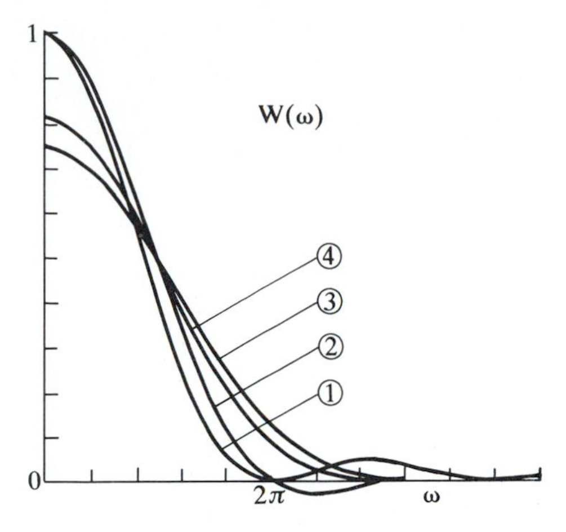

FIGURE 13-9

unknown $S(\omega)$ are difficult to generate. However, as we show next, for high-resolution estimates (large T) the last example of Table 13-1 minimizes the bias. In this table and in Fig. 13-9, we list the most common window pairs $w(t) \leftrightarrow W(\omega)$. We also show the values of the second moment m_2, the energy E_w, and the exponent n of the asymptotic attenuation A/ω^n of $W(\omega)$. In all cases, $w(t) = 0$ for $|t| > 1$.

OPTIMUM WINDOWS. We introduce next three classes of windows. In all cases, we assume that the data size T and the scaling factor M are large (high-resolution estimates) in the sense that we can use the parabolic approximation of $S(\omega - \alpha)$ in the evaluation of the bias. This yields [see (11-168)]

$$\frac{1}{2\pi}\int_{-\infty}^{\infty} S(\omega - \alpha)W(\alpha)\,d\alpha \simeq S(\omega) + \frac{S''(\omega)}{4\pi}\int_{-\infty}^{\infty} \alpha^2 W(\alpha)\,d\alpha \quad (13\text{-}62)$$

Note that since $W(\omega) > 0$, the above is an equality if we replace the term $S''(\omega)$ by $S''(\omega + \delta)$ where δ is a constant in the region where $W(\omega)$ takes significant values.

Minimum bias data window. The modified periodogram $\mathbf{S}_c(\omega)$ obtained with the data window $c(t)$ is a biased estimator of $S(\omega)$. Inserting (13-62) into (13-45), we conclude that the bias equals

$$B_c(\omega) = \frac{1}{2\pi}\int_{-\infty}^{\infty} S(\omega - \alpha)C^2(\alpha)\,d\alpha - S(\omega)$$

$$\simeq \frac{S''(\omega)}{4\pi}\int_{-\infty}^{\infty} \alpha^2 C^2(\alpha)\,d\alpha \quad (13\text{-}63)$$

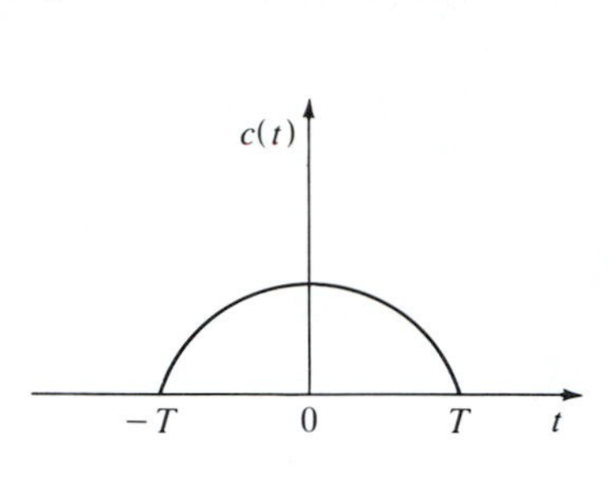

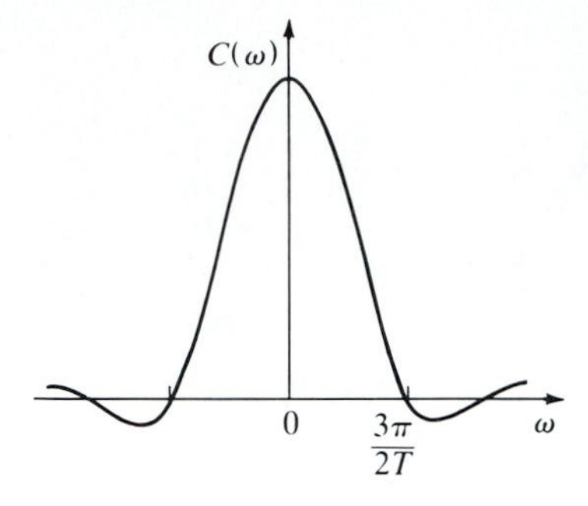

FIGURE 13-10

We have thus expressed the bias as a product where the first factor depends only on $S(\omega)$ and the second depends only on $C(\omega)$. This separation permits us to find $C(\omega)$ so as to minimize $B_c(\omega)$. To do so, it suffices to minimize the second moment

$$M_2 = \frac{1}{2\pi}\int_{-\infty}^{\infty} \omega^2 C^2(\omega)\, d\omega = \int_{-T}^{T} |c'(t)|^2\, dt \tag{13-64}$$

of $C^2(\omega)$ subject to the constraints

$$\frac{1}{2\pi}\int_{-\infty}^{\infty} C^2(\omega)\, d\omega = 1 \qquad C(-\omega) = C(\omega)$$

It can be shown that† the optimum data window is a truncated cosine (Fig. 13-10):

$$c(t) = \begin{cases} \dfrac{1}{\sqrt{T}} \cos \dfrac{\pi}{2T} t & |t| < T \\ 0 & |t| > T \end{cases} \leftrightarrow C(\omega) = 4\pi\sqrt{T}\, \frac{\cos T\omega}{\pi^2 - 4T^2\omega^2} \tag{13-65}$$

The resulting second moment M_2 equals 1. Note that if no data window is used, then $c(t) = 1$ and $M_2 = 2$. Thus the optimum data window yields a 50 percent reduction of the bias.

Minimum bias spectral window. From (13-53) and (13-62) it follows that the bias of $\mathbf{S}_w(\omega)$ equals

$$B(\omega) = \frac{1}{2\pi}\int_{-\infty}^{\infty} S(\omega - \alpha) W(\alpha)\, d\alpha - S(\omega) \simeq \frac{m_2}{2} S''(\omega) \tag{13-66}$$

where m_2 is the second moment of $W(\omega)/2\pi$. To minimize $B(\omega)$, it suffices,

†A. Papoulis: "Apodization for Optimum Imaging of Smooth Objects", *J. Opt. Soc. Am.*, Vol. 62, December, 1972.

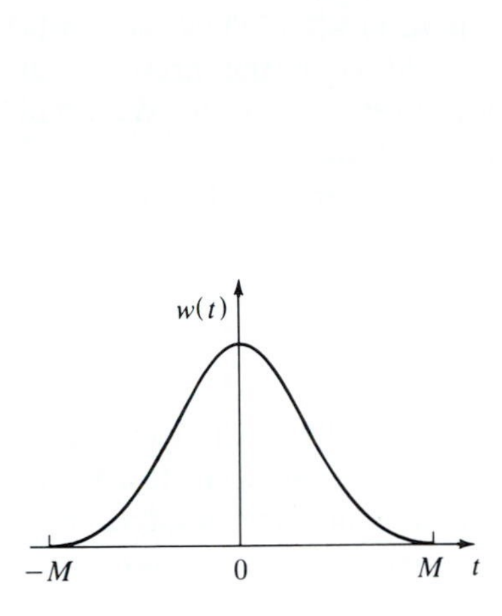

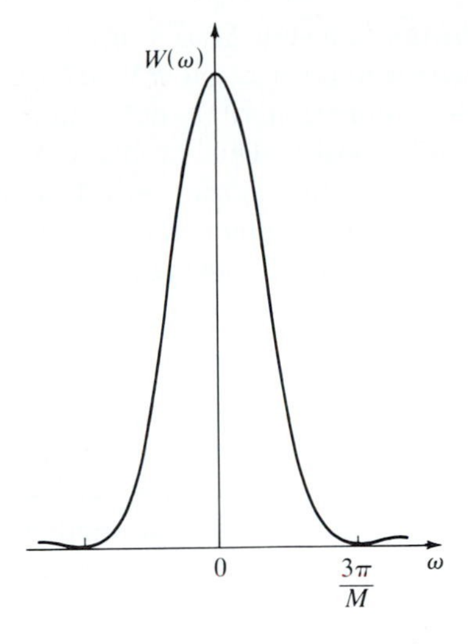

FIGURE 13-11

therefore, to minimize m_2 subject to the constraints

$$W(\omega) \geq 0 \qquad W(-\omega) = W(\omega) \qquad \frac{1}{2\pi}\int_{-\infty}^{\infty} W(\omega)\, d\omega = 1 \tag{13-67}$$

This is the same as the problem just considered if we replace $2T$ by M and we set

$$W(\omega) = C^2(\omega) \qquad w(t) = c(t) * c(-t)$$

This yields the pair (Fig. 13-11)

$$w(t) = \begin{cases} \dfrac{1}{\pi}\left|\sin\dfrac{\pi}{M}t\right| + \left(1 - \dfrac{|t|}{M}\right)\cos\dfrac{\pi}{M}t & |t| \leq M \\ 0 & |t| > M \end{cases} \tag{13-68}$$

$$W(\omega) = 8M\pi^2 \frac{\cos^2(M\omega/2)}{\left(\pi^2 - M^2\omega^2\right)^2} \tag{13-69}$$

Thus the last window in Table 13-1 minimizes the bias in high-resolution spectral estimates.

LMS spectral window. We shall finally select the spectral window $W(\omega)$ so as to minimize the MS estimation error

$$e = B^2(\omega) + \operatorname{Var} \mathbf{S}_w(\omega) \tag{13-70}$$

We have shown that for sufficiently large values of T, the periodogram $\mathbf{S}_T(\omega)$

can be written as a sum $S(\omega) + \mathbf{v}(\omega)$ where $\mathbf{v}(\omega)$ is a nonstationary white noise process with autocorrelation $\pi S^2(u)\delta(u - v)/T$ as in (13-57). Thus our problem is the estimation of a deterministic function $S(\omega)$ in the presence of additive noise $\mathbf{v}(\omega)$. This problem was considered in Sec. 11-6. We shall reestablish the results in the context of spectral estimation.

We start with a rectangular window of size 2Δ and area 1. The resulting estimate of $S(\omega)$ is the moving average

$$\mathbf{S}_\Delta(\omega) = \frac{1}{2\Delta}\int_{-\Delta}^{\Delta} \mathbf{S}_T(\omega - \alpha)\, d\alpha \tag{13-71}$$

of $\mathbf{S}_T(\omega)$. The rectangular window was used first by Daniell† in the early years of spectral estimation. It is a special case of the spectral window $W(\omega)/2\pi$. Note that the corresponding lag window $\sin \Delta t/2\pi\Delta t$ is not time-limited.

With the familiar large-T assumption, the periodogram $\mathbf{S}_T(\omega)$ is an unbiased estimator of $S(\omega)$. Hence the bias of $\mathbf{S}_\Delta(\omega)$ equals

$$\frac{1}{2\Delta}\int_{-\Delta}^{\Delta} S(\omega - y)\, dy - S(\omega) \simeq \frac{S''(\omega)}{2\Delta}\int_{-\Delta}^{\Delta} y^2\, dy = S''(\omega)\frac{\Delta^2}{6}$$

and its variance equals

$$\frac{\pi S^2(\omega)}{4\Delta^2 T}\int_{-\Delta}^{\Delta} d\omega = \frac{\pi S^2(\omega)}{2\Delta T}$$

This follows from (13-71) because $\mathbf{S}_T(\omega)$ is white noise as in (13-57) with $q(u) = \pi S^2(u)/T$ [see also (11-172)]. The above leads to the conclusion that

$$\operatorname{Var} \mathbf{S}_\Delta(\omega) = \frac{\pi S^2(\omega)}{2\Delta T} \qquad e = \frac{\Delta^4}{36}[S''(\omega)]^2 + \frac{\pi S^2(\omega)}{2\Delta T} \tag{13-72}$$

Proceeding as in (11-176), we conclude that e is minimum if

$$\Delta = \left(\frac{9\pi}{2T}\right)^{0.2}\left(\frac{S(\omega)}{S''(\omega)}\right)^{0.4}$$

The resulting bias equals twice the standard deviation of $\mathbf{S}_\Delta(\omega)$ (see two-to-one rule).

Suppose finally that the spectral window is a function of unknown form. We wish to determine is shape so as to minimize the MS error e. Proceeding as in (11-177), we can show that e is minimum if the window is a truncated

†P. J. Daniell: Discussion on "Symposium on Autocorrelation in Time Series," *J. Roy. Statist. Soc. Suppl.*, **8**, 1946.

parabola:

$$\mathbf{S}_w(\omega) = \frac{3}{4\Delta}\int_{-\Delta}^{\Delta} \mathbf{S}_T(\omega - y)\left(1 - \frac{y^2}{\Delta^2}\right) dy \qquad \Delta = \left(\frac{15\pi}{T}\right)^{0.2}\left(\frac{S(\omega)}{S''(\omega)}\right)^{0.4} \tag{13-73}$$

This window was first suggested by Priestley.† Note that unlike the earlier windows, it is frequency dependent and its size is a function of the unknown spectrum $S(\omega)$ and its second derivative. To determine $\mathbf{S}_w(\omega + \delta)$ we must therefore estimate first not only $S(\omega)$ but also $S''(\omega)$. Using these estimates we determine Δ for the next step.

13-3 EXTRAPOLATION AND SYSTEM IDENTIFICATION

In the preceding discussion, we computed the estimate $\mathbf{R}_T(\tau)$ of $R(\tau)$ for $|\tau| < M$ and used as the estimate of $S(\omega)$ the Fourier transform $\mathbf{S}_w(\omega)$ of the product $w(t)\mathbf{R}_T(t)$. The portion of $\mathbf{R}_T(\tau)$ for $|\tau| > M$ was not used. In this section, we shall assume that $S(\omega)$ belongs to a class of functions that can be specified in terms of certain parameters, and we shall use the estimated part of $R(\tau)$ to determine these parameters. In our development, we shall not consider the variance problem. We shall assume that the portion of $R(\tau)$ for $|\tau| < M$ is known exactly. This is a realistic assumption if $T \gg M$ because $\mathbf{R}_T(\tau) \to R(\tau)$ for $|\tau| < M$ as $T \to \infty$. A physical problem leading to the assumption that $R(\tau)$ is known exactly but only for $|\tau| < M$ is the Michelson interferometer. In this example, the time of observation is arbitrarily large; however, $R(\tau)$ can be determined only for $|\tau| < M$ where M is a constant proportional to the maximum displacement of the moving mirror (Fig. 13-5).

Our problem can thus be phrased as follows: We are given a finite segment

$$R_M(\tau) = \begin{cases} R(\tau) & |\tau| < M \\ 0 & |\tau| > M \end{cases}$$

of the autocorrelation $R(\tau)$ of a process $\mathbf{x}(t)$ and we wish to estimate its power spectrum $S(\omega)$. This is essentially a deterministic problem: We wish to find the Fourier transform $S(\omega)$ of a function $R(\tau)$ knowing only the segment $R_M(\tau)$ of $R(\tau)$ and the fact that $S(\omega) \geq 0$. This problem does not have a unique solution. Our task then is to find a particular $S(\omega)$ that is close in some sense to the unknown spectrum. In the early years of spectral estimation, the function $S(\omega)$

†M. B. Priestley: "Basic Considerations in the Estimation of Power Spectra," *Technometrics*, **4**, 1962.

was estimated with the method of windows (Blackman and Tukey†). In this method, the unknown $R(\tau)$ is replaced by 0 and the known or estimated part is tapered by a suitable factor $w(\tau)$. In recent years, a different approach has been used: It is assumed that $S(\omega)$ can be specified in terms of a finite number of parameters (parametric extrapolation) and the problem is reduced to the estimation of these parameters. In this section we concentrate on the extrapolation method starting with brief coverage of the method of windows.

Method of windows. The continuous-time version of this method is treated in the last section in the context of the bias reduction problem: We use as the estimate of $S(\omega)$ the integral

$$S_w(\omega) = \int_{-M}^{M} w(\tau)R(\tau)e^{-j\omega\tau}\,d\tau = \frac{1}{2\pi}\int_{-\infty}^{\infty} S(\omega-\alpha)W(\alpha)\,d\alpha \quad (13\text{-}74)$$

and we select $w(t)$ so as to minimize in some sense the estimation error $S_w(\omega) - S(\omega)$. If M is large in the sense that $S(\omega - \alpha) \simeq S(\omega)$ for $|\alpha| \le 1/M$, we can use the approximation [see (13-62)]

$$S_w(\omega) - S(\omega) \simeq \frac{S''(\omega)}{4\pi}\int_{-\infty}^{\infty} \alpha^2 W(\alpha)\,d\alpha$$

This is minimum if

$$w(\tau) = \frac{1}{\pi}\left|\sin\frac{\pi}{M}\tau\right| + \left(1 - \frac{|\tau|}{M}\right)\cos\frac{\pi}{M}\tau \qquad |\tau| < M$$

The discrete-time version of this method is similar: We are given a finite segment

$$R_L[m] = \begin{cases} R[m] & |m| \le L \\ 0 & |m| > L \end{cases} \quad (13\text{-}75)$$

of the autocorrelation $R[m] = E\{\mathbf{x}[n+m]\mathbf{x}[n]\}$ of a process $\mathbf{x}[n]$ and we wish to estimate its power spectrum

$$S(\omega) = \sum_{m=-\infty}^{\infty} R[m]e^{-jm\omega}$$

We use as the estimate of $S(\omega)$ the DFT

$$S_w(\omega) = \sum_{m=-L}^{L} w[m]R[m]e^{-jm\omega} = \frac{1}{2\pi}\int_{-\pi}^{\pi} S(\omega-\alpha)W(\alpha)\,d\alpha \quad (13\text{-}76)$$

of the product $w[m]R[m]$ where $w[m] \leftrightarrow W(\omega)$ is a DFT pair. The criteria for selecting $w[m]$ are the same as in the continuous-time case. In fact, if M is

†R. B. Blackman and J.W. Tukey: *The Measurement of Power Spectra*, Dover, New York, 1959.

large, we can choose for $w[m]$ the samples

$$w[m] = w(Mm/L) \qquad m = 0, \ldots, L \tag{13-77}$$

of an analog window $w(t)$ where M is the size of $w(t)$.

In a real problem, the data $R_L[m]$ are not known exactly. They are estimated in terms of the J samples of $\mathbf{x}[n]$:

$$\mathbf{R}_L[m] = \frac{1}{J}\sum_n \mathbf{x}[n+m]\mathbf{x}[n] \tag{13-78}$$

The mean and variance of $\mathbf{R}_L[m]$ can be determined as in the analog case. The details, however, will not be given. In the following, we assume that $R_L[m]$ is known exactly. This assumption is satisfactory if $J \gg L$.

Extrapolation Method

The spectral estimation problem is essentially numerical. This involves digital data even if the given process is analog. We shall, therefore, carry out the analysis in digital form. In the extrapolation method we assume that $S(z)$ is of known form. We shall assume that it is rational

$$\mathsf{S}(z) = \mathsf{L}(z)\mathsf{L}(1/z) \qquad \mathsf{L}(z) = \frac{b_0 + b_1 z^{-1} + \cdots + b_M z^{-M}}{1 + a_1 z^{-1} + \cdots + a_N z^{-N}} = \frac{N(z)}{D(z)} \tag{13-79}$$

We select the rational model for the following reasons: The numerical evaluation of its unknown parameters is relatively simple. An arbitrary spectrum can be closely approximated by a rational model of sufficiently large order. Spectra involving responses of dynamic systems are often rational.

System identification. The rational model leads directly to the solution of the identification problem (see also Sec. 11-7): We wish to determine the system function $\mathsf{H}(z)$ of a system driven by white noise in terms of the measurements of its output $\mathbf{x}[n]$. As we know, the power spectrum of the output is proportional to $\mathsf{H}(z)\mathsf{H}(1/z)$. If, therefore, the system is of finite order and minimum phase, then $\mathsf{H}(z)$ is proportional to $\mathsf{L}(z)$. To determine $\mathsf{H}(z)$, it suffices, therefore, to determine the $M+N+1$ parameters of $\mathsf{L}(z)$. We shall do so under the assumption that $R_L[m]$ is known exactly for $|m| \le M+N+1$.

We should stress that the proposed model is only an abstraction. In a real problem, $R[m]$ is not known exactly. Furthermore, $\mathsf{S}(z)$ might not be rational; even if it is, the constants M and N might not be known. However, the method leads to reasonable approximations if $R_L[m]$ is replaced by its time-average estimate $\mathbf{R}_L[m]$ and L is large.

Autoregressive processes. Our objective is to determine the $M+N+1$ coefficients b_i and a_k specifying the spectrum $\mathsf{S}(z)$ in terms of the first $M+N+1$

values $R_L[m]$ of $R[m]$. We start with the assumption that

$$\mathsf{L}(z) = \frac{\sqrt{P_N}}{1 + a_1 z^{-1} + \cdots + a_N z^{-N}} = \frac{\sqrt{P_N}}{D(z)} \tag{13-80}$$

This is a special case of (12-36) with $M = 0$ and $b_0 = \sqrt{P_N}$. As we know, the process $\mathbf{x}[n]$ satisfies the equation

$$\mathbf{x}[n] + a_1\mathbf{x}[n-1] + \cdots + a_N\mathbf{x}[n-N] = \boldsymbol{\varepsilon}[n] \tag{13-81}$$

where $\boldsymbol{\varepsilon}[n]$ is white noise with average power P_N. Our problem is to find the $N+1$ coefficients a_k and P_N. To do so, we multiply (13-81) by $\mathbf{x}[n-m]$ and take expected values. With $m = 0, \ldots, N$, this yields the *Yule–Walker* equations

$$\begin{aligned} R[0] + a_1R[1] + \cdots + a_NR[N] &= P_N \\ R[1] + a_1R[0] + \cdots + a_NR[N-1] &= 0 \\ \cdots\cdots\cdots\cdots\cdots\cdots\cdots&\cdots \\ R[N] + a_1R[N-1] + \cdots + a_NR[0] &= 0 \end{aligned} \tag{13-82}$$

This is a system of $N+1$ equations involving the $N+1$ unknowns a_k and P_N, and it has a unique solution if the determinant Δ_N of the correlation matrix D_N of $\mathbf{x}[n]$ is strictly positive. We note, in particular, that

$$P_N = \frac{\Delta_{N+1}}{\Delta_N} \qquad \Delta_N > 0 \tag{13-83}$$

If $\Delta_{N+1} = 0$, then $P_N = 0$ and $\boldsymbol{\varepsilon}_N[m] = 0$. In this case, the unknown $S(\omega)$ consists of lines [see (12-44)].

To find $\mathsf{L}(z)$, it suffices, therefore, to solve the system (13-82). This involves the inversion of the matrix D_N. The problem of inversion can be simplified because the matrix D_N is Toeplitz; that is, it is symmetrical with respect to its diagonal. We give later a simple method for determining a_k and P_N based on this property (Levinson's algorithm).

Moving average processes. If $\mathbf{x}[n]$ is an MA process, then

$$\mathsf{S}(z) = \mathsf{L}(z)\mathsf{L}(1/z) \qquad \mathsf{L}(z) = b_0 + b_1z^{-1} + \cdots + b_Mz^{-M} \tag{13-84}$$

In this case, $R[m] = 0$ for $|m| > M$ [see (12-47)]; hence $\mathsf{S}(z)$ can be expressed directly in terms of $R[m]$:

$$\mathsf{S}(z) = \sum_{m=-M}^{M} R[m]z^{-m} \qquad \mathsf{S}(e^{j\omega}) = \left|\sum_{m=0}^{M} b_m e^{-jm\omega}\right|^2 \tag{13-85}$$

In the identification problem, our objective is to find not the function $\mathsf{S}(z)$, but the $M+1$ coefficients b_m of $\mathsf{L}(z)$. One method for doing so is the factorization $\mathsf{S}(z) = \mathsf{L}(z)\mathsf{L}(1/z)$ of $\mathsf{S}(z)$ as in Sec. 12-1. This method involves the determination of the roots of $\mathsf{S}(z)$. We discuss later a method that avoids factorization (see page 470).

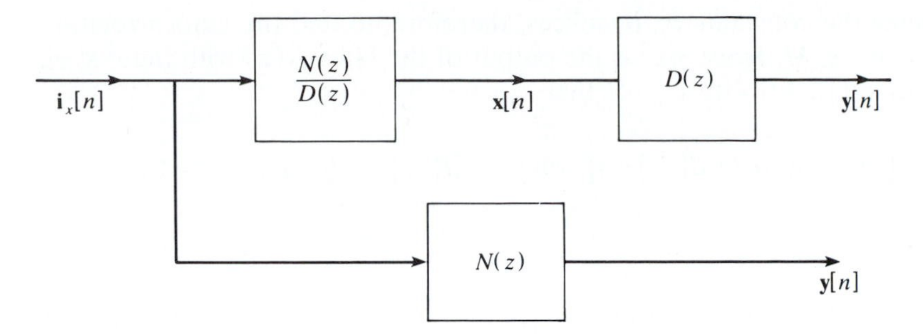

FIGURE 13-12

ARMA processes.† We assume now that $\mathbf{x}[n]$ is an ARMA process:

$$\mathsf{L}(z) = \frac{b_0 + b_1 z^{-1} + \cdots + b_M z^{-M}}{1 + a_1 z^{-1} + \cdots + a_N z^{-N}} = \frac{N(z)}{D(z)} \tag{13-86}$$

In this case, $\mathbf{x}[n]$ satisfies the equation

$$\mathbf{x}[n] + a_1\mathbf{x}[n-1] + \cdots + a_N\mathbf{x}[n-N] = b_0\mathbf{i}[n] + \cdots + b_M\mathbf{i}[n-M] \tag{13-87}$$

where $\mathbf{i}[n]$ is its innovations. Multiplying both sides of (13-87) by $\mathbf{x}[n-m]$ and taking expected values, we conclude as in (12-49) that

$$R[m] + a_1 R[m-1] + \cdots + a_N R[m-N] = 0 \qquad m > M \tag{13-88}$$

Setting $m = M+1, M+2, \ldots, M+N$ into (13-88), we obtain a system of N equations. The solution of this system yields the N unknowns $a_1, \ldots, a_N$.

To complete the specification of $\mathsf{L}(z)$, it suffices to find the $M+1$ constants $b_0, \ldots, b_M$. To do so, we form a filter with input $\mathbf{x}[n]$, and system function (Fig. 13-12)

$$D(z) = 1 + a_1 z^{-1} + \cdots + a_N z^{-N}$$

The resulting output $\mathbf{y}[n]$ is called the *residual sequence*. Inserting into (10-183), we obtain

$$\mathsf{S}_{yy}(z) = \mathsf{S}(z)D(z)D(1/z) = N(z)N(1/z)$$

From this it follows that $\mathbf{y}[n]$ is an MA process, and its innovations filter equals

$$\mathsf{L}_y(z) = N(z) = b_0 + b_1 z^{-1} + \cdots + b_M z^{-M} \tag{13-89}$$

†M. Kaveh: "High Resolution Spectral Estimation for Noisy Signals," *IEEE Transactions on Acoustics, Speech, and Signal Processing*, vol. **ASSP-27**, 1979. See also J. A. Cadzow: "Spectral Estimation: An Overdetermined Rational Model Equation Approach," *IEEE Proceedings*, vol. **70**, 1982.

To determine the constants b_i, it suffices, therefore, to find the autocorrelation $R_{yy}[m]$ for $|m| \le M$. Since $\mathbf{y}[n]$ is the output of the filter $D(z)$ with input $\mathbf{x}[n]$, it follows from (12-47) with $a_0 = 1$ that

$$R_{yy}[m] = R[m] * d[m] * d[-m] \qquad d[m] = \sum_{k=0}^{N} a_k \delta[m-k]$$

This yields

$$R_{yy}[m] = \sum_{i=-N}^{N} R[m-i]\rho[i] \qquad \rho[m] = \sum_{k=m}^{N} a_{k-m} a_k = \rho[-m] \quad (13\text{-}90)$$

for $0 \le m \le M$ and 0 for $m > M$. With $R_{yy}[m]$ so determined, we proceed as in the MA case.

The determination of the ARMA model involves thus the following steps:

Find the constants a_k from (13-88); this yields $D(z)$.
Find $R_{yy}[m]$ from (13-90).
Find the roots of the polynomial

$$\mathsf{S}_{yy}(z) = \sum_{m=-M}^{M} R_{yy}[m] z^{-m} = N(z)N(1/z)$$

Form the Hurwitz factor $N(z)$ of $\mathsf{S}_{yy}(z)$.

Lattice filters and Levinson's algorithm. An MA filter is a polynomial in z^{-1}. Such a filter is usually realized by a ladder structure as in Fig. 13-14*a*. A lattice filter is an alternate realization of an MA filter in the form of Fig. 13-14*b*. In the context of spectral estimation, lattice filters are used to simplify the solution of the Yule–Walker equations and the factorization of polynomials. Furthermore, as we show later, they are also used to give a convenient description of the properties of extrapolating spectra. Related applications are developed in the next chapter in the solution of the prediction problem.

The polynomial

$$D(z) = 1 - a_1^N z^{-1} - \cdots - a_N^N z^{-N} = 1 - \sum_{k=1}^{N} a_k^N z^{-k}$$

specifies an MA filter with $\mathsf{H}(z) = D(z)$. The superscript in a_k^N identifies the order of the filter. If the input to this filter is an AR process $\mathbf{x}[n]$ with $\mathsf{L}(z)$ as in (13-80) and $a_k^N = -a_k$, then the resulting output

$$\boldsymbol{\varepsilon}[n] = \mathbf{x}[n] - a_1^N \mathbf{x}[n-1] - \cdots - a_N^N \mathbf{x}[n-N] \qquad (13\text{-}91)$$

is white noise as in (13-81). The filter $D(z)$ is usually realized by the ladder structure of Fig. 13-14*a*. We shall show that the lattice filter of Fig. 13-14*b* is an equivalent realization. We start with $N = 1$.

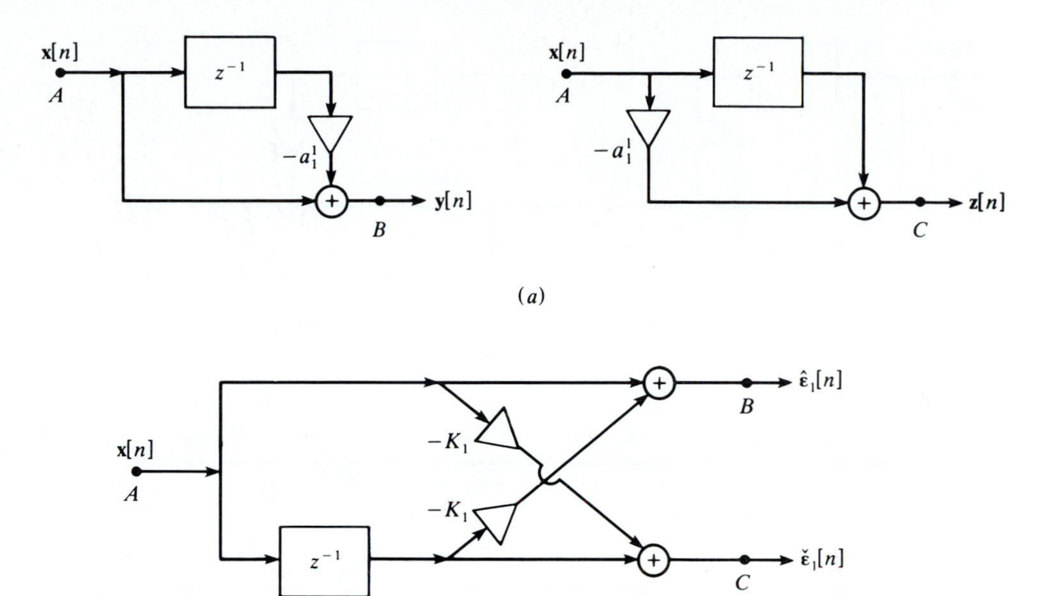

FIGURE 13-13

In Fig. 13-13*a* we show the ladder realization of an MA filter of order 1 and its mirror image. The input to both systems is the process $\mathbf{x}[n]$; the outputs equal

$$\mathbf{y}[n] = \mathbf{x}[n] - a_1^1\mathbf{x}[n-1] \qquad \mathbf{z}[n] = -a_1^1\mathbf{x}[n] + \mathbf{x}[n-1]$$

The corresponding system functions equal

$$1 - a_1^1 z^{-1} \qquad -a_1^1 + z^{-1}$$

In Fig. 13-13*b* we show a lattice filter of order 1. It has a single input $\mathbf{x}[n]$ and two outputs

$$\hat{\boldsymbol{\varepsilon}}_1[n] = \mathbf{x}[n] - K_1\mathbf{x}[n-1] \qquad \check{\boldsymbol{\varepsilon}}_1[n] = -K_1\mathbf{x}[n] + \mathbf{x}[n-1]$$

The corresponding system functions are

$$\hat{\mathsf{E}}_1(z) = 1 - K_1 z^{-1} \qquad \check{\mathsf{E}}_1(z) = -K_1 + z^{-1} = z^{-1}\hat{\mathsf{E}}_1(1/z)$$

If $K_1 = a_1^1$ then the lattice filter of Fig. 13-13*b* is equivalent to the two MA filters of Fig. 13-13*a*.

In Fig. 13-14*b* we show a lattice filter of order N formed by cascading N first-order filters. The input to this filter is the process $\mathbf{x}[n]$. The resulting

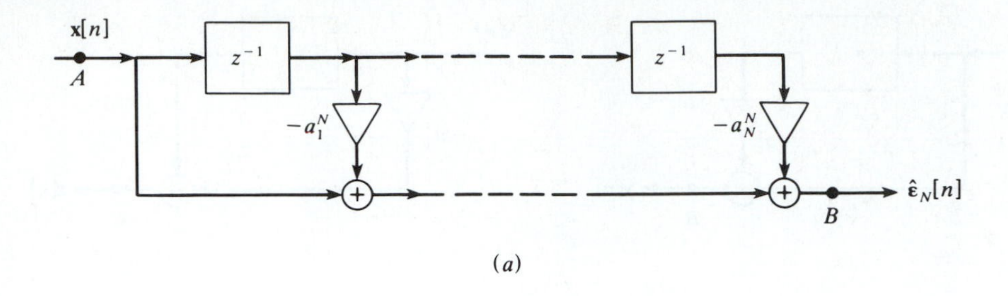

(a)

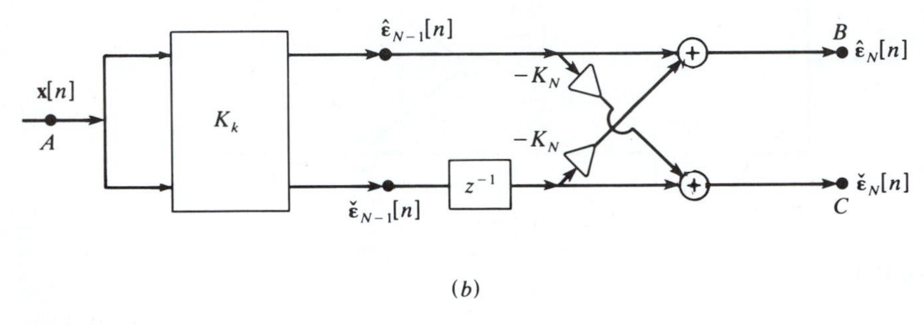

(b)

FIGURE 13-14

outputs are denoted by $\hat{\varepsilon}_N[n]$ and $\check{\varepsilon}_N[n]$ and are called *forward* and *backward* respectively. As we see from the diagram these signals satisfy the equations

$$\hat{\varepsilon}_N[n] = \hat{\varepsilon}_{N-1}[n] - K_N \check{\varepsilon}_{N-1}[n-1] \tag{13-92a}$$

$$\check{\varepsilon}_N[n] = \check{\varepsilon}_{N-1}[n-1] - K_N \hat{\varepsilon}_{N-1}[n] \tag{13-92b}$$

Denoting by $\hat{\mathrm{E}}_N(z)$ and $\check{\mathrm{E}}_N(z)$ the system functions from the input A to the upper output B and lower output C respectively, we conclude that

$$\hat{\mathrm{E}}_N(z) = \hat{\mathrm{E}}_{N-1}(z) - K_N z^{-1} \check{\mathrm{E}}_{N-1}(z) \tag{13-93a}$$

$$\check{\mathrm{E}}_N(z) = z^{-1} \check{\mathrm{E}}_{N-1}(z) - K_N \hat{\mathrm{E}}_{N-1}(z) \tag{13-93b}$$

where $\hat{\mathrm{E}}_{N-1}(z)$ and $\check{\mathrm{E}}_{N-1}(z)$ are the forward and backward system functions of the lattice of the first $N-1$ sections. From (13-93) it follows by a simple induction that

$$\check{\mathrm{E}}_N(z) = z^{-N} \hat{\mathrm{E}}_N(1/z) \tag{13-94}$$

The lattice filter is thus specified in terms of the N constants K_k. These constants are called *reflection coefficients*.

Since $\hat{\mathsf{E}}_1(z) = 1 - K_1 z^{-1}$ and $\check{\mathsf{E}}_1(z) = -K_1 + z^{-1}$, we conclude from (13-93) that the functions $\hat{\mathsf{E}}_1(z)$ and $\check{\mathsf{E}}_N(z)$ are polynomials in z^{-1} of the form

$$\hat{\mathsf{E}}_N(z) = 1 - a_1^N z^{-1} - \cdots - a_N^N z^{-N} \tag{13-95}$$

$$\check{\mathsf{E}}_N(z) = z^{-N} - a_1^N z^{-N+1} - \cdots - a_N^N \tag{13-96}$$

where a_k^N are N constants that are specified in terms of the reflection coefficients K_k.

LEVINSON'S ALGORITHM.† We denote by a_k^{N-1} the coefficients of the lattice filter of the first $N - 1$ sections:

$$\hat{\mathsf{E}}_{N-1}(z) = 1 - a_1^{N-1} z^{-1} - \cdots - a_{N-1}^{N-1} z^{-(N-1)}$$

From (13-94) it follows that

$$z^{-1}\check{\mathsf{E}}_{N-1}(z) = z^{-N}\hat{\mathsf{E}}_{N-1}(1/z)$$

Inserting into (13-93*a*) and equating coefficients of equal powers of z, we obtain

$$\begin{aligned} a_k^N &= a_k^{N-1} - K_N a_{N-k}^{N-1} \qquad k = 1, \ldots, N-1 \\ a_N^N &= K_N \end{aligned} \tag{13-97}$$

We have thus expressed the coefficients a_k^N of a lattice of order N in terms of the coefficients a_k^{N-1} and the last reflection coefficient K_N. Starting with $a_1^1 = K_1$, we can express recursively the N parameters a_k^N in terms of the N reflection coefficients K_k.

Conversely, if we know a_k^N, we find K_k using inverse recursion: The coefficient K_N equals a_N^N. To find K_{N-1}, it suffices to find the polynomial $\hat{\mathsf{E}}_{N-1}(z)$. Multiplying (13-93*b*) by K_N and adding to (13-93*a*), we obtain

$$(1 - K_N^2)\hat{\mathsf{E}}_{N-1}(z) = \hat{\mathsf{E}}_N(z) + K_N z^{-N}\hat{\mathsf{E}}_N(1/z) \tag{13-98}$$

This expresses $\hat{\mathsf{E}}_{N-1}(z)$ in terms of $\hat{\mathsf{E}}_N(z)$ because $K_N = a_N^N$. With $\hat{\mathsf{E}}_{N-1}(z)$ so determined, we set $K_{N-1} = a_{N-1}^{N-1}$. Continuing this process, we find $\hat{\mathsf{E}}_{N-k}(z)$ and K_{N-k} for every $k < N$.

Minimum-phase properties. We shall relate the location of the roots z_i^N of the polynomial $\hat{\mathsf{E}}_N(z)$ to the magnitude of the reflection coefficients K_k.

THEOREM. If

$$|K_k| < 1 \quad \text{for all } k \le N \qquad \text{then} \quad |z_i^N| < 1 \quad \text{for all } i \le N \tag{13-99}$$

†N. Levinson: "The Wiener RMS Error Criterion in Filter Design and Prediction," *Journal of Mathematics and Physics*, vol. **25**, 1947. See also J. Durbin: "The Fitting of Time Series Models," *Revue L'Institut Internationale de Statisque*, vol. 28, 1960.

Proof. By induction. The theorem is true for $N = 1$ because $\hat{E}_1(z) = 1 - K_1 z^{-1}$; hence $|z_1^1| = |K_1| < 1$. Suppose that $|z_j^{N-1}| \le 1$ for all $j \le N-1$ where z_j^{N-1} are the roots of $\hat{E}_{N-1}(z)$. From this it follows that the function

$$A_{N-1}(z) = \frac{z^{-N}\hat{E}_{N-1}(1/z)}{\hat{E}_{N-1}(z)} \tag{13-100}$$

is all-pass. Since $\hat{E}_N(z_i^N) = 0$ by assumption, we conclude from (13-93*a*) and (13-94) that

$$\hat{E}_N(z_i) = \hat{E}_{N-1}(z_i) - K_N z^{-N}\hat{E}_{N-1}(1/z_i) = 0$$

Hence

$$\left|A_{N-1}(z_i^N)\right| = \frac{1}{|K_N|} > 1$$

This shows that $|z_i^N| < 1$ [see (13B-2)].

CONVERSE THEOREM. If

$$|z_i^N| < 1 \quad \text{for all } i \le N \qquad \text{then} \quad |K_k| < 1 \quad \text{for all } k \le N \tag{13-101}$$

Proof. The product of the roots of the polynomial $\hat{E}_N(z)$ equals the last coefficient a_N^N. Hence

$$K_N = a_N^N = z_1^N \cdots z_N^N \qquad |K_N| < 1$$

Thus (13-100) is true for $k = N$. To show that it is true for $k = N-1$, it suffices to show that $|z_j^{N-1}| < 1$ for $j \le N-1$. To do so, we form the all-pass function

$$A_N(z) = \frac{z^{-N}\hat{E}_N(1/z)}{\hat{E}_N(z)} \tag{13-102}$$

Since $\hat{E}_{N-1}(z_j^{N-1}) = 0$ it follows from (13-98) that

$$\left|A_N\left(z_j^{N-1}\right)\right| = \frac{1}{|K_N|} > 1$$

Hence $|z_j^{N-1}| < 1$ and $|K_{N-1}| = |a_{N-1}^{N-1}| = |z_1^{N-1} \cdots z_{N-1}^{N-1}| < 1$. Proceeding similarly, we conclude that $|K_k| < 1$ for all $k \le N$.

COROLLARY. If $|K_k| < 1$ for $k \le N-1$ and $|K_N| = 1$, then

$$|z_i^N| = 1 \qquad \text{for all} \quad i \le N \tag{13-103}$$

Proof. From the theorem it follows that $|z_j^{N-1}| < 1$ because $|K_k| < 1$ for all $k \le N-1$. Hence the function $A_{N-1}(z)$ in (13-100) is all-pass and $|A_{N-1}(z_i^N)| = 1/\,|K_N| = 1$. This leads to the conclusion that $|z_i^N| = 1$ [see (13B-2)].

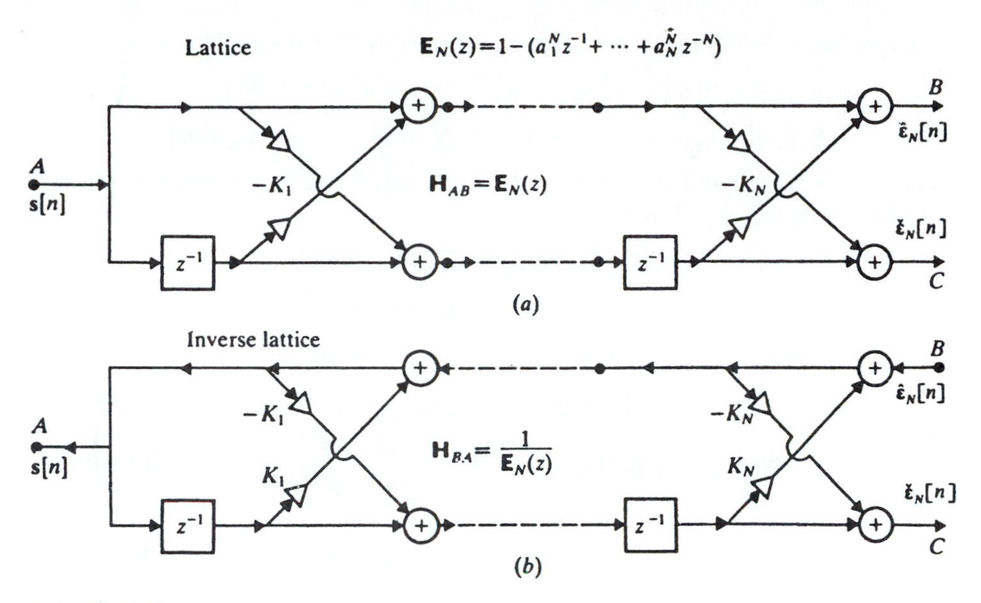

FIGURE 13-15

We have thus established the equivalence between a polynomial $\hat{E}_N(z)$ and a set of N constants K_k. We have shown further that the polynomial is strictly Hurwitz, iff $|K_k| < 1$ for all k.

Inverse lattice realization of AR systems. An *inverse lattice* is a modification of a lattice as in Fig. 13-15. In this modification, the input is at point B and the outputs are at points A and C. Furthermore, the multipliers from the lower to the upper line are changed from $-K_k$ to K_k. Denoting by $\hat{\boldsymbol{\varepsilon}}_N[n]$ the input at point B and by $\check{\boldsymbol{\varepsilon}}_{N-1}[n]$ the resulting output at C, we observe from the figure that

$$\hat{\boldsymbol{\varepsilon}}_{N-1}[n] = \hat{\boldsymbol{\varepsilon}}_N[n] + K_N \check{\boldsymbol{\varepsilon}}_{N-1}[n-1] \tag{13-104a}$$

$$\check{\varepsilon}_N[n] = \check{\varepsilon}_{N-1}[n-1] - K_N \hat{\boldsymbol{\varepsilon}}_{N-1}[n] \tag{13-104b}$$

These equations are identical with the two equations in (13-92). From this it follows that the system function from B to A equals

$$\frac{1}{\hat{E}_N(z)} = \frac{1}{1 - a_1^N z^{-1} - \cdots - a_N^N z^{-N}}$$

We have thus shown that an AR system can be realized by an inverse lattice. The coefficients a_k^N and K_k satisfy Levinson's algorithm (13-97).

Iterative solution of the Yule–Walker equations. Consider an AR process $\mathbf{x}[n]$ with innovations filter $L(z) = \sqrt{P_N}/D(z)$ as in (13-80). We form the lattice equivalent of the MA system $D(z)$ with $a_k^N = -a_k$, and use $\mathbf{x}[n]$ as its input. As we know [see (13-95)] the forward and backward responses are given by

$$\begin{aligned}\hat{\boldsymbol{\varepsilon}}_N[n] &= \mathbf{x}[n] - a_1^N\mathbf{x}[n-1] - \cdots - a_N^N\mathbf{x}[n-N] \\ \check{\boldsymbol{\varepsilon}}_N[n] &= \mathbf{x}[n-N] - a_1^N x[n-N+1] - \cdots - a_N^N\mathbf{x}[n]\end{aligned} \tag{13-105}$$

Denoting by $\hat{\mathbf{S}}_N(z)$ and $\check{\mathbf{S}}_N(z)$ the spectra of $\hat{\boldsymbol{\varepsilon}}_N[n]$ and $\check{\boldsymbol{\varepsilon}}_N[n]$ respectively, we conclude from (13-105) that

$$\hat{\mathbf{S}}_N(z) = \mathbf{S}(z)\hat{\mathbf{E}}_N(z)\hat{\mathbf{E}}_N(1/z) = P_N$$

$$\check{\mathbf{S}}_N(z) = \mathbf{S}(z)\check{\mathbf{E}}_N(z)\check{\mathbf{E}}_N(1/z) = P_N$$

From this it follows that $\hat{\boldsymbol{\varepsilon}}_N[n]$ and $\check{\boldsymbol{\varepsilon}}_N[n]$ are two white-noise processes and

$$E\{\hat{\boldsymbol{\varepsilon}}_N^2[n]\} = E\{\check{\boldsymbol{\varepsilon}}_N^2[n]\} = P_N \tag{13-106a}$$

$$E\{\mathbf{x}[n-m]\hat{\boldsymbol{\varepsilon}}_N[n]\} = \begin{cases} P_N & m = 0 \\ 0 & 1 \le m \le N \end{cases} \tag{13-106b}$$

$$E\{\mathbf{x}[n-m]\check{\boldsymbol{\varepsilon}}_N[n]\} = \begin{cases} 0 & 0 \le m \le N-1 \\ P_N & m = N \end{cases} \tag{13-106c}$$

These equations also hold for all filters of lower order. We shall use them to express recursively the parameters a_k^N, K_N, and P_N in terms of the $N+1$ constants $R[0], \ldots, R[N]$.

For $N = 1$ (13-82) yields

$$R[0] - a_1^1 R[1] = P_1 \qquad R[1] - a_1^1 R[0] = 0$$

Setting $P_0 = R[0]$, we obtain

$$a_1^1 = K_1 = \frac{R[1]}{R[0]} \qquad P_1 = (1 - K_1^2)P_0$$

Suppose now that we know the $N+1$ parameters a_k^{N-1}, K_{N-1}, and P_N. From Levinson's algorithm (13-97) it follows that we can determine a_k^N if K_N is known. To complete the iteration, it suffices, therefore, to find K_N and P_N.

We maintain that

$$P_{N-1}K_N = R[N] - \sum_{k=1}^{N-1} a_k^{N-1} R[N-k] \tag{13-107}$$

$$P_N = (1 - K_N^2)P_{N-1} \tag{13-108}$$

The first equation yields K_N in terms of the known parameters a_k^{N-1}, $R[m]$, and P_{N-1}. With K_N so determined, P_N is determined from the second equation.

Proof. Multiplying (13-92a) by $\mathbf{x}[n-N]$ and using the identities

$$E\{\hat{\boldsymbol{\varepsilon}}_{N-1}[n]\mathbf{x}[n-N]\} = R[N] - \sum_{k=1}^{N-1} a_k^{N-1} R[k]$$

$$E\{\hat{\boldsymbol{\varepsilon}}_N[n]\mathbf{x}[n]\} = P_N \qquad E\{\hat{\boldsymbol{\varepsilon}}_{N-1}[n-1]\mathbf{x}[n]\} = P_{N-1}$$

we obtain (13-107). From (13-92a) and the identities

$$E\{\hat{\varepsilon}_N[n]\mathbf{x}[n]\} = P_N \qquad E\{\hat{\varepsilon}_{N-1}[n-1]\mathbf{x}[n]\} = P_{N-1}$$

$$E\{\check{\varepsilon}_{N-1}[n-1]\mathbf{x}[n]\} = R[N] - \sum_{k=1}^{N-1} a_k^{N-1} R[N-k] = P_{N-1}K_N$$

it follows similarly that $P_N = P_{N-1} - K_N^2 P_{N-1}$ and (13-108) results.

Since $P_k \geq 0$ for every k, it follows from (13-108) that

$$|K_k| \leq 1 \qquad \text{and} \qquad P_0 \geq P_1 \geq \cdots \geq P_N \geq 0 \tag{13-109}$$

If $|K_N| = 1$ but $|K_k| < 1$ for all $k < N$, then

$$P_0 > P_1 > \cdots > P_N = 0 \tag{13-110}$$

As we show next this is the case if $S(\omega)$ consists of lines.

Line spectra and hidden periodicities. If $P_N = 0$, then $\hat{\varepsilon}_N[n] = 0$; hence the process $\mathbf{x}[n]$ satisfies the homogeneous recursion equation

$$\mathbf{x}[n] = a_1^N \mathbf{x}[n-1] + \cdots + a_N^N \mathbf{x}[n-N] \tag{13-111}$$

This shows that $\mathbf{x}[n]$ is a predictable process, that is, it can be expressed in terms of its N past values. Furthermore,

$$R[m] - a_1^N R[m-1] - \cdots - a_N^N R[m-N] = 0 \tag{13-112}$$

As we know [see (13-103)] the roots z_i^N of the characteristic polynomial $\hat{\mathsf{E}}_N(z)$ of this equation are on the unit circle: $z_i^N = e^{j\omega_i}$. From this it follows that

$$R[m] = \sum_{i=1}^{N} \alpha_i e^{j\omega_i m} \qquad S(\omega) = 2\pi \sum_{i=1}^{N} \alpha_i \delta(\omega - \omega_i) \tag{13-113}$$

And since $S(\omega) \geq 0$, we conclude that $\alpha_i \geq 0$.

Solving (13-111), we obtain

$$\mathbf{x}[n] = \sum_{i=1}^{N} \mathbf{c}_i e^{j\omega_i n} \qquad E\{\mathbf{c}_i\} = 0 \qquad E\{\mathbf{c}_i \mathbf{c}_k\} = \begin{cases} \alpha_i & i = k \\ 0 & i \neq k \end{cases} \tag{13-114}$$

CARATHEODORY'S THEOREM. We show next that if $R[m]$ is a p.d. sequence and its correlation matrix is of rank N, that is, if

$$\Delta_N > 0 \qquad \Delta_{N+1} = 0 \tag{13-115}$$

then $R[m]$ is a sum of exponentials with positive coefficients:

$$R[m] = \sum_{i=1}^{N} \alpha_i e^{j\omega_i m} \qquad \alpha_i > 0 \tag{13-116}$$

Proof. Since $R[m]$ is a p.d. sequence, we can construct a process $\mathbf{x}[n]$ with autocorrelation $R[m]$. Applying Levinson's algorithm, we obtain a sequence of constants K_k and P_k. The iteration stops at the Nth step because $P_N = \Delta_{N+1}/\Delta_N = 0$. This shows that the process $\mathbf{x}[n]$ satisfies the recursion equation (13-111).

Detection of hidden periodicities.† We shall use the above to solve the following problem: We wish to determine the frequencies ω_i of a process $\mathbf{x}[n]$ consisting of at most N exponentials as in (13-114). The available information is the sum

$$\mathbf{y}[n] = \mathbf{x}[n] + \boldsymbol{\nu}[n] \qquad E\{\boldsymbol{\nu}^2[n]\} = q \tag{13-117}$$

where $\boldsymbol{\nu}[n]$ is white noise independent of $\mathbf{x}[n]$.

Using J samples of $\mathbf{y}[n]$, we estimate its autocorrelation

$$R_{yy}[m] = R_{xx}[m] + q\delta[m] \tag{13-118}$$

as in (13-78). The correlation matrix D_{N+1} of $\mathbf{x}[n]$ is thus given by

$$D_{N+1} = \begin{bmatrix} R_{yy}[0] - q & R_{yy}[1] & \cdots & R_{yy}[N] \\ R_{yy}[1] & R_{yy}[0] - q & \cdots & R_{yy}[N-1] \\ \cdots & \cdots & \cdots & \cdots \\ R_{yy}[N] & R_{yy}[N-1] & \cdots & R_{yy}[0] - q \end{bmatrix} \tag{13-119}$$

In this expression, $R_{yy}[m]$ is known but q is unknown. We know, however, that $\Delta_{N+1} = 0$ because $\mathbf{x}[n]$ consists of N lines. Hence q is an eigenvalue of D_{N+1}. It is, in fact, the smallest eigenvalue q_0 because $D_{N+1} > 0$ for $q < q_0$. With $R_{xx}[m]$ so determined, we proceed as before: Using Levinson's algorithm, we find the coefficients a_k^N and the roots $e^{j\omega_i}$ of the resulting polynomial $\hat{\mathsf{E}}_N(z)$. If q_0 is a simple eigenvalue, then all roots are distinct and $\mathbf{x}[n]$ is a sum of N exponentials. If, however, q_0 is a multiple root with multiplicity N_0 then $\mathbf{x}[n]$ consists of $N - N_0 + 1$ exponentials.

This analysis leads to the following extension of Caratheodory's theorem: The $N + 1$ values $R[0], \ldots, R[N]$ of a strictly p.d. sequence $R[m]$ can be expressed in the form

$$R[m] = q_0\delta[m] + \sum_{i=1}^{N} \alpha_i e^{j\omega_i m} \tag{13-120}$$

where q_0 and α_i are positive constants and ω_i are real frequencies.

Burg's iteration.‡ Levinson's algorithm is used to determine recursively the coefficients a_k^N of the innovations filter $\mathsf{L}(z)$ of an AR process $\mathbf{x}[n]$ in terms of $R[m]$. In a real problem the data $R[m]$ are not known exactly. They are estimated from the J samples of $\mathbf{x}[n]$ and these estimates are inserted into (13-107) and (13-108) yielding the estimates of K_N and P_N. The results are then used to estimate a_k^N from (13-97). A more direct approach, suggested by Burg, avoids the estimation of $R[m]$. It is based on the observation that Levinson's

†V. F. Pisarenko: "The Retrieval of Harmonics," *Geophysical Journal of the Royal Astronomical Society*, 1973.

‡J. P. Burg: Maximum entropy spectral analysis, presented at the International Meeting of the Society for the Exploration of Geophysics, Orlando, FL, 1967.

algorithm expresses recursively the coefficients a_k^N in terms of K_N and P_N. The estimates of these coefficients can, therefore, be obtained directly in terms of the estimates of K_N and P_N. These estimates are based on the following identities [see (13-106)]:

$$P_{N-1}K_N = E\{\hat{\boldsymbol{\varepsilon}}_{N-1}[n]\check{\boldsymbol{\varepsilon}}_{N-1}[n-1]\}$$
$$P_N = \tfrac{1}{2}E\{\hat{\boldsymbol{\varepsilon}}_N^2[n] + \check{\boldsymbol{\varepsilon}}_N^2[n]\} \tag{13-121}$$

Replacing expected values by time averages, we obtain the following iteration: Start with

$$\mathbf{P}_0 = \frac{1}{J}\sum_{n=1}^{J}\mathbf{x}^2[n] \qquad \hat{\boldsymbol{\varepsilon}}_0[n] = \check{\boldsymbol{\varepsilon}}_0[n] = \mathbf{x}[n]$$

Find $\mathbf{K}_{N-1}$, $\mathbf{P}_{N-1}$, $\mathbf{a}_k^{N-1}$, $\hat{\boldsymbol{\varepsilon}}_{N-1}[n]$, $\check{\boldsymbol{\varepsilon}}_{N-1}$. Set

$$\mathbf{K}_N = \frac{\sum_{n=N+1}^{J}\hat{\boldsymbol{\varepsilon}}_{N-1}[n]\check{\boldsymbol{\varepsilon}}_{N-1}[n-1]}{\frac{1}{2}\sum_{n=N+1}^{J}\left(\hat{\boldsymbol{\varepsilon}}_{N-1}^2[n] + \check{\boldsymbol{\varepsilon}}_{N-1}^2[n-1]\right)} \tag{13-122}$$

$$\mathbf{P}_N = \left(1 - \mathbf{K}_N^2\right)P_{N-1} \tag{13-123}$$

$$\mathbf{a}_k^N = \mathbf{a}_k^{N-1} - \mathbf{K}_N\mathbf{a}_{N-k}^{N-1} \qquad k = 1,\ldots,N-1$$
$$\mathbf{a}_N^N = K_N \tag{13-124}$$

$$\hat{\boldsymbol{\varepsilon}}_N[n] = \mathbf{x}[n] - \sum_{k=1}^{N-1}\mathbf{a}_k^N\mathbf{x}[n-k]$$
$$\check{\boldsymbol{\varepsilon}}_N[n] = \mathbf{x}[n-N] - \sum_{k=1}^{N}\mathbf{a}_{N-k}^N\mathbf{x}[n-N+k] \tag{13-125}$$

This completes the Nth iteration step. Note that

$$|\mathbf{K}_N| \le 1 \qquad \mathbf{P}_N \ge 0$$

This follows readily if we apply Cauchy's inequality (see Prob. 11-23) to the numerator of (13-122).

Levinson's algorithm yields the correct spectrum $\mathsf{S}(z)$ only if $\mathbf{x}[n]$ is an AR process. If it is not, the result is only an approximation. If $R[m]$ is known exactly, the approximation improves as N increases. However, if $R[m]$ is estimated as above, the error might increase because the number of terms in (13-49) equals $J - N - 1$ and it decreases as N increases. The determination of an optimum N is in general difficult.

FEJÉR–RIESZ THEOREM AND LEVINSON'S ALGORITHM. Given a positive trigonometric polynomial

$$W(e^{j\omega}) = \sum_{n=-N}^{N} w_n e^{-jn\omega} \ge 0 \tag{13-126}$$

we can find a Hurwitz polynomial

$$Y(z) = \sum_{n=0}^{N} y_n z^{-n} \tag{13-127}$$

such that $W(e^{j\omega}) = |Y(e^{j\omega})|^2$. This theorem has extensive applications. We used it in Sec. 12-1 (spectral factorization) and in the estimation of the spectrum of an MA and an ARMA process. The construction of the polynomial $Y(z)$ involves the determination of the roots of $W(z)$. This is not a simple problem particularly if $W(e^{j\omega})$ is known only as a function of ω. We discuss next a method for determining $Y(z)$ involving Levinson's algorithm and Fourier series.

We compute, first, the Fourier series coefficients

$$R[m] = \frac{1}{2\pi}\int_{-\pi}^{\pi} \frac{1}{W(e^{j\omega})} e^{-jm\omega}\, d\omega \qquad 0 \le m \le N \tag{13-128}$$

of the function $S(e^{j\omega}) = 1/W(e^{j\omega})$. The numbers $R[m]$ so obtained are the values of a p.d. sequence because $S(e^{j\omega}) \ge 0$. Applying Levinson's algorithm to the numbers $R[m]$ so computed, we obtain $N+1$ constants a_k^N and P_N. This yields

$$S(e^{j\omega}) = \frac{1}{W(e^{j\omega})} = \frac{P_N}{|1 - \sum_{n=1}^{N} a_n^N e^{-jn\omega}|^2}$$

Hence

$$Y(z) = \frac{1}{\sqrt{P_N}}\left(1 - \sum_{n=1}^{N} a_n^N z^{-n}\right)$$

as in (13-127). This method thus avoids the factorization problem.

The General Class of Extrapolating Spectra†

We consider now the following problem: We are given the $N+1$ values (data)

$$R[0], \ldots, R[N]$$

of the autocorrelation $R[m]$ of a process $\mathbf{x}[n]$ and we wish to find all its p.d. extrapolations, that is, we wish to find the family C_N of spectra $\mathsf{S}(e^{j\omega}) \ge 0$ such that the first $N+1$ coefficients of their Fourier series expansion equal the given data. The sequences $R[m]$ of the class C_N and their spectra will be called admissible.

A member of the class C_N is the AR spectrum

$$\mathsf{S}(z) = \mathsf{L}(z)\mathsf{L}(1/z) \qquad \mathsf{L}(z) = \sqrt{P_N}/\mathsf{E}_N(z)$$

†A. Papoulis: "Levinson's Algorithm, Wold's Decomposition, and Spectral Estimation," *SIAM Review*, vol. **27**, 1985.

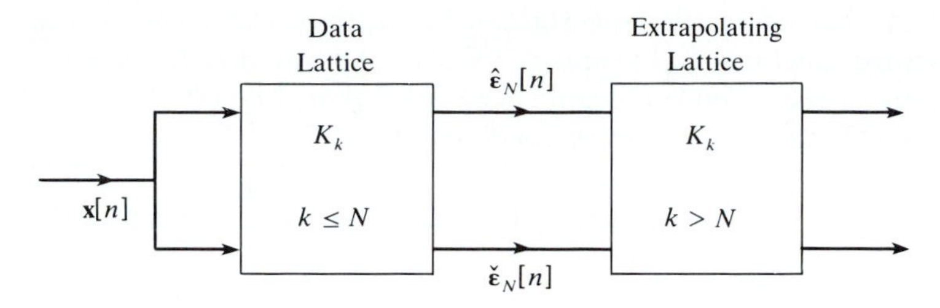

FIGURE 13-16

where $\mathsf{E}_N(z) = \hat{\mathsf{E}}_N(z)$ is the forward filter of order N obtained from an N-step Levinson algorithm. The continuation of the corresponding $R[m]$ is obtained from (12-41b):

$$R[m] = \sum_{k=1}^{N} a_k^N R[m-k] \qquad m > N$$

To find all members of the class C_N, we continue the algorithm assigning arbitrary values

$$|K_k| \le 1 \qquad k = N+1, N+2, \ldots$$

to the reflection coefficients. The resulting values of $R[m]$ are determined recursively [see (13-107)]

$$R[m] = \sum_{k=1}^{m-1} a_k^{m-1} R[m-k] + P_{m-1} K_m \tag{13-129}$$

This shows that the admissible values of $R[m]$ at the mth iteration are in an interval of length $2P_{m-1}$:

$$\sum_{k=1}^{m-1} a_k^{m-1} R[m-k] - P_{m-1} \le R[m] \le \sum_{k=1}^{m-1} a_k^{m-1} R[m-k] + P_{m-1} \tag{13-130}$$

because $|K_m| \le 1$. At the endpoints of the interval, $|K_m| = 1$; in this case, $P_m = 0$ and $\Delta_{m+1} = 0$. As we have shown, the corresponding spectrum $S(\omega)$ consists of lines. If $|K_{m_0}| < 1$ and $K_m = 0$ for $m > m_0$, then $\mathsf{S}(z)$ is an AR spectrum of order m_0. In Fig. 13-16, we show the iteration lattice. The first N sections are uniquely determined in terms of the data. The remaining sections form a four-terminal lattice specified in terms of the arbitrarily chosen reflection coefficients $K_{N+1}, K_{N+2}, \ldots$.

Admissible spectra. The DFTs of the sequences generated by the preceding iteration form the class C_N of admissible spectra. We give next a simple characterization of this class starting with regular spectra. Such spectra are the transforms of all admissible sequences obtained with $|K_m| < 1$ for all m.

We shall show that all regular spectra can be expressed in terms of the forward and backward functions $\mathsf{E}_N(z)$ and $z^{-N}\mathsf{E}_N(1/z)$ of the first N sections, the constant P_N, and a reflection coefficient $\rho(z)$ defined as follows:

A function $\rho(z)$ is called a *reflection coefficient* if

$$\rho(z) = \frac{b(z)}{a(z)} \qquad |\rho(z)| < 1 \qquad \text{for} \quad |z| \geq 1 \tag{13-131}$$

where $a(z)$ and $b(z)$ are two power series in z^{-1} analytic for $|z| \geq 1$ and such that $a(\infty) = 1$, $b(\infty) = 0$.

It can be shown that the functions

$$\mathsf{S}(e^{j\omega}) = P_N \frac{1 - |\rho(e^{j\omega})|^2}{|\mathsf{E}_N(e^{j\omega}) + e^{-jN\omega}\mathsf{E}_N(e^{-j\omega})\rho(e^{j\omega})|^2} \tag{13-132}$$

generate the class of all regular spectra of the class C_N where $\rho(z)$ is an arbitrary reflection coefficient. The proof of this is based on the properties of four-terminal lattices. The details, however, are involved (see page 470*n*).

We shall determine the innovations filter $\mathsf{L}(z)$ of a process with the above spectrum. To do so, we must factor $\mathsf{S}(z)$. The denominator of $\mathsf{S}(z)$ is factored readily. To factor the numerator, we observe that

$$1 - \rho(z)\rho(1/z) = \frac{a(z)a(1/z) - b(z)b(1/z)}{a(z)a(1/z)}$$

It suffices, therefore, to find the numerator. To do so, we determine a Hurwitz polynomial $y(z)$ such that

$$y(z)y(1/z) = a(z)a(1/z) - b(z)b(1/z)$$

With $y(z)$ so determined, (13-132) yields

$$\mathsf{L}(z) = \frac{\sqrt{P_N}\, y(z)}{\mathsf{E}_N(z)a(z) - z^{-N}\mathsf{E}_N(1/z)b(z)} \tag{13-133}$$

If $\rho(z)$ is a rational function, $\mathsf{S}(z)$ is an ARMA spectrum. If $\rho(z) = 0$, it is an AR spectrum of order N. If $y[z] =$ constant, $\mathsf{S}(z)$ is an AR spectrum of order higher than N.

Line spectra. If in the preceding algorithm $|K_m| = 1$ for $m = m_0 > N$, then the iteration terminates and the resulting spectrum consists of m_0 lines. We can, therefore, fit the $N + 1$ values $R[m]$ of a p.d. sequence with the sum of m_0 exponentials where m_0 is any number larger than N. We can do so with N lines only if $\Delta_{N+1} = 0$. If $\Delta_{N+1} > 0$, we obtain a spectrum consisting of N lines and a constant:

$$S(\omega) = q_0 + 2\pi \sum_{i=1}^{N} \alpha_i \delta(\omega - \omega_i) \tag{13-134}$$

where q_0 is the smallest eigenvalue of the matrix D_{N+1}. This is a consequence of the modified form (13-120) of Caratheodory's theorem.

Maximum Entropy and Smoothness Conditions

In the preceding discussion we used as the parametric form of $\mathbf{S}(z)$ a rational function of z. In the following we determine the parametric form of $\mathbf{S}(z)$ in terms of certain smoothness conditions leading to the maximization of the integral of some function of $S(\omega)$. The method of maximum entropy is a special case. We repeat the problem: We are given the first $N+1$ values of the p.d. sequence $R[m]$ and we wish to determine its spectrum

$$S(\omega) = \sum_{m=-\infty}^{\infty} R[m]e^{-jm\omega}$$

To solve this problem, we introduce a nonlinear function $G(S(\omega))$ of $S(\omega)$ and we determine the unknown values of $R[m]$ so as to maximize the integral

$$H = \int_{-\pi}^{\pi} G(S(\omega))\, d\omega \tag{13-135}$$

subject to the constraints

$$R[m] = \frac{1}{2\pi}\int_{-\pi}^{\pi} S(\omega)e^{jm\omega}\, d\omega \qquad |m| \le N \tag{13-136}$$

where $R[m]$ are the given data.

The integral H depends on the unknown values of $R[m]$. It is, therefore, maximum if

$$\frac{\partial H}{\partial R[m]} = \int_{-\pi}^{\pi} \frac{d}{dS} G(S(\omega)) \frac{\partial S(\omega)}{\partial R[m]}\, d\omega = 0 \qquad |m| > N$$

With $F(S(\omega)) = G'(S(\omega))$ this yields

$$\int_{-\pi}^{\pi} F(S(\omega))e^{-jm\omega}\, d\omega = 0 \qquad |m| > N \qquad \text{because} \quad \frac{\partial S(\omega)}{\partial R[m]} = e^{-jm\omega}$$

From this it follows that the Fourier series coefficient of the function $F(S(\omega))$ must be 0 for $|m| > M$. In other words

$$F(S(\omega)) = \sum_{k=-N}^{N} c_k e^{-jk\omega} \tag{13-137}$$

The constants c_k can, in principle, be determined in terms of the data $R[m]$. Indeed, from (13-136) and (13-137) it follows that

$$R[m] = \frac{1}{2\pi}\int_{-\pi}^{\pi} F^{(-1)}\left(\sum_{k=-N}^{N} c_k e^{-jk\omega}\right) e^{jm\omega}\, d\omega \qquad |m| \le N \tag{13-138}$$

where $F^{(-1)}$ is the inverse of $F(s)$. This is a nonlinear system of $2N+1$ equations involving the $2N+1$ unknowns c_k. Its solution is in general difficult.

The selection of the function $G(S)$ depends on the applications. It might be selected, for example, to emphasize the high or low values of $S(\omega)$. The following special case is of particular interest. It leads to a system that can be simply solved and the result maximizes the uncertainty about the unknown spectrum.

The method of maximum entropy.† We now assume that

$$G(S(\omega)) = \ln S(\omega)$$

In this case,

$$H = \int_{-\pi}^{\pi} \ln S(\omega)\, d\omega \tag{13-139}$$

If $S(\omega)$ is the power spectrum of a process $\mathbf{x}[n]$, then H is the *entropy rate* of $\mathbf{x}[n]$ [see (15-130)].

From (13-135) it follows that $G(S(\omega)) = \ln S(\omega)$; hence

$$F(S(\omega)) = \frac{1}{S(\omega)} = \sum_{k=-N}^{N} c_k e^{-jk\omega} \geq 0 \tag{13-140}$$

This shows that the spectrum $S(\omega)$ is ARMA. It can, therefore, be written in the form

$$S(\omega) = \frac{P_N}{|1 + \sum_{k=1}^{N} a_k e^{-j\omega k}|^2} \tag{13-141}$$

Hence its coefficients a_k and P_N can be determined recursively from Levinson's algorithm.

We have thus shown that the estimation of $S(\omega)$ based on the principle of maximum entropy rate is equivalent to the assumption that the unknown $S(\omega)$ is AR.

APPENDIX 13A MINIMUM-PHASE FUNCTIONS

A function

$$\mathbf{H}(z) = \sum_{n=0}^{\infty} h_n z^{-n}$$

is called minimum-phase, if it is analytic and its inverse $1/\mathbf{H}(z)$ is also analytic

†A. Papoulis: "Maximum Entropy and Spectral Estimation: A Review," *IEEE Transactions on Acoustics, Speech, and Signal Processing*, vol. ASSP-29, 1981.

for $|z| \geq 1$. We shall show that if $\mathbf{H}(z)$ is minimum-phase, then

$$\ln h_0^2 = \frac{1}{2\pi}\int_{-\pi}^{\pi} \ln |\mathbf{H}(e^{j\varphi})|^2 \, d\varphi \tag{13A-1}$$

Proof. Using the identity $|\mathbf{H}(e^{j\varphi})|^2 = \mathbf{H}(e^{j\varphi})\mathbf{H}(e^{-j\varphi})$, we conclude with $e^{j\varphi} = z$, that

$$\int_{-\pi}^{\pi} \ln |\mathbf{H}(e^{j\varphi})|^2 \, d\varphi = \oint \frac{1}{jz} \ln[\mathbf{H}(z)\mathbf{H}(z^{-1})] \, dz$$

where the path of integration is the unit circle. We note further, changing z to $1/z$, that

$$\oint \frac{1}{z} \ln \mathbf{H}(z) \, dz = \oint \frac{1}{z} \ln \mathbf{H}(z^{-1}) \, dz$$

To prove (13A-1), it suffices, therefore, to show that

$$\ln |h_0| = \frac{1}{2\pi j}\oint \frac{1}{z} \ln \mathbf{H}(z) \, dz$$

This follows readily because $\mathbf{H}(z)$ tends to h_0 as $z \to \infty$ and the function $\ln \mathbf{H}(z)$ is analytic for $|z| \geq 1$ by assumption.

APPENDIX 13B ALL-PASS FUNCTIONS

The unit circle is the locus of points N such that (see Fig. 13-17a)

$$\frac{(NA)}{(NB)} = \frac{|e^{j\varphi} - 1/z_i^*|}{|e^{j\varphi} - z_i|} = \frac{1}{|z_i|} \qquad |z_i| < 1$$

From this it follows that, if

$$\mathbf{F}(z) = \frac{zz_i^* - 1}{z - z_i} \qquad |z_i| < 1$$

then $|\mathbf{F}(e^{j\varphi})| = 1$. Furthermore, $|\mathbf{F}(z)| > 1$ for $|z| < 1$ and $|\mathbf{F}(z)| < 1$ for $|z| > 1$ because $\mathbf{F}(z)$ is continuous and

$$|\mathbf{F}(0)| = \frac{1}{|z_i|} > 1 \qquad |\mathbf{F}(\infty)| = |z_i^*| < 1$$

Multiplying N bilinear fractions of the above form, we conclude that, if

$$\mathbf{H}(z) = \prod_{i=1}^{N} \frac{zz_i^* - 1}{z - z_i} \qquad |z_i| < 1 \tag{13B-1}$$

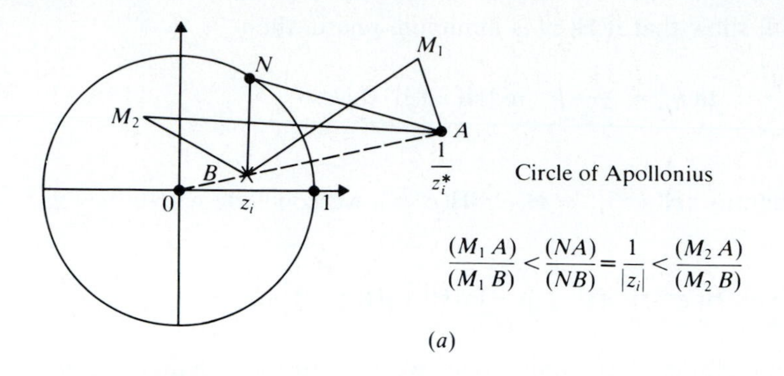

(a)

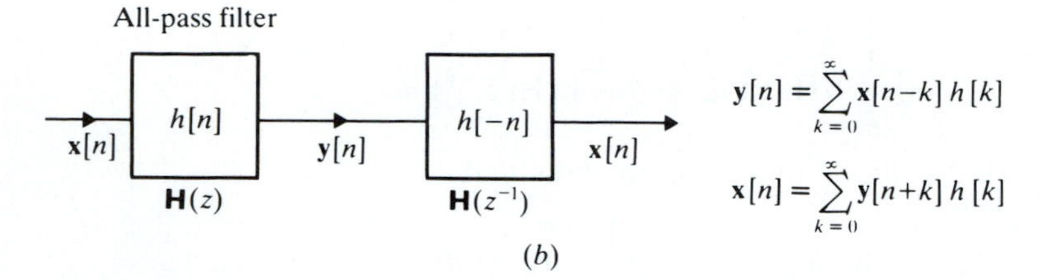

(b)

FIGURE 13-17

then

$$|\mathbf{H}(z)| \begin{cases} > 1 & |z| < 1 \\ = 1 & |z| = 1 \\ < 1 & |z| > 1 \end{cases} \tag{13B-2}$$

A system with system function $\mathbf{H}(z)$ as in (13B-1) is called *all-pass*. Thus an all-pass system is stable, causal, and

$$|\mathbf{H}(e^{j\omega T})| = 1$$

Furthermore,

$$\frac{1}{\mathbf{H}(z)} = \prod_{i=1}^{N} \frac{z - z_i}{zz_i^* - 1} = \prod_{i=1}^{N} \frac{1 - z_i/z}{z_i^* - 1/z} = \mathbf{H}\left(\frac{1}{z}\right) \tag{13B-3}$$

because if z_i is a pole of $\mathbf{H}(z)$, then z_i^* is also a pole.

From the above it follows that if $h[n]$ is the delta response of an all-pass system, then the delta response of its inverse is $h[-n]$:

$$\mathbf{H}(z) = \sum_{n=0}^{\infty} h[n] z^{-n} \qquad \frac{1}{\mathbf{H}(z)} = \sum_{n=0}^{\infty} h[n] z^{n} \tag{13B-4}$$

where both series converge in a ring containing the unit circle.

PROBLEMS

13-1. Find the mean and variance of the RV

$$\mathbf{n}_T = \frac{1}{2T}\int_{-T}^{T} \mathbf{x}(t)\, dt \qquad \text{where} \quad \mathbf{x}(t) = 10 + \boldsymbol{\nu}(t)$$

for $T = 5$ and for $T = 100$. Assume that $E\{\boldsymbol{\nu}(t)\} = 0$, $R_\nu(\tau) = 2\delta(\tau)$.

13-2. Show that if a process is normal and distribution-ergodic as in (13-35), then it is also mean-ergodic.

13-3. Show that if $\mathbf{x}(t)$ is normal with $\eta_x = 0$ and $R_x(\tau) = 0$ for $|\tau| > a$, then it is correlation-ergodic.

13-4. Show that the process $\mathbf{a}e^{j(\omega t + \varphi)}$ is not correlation-ergodic.

13-5. Show that

$$R_{xy}(\lambda) = \lim_{T\to\infty} \frac{1}{2T}\int_{-T}^{T} \mathbf{x}(t+\lambda)\mathbf{y}(t)\, dt$$

iff

$$\lim_{T\to\infty} \frac{1}{2T}\int_{-2T}^{2T}\left(1 - \frac{|\tau|}{2T}\right) E\{\mathbf{x}(t+\lambda+\tau)\mathbf{y}(t+\tau)\mathbf{x}(t+\lambda)\mathbf{y}(t)\}\, d\tau = R_{xy}^2(\lambda)$$

13-6. The process $\mathbf{x}(t)$ is cyclostationary with period T, mean $\eta(t)$, and correlation $R(t_1, t_2)$. Show that if $R(t+\tau, t) \to \eta^2(t)$ as $|\tau| \to \infty$, then

$$\lim_{c\to\infty} \frac{1}{2c}\int_{-c}^{c} \mathbf{x}(t)\, dt = \frac{1}{T}\int_0^T \eta(t)\, dt$$

Hint: The process $\bar{\mathbf{x}}(t) = \mathbf{x}(t - \boldsymbol{\theta})$ is mean-ergodic.

13-7. Show that if

$$C(t+\tau, t) \xrightarrow[t\to\infty]{} 0$$

uniformly in t; then $\mathbf{x}(t)$ is mean-ergodic.

13-8. The process $\mathbf{x}(t)$ is normal with 0 mean and WSS. (a) Show that (Fig. P13-8a)

$$E\{\mathbf{x}(t+\lambda)|\mathbf{x}(t) = x\} = \frac{R(\lambda)}{R(0)}x$$

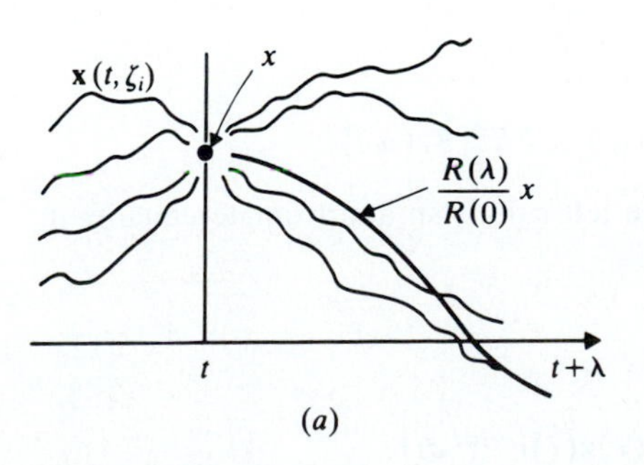

(a)

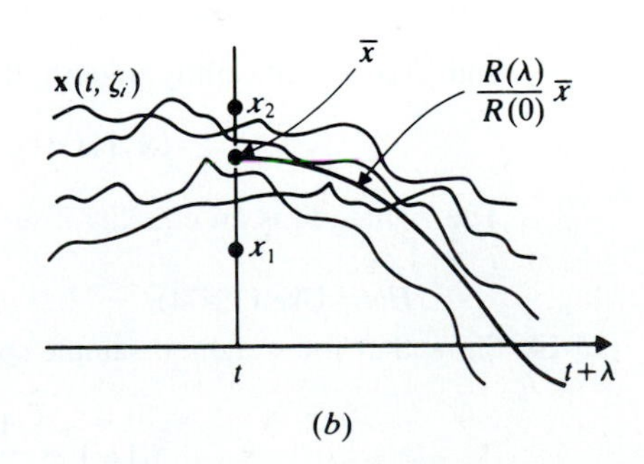

(b)

FIGURE P13-8

(b) Show that if D is an arbitrary set of real numbers x_i and $\bar{x} = E\{\mathbf{x}(t)|\mathbf{x}(t) \in D\}$, then (Fig. P13-8$b$)

$$E\{\mathbf{x}(t+\lambda)|\mathbf{x}(t) \in D\} = \frac{R(\lambda)}{R(0)}\bar{x}$$

(c) Using the above, design an analog correlometer for normal processes

13-9. The processes $\mathbf{x}(t)$ and $\mathbf{y}(t)$ are jointly normal with zero mean. Show that: (a) If $\mathbf{w}(t) = \mathbf{x}(t+\lambda)\mathbf{y}(t)$, then

$$C_{ww}(\tau) = C_{xy}(\lambda+\tau)C_{xy}(\lambda-\tau) + C_{xx}(\tau)C_{yy}(\tau)$$

(b) If the functions $C_{xx}(\tau), C_{yy}(\tau)$, and $C_{xy}(\tau)$ tend to 0 as $\tau \to \infty$ then the processes $\mathbf{x}(t)$ and $\mathbf{y}(t)$ are cross-variance ergodic.

13-10. Using Schwarz's inequality (11B-1), show that

$$\left|\int_a^b f(x)\,dx\right|^2 \le (b-a)\int_a^b |f(x)|^2\,dx$$

13-11. We wish to estimate the mean η of a process $\mathbf{x}(t) = \eta + \boldsymbol{\nu}(t)$ where $R_{\nu\nu}(\tau) = 5\delta(\tau)$. ($a$) Using (5-57), find the 0.95 confidence interval of η. (b) Improve the estimate if $\boldsymbol{\nu}(t)$ is a normal process.

13-12. (a) Show that if we use as estimate of the power spectrum $S(\omega)$ of a discrete-time process $\mathbf{x}[n]$ the function

$$S_w(\omega) = \sum_{m=-N}^{N} w_m R[m] e^{-jm\omega T}$$

then

$$S_w(\omega) = \frac{1}{2\sigma}\int_{-\sigma}^{\sigma} S(y)W(\omega-y)\,dy \qquad W(\omega) = \sum_{-N}^{N} w_n e^{-jn\omega T}$$

(b) Find $W(\omega)$ if $N = 10$ and $w_n = 1 - |n|/11$.

13-13. Show that if $\mathbf{x}(t)$ is zero-mean normal process with sample spectrum

$$\mathbf{S}_T(\omega) = \frac{1}{2T}\left|\int_{-T}^{T} \mathbf{x}(t)e^{-j\omega t}\,dt\right|^2$$

and $S(\omega)$ is sufficiently smooth, then

$$E^2\{\mathbf{S}_T(\omega)\} \le \operatorname{Var}\mathbf{S}_T(\omega) \le 2E^2\{\mathbf{S}_T(\omega)\}$$

The right side is an equality if $\omega = 0$. The left side is an approximate equality if $T \gg 1/\omega$.

Hint: Use (12-74).

13-14. Show that the weighted sample spectrum

$$\mathbf{S}_c(\omega) = \frac{1}{2T}\left|\int_{-T}^{T} c(t)\mathbf{x}(t)e^{-j\omega t}\,dt\right|^2$$

of a process $\mathbf{x}(t)$ is the Fourier transform of the function

$$\mathbf{R}_c(\tau) = \frac{1}{2T}\int_{-T+|\tau|/2}^{T-|\tau|/2} c\left(t+\frac{\tau}{2}\right)c\left(t-\frac{\tau}{2}\right)\mathbf{x}\left(t+\frac{\tau}{2}\right)\mathbf{x}\left(t-\frac{\tau}{2}\right)dt$$

13-15. Given a normal process $\mathbf{x}(t)$ with zero mean and power spectrum $S(\omega)$, we form its sample autocorrelation $\mathbf{R}_T(\tau)$ as in (13-38). Show that for large T,

$$\operatorname{Var}\mathbf{R}_T(\lambda) \simeq \frac{1}{4\pi T}\int_{-\infty}^{\infty}(1+e^{j2\lambda\omega})S^2(\omega)\,d\omega$$

13-16. Show that if

$$\mathbf{R}_T(\tau) = \frac{1}{2T}\int_{-T+|\tau|/2}^{T-|\tau|/2}\mathbf{x}\left(t+\frac{\tau}{2}\right)\mathbf{x}\left(t-\frac{\tau}{2}\right)dt$$

is the estimate of the autocorrelation $R(\tau)$ of a zero-mean normal process, then

$$\sigma_{R_T}^2 = \frac{1}{2T}\int_{-2T+|\tau|}^{2T-|\tau|}\left[R^2(\alpha)+R(\alpha+\tau)R(\alpha-\tau)\right]\left(1-\frac{|\tau|+|\alpha|}{2T}\right)d\alpha$$

13-17. Show that in Levinson's algorithm,

$$a_k^{N-1} = \frac{a_k^N + K_N a_{N-K}^N}{1-K_N^2}$$

13-18. Show that if $R[0]=8$ and $R[1]=4$, then the MEM estimate of $S(\omega)$ equals

$$S_{\mathrm{MEM}}(\omega) = \frac{6}{|1-0.5e^{-j\omega}|^2}$$

13-19. Find the maximum entropy estimate $S_{\mathrm{MEM}}(\omega)$ and the line-spectral estimate (13–111) of a process $\mathbf{x}[n]$ if

$$R[0]=13 \qquad R[1]=5 \qquad R[2]=2$$

CHAPTER 14

MEAN SQUARE ESTIMATION

14-1 INTRODUCTION†

In this chapter, we consider the problem of estimating the value of a stochastic process $\mathbf{s}(t)$ at a specific time in terms of the values (data) of another process $\mathbf{x}(\xi)$ specified for every ξ in an interval $a \le \xi \le b$ of finite or infinite length. In the digital case, the solution of this problem is a direct application of the orthogonality principle (see Sec. 8-4). In the analog case, the linear estimator $\hat{\mathbf{s}}(t)$ of $\mathbf{s}(t)$ is not a sum. It is an integral

$$\hat{\mathbf{s}}(t) \triangleq \hat{E}\{\mathbf{s}(t)|\mathbf{x}(\xi), a \le \xi \le b\} = \int_a^b h(\alpha)\mathbf{x}(\alpha)\,d\alpha \tag{14-1}$$

and our objective is to find $h(\alpha)$ so as to minimize the MS error

$$P = E\{[\mathbf{s}(t) - \hat{\mathbf{s}}(t)]^2\} = E\left\{\left[\mathbf{s}(t) - \int_a^b h(\alpha)\mathbf{x}(\alpha)\,d\alpha\right]^2\right\} \tag{14-2}$$

The function $h(\alpha)$ involves a noncountable number of unknowns, namely, its

†N. Wiener: *Extrapolation, Interpolation, and Smoothing of Stationary Time Series*, MIT Press, 1950. J. Makhoul: "Linear Prediction: A Tutorial Review," *Proceedings of the IEEE*, vol. 63, 1975. T. Kailath: "A View of Three Decades of Linear Filtering Theory," *IEEE Transactions Information Theory*, vol. IT-20, 1974.

values for every α in the interval (a, b). To determine $h(\alpha)$, we shall use the following extension of the orthogonality principle:

THEOREM. The MS error P of the estimation of a process $\mathbf{s}(t)$ by the integral in (14-1) is minimum if the data $\mathbf{x}(\xi)$ are orthogonal to the error $\mathbf{s}(t) - \hat{\mathbf{s}}(t)$:

$$E\left\{\left[\mathbf{s}(t) - \int_a^b h(\alpha)\mathbf{x}(\alpha)\,d\alpha\right]\mathbf{x}(\xi)\right\} = 0 \qquad a \le \xi \le b \tag{14-3}$$

or, equivalently, if $h(\alpha)$ is the solution of the integral equation

$$R_{sx}(t, \xi) = \int_a^b h(\alpha) R_{xx}(\alpha, \xi)\,d\alpha \qquad a \le \xi \le b \tag{14-4}$$

Proof. We shall give a formal proof based on the approximation of the integral in (14-1) by its Riemann sum. Dividing the interval (a, b) into m segments $(\alpha_k, \alpha_k + \Delta\alpha)$, we obtain

$$\hat{\mathbf{s}}(t) \simeq \sum_{k=1}^{m} h(\alpha_k)\mathbf{x}(\alpha_k)\,\Delta\alpha \qquad \Delta\alpha = \frac{b-a}{m}$$

Applying (8-70) with $a_k = h(\alpha_k)\,\Delta\alpha$, we conclude that the resulting MS error P is minimum if

$$E\left\{\left[\mathbf{s}(t) - \sum_{k=1}^{m} h(\alpha_k)\mathbf{x}(\alpha_k)\,\Delta\alpha\right]\mathbf{x}(\xi_j)\right\} = 0 \qquad 1 \le j \le m$$

where ξ_j is a point in the interval $(\alpha_j, \alpha_j + \Delta\alpha)$. This yields the system

$$R_{sx}(t, \xi_j) = \sum_{k=1}^{m} h(\alpha_k) R_{xx}(\alpha_k, \xi_j)\,\Delta\alpha \qquad j = 1, \dots, m \tag{14-5}$$

The integral equation (14-4) is the limit of (14-5) as $\Delta\alpha \to 0$.

From (8-73) it follows that the LMS error of the estimation of $\mathbf{s}(t)$ by the integral in (14-1) equals

$$P = E\left\{\left[\mathbf{s}(t) - \int_a^b h(\alpha)\mathbf{x}(\alpha)\,d\alpha\right]\mathbf{s}(t)\right\} = R_{ss}(0) - \int_a^b h(\alpha) R_{sx}(t, \alpha)\,d\alpha \tag{14-6}$$

In general, the integral equation (14-4) can only be solved numerically. In fact, if we assign to the variable ξ the values ξ_j and we approximate the integral by a sum, we obtain the system (14-5). In this chapter, we consider various special cases that lead to explicit solutions. Unless stated otherwise, it will be assumed that all processes are WSS and real.

We shall use the following terminology:

If the time t in (14-1) is in the interior of the data interval (a, b), then the estimate $\hat{\mathbf{s}}(t)$ of $\mathbf{s}(t)$ will be called *smoothing*.

If t is outside this interval and $\mathbf{x}(t) = \mathbf{s}(t)$ (no noise), then $\hat{\mathbf{s}}(t)$ is a *predictor* of $\mathbf{s}(t)$. If $t > b$, then $\hat{\mathbf{s}}(t)$ is a "forward predictor"; if $t < a$, it is a "backward predictor."

If t is outside the data interval and $\mathbf{x}(t) \neq \mathbf{s}(t)$, then the estimate is called *filtering and prediction*.

Simple Illustrations

In this section, we present a number of simple estimation problems involving a finite number of data and we conclude with the smoothing problem when the data $\mathbf{x}(\xi)$ are available from $(-\infty, \infty)$. In this case, the solution of the integral equation (14-4) is readily obtained in terms of Fourier transforms.

Prediction. We wish to estimate the future value $\mathbf{s}(t+\lambda)$ of a stationary process $\mathbf{s}(t)$ in terms of its present value

$$\hat{\mathbf{s}}(t+\lambda) = \hat{E}\{\mathbf{s}(t+\lambda) | \mathbf{s}(t)\} = a\mathbf{s}(t)$$

From (7-71) and (7-72) it follows with $n = 1$ that

$$E\{[\mathbf{s}(t+\lambda) - a\mathbf{s}(t)]\mathbf{s}(t)\} = 0 \qquad a = \frac{R(\lambda)}{R(0)}$$

$$P = E\{[\mathbf{s}(t+\lambda) - a\mathbf{s}(t)]\mathbf{s}(t+\lambda)\} = R(0) - aR(\lambda)$$

Special case If

$$R(\tau) = Ae^{-\alpha|\tau|} \qquad \text{then} \qquad a = e^{-\alpha\lambda}$$

In this case, the difference $\mathbf{s}(t+\lambda) - a\mathbf{s}(t)$ is orthogonal to $\mathbf{s}(t-\xi)$ for every $\xi \geq 0$:

$$E\{[\mathbf{s}(t+\lambda) - a\mathbf{s}(t)]\mathbf{s}(t-\xi)\} = R(\lambda+\xi) - aR(\xi)$$
$$= Ae^{-\alpha(\lambda+\xi)} - Ae^{-\alpha\lambda}e^{-\alpha\xi} = 0$$

This shows that $a\mathbf{s}(t)$ is the estimate of $\mathbf{s}(t+\lambda)$ in terms of its entire past. Such a process is called *wide-sense Markoff* of order 1.

We shall now find the estimate of $\mathbf{s}(t+\lambda)$ in terms of $\mathbf{s}(t)$ and $\mathbf{s}'(t)$:

$$\hat{\mathbf{s}}(t+\lambda) = a_1\mathbf{s}(t) + a_2\mathbf{s}'(t)$$

The orthogonality condition (8-70) yields

$$\mathbf{s}(t+\lambda) - \hat{\mathbf{s}}(t+\lambda) \perp \mathbf{s}(t), \mathbf{s}'(t)$$

Using the identities

$$R'(0) = 0 \qquad R_{ss'}(\tau) = -R'(\tau) \qquad R_{s's'}(\tau) = -R''(\tau)$$

we obtain

$$a_1 = R(\lambda)/R(0) \qquad a_2 = R'(\lambda)/R''(0)$$

$$P = E\{[\mathbf{s}(t+\lambda) - a_1\mathbf{s}(t) - a_2\mathbf{s}'(t)]\mathbf{s}(t+\lambda)\} = R(0) - a_1R(\lambda) + a_2R'(\lambda)$$

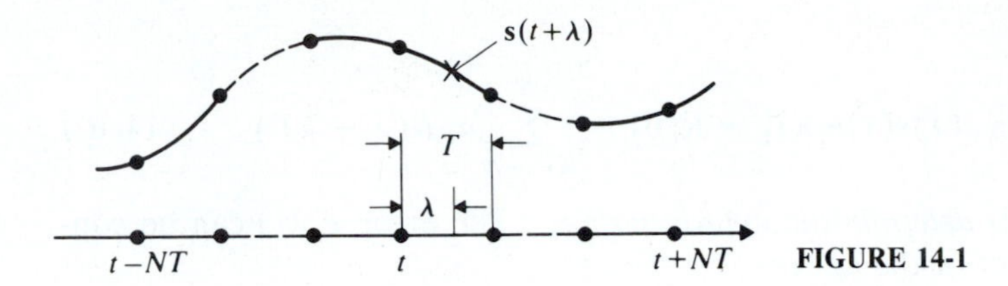

FIGURE 14-1

If λ is small, then

$$R(\lambda) \simeq R(0) \qquad R'(\lambda) \simeq R'(0) + R''(0)\lambda \simeq R''(0)\lambda$$
$$a_1 \simeq 1 \qquad a_2 \simeq \lambda \qquad \hat{\mathbf{s}}(t+\lambda) \simeq \mathbf{s}(t) + \lambda\mathbf{s}'(t)$$

Filtering We shall estimate the present value of a process $\mathbf{s}(t)$ in terms of the present value of another process $\mathbf{x}(t)$:

$$\hat{\mathbf{s}}(t) = \hat{E}\{\mathbf{s}(t)|\mathbf{x}(t)\} = a\mathbf{x}(t)$$

From (7-71) and (7-72) it follows that

$$E\{[\mathbf{s}(t) - a\mathbf{x}(t)]\mathbf{x}(t)\} = 0 \qquad a = R_{sx}(0)/R_{xx}(0)$$
$$P = E\{[\mathbf{s}(t) - a\mathbf{x}(t)]\mathbf{s}(t)\} = R_{ss}(0) - aR_{sx}(0)$$

Interpolation We wish to estimate the value $\mathbf{s}(t+\lambda)$ of a process $\mathbf{s}(t)$ at a point $t+\lambda$ in the interval $(t, t+T)$, in terms of its $2N+1$ samples $\mathbf{s}(t+kT)$ that are nearest to t (Fig. 14-1)

$$\hat{\mathbf{s}}(t+\lambda) = \sum_{k=-N}^{N} a_k \mathbf{s}(t+kT) \qquad 0 < \lambda < T \tag{14-7}$$

The orthogonality principle now yields

$$E\left\{\left[\mathbf{s}(t+\lambda) - \sum_{k=-N}^{N} a_k \mathbf{s}(t+kT)\right]\mathbf{s}(t+nT)\right\} = 0 \qquad |n| \le N$$

from which it follows that

$$\sum_{k=-N}^{N} a_k R(kT - nT) = R(\lambda - nT) \qquad -N \le n \le N \tag{14-8}$$

This is a system of $2N+1$ equations and its solution yields the $2N+1$ unknowns a_k. The MS value P of the estimation error

$$\boldsymbol{\varepsilon}_N(t) = \mathbf{s}(t+\lambda) - \sum_{k=-N}^{N} a_k \mathbf{s}(t+kT) \tag{14-9}$$

equals

$$P = E\{\boldsymbol{\varepsilon}_N(t)\mathbf{s}(t+\lambda)\} = R(0) - \sum_{k=-N}^{N} a_k R(\lambda - kT) \tag{14-10}$$

Interpolation as deterministic approximation The error $\boldsymbol{\varepsilon}_N(t)$ can be considered as the output of the system

$$E_N(\omega) = e^{j\omega\lambda} - \sum_{k=-N}^{N} a_k e^{jkT\omega}$$

(error filter) with input $\mathbf{s}(t)$. Denoting by $S(\omega)$ the power spectrum of $\mathbf{s}(t)$, we conclude from (10-139) that

$$P = E\{\boldsymbol{\varepsilon}_N^2(t)\} = \frac{1}{2\pi}\int_{-\infty}^{\infty} S(\omega)\left| e^{j\omega\lambda} - \sum_{k=-N}^{N} a_k e^{jkT\omega}\right|^2 d\omega \tag{14-11}$$

This shows that the minimization of P is equivalent to the deterministic problem of minimizing the weighted mean square error of the approximation of the exponential $e^{j\omega\lambda}$ by a trigonometric polynomial (truncated Fourier series).

Quadrature We shall estimate the integral

$$\mathbf{z} = \int_0^b \mathbf{s}(t)\,dt$$

of a process $\mathbf{s}(t)$ in terms of its $N+1$ samples $\mathbf{s}(nT)$:

$$\hat{\mathbf{z}} = a_0\mathbf{s}(0) + a_1\mathbf{s}(T) + \cdots + a_N\mathbf{s}(NT) \qquad T = \frac{b}{N}$$

Applying (8-70), we obtain

$$E\left\{\left[\int_0^b \mathbf{s}(t)\,dt - \hat{\mathbf{z}}\right]\mathbf{s}(kT)\right\} = 0 \qquad 0 \le k \le N$$

Hence

$$\int_0^b R(t-kT)\,dt = a_0R(kT) + \cdots + a_NR(kT - NT) \qquad 0 \le k \le N$$

This is a system of $N+1$ equations and its solution yields the coefficients a_k.

Smoothing

We wish to estimate the present value of a process $\mathbf{s}(t)$ in terms of the values $\mathbf{x}(\xi)$ of the sum

$$\mathbf{x}(t) = \mathbf{s}(t) + \boldsymbol{\nu}(t)$$

available for every ξ from $-\infty$ to ∞. The desirable estimate

$$\hat{\mathbf{s}}(t) = \hat{E}\{\mathbf{s}(t)\,|\,\mathbf{x}(\xi),\ -\infty < \xi < \infty\}$$

will be written in the form

$$\hat{\mathbf{s}}(t) = \int_{-\infty}^{\infty} h(\alpha)\mathbf{x}(t-\alpha)\,d\alpha \tag{14-12}$$

In this notation, $h(\alpha)$ is independent of t and $\hat{\mathbf{s}}(t)$ can be considered as the output of a linear time-invariant noncausal system with input $\mathbf{x}(t)$ and impulse response $h(t)$. Our problem is to find $h(t)$.

Clearly,

$$\mathbf{s}(t) - \hat{\mathbf{s}}(t) \perp \mathbf{x}(\xi) \qquad \text{all } \xi$$

Setting $\xi = t - \tau$, we obtain

$$E\left\{\left[\mathbf{s}(t) - \int_{-\infty}^{\infty} h(\alpha)\mathbf{x}(t-\alpha)\,d\alpha\right]\mathbf{x}(t-\tau)\right\} = 0 \qquad \text{all } \tau$$

This yields

$$R_{sx}(\tau) = \int_{-\infty}^{\infty} h(\alpha)R_{xx}(\tau-\alpha)\,d\alpha \qquad \text{all } \tau \tag{14-13}$$

Thus, to determine $h(t)$, we must solve the above integral equation. This equation can be solved easily because it holds for all τ and the integral is a convolution of $h(\tau)$ with $R_{xx}(\tau)$. Taking transforms of both sides, we obtain $S_{sx}(\omega) = H(\omega)S_{xx}(\omega)$. Hence

$$H(\omega) = \frac{S_{sx}(\omega)}{S_{xx}(\omega)} \tag{14-14}$$

The resulting system is called the *noncausal Wiener filter*.

The MS estimation error P equals

$$P = E\left\{\left[\mathbf{s}(t) - \int_{-\infty}^{\infty} h(\alpha)\mathbf{x}(t-\alpha)\,d\alpha\right]\mathbf{s}(t)\right\}$$

$$= R_{ss}(0) - \int_{-\infty}^{\infty} h(\alpha)R_{sx}(\alpha)\,d\alpha = \frac{1}{2\pi}\int_{-\infty}^{\infty}\left[S_{ss}(\omega) - H^*(\omega)S_{sx}(\omega)\right]d\omega \tag{14-15}$$

If the signal $\mathbf{s}(t)$ and the noise $\boldsymbol{\nu}(t)$ are *orthogonal*, then

$$S_{sx}(\omega) = S_{ss}(\omega) \qquad S_{xx}(\omega) = S_{ss}(\omega) + S_{\nu\nu}(\omega)$$

Hence (Fig. 14-2)

$$H(\omega) = \frac{S_{ss}(\omega)}{S_{ss}(\omega) + S_{\nu\nu}(\omega)} \qquad P = \frac{1}{2\pi}\int_{-\infty}^{\infty} \frac{S_{ss}(\omega)S_{\nu\nu}(\omega)}{S_{ss}(\omega) + S_{\nu\nu}(\omega)}\,d\omega \tag{14-16}$$

If the spectra $S_{ss}(\omega)$ and $S_{\nu\nu}(\omega)$ do not overlap, then $H(\omega) = 1$ in the band of the signal and $H(\omega) = 0$ in the band of the noise. In this case, $P = 0$.

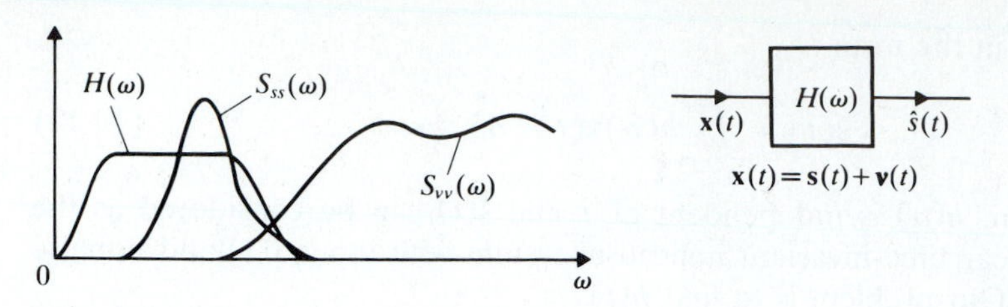

FIGURE 14-2

Example 14-1. If

$$S_{ss}(\omega) = \frac{N_0}{\alpha^2 + \omega^2} \qquad S_{\nu\nu}(\omega) = N \qquad S_{s\nu}(\omega) = 0$$

then (14-16) yields

$$H(\omega) = \frac{N_0}{N_0 + N(\alpha^2 + \omega^2)} \qquad h(t) = \frac{N_0}{2\beta N} e^{-\beta|t|}$$

$$P = \frac{1}{2\pi} \int_{-\infty}^{\infty} \frac{N_0}{\beta^2 + \omega^2} \, d\omega = \frac{N_0}{2\beta} \qquad \beta^2 = \alpha^2 + \frac{N_0}{N}$$

DISCRETE-TIME PROCESSES. The noncausal estimate $\hat{\mathbf{s}}[n]$ of a discrete-time process in terms of the data

$$\mathbf{x}[n] = \mathbf{s}[n] + \boldsymbol{\nu}[n]$$

is the output

$$\hat{\mathbf{s}}[n] = \sum_{k=-\infty}^{\infty} h[k]\mathbf{x}[n-k]$$

of a linear time-invariant noncausal system with input $\mathbf{x}[n]$ and delta response $h[n]$. The orthogonality principle yields

$$E\left\{\left(\mathbf{s}[n] - \sum_{k=-\infty}^{\infty} h[k]\mathbf{x}[n-k]\right)\mathbf{x}[n-m]\right\} = 0 \qquad \text{all } m$$

Hence

$$R_{sx}[m] = \sum_{k=-\infty}^{\infty} h[k]R_{xx}[m-k] \qquad \text{all } m \tag{14-17}$$

Taking transforms of both sides, we obtain

$$\mathbf{H}(z) = \frac{\mathbf{S}_{sx}(z)}{\mathbf{S}_{xx}(z)} \tag{14-18}$$

The resulting MS error equals

$$P = E\left\{\left[\mathbf{s}[n] - \sum_{k=-\infty}^{\infty} h[k]\mathbf{x}[n-k]\right]\mathbf{s}[n]\right\}$$

$$= R_{ss}(0) - \sum_{k=-\infty}^{\infty} h[k]R_{sx}[k] = \frac{1}{2\pi}\int_{-\pi}^{\pi}\left[S_{ss}(\omega) - \mathbf{H}(e^{-j\omega T})S_{sx}(\omega)\right]d\omega$$

Example 14-2. Suppose that $\mathbf{s}[n]$ is a first-order AR process and $\boldsymbol{\nu}[n]$ is white noise orthogonal to $\mathbf{s}[n]$:

$$\mathbf{S}_{ss}(z) = \frac{N_0}{(1-az^{-1})(1-az)} \qquad \mathbf{S}_{\nu\nu}(z) = N \qquad \mathbf{S}_{s\nu}(z) = 0$$

In this case,

$$\mathbf{S}_{xx}(z) = \mathbf{S}_{ss}(z) + N = \frac{aN(1-bz^{-1})(1-bz)}{b(1-az^{-1})(1-az)}$$

where

$$0 < b < a < 1 \qquad b + b^{-1} = a + a^{-1} + \frac{N_0}{aN}$$

Hence

$$\mathbf{H}(z) = \frac{bN_0}{aN(1-bz^{-1})(1-bz)} \qquad h[n] = cb^{|n|} \qquad c = \frac{bN_0}{aN(1-b^2)}$$

$$P = \frac{N_0}{1-a^2}\left[1 - c\sum_{k=-\infty}^{\infty}(ab)^{|k|}\right] = \frac{bN_0}{a(1-b^2)}$$

14-2 PREDICTION

Prediction is the estimation of the future $\mathbf{s}(t+\lambda)$ of a process $\mathbf{s}(t)$ in terms of its past $\mathbf{s}(t-\tau)$, $\tau > 0$. This problem has three parts: The past (data) is known in the interval $(-\infty, t)$; it is known in the interval $(t-T, t)$ of finite length T; it is known in the interval $(0, t)$ of variable length t. We shall develop all three parts for digital processes only. The discussion of analog predictors will be limited to the first part. In the digital case, we find it more convenient to predict the present $\mathbf{s}[n]$ of the given process in terms of its past $\mathbf{s}[n-k]$, $k \geq r$.

Infinite Past

We start with the estimation of a process $\mathbf{s}[n]$ is terms of its entire past $\mathbf{s}[n-k]$, $k \geq 1$:

$$\hat{\mathbf{s}}[n] = \hat{E}\{\mathbf{s}[n] | \mathbf{s}[n-k], k \geq 1\} = \sum_{k=1}^{\infty} h[k]\mathbf{s}[n-k] \tag{14-19}$$

This estimator will be called the one-step predictor of $\mathbf{s}[n]$. Thus $\hat{\mathbf{s}}[n]$ is the response of the *predictor filter*

$$\mathbf{H}(z) = h[1]z^{-1} + \cdots + h[k]z^{-k} + \cdots \tag{14-20}$$

to the input $\mathbf{s}[n]$ and our objective is to find the constants $h[k]$ so as to minimize the MS estimation error. From the orthogonality principle it follows that the error $\boldsymbol{\varepsilon}[n] = \mathbf{s}[n] - \hat{\mathbf{s}}[n]$ must be orthogonal to the data $\mathbf{s}[n-m]$:

$$E\left\{\left(\mathbf{s}[n] - \sum_{k=1}^{\infty} h[k]\mathbf{s}[n-k]\right)\mathbf{s}[n-m]\right\} = 0 \qquad m \geq 1 \tag{14-21}$$

This yields

$$R[m] - \sum_{k=1}^{\infty} h[k]R[m-k] = 0 \qquad m \geq 1 \tag{14-22}$$

We have thus obtained a system of infinitely many equations expressing the unknowns $h[k]$ in terms of the autocorrelation $R[m]$ of $\mathbf{s}[n]$. These equations are called Wiener–Hopf (digital form).

The Wiener–Hopf equations cannot be solved directly with z transforms even though the right side equals the convolution of $h[m]$ with $R[m]$. The reason is that, unlike (14-17), the two sides of (14-22) are not equal for every m. A solution based on the analytic properties of the z transforms of causal and anticausal sequences can be found (see Prob. 14-12); however, the underlying theory is not simple. We shall give presently a very simple solution based on the concept of innovations. We comment first on a basic property of the estimation error $\boldsymbol{\varepsilon}[n]$ and of the error filter

$$\mathbf{E}(z) = 1 - \mathbf{H}(z) = 1 - \sum_{k=1}^{\infty} h[n]z^{-k} \tag{14-23}$$

The error $\boldsymbol{\varepsilon}[n]$ is orthogonal to the data $\mathbf{s}[n-m]$ for every $m \geq 1$; furthermore, $\boldsymbol{\varepsilon}[n-m]$ is a linear function of $\mathbf{s}[n-m]$ and its past because $\boldsymbol{\varepsilon}[n]$ is the response of the causal system $\mathbf{E}(z)$ to the input $\mathbf{s}[n]$. From this it follows that $\boldsymbol{\varepsilon}[n]$ is orthogonal to $\boldsymbol{\varepsilon}[n-m]$ for every $m \geq 1$ and every n. Hence $\boldsymbol{\varepsilon}[n]$ is white noise:

$$R_{\varepsilon\varepsilon}[m] = E\{\boldsymbol{\varepsilon}[n]\boldsymbol{\varepsilon}[n-m]\} = P\delta[m] \tag{14-24}$$

where

$$P = E\{\boldsymbol{\varepsilon}^2[n]\} = E\{(\mathbf{s}[n] - \hat{\mathbf{s}}[n])\mathbf{s}[n]\} = R[0] - \sum_{k=1}^{\infty} h[k]R[k]$$

is the LMS error. This error can be expressed in terms of the power spectrum $S(\omega)$ of $\mathbf{s}[n]$; as we see from (10-139),

$$P = \frac{1}{2\pi}\int_{-\pi}^{\pi} |\mathbf{E}(e^{j\omega})|^2 S(\omega)\, d\omega \tag{14-25}$$

Using the above, we shall show that the function $\mathbf{E}(z)$ has no 0's outside the unit circle.

THEOREM. If

$$E(z_i) = 0 \qquad \text{then} \quad |z_i| \le 1 \tag{14-26}$$

Proof. We form the function

$$\mathbf{E}_0(z) = \mathbf{E}(z)\frac{1 - z^{-1}/z_i^*}{1 - z_i z^{-1}}$$

This function is an error filter because it is causal and $\mathbf{E}_0(\infty) = \mathbf{E}(\infty) = 1$. Furthermore, if $|z_i| > 1$, then [see (13B-2)]

$$|\mathbf{E}_0(e^{j\omega})| = \frac{1}{|z_i|}|\mathbf{E}(e^{j\omega})| < |\mathbf{E}(e^{j\omega})|$$

Inserting into (14-25), we conclude that if we use as the estimator filter the function 1-$\mathbf{E}_0(z)$, the resulting MS error will be smaller than P. This, however, is impossible because P is minimum; hence $|z_i| \le 1$.

Regular Processes

We shall solve the Wiener–Hopf equations (14-22) under the assumption that the process $\mathbf{s}[n]$ is regular. As we have shown in Sec. 12-1, such a process is linearly equivalent to a white-noise process $\mathbf{i}[n]$ in the sense that

$$\mathbf{s}[n] = \sum_{k=0}^{\infty} l[k]\mathbf{i}[n-k] \tag{14-27}$$

$$\mathbf{i}[n] = \sum_{k=0}^{\infty} \gamma[k]\mathbf{s}[n-k] \tag{14-28}$$

From this it follows that the predictor $\hat{\mathbf{s}}[n]$ of $\mathbf{s}[n]$ can be written as a linear sum involving the past of $\mathbf{i}[n]$:

$$\hat{\mathbf{s}}[n] = \sum_{k=1}^{\infty} h_i[k]\mathbf{i}[n-k] \tag{14-29}$$

To find $\hat{\mathbf{s}}[n]$, it suffices, therefore, to find the constants $h_i[k]$ and to express $\mathbf{i}[n]$ in terms of $\mathbf{s}[n]$ using (14-28). To do so, we shall determine first the cross-correlation of $\mathbf{s}[n]$ and $\mathbf{i}[n]$. We maintain that

$$R_{si}[m] = l[m] \tag{14-30}$$

Proof. We multiply (14-27) by $\mathbf{i}[n-m]$ and take expected values. This yields

$$E\{\mathbf{s}[n]\mathbf{i}[n-m]\} = \sum_{k=0}^{\infty} l[k]E\{\mathbf{i}[n-k]\mathbf{i}[n-m]\} = \sum_{k=0}^{\infty} l[k]\delta[m-k]$$

because $R_{ii}[m] = \delta[m]$, and (14-30) results.

To find $h_i[k]$, we apply the orthogonality principle:

$$E\left\{\left(\mathbf{s}[n] - \sum_{k=1}^{\infty} h_i[k]\mathbf{i}[n-k]\right)\mathbf{i}[n-m]\right\} = 0 \qquad m \geq 1$$

This yields

$$R_{si}[m] - \sum_{k=1}^{\infty} h_i[k]R_{ii}[m-k] = R_{si}[m] - \sum_{k=1}^{\infty} h_i[k]\delta[m-k] = 0$$

and since the last sum equals $h_i[m]$, we conclude that $h_i[m] = R_{si}[m]$. From this and (14-30) it follows that the predictor $\hat{\mathbf{s}}[n]$, expressed in terms of its innovations, equals

$$\hat{\mathbf{s}}[n] = \sum_{k=1}^{\infty} l[k]\mathbf{i}[n-k] \tag{14-31}$$

We shall rederive this important result using (14-27). To do so, it suffices to show that the difference $\mathbf{s}[n] - \hat{\mathbf{s}}[n]$ is orthogonal to $\mathbf{i}[n-m]$ for every $m \geq 1$. This is indeed the case because

$$\boldsymbol{\varepsilon}[n] = \sum_{k=0}^{\infty} l[k]\mathbf{i}[n-k] - \sum_{k=1}^{\infty} l[k]\mathbf{i}[n-k] = l[0]\mathbf{i}[n] \tag{14-32}$$

and $\mathbf{i}[n]$ is white noise.

The sum in (14-31) is the response of the filter

$$\sum_{k=1}^{\infty} l[k]z^{-k} = \mathbf{L}(z) - l[0]$$

to the input $\mathbf{i}[n]$. To complete the specification of $\hat{\mathbf{s}}[n]$, we must express $\mathbf{i}[n]$ in terms of $\mathbf{s}[n]$. Since $\mathbf{i}[n]$ is the response of the filter $1/\mathbf{L}(z)$ to the input $\mathbf{s}[n]$, we conclude, cascading as in Fig. 14-3, that the predictor filter of $\mathbf{s}[n]$ is the product

$$\mathbf{H}(z) = \frac{1}{\mathbf{L}(z)}(\mathbf{L}(z) - l[0]) = 1 - \frac{l[0]}{\mathbf{L}(z)} \tag{14-33}$$

shown in Fig. 14-4. Thus, to obtain $\mathbf{H}(z)$, it suffices to factor $\mathbf{S}(z)$ as in (12-6). The constant $l[0]$ is determined from the initial value theorem:

$$l[0] = \lim_{z\to\infty} \mathbf{L}(z)$$

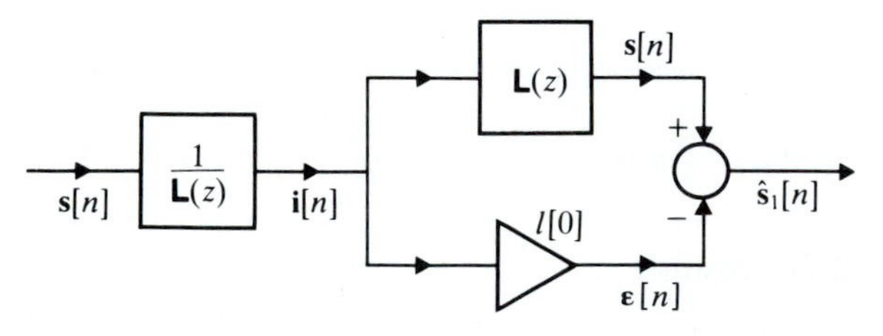

FIGURE 14-3

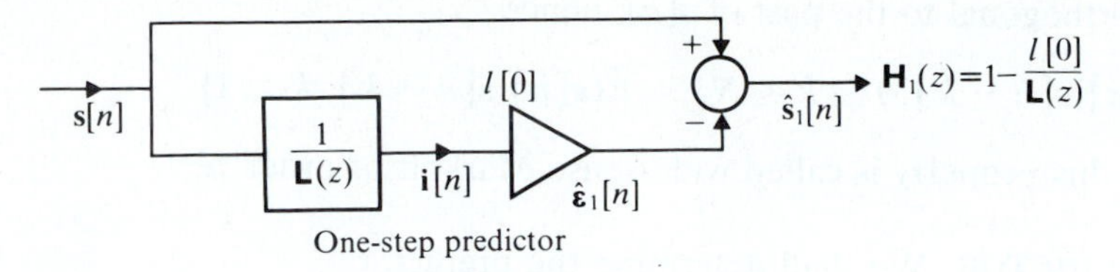

FIGURE 14-4

Example 14-3. Suppose that

$$S(\omega) = \frac{5 - 4\cos\omega}{10 - 6\cos\omega} \qquad \mathbf{L}(z) = \frac{2z - 1}{3z - 1} \qquad l[0] = \frac{2}{3}$$

as in Example 12-4. In this case, (14-33) yields

$$\mathbf{H}(z) = 1 - \frac{2}{3} \times \frac{3z - 1}{2z - 1} = \frac{-z^{-1}}{6(1 - z^{-1}/2)}$$

Note that $\hat{\mathbf{s}}[n]$ can be determined recursively:

$$\hat{\mathbf{s}}[n] - \tfrac{1}{2}\hat{\mathbf{s}}[n-1] = -\tfrac{1}{6}\mathbf{s}[n-1]$$

The Kolmogoroff–Szego MS error formula† As we have seen from (14-32), the MS estimation error equals

$$P = E\{\boldsymbol{\varepsilon}^2[n]\} = l^2[0]$$

Furthermore [see (13A-1)]

$$\ln l^2[0] = \frac{1}{2\pi}\int_{-\pi}^{\pi} \ln|\mathbf{L}(e^{i\omega})|^2\,d\omega$$

Since $S(\omega) = |\mathbf{L}(e^{j\omega})|^2$, this yields the identity

$$P = \exp\left\{\frac{1}{2\pi}\int_{-\pi}^{\pi} \ln S(\omega)\,d\omega\right\} \tag{14-34}$$

expressing P directly in terms of $S(\omega)$.

Autoregressive processes. If $\mathbf{s}[n]$ is an AR process as in (12-39), then $l[0] = b_0$ and

$$\begin{aligned} \mathbf{H}(z) &= -a_1 z^{-1} - \cdots - a_N z^{-N} \\ \hat{\mathbf{s}}[n] &= -a_1\mathbf{s}[n-1] - \cdots - a_N\mathbf{s}[n-N] \qquad P = b_0^2 \end{aligned} \tag{14-35}$$

The above shows that the predictor $\hat{\mathbf{s}}[n]$ of $\mathbf{s}[n]$ in terms of its entire past is the same as the predictor in terms of the N most recent past values. This result can be established directly: From (12-39) and (14-35) it follows that $\mathbf{s}[n] - \hat{\mathbf{s}}[n] =$

†U. Grenander and G. Szego: *Toeplitz Forms and Their Applications*, Berkeley University Press, 1958.

$b_0\mathbf{i}[n]$. This is orthogonal to the past of $\mathbf{s}[n]$; hence

$$\hat{E}\{\mathbf{s}[n]|\mathbf{s}[n-k], 1 \le k \le N\} = \hat{E}\{\mathbf{s}[n]|\mathbf{s}[n-k], k \ge 1\}$$

A process with this property is called wide-sense Markoff of order N.

THE *r*-STEP PREDICTOR. We shall determine the predictor

$$\hat{\mathbf{s}}_r[n] = \hat{E}\{\mathbf{s}[n]|\mathbf{s}[n-k], k \ge r\}$$

of $\mathbf{s}[n]$ in terms of $\mathbf{s}[n-r]$ and its past using innovations. We maintain that

$$\hat{\mathbf{s}}_r[n] = \sum_{k=r}^{\infty} l[k]\mathbf{i}[n-k] \tag{14-36}$$

Proof. It suffices to show that the difference

$$\hat{\boldsymbol{\varepsilon}}_r[n] = \mathbf{s}[n] - \hat{\mathbf{s}}_r[n] = \sum_{k=0}^{r-1} l[k]\mathbf{i}[n-k]$$

is orthogonal to the data $\mathbf{s}[n-k]$, $k \ge r$. This is a consequence of the fact that $\mathbf{s}[n-k]$ is linearly equivalent to $\mathbf{i}[n-k]$ and its past for $k \ge r$; hence it is orthogonal to $\mathbf{i}[n], \mathbf{i}[n-1], \ldots, \mathbf{i}[n-r+1]$.

The prediction error $\hat{\boldsymbol{\varepsilon}}_r[n]$ is the response of the MA filter $l[0] + l[1]z^{-1} + \cdots + l[r-1]z^{-r+1}$ of Fig. 14-5 to the input $\mathbf{i}[n]$. Cascading this filter with $1/\mathbf{L}(z)$ as in Fig. 14-5, we conclude that the process $\hat{\mathbf{s}}_r[n] = \mathbf{s}[n] - \hat{\boldsymbol{\varepsilon}}_r[n]$ is the response of the system

$$\mathbf{H}_r(z) = 1 - \frac{1}{\mathbf{L}(z)} \sum_{k=0}^{r-1} l[k]z^{-k} \tag{14-37}$$

to the input $\mathbf{s}[n]$. This is the *r-step predictor filter* of $\mathbf{s}[n]$. The resulting MS error

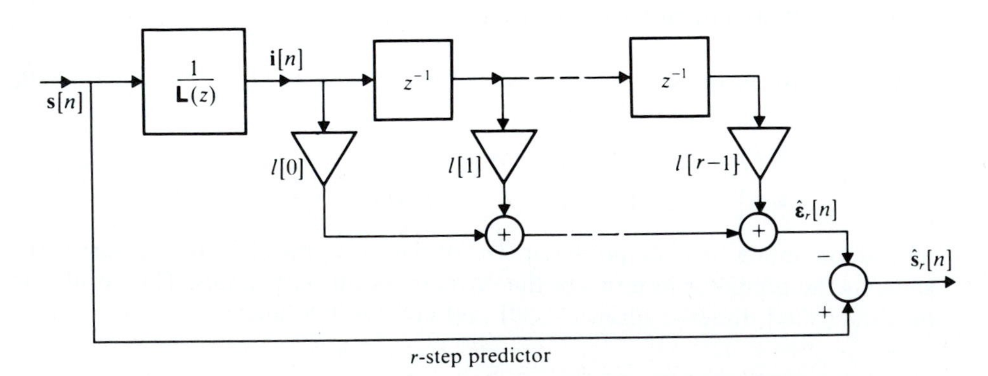

FIGURE 14-5

equals

$$P_r = E\{\varepsilon_r^2[n]\} = \sum_{k=0}^{r-1} l^2[k] \tag{14-38}$$

Example 14-4. We are given a process $s[n]$ with autocorrelation $R[m] = a^{|m|}$ and we wish to determine its r-step predictor. In this case (see Example 10-30)

$$\mathbf{S}(z) = \frac{a^{-1} - a}{(a^{-1} + a) - (z^{-1} + z)} = \frac{b^2}{(1 - az^{-1})(1 - az)} \qquad b^2 = 1 - a^2$$

$$\mathbf{L}(z) = \frac{b}{1 - az^{-1}} \qquad l[n] = ba^n U[n]$$

Hence

$$\mathbf{H}_r(z) = 1 - \frac{1 - az^{-1}}{b} \sum_{k=0}^{r-1} ba^k z^{-k} = a^r z^{-r}$$

$$\hat{s}_r[n] = a^r s[n - r] \qquad P_r = b^2 \sum_{k=0}^{r-1} a^{2k} = 1 - a^{2r}$$

ANALOG PROCESSES. We consider now the problem of predicting the future value $s(t + \lambda)$ of a process $s(t)$ in terms of its entire past $s(t - \tau)$, $\tau \geq 0$. In this problem, our estimator is an integral:

$$\hat{s}(t + \lambda) = \hat{E}\{s(t + \lambda) | s(t - \tau), \tau \geq 0\} = \int_0^\infty h(\alpha) s(t - \alpha)\, d\alpha \tag{14-39}$$

and the problem is to find the function $h(\alpha)$. From the analog form (14-4) of the orthogonality principle, it follows that

$$E\left\{\left[s(t + \lambda) - \int_0^\infty h(\alpha) s(t - \alpha)\right] s(t - \tau)\right\} = 0 \qquad \tau \geq 0$$

This yields the Wiener–Hopf integral equation

$$R(\tau + \lambda) = \int_0^\infty h(\alpha) R(\tau - \alpha)\, d\alpha \qquad \tau \geq 0 \tag{14-40}$$

The solution of this equation is the impulse response of the *causal Wiener filter*

$$\mathbf{H}(s) = \int_0^\infty h(t) e^{-st}\, dt$$

The corresponding MS error equals

$$P = E\{[s(t + \lambda) - \hat{s}(t + \lambda)] s(t + \lambda)\} = R(0) - \int_0^\infty h(\alpha) R(\lambda + \alpha)\, d\alpha \tag{14-41}$$

Equation (14-40) cannot be solved directly with transforms because the two sides are equal for $\tau \geq 0$ only. A solution based on the analytic properties of Laplace transforms is outlined in Prob. 14-11. We give next a solution using innovations.

As we have shown in (12-8), the process $\mathbf{s}(t)$ is the response of its innovations filter $\mathbf{L}(s)$ to the white-noise process $\mathbf{i}(t)$. From this it follows that

$$\mathbf{s}(t+\lambda) = \int_0^\infty l(\alpha)\mathbf{i}(t+\lambda-\alpha)\,d\alpha \tag{14-42}$$

We maintain that $\hat{\mathbf{s}}(t+\lambda)$ is the part of the above integral involving only the past of $\mathbf{i}(t)$:

$$\hat{\mathbf{s}}(t+\lambda) = \int_\lambda^\infty l(\alpha)\mathbf{i}(t+\lambda-\alpha)\,d\alpha = \int_0^\infty l(\beta+\lambda)\mathbf{i}(t-\beta)\,d\beta \tag{14-43}$$

Proof. The difference

$$\mathbf{s}(t+\lambda) - \hat{\mathbf{s}}(t+\lambda) = \int_0^\lambda l(\alpha)\mathbf{i}(t+\lambda-\alpha)\,d\alpha \tag{14-44}$$

depends only on the values of $\mathbf{i}(t)$ in the interval $(t, t+\lambda)$; hence it is orthogonal to the past of $\mathbf{i}(t)$ and, therefore, it is also orthogonal to the past of $\mathbf{s}(t)$.

The predictor $\hat{\mathbf{s}}(t+\lambda)$ of $\mathbf{s}(t)$ is the response of the system

$$\mathbf{H}_i(s) = \int_0^\infty h_i(t)e^{-st}\,dt \qquad h_i(t) = l(t+\lambda)U(t) \tag{14-45}$$

(Fig. 14-6) to the input $\mathbf{i}(t)$. Cascading with $1/\mathbf{L}(s)$, we conclude that $\hat{\mathbf{s}}(t+\lambda)$ is

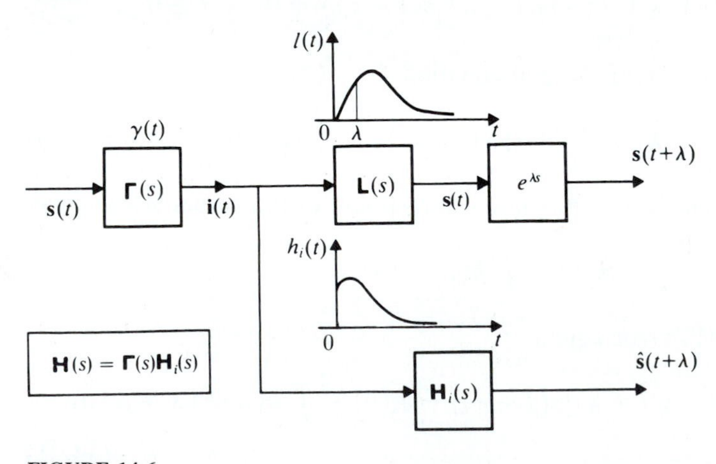

FIGURE 14-6

the response of the system

$$\mathbf{H}(s) = \frac{\mathbf{H}_i(s)}{\mathbf{L}(s)} \tag{14-46}$$

to the input $\mathbf{s}(t)$. Thus, to determine the predictor filter $\mathbf{H}(s)$ of $\mathbf{s}(t)$, proceed as follows:

Factor the spectrum of $\mathbf{s}(t)$ as in (12-3): $\mathbf{S}(s) = \mathbf{L}(s)\mathbf{L}(-s)$.

Find the inverse transform $l(t)$ of $\mathbf{L}(s)$ and form the function $h_i(t) = l(t + \lambda)U(t)$.

Find the transform $\mathbf{H}_i(s)$ of $h_i(t)$ and determine $\mathbf{H}(s)$ from (14-46).

The MS estimation error is determined from (14-44):

$$P = E\left\{\left|\int_0^\lambda l(\alpha)\mathbf{i}(t + \lambda - \alpha)\,d\alpha\right|^2\right\} = \int_0^\lambda l^2(\alpha)\,d\alpha \tag{14-47}$$

Example 14-5. We are given a process $\mathbf{s}(t)$ with autocorrelation $R(\tau) = 2\alpha e^{-\alpha|\tau|}$ and we wish to determine its predictor. In this problem,

$$\mathbf{S}(s) = \frac{1}{\alpha^2 - s^2} \qquad \mathbf{L}(s) = \frac{1}{\alpha + s} \qquad l(t) = e^{-\alpha t}U(t)$$

$$h_i(t) = e^{-\alpha\lambda}e^{-\alpha t}U(t) \qquad \mathbf{H}_i(s) = \frac{e^{-\alpha\lambda}}{\alpha + s}$$

$$\mathbf{H}(s) = e^{-\alpha\lambda} \qquad \hat{\mathbf{s}}(t + \lambda) = e^{-\alpha\lambda}\mathbf{s}(t)$$

This shows that the predictor of $\mathbf{s}(t + \lambda)$ in terms of its entire past is the same as the predictor in terms of its present $\mathbf{s}(t)$. In other words, if $\mathbf{s}(t)$ is specified, the past has no effect on the linear prediction of the future.

The determination of $\mathbf{H}(s)$ is simple if $\mathbf{s}(t)$ has a rational spectrum. Assuming that the poles of $\mathbf{H}(s)$ are simple, we obtain

$$\mathbf{L}(s) = \frac{N(s)}{D(s)} = \sum_i \frac{c_i}{s - s_i} \qquad l(t) = \sum_i c_i e^{s_i t}U(t)$$

$$h_i(t) = \sum_i c_i e^{s_i\lambda}e^{s_i t}U(t) \qquad \mathbf{H}_i(s) = \sum_i \frac{c_i e^{s_i\lambda}}{s - s_i} = \frac{N_1(s)}{D(s)} \tag{14-48}$$

and (14-46) yields $\mathbf{H}(s) = N_1(s)/N(s)$.

If $N(s) = 1$, then $\mathbf{H}(s)$ is a polynomial:

$$\mathbf{H}(s) = N_1(s) = b_0 + b_1 s + \cdots + b_n s^n$$

and $\hat{\mathbf{s}}(t + \lambda)$ is a linear sum of $\mathbf{s}(t)$ and its first n derivatives:

$$\hat{\mathbf{s}}(t + \lambda) = b_0\mathbf{s}(t) + b_1\mathbf{s}'(t) + \cdots + b_n\mathbf{s}^{(n)}(t)$$

Example 14-6. We are given a process $\mathbf{s}(t)$ with

$$\mathbf{S}(s) = \frac{49 - 25s^2}{(1-s^2)(9-s^2)} \qquad \mathbf{L}(s) = \frac{7+5s}{(1+s)(3+s)}$$

and we wish to estimate its future $s(t+\lambda)$ for $\lambda = \ln 2$. In this problem, $e^{\lambda} = 2$:

$$\mathbf{L}(s) = \frac{1}{s+1} + \frac{4}{s+3} \qquad \mathbf{H}_i(s) = \frac{e^{-\lambda}}{s+1} + \frac{4e^{-3\lambda}}{s+3} = \frac{s+2}{(s+1)(s+3)}$$

$$\mathbf{H}(s) = \frac{s+2}{5s+7} \qquad h(t) = \frac{1}{5}\delta(t) + \frac{3}{25}e^{-1.4t}U(t)$$

Hence

$$E\{\mathbf{s}(t+\lambda)|\mathbf{s}(t-\tau), \tau \geq 0\} = 0.2\mathbf{s}(t) + \hat{E}\{\mathbf{s}(t+\lambda)|\mathbf{s}(t-\tau), \tau > 0\}$$

Notes 1. The integral

$$y(\tau) = \int_0^{\infty} h(\alpha)R(\tau - \alpha)\, d\alpha$$

in (14-40) is the response of the Wiener filter $\mathbf{H}(s)$ to the input $R(\tau)$. From (14-40) and (14-41) it follows that

$$y(\tau) = R(\tau + \lambda) \quad \text{for } \tau \geq 0 \qquad \text{and} \qquad y(-\lambda) = R(0) - P$$

2. In all MS estimation problems, only second-order moments are used. If, therefore, two processes have the same autocorrelation, then their predictors are identical. This suggests the following derivation of the Wiener–Hopf equation: Suppose that $\boldsymbol{\omega}$ is an RV with density $f(\omega)$ and $\mathbf{z}(t) = e^{j\boldsymbol{\omega} t}$. Clearly,

$$R_{zz}(\tau) = E\{e^{j\boldsymbol{\omega}(t+\tau)}e^{-j\boldsymbol{\omega} t}\} = \int_{-\infty}^{\infty} f(\omega)e^{j\omega\tau}\, d\omega$$

From this it follows that the power spectrum of $\mathbf{z}(t)$ equals $2\pi f(\omega)$ [see also (10-127)]. If, therefore, $\mathbf{s}(t)$ is a process with power spectrum $S(\omega) = 2\pi f(\omega)$, then its predictor $h(t)$ will equal the predictor of $\mathbf{z}(t)$:

$$\hat{\mathbf{z}}(t+\lambda) = \hat{E}\{e^{j\boldsymbol{\omega}(t+\lambda)}|e^{j\boldsymbol{\omega}(t-\alpha)}, \alpha \geq 0\} = \int_0^{\infty} h(\alpha)e^{j\boldsymbol{\omega}(t-\alpha)}\, d\alpha$$

$$= e^{j\boldsymbol{\omega} t}\int_0^{\infty} h(\alpha)e^{-j\boldsymbol{\omega}\alpha}\, d\alpha = e^{j\boldsymbol{\omega} t}H(\boldsymbol{\omega})$$

And since $\mathbf{z}(t+\lambda) - \hat{\mathbf{z}}(t+\lambda) \perp \mathbf{z}(t-\tau)$, for $\tau \geq 0$, we conclude from the above that

$$E\{[e^{j\boldsymbol{\omega}(t+\lambda)} - e^{j\boldsymbol{\omega} t}H(\boldsymbol{\omega})]e^{-j\boldsymbol{\omega}(t-\tau)}\} = 0 \qquad \tau \geq 0$$

Hence

$$\int_{-\infty}^{\infty} f(\omega)[e^{j\omega(\tau+\lambda)} - e^{j\omega\tau}H(\omega)]\, d\omega = 0 \qquad \tau \geq 0$$

This yields (14-40) because the inverse transform of $f(\omega)e^{j\omega(\tau+\lambda)}$ equals $R(\tau+\lambda)$ and the inverse transform of $f(\omega)e^{j\omega\tau}H(\omega)$ equals the integral in (14-40).

Predictable processes. We shall say that a process $\mathbf{s}[n]$ is *predictable* if it equals its predictor:

$$\mathbf{s}[n] = \sum_{k=1}^{\infty} h[k]\mathbf{s}[n-k] \tag{14-49}$$

In this case [see (14-25)]

$$P = \frac{1}{2\pi}\int_{-\pi}^{\pi} |\mathbf{E}(e^{j\omega})|^2 S(\omega)\, d\omega = 0 \tag{14-50}$$

Since $S(\omega) \geq 0$, the above integral is 0 if $S(\omega) \neq 0$ only in a region R of the ω axis where $\mathbf{E}(e^{j\omega}) = 0$. It can be shown that this region consists of a countable number of points ω_i—the proof is based on the Paley–Wiener condition (12-9). From this it follows that

$$S(\omega) = 2\pi \sum_{i=1}^{m} \alpha_i \delta(\omega - \omega_i) \qquad \mathbf{E}(e^{j\omega_i}) = 0 \tag{14-51}$$

Thus a process $\mathbf{s}[n]$ is predictable if it is a sum of exponentials as in (12-9):

$$\mathbf{s}[n] = \sum_{i=1}^{m} \mathbf{c}_i e^{j\omega_i n} \qquad E\{\mathbf{c}_i^2\} = \alpha_i \tag{14-52}$$

We maintain that the converse is also true: If $\mathbf{s}[n]$ is a sum of m exponentials as in (14-52), then it is predictable and its predictor filter equals $1 - D(z)$ where

$$D(z) = (1 - e^{j\omega_1}z^{-1}) \cdots (1 - e^{j\omega_m}z^{-1}) \tag{14-53}$$

Proof. In this case, $\mathbf{E}(z) = D(z)$ and $\mathbf{E}(e^{j\omega_i}) = 0$; hence $\mathbf{E}(e^{j\omega})S(\omega) = 0$ because $\mathbf{E}(e^{j\omega})\delta(\omega - \omega_i) = \mathbf{E}(e^{j\omega_i})\delta(\omega - \omega_i) = 0$. From this it follows that $P = 0$.

Note The preceding result seems to be in conflict with the sampling expansion (11-138) of a BL process $\mathbf{s}(t)$: This expansion shows that $\mathbf{s}(t)$ is predictable in the sense that it can be approximated within an arbitrary error ε by a linear sum involving only its past samples $\mathbf{s}(nT_0)$. From this it follows that the digital process $\mathbf{s}[n] = \mathbf{s}(nT_0)$ is predictable in the same sense. Such an expansion, however, does not violate (14-50). It is only an approximation and its coefficients tend to ∞ as $\varepsilon \to 0$.

GENERAL PROCESSES AND WOLD'S DECOMPOSITION† We show finally that an arbitrary process $\mathbf{s}[n]$ can be written as a sum

$$\mathbf{s}[n] = \mathbf{s}_1[n] + \mathbf{s}_2[n] \tag{14-54}$$

of a regular process $\mathbf{s}_1[n]$ and a predictable process $\mathbf{s}_2[n]$, that these processes are

†A. Papoulis: Predictable Processes and Wold's Decomposition: A Review. *IEEE Transactions on Acoustics, Speech, and Signal Processing*, Vol. 22, 1985.

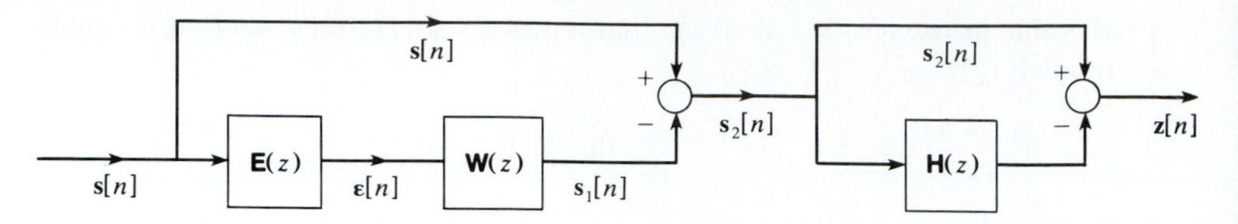

FIGURE 14-7

orthogonal, and that they have the same predictor filter. We thus reestablish constructively Wold's decomposition (12-89) in the context of MS estimation.

As we know [see (14-24)], the error $\varepsilon[n]$ of the one-step estimate of $s[n]$ is a white-noise process. We form the estimator $s_1[n]$ of $s[n]$ in terms of $\varepsilon[n]$ and its past:

$$s_1[n] = \hat{E}\{s[n] \mid \varepsilon[n-k], k \geq 0\} = \sum_{k=0}^{\infty} w_k \varepsilon[n-k] \tag{14-55}$$

Thus $s_1[n]$ is the response of the system (Fig. 14-7)

$$\mathbf{W}(z) = \sum_{k=0}^{\infty} w_k z^{-k}$$

to the input $\varepsilon[n]$. The difference $s_2[n] = s[n] - s_1[n]$ is the estimation error (Fig. 14-7). Clearly (orthogonality principle)

$$s_2[n] \perp \varepsilon[n-k] \qquad k \geq 0 \tag{14-56}$$

Note that if $s[n]$ is a regular process, then [see (14-32)] $\varepsilon[n] = l[0]\mathbf{i}[n]$; in this case, $s_1[n] = s[n]$.

THEOREM. (a) The processes $s_1[n]$ and $s_2[n]$ are orthogonal:

$$s_1[n] \perp s_2[n-k] \qquad \text{all} \quad k \tag{14-57}$$

(b) $s_1[n]$ is a regular process.

(c) $s_2[n]$ is a predictable process and its predictor filter is the sum in (14-19):

$$s_2[n] = \sum_{k=1}^{\infty} h[k] s_2[n-k] \tag{14-58}$$

Proof. (a) The process $\varepsilon[n]$ is orthogonal to $s[n-k]$ for every $k > 0$. Furthermore, $s_2[n-k]$ is a linear function of $s[n-k]$ and its past; hence $s_2[n-k] \perp \varepsilon[n]$ for $k > 0$. Combining with (14-56), we conclude that

$$s_2[n-k] \perp \varepsilon[n] \qquad \text{all} \quad k \tag{14-59}$$

And since $s_1[n]$ depends linearly on $\varepsilon[n]$ and its past, (14-57) follows.

(*b*) The process $\mathbf{s}_1[n]$ is the response of the system $\mathbf{W}(z)$ to the white noise $\boldsymbol{\varepsilon}[n]$. To prove that it is regular, it suffices to show that

$$\sum_{k=0}^{\infty} w_k^2 < \infty \tag{14-60}$$

From (14-54) and (14-55) it follows that

$$E\{\mathbf{s}^2[n]\} = E\{\mathbf{s}_1^2[n]\} + E\{\mathbf{s}_2^2[n]\} \geq E\{\mathbf{s}_1^2[n]\} = \sum_{k=0}^{\infty} w_k^2$$

This yields (14-60) because $E\{\mathbf{s}^2[n]\} = R(0) < \infty$.

(*c*) To prove (14-58), it suffices to show that the difference

$$\mathbf{z}[n] = \mathbf{s}_2[n] - \sum_{k=1}^{\infty} h[k]\mathbf{s}_2[n-k]$$

equals 0. From (14-59) it follows that $\mathbf{z}[n] \perp \boldsymbol{\varepsilon}[n-k]$ for all k. But $\mathbf{z}[n]$ is the response of the system $1 - \mathbf{H}(z) = \mathbf{E}(z)$ to the input $\mathbf{s}_2[n] = \mathbf{s}[n] - \mathbf{s}_1[n]$; hence (see Fig. 14-8)

$$\mathbf{z}[n] = \boldsymbol{\varepsilon}[n] - \mathbf{s}_1[n] + \sum_{k=1}^{\infty} h[k]\mathbf{s}_1[n-k] \tag{14-61}$$

This shows that $\mathbf{z}[n]$ is a linear function of $\boldsymbol{\varepsilon}[n]$ and its past. And since it is also orthogonal to $\boldsymbol{\varepsilon}[n]$, we conclude that $\mathbf{z}[n] = 0$.

Note finally that [see (14-61)]

$$\mathbf{s}_1[n] - \sum_{k=1}^{\infty} h[k]\mathbf{s}_1[n-k] = \boldsymbol{\varepsilon}[n] \perp \mathbf{s}_1[n-m] \qquad m \geq 1$$

Hence the above sum is the predictor of $\mathbf{s}_1[n]$. We thus conclude that the sum $\mathbf{H}(z)$ in (14-20) is the predictor filter of the processes $\mathbf{s}[n]$, $\mathbf{s}_1[n]$, and $\mathbf{s}_2[n]$.

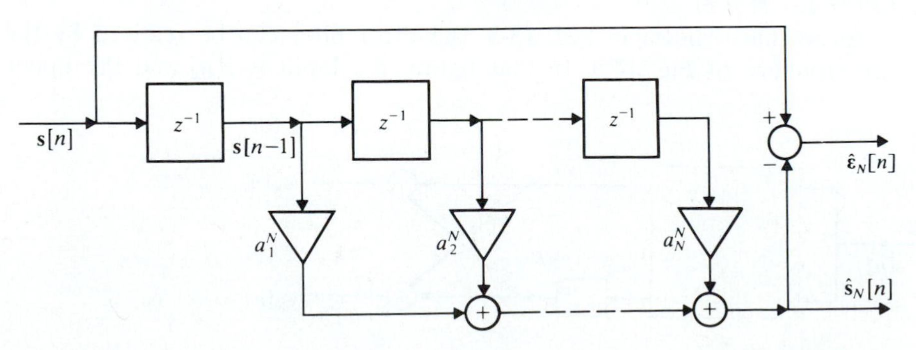

FIGURE 14-8

FIR PREDICTORS. We shall find the estimate $\hat{\mathbf{s}}_N[n]$ of a process $\mathbf{s}[n]$ in terms of its N most recent past values:

$$\hat{\mathbf{s}}_N[n] = \hat{E}\{\mathbf{s}[n] | \mathbf{s}[n-k], 1 \le k \le N\} = \sum_{k=1}^{N} a_k^N \mathbf{s}[n-k] \quad (14\text{-}62)$$

This estimate will be called the *forward* predictor of order N. The superscript in a_k^N identifies the order. The process $\hat{\mathbf{s}}_N[n]$ is the response of the *forward predictor filter*

$$\hat{\mathbf{H}}_N(z) = \sum_{k=1}^{N} a_k^N z^{-k} \quad (14\text{-}63)$$

to the input $\mathbf{s}[n]$. Our objective is to determine the constants a_k^N so as to minimize the MS value

$$P_N = E\{\hat{\boldsymbol{\varepsilon}}_N^2[n]\} = E\{(\mathbf{s}[n] - \hat{\mathbf{s}}_N[n])\mathbf{s}[n]\} \quad (14\text{-}64)$$

of the forward prediction error $\hat{\boldsymbol{\varepsilon}}_N[n] = \mathbf{s}[n] - \hat{\mathbf{s}}_N[n]$

The Yule–Walker equations. From the orthogonality principle it follows that

$$E\left\{\left(\mathbf{s}[n] - \sum_{k=1}^{N} a_k^N \mathbf{s}[n-k]\right)\mathbf{s}[n-m]\right\} = 0 \qquad 1 \le m \le N$$

This yields the system

$$R[m] - \sum_{k=1}^{N} a_k^N R[m-k] = 0 \qquad 1 \le m \le N \quad (14\text{-}65)$$

Solving, we obtain the coefficients a_k^N of the predictor filter $\hat{\mathbf{H}}_N(z)$. The resulting MS error equals [see (13-83)]

$$P_N = R[0] - \sum_{k=1}^{N} a_k^N R[k] = \frac{\Delta_{N+1}}{\Delta_N} \quad (14\text{-}66)$$

In Fig. 14-8 we show the ladder realization of $\hat{\mathbf{H}}_N(z)$ and the *forward error filter* $\hat{\mathbf{E}}_N(z) = 1 - \hat{\mathbf{H}}_N(z)$.

As we have shown in Sec. 13-3, the error filter can be realized by the lattice structure of Fig. 14-9. In that figure, the input is $\mathbf{s}[n]$ and the upper

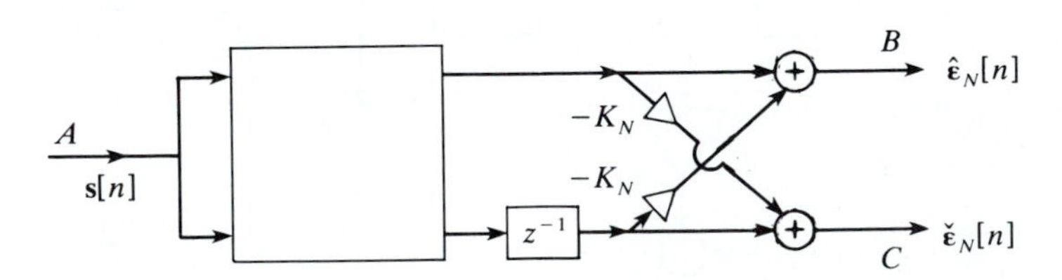

FIGURE 14-9

output $\hat{\boldsymbol{\varepsilon}}_N[N]$. The lower output $\check{\boldsymbol{\varepsilon}}_N[n]$ is the *backward prediction error* defined as follows: The processes $\mathbf{s}[n]$ and $\mathbf{s}[-n]$ have the same autocorrelation; hence their predictor filters are identical. From this it follows that the backward predictor $\check{\mathbf{s}}_N[n]$, that is, the predictor of $\mathbf{s}[n]$ in terms of its N most recent future values, equals

$$\check{\mathbf{s}}_N[n] = \hat{E}\{\mathbf{s}[n]|\mathbf{s}[n+k], 1 \le k \le N\} = \sum_{k=1}^{N} a_k^N \mathbf{s}[n+k]$$

The backward error

$$\check{\boldsymbol{\varepsilon}}_N[n] = \mathbf{s}[n-N] - \check{\mathbf{s}}_N[n-N]$$

is the response of the filter

$$\check{\mathbf{E}}_N(z) = z^{-N}\left(1 - a_1^N z - \cdots - a_N^N z^N\right) = z^{-N}\hat{\mathbf{E}}_N(1/z)$$

with input $\mathbf{s}[n]$. From this and (13-94) it follows that the lower output of the lattice of Fig. 14-8 is $\check{\boldsymbol{\varepsilon}}_N[n]$.

In Sec. 13-3, we used the ladder–lattice equivalence to simplify the solution of the Yule–Walker equations. We summarize next the main results in the context of the prediction problem. We note that the lattice realization also has the following advantage. Suppose that we have a predictor of order N and we wish to find the predictor of order $N+1$. In the ladder realization, we must find a new set of $N+1$ coefficients a_k^{N+1}. In the lattice realization, we need only the new reflection coefficient K_{N+1}; the first N reflection coefficients K_k do not change.

Levinson's algorithm. We shall determine the constants a_k^N, K_N, and P_N recursively. This involves the following steps: Start with

$$a_1^1 = K_1 = R[1]/R[0] \qquad P_1 = \left(1 - K_1^2\right)R[0]$$

Assume that the $N+1$ constants a_k^{N-1}, K_{N-1}, and P_{N-1} are known.
Find K_N and P_N from (13-107) and (13-108):

$$P_{N-1}K_N = R[N] - \sum_{k=1}^{N-1} a_k^{N-1}R[N-k] \qquad P_N = \left(1 - K_N^2\right)P_{N-1} \quad (14\text{-}67)$$

Find a_k^N from (13-97)

$$a_N^N = K_N \qquad a_k^N = a_k^{N-1} - K_N a_{N-k}^{N-1} \qquad 1 \le k \le N-1 \quad (14\text{-}68)$$

In Levinson's algorithm, the order N of the iteration is finite but it can continue indefinitely. We shall examine the properties of the predictor and of the MS error P_N as $N \to \infty$. It is obvious that P_N is a nonincreasing sequence of positive numbers; hence it tends to a positive limit:

$$P_1 \ge P_2 \cdots \ge P_N \xrightarrow[N\to\infty]{} P \ge 0 \quad (14\text{-}69)$$

As we have shown in Sec. 12-3, the zeros z_i of the error filter

$$\hat{\mathbf{E}}_N(z) = 1 - \sum_{k=1}^{N} a_k^N z^{-k}$$

are either all inside the unit circle or they are all on the unit circle:

If $P_N > 0$, then $|K_k| < 1$ for every $k \le N$ and $|z_i| < 1$ for every i [see (13-99)].

If $P_{N-1} > 0$ and $P_N = 0$, then $|K_k| < 1$ for every $k \le N - 1$, $|K_N| = 1$, and $|z_i| = 1$ for every i [see (13-101)]. In this case, the process $\mathbf{s}[n]$ is predictable and its spectrum consists of lines.

If $P > 0$, then $|z_i| \le 1$ for every i [see (14-26)]. In this case, the predictor $\hat{\mathbf{s}}_N[n]$ of $\mathbf{s}[n]$ tends to the Wiener predictor $\hat{\mathbf{s}}[n]$ as in (14-19). From this and (14-34) it follows that

$$P = \exp\left\{\frac{1}{2\pi}\int_{-\pi}^{\pi} \ln S(\omega)\, d\omega\right\} = l[0] = \lim_{N\to\infty} \frac{\Delta_{N+1}}{\Delta_N} \qquad (14\text{-}70)$$

This shows the connection between the LMS error P of the prediction of $\mathbf{s}[n]$ in terms of its entire past, the power spectrum $S(\omega)$ of $\mathbf{s}[n]$, the initial value $l[0]$ of the delta response $l[n]$ of its innovations filter, and the correlation determinant Δ_N.

Suppose, finally, that $P_{M-1} > P_M$ and

$$P_M = P_{M-1} = \cdots = P \qquad (14\text{-}71)$$

In this case, $K_k = 0$ for $|k| > M$; hence the algorithm terminates at the Mth step. From this it follows that the Mth order predictor $\hat{\mathbf{s}}_M[n]$ of $\mathbf{s}[n]$ equals its Wiener predictor:

$$\hat{\mathbf{s}}_M[n] = \hat{E}\{\mathbf{s}[n]|\mathbf{s}[n-k], 1 \le k \le M\} = \hat{E}\{\mathbf{s}[n]|\mathbf{s}[n-k], k \ge 1\}$$

In other words, the process $\mathbf{s}[n]$ is *wide-sense Markoff* of order M. This leads to the conclusion that the prediction error $\hat{\boldsymbol{\varepsilon}}_M[n] = \mathbf{s}[n] - \hat{\mathbf{s}}_M[n]$ is white noise with average power P [see (14-24)]:

$$\mathbf{s}[n] - \sum_{k=1}^{M} a_k^N \mathbf{s}[n-k] = \hat{\boldsymbol{\varepsilon}}_M[n] \qquad E\{\hat{\boldsymbol{\varepsilon}}_M^2[n]\} = P$$

and it shows that $\mathbf{s}[n]$ is an AR process. Conversely, if $\mathbf{s}[n]$ is AR, then it is also wide-sense Markoff.

Autoregressive processes and maximum entropy. Suppose that $\mathbf{s}[n]$ is an AR process of order M with autocorrelation $R[m]$ and $\bar{\mathbf{s}}[n]$ is a general process with autocorrelation $\bar{R}[m]$ such that

$$\bar{R}[m] = R[m] \quad \text{for } |m| \le M$$

The predictors of these processes of order M are identical because they depend on the values of $R[m]$ for $|m| \leq M$ only. From this it follows that the corresponding prediction errors P_M and $\bar{P}_M$ are equal. As we have noted, $P_M = P$ for the AR process $\mathbf{s}[n]$ and $\bar{P}_M \geq P$ for the general process $\bar{\mathbf{s}}[n]$.

Consider now the class C_M of processes with identical autocorrelations (data) for $|m| \leq M$. Each $R[m]$ is a p.d. extrapolation of the given data. We have shown in Sec. 13-3 that the extrapolating sequence obtained with the maximum entropy (ME) method is the autocorrelation of an AR process [see (13-141)]. This leads to the following relationship between MS estimation and maximum entropy: The ME extrapolation is the autocorrelation of a process $\mathbf{s}[n]$ in the class C_M, the predictor of which maximizes the minimum MS error P. In this sense, the ME method maximizes our uncertainty about the values of $R[m]$ for $|m| > M$.

Causal Data

We wish to estimate the present value of a regular process $\mathbf{s}[n]$ in terms of its finite past, starting from some origin. The data are now available from 0 to $n - 1$ and the desired estimate is given by

$$\hat{\mathbf{s}}_n[n] = \hat{E}\{\mathbf{s}[n] | \mathbf{s}[n-k], 1 \leq k \leq n\} = \sum_{k=1}^{n} a_k^n \mathbf{s}[n-k] \qquad (14\text{-}72)$$

Unlike the fixed length N of the FIR predictor $\hat{\mathbf{s}}_N[n]$ considered in (14-62), the length n of this estimate is not constant. Furthermore, the values a_k^n of the coefficients of the filter specified by (14-72) depend on n. Thus the estimator of the process $\mathbf{s}[n]$ in terms of its causal past is a linear *time-varying filter*. If it is realized by a tapped-delay line as in Fig. 14-8, the number of the taps increases and the values of the weights change as n increases.

The coefficients a_k^n of $\hat{\mathbf{s}}_n[n]$ can be determined recursively from Levinson's algorithm where now $N = n$. Introducing the backward estimate $\check{\mathbf{s}}[n]$ of $\mathbf{s}[n]$ in terms of its n most recent future values, we conclude from (13-92) that

$$\begin{aligned} \hat{\mathbf{s}}_n[n] &= \hat{\mathbf{s}}_{n-1}[n] + K_n(\mathbf{s}[0] - \check{\mathbf{s}}_{n-1}[0]) \\ \check{\mathbf{s}}_n[0] &= \check{\mathbf{s}}_{n-1}[0] + K_n(\mathbf{s}[n] - \hat{\mathbf{s}}_{n-1}[n]) \end{aligned} \qquad (14\text{-}73)$$

In Fig. 14-10, we show the *normalized lattice* realization of the error filter $\mathbf{E}_n(z)$ where we use as upper output the process

$$\mathbf{i}[n] = \frac{1}{\sqrt{P_n}} \hat{\boldsymbol{\varepsilon}}_n[n] \qquad E\{\mathbf{i}^2[n]\} = 1 \qquad (14\text{-}74)$$

The filter is formed by switching "on" successively a new lattice section starting from the left. This filter is again time-varying; however, unlike the tapped-delay line realization, the elements of each section remain unchanged as n increases. We should point out that whereas $\hat{\boldsymbol{\varepsilon}}_k[n]$ is the value of the upper response of

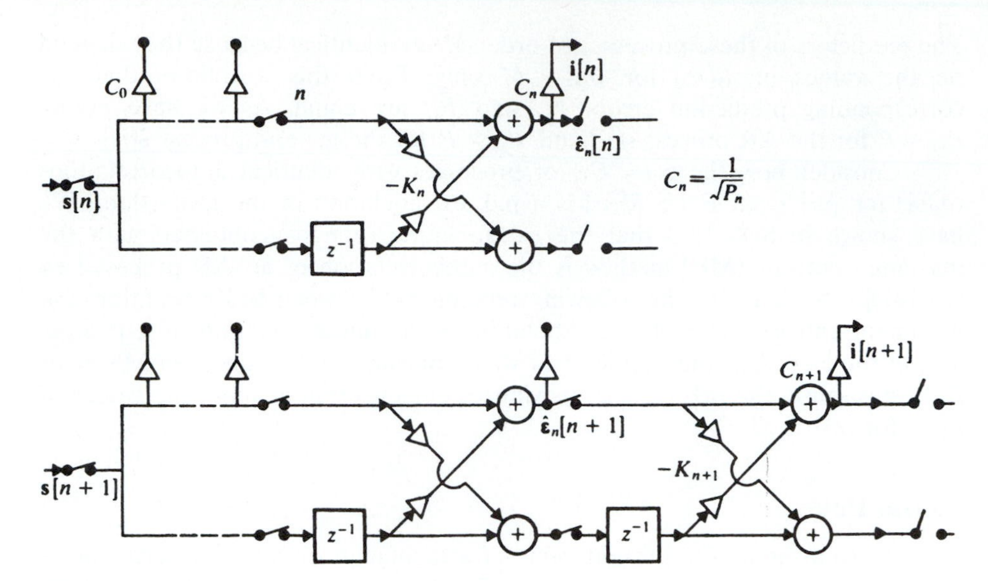

FIGURE 14-10

the kth section at time n, the process $\mathbf{i}[n]$ does not appear at a fixed position. It is the output of the last section that is switched "on" and as n increases, the point where $\mathbf{i}[n]$ is observed changes.

We conclude with the observation that if the process $\mathbf{s}[n]$ is AR of order M [see (13-81)], then the lattice stops increasing for $n > M$, realizing, thus, the time invariant system $\mathbf{E}_M(z)/\sqrt{P_M}$. The corresponding inverse lattice (see Fig. 13-15) realizes the all-pole system

$$\frac{\sqrt{P_M}}{\mathbf{E}_M(z)}$$

We shall now show that the output $\mathbf{i}[n]$ of the normalized lattice is white noise

$$R_{ii}[m] = \delta[m] \tag{14-75}$$

Indeed, as we know, $\hat{\boldsymbol{\varepsilon}}_n[n] \perp \mathbf{s}[n-k]$ for $1 \le k \le n$. Furthermore, $\hat{\boldsymbol{\varepsilon}}_{n-k}[n-r]$ depends linearly only on $\mathbf{s}[n-r]$ and its past values. Hence

$$\hat{\boldsymbol{\varepsilon}}_n[n] \perp \hat{\boldsymbol{\varepsilon}}_{n-1}[n-1] \tag{14-76}$$

This yields (14-75) because $P_n = E\{\boldsymbol{\varepsilon}_n^2[n]\}$.

Note In a lattice of fixed length, the output $\hat{\boldsymbol{\varepsilon}}_N[n]$ is not white noise and it is not orthogonal to $\hat{\boldsymbol{\varepsilon}}_{N-1}[n]$. However for a specific n, the random variables $\hat{\boldsymbol{\varepsilon}}_N[n]$ and $\hat{\boldsymbol{\varepsilon}}_{N-1}[n-1]$ are orthogonal.

KALMAN INNOVATIONS†. The output $\mathbf{i}[n]$ of the time-varying lattice of Fig. 14-10 is an orthonormal process that depends linearly on $\mathbf{s}[n-k]$. Denoting by γ_k^n the response of the lattice at time n to the input $\mathbf{s}[n] = \delta[n-k]$, we obtain

$$\begin{aligned} \mathbf{i}[0] &= \gamma_0^0 \mathbf{s}[0] \\ \mathbf{i}[1] &= \gamma_0^1 \mathbf{s}[0] + \gamma_1^1 \mathbf{s}[1] \\ &\cdots\cdots\cdots\cdots\cdots\cdots \\ \mathbf{i}[n] &= \gamma_0^n \mathbf{s}[0] + \cdots + \gamma_k^n \mathbf{s}[k] + \cdots + \gamma_n^n \mathbf{s}[n] \end{aligned} \tag{14-77}$$

or in vector form

$$\mathbf{I}_{n+1} = \mathbf{S}_{n+1}\Gamma_{n+1} \qquad \Gamma_{n+1} = \begin{bmatrix} \gamma_0^0 & \gamma_0^1 & \cdots & \gamma_0^n \\ & \gamma_1^1 & \cdots & \gamma_1^n \\ & 0 & & \cdots\cdots \\ & & & \gamma_n^n \end{bmatrix}$$

where $\mathbf{S}_{n+1}$ and $\mathbf{I}_{n+1}$ are row vectors with components

$$\mathbf{s}[0], \ldots, \mathbf{s}[n] \qquad \text{and} \qquad \mathbf{i}[0], \ldots, \mathbf{i}[n]$$

respectively.

From the above it follows that if

$$\mathbf{s}[n] = \delta[n-k] \qquad \text{then} \quad \mathbf{i}[n] = \gamma_k^n \qquad n \geq k$$

This shows that to determine the delta response of the lattice of Fig. 14-10, we use as input the delta sequence $\delta[n-k]$ and we observe the moving output $\mathbf{i}[n]$ for $n \geq k$.

The elements γ_k^n of the triangular matrix Γ_{n+1} can be expressed in terms of the weights a_k^n of the causal predictor $\hat{\mathbf{s}}_n[n]$. Since

$$\hat{\boldsymbol{\varepsilon}}_n[n] = \mathbf{s}[n] - \hat{\mathbf{s}}_n[n] = \sqrt{P_n}\,\mathbf{i}[n]$$

it follows from (14-72) that

$$\gamma_n^n = \frac{1}{\sqrt{P_n}} \qquad \gamma_{n-k}^n = \frac{-1}{\sqrt{P_n}} a_k^n \qquad k \geq 1$$

The inverse of the lattice of Fig. 14-10 is obtained by reversing the flow direction of the upper line and the sign of the upward weights $-K_n$ as in Fig. 13-15. The turn-on switches close again in succession starting from the left, and the input $\mathbf{i}[n]$ is applied at the terminal of the section that is connected last. The

†T. Kailath, A. Vieira, and M. Morf: "Inverses of Toeplitz Operators, Innovations, and Orthogonal Polynomials," *SIAM Review*, vol. 20, no. 1, 1978.

output at A is thus given by

$$\begin{aligned} \mathbf{s}[0] &= l_0^0\mathbf{i}[0] \\ \mathbf{s}[1] &= l_0^1\mathbf{i}[0] + l_1^1\mathbf{i}[1] \qquad\qquad \mathbf{S}_{n+1} = \mathbf{I}_{n+1}L_{n+1} \\ &\cdots\cdots\cdots\cdots\cdots \\ \mathbf{s}[n] &= l_0^n\mathbf{i}[0] + \cdots + l_n^n\mathbf{i}[n] \qquad L_n = \Gamma_n^{-1} \end{aligned} \tag{14-78}$$

From this it follows that if

$$\mathbf{i}[n] = \delta[n-k] \qquad \text{then} \quad \mathbf{s}[n] = l_k^n \qquad n \ge k$$

Thus, to determine the delta response l_k^n of the inverse lattice, we use as moving input the delta sequence $\delta[n-k]$ and we observe the left output $\mathbf{s}[n]$ for $n \ge k$.

From the preceding discussion it follows that the random vector $\mathbf{S}_n$ is linearly equivalent to the orthonormal vector $\mathbf{I}_n$. Thus Eqs. (14-77) and (14-78) correspond to the Gram–Schmidt orthonormalization equations (8-88) and (8-91) of Sec. 8-3. Applying the terminology of Sec. 12-1 to causal signals, we shall call the process $\mathbf{i}[n]$ the *Kalman innovations* of $\mathbf{s}[n]$ and the lattice filter and its inverse *Kalman whitening and Kalman innovations* filters respectively. These filters are *time-varying* and their transition matrices equal Γ_n and L_n respectively. Their elements can be expressed in terms of the parameters K_n and P_n of Levinson's algorithm because these parameters specify completely the filters.

Cholesky factorization We maintain that the correlation matrix R_n and its inverse can be written as products

$$R_n = L_n^t L_n \qquad R_n^{-1} = \Gamma_n \Gamma_n^t \tag{14-79}$$

where Γ_n and L_n are the triangular matrices introduced earlier. Indeed, from the orthonormality of $\mathbf{I}_n$ and the definition of R_n, it follows that

$$E\{\mathbf{I}_n^t\mathbf{I}_n\} = 1_n \qquad E\{\mathbf{S}_n^t\mathbf{S}_n\} = R_n$$

where 1_n is the identity matrix. Since $\mathbf{I}_n = \mathbf{S}_n\Gamma_n$ and $\mathbf{S}_n = \mathbf{I}_n L_n$, the above yields

$$\Gamma_n^t R_n \Gamma_n = 1_n \qquad L_n^t 1_n L_n = R_n$$

and (14-79) results.

Autocorrelation as lattice response. We shall determine the autocorrelation $R[m]$ of the process $\mathbf{s}[n]$ in terms of the Levinson parameters K_N and P_N. For this purpose, we form a lattice of order N_0 and we denote by $\hat{q}_N[m]$ and $\check{q}_N[m]$ respectively its upper and lower responses (14-11a) to the input $R[m]$. As we see from the figure

$$\hat{q}_{N-1}[m] = \hat{q}_N[m] + K_N\check{q}_{N-1}[m-1] \tag{14-80a}$$

$$\check{q}_N[m] = \check{q}_{N-1}[m-1] - K_N\hat{q}_{N-1}[m] \tag{14-80b}$$

$$\hat{q}_0[m] = \check{q}_0[m] = R[m] \tag{14-80c}$$

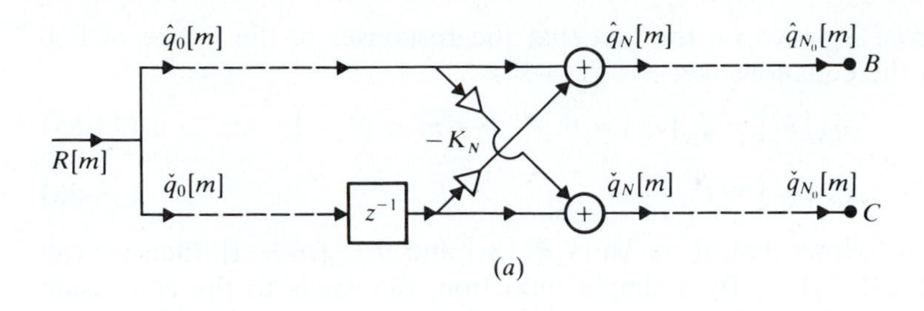

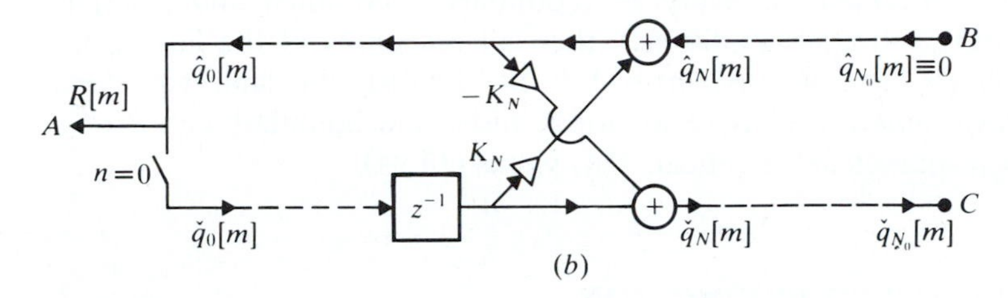

FIGURE 14-11

Using the above, we shall show that $R[m]$ can be determined as the response of the inverse lattice† of Fig. 14-11*b* provided that the following boundary and initial conditions are satisfied: The input to the system (point B) is identically 0:

$$\hat{q}_{N_0}[m] = 0 \qquad \text{all} \quad m \tag{14-81}$$

The initial conditions of all delay elements except the first are 0:

$$\check{q}_N[0] = 0 \qquad N > 0 \tag{14-82}$$

The first delay element is connected to the system at $m = 0$ and its initial condition equals $R[0]$:

$$\check{q}_0[0] = R[0] \tag{14-83}$$

From the above and (14-81) it follows that

$$\check{q}_N[1] = 0 \qquad N > 1$$

We maintain that under the stated conditions, the left output of the inverse lattice (point A) equals $R[m]$ and the right output of the mth section equals the MS error P_m:

$$\hat{q}_0[m] = R[m] \qquad \check{q}_m[m] = P_m \tag{14-84}$$

†E. A. Robinson and S. Treitel: "Maximum Entropy and the Relationship of the Partial Autocorrelation to the Reflection Coefficients of a Layered System," *IEEE Transactions on Acoustics, Speech, and Signal Process*, vol. ASSP-28, no. 2, 1980.

Proof. The proof is based on the fact that the responses of the lattice of Fig. 14-10a satisfy the equations (see Prob. 14-24)

$$\hat{q}_N[m] = \check{q}_N[m] = 0 \qquad 1 \le m \le N - 1 \tag{14-85}$$

$$\check{q}_N[N] = P_N \tag{14-86}$$

From (14-80) it follows that, if we know $\hat{q}_N[m]$ and $\check{q}_{N-1}[m-1]$, then we can find $\hat{q}_{N-1}[m]$ and $\check{q}_N[m]$. By a simple induction, this leads to the conclusion that if $\hat{q}_{N_0}[m]$ is specified for every m (boundary conditions) and $\check{q}_N[1]$ is specified for every N (initial conditions), then all responses of the lattice are determined uniquely. The two systems of Fig. 14-11 satisfy the same equations (14-80) and, as we noted, they have identical initial and boundary conditions. Hence all their responses are identical. This yields (14-84).

14-3 FILTERING AND PREDICTION

In this section, we consider the problem of estimating the future value $\mathbf{s}(t+\lambda)$ of a stochastic process $\mathbf{s}(t)$ (signal) in terms of the present and past values of a regular process $\mathbf{x}(t)$ (signal plus noise)

$$\hat{\mathbf{s}}(t+\lambda) = \hat{E}\{\mathbf{s}(t+\lambda)|\mathbf{x}(t-\tau), \tau \ge 0\} = \int_0^\infty h_x(\alpha)\mathbf{x}(t-\alpha)\,d\alpha \tag{14-87}$$

Thus $\hat{\mathbf{s}}(t+\lambda)$ is the output of a linear time-invariant causal system $\mathbf{H}_x(s)$ with input $\mathbf{x}(t)$. To determine $\mathbf{H}_x(s)$, we use the orthogonality principle

$$E\left\{\left[\mathbf{s}(t+\lambda) - \int_0^\infty h_x(\alpha)\mathbf{x}(t-\alpha)\,d\alpha\right]\mathbf{x}(t-\tau)\right\} = 0 \qquad \tau \ge 0$$

This yields the *Wiener–Hopf* equation

$$R_{sx}(\tau+\lambda) = \int_0^\infty h_x(\alpha)R_{xx}(\tau-\alpha)\,d\alpha \qquad \tau \ge 0 \tag{14-88}$$

The solution $h_x(t)$ of (14-88) is the impulse response of the prediction and filtering system known as the *Wiener filter*. If $\mathbf{x}(t) = \mathbf{s}(t)$, then $h_x(t)$ is a pure predictor as in (14-39). If $\lambda = 0$, then $h_x(t)$ is a pure filter.

To solve (14-88), we express $\mathbf{x}(t)$ in terms of its innovations $\mathbf{i}_x(t)$ (Fig. 14-12)

$$\mathbf{x}(t) = \int_0^\infty l_x(\alpha)\mathbf{i}_x(t-\alpha)\,d\alpha \qquad R_{ii}(\tau) = \delta(\tau) \tag{14-89}$$

where $l_x(t)$ is the impulse response of the innovations filter $\mathbf{L}_x(s)$ obtained by factoring the spectrum of $\mathbf{x}(t)$ as in (12-3):

$$\mathbf{S}_{xx}(s) = \mathbf{L}_x(s)\mathbf{L}_x(-s) \tag{14-90}$$

As we know, the processes $\mathbf{i}_x(t)$ and $\mathbf{x}(t)$ are linearly equivalent; hence the estimate $\hat{\mathbf{s}}(t+\lambda)$ can be expressed as the output of a causal filter $\mathbf{H}_{i_x}(s)$ with

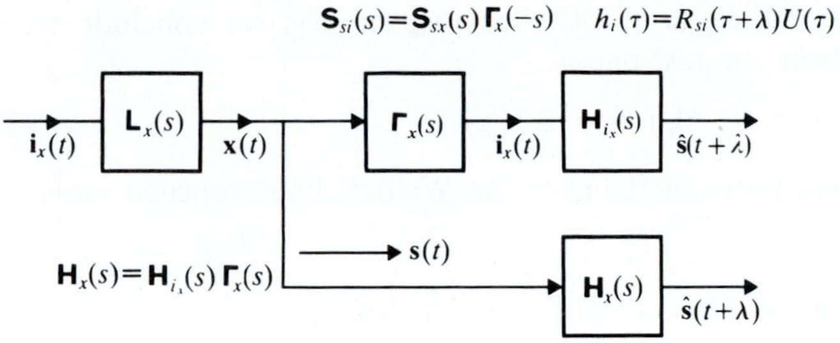

$\mathbf{S}_{si}(s) = \mathbf{S}_{sx}(s)\boldsymbol{\Gamma}_x(-s)$ $h_i(\tau) = R_{si}(\tau+\lambda)U(\tau)$

$\mathbf{i}_x(t)$ $\mathbf{L}_x(s)$ $\mathbf{x}(t)$ $\boldsymbol{\Gamma}_x(s)$ $\mathbf{i}_x(t)$ $\mathbf{H}_{i_x}(s)$ $\hat{\mathbf{s}}(t+\lambda)$

$\mathbf{s}(t)$

$\mathbf{H}_x(s) = \mathbf{H}_{i_x}(s)\boldsymbol{\Gamma}_x(s)$ $\mathbf{H}_x(s)$ $\hat{\mathbf{s}}(t+\lambda)$

FIGURE 14-12

input $\mathbf{i}_x(t)$:

$$\hat{\mathbf{s}}(t+\lambda) = \int_0^\infty h_{i_x}(\alpha)\mathbf{i}_x(t-\alpha)\,d\alpha \tag{14-91}$$

To determine $h_{i_x}(t)$, we use the orthogonality principle

$$E\left\{\left[\mathbf{s}(t+\lambda) - \int_0^\infty h_{i_x}(\alpha)\mathbf{i}_x(t-\alpha)\,d\alpha\right]\mathbf{i}_x(t-\tau)\right\} = 0 \qquad \tau \ge 0$$

Since $\mathbf{i}_x(t)$ is white noise, the above yields

$$R_{si_x}(\tau+\lambda) = \int_0^\infty h_{i_x}(\alpha)\delta(\tau-\alpha)\,d\alpha = h_{i_x}(\tau) \qquad \tau \ge 0 \tag{14-92}$$

This determines $h_{i_x}(\tau)$ for all τ because $h_{i_x}(\tau) = 0$ for $\tau < 0$:

$$h_{i_x}(\tau) = R_{si_x}(\tau+\lambda)U(\tau) \tag{14-93}$$

In the above, $R_{si_x}(\tau)$ is the cross-correlation between the signal $\mathbf{s}(t)$ and the process $\mathbf{i}_x(t)$. The function $R_{si_x}(\tau)$ can be expressed in terms of the cross-correlation $R_{sx}(\tau)$ between $\mathbf{s}(t)$ and $\mathbf{x}(t)$. Indeed, since $\mathbf{i}_x(t)$ is the output of the whitening filter $\boldsymbol{\Gamma}_x(s)$ with input $\mathbf{x}(t)$, we can show as in (10-118) and (10-157) that

$$\mathbf{S}_{si_x}(s) = \mathbf{S}_{sx}(s)\boldsymbol{\Gamma}_x(-s) \tag{14-94}$$

Thus, since $\mathbf{S}_{sx}(s)$ is assumed known, (14-94) yields $R_{si_x}(\tau)$. Shifting to the left and truncating as in (14-93), we obtain $h_{i_x}(\tau)$.

To complete the specification of $\mathbf{H}_x(s)$, we multiply the transform $\mathbf{H}_{i_x}(s)$ of the function $h_{i_x}(t)$ so obtained with $\boldsymbol{\Gamma}_x(s)$ (see Fig. 14-12)

$$\mathbf{H}_x(s) = \mathbf{H}_{i_x}(s)\boldsymbol{\Gamma}_x(s) \tag{14-95}$$

The function $\mathbf{H}_{i_x}(s)$ can be determined directly from (14-94): As we know (shifting theorem) the transform of $R_{si_x}(\tau+\lambda)$ equals

$$\mathbf{S}_\lambda(s) = \mathbf{S}_{si_x}(s)e^{\lambda s} = \mathbf{S}_{sx}(s)\boldsymbol{\Gamma}_x(-s)e^{\lambda s} \tag{14-96}$$

To find $\mathbf{H}_{i_x}(s)$, it suffices to write $\mathbf{S}_\lambda(s)$ as a sum

$$\mathbf{S}_\lambda(s) = \mathbf{S}_\lambda^+(s) + \mathbf{S}_\lambda^-(s) \tag{14-97}$$

where $\mathbf{S}_\lambda^+(s)$ is analytic in the right-hand s plane and $\mathbf{S}_\lambda^-(s)$ is analytic in the left-hand s plane. Since the inverse transforms of the function $\mathbf{S}_\lambda^+(s)$ and $\mathbf{S}_\lambda^-(s)$

equal $R_{si_x}(\tau+\lambda)U(\tau)$ and $R_{si_x}(\tau+\lambda)U(-\tau)$ respectively, we conclude from (14-93) that (see also Note on next page)

$$\mathbf{H}_{i_x}(s) = \mathbf{S}_\lambda^+(s) \tag{14-98}$$

To determine the system function $\mathbf{H}_x(s)$ of the Wiener filter, proceed, thus, as follows:

Factor $\mathbf{S}_{xx}(s)$ as in (14-90) and set $\mathbf{\Gamma}_x(s) = 1/\mathbf{L}_x(s)$.
Evaluate $\mathbf{S}_{si_x}(s)$ from (14-94) and form the function $\mathbf{S}_\lambda(s)$ using (14-96).
Decompose $\mathbf{S}_\lambda(s)$ as in (14-97) and form the function $\mathbf{H}_{i_x}(s)$ using (4-98).
Determine $\mathbf{H}_x(s)$ from (14-95).

If the function $\mathbf{S}_\lambda(s)$ is rational, then the decomposition (14-97) can be accomplished by expanding $\mathbf{S}_{si_x}(s)$ into partial fractions. Assuming that $\mathbf{S}_{si_x}(s)$ is a proper fraction with simple poles, we obtain

$$\mathbf{S}_{si_x}(s) = \sum_i \frac{a_i}{s-s_i} + \sum_k \frac{b_k}{s-z_k} \qquad \begin{matrix} \operatorname{Re} s_i < 0 \\ \operatorname{Re} z_k > 0 \end{matrix} \tag{14-99}$$

The inverse of the second sum is 0 for $\tau > 0$. If, therefore, it is shifted to the left, it will remain 0 for $\tau > 0$. This shows that only the first sum will contribute to the term $R_{si_x}(\tau+\lambda)U(\tau)$. In other words,

$$R_{si_x}(\tau+\lambda)U(\tau) = \left[a_1 e^{s_1(\tau+\lambda)} + \cdots + a_n e^{s_n(\tau+\lambda)}\right]U(\tau)$$

The transform of the above yields

$$\mathbf{S}_\lambda^+(s) = \frac{a_1 e^{s_1\lambda}}{s-s_1} + \cdots + \frac{a_n e^{s_n\lambda}}{s-s_n} \tag{14-100}$$

Example 14-7. Suppose that $\mathbf{x}(t) = \mathbf{s}(t) + \boldsymbol{\nu}(t)$ and

$$S_{ss}(\omega) = \frac{N_0}{\alpha^2+\omega^2} \qquad S_{\nu\nu}(\omega) = N \qquad S_{s\nu}(\omega) = 0 \tag{14-101}$$

as in Example 14-1. In this case, $\mathbf{S}_{sx}(s) = \mathbf{S}_{ss}(s)$ and

$$\mathbf{S}_{xx}(s) = \frac{N_0}{\alpha^2-s^2} + N = N\frac{\beta^2-s^2}{\alpha^2-s^2} \qquad \beta^2 = \alpha^2 + \frac{N_0}{N}$$

Hence

$$\mathbf{L}_x(s) = \sqrt{N}\frac{s+\beta}{s+\alpha} \qquad \mathbf{\Gamma}_x(-s) = \frac{1}{\sqrt{N}}\frac{\alpha-s}{\beta-s} \tag{14-102}$$

Inserting into (14-94) and expanding into partial fractions, we obtain

$$\mathbf{S}_{si_x}(s) = \frac{N_0}{\alpha^2-s^2}\frac{\alpha-s}{(\beta-s)\sqrt{N}} = \frac{A}{s+\alpha} - \frac{A}{s-\beta} \qquad A = \frac{N_0}{(\alpha+\beta)\sqrt{N}}$$

and with $s_1 = -\alpha$, (14-100) yields

$$\mathbf{S}_\lambda^+(s) = \frac{A}{s+\alpha} e^{-\alpha\lambda}$$

Hence

$$\mathbf{H}_x(s) = \mathbf{S}_\lambda^+(s)\mathbf{\Gamma}_x(s) = \frac{\beta - \alpha}{s+\beta} e^{-\alpha\lambda} \tag{14-103}$$

Note In the decomposition (14-97) of $\mathbf{S}_\lambda(s)$, the functions $\mathbf{S}_\lambda^+(s)$ and $\mathbf{S}_\lambda^-(s)$ are unique within an additive constant. This causes an ambiguity in the determination of $h_{i_x}(t)$. The ambiguity is removed if we impose the condition that

$$\mathbf{S}_\lambda^-(\infty) = 0$$

In the pure filtering case ($\lambda = 0$); the resulting $h_x(t)$ might contain impulses at the origin. This is acceptable because, by assumption the estimate $\hat{\mathbf{s}}(t)$ of $\mathbf{s}(t)$ is a functional of the past *and* the present value of the data $\mathbf{x}(t)$.

Filtering white noise. In the pure filtering problem, the determination of the estimator $\mathbf{H}_x(s)$ can be simplified if $R_{ss}(0) < \infty$ and $\boldsymbol{\nu}(t)$ is white noise orthogonal to the signal as in (14-101). We maintain, in fact, that in this case

$$\mathbf{H}_x(s) = 1 - \sqrt{N}\,\mathbf{\Gamma}_x(s) \tag{14-104}$$

where $\mathbf{\Gamma}_x(s)$ is the whitening filter of $\mathbf{x}(t)$.

Proof. From the above assumptions it follows that $\mathbf{S}_{ss}(\infty) = 0$; hence

$$\mathbf{S}_{sx}(s) = \mathbf{S}_{ss}(s) = \mathbf{S}_{xx}(s) - N = \mathbf{L}_x(s)\mathbf{L}_x(-s) - N$$

$$\mathbf{S}_{sx}(\infty) = 0 \qquad \mathbf{S}_{xx}(\infty) = N \qquad \mathbf{L}_x(\pm\infty) = \sqrt{N}$$

Inserting into (14-94), we obtain

$$\mathbf{S}_{si_x}(s) = \mathbf{L}_x(s) - N\mathbf{\Gamma}_x(-s) = \mathbf{L}_x(s) + K - N\mathbf{\Gamma}_x(-s) - K$$

From the preceding note it follows that the constant K must be such that the noncausal component of $\mathbf{S}_{si_x}(s)$ satisfies the infinity condition $-N\mathbf{\Gamma}_x(-\infty) - K = 0$. And since $\mathbf{\Gamma}_x(-\infty) = 1/\mathbf{L}_x(-\infty) = 1/\sqrt{N}$, (14-104) follows from (14-95).

Example 14-8. We shall determine the pure filter of the process in Example 14-7. From (14-102) and (14-104) it follows that

$$\mathbf{H}_x(s) = 1 - \frac{\alpha + s}{\beta + s} = \frac{\beta - \alpha}{s + \beta} \qquad h_x(t) = (\beta - \alpha)e^{-\beta t}U(t)$$

in agreement with (14-103). Note that the resulting MS error equals

$$P = E\left\{\left[\mathbf{s}(t) - \int_0^\infty h_x(\alpha)\mathbf{x}(t-\alpha)\,d\alpha\right]\mathbf{s}(t)\right\} = \frac{N_0}{\alpha + \beta}$$

Digital Processes

We shall state briefly the discrete-time version of the preceding results. Our problem now is the determination of the future value $\mathbf{s}[n+r]$ of a stochastic process in terms of the present and past values of another process $\mathbf{x}[n]$:

$$\hat{\mathbf{s}}_r[n+r] = \sum_{k=0}^{\infty} h_x^r[k]\mathbf{x}[n-k] \tag{14-105}$$

In this case,

$$\mathbf{s}[n+r] - \hat{\mathbf{s}}_r[n+r] \perp \mathbf{x}[n-m] \qquad m \geq 0$$

hence

$$R_{sx}[m+r] = \sum_{k=0}^{\infty} h_x^r[k]R_{xx}[m-k] \qquad m \geq 0 \tag{14-106}$$

This is the discrete version of the Wiener–Hopf equation (14-88).

To determine $h_x^r[n]$, we proceed as in the analog case: We express $\hat{\mathbf{s}}_r[n+r]$ in terms of the innovations $\mathbf{i}_x[n]$ of $\mathbf{x}[n]$ (Fig. 14-13)

$$\hat{\mathbf{s}}_r[n+r] = \sum_{k=0}^{\infty} h_{i_x}^r[k]\mathbf{i}_x[n-k] \tag{14-107}$$

From this and (8-70) it follows that

$$R_{si_x}[m+r] = \sum_{k=0}^{\infty} h_{i_x}^r[k]\delta[m-k] = h_{i_x}^r[m] \qquad m \geq 0$$

because $R_{i_x}[m] = \delta[m]$. Hence

$$h_{i_x}^r[m] = R_{si_x}[m+r]U[m] \qquad \text{all} \quad m \tag{14-108}$$

The function $R_{si_x}[m]$ can be expressed in terms of $R_{sx}[m]$ as in (14-94)

$$\mathbf{S}_{si_x}(z) = \mathbf{S}_{sx}(z)\boldsymbol{\Gamma}_x(z^{-1}) \tag{14-109}$$

Thus the transform of $R_{si_x}[m+r]$ equals

$$\mathbf{S}_r(z) = z^r\mathbf{S}_{si_x}(z) = z^r\mathbf{S}_{sx}(z)\boldsymbol{\Gamma}_x(z^{-1}) \tag{14-110}$$

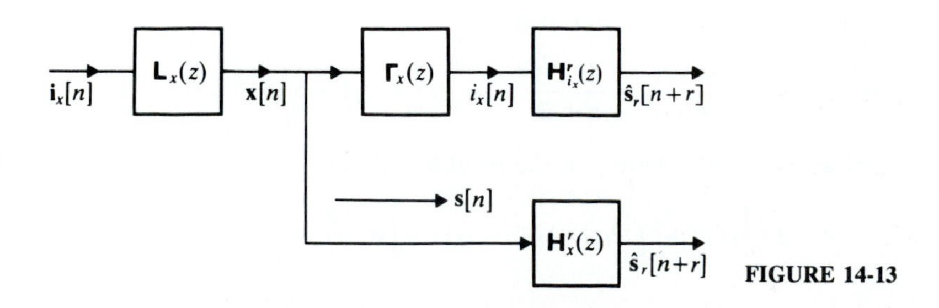

FIGURE 14-13

The function $\mathbf{S}_r(z)$ is then written as a sum

$$\mathbf{S}_r(z) = \mathbf{S}_r^+(z) + \mathbf{S}_r^-(z) \tag{14-111}$$

where $\mathbf{S}_r^+(z)$ is analytic for $|z| > 1$ and $\mathbf{S}_r^-(z)$ is analytic for $|z| < 1$. Furthermore, the inverse of $\mathbf{S}_r^-(z)$ at the origin is 0. Thus $\mathbf{S}_r^+(z)$ is the transform of the causal function $R_{si_x}[m + r]U[m]$. And since $\mathbf{i}_x[n]$ is the response of the whitening filter $\mathbf{\Gamma}_x(z)$ with input $\mathbf{x}[n]$, we conclude from (13-108) that

$$\mathbf{H}_x^r(z) = \mathbf{H}_{i_x}^r(z)\mathbf{\Gamma}_x(z) = \mathbf{S}_r^+(z)\mathbf{\Gamma}_x(z) \tag{14-112}$$

Example 14-9. We shall determine the one-step predictor $\hat{\mathbf{s}}_1[n+1]$ of the process $\mathbf{s}[n]$ where

$$\mathbf{S}_{ss}(z) = \frac{N_0}{(1 - az^{-1})(1 - az)} \qquad \mathbf{S}_{\nu\nu}(z) = N \qquad \mathbf{S}_{s\nu}(z) = 0$$

In this case (see Example 14-2)

$$\mathbf{L}_x(z) = \sqrt{\frac{Na}{b}}\,\frac{1 - bz^{-1}}{1 - az^{-1}}$$

From (14-110) it follows with $r = 1$ that

$$z\mathbf{S}_{si_x}(z) = \frac{zN_0\sqrt{b/Na}}{(1 - az^{-1})(1 - bz)} = \frac{Aaz}{z - a} - \frac{Az/b}{z - 1/b} \qquad A = (a - b)\sqrt{\frac{N}{ab}}$$

Since $0 < a < 1$ and $1/b > 1$, we conclude from the above that $\mathbf{S}_1^+(z) = Aaz/(z - a)$ and (14-112) yields

$$\mathbf{H}_x^1(z) = (a - b)\frac{z}{z - b} \qquad h_x^1[n] = (a - b)b^n U[n]$$

We discuss presently a more direct method for determining $\mathbf{H}_x^r(z)$ [see (14-118) below].

White noise. We shall examine the nature of the predictor $\mathbf{H}_x^r(z)$ of $\mathbf{s}[n + r]$ under the assumption that the noise is white and orthogonal to the signal

$$R_{\nu\nu}[m] = N\delta[m] \qquad R_{s\nu}[m] = 0 \tag{14-113}$$

Pure filter Suppose first that $r = 0$. In this case, $\mathbf{H}_x^0(z)$ is a pure filter and $\hat{\mathbf{s}}_0[n]$ is the estimate of the signal $\mathbf{s}[n]$ in terms of $\mathbf{x}[n]$ and its past.

We maintain that (Fig. 14-14)

$$\mathbf{H}_x^0(z) = 1 - \frac{D}{\mathbf{L}_x(z)} \qquad D = \frac{N}{l_x[0]} \tag{14-114}$$

Proof. From (14-113) it follows that

$$\mathbf{S}_{sx}(z) = \mathbf{S}_{ss}(z) = \mathbf{S}_{xx}(z) - N = \mathbf{L}_x(z)\mathbf{L}_x(z^{-1}) - N$$

Inserting into (14-109), we obtain

$$\mathbf{S}_{si_x}(z) = \mathbf{L}_x(z) - N\mathbf{\Gamma}_x(z^{-1}) \tag{14-115}$$

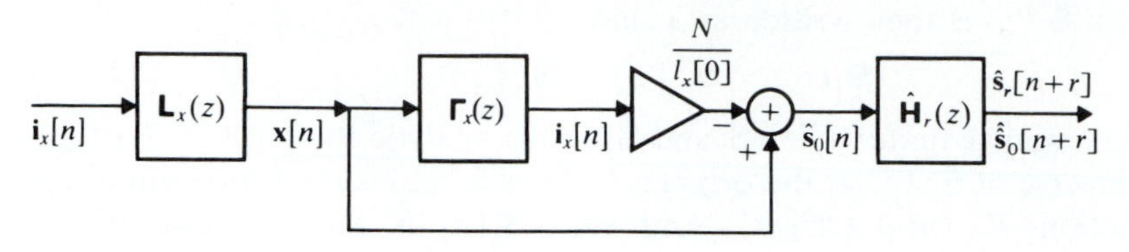

FIGURE 14-14

We wish to find the causal part of the above, including the value of its inverse at $n = 0$. Since the inverse z transform of $\mathbf{\Gamma}_x(1/z)$ is 0 for $n > 0$ and for $n = 0$ it equals $\mathbf{\Gamma}_x(\infty)$, we conclude that

$$\mathbf{H}^0_{i_x}(z) = \mathbf{L}_x(z) - N\mathbf{\Gamma}_x(\infty) \tag{14-116}$$

Multiplying by $\mathbf{\Gamma}_x(z)$, we obtain (14-114) because $\mathbf{\Gamma}_x(\infty) = 1/l_x[0]$.

Filtering and prediction We shall now show that the estimate $\hat{\mathbf{s}}_r[n+r]$ of $\mathbf{s}[n+r]$ equals the pure predictor $\hat{\hat{\mathbf{s}}}_0[n+r]$ of the estimate $\hat{\mathbf{s}}_0[n]$ of $\mathbf{s}[n]$ (Fig. 14-14)

$$\hat{\mathbf{s}}_r[n+r] = \hat{\hat{\mathbf{s}}}_0[n+r] = \hat{E}\{\hat{\mathbf{s}}_0[n+r] | \hat{\mathbf{s}}_0[n-k], k \geq 0\} \tag{14-117}$$

Proof. From (14-110) and (14-115) it follows that

$$\mathbf{S}_r(z) = z^r\left[\mathbf{L}_x(z) - N\mathbf{\Gamma}_x(z^{-1})\right]$$

But the inverse of $z^r\mathbf{\Gamma}_x(1/z)$ is 0 for $n \geq 0$. Hence $\mathbf{S}_r^+(z)$ is the causal part of $z^r\mathbf{L}_x(z)$. Inserting into (14-112), we obtain

$$\mathbf{H}^r_x(z) = z^r\left(\mathbf{L}_x(z) - \sum_{k=0}^{r-1} l_x[k]z^{-k}\right)\mathbf{\Gamma}_x(z)$$

$$= z^r\left(1 - \frac{\sum_{k=0}^{r-1} l_x[k]z^{-k}}{\mathbf{L}_x(z)}\right) \tag{14-118}$$

As we see from Fig. 14-14, the innovations filter of $\hat{\mathbf{s}}_0[n]$ equals $\mathbf{L}_x(z)\mathbf{H}^0_x(z)$. To determine the pure predictor $\hat{\mathbf{H}}_r(z)$ of $\hat{\mathbf{s}}_0[n+r]$, it suffices, therefore, to multiply (14-37) by z^r (we are predicting now the future) and to replace the function $\mathbf{L}(z)$ by $\mathbf{L}_x(z)\mathbf{H}^0_x(z)$. This yields

$$\hat{\mathbf{H}}_r(z) = z^r\left(1 - \frac{\sum_{k=0}^{r-1} l_x[k]z^{-k} - D}{\mathbf{L}_x(z) - D}\right)$$

because the inverse of $\mathbf{L}_x(z) - D$ equals $l_x[n] - D\delta[n]$. Comparing with (14-118), we conclude that

$$\mathbf{H}^r_x(z) = \mathbf{H}^0_x(z)\hat{\mathbf{H}}_r(z)$$

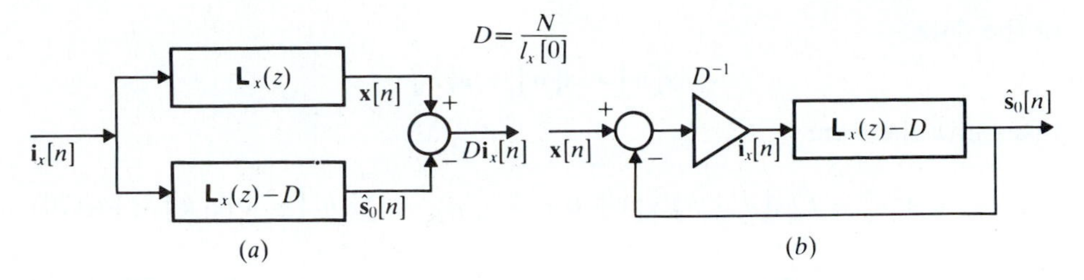

FIGURE 14-15

The preceding discussion leads to the following important consequences of the white-noise assumption (14-113):

1. The innovations $\mathbf{i}_x[n]$ of $\mathbf{x}[n]$ are proportional to the difference $\mathbf{x}[n] - \hat{\mathbf{s}}_0[n]$:

$$\mathbf{x}[n] - \hat{\mathbf{s}}_0[n] = D\mathbf{i}_x[n] \qquad D = \frac{N}{l_x[0]} \tag{14-119}$$

   Indeed, $\mathbf{x}[n] - \hat{\mathbf{s}}_0[n]$ is the output of the filter

$$\mathbf{L}_x(z) - [\mathbf{L}_x(z) - D] = D$$

   with input $\mathbf{i}_x[n]$ (Fig. 14-15*a*). Thus the process $\mathbf{i}_x[n]$ can be realized simply by a feedback system (Fig. 14-15*b*) involving merely the filter $\mathbf{H}^0_{i_x}(z)$.
2. The r-step filtering and prediction estimate $\hat{\mathbf{s}}_0[n+r]$ can be obtained by cascading the pure filter $\mathbf{H}^0_x(z)$ of $\mathbf{s}[n]$ with the pure predictor $\hat{\mathbf{H}}_r(z)$ of $\hat{\mathbf{s}}_0[n+r]$.
3. If the signal $\mathbf{s}[n]$ is an ARMA process, then its estimate $\hat{\mathbf{s}}_0[n]$ is also an ARMA process.

   Indeed, if $\mathbf{L}_x(z) = A(z)/B(z)$ is rational, then [see (14-114)], the filter $\mathbf{H}^0_x(z)$ is also rational. Furthermore, the denominator $B(z)$ of $\mathbf{L}_x(z)$ is the same as the denominator of the forward component $\mathbf{L}_x(z) - D$ of the feedback realization of $\mathbf{H}^0_x(z)$ shown in Fig. 14-15*b*.

As we shall presently see, these results are central in the development of Kalman filters.

14-4 KALMAN FILTERS†

In this section we extend the preceding results to nonstationary processes with causal data and we show that the results can be simplified if the noise is white and the signal is an ARMA process. The estimate $\hat{\mathbf{s}}_r[n+r]$ of $\mathbf{s}[n+r]$ in terms

†R. E. Kalman: "A New Approach to Linear Filtering and Prediction Problems," *ASME Transactions*, vol. 82D, 1960.

of the data

$$\mathbf{x}[n] = \mathbf{s}[n] + \boldsymbol{\nu}[n]$$

takes now the form

$$\hat{\mathbf{s}}_r[n+r] = \hat{E}\{\mathbf{s}[n+r]|\mathbf{x}[k], 0 \le k \le n\} = \sum_{k=0}^{n} h_x^r[n,k]\mathbf{x}[k] \quad (14\text{-}120)$$

Thus $\hat{\mathbf{s}}_r[n+r]$ is the output of a causal, time-varying system with input $\mathbf{x}[n]U[n]$, and our problem is to find its delta response $h_x^r[n,k]$.

As we know,

$$\mathbf{s}[n+r] - \hat{\mathbf{s}}_r[n+r] \perp \mathbf{x}[m] \qquad 0 \le m \le n$$

This yields

$$R_{sx}[n+r,m] = \sum_{k=0}^{n} h_x^r[n,k]R_{xx}[k,m] \qquad 0 \le m \le n \quad (14\text{-}121)$$

Thus $h_x^r[n,k]$ must be such that its response to $R_{xx}[n,m]$ (the time variable is n) equals $R_{sx}[n+r,m]$ for every $0 \le m \le n$. For a specific n, this yields $n+1$ equations for the $n+1$ unknowns $h_x^r[n,k]$.

To simplify the determination of $h_x^r[n,k]$, we shall express the desired estimates $\hat{\mathbf{s}}_r[n+r]$ in terms of the *Kalman innovations* [see (14-77)]

$$\mathbf{i}_x[n] = \sum_{k=0}^{n} \gamma_x[n,k]\mathbf{x}[k] \quad (14\text{-}122)$$

of the process $\mathbf{x}[n]U[n]$ where $\gamma_x[n,k]$ is the Kalman whitening filter. The process $\mathbf{i}_x[n]$ is orthonormal and, if the data are linearly independent, then the processes $\mathbf{x}[n]$ and $\mathbf{i}_x[n]$ are linearly equivalent. This leads to the conclusion that $\hat{\mathbf{s}}_r[n+r]$ can be expressed in terms of $\mathbf{i}_x[n]$ and its past (Fig. 14-16)

$$\hat{\mathbf{s}}_r[n+r] = \sum_{k=0}^{n} h_{i_x}^r[n,k]\mathbf{i}_x[k] \quad (14\text{-}123)$$

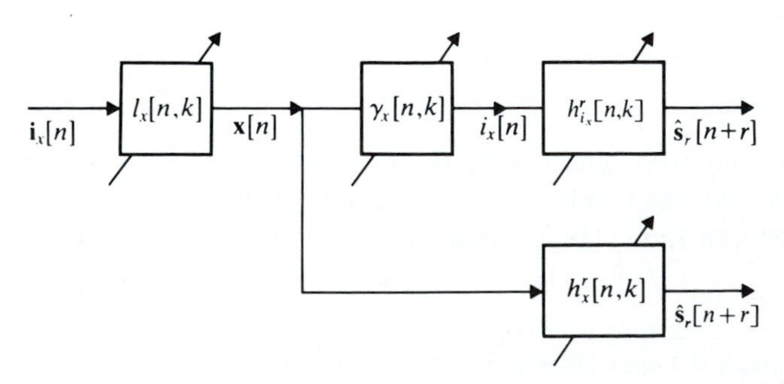

FIGURE 14-16

To determine $h_{i_x}^r[n,k]$, we apply the orthogonality principle. Since

$$R_{i_x}[m,n] = \delta[m-n]$$

this yields

$$R_{si_x}[n+r,m] = \sum_{k=0}^{n} h_{i_x}^r[n,k]\delta[k-m]$$

Hence

$$h_{i_x}^r[n,m] = R_{si_x}[n+r,m] \qquad 0 \le m \le n \tag{14-124}$$

This function can be expressed in terms of the cross-correlation $R_{sx}[m,n]$. Multiplying (14-122) by $\mathbf{s}[m]$, we obtain

$$R_{si_x}[m,n] = \sum_{k=0}^{n} \gamma_k[n,k]R_{sx}[m,k] \tag{14-125}$$

Thus, for a specific m, $R_{si_x}[m,n]$ is the response of the Kalman whitening filter of $\mathbf{x}[n]$ to the function $R_{sx}[m,n]$ where n is the variable. To complete the specification of $\hat{\mathbf{s}}_r[n+r]$, we cascade the filter $h_{i_x}^r[n,m]$ with the whitening filter $\gamma_x[n,k]$ as in Fig. 14-16.

ARMA Signals in White Noise

In the numerical implementation of the above, we are faced with two problems: (1) the realization of the Kalman innovations process $\mathbf{i}_x[n]$: (2) the determination of the sum in (14-123). In general, these problems are complex, involving storage capacity and number of computations proportional to n. However, as we show next, under certain realistic assumptions the problem can be simplified drastically.

ASSUMPTION 1. The noise is white and orthogonal to the signal:

$$R_{\nu\nu}[m,n] = N_n\delta[m-n] \qquad R_{s\nu}[m,n] = 0 \tag{14-126}$$

This leads to the following conclusions.

Property 1 If $\hat{\mathbf{s}}_0[n]$ is the estimate of $\mathbf{s}[n]$ in terms of $\mathbf{x}[n]$ and its past and D_n^2 is the MS estimation error, then the difference $\mathbf{x}[n] - \hat{\mathbf{s}}_0[n]$ is proportional to the Kalman innovations $\mathbf{i}_x[n]$ of the data $\mathbf{x}[n]$:

$$\mathbf{x}[n] - \hat{\mathbf{s}}_0[n] = D_n\mathbf{i}_x[n] \tag{14-127a}$$

$$D_n^2 = E\{|\mathbf{x}[n] - \hat{\mathbf{s}}_0[n]|^2\} \tag{14-127b}$$

Proof. The difference $\mathbf{x}[n] - \hat{\mathbf{s}}_0[n]$ depends linearly on $\mathbf{x}[n]$ and its past. Furthermore, the processes $\boldsymbol{\nu}[n]$ and $\mathbf{s}[n] - \hat{\mathbf{s}}_0[n]$ are orthogonal to the past of $\mathbf{x}[n]$. Hence

$$\mathbf{x}[n] - \hat{\mathbf{s}}_0[n] = \mathbf{s}[n] - \hat{\mathbf{s}}_0[n] + \boldsymbol{\nu}[n] \perp \mathbf{x}[k] \qquad k < n$$

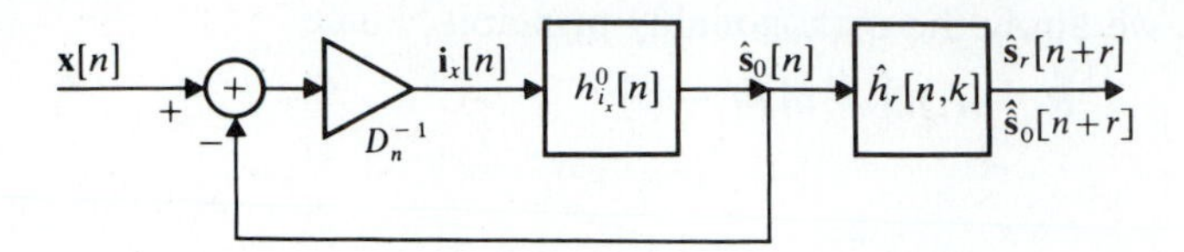

FIGURE 14-17

From this it follows that the process $\mathbf{x}[n] - \hat{\mathbf{s}}_0[n]$ is white noise and

$$\mathbf{x}[n] - \hat{\mathbf{s}}_0[n] \perp \mathbf{i}_x[k] \qquad 0 \le k \le n-1$$

because the processes $\mathbf{x}[k]$ and $\mathbf{i}_x[k]$ are linearly equivalent. And since $\mathbf{x}[n] - \hat{\mathbf{s}}_0[n]$ depends linearly on $\mathbf{i}_x[k]$ for $0 \le k \le n$, (14-127*a*) results. Equation (14-127*b*) is a consequence of the requirement that $E\{\mathbf{i}_x^2[n]\} = 1$.

Property 1 shows that the process $\mathbf{i}_x[n]$ can be realized simply by the feedback system of Fig. 14-17. This eliminates the need for designing the whitening filter $\gamma_x[n, k]$.

Property 2 The estimate $\hat{\mathbf{s}}_r[n+r]$ of $\mathbf{s}[n+r]$ equals the pure predictor $\hat{\hat{\mathbf{s}}}_0[n+r]$ of the estimate $\hat{\mathbf{s}}_0[n]$ of $\mathbf{s}[n]$ (Fig. 14-17)

$$\hat{\mathbf{s}}_r[n+r] = \hat{\hat{\mathbf{s}}}_0[n+r] = \sum_{k=0}^{n} \hat{h}_r[n,k]\hat{\mathbf{s}}_0[k] \tag{14-128}$$

provided that, for every $n \ge 0$,

$$E\{\hat{\mathbf{s}}_0[n]\mathbf{i}_x[n]\} = E\{\mathbf{x}[n]\mathbf{i}_x[n]\} - D_n \ne 0 \tag{14-129}$$

Proof. The process $\hat{\mathbf{s}}_0[n]$ is linearly dependent on $\mathbf{x}[n]$ and its past. Condition (14-129) means that the component of $\hat{\mathbf{s}}_0[n]$ in the $\mathbf{i}_x[n]$ direction is not 0. Hence the processes $\hat{\mathbf{s}}_0[n]$ and $\mathbf{x}[n]$ are linearly equivalent. And since

$$\hat{\mathbf{s}}_0[n+r] - \hat{\hat{\mathbf{s}}}_0[n+r] \perp \hat{\mathbf{s}}_0[k] \qquad 0 \le k \le n$$

we conclude that

$$\hat{\hat{\mathbf{s}}}_0[n+r] - \hat{\mathbf{s}}_0[n+r] \perp \mathbf{x}[k] \qquad 0 \le k \le n$$

Furthermore,

$$\mathbf{s}[n+r] - \hat{\mathbf{s}}_0[n+r] \perp \mathbf{x}[k] \qquad 0 \le k \le n+r$$

because $\hat{\mathbf{s}}_0[n+r]$ is the estimate of $\mathbf{s}[n+r]$ in terms of $\mathbf{x}[k]$ for $0 \le k \le n+r$. Finally,

$$\mathbf{s}[n+r] - \hat{\hat{\mathbf{s}}}_0[n+r] = (\mathbf{s}[n+r] - \hat{\mathbf{s}}_0[n+r]) + (\hat{\mathbf{s}}_0[n+r] - \hat{\hat{\mathbf{s}}}_0[n+r])$$

Hence

$$\mathbf{s}[n+r] - \hat{\hat{\mathbf{s}}}_0[n+r] \perp \mathbf{x}[k] \qquad 0 \le k \le n$$

and (14-128) results.

This property shows that filtering and prediction can be reduced to a cascade of a pure filter and a pure predictor.

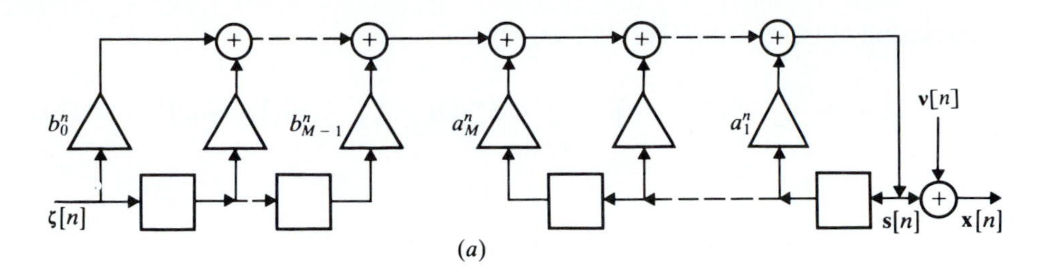

(a)

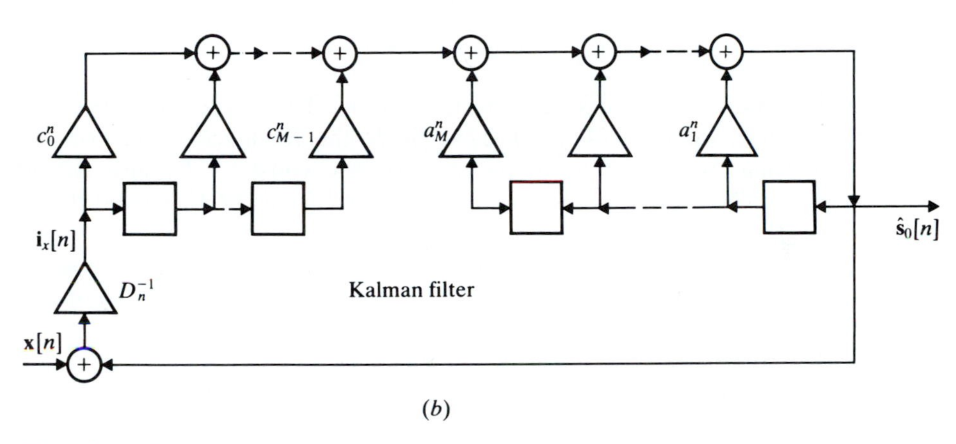

(b)

FIGURE 14-18

ASSUMPTION 2. The signal $\mathbf{s}[n]$ is a time-varying ARMA process (Fig. 14-18a)

$$\mathbf{s}[n] - a_1^n\mathbf{s}[n-1] - \cdots - a_M^n\mathbf{s}[n-M] = \sum_{k=0}^{M-1} b_k^n \boldsymbol{\zeta}[n-k] \qquad (14\text{-}130)$$

$$R_{\zeta\zeta}[m,n] = V_n\delta[m-n]$$

Property 3 The estimate $\hat{\mathbf{s}}_0[n]$ is also an ARMA process

$$\hat{\mathbf{s}}_0[n] - a_1^n\hat{\mathbf{s}}_0[n-1] - \cdots - a_M^n\hat{\mathbf{s}}_0[n-M] = \sum_{k=0}^{M-1} c_k^n \mathbf{i}_x[n-k] \qquad (14\text{-}131)$$

where the coefficients a_k^n are the same as in (14-130) and the coefficients c_k^n are M constants to be determined.

Proof. We assume that the above is true for all past estimates $\hat{\mathbf{s}}_0[n-k]$ and we shall prove that if $\hat{\mathbf{s}}_0[n]$ is given by (14-131), then it is the estimate of $\mathbf{s}[n]$. It suffices to show that if the constants c_k^n are suitably chosen, then the resulting error satisfies the orthogonality principle

$$\boldsymbol{\varepsilon}[n] = \mathbf{s}[n] - \hat{\mathbf{s}}_0[n] \perp \mathbf{x}[r] \qquad 0 \le r \le n \qquad (14\text{-}132)$$

Subtracting (14-131) from (14-130), we obtain

$$\boldsymbol{\varepsilon}[n] = \sum_{k=1}^{M} a_k^n \boldsymbol{\varepsilon}[n-k] + \sum_{k=0}^{M-1} \left(b_k^n \boldsymbol{\zeta}[n-k] - c_k^n \mathbf{i}_x[n-k]\right) \tag{14-133}$$

But

$$\boldsymbol{\zeta}[n_1], \mathbf{i}_x[n_1] \perp \mathbf{x}[r] \qquad \text{for} \quad r < n_1$$

and $\boldsymbol{\varepsilon}[n-k] \perp \mathbf{x}[r]$ for $r \le n-k$ (induction hypothesis). Hence (14-132) is true for $r \le n - M$. It suffices, therefore, to select the M constants c_k^n such that

$$E\{\boldsymbol{\varepsilon}[n]\mathbf{x}[r]\} = 0 \qquad n - M + 1 \le r \le n \tag{14-134}$$

We have thus expressed $\hat{\mathbf{s}}_0[n]$ in terms of $\mathbf{i}_x[n]$. To complete the specification of the filter, we use (14-127*a*). This yields the feedback system of Fig. 14-18*b* involving $M+1$ unknown parameters: the constant D_n and the M coefficients c_k^n. These parameters can be determined from (14-127*b*) and the M equations in (14-134).

The recursion equation (14-131) can be written as a system of M first-order equations (state equations) or, equivalently, as a first-order vector equation (see Sec. 12-2). The unknowns are then the scalar D_n and the coefficients c_k^n. To simplify the analysis, we shall carry out the determination of the unknown parameters for the first-order scalar case only. The results hold also for the vector case mutatis mutandis.

FIRST-ORDER. If

$$\mathbf{s}[n] - A_n \mathbf{s}[n-1] = \boldsymbol{\zeta}[n] \qquad E\{\boldsymbol{\zeta}^2[n]\} = V_n \tag{14-135}$$

then (14-131) yields

$$\hat{\mathbf{s}}_0[n] - A_n \hat{\mathbf{s}}_0[n-1] = K_n(\mathbf{x}[n] - \hat{\mathbf{s}}_0[n]) \tag{14-136}$$

where $K_n = c_0^n / D_n$. This is a first-order system as in Fig. 14-19*a*. To complete

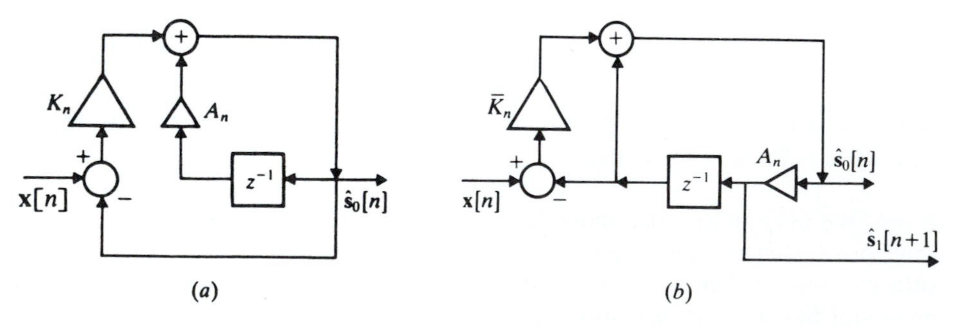

FIGURE 14-19

its specification, we must find the constant K_n. We maintain that

$$K_n = \frac{P_n}{N_n - P_n} \qquad P_n = E\{\boldsymbol{\varepsilon}^2[n]\} \tag{14-137}$$

In the above, N_n is the average intensity of $\boldsymbol{\nu}[n]$, which we assume known.

The MS error P_n can be determined recursively

$$\frac{P_n}{N_n - P_n} = \frac{A_n^2 P_{n-1} + V_n}{N_n} \tag{14-138}$$

Proof. Multiplying the data $\mathbf{x}[n] = \mathbf{s}[n] + \boldsymbol{\nu}[n]$ by the error

$$\boldsymbol{\varepsilon}[n] = \mathbf{s}[n] - \hat{\mathbf{s}}_0[n] = \mathbf{x}[n] - \hat{\mathbf{s}}_0[n] - \boldsymbol{\nu}[n]$$

and using the orthogonality condition (14-132), we obtain

$$E\{\boldsymbol{\varepsilon}[n]\mathbf{x}[n]\} = 0 = P_n + E\{\boldsymbol{\varepsilon}[n]\boldsymbol{\nu}[n]\}$$

From (14-135) and (14-136) it follows that

$$\begin{aligned} \boldsymbol{\varepsilon}[n] &= A_n\boldsymbol{\varepsilon}[n-1] + \boldsymbol{\zeta}[n] - K_n(\boldsymbol{\varepsilon}[n] + \boldsymbol{\nu}[n]) \\ (1 + K_n)\boldsymbol{\varepsilon}[n] &= A_n\boldsymbol{\varepsilon}[n-1] + \boldsymbol{\zeta}[n] - K_n\boldsymbol{\nu}[n] \end{aligned} \tag{14-139}$$

Hence

$$(1 + K_n)E\{\boldsymbol{\varepsilon}[n]\boldsymbol{\nu}[n]\} = -K_nE\{\boldsymbol{\nu}^2[n]\}$$

and (14-137) results.

To prove (14-138), we multiply each side of (14-139) by each side of the identity $\mathbf{s}[n] = A_n\mathbf{s}[n-1] + \boldsymbol{\zeta}[n]$. This yields

$$(1 + K_n)P_n = A_n^2 P_{n-1} + V_n$$

Since $1 + K_N = N_n/(N_n - P_n)$, the above yields (14-138).

Note Using (14-135), we can readily show that

$$\hat{\mathbf{s}}_0[n] - A_n\hat{\mathbf{s}}_0[n-1] = \overline{K}_n(\mathbf{x}[n] - A_n\hat{\mathbf{s}}_0[n-1]) \tag{14-140}$$

where

$$\overline{K}_n = \frac{P_n}{N_n} = \frac{A_n^2 P_{n-1} + V_n}{A_n^2 P_{n-1} + V_n + N_n} \tag{14-141}$$

The corresponding system is shown in Fig. 14-19*b*. In the same diagram, we also show the realization of the one-step predictor

$$\hat{\mathbf{s}}_1[n+1] = \hat{\hat{\mathbf{s}}}_0[n+1] = A_n\hat{\mathbf{s}}_0[n]$$

of $\mathbf{s}[n+1]$. This follows readily from (14-128) because the process $\hat{\mathbf{s}}_0[n]$ is AR; hence its pure predictor equals $A_n\hat{\mathbf{s}}_0[n]$.

The iteration The estimate $\hat{\mathbf{s}}_0[n]$ of $\mathbf{s}[n]$ is determined recursively: If $\overline{K}_{n-1}$ and $\hat{\mathbf{s}}_0[n-1]$ are known, then $\overline{K}_n$ is determined from (14-141) and $\hat{\mathbf{s}}_0[n]$ from (14-140). To start the iteration, we must specify the initial conditions of

(14-135). We shall assume that

$$\mathbf{s}[0] = \boldsymbol{\zeta}[0]$$

This leads to the initial estimate

$$\hat{\mathbf{s}}_0[0] = \overline{K}_0 \mathbf{x}[0]$$

from which it follows that

$$\overline{K}_0 = \frac{E\{\mathbf{s}[0]\mathbf{x}[0]\}}{E\{\mathbf{x}^2[0]\}} = \frac{E\{\boldsymbol{\zeta}^2[0]\}}{E\{\boldsymbol{\zeta}^2[0]\} + E\{\boldsymbol{\nu}^2[0]\}}$$

Hence

$$\overline{K}_0 = \frac{V_0}{V_0 + N_0} \qquad P_0 = \frac{V_0 N_0}{V_0 + N_0} \tag{14-142}$$

Linearization Equation (14-138) and its equivalent (14-141) are nonlinear. However, each can be replaced by two linear equations. Indeed, if F_n and G_n are two sequences such that

$$\begin{aligned} F_n &= A_n^2 F_{n-1} + V_n G_{n-1} & F_0 &= V_0 N_0 \\ N_n G_n &= A_n^2 F_{n-1} + (V_n + N_n) G_{n-1} & G_0 &= V_0 + N_0 \end{aligned} \tag{14-143}$$

then

$$P_n = \frac{F_n}{G_n}$$

Example 14-10. We shall determine the noncausal, the causal, and the Kalman estimate of a process $\mathbf{s}[n]$ in terms of the data $\mathbf{x}[k] = \mathbf{s}[k] + \boldsymbol{\nu}[k]$, and the corresponding MS error P. We assume that the process $\mathbf{s}[n]$ satisfies the equation

$$\mathbf{s}[n] - 0.8\mathbf{s}[n-1] = \boldsymbol{\zeta}[n]$$

and that

$$R_{\zeta\zeta}[m] = 0.36\delta[m] \qquad R_{\zeta\nu}[m] = 0 \qquad R_{\nu\nu}[m] = \delta[n-m]$$

This is a special case of the process considered in Example 14-2 with

$$a = 0.8 \qquad N = 1 \qquad N_0 = 0.36 \qquad b = 0.5$$

Hence

$$\mathbf{S}_{ss}(z) = \frac{0.36}{(1 - 0.8z^{-1})(1 - 0.8z)} \qquad R_{ss}[m] = 0.8^{|m|}$$

$$\mathbf{S}_{xx}(z) = \mathbf{L}_x(z)\mathbf{L}_x(z^{-1}) \qquad \mathbf{L}_x(z) = \sqrt{1.6}\,\frac{z - 0.5}{z - 0.8}$$

(*a*) *Smoothing:* $\mathbf{x}[k]$ is available for all k. In this case the solution is obtained from Example 14-2 with $b = 0.5$ and $c = 0.375$:

$$h[n] = 0.3 \times 0.5^{|n|} \qquad P = 0.3$$

(*b*) *Causal filter:* $\mathbf{x}[k]$ is available for $k \leq n$. The unknown filter is determined from (14-114) where now $l_x[0] = \sqrt{1.6}$:

$$\mathbf{H}_x^0(z) = 1 - \frac{z - 0.8}{1.6(z - 0.5)} = \frac{0.375z}{z - 0.5} \qquad h[n] = 0.375 \times 0.5^n U[n]$$

This shows that the estimate $\hat{\mathbf{s}}[n]$ of $\mathbf{s}[n]$ satisfies the recursion equation

$$\hat{\mathbf{s}}[n] - 0.5\hat{\mathbf{s}}[n-1] = 0.375\mathbf{x}[n] \qquad n \geq 0$$

The resulting MS error equals

$$P = R_{ss}[0] - \sum_{k=0}^{\infty} R_{ss}[k]h[k] = 0.375$$

(c) *Kalman filter:* $\mathbf{x}[k]$ is available for $0 \leq k \leq n$. Our process is a special case of (14-135) with

$$A_n = 0.8 \qquad V_n = E\{\boldsymbol{\zeta}^2[n]\} = 0.36 \qquad N_n = E\{\boldsymbol{\nu}^2[n]\} = 1$$

Inserting into (14-143), we obtain

$$F_n = 0.64F_{n-1} + 0.36G_{n-1} \qquad F_0 = 0.36$$
$$G_n = 0.64F_{n-1} + 1.36G_{n-1} \qquad G_0 = 1.36$$

This is a system of linear recursion equations and can be readily solved with z transforms. Since

$$\overline{K}_n = \frac{P_n}{N} = \frac{F_n}{G_n}$$

and $N = 1$, the solution yields

$$\overline{K}_n = P_n = \frac{0.48z_1^n - 0.12z_2^n}{1.28z_1^n + 0.08z_2^n} \qquad \begin{matrix} z_1 = 1.6 \\ z_2 = 0.4 \end{matrix}$$

In particular,

$n =$	0	1	2	3	4
$P_n \simeq$	0.3	0.357	0.371	0.374	0.375

Thus, although the number of the available data increases as n increases, the MS error P_n also increases. The reason is that $\mathbf{s}[n]$ is a nonstationary process with initial second moment $V_0 = 0.36$ because $\mathbf{s}[0] = \boldsymbol{\zeta}[0]$, and, as n increases, $E\{\mathbf{s}^2[n]\}$ approaches the value 1.

We note, finally, that

$$\overline{K}_n = P_n \xrightarrow[n \to \infty]{} \frac{0.48}{1.28} = 0.375$$

and (14-140) yields

$$\hat{\mathbf{s}}_0[n] - 0.8\hat{\mathbf{s}}_0[n-1] = 0.375\mathbf{x}[n] - 0.3\hat{\mathbf{s}}_0[n-1]$$

The above shows that, if the process $\mathbf{s}[n]$ is WSS, then its Kalman filter approaches the causal Wiener filter as $n \to \infty$. This is the case for any P_0 because the limit of $F_n G_n$ as $n \to \infty$ equals 0.375 regardless of the initial conditions.

Example 14-11. We wish to estimate the RV $\mathbf{s}$ in terms of the sum

$$\mathbf{x}[n] = \mathbf{s} + \boldsymbol{\nu}[n] \qquad \text{where} \quad E\{\mathbf{s}\boldsymbol{\nu}[n]\} = 0 \qquad R_{\nu\nu}[m,n] = N\delta[m-n]$$

The estimate $\hat{\mathbf{s}}_0[n]$ in terms of the data $\mathbf{x}[n]$ can be obtained as the output of a Kalman filter if we consider the RV $\mathbf{s}$ as a stochastic process satisfying trivially (14-135)

$$\mathbf{s}[n] = \mathbf{s}[n-1] + \boldsymbol{\zeta}[n] \qquad \mathbf{s}[-1] = 0$$

$$\boldsymbol{\zeta}[n] = \begin{cases} \mathbf{s} & n = 0 \\ 0 & n > 0 \end{cases} \qquad V_n = \begin{cases} E\{\mathbf{s}^2\} = M & n = 0 \\ 0 & n > 0 \end{cases}$$

In this case, $A_n = 1$, $N_n = N$, and (14-143) yields

$$F_n = F_{n-1} \qquad NG_n = F_{n-1} + NG_{n-1} \qquad F_0 = MN \qquad G_0 = M + N$$

Solving, we obtain

$$F_n = MN \qquad G_n = M + N + Mn$$

Hence

$$\hat{\mathbf{s}}_0[n] = \frac{N + Mn}{M + N + Mn}\hat{\mathbf{s}}_0[n-1] + \frac{M}{M + N + Mn}\mathbf{x}[n]$$

Continuous-Time Processes

We wish, finally, to determine the estimate

$$\hat{\mathbf{s}}_0(t) = \hat{E}\{\mathbf{s}(t)|\mathbf{x}(\tau), 0 \le \tau \le t\} \tag{14-144}$$

of a continuous-time process $\mathbf{s}(t)$ in terms of the data

$$\mathbf{x}(t) = \mathbf{s}(t) + \boldsymbol{\nu}(t) \tag{14-145}$$

The solution of this problem parallels the discrete-time solution if recursion equations are replaced by differential equations and sums by integrals. It might be instructive, however, to rederive the principal results using a different approach.

To avoid repetition, we start directly with the white-noise assumption

$$R_{\nu\nu}(t,\tau) = N(\tau)\delta(t-\tau) \qquad N(\tau) > 0 \qquad R_{s\nu}(t,\tau) = 0 \tag{14-146}$$

and we show that the process

$$\mathbf{w}(t) = \mathbf{x}(t) - \hat{\mathbf{s}}_0(t)$$

is white noise with autocorrelation

$$R_{ww}(t,\tau) = N(\tau)\delta(t-\tau) \tag{14-147}$$

Proof. As we know

$$\boldsymbol{\varepsilon}(t) = \mathbf{s}(t) - \hat{\mathbf{s}}_0(t) \perp \mathbf{x}(\tau) \qquad \boldsymbol{\nu}(t) \perp \mathbf{x}(\tau)$$

for $\tau < t$. Furthermore, $\mathbf{w}(\tau)$ depends linearly on $\mathbf{x}(\tau)$ and its past. Hence

$$\mathbf{w}(t) = \boldsymbol{\varepsilon}(t) + \boldsymbol{\nu}(t) \perp \mathbf{w}(\tau) \qquad \tau < t \tag{14-148}$$

To complete the proof of (14-147), we shall assume that $\mathbf{s}_0(t)$ is continuous from the left

$$\hat{\mathbf{s}}_0(t^-) = \hat{\mathbf{s}}_0(t)$$

This is not true at the origin if $\mathbf{s}(0) \neq 0$. However, for sufficiently large t, the effect of the initial condition can be neglected. From the above it follows that

$$P(t) = E\{\boldsymbol{\varepsilon}^2(t)\} < \infty$$

and since $\boldsymbol{\varepsilon}(t) \perp \boldsymbol{\nu}(\tau)$ for $\tau > t$, we conclude that

$$R_{ww}(t,\tau) = R_{\nu w}(t,\tau) = R_{\nu\nu}(t,\tau) = N(\tau)\delta(t-\tau)$$

Using a limit argument, we can show that, as in the discrete-time case, the normalized process $\mathbf{w}(t)/\sqrt{N(t)}$ is the Kalman innovations of $\mathbf{x}(t)$. The details, however, will be omitted. This leads to the conclusion that $\hat{\mathbf{s}}_0(t)$ can be expressed in terms of $\mathbf{w}(t)$ [see also (14-123)]

$$\hat{\mathbf{s}}_0(t) = \int_0^t h_w(t,\alpha)\mathbf{w}(\alpha)\,d\alpha \tag{14-149}$$

Since $\mathbf{s}(t) - \hat{\mathbf{s}}_0(t) \perp \mathbf{w}(\tau)$ for $\tau \le t$, we conclude from the above and (14-147) that

$$R_{sw}(t,\tau) = \int_0^t h_w(t,\alpha)N(\alpha)\delta(\tau-\alpha)\,d\alpha = h_w(t,\tau)N(\tau)$$

and (14-149) yields

$$\hat{\mathbf{s}}_0(t) = \int_0^t \frac{1}{N(\alpha)} R_{sw}(t,\alpha)\mathbf{w}(\alpha)\,d\alpha \tag{14-150}$$

We note that [see (14-148) and (14-146)]

$$R_{sw}(t,t) = E\{\mathbf{s}(t)[\boldsymbol{\varepsilon}(t) + \boldsymbol{\nu}(t)]\} = P(t)$$

WIDE-SENSE MARKOFF PROCESSES. Using the above, we shall show that, if the signal $\mathbf{s}(t)$ is WS Markoff, that is, if it satisfies a differential equation driven by white noise, then its estimate $\hat{\mathbf{s}}_0(t)$ satisfies a similar equation. For simplicity, we consider the first-order case

$$\mathbf{s}'(t) + A(t)\mathbf{s}(t) = \boldsymbol{\zeta}(t) \qquad R_{\zeta\zeta}(t,\tau) = V(\tau)\delta(t-\tau) \tag{14-151}$$

The Kalman–Bucy equations†. We maintain that

$$\hat{\mathbf{s}}_0'(t) + A(t)\hat{\mathbf{s}}_0(t) = K(t)[\mathbf{x}(t) - \hat{\mathbf{s}}_0(t)] \tag{14-152}$$

†R. E. Kalman and R. C. Bucy: "New Results in Linear Filtering and Prediction Theory," *ASME Transactions*, vol. 83D, 1961.

where

$$K(t) = \frac{P(t)}{N(t)} \tag{14-153}$$

Furthermore, the MS error $P(t)$ satisfies the *Riccati equation*

$$P'(t) + 2A(t)P(t) = V(t) - \frac{1}{N(t)}P^2(t) \tag{14-154}$$

Proof. Multiplying the differential equation in (14-151) by $\mathbf{w}(\tau)$, we obtain

$$\frac{\partial}{\partial t}R_{sw}(t,\tau) + A(t)R_{sw}(t,\tau) = 0 \qquad \tau < t \tag{14-155}$$

We next equate the derivatives of both sides of (14-150)

$$\hat{\mathbf{s}}_0'(t) = \frac{1}{N(t)}R_{sw}(t,t)\mathbf{w}(t) + \int_0^t \frac{1}{N(\alpha)}\frac{\partial}{\partial t}R_{sw}(t,\alpha)\,d\alpha$$

Finally, we multiply (14-150) by $A(t)$ and add with the above. This yields (15-152) because, as we see from (14-155), the sum of the two integrals is 0.

To prove (14-154), we use the following version of (10-90): If $\mathbf{z}(t)$ is a process with $E\{\mathbf{z}^2(t)\} = I(t)$ and such that

$$\mathbf{z}'(t) + B(t)\mathbf{z}(t) = \boldsymbol{\xi}(t) \qquad R_{\xi\xi}(t,\tau) = Q(\tau)\delta(t-\tau) \tag{14-156}$$

then (see Prob. 10-28*b*)

$$I'(t) + 2B(t)I(t) = Q(t) \tag{14-157}$$

Returning to (14-152), we observe, subtracting from (14-151), that the estimation error $\boldsymbol{\varepsilon}(t)$ satisfies the equation

$$\boldsymbol{\varepsilon}'(t) + [A(t) + K(t)]\boldsymbol{\varepsilon}(t) = \boldsymbol{\zeta}(t) - K(t)\boldsymbol{\nu}(t)$$

In the above, the right side $\boldsymbol{\xi}(t) = \boldsymbol{\zeta}(t) - K(t)\boldsymbol{\nu}(t)$ is white noise as in (14-156) with

$$Q(\tau) = V(\tau) + K^2(\tau)N(\tau)$$

Hence the function $P(t) = E\{\boldsymbol{\varepsilon}^2(t)\}$ satisfies (14-157) where $B(t) = A(t) + K(t)$. This yields

$$P'(t) + 2[A(t) + K(t)]P(t) = V(t) + K^2(t)N(t)$$

and (14-154) results.

Linearization We shall now show that the nonlinear equation (14-154) is equivalent to two linear equations. For this purpose, we introduce the functions $F(t)$ and $G(t)$ such that

$$P(t) = \frac{F(t)}{G(t)} \tag{14-158}$$

Clearly,

$$F'(t) = P'(t)G(t) + P(t)G'(t)$$

and (14-154) yields

$$F'(t) + A(t)F(t) - V(t)G(t) = P(t)\left[G'(t) - A(t)G(t) - \frac{F(t)}{N(t)}\right]$$

This is satisfied if

$$\begin{aligned} F'(t) &= -A(t)F(t) + V(t)G(t) \\ G'(t) &= \frac{F(t)}{N(t)} + A(t)G(t) \end{aligned} \tag{14-159}$$

To solve the above system, we must specify $F(0)$ and $G(0)$. Setting arbitrarily $G(0) = 1$, we obtain $F(0) = P(0)$ where

$$P(0) = E\{\mathbf{s}^2(0)\}$$

is the initial value of the MS error $P(t)$. The determination of the Kalman filter thus depends on the second moment of $\mathbf{s}(0)$.

Example 14-12. We shall determine the noncausal, the causal, and the Kalman estimate of a process $\mathbf{s}(t)$ in terms of the data $\mathbf{x}(t) = \mathbf{s}(t) + \boldsymbol{\nu}(t)$, and the corresponding MS error P. We assume that $\mathbf{s}(t)$ satisfies the equation

$$\mathbf{s}'(t) + 2\mathbf{s}(t) = \boldsymbol{\zeta}(t)$$

and that

$$R_{\zeta\zeta}(\tau) = 12\delta(\tau) \qquad R_{s\nu}(\tau) = 0 \qquad R_{\nu\nu}(\tau) = \delta(\tau)$$

This is a special case of the process considered in Example 14-7 with

$$\alpha = 2 \qquad N = 1 \qquad N_0 = 12 \qquad \beta = 4$$

Hence

$$S_{ss}(\omega) = \frac{12}{4 + \omega^2} \qquad R_{ss}(\tau) = 3e^{-2|\tau|}$$

$$S_{xx}(\omega) = \frac{16 + \omega^2}{4 + \omega^2} \qquad \mathbf{L}_x(s) = \frac{s + 4}{s + 2}$$

(*a*) *Smoothing:* $\mathbf{x}(\xi)$ is available for all ξ. In this case, (14-16) yields

$$H(\omega) = \frac{12}{16 + \omega^2} \qquad h(t) = \frac{3}{2}e^{-4|\tau|}$$

The MS error is obtained from (14-15)

$$P = 3 - \frac{9}{2}\int_{-\infty}^{\infty} e^{-4|\tau|}e^{-2|\tau|}\, d\tau = 1.5$$

(*b*) *Causal filter:* $\mathbf{x}(\xi)$ is available for $\xi \le t$. The unknown filter is specified in Example 14-8 with

$$\alpha = 2 \qquad \beta = 4 \qquad N = 12$$

Thus

$$\mathbf{H}_x(s) = \frac{2}{s+4} \qquad h_x(t) = 2e^{-4t}U(t) \qquad P = 2$$

This shows that the estimate $\hat{\mathbf{s}}(t)$ of $\mathbf{s}(t)$ satisfies the differential equation

$$\hat{\mathbf{s}}'(t) + 4\hat{\mathbf{s}}(t) = 2\mathbf{x}(t)$$

(*c*) *Kalman filter:* $\mathbf{x}(\xi)$ is available for $0 \le \xi \le t$. Our problem is a special case of (14-151) with

$$A(t) = 2 \qquad V(t) = 12 \qquad N(t) = 1$$

Hence [see (14-159)]

$$F'(t) = -2F(t) + 12G(t) \qquad G'(t) = F(t) + 2G(t)$$

To solve this system, we must know $P(0)$.

Case 1 If $\mathbf{s}(0) = 0$, then $P(0) = 0$. In this case, $F(0) = 0$, $G(0) = 1$. Inserting the solution of the above system into (14-153), we obtain

$$K(t) = P(t) = \frac{6e^{4t} - 6e^{-4t}}{3e^{4t} + e^{-4t}} \xrightarrow[t\to\infty]{} 2$$

Case 2 We now assume that $\mathbf{s}(t)$ is the stationary solution of the differential equation specifying $\mathbf{s}(t)$. In this case, $E\{\mathbf{s}^2(0)\} = 3$; hence $P(0) = F(0) = 3$ and

$$K(t) = P(t) = \frac{18e^{4t} + 6e^{-4t}}{9e^{4t} - e^{-4t}} \xrightarrow[t\to\infty]{} 2$$

Thus, in both cases, the solution $\hat{\mathbf{s}}_0(t)$ of the Kalman–Bucy equation (14-152) tends to the solution of the causal Wiener filter

$$\hat{\mathbf{s}}_0'(t) + 2\hat{\mathbf{s}}_0(t) = 2\mathbf{x}(t) - 2\hat{\mathbf{s}}_0(t)$$

as $t \to \infty$.

Example 14-13. We wish to estimate the RV $\mathbf{s}$ in terms of the sum

$$\mathbf{x}(t) = \mathbf{s} + \boldsymbol{\nu}(t) \qquad E\{\mathbf{s}\boldsymbol{\nu}(t)\} = 0 \qquad R_{\nu\nu}(\tau) = N\delta(\tau)$$

This is a special case of (14-151) if

$$A(t) = 0 \qquad \mathbf{s}(t) = \mathbf{s} \qquad \zeta(t) = 0 \qquad N(t) = N$$

In this case, $V(t) = 0$, $P(0) = E\{\mathbf{s}^2\} \equiv M$, and (14-159) yields

$$F'(t) = 0 \qquad G'(t) = \frac{F(t)}{N} \qquad F(0) = M \qquad G(0) = 1$$

Hence

$$F(t) = M \qquad G(t) = 1 + \frac{Mt}{N}$$

Inserting into (14-152), we obtain

$$\hat{\mathbf{s}}_0'(t) + \frac{M}{N + Mt}\hat{\mathbf{s}}_0(t) = \frac{M}{N + Mt}\mathbf{x}(t)$$

PROBLEMS

14-1. If $R_s(\tau) = Ie^{-|\tau|/T}$ and

$$\hat{E}\{\mathbf{s}(t-T/2)|\mathbf{s}(t), \mathbf{s}(t-T)\} = a\mathbf{s}(t) + b\mathbf{s}(t-T)$$

find the constants a, b and the MS error.

14-2. Show that if $\hat{\mathbf{z}} = a\mathbf{s}(0) + b\mathbf{s}(T)$ is the MS estimate of

$$\mathbf{z} = \int_0^T \mathbf{s}(t)\,dt \qquad \text{then} \quad a = b = \frac{\int_0^T R_s(\tau)\,d\tau}{R_s(0) + R_s(T)}$$

14-3. Show that if $\mathbf{x}(t) = \mathbf{s}(t) + \boldsymbol{\nu}(t)$, $R_{s\nu}(\tau) = 0$ and

$$\hat{E}\{\mathbf{s}'(t)|\mathbf{x}(t), \mathbf{x}(t-\tau)\} = a\mathbf{x}(t) + b\mathbf{x}(t-\tau)$$

then for small τ, $a = -b \simeq R''_{ss}(0)/\tau R''_{xx}(0)$.

14-4. Show that, if $S_x(\omega) = 0$ for $|\omega| > \sigma = \pi/T$, then the linear MS estimate of $\mathbf{x}(t)$ in terms of its samples $\mathbf{x}(nT)$ equals

$$\hat{E}\{\mathbf{x}(t)|\mathbf{x}(nT), n = -\infty, \dots, \infty\} = \sum_{n=-\infty}^{\infty} \frac{\sin(\sigma t - nT)}{\sigma t - n\pi}\mathbf{x}(nT)$$

and the MS error equals 0.

14-5. Show that if

$$\hat{E}\{\mathbf{s}(t+\lambda)|\mathbf{s}(t), \mathbf{s}(t-\tau)\} = \hat{E}\{\mathbf{s}(t+\lambda)|\mathbf{s}(t)\}$$

then $R_s(\tau) = Ie^{-\alpha|\tau|}$.

14-6. A random sequence $\mathbf{x}_n$ is called a *martingale* if $E\{\mathbf{x}_n = 0\}$ and

$$E\{\mathbf{x}_n|\mathbf{x}_{n-1}, \dots, \mathbf{x}_1\} = \mathbf{x}_{n-1}$$

Show that if the RVs $\mathbf{y}_n$ are *independent*, then their sum $\mathbf{x}_n = \mathbf{y}_1 + \cdots + \mathbf{y}_n$ is a martingale.

14-7. A random sequence $\mathbf{x}_n$ is called *wide-sense martingale* if

$$\hat{E}\{\mathbf{x}_n|\mathbf{x}_{n-1}, \dots, \mathbf{x}_1\} = \mathbf{x}_{n-1}$$

(*a*) Show that a sequence $\mathbf{x}_n$ is WS martingale if it can be written as a sum $\mathbf{x}_n = \mathbf{y}_1 + \cdots + \mathbf{y}_n$ where the RVs $\mathbf{y}_n$ are *orthogonal*.

(*b*) Show that if the sequence $\mathbf{x}_n$ is WS martingale, then

$$E\{\mathbf{x}_n^2\} \ge E\{\mathbf{x}_{n-1}^2\} \ge \cdots \ge E\{\mathbf{x}_1^2\}$$

Hint: $\mathbf{x}_n = \mathbf{x}_n - \mathbf{x}_{n-1} + \mathbf{x}_{n-1}$ and $\mathbf{x}_n - \mathbf{x}_{n-1} \perp \mathbf{x}_{n-1}$.

14-8. Find the noncausal estimators $H_1(\omega)$ and $H_2(\omega)$ respectively of a process $\mathbf{s}(t)$ and its derivative $\mathbf{s}'(t)$ in terms of the data $\mathbf{x}(t) = \mathbf{s}(t) + \boldsymbol{\nu}(t)$ where

$$R_s(\tau) = A\frac{\sin^2 a\tau}{\tau^2} \qquad R_\nu(\tau) = N\delta(\tau) \qquad R_{s\nu}(\tau) = 0$$

14-9. We denote by $H_s(\omega)$ and $H_y(\omega)$ respectively the noncausal estimators of the input $\mathbf{s}(t)$ and the output $\mathbf{y}(t)$ of the system $T(\omega)$ in terms of the data $\mathbf{x}(t)$ (Fig. P14-9). Show that $H_y(\omega) = H_s(\omega)T(\omega)$.

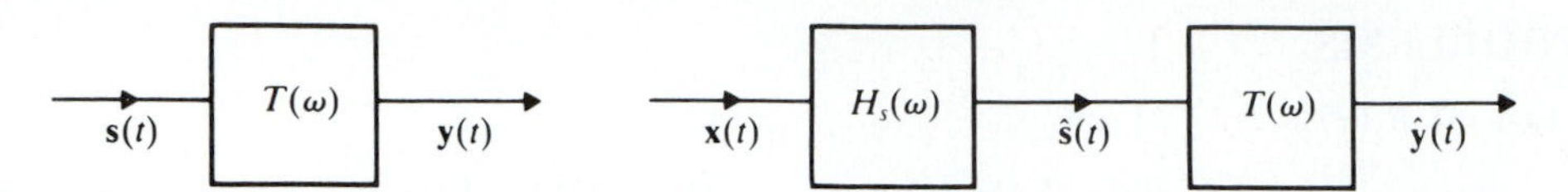

FIGURE P14-9

14-10. Show that if $S(\omega) = 1/(1+\omega^4)$, then the predictor of $\mathbf{s}(t)$ in terms of its entire past equals $\hat{\mathbf{s}}(t+\lambda) = b_0\mathbf{s}(t) + b_1\mathbf{s}'(t)$ where

$$b_0 = e^{-\lambda/\sqrt{2}}\left(\cos\frac{\lambda}{\sqrt{2}} + \sin\frac{\lambda}{\sqrt{2}}\right) \qquad b_1 = \sqrt{2}e^{-\lambda/\sqrt{2}}\sin\frac{\lambda}{\sqrt{2}}$$

14-11. (*a*) Find a function $h(t)$ satisfying the integral equation (Wiener–Hopf)

$$\int_0^\infty h(\alpha)R(\tau-\alpha)\,d\alpha = R(\tau+\ln 2) \qquad t \ge 0 \qquad R(\tau) = \tfrac{3}{2}e^{-\tau} + \tfrac{11}{3}e^{-3\tau}$$

(*b*) The function $\mathbf{H}(s)$ is rational with poles in the left-hand plane. The function $\mathbf{Y}(s)$ is analytic in the left-hand plane. Find $\mathbf{H}(s)$ and $\mathbf{Y}(s)$ if

$$[\mathbf{H}(s) - 2^s]\frac{49-25s^2}{9-10s^2+s^4} = \mathbf{Y}(s)$$

(*c*) Discuss the relationship between (*a*) and (*b*).

14-12. (*a*) Find a sequence h_n satisfying the system

$$\sum_{k=0}^{\infty} h_k R_{m-k} = R_{m+1} \qquad m \ge 0 \qquad R_m = \frac{1}{2^m} + \frac{1}{3^m}$$

(*b*) The function $\mathbf{H}(z)$ is rational with poles in the unit circle. The function $\mathbf{Y}(z)$ is rational with poles outside the unit circle. Find $\mathbf{H}(z)$ and $\mathbf{Y}(z)$ if

$$[\mathbf{H}(z) - z]\frac{70-25(z+z^{-1})}{6(z+z^{-1})^2 - 35(z+z^{-1}) + 50} = \mathbf{Y}(z)$$

(*c*) Discuss the relationship between (*a*) and (*b*).

14-13. Show that if $\mathbf{H}(z)$ is a predictor of a process $\mathbf{s}[n]$ and $\mathbf{H}_a(z)$ is an all-pass function such that $|\mathbf{H}_a(e^{j\omega})| = 1$, then the function $1 - (1 - \mathbf{H}(z))\mathbf{H}_a(z)$ is also a predictor with the same MS error P.

14-14. We have shown that the one-step predictor $\hat{\mathbf{s}}_1[n]$ of an AR process of order m in terms of its entire past equals [see (14-35)]

$$\hat{E}\{\mathbf{s}[n]|\mathbf{s}[n-k], k \ge 1\} = -\sum_{k=1}^{m} a_k\mathbf{s}[n-k]$$

Show that its two-step predictor $\hat{\mathbf{s}}_2[n]$ is given by

$$\hat{E}\{\mathbf{s}[n]|\mathbf{s}[n-k], k \ge 2\} = -a_1\hat{\mathbf{s}}_1[n-1] - \sum_{k=2}^{m} a_k\mathbf{s}[n-k]$$

14-15. Using (14-70) show that

$$\lim_{N\to\infty} \ln\frac{\Delta_{N+1}}{\Delta_N} = \lim_{N\to\infty} \frac{\ln \Delta_N}{N} = \frac{1}{2\pi}\int_{-\pi}^{\pi} \ln S_s(\omega)\, d\omega$$

Hint:

$$\frac{1}{N}\sum_{n=1}^{N} \ln\frac{\Delta_{n+1}}{\Delta_n} = \frac{1}{N}\ln \Delta_{N+1} - \frac{1}{N}\ln \Delta_1 \to \lim_{N\to\infty} \ln\frac{\Delta_{N+1}}{\Delta_N}$$

14-16. Find the predictor

$$\hat{\mathbf{s}}_N[n] = \hat{E}\{\mathbf{s}[n] | \mathbf{s}[n-k], 1 \le k \le N]\}$$

of a process $\mathbf{s}[n]$ and realize the error filter $\mathbf{E}_N(z)$ as an FIR filter (Fig. 14-8) and as a lattice filter (Fig. 13-15) for $N = 1$, 2, and 3 if

$$R_s[m] = \begin{cases} 5(3 - |m|) & |m| < 3 \\ 0 & |m| \ge 3 \end{cases}$$

14-17. The lattice filter of a process $\mathbf{s}[n]$ is shown in Fig. P14-17 for $N = 3$. Find the corresponding FIR filter for $N = 1$, 2, and 3 and the values of $R[m]$ for $|m| \le 3$ if $R[0] = 5$.

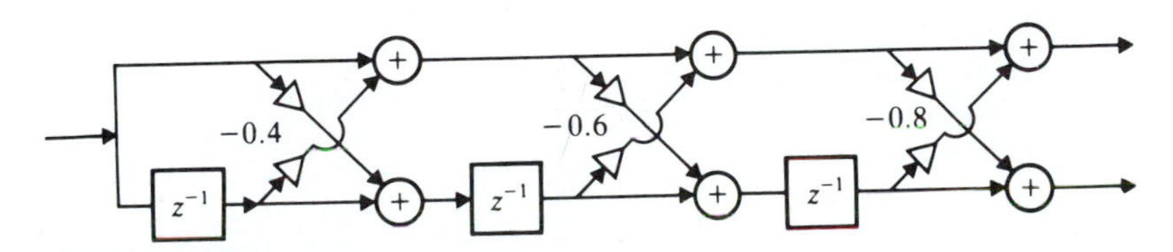

FIGURE P14-17

14-18. We wish to find the estimate $\hat{\mathbf{s}}(t)$ of the random telegraph signal $\mathbf{s}(t)$ in terms of the sum $\mathbf{x}(t) = \mathbf{s}(t) + \boldsymbol{\nu}(t)$ and its past, where

$$R_s(\tau) = e^{-2\lambda|\tau|} \qquad R_\nu(\tau) = N\delta(\tau) \qquad R_{s\nu}(\tau) = 0$$

Show that

$$\hat{\mathbf{s}}(t) = (c - 2\lambda)\int_0^\infty \mathbf{x}(t-\alpha)e^{-c\alpha}\, d\alpha \qquad c = 2\lambda\sqrt{1 + \frac{1}{\lambda N}}$$

14-19. Show that if $\hat{\boldsymbol{\varepsilon}}_N[n]$ is the forward prediction error and $\check{\boldsymbol{\varepsilon}}_N[n]$ is the backward prediction error of a process $\mathbf{s}[n]$, then (*a*) $\hat{\boldsymbol{\varepsilon}}_N[n] \perp \hat{\boldsymbol{\varepsilon}}_{N+m}[n+m]$, (*b*) $\check{\boldsymbol{\varepsilon}}_N[n] \perp \check{\boldsymbol{\varepsilon}}_{N+m}[n-m]$, (*c*) $\hat{\boldsymbol{\varepsilon}}_N[n] \perp \check{\boldsymbol{\varepsilon}}_{N+m}[n-N-m]$.

14-20. If $\mathbf{x}(t) = \mathbf{s}(t) + \boldsymbol{\nu}(t)$ and

$$R_s(\tau) = 5e^{-0.2|\tau|} \qquad R_\nu(\tau) = 5\delta(\tau) \qquad R_{s\nu}(\tau) = 0$$

find the following MS estimates and the corresponding MS errors: (*a*) the noncausal filter of $\mathbf{s}(t)$; (*b*) the causal filter of $\mathbf{s}(t)$; (*c*) the estimate of $\mathbf{s}(t+2)$ in terms of $\mathbf{s}(t)$ and its past; (*d*) the estimate of $\mathbf{s}(t+2)$ in terms of $\mathbf{x}(t)$ and its past.

14-21. If $\mathbf{x}[n] = \mathbf{s}[n] + \boldsymbol{\nu}[n]$:

$$R_s[m] = 5 \times 0.8^{|m|} \qquad R_\nu[m] = 5\delta[m] \qquad R_{s\nu}[m] = 0$$

Find the following MS estimates and the corresponding MS errors: (*a*) the noncausal filter of $\mathbf{s}[n]$; (*b*) the causal filter of $\mathbf{s}[n]$; (*c*) the estimate of $\mathbf{s}[n+1]$ in terms of $\mathbf{s}[n]$ and its past; (*d*) the estimate of $\mathbf{s}[n+1]$ in terms of $\mathbf{x}[n]$ and its past.

14-22. Find the Kalman estimate

$$\hat{\mathbf{s}}_0[n] = E\{\mathbf{s}[n] \mid \mathbf{s}[k] + \boldsymbol{\nu}[k], 0 \le k \le n\}$$

of $\mathbf{s}[n]$ and the MS error $P_n = E\{(\mathbf{s}[n] - \hat{\mathbf{s}}_0[n])^2\}$ if

$$R_s[m] = 5 \times 0.8^{|m|} \qquad R_\nu[m] = 5\delta[m] \qquad R_{s\nu}[m] = 0$$

14-23. Find the Kalman estimate

$$\hat{\mathbf{s}}_0(t) = E\{\mathbf{s}(t) \mid \mathbf{s}(\tau) + \boldsymbol{\nu}(\tau), 0 \le \tau \le t\}$$

of $\mathbf{s}(t)$ and the MS error $P(t) = E\{[\mathbf{s}(t) - \hat{\mathbf{s}}_0(t)]^2\}$ if

$$R_s(\tau) = 5e^{-0.2|\tau|} \qquad R_\nu(\tau) = \tfrac{10}{3}\delta(\tau) \qquad R_{s\nu}(\tau) = 0$$

14-24. Show that the sequences $\hat{\mathbf{q}}_N[m]$ and $\check{\mathbf{q}}_N[m]$ of the inverse lattice of Fig. 14-11*b* satisfy (14-85) and (14-86) (see Note 1 page 496).

CHAPTER 15

ENTROPY

15-1 INTRODUCTION

As we have noted in Chap. 1, the probability $P(\mathscr{A})$ of an event $\mathscr{A}$ can be interpreted as a measure of our uncertainty about the occurrence or nonoccurrence of $\mathscr{A}$ in a single performance of the underlying experiment $\mathscr{S}$. If $P(\mathscr{A}) \simeq 0.999$, then we are almost certain that $\mathscr{A}$ will occur; if $P(\mathscr{A}) = 0.1$, then we are reasonably certain that $\mathscr{A}$ will not occur; our uncertainty is maximum if $P(\mathscr{A}) = 0.5$. In this chapter, we consider the problem of assigning a measure of uncertainty to the occurrence or nonoccurrence not of a single event of $\mathscr{S}$, but of any event $\mathscr{A}_i$ of a partition $\mathfrak{A}$ of $\mathscr{S}$ where, as we recall, a partition is a collection of mutually exclusive events whose union equals $\mathscr{S}$ (Fig. 15-1). The measure of uncertainty about $\mathfrak{A}$ will be denoted by $H(\mathfrak{A})$ and will be called the *entropy of the partitioning* $\mathfrak{A}$.

Historically, the functional $H(\mathfrak{A})$ was derived from a number of postulates based on our heuristic understanding of uncertainty. The following is a typical set of such postulates†:

1. $H(\mathfrak{A})$ is a continuous function of $p_i = P(\mathscr{A}_i)$.

2. If $p_1 = \cdots = p_N = 1/N$, then $H(\mathfrak{A})$ is an increasing function of N.

3. If a new partition $\mathfrak{B}$ is formed by subdividing one of the sets of $\mathfrak{A}$, then $H(\mathfrak{B}) \geq H(\mathfrak{A})$.

†C. E. Shannon and W. Weaver: *The Mathematical Theory of Communication*, University of Illinois Press, 1949.

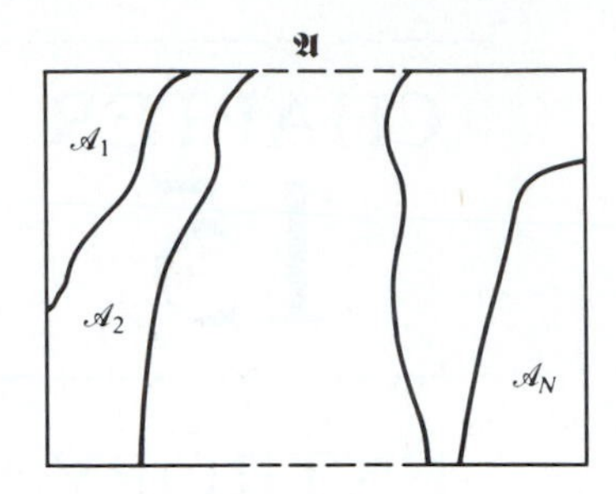

FIGURE 15-1

It can be shown that the sum

$$H(\mathfrak{A}) = -p_1 \log p_1 - \cdots - p_N \log p_N \tag{15-1}$$†

satisfies these postulates and it is unique within a constant factor. The proof of this assertion is not difficult but we choose not to reproduce it. We propose, instead, to introduce (15-1) as the *definition* of entropy and to develop axiomatically all its properties within the framework of probability. It is true that the introduction of entropy in terms of postulates establishes a link between the sum in (15-1) and our heuristic understanding of uncertainty. However, for our purposes, this is only incidental. In the last analysis, the justification of the concept must ultimately rely on the usefulness of the resulting theory.

The applications of entropy can be divided into two categories. The first deals with problems involving the determination of unknown distributions (Sec. 15-4). The available information is in the form of known expected values or other statistical functionals, and the solution is based on the principle of maximum entropy: We determine the unknown distributions so as to *maximize* the entropy $H(\mathfrak{A})$ of some partition $\mathfrak{A}$ subject to the given constraints (statistical mechanics). In the second category (coding theory), we are given $H(\mathfrak{A})$ (source entropy) and we wish to construct various random variables (code lengths) so as to *minimize* their expected values (Sec. 15-5). The solution involves the construction of optimum mappings (codes) of the random variables under consideration, into the given probability space.

Uncertainty and information In the heuristic interpretation of entropy, the number $H(\mathfrak{A})$ is a measure of our uncertainty about the events $\mathscr{A}_i$ of the partition $\mathfrak{A}$ prior to the performance of the underlying experiment. If the experiment is performed and the results concerning $\mathscr{A}_i$ become known, the uncertainty is removed. We can thus say that the experiment provides *information* about the events $\mathscr{A}_i$ equal to the *entropy* of their partition. Thus uncertainty equals information and both are measured by the sum in (15-1).

---

†We shall use as logarithmic base either the number 2 or the number e. In the first case, the unit of entropy is the *bit*.

Example 15-1. (*a*) We shall determine the entropy of the partition $\mathfrak{A}$ = [even, odd] in the fair-die experiment. Clearly, $P\{\text{even}\} = P\{\text{odd}\} = 1/2$. Hence

$$H(\mathfrak{A}) = -\tfrac{1}{2}\log\tfrac{1}{2} - \tfrac{1}{2}\log\tfrac{1}{2} = \log 2$$

(*b*) In the same experiment, $\mathfrak{S}$ is the partition consisting of the elementary events $\{f_i\}$. In this case, $P\{f_i\} = 1/6$; hence

$$H(\mathfrak{S}) = -\tfrac{1}{6}\log\tfrac{1}{6} - \cdots - \tfrac{1}{6}\log\tfrac{1}{6} = \log 6$$

If the die is rolled and we are told which face showed, then we gain information about the partition $\mathfrak{S}$ equal to its entropy $\log 6$. If we are told merely that "even" or "odd" showed, then we gain information about the partition $\mathfrak{A}$ equal to its entropy $\log 2$. In this case, the information gained about the partition $\mathfrak{S}$ equals again $\log 2$. As we shall see, the difference $\log 6 - \log 2 = \log 3$ is the uncertainty about $\mathfrak{S}$ assuming $\mathfrak{A}$ (conditional entropy).

Example 15-2. We consider now the coin experiment where $P\{h\} = p$. In this case, the entropy of $\mathfrak{S}$ equals

$$H(\mathfrak{S}) = -p\log p - (1-p)\log(-p) \equiv r(p) \qquad (15\text{-}2)$$

The function $r(p)$ is shown in Fig. 15-2 for $0 \le p \le 1$. This function is symmetrical, convex, even about the point $p = 0.5$, and it reaches its maximum at that point. Furthermore, $r(0) = r(1) = 0$.

Historical note The term *entropy* as a scientific concept was first used in thermodynamics (Clausius 1850). Its probabilistic interpretation in the context of statistical mechanics is attributed to Boltzmann (1877). However, the explicit relationship between entropy and probability was recorded several years later (Planck, 1906). Shannon, in his celebrated paper (1948), used the concept to give an economical description of the properties of long sequences of symbols, and applied the results to a number of basic problems in coding theory and data transmission. His remarkable contributions form the basis of modern information theory. Jaynes† (1957) reexamined the method of maximum entropy and applied it to a variety of problems involving the determination of unknown parameters from incomplete data.

Maximum entropy and classical definition. An important application of entropy is the determination of the probabilities p_i of the events of a partition $\mathfrak{A}$, subject to various constraints, with the method of maximum entropy (MEM). The method states that the unknown p_i's must be so chosen as to maximize the entropy of $\mathfrak{A}$ subject to the given constraints. This topic is considered in Sec. 15-4. In the following we introduce the main idea and we show the equivalence between the MEM and the classical definition of probability (principle of insufficient reason), using as illustration the die experiment.

†E. T. Jaynes: *Physical Review*, vols. 106–107, 1957.

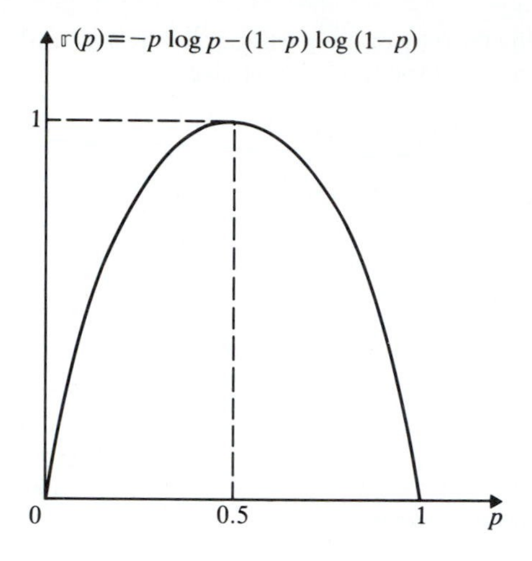

FIGURE 15-2

Example 15-3. (*a*) We wish to determine the probabilities p_i of the six faces of a die, having access to no prior information. The MEM states that the p_i's must be such as to maximize the sum

$$H(\mathfrak{S}) = -p_1 \log p_1 - \cdots - p_6 \log p_6$$

Since $p_1 + \cdots + p_6 = 1$, this yields

$$p_1 = \cdots = p_6 = \tfrac{1}{6}$$

in agreement with the classical definition.

(*b*) Suppose now that we are given the following information: A player places a bet of one dollar on "odd" and he wins, on the average, 20 cents per game. We wish again to determine the p_i's using the MEM; however, now we must satisfy the constraints

$$p_1 + p_3 + p_5 = 0.6 \qquad p_2 + p_4 + p_6 = 0.4$$

This is a consequence of the available information because an average gain of 20 cents means that $P\{\text{odd}\} - P\{\text{even}\} = 0.2$. Maximizing $H(\mathfrak{S})$ subject to the above constraints, we obtain

$$p_1 = p_3 = p_5 = 0.2 \qquad p_2 = p_4 = p_6 = 0.133\ldots$$

This agrees again with the classical definition if we apply the principle of insufficient reason to the outcomes of the events {odd} and {even} separately.

Although conceptually the ME principle is equivalent to the principle of insufficient reason, operationally the MEM simplifies the analysis drastically when, as is the case in most applications, the constraints are phrased in terms of probabilities in the space $\mathscr{S}^n$ of repeated trials. In such cases the equivalence still holds, although it is less obvious, but the reasoning is involved and rather forced if we derive the unknown probabilities starting from the classical definition.

The MEM is thus a valuable tool in the solution of applied problems. It is used, in fact, even in deterministic problems involving the estimation of unknown parameters from insufficient data. The ME principle is then accepted as a smoothness criterion. We should emphasize, however, that as in the case of the classical definition, the conclusions drawn from the ME principle must be accepted with skepticism particularly when they involve elaborate constraints. This is evident even in the interpretation of the results in Example 15-3: In the absence of prior constraints, we conclude that all p_i's must be equal. This conclusion we accept readily because it is not in conflict with our experience concerning dice. The second conclusion, however, that $p_2 = p_4 = p_6 = 0.133\ldots$ and $p_1 = p_3 = p_5 = 0.2$ is not as convincing, we would think, even though we have no basis for any other conclusion. In our experience, no crooked dice exhibit such symmetries.

One might argue that this apparent conflict between the MEM and our experience is due to the fact that we did not make total use of our prior knowledge. Had we included among the constraints everything we know about dice, there would be no conflict. This might be true; however, it is not always clear how such constraints can be phrased analytically and, even if they can, how complex the required computations might be.

Typical Sequences and Relative Frequency

Suppose that $\mathfrak{A} = [\mathscr{A}_1, \ldots, \mathscr{A}_N]$ is an N-element partition of an experiment $\mathscr{S}$. In the space $\mathscr{S}^n$ of repeated trials, the elements $\mathscr{A}_i$ of $\mathfrak{A}$ form N^n sequences of the form

$$\{\mathscr{A}_i \text{ occurs } n_i \text{ times in a specific order}\} \tag{15-3}$$

and the probability of each sequence equals

$$p_1^{n_1} \cdots p_i^{n_i} \cdots p_N^{n_N} \tag{15-4}$$

where $p_i = P(\mathscr{A}_i)$. The numbers n_i are arbitrary subject only to the constraint $n_1 + \cdots + n_N = n$. However, according to the relative frequency interpretation of probability, if n is "sufficiently large," then "almost certainly"

$$n_i \simeq np_i \qquad i = 1, \ldots, N \tag{15-5}$$

This is, of course, only a heuristic statement; hence the resulting consequences must be interpreted accordingly. However, as we know, the approximation (15-5) can be given a precise interpretation in the form of the law of large numbers. Following a similar approach, we prove at the end of the section the main consequence [Eq. (15-10)] of (15-5) in the context of entropy.

Guided by (15-5), we shall separate the N^n sequences of the form (15-3) into two groups: (a) typical and (b) rare. We shall say that a sequence is *typical*, if $n_i \simeq np_i$. All other sequences will be called *rare*. A typical sequence will be identified with the letter $\mathfrak{t}$:

$$\mathfrak{t} = \{\mathscr{A}_i \text{ occurs } n_i \simeq np_i \text{ times in a specific order}\} \tag{15-6}$$

From the definition it follows that to each set of numbers $n_1, \ldots, n_N$ "close" to the numbers $np_1, \ldots, np_N$ there corresponds one typical sequence. The union of all typical sequences will be denoted by $\mathbb{T}$. Thus $\mathbb{T}$ is the totality of all sequences of the form (15-3) where $n_i \simeq np$. As we noted, it is almost certain that for large n, each observed sequence is typical. This leads to the conclusion that

$$P(\mathbb{T}) \simeq 1 \tag{15-7}$$

The complement $\overline{\mathbb{T}}$ of $\mathbb{T}$ is the union of all rare sequences and its probability is negligible for large n:

$$P(\overline{\mathbb{T}}) \simeq 0 \tag{15-8}$$

Since $n_i \simeq np_i$ for all typical sequences, (15-4) yields

$$P(\mathfrak{t}) = p_1^{n_1} \cdots p_N^{n_N} \simeq e^{np_1 \ln p_1 + \cdots + np_N \ln p_N}$$

Hence the probability of each typical sequence equals

$$P(\mathfrak{t}) = e^{-nH(\mathfrak{A})} \tag{15-9}$$

where $H(\mathfrak{A})$ is the entropy of the partition $\mathfrak{A}$. Denoting by $n_{\mathbb{T}}$ the number of typical sequences, we conclude from (15-7) and the above that

$$n_{\mathbb{T}} = \frac{P(\mathbb{T})}{P(\mathfrak{t})} \simeq e^{nH(\mathfrak{A})} \tag{15-10}$$

We have thus expressed the number of typical sequences in terms of the entropy of $\mathfrak{A}$. If all the events of $\mathfrak{A}$ are equally likely, then $H(\mathfrak{A}) = \ln N$ and $n_{\mathbb{T}} = N^n$. In all other cases, $H(\mathfrak{A}) < \ln N$ [see (15-38)]. Hence

$$n_{\mathbb{T}} \simeq e^{nH(\mathfrak{A})} \ll N^n \qquad \text{for} \quad n \gg 1 \tag{15-11}$$

This leads to the important conclusion that, if n is sufficiently large, then *most* sequences are rare even though "almost certainly" none will occur.

Notes 1. We should point out that each typical sequence is not more likely than each rare sequence. In fact, the sequence with the largest probability is the rare sequence $\{\mathscr{A}_m$ occurs n times$\}$, where $\mathscr{A}_m$ is the event with the largest probability. As we presently show, the distinction between typical and rare sequences is best expressed in terms of the events

$$\{\mathscr{A}_i \text{ occurs } n_i \text{ times in } \textit{any} \text{ order}\}$$

As we know [see (3-38)], the probability of these events equals

$$\frac{n!}{k_1! \cdots k_N!} p_1^{k_1} \cdots p_N^{k_N}$$

and for large n, it takes significant values only in a small vicinity of the point $(k_1 = n_1 p_1, \ldots, k_N = n_N p_N)$. This follows by repeating the argument leading to (3-17) or, from the DeMoivre–Laplace approximation (3-39).

2. On page 1 of Chap. 1 we noted that the theory of probability applied to averages of mass phenomena leads to useful results only if the ratio k/n approaches a constant as n increases and this constant is the same for any subsequence. This apparently mild requirement results in severe restrictions on the properties of the resulting sequences. It leads to the conclusion that of all possible N^n sequences formed with the N elements of a partition $\mathfrak{A}$, only the $e^{nH(\mathfrak{A})}$ typical sequences are likely to occur; all other sequences are nearly impossible.

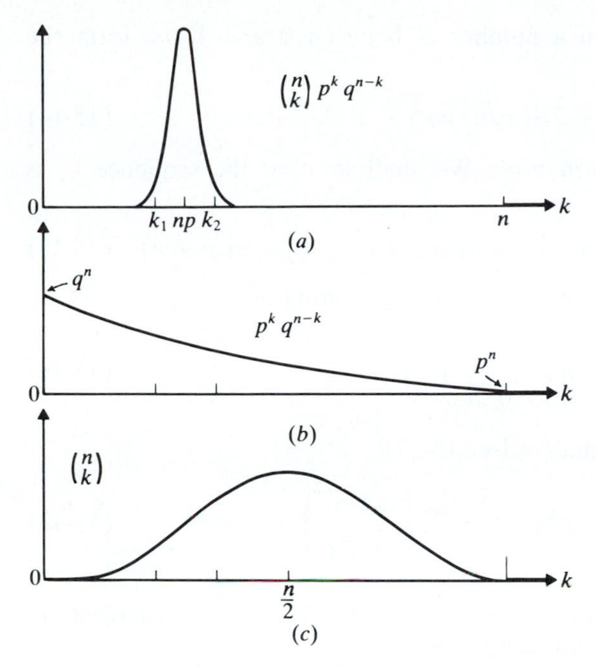

FIGURE 15-3

Typical Sequences and the Law of Large Numbers

We show next that the preceding results can be reestablished rigorously as consequences of the law of large numbers. For simplicity, we consider only two-element partitions and, to be concrete, we assume that $\mathscr{A}$ and $\overline{\mathscr{A}}$ are the events "heads" and "tails" respectively in the coin experiment. In the space $\mathscr{S}^n$, the probability of the elementary event $\{\zeta_k\} = \{k \text{ heads in a specific order}\}$ equals

$$P\{\zeta_k\} = p^k q^{n-k}$$

and the probability of the event†

$$\mathscr{A}_k = \{k \text{ heads in any order}\}$$

equals

$$P(\mathscr{A}_k) = \binom{n}{k} p^k q^{n-k} \simeq \frac{1}{\sqrt{2\pi npq}} e^{-(k-np)^2/2npq} \tag{15-12}$$

In Fig. 15-3 we plot the probability $P(\mathscr{A}_k)$, the geometric progression $q^n(p/q)^k$, and the binomial coefficients

$$\binom{n}{k} = p^{-k} q^{-(n-k)} P(\mathscr{A}_k) \tag{15-13}$$

as functions of k.

†The event $\mathscr{A}_k$ is not, of course, an element of the partition $\mathfrak{A} = [\mathscr{A}, \overline{\mathscr{A}}]$.

α-TYPICAL SEQUENCES. Given a number α between 0 and 1, we form the number ε such that

$$\alpha = 2\mathbb{G}\left(\varepsilon\sqrt{n/pq}\right) - 1 \tag{15-14}$$

where $\mathbb{G}(x)$ is the normal distribution. We shall say that the sequence ζ_k is *α-typical* if k is such that

$$k_1 \le k \le k_2 \qquad \text{where} \quad k_1 = n(p - \varepsilon) \qquad k_2 = n(p + \varepsilon) \tag{15-15}$$

The union of all α-typical sequences is a set $\mathbb{T}$ consisting of

$$n_{\mathbb{T}} = \sum_{k=k_1}^{k_2} \binom{n}{k} \tag{15-16}$$

elements and its probability equals α [see (3-37)]

$$P(\mathbb{T}) = \sum_{k=k_1}^{k_2} \binom{n}{k} p^k q^{n-k} \simeq 2\mathbb{G}\left(\varepsilon\sqrt{\frac{n}{pq}}\right) - 1 = \alpha \tag{15-17}$$

FUNDAMENTAL THEOREM. For any $\alpha < 1$, the number $n_{\mathbb{T}}$ of α-typical sequences tends to $e^{nH(\mathfrak{A})}$ in the following sense

$$\frac{\ln n_{\mathbb{T}}}{n} \xrightarrow[n\to\infty]{} H(\mathfrak{A}) \tag{15-18}$$

Proof. If $p = q = 0.5$, then the DeMoivre–Laplace approximation yields

$$\binom{n}{k} \simeq \frac{2^n}{\sqrt{\pi n/2}} e^{-2(k-n/2)^2/n}$$

for k in the $\sqrt{n}$ vicinity of $n/2$. This approximation cannot be used to evaluate the sum in (15-16) for $p \ne 0.5$ because then the center np of the interval (k_1, k_2) is not $n/2$. We shall bound $n_{\mathbb{T}}$ using (15-13) and (15-16). Clearly,

$$n_{\mathbb{T}} = \sum_{k=k_1}^{k_2} p^{-k} q^{k-n} P(\mathscr{A}_k) \tag{15-19}$$

where we assume that $p < q$. As k increases, the term $p^{-k}q^{k-n}$ increases monotonically. Hence

$$q^{-n}\left(\frac{q}{p}\right)^{k_1} \sum_{k=k_1}^{k_2} P(\mathscr{A}_k) < n_{\mathbb{T}} < q^{-n}\left(\frac{q}{p}\right)^{k_2} \sum_{k=k_1}^{k_2} P(\mathscr{A}_k) \tag{15-20}$$

And since [see (15-17)]

$$\sum_{k=k_1}^{k_2} P(\mathscr{A}_k) = P(\mathbb{T}) = \alpha$$

(15-20) yields

$$\frac{\alpha}{q^n}\left(\frac{q}{p}\right)^{k_1} < \sum_{k=k_1}^{k_2}\binom{n}{k} < \frac{\alpha}{q^n}\left(\frac{q}{p}\right)^{k_2} \tag{15-21}$$

Setting $k_1 = np - n\varepsilon$ and $k_2 = np + n\varepsilon$ in the above and using the identity

$$p^{-np}q^{-nq} = e^{-n(p\ln p + q\ln q)} = e^{nH(\mathfrak{A})}$$

we conclude from (15-21) that

$$\alpha e^{nH(\mathfrak{A})}\left(\frac{q}{p}\right)^{-n\varepsilon} < n_{\mathbb{T}} < \alpha e^{nH(\mathfrak{A})}\left(\frac{q}{p}\right)^{n\varepsilon}$$

Hence

$$nH(\mathfrak{A}) + \ln\alpha - n\varepsilon\ln\frac{q}{p} < \ln n_{\mathbb{T}} < nH(\mathfrak{A}) + \ln\alpha + n\varepsilon\ln\frac{q}{p}$$

Dividing by n, we obtain (15-18) because α is constant and, as we see from(15-14), $\varepsilon \to 0$ as $n \to \infty$.

Important conclusion Theorem (15-18) holds for any $\alpha < 1$; it will be assumed, however, that $\alpha \simeq 1$ and the corresponding sequences will be called typical. With this assumption

$$P(\mathbb{T}) = \alpha \simeq 1 \qquad P(\overline{\mathbb{T}}) = 1 - \alpha \simeq 0 \tag{15-22}$$

The probability of an arbitrary event $\mathscr{M}$ equals, therefore, its conditional probability

$$P(\mathscr{M}) = P(\mathscr{M}|\mathbb{T})P(\mathbb{T}) + P(\mathscr{M}|\overline{\mathbb{T}})P(\overline{\mathbb{T}}) \simeq P(\mathscr{M}|\mathbb{T}) \tag{15-23}$$

In other words, in any conclusions concerning probabilities in the space $\mathscr{S}^n$, it suffices to consider the subspace of $\mathscr{S}^n$ consisting of typical sequences only. This is, of course, only approximately true for finite n. It is, however, exact in the limit as $n \to \infty$.

CONCLUDING REMARKS. In Chap. 1, we presented the following interpretations of the probability $P(\mathscr{A})$ of an event $\mathscr{A}$.

Axiomatic. $P(\mathscr{A})$ is a number assigned to the event $\mathscr{A}$. This number satisfies three axioms but is otherwise arbitrary.

Empirical. For large n,

$$P(\mathscr{A}) \simeq \frac{k}{n}$$

where k is the number of times $\mathscr{A}$ occurs in n repetitions of the underlying experiment $\mathscr{S}$.

Subjective. $P(\mathscr{A})$ is a measure of our uncertainty about the occurrence of $\mathscr{A}$ in a single performance of $\mathscr{S}$.

Principle of insufficient reason. If $\mathscr{A}_i$ are N events of a partition $\mathfrak{A}$ of $\mathscr{S}$ and nothing is known about their probabilities, then $P(\mathscr{A}_i) = 1/N$.

We give next four related interpretations of the entropy $H(\mathfrak{A})$ of $\mathfrak{A}$.

Axiomatic. $H(\mathfrak{A})$ is a number assigned to each partition of $\mathscr{S}$. This number equals the sum $-\Sigma p_i \ln p_i$ where $p_i = P(\mathscr{A}_i)$.

Empirical. This interpretation involves the repeated performance not of the experiment $\mathscr{S}$, but of the experiment $\mathscr{S}^n$ of repeated trials. In this experiment, a specific typical sequence $\mathfrak{t}_j$ is an event with probability $e^{-nH(\mathfrak{A})}$. Applying the relative frequency interpretation of probability to this event, we conclude that if the experiment $\mathscr{S}^n$ is repeated m times and the event $\mathfrak{t}_j$ occurs m_j times, then for sufficiently large m,

$$P(\mathfrak{t}_j) = e^{-nH(\mathfrak{A})} \simeq \frac{m_j}{m} \quad \text{hence} \quad H(\mathfrak{A}) \simeq -\frac{1}{n} \ln \frac{m_j}{m}$$

This relates the theoretical quantity $H(\mathfrak{A})$ to the experimental numbers m_j and m.

Subjective. $H(\mathfrak{A})$ is a measure of our uncertainty about the occurrence of the events $\mathscr{A}_i$ of the partition $\mathfrak{A}$ in a single performance of $\mathscr{S}$.

Principle of maximum entropy. The probabilities $p_i = P(\mathscr{A}_i)$ must be such as to maximize $H(\mathfrak{A})$ subject to the given constraints. Since $n_t = e^{nH(\mathfrak{A})}$, the ME principle is equivalent to the principle of maximizing the number of typical sequences. If there are no constraints, that is, if nothing is known about the probabilities p_i, then the ME principle leads to the estimates $p_i = 1/N$, $H(\mathfrak{A}) = \ln N$, and $n_t = N^n$.

15-2 BASIC CONCEPTS

In this section, we develop deductively the properties of entropy starting with various notations and set operations. At the end of the section, we reexamine the results in terms of the heuristic notion of entropy as a measure of uncertainty, and we conclude with a typical sequence interpretation of the main theorems.

DEFINITIONS. The notation

$$\mathfrak{A} = [\mathscr{A}_1, \ldots, \mathscr{A}_k] \qquad \text{or simply} \qquad \mathfrak{A} = [\mathscr{A}_i]$$

will mean that $\mathfrak{A}$ is a partition consisting of the events $\mathscr{A}_i$. These events will be called elements† of $\mathfrak{A}$.

†It will be clear from the context whether the word *element* means an event $\mathscr{A}_i$ of a partition $\mathfrak{A}$ or an element ζ_i of the space $\mathscr{S}$.

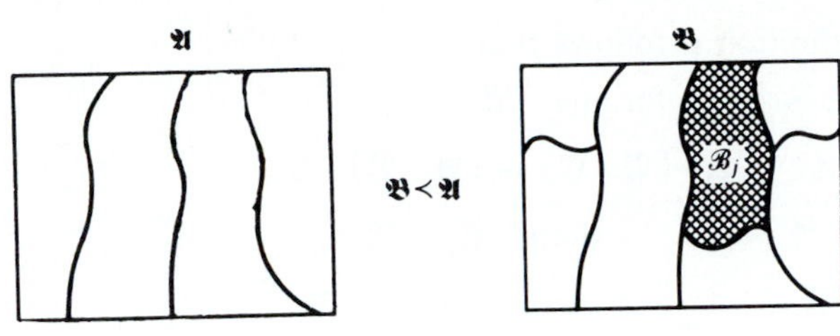

FIGURE 15-4

I. A partition with only two elements will be called *binary*. Thus

$$\mathfrak{A} = [\mathscr{A}, \overline{\mathscr{A}}]$$

is a binary partition consisting of the event $\mathscr{A}$ and its complement $\overline{\mathscr{A}}$.

II. A partition whose elements are the elementary events $\{\zeta_i\}$ of the space $\mathscr{S}$ will be denoted by $\mathfrak{S}$ and will be called the *element partition*.

III. A *refinement* of a partition $\mathfrak{A}$ is a partition $\mathfrak{B}$ such that each element $\mathscr{B}_j$ of $\mathfrak{B}$ is a subset of some element $\mathscr{A}_i$ of $\mathfrak{A}$ (Fig. 15-4). We shall use the notation $\mathfrak{B} \prec \mathfrak{A}$ to indicate that $\mathfrak{B}$ is a refinement of $\mathfrak{A}$ and we shall say that $\mathfrak{A}$ is larger† than $\mathfrak{B}$. Thus

$$\mathfrak{B} \prec \mathfrak{A} \qquad \text{iff} \quad \mathscr{B}_j \subset \mathscr{A}_i \tag{15-24}$$

A *common refinement* of two partitions is a refinement of both.
The partition $\mathfrak{D}$ in Fig. 15-5 is a common refinement of the partitions $\mathfrak{A}$ and $\mathfrak{B}$.

IV. The *product*‡ of two partitions $\mathfrak{A} = [\mathscr{A}_i]$ and $\mathfrak{B} = [\mathscr{B}_j]$ is a partition whose elements are all intersections $\mathscr{A}_i\mathscr{B}_j$ of the elements of $\mathfrak{A}$ and $\mathfrak{B}$. This partition will be denoted by

$$\mathfrak{A} \cdot \mathfrak{B}$$

Clearly, $\mathfrak{A} \cdot \mathfrak{B}$ is the largest common refinement of $\mathfrak{A}$ and $\mathfrak{B}$.

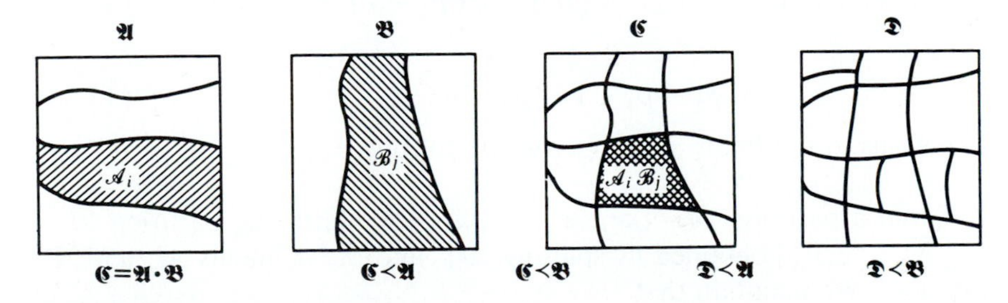

FIGURE 15-5

†The symbol $\prec$ is *not* an ordering of two arbitrary partitions. It has a meaning only if $\mathfrak{B}$ is a refinement of $\mathfrak{A}$.

‡We should emphasize that partition product is *not* a set operation.

Properties From the definition it follows that

$$\mathfrak{S} \prec \mathfrak{A} \qquad \text{for any} \quad \mathfrak{A}$$

$$\mathfrak{A} \cdot \mathfrak{B} = \mathfrak{B} \cdot \mathfrak{A} \qquad \mathfrak{A} \cdot (\mathfrak{B} \cdot \mathfrak{C}) = (\mathfrak{A} \cdot \mathfrak{B}) \cdot \mathfrak{C}$$

$$\text{If} \quad \mathfrak{A}_1 \prec \mathfrak{A}_2 \prec \mathfrak{A}_3 \qquad \text{then} \quad \mathfrak{A}_1 \prec \mathfrak{A}_3$$

$$\text{If} \quad \mathfrak{B} \prec \mathfrak{A} \qquad \text{then} \quad \mathfrak{A} \cdot \mathfrak{B} = \mathfrak{B}$$

ENTROPY. The entropy of a partition $\mathfrak{A}$ is by definition the sum

$$H(\mathfrak{A}) = -(p_1 \log p_1 + \cdots + p_N \log p_N) = \sum_{i=1}^{N} \varphi(p_i) \qquad (15\text{-}25)$$

where $p_i = P(\mathscr{A}_i)$ and $\varphi(p) = -p \log p$.

Since $\varphi(p) \geq 0$ for $0 \leq p \leq 1$, it follows from (15-25) that

$$H(\mathfrak{A}) \geq 0 \qquad (15\text{-}26)$$

where $H(\mathfrak{A}) = 0$ iff one of the p_i's equals 1; all others are then equal to 0.

Binary partitions If $\mathfrak{A} = [\mathscr{A}, \overline{\mathscr{A}}]$ and $P(\mathscr{A}) = p$, then (Fig. 15-2)

$$H(\mathfrak{A}) = -p \log p - (1-p)\log(1-p) = \mathfrak{r}(p) \qquad (15\text{-}27)$$

Equally likely events If

$$p_1 = p_2 = \cdots = p_N$$

then

$$H(\mathfrak{A}) = -\frac{1}{N}\log\frac{1}{N} - \cdots - \frac{1}{N}\log\frac{1}{N} = \log N \qquad (15\text{-}28)$$

If, in particular, $N = 2^m$, then $H(\mathfrak{A}) = m$.

INEQUALITIES. The function $\varphi(p) = -p \log p$ is convex. Therefore (see Fig. 15-6 and Prob. 15-2)

$$\varphi(p_1 + p_2) < \varphi(p_1) + \varphi(p_2) < \varphi(p_1 + \varepsilon) + \varphi(p_2 - \varepsilon) \qquad (15\text{-}29)$$

where

$$p_1 < p_1 + \varepsilon \leq p_2 - \varepsilon < p_2 \qquad (15\text{-}30)$$

This leads to the following properties of entropy:

1. Given a partition $\mathfrak{A} = [\mathscr{A}_1, \mathscr{A}_2, \ldots, \mathscr{A}_N]$, we form the partition $\mathfrak{B} = [\mathscr{B}_a, \mathscr{B}_b, \mathscr{A}_2, \ldots, \mathscr{A}_N]$ obtained by splitting $\mathscr{A}_1$ into the elements $\mathscr{B}_a$ and $\mathscr{B}_b$ as in Fig. 15-7. We maintain that

$$H(\mathfrak{A}) \leq H(\mathfrak{B}) \qquad (15\text{-}31)$$

Proof. Clearly,

$$H(\mathfrak{A}) - \varphi(p_a + p_b) = H(\mathfrak{B}) - \varphi(p_a) - \varphi(p_b)$$

because each side equals the contribution to $H(\mathfrak{A})$ and $H(\mathfrak{B})$ respectively due

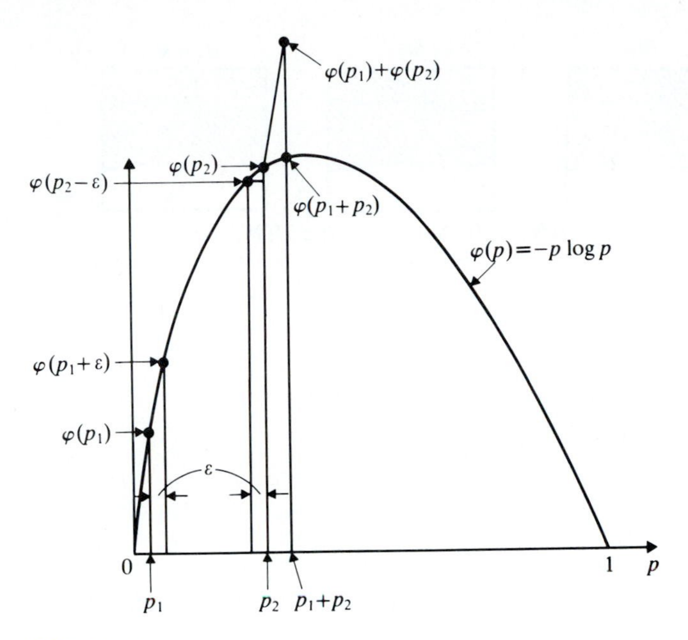

FIGURE 15-6

to the common elements of $\mathfrak{A}$ and $\mathfrak{B}$. Hence (15-31) follows from the first inequality in (15-29).

Example 15-4. In the next table we list the probabilities of the events of a partition $\mathfrak{A}$ and of its refinement $\mathfrak{B}$ obtained as above.

$\mathfrak{A}$	$p = 0.4$		0.35	0.25
$\mathfrak{B}$	$p_a = 0.22$	$p_b = 0.18$	0.35	0.25

In this case,

$$H(\mathfrak{A}) = -(0.4 \log 0.4 + 0.35 \log 0.35 + 0.25 \log 0.25) = 1.559$$

$$H(\mathfrak{B}) = -(0.22 \log 0.22 + 0.18 \log 0.18 + 0.35 \log 0.35 + 0.25 \log 0.25) = 1.956$$

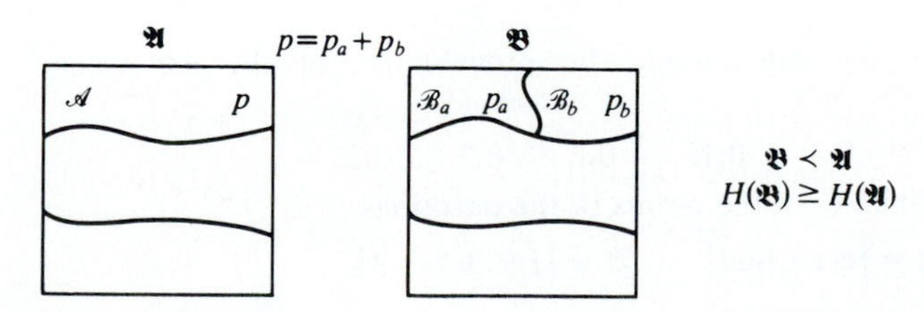

FIGURE 15-7

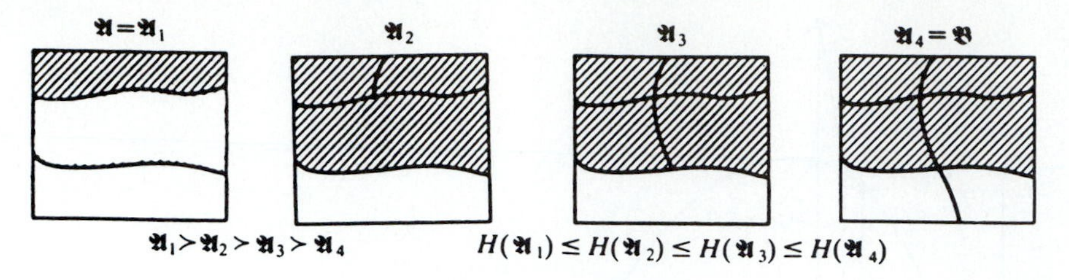

FIGURE 15-8

Thus

$$H(\mathfrak{A}) = 1.559 < 1.956 = H(\mathfrak{B})$$

in agreement with (15-31).

2. If

$$\mathfrak{B} \prec \mathfrak{A} \qquad \text{then} \quad H(\mathfrak{B}) \ge H(\mathfrak{A}) \tag{15-32}$$

Proof. Repeating the construction of Fig. 15-7, we form a chain of refinements

$$\mathfrak{A} = \mathfrak{A}_1 \prec \cdots \prec \mathfrak{A}_{m-1} \prec \mathfrak{A}_m \prec \cdots \prec \mathfrak{A}_n = \mathfrak{B}$$

where $\mathfrak{A}_m$ is obtained by splitting one of the elements of $\mathfrak{A}_{m-1}$ as in Fig. 15-8. From this and (15-31) it follows that

$$H(\mathfrak{A}) = H(\mathfrak{A}_1) \le \cdots \le H(\mathfrak{A}_n) = H(\mathfrak{B})$$

and (15-32) results.

3. For any $\mathfrak{A}$:

$$H(\mathfrak{A}) \le H(\mathfrak{S}) \tag{15-33}$$

where $\mathfrak{S}$ is the element partition.

Proof. It follows from (15-31) because $\mathfrak{S}$ is a refinement of $\mathfrak{A}$.

4. For any $\mathfrak{A}$ and $\mathfrak{B}$:

$$H(\mathfrak{A} \cdot \mathfrak{B}) \ge H(\mathfrak{A}) \qquad H(\mathfrak{A} \cdot \mathfrak{B}) \ge H(\mathfrak{B}) \tag{15-34}$$

Proof. It follows from (15-31) because $\mathfrak{A} \cdot \mathfrak{B}$ is a refinement of $\mathfrak{A}$ and of $\mathfrak{B}$.

Example 15-5. In the die experiment, the probabilities of the six events $\{f_1\}, \ldots, \{f_6\}$ equal

$$0.1 \qquad 0.1 \qquad 0.15 \qquad 0.2 \qquad 0.2 \qquad 0.25$$

respectively. The probabilities of the events of the partitions

$$\mathfrak{A} = [\text{even}, \text{odd}] \qquad \mathfrak{B} = [i \le 3, i > 3]$$

are given by

$$P\{\text{even}\} = 0.55 \qquad P\{\text{odd}\} = 0.45 \qquad P\{i \le 3\} = 0.35 \qquad P\{i > 3\} = 0.65$$

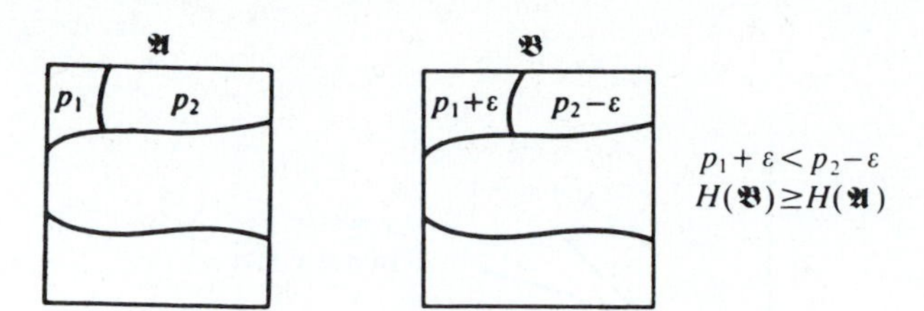

FIGURE 15-9

The product $\mathfrak{A} \cdot \mathfrak{B}$ is a partition consisting of the four elements

$$\{f_2\} \qquad \{f_1 f_3\} \qquad \{f_4 f_6\} \qquad \{f_5\}$$

with respective probabilities

$$0.1 \qquad 0.25 \qquad 0.45 \qquad 0.2$$

From the above it follows that

$$H(\mathfrak{A}) = 0.993 \qquad H(\mathfrak{B}) = 0.934 \qquad H(\mathfrak{A} \cdot \mathfrak{B}) = 1.815$$

in agreement with (15-34).

5. Suppose that $\mathfrak{A}$ and $\mathfrak{B}$ are two partitions that have the same elements except the first two (Fig. 15-9)

$$\mathfrak{A} = [\mathscr{A}_1, \mathscr{A}_2, \mathscr{A}_3, \ldots, \mathscr{A}_N] \qquad \mathfrak{B} = [\mathscr{B}_1, \mathscr{B}_2, \mathscr{A}_3, \ldots, \mathscr{A}_N]$$

We maintain that if

$$P(\mathscr{A}_1) = p_1 \qquad P(\mathscr{A}_2) = p_2 \qquad P(\mathscr{B}_1) = p_1 + \varepsilon \leq p_2 - \varepsilon = P(\mathscr{B}_2)$$

as in (15-30), then

$$H(\mathfrak{A}) \leq H(\mathfrak{B}) \tag{15-35}$$

Proof. Clearly,

$$H(\mathfrak{A}) - \varphi(p_1) - \varphi(p_2) = H(\mathfrak{B}) - \varphi(p_1 + \varepsilon) - \varphi(p_2 + \varepsilon)$$

because each side equals the contribution to $H(\mathfrak{A})$ and $H(\mathfrak{B})$ respectively due to the common elements of $\mathfrak{A}$ and $\mathfrak{B}$. Hence (15-35) follows from the second inequality in (15-29).

Example 15-6. In the next table we list the probabilities of the events of the partitions $\mathfrak{A}$ and $\mathfrak{B}$.

$\mathfrak{A}$	0.1	0.3	0.35	0.25
$\mathfrak{B}$	0.18	0.22	0.35	0.25

$p_1 = 0.1$ $\quad p_2 = 0.3$ $\qquad \varepsilon = 0.08$

In this case,

$$H(\mathfrak{A}) = 1.883 \qquad H(\mathfrak{B}) = 1.956$$

in agreement with (15-35).

6. If we equalize the entropies of two elements of a partition, leaving all others unchanged, its entropy increases.

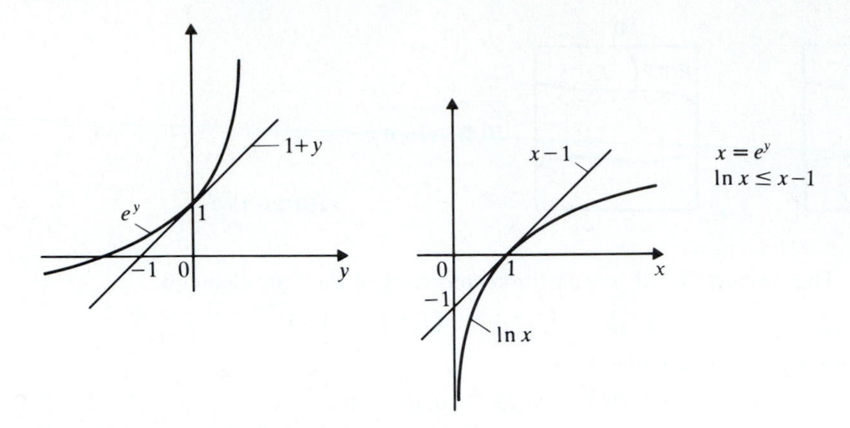

FIGURE 15-10

Proof. It follows from the above with $\varepsilon = (p_2 - p_1)/2$.

7. The entropy of a partition is maximum if all its elements are equally likely as in (15-28).

Proof. Suppose that $\mathfrak{A}$ is a partition such that $H(\mathfrak{A}) = H_m$ is maximum and two of its elements have unequal probabilities. If they are made equal, then (property 6) $H(\mathfrak{A})$ increases. But this is impossible because H_m is maximum by assumption.

A useful inequality. If a_i and b_i are N positive numbers such that

$$a_1 + \cdots + a_N = 1 \qquad b_1 + \cdots + b_N \le 1 \tag{15-36}$$

then

$$-\sum_i a_i \log a_i \le -\sum_i a_i \log b_i \tag{15-37}$$

with equality iff $a_i = b_i$.

Proof. From the inequality $e^y \ge 1 + y$ it follows that $\ln x \le x - 1$ (Fig. 15-10). With $x = b_i/a_i$, this yields

$$\ln b_i - \ln a_i = \ln \frac{b_i}{a_i} \le \frac{b_i}{a_i} - 1$$

Multiplying by a_i and adding, we obtain

$$\sum_i a_i(\ln b_i - \ln a_i) \le \sum_i a_i\left(\frac{b_i}{a_i} - 1\right) = \sum_i (b_i - a_i) \le 0$$

and (15-37) results.

Maximum entropy. Using (15-37), we shall rederive property 7. It suffices to show that

$$-\sum_i p_i \log p_i \leq \log N \tag{15-38}$$

Proof. The numbers $a_i = p_i$ and $b_i = 1/N$ satisfy (15-36). Inserting into (15-37), we conclude that

$$-\sum_i p_i \log p_i \leq -\sum_i p_i \log \frac{1}{N} = \log N \sum_i p_i = \log N$$

Conditional Entropy and Mutual Information

The conditional entropy of a partition $\mathfrak{A}$ assuming $\mathscr{M}$ is by definition the sum

$$H(\mathfrak{A}|\mathscr{M}) = -\sum_{i=1}^{N_{\mathfrak{A}}} P(\mathscr{A}_i|\mathscr{M}) \log P(\mathscr{A}_i|\mathscr{M}) \tag{15-39}$$

where $P(\mathscr{M}) \neq 0$, $N_{\mathfrak{A}}$ is the number of elements $\mathscr{A}_i$ of $\mathfrak{A}$, and

$$P(\mathscr{A}_i|\mathscr{M}) = \frac{P(\mathscr{A}_i \mathscr{M})}{P(\mathscr{M})}$$

As we explain later, $H(\mathfrak{A}|\mathscr{M})$ is the uncertainty about $\mathfrak{A}$ in the subsequence of trials in which $\mathscr{M}$ occurs.

Suppose now that $\mathfrak{B}$ is a partition consisting of the $N_{\mathfrak{B}}$ elements $\mathscr{B}_j$. Clearly,

$$H(\mathfrak{A}|\mathscr{B}_j) = -\sum_{i=1}^{N_{\mathfrak{A}}} P(\mathscr{A}_i|\mathscr{B}_j) \log P(\mathscr{A}_i|\mathscr{B}_j) \tag{15-40}$$

is the conditional entropy of $\mathfrak{A}$ assuming $\mathscr{B}_j$ defined as in (15-39). The conditional entropy of $\mathfrak{A}$ assuming $\mathfrak{B}$ is the weighted average of $H(\mathfrak{A}|\mathscr{B}_j)$:

$$H(\mathfrak{A}|\mathfrak{B}) = \sum_{j=1}^{N_{\mathfrak{B}}} P(\mathscr{B}_j) H(\mathfrak{A}|\mathscr{B}_j) \tag{15-41}$$

This equals the uncertainty about $\mathfrak{A}$ if at each trial we know which of the events $\mathscr{B}_j$ of $\mathfrak{B}$ has occurred.

Example 15-7. We shall determine the conditional entropy $H(\mathfrak{S}|\mathfrak{B})$ of the element partition $\mathfrak{S}$ in the fair-die experiment where $\mathfrak{B} = [\text{even}, \text{odd}]$.

Clearly, $P\{f_i|\text{even}\} = \frac{1}{3}$ if i is even and $P\{f_i|\text{even}\} = 0$ if i is odd. Similarly, $P\{f_i|\text{odd}\} = \frac{1}{3}$ if i is odd and $P\{f_i|\text{odd}\} = 0$ if i is even. Hence

$$H(\mathfrak{S}|\text{even}) = -\left(\tfrac{1}{3}\log\tfrac{1}{3} + \tfrac{1}{3}\log\tfrac{1}{3} + \tfrac{1}{3}\log\tfrac{1}{3}\right) = \log 3 = H(\mathfrak{S}|\text{odd})$$

And since $P\{\text{even}\} = P\{\text{odd}\} = 0.5$, we conclude from (15-41) that

$$H(\mathfrak{S}|\mathfrak{B}) = 0.5 \log 3 + 0.5 \log 3 = \log 3$$

Thus, in the absence of any information, our uncertainty about $\mathfrak{S}$ equals $H(\mathfrak{S}) = \log 6$. If we know, however, whether at each trial "even" or "odd" showed, then our uncertainty is reduced to $H(\mathfrak{S}|\mathfrak{B}) = \log 3$.

THEOREM 1. If

$$\mathfrak{B} \prec \mathfrak{A} \qquad \text{then} \quad H(\mathfrak{A}|\mathfrak{B}) = 0 \tag{15-42}$$

Proof. Since $\mathfrak{B}$ is a refinement of $\mathfrak{A}$, each element $\mathscr{B}_j$ of $\mathfrak{B}$ is a subset of some element $\mathscr{A}_k$ of $\mathfrak{A}$ and, therefore, it is disjoint with all other elements of $\mathfrak{A}$. Hence $\mathscr{A}_i\mathscr{B}_j = \mathscr{B}_j$ if $i = k$ and $\mathscr{A}_i\mathscr{B}_j = 0$ otherwise. This leads to the conclusion that

$$P(\mathscr{A}_i|\mathscr{B}_j) = \frac{P(\mathscr{A}_i\mathscr{B}_j)}{P(\mathscr{B}_j)} = \begin{cases} 1 & i = k \\ 0 & i \neq k \end{cases}$$

And since $p \log p = 0$ for $p = 0$ and $p = 1$, we conclude that all terms in (15-40) equal 0; hence

$$H(\mathfrak{A}|\mathscr{B}_j) = 0$$

for every j. From this and (15-41) it follows that $H(\mathfrak{A}|\mathfrak{B}) = 0$.

Independent partitions Two partitions $\mathfrak{A} = [\mathscr{A}_i]$ and $\mathfrak{B} = [\mathscr{B}_j]$ are called independent if the events $\mathscr{A}_i$ and $\mathscr{B}_j$ are independent for every i and j:

$$P(\mathscr{A}_i\mathscr{B}_j) = P(\mathscr{A}_i)P(\mathscr{B}_j) \tag{15-43}$$

THEOREM 2. If the partitions $\mathfrak{A}$ and $\mathfrak{B}$ are independent, then

$$H(\mathfrak{A}|\mathfrak{B}) = H(\mathfrak{A}) \qquad H(\mathfrak{B}|\mathfrak{A}) = H(\mathfrak{B}) \tag{15-44}$$

Proof. Clearly, $P(\mathscr{A}_i|\mathscr{B}_j) = P(\mathscr{A}_i)$; hence [see (15-40)]

$$H(\mathfrak{A}|\mathscr{B}_j) = -\sum_i P(\mathscr{A}_i)\log P(\mathscr{A}_i) = H(\mathfrak{A})$$

Inserting into (15-41), we obtain

$$H(\mathfrak{U}|\mathfrak{B}) = H(\mathfrak{A})\sum_j P(\mathscr{B}_j) = H(\mathfrak{A})$$

and (15-43) results. We can show similarly that $H(\mathfrak{B}|\mathfrak{A}) = H(\mathfrak{B})$.

THEOREM 3. For any $\mathfrak{A}$ and $\mathfrak{B}$:

$$H(\mathfrak{A} \cdot \mathfrak{B}) \leq H(\mathfrak{A}) + H(\mathfrak{B}) \tag{15-45}$$

Proof. As we know [see (2-36)]

$$P(\mathscr{A}_i) = \sum_j P(\mathscr{A}_i\mathscr{B}_j)$$

Hence

$$H(\mathfrak{A}) - \sum_i P(\mathscr{A}_i)\log P(\mathscr{A}_i) = -\sum_{i,j} P(\mathscr{A}_i\mathscr{B}_j)\log P(\mathscr{A}_i)$$

Writing a similar equation for $H(\mathfrak{B})$ and adding, we obtain

$$H(\mathfrak{A}) + H(\mathfrak{B}) = -\sum_{i,j} P(\mathscr{A}_i\mathscr{B}_j)\log\left[P(\mathscr{A}_i)P(\mathscr{B}_j)\right] \tag{15-46}$$

Clearly, $H(\mathfrak{A}\cdot\mathfrak{B})$ is a partition with elements $\mathscr{A}_i\mathscr{B}_j$. Hence

$$H(\mathfrak{A}\cdot\mathfrak{B}) = -\sum_{i,j} P(\mathscr{A}_i\mathscr{B}_j)\log P(\mathscr{A}_i\mathscr{B}_j) \tag{15-47}$$

To prove (15-45), we shall apply (15-37) identifying the numbers a_i and b_i with the numbers $P(\mathscr{A}_i\mathscr{B}_j)$ and $P(\mathscr{A}_i)P(\mathscr{B}_j)$ respectively. We can do so because

$$\sum_{i,j} P(\mathscr{A}_i\mathscr{B}_j) = 1 \qquad \sum_{i,j} P(\mathscr{A}_i)P(\mathscr{B}_j) = 1$$

From (15-37) it follows that the sum in (15-47) cannot exceed the sum in (15-46); hence (15-45) must be true.

COROLLARY.

$$H(\mathfrak{A}\cdot\mathfrak{B}) = H(\mathfrak{A}) + H(\mathfrak{B}) \tag{15-48}$$

iff the partitions $\mathfrak{A}$ and $\mathfrak{B}$ are independent.

Proof. This follows from (15-45) because (15-37) is an equality iff $a_i = b_i$ for every i. Hence (15-45) is an equality iff

$$P(\mathscr{A}_i\mathscr{B}_j) = P(\mathscr{A}_i)P(\mathscr{B}_j)$$

for every i and j.

THEOREM 4. For any $\mathfrak{A}$ and $\mathfrak{B}$:

$$H(\mathfrak{A}\cdot\mathfrak{B}) = H(\mathfrak{B}) + H(\mathfrak{A}|\mathfrak{B}) = H(\mathfrak{A}) + H(\mathfrak{B}|\mathfrak{A}) \tag{15-49}$$

Proof. Since

$$P(\mathscr{A}_i\mathscr{B}_j) = P(\mathscr{B}_j)P(\mathscr{A}_i|\mathscr{B}_j)$$

we conclude from (15-40) that

$$\begin{aligned}
P(\mathscr{B}_j)H(\mathfrak{A}|\mathscr{B}_j) &= -\sum_i P(\mathscr{B}_j)P(\mathscr{A}_i|\mathscr{B}_j)\log P(\mathscr{A}_i|\mathscr{B}_j) \\
&= -\sum_i P(\mathscr{A}_i\mathscr{B}_j)\left[\log P(\mathscr{A}_i\mathscr{B}_j) - \log P(\mathscr{B}_j)\right] \\
&= -\sum_i P(\mathscr{A}_i\mathscr{B}_j)\log P(\mathscr{A}_i\mathscr{B}_j) + P(\mathscr{B}_j)\log P(\mathscr{B}_j)
\end{aligned}$$

Summing over all j, we obtain

$$\sum_j P(\mathscr{B}_j)H(\mathfrak{A}|\mathscr{B}_j) = -\sum_{i,j} P(\mathscr{A}_i\mathscr{B}_j)\log P(\mathscr{A}_i\mathscr{B}_j) + \sum_j P(\mathscr{B}_j)\log P(\mathscr{B}_j)$$

and the first equation in (15-49) follows because the above three sums equal $H(\mathfrak{A}|\mathfrak{B})$, $H(\mathfrak{A}\cdot\mathfrak{B})$, and $-H(\mathfrak{B})$ respectively. The second equation follows because $\mathfrak{A}\cdot\mathfrak{B} = \mathfrak{B}\cdot\mathfrak{A}$.

COROLLARIES. The following relationships follow readily from the last two theorems: For any $\mathfrak{A}$ and $\mathfrak{B}$:

$$H(\mathfrak{B}) \le H(\mathfrak{A}\cdot\mathfrak{B}) \le H(\mathfrak{A}) + H(\mathfrak{B}) \tag{15-50}$$

$$H(\mathfrak{A}|\mathfrak{B}) \le H(\mathfrak{A}) \tag{15-51}$$

$$H(\mathfrak{A}) - H(\mathfrak{A}|\mathfrak{B}) = H(\mathfrak{B}) - H(\mathfrak{B}|\mathfrak{A}) \tag{15-52}$$

Mutual information. The function

$$I(\mathfrak{A},\mathfrak{B}) = H(\mathfrak{A}) + H(\mathfrak{B}) - H(\mathfrak{A}\cdot\mathfrak{B}) \tag{15-53}$$

is called the mutual information of the partitions $\mathfrak{A}$ and $\mathfrak{B}$. From (15-49) it follows that

$$I(\mathfrak{A},\mathfrak{B}) = H(\mathfrak{A}) - H(\mathfrak{A}|\mathfrak{B}) = H(\mathfrak{B}) - H(\mathfrak{B}|\mathfrak{A}) \tag{15-54}$$

Clearly [see (15-51)]

$$I(\mathfrak{A},\mathfrak{B}) \ge 0 \tag{15-55}$$

As we shall presently see, $I(\mathfrak{A},\mathfrak{B})$ can be interpreted as the "information about $\mathfrak{A}$ contained in $\mathfrak{B}$" and it equals the "information about $\mathfrak{B}$ contained in $\mathfrak{A}$."

Example 15-8. In the fair-die experiment of Example 15-7,

$$H(\mathfrak{S}) = \log 6 \qquad H(\mathfrak{S}|\mathfrak{B}) = \log 3 \qquad H(\mathfrak{B}) = \log 2 \qquad H(\mathfrak{B}|\mathfrak{S}) = 0$$

Hence

$$I(\mathfrak{S},\mathfrak{B}) = \log 2$$

Thus the information about the element partition $\mathfrak{S}$ resulting from the observation of the even–odd partition $\mathfrak{B}$ equals log 2.

Generalizations. The preceding results can be readily generalized to an arbitrary number of partitions. We list below several special cases leaving the simple proofs as problems:

(a) If

$$\mathfrak{B} \prec \mathfrak{C} \qquad \text{then} \qquad H(\mathfrak{A}|\mathfrak{B}) \le H(\mathfrak{A}|\mathfrak{C}) \tag{15-56}$$

(b) If the partitions $\mathfrak{A}$, $\mathfrak{B}$ and $\mathfrak{C}$ are independent, then

$$H(\mathfrak{A}\cdot\mathfrak{B}\cdot\mathfrak{C}) = H(\mathfrak{A}) + H(\mathfrak{B}\cdot\mathfrak{C}) = H(\mathfrak{A}) + H(\mathfrak{B}) + H(\mathfrak{C}) \tag{15-57}$$

(*c*) *Chain rule* For any $\mathfrak{A}$, $\mathfrak{B}$, and $\mathfrak{C}$:

$$H(\mathfrak{B}\cdot\mathfrak{C}|\mathfrak{A}) = H(\mathfrak{B}|\mathfrak{A}) + H(\mathfrak{C}|\mathfrak{A}\cdot\mathfrak{B}) \tag{15-58}$$

$$H(\mathfrak{A}\cdot\mathfrak{B}\cdot\mathfrak{C}) = H(\mathfrak{A}) + H(\mathfrak{B}\cdot\mathfrak{C}|\mathfrak{A}) = H(\mathfrak{A}) + H(\mathfrak{B}|\mathfrak{A}) + H(\mathfrak{C}|\mathfrak{A}\cdot\mathfrak{B}) \tag{15-59}$$

Repeated trials. In the space $\mathscr{S}^n$ of repeated trials all outcomes are sequences of the form

$$\zeta_{i_1}\cdots\zeta_{i_k}\cdots\zeta_{i_n} \tag{15-60}$$

where each ζ_{i_k} is an element of $\mathscr{S}$. Consider a partition $\mathfrak{A}$ of $\mathscr{S}$ consisting of N events. At the kth trial, one and only one of these events will occur, namely the event $\mathscr{A}_{j_k}$ that contains the element ζ_{i_k}. The cartesian product

$$\mathscr{A}_{j,k} = \mathscr{S}\times\cdots\mathscr{S}\times\mathscr{A}_{j_k}\times\mathscr{S}\cdots\times\mathscr{S} \qquad \zeta_{i_k}\in\mathscr{A}_{j_k} \tag{15-61}$$

is an event in $\mathscr{S}^n$ with probability

$$P(\mathscr{A}_{j,k}) = P(\mathscr{A}_{j_k}) \tag{15-62}$$

because it occurs iff the event $\mathscr{A}_{j_k}$ occurs at the kth trial. For specific k, the events $\mathscr{A}_{j,k}$ form an N element partition of the space $\mathscr{S}^n$. This partition will be denoted by $\mathfrak{A}_k$. From (15-62) it follows readily that

$$H(\mathfrak{A}_k) = H(\mathfrak{A}) \tag{15-63}$$

We can define similarly the partition $\mathfrak{B}_k$ of $\mathscr{S}^n$ formed with the elements of another partition $\mathfrak{B}$ of $\mathscr{S}$. Reasoning as in (15-63), we conclude that $H(\mathfrak{B}_k) = H(\mathfrak{B})$ and

$$H(\mathfrak{A}_k|\mathfrak{B}_k) = H(\mathfrak{A}|\mathfrak{B}) \qquad I(\mathfrak{A}_k,\mathfrak{B}_k) = I(\mathfrak{A},\mathfrak{B}) \tag{15-64}$$

We next form the product of the n partitions $\mathfrak{A}_k$:

$$\mathfrak{A}^n = \mathfrak{A}_1\cdot\mathfrak{A}_2\cdots\mathfrak{A}_n \tag{15-65}$$

The elements of this partition are cartesian products of the form

$$\mathscr{A}_{j_1}\times\cdots\times\mathscr{A}_{j_k}\times\cdots\times\mathscr{A}_{j_n} \tag{15-66}$$

If $\mathfrak{A}$ is the element partition of $\mathscr{S}$, then $\mathfrak{A}^n$ is the element partition of $\mathscr{S}^n$. In general, however, the elements of $\mathfrak{A}^n$ are events consisting of a large number of sequences of the form (15-60). If we picture these sequences as wires, then the elements (15-66) of the partition $\mathfrak{A}^n$ can be viewed as cables and their union as a collection of such cables (Fig. 15-11).

From the independence of the trials, it follows that the n partitions $\mathfrak{A}_1,\ldots,\mathfrak{A}_n$ of $\mathscr{S}^n$ are independent. Hence [see (15-57) and (15-63)]

$$H(\mathfrak{A}^n) = H(\mathfrak{A}_1) + \cdots + H(\mathfrak{A}_n) = nH(\mathfrak{A}) \tag{15-67}$$

Defining similarly the partition $\mathfrak{B}^n$, we conclude as in (15-64) that

$$H(\mathfrak{A}^n|\mathfrak{B}^n) = nH(\mathfrak{B}|\mathfrak{A}) \qquad I(\mathfrak{A}^n,\mathfrak{B}^n) = nI(\mathfrak{A},\mathfrak{B}) \tag{15-68}$$

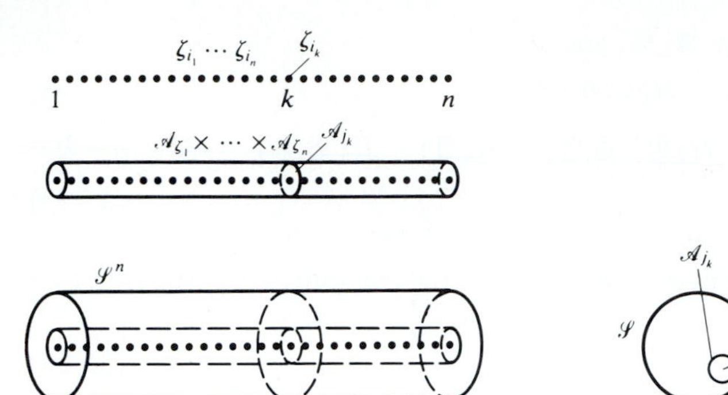

FIGURE 15-11

Example 15-9. In the coin experiment, the entropy of the element partition equals

$$H(\mathfrak{S}) = -p \log p - q \log q$$

In the space $\mathscr{S}^2$, the element partition consists of four events with

$$P\{hh\} = p^2 \qquad P\{ht\} = P\{th\} = pq \qquad P\{tt\} = q^2$$

Hence

$$H(\mathfrak{S}^2) = -p^2 \log p^2 - 2pq \log pq - q^2 \log q^2 = -2p \log p - 2q \log q$$

Thus

$$H(\mathfrak{S}^2) = 2H(\mathfrak{S})$$

in agreement with (15-67).

CONDITIONAL ENTROPY AND UNCERTAINTY. As we have noted, the entropy $H(\mathfrak{A})$ of a partition $\mathfrak{A} = [\mathscr{A}_i]$ gives us a measure of uncertainty about the occurrence of the events $\mathscr{A}_i$ at a given trial. Once the trial is performed and the events $\mathscr{A}_i$ are observed, the uncertainty is removed. We give next a similar interpretation to the conditional entropy $H(\mathfrak{A}|\mathscr{M})$ of $\mathfrak{A}$ assuming that the event $\mathscr{M}$ has been observed, and of the conditional entropy $H(\mathfrak{A}|\mathfrak{B})$ of $\mathfrak{A}$ assuming that the partitioning $\mathfrak{B}$ has been observed.†

If in the definition (15-25) of entropy we replace the probabilities $P(\mathscr{A}_i)$ by the conditional probabilities $P(\mathscr{A}_i|\mathscr{M})$, we obtain the conditional entropy $H(\mathfrak{A}|\mathscr{M})$ of $\mathfrak{A}$ assuming $\mathscr{M}$ [see (15-39)]. The relative frequency interpretation of $P(\mathscr{A}_i|\mathscr{M})$ is the same as that of $P(\mathscr{A}_i)$ if we consider not the entire sequence of n trials but only the subsequence of trials in which the event $\mathscr{M}$ occurs. From this it follows that $H(\mathfrak{A}|\mathscr{M})$ is the uncertainty about $\mathfrak{A}$ per trial in that subsequence. In other words, if at a given trial we know that $\mathscr{M}$ occurs, then our uncertainty about $\mathfrak{A}$ equals $H(\mathfrak{A}|\mathscr{M})$; if we know that $\overline{\mathscr{M}}$ occurs, then our

†The expression *a partition* $\mathfrak{B}$ *is observed* will mean that we know which of the events of $\mathfrak{B}$ has occurred.

uncertainty equals $H(\mathfrak{A}|\overline{\mathscr{M}})$. The weighted sum

$$P(\mathscr{M})H(\mathfrak{A}|\mathscr{M}) + P(\overline{\mathscr{M}})H(\mathfrak{A}|\overline{\mathscr{M}})$$

is the uncertainty about $\mathfrak{A}$ assuming that the binary partition $[\mathscr{M}, \overline{\mathscr{M}}]$ is observed.

Suppose now that at each trial we observe the partition $\mathfrak{B} = [\mathscr{B}_j]$. We maintain that, under this assumption, the uncertainty per trial about $\mathfrak{A}$ equals $H(\mathfrak{A}|\mathfrak{B})$. Indeed, in a sequence of n trials, the number of times the event $\mathscr{B}_j$ occurs equals

$$n_j \simeq nP(\mathscr{B}_j)$$

In this subsequence, the uncertainty about $\mathfrak{A}$ equals $H(\mathfrak{A}|\mathscr{B}_j)$ per trial. Hence the total uncertainty about $\mathfrak{A}$ equals

$$\sum_j n_j H(\mathfrak{A}|\mathscr{B}_j) \simeq \sum_j nP(\mathscr{B}_j)H(\mathfrak{A}|\mathscr{B}_j) = nH(\mathfrak{A}|\mathfrak{B})$$

and the uncertainty per trial equals $H(\mathfrak{A}|\mathfrak{B})$.

Thus the observation of $\mathfrak{B}$ reduces the uncertainty about $\mathfrak{A}$ from $H(\mathfrak{A})$ to $H(\mathfrak{A}|\mathfrak{B})$. The difference

$$I(\mathfrak{A}, \mathfrak{B}) = H(\mathfrak{A}) - H(\mathfrak{A}|\mathfrak{B})$$

is the reduction of the uncertainty about $\mathfrak{B}$ resulting from the observation of $\mathfrak{B}$. This justifies the statement that the mutual information $I(\mathfrak{A}, \mathfrak{B})$ equals the *information* about $\mathfrak{A}$ contained in $\mathfrak{B}$.

We show next the consistency between the properties of entropy developed earlier and the subjective notion of uncertainty.

1. If $\mathfrak{B}$ is a refinement of $\mathfrak{A}$ and $\mathfrak{B}$ is observed, then we know which of the events of $\mathfrak{A}$ occurred. Hence $H(\mathfrak{A}|\mathfrak{B}) = 0$ in agreement with (15-42).
2. If the partitions $\mathfrak{A}$ and $\mathfrak{B}$ are independent and $\mathfrak{B}$ is observed, no information about $\mathfrak{A}$ is gained. Hence $H(\mathfrak{A}|\mathfrak{B}) = H(\mathfrak{A})$ in agreement with (15-44).
3. If we observe $\mathfrak{B}$, our uncertainty about $\mathfrak{A}$ can only decrease. Hence $H(\mathfrak{A}|\mathfrak{B}) \le H(\mathfrak{A})$ in agreement with (15-51).
4. To observe $\mathfrak{A} \cdot \mathfrak{B}$, we must observe $\mathfrak{A}$ and $\mathfrak{B}$. If only $\mathfrak{B}$ is observed, the information gained equals $H(\mathfrak{B})$. The uncertainty about $\mathfrak{A}$ assuming $\mathfrak{B}$ equals, therefore, the remaining uncertainty $H(\mathfrak{A}|\mathfrak{B})$ about $\mathfrak{B}$. Hence $H(\mathfrak{A} \cdot \mathfrak{B}) - H(\mathfrak{B}) = H(\mathfrak{A}|\mathfrak{B})$ in agreement with (15-49).
5. Combining 3 and 4, we conclude that $H(\mathfrak{A} \cdot \mathfrak{B}) - H(\mathfrak{B}) \le H(\mathfrak{A})$ in agreement with (15-45).
6. If $\mathfrak{B}$ is observed, then the information that is gained about $\mathfrak{A}$ equals $I(\mathfrak{A}, \mathfrak{B})$. If $\mathfrak{B} \prec \mathfrak{C}$ and $\mathfrak{B}$ is observed, then $\mathfrak{C}$ is known. But knowledge of $\mathfrak{C}$ yields information about $\mathfrak{A}$ equal to $I(\mathfrak{A}, \mathfrak{C})$. Hence, if $\mathfrak{B} \prec \mathfrak{C}$, then $I(\mathfrak{A}, \mathfrak{B}) \ge I(\mathfrak{A}, \mathfrak{C})$ or, equivalently, $H(\mathfrak{A}|\mathfrak{B}) \le H(\mathfrak{A}|\mathfrak{C})$ in agreement with (15-56).

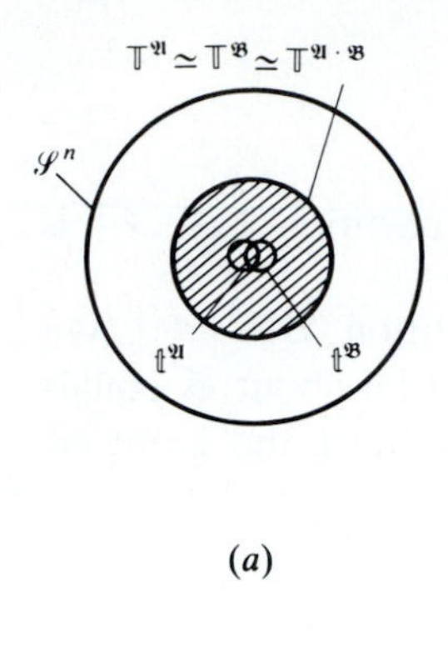

(a)

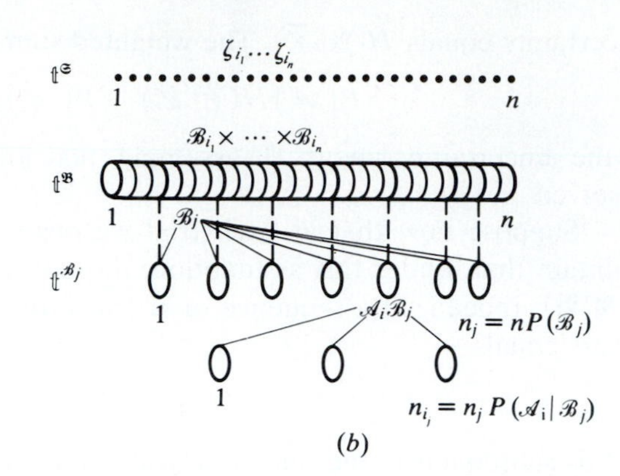

(b)

FIGURE 15-12

CONDITIONAL ENTROPY AND TYPICAL SEQUENCES. We give next a typical sequence interpretation of the properties of conditional entropy limiting the discussion to (15-45) and (15-49). The underlying reasoning is used in the proof of the channel capacity theorem (Sec. 15-6).

We denote by $\mathfrak{t}^{\mathfrak{A}}$, $\mathfrak{t}^{\mathfrak{B}}$, and $\mathfrak{t}^{\mathfrak{A}\cdot\mathfrak{B}}$ the typical sequences of the partition $\mathfrak{A}$, $\mathfrak{B}$, and $\mathfrak{A} \cdot \mathfrak{B}$ respectively, and by $\mathbb{T}^{\mathfrak{A}}$, $\mathbb{T}^{\mathfrak{B}}$, and $\mathbb{T}^{\mathfrak{A}\cdot\mathfrak{B}}$ their unions (Fig. 15-12a). As we know [see (15-7)]

$$P(\mathbb{T}^{\mathfrak{A}}) \simeq P(\mathbb{T}^{\mathfrak{B}}) \simeq P(\mathbb{T}^{\mathfrak{A}\cdot\mathfrak{B}}) \simeq 1$$

Furthermore, the number of typical sequences in each of the above three sets equals [see (15-10)]

$$n_{\mathbb{T}^{\mathfrak{A}}} \simeq e^{nH(\mathfrak{A})} \qquad n_{\mathbb{T}^{\mathfrak{B}}} \simeq e^{nH(\mathfrak{B})} \qquad n_{\mathbb{T}^{\mathfrak{A}\cdot\mathfrak{B}}} \simeq e^{nH(\mathfrak{A}\cdot\mathfrak{B})} \tag{15-69}$$

I. We maintain that

$$H(\mathfrak{A} \cdot \mathfrak{B}) \le H(\mathfrak{A}) + H(\mathfrak{B}) \tag{15-70}$$

Proof. Each $\mathfrak{t}^{\mathfrak{A}\cdot\mathfrak{B}}$ sequence specifies a pair $(\mathfrak{t}^{\mathfrak{A}}, \mathfrak{t}^{\mathfrak{B}})$. The total number of such pairs formed with all the elements of $\mathbb{T}^{\mathfrak{A}}$ and $\mathbb{T}^{\mathfrak{B}}$ equals $n_{\mathbb{T}^{\mathfrak{A}}} \cdot n_{\mathbb{T}^{\mathfrak{B}}}$. However, not all such pairs generate $\mathfrak{t}^{\mathfrak{A}\cdot\mathfrak{B}}$ sequences because, if the partitions $\mathfrak{A}$ and $\mathfrak{B}$ are not independent, then not all pairs can occur. For example, if $\mathscr{A}_i = \mathscr{B}_j$ for some i and j and $\mathscr{A}_i$ occurs at the kth trial, then B_j must also occur at this trial. From the above it follows that

$$n_{\mathbb{T}^{\mathfrak{A}\cdot\mathfrak{B}}} \le n_{\mathbb{T}^{\mathfrak{A}}} \cdot n_{\mathbb{T}^{\mathfrak{B}}}$$

and (15-70) follows from (15-69).

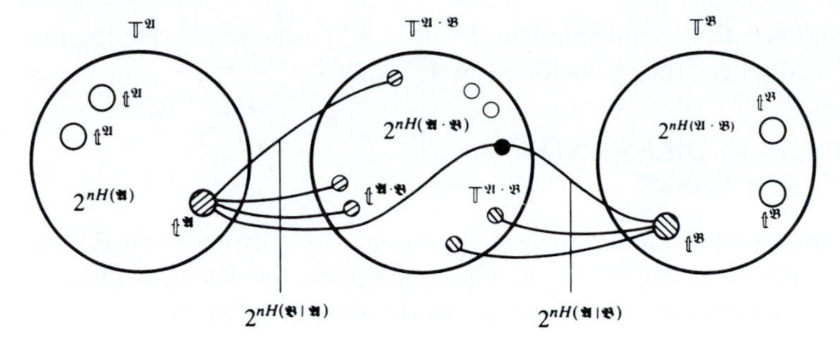

FIGURE 15-13

II. We shall show, finally, that

$$H(\mathfrak{A}\cdot\mathfrak{B}) = H(\mathfrak{B}) + H(\mathfrak{A}|\mathfrak{B}) \tag{15-71}$$

Proof. There are $n_{\mathbb{T}^{\mathfrak{B}}}$ sequences in the set $\mathbb{T}^{\mathfrak{B}}$ and $n_{\mathbb{T}^{\mathfrak{A}\cdot\mathfrak{B}}}$ sequences in the set $\mathbb{T}^{\mathfrak{A}\cdot\mathfrak{B}}$. The ratio

$$\frac{n_{\mathbb{T}^{\mathfrak{A}\cdot\mathfrak{B}}}}{n_{\mathbb{T}^{\mathfrak{B}}}} \simeq e^{n[H(\mathfrak{A}\cdot\mathfrak{B})-H(\mathfrak{B})]}$$

equals, therefore, the number of $\mathfrak{t}^{\mathfrak{A}\cdot\mathfrak{B}}$ sequences contained in a single $\mathfrak{t}^{\mathfrak{B}}$ sequence *on the average*. To prove (15-71), we must prove, therefore, that this number equals $e^{nH(\mathfrak{A}|\mathfrak{B})}$. We shall prove a stronger statement: The number of $\mathfrak{t}^{\mathfrak{A}\cdot\mathfrak{B}}$ sequences contained in a *single* $\mathfrak{t}^{\mathfrak{B}}$ sequence (Fig. 15-13) equals $e^{nH(\mathfrak{A}|\mathfrak{B})}$.

As we know [see (1-1)], the number of times the event $\mathscr{B}_j$ occurs in a $\mathfrak{t}^{\mathfrak{B}}$ sequence "almost certainly" equals

$$n_j \simeq nP(\mathscr{B}_j) \tag{15-72}$$

We denote by $\mathfrak{t}^{\mathscr{B}_j}$ a subsequence (Fig. 15-12*b*) of $\mathfrak{t}^{\mathfrak{B}}$ in which the number of occurrences of $\mathscr{B}_j$ satisfies (15-72). In this subsequence, the relative frequency of the occurrence of an event $\mathscr{A}_i$ equals $P(\mathscr{A}_i|\mathscr{B}_j)$ [see (2-32)].

We shall use (15-10) to show that the number of typical sequences formed with the elements $\mathscr{A}_i$ of $\mathfrak{A}$ that are included in a $\mathfrak{t}^{\mathscr{B}_j}$ sequence equals

$$e^{n_j H(\mathfrak{A}|\mathscr{B}_j)} \simeq e^{nP(\mathscr{B}_j)H(\mathfrak{A}|\mathscr{B}_j)} \tag{15-73}$$

Indeed, this follows from (15-10) if we introduce the following changes: We replace $P(\mathscr{A}_i)$ by $P(\mathscr{A}_i|\mathscr{B}_j)$, the length n of the original sequences with the length $n_j \simeq nP(\mathscr{B}_j)$, and the entropy $H(\mathfrak{A})$ of $\mathfrak{A}$ with the conditional entropy $H(\mathfrak{A}|\mathscr{B}_j)$.

Returning to the original $\mathfrak{t}^{\mathfrak{B}}$ sequence, we note that it is formed by combining the $\mathfrak{t}^{\mathscr{B}_j}$ sequences that are included in $\mathfrak{t}^{\mathfrak{B}}$. This shows that the total number of $\mathfrak{t}^{\mathfrak{A}}$ sequences that are included in $\mathfrak{t}^{\mathfrak{B}}$ equals the product

$$\prod_j e^{nP(\mathscr{B}_j)H(\mathfrak{A}|\mathscr{B}_j)} = e^{nH(\mathfrak{A}|\mathfrak{B})} \tag{15-74}$$

But each $\mathfrak{t}^{\mathfrak{A}}$ sequence that is included in $\mathfrak{t}^{\mathfrak{B}}$ is a $\mathfrak{t}^{\mathfrak{A}\cdot\mathfrak{B}}$ sequence. Hence the number of $\mathfrak{t}^{\mathfrak{A}\cdot\mathfrak{B}}$ sequences that is included in $\mathfrak{t}^{\mathfrak{B}}$ equals $e^{nH(\mathfrak{A}|\mathfrak{B})}$.

15-3 RANDOM VARIABLES AND STOCHASTIC PROCESSES

Entropy is a number assigned to a partition. To define the entropy of an RV we must, therefore, form a suitable partition. This is simple if the RV is of discrete type. However, for continuous-type RVs we can do so only indirectly.

Discrete type. Suppose that $\mathbf{x}$ is an RV taking the values x_i with

$$P\{\mathbf{x} = x_i\} = p_i$$

The events $\{\mathbf{x} = x_i\}$ are mutually exclusive and their union is the certain event; hence they form a partition. This partition will be denoted by $\mathfrak{A}_x$ and will be called the partition of $\mathbf{x}$.

Definition The entropy $H(\mathbf{x})$ of a discrete-type RV $\mathbf{x}$ is the entropy $H(\mathfrak{A}_x)$ of its partition $\mathfrak{A}_x$:

$$H(\mathbf{x}) = H(\mathfrak{A}_x) = -\sum_i p_i \ln p_i \tag{15-75}$$

Continuous type. The entropy of a continuous-type RV cannot be so defined because the events $\{\mathbf{x} = x_i\}$ do not form a partition (they are not countable). To define $H(\mathbf{x})$, we form, first, the discrete-type RV $\mathbf{x}_\delta$ obtained by rounding off $\mathbf{x}$ as in Fig. 15-14:

$$\mathbf{x}_\delta = n\delta \qquad \text{if} \quad n\delta - \delta < \mathbf{x} \le n\delta \tag{15-76}$$

Clearly,

$$P\{\mathbf{x}_\delta = n\delta\} = P\{n\delta - \delta < \mathbf{x} \le \delta\} = \int_{n\delta-\delta}^{n\delta} f(x)\,dx = \delta \bar{f}(n\delta)$$

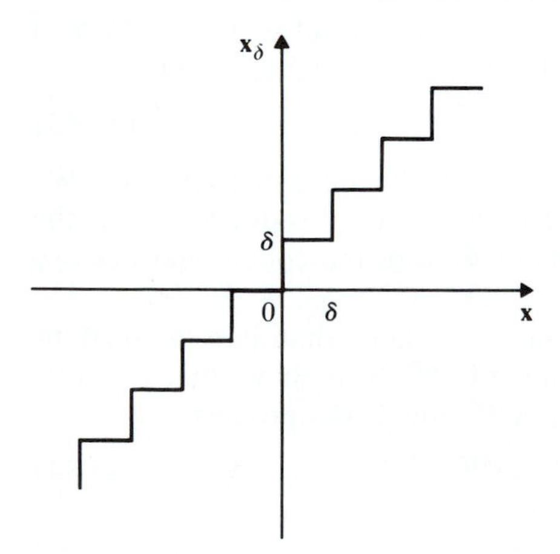

FIGURE 15-14

where $\bar{f}(n\delta)$ is a number between the maximum and the minimum of $f(x)$ in the interval $(n\delta - \delta, n\delta)$. Applying (15-75) to the RV $\mathbf{x}_\delta$, we obtain

$$H(\mathbf{x}_\delta) = -\sum_{n=-\infty}^{\infty} \delta\bar{f}(n\delta)\ln[\delta\bar{f}(n\delta)]$$

and since

$$\sum_{n=-\infty}^{\infty} \delta\bar{f}(n\delta) = \int_{-\infty}^{\infty} f(x)\,dx = 1$$

we conclude that

$$H(\mathbf{x}_\delta) = -\ln\delta - \sum_{n=-\infty}^{\infty} \delta\bar{f}(n\delta)\ln\bar{f}(n\delta) \tag{15-77}$$

As $\delta \to 0$, the RV $\mathbf{x}_\delta$ tends to $\mathbf{x}$; however, its entropy $H(\mathbf{x}_\delta)$ tends to ∞ because $-\ln\delta \to \infty$. For this reason, we define the entropy $H(\mathbf{x})$ of $\mathbf{x}$ not as the limit of $H(\mathbf{x}_\delta)$ but as the limit of the sum $H(\mathbf{x}_\delta) + \ln\delta$ as $\delta \to 0$. This yields

$$H(\mathbf{x}_\delta) + \ln\delta \xrightarrow[\delta\to 0]{} -\int_{-\infty}^{\infty} f(x)\ln f(x)\,dx \tag{15-78}$$

Definition The entropy of a continuous-type RV $\mathbf{x}$ is by definition the integral

$$H(\mathbf{x}) = -\int_{-\infty}^{\infty} f(x)\ln f(x)\,dx \tag{15-79}$$

The integration extends only over the region where $f(x) \neq 0$ because $f(x)\ln f(x) = 0$ if $f(x) = 0$.

Example 15-10. If $\mathbf{x}$ is uniform in the interval $(0, a)$, then

$$H(\mathbf{x}) = -\frac{1}{a}\int_0^a \ln\frac{1}{a}\,dx = \ln a \tag{15-80}$$

Notes 1. The entropy $H(\mathbf{x}_\delta)$ of $\mathbf{x}_\delta$ is a measure of our uncertainty about the RV $\mathbf{x}$ rounded off to the nearest $n\delta$. If δ is small, the resulting uncertainty is large and it tends to ∞ as $\delta \to 0$. This conclusion is based on the assumption that $\mathbf{x}$ can be *observed perfectly*; that is, its various values can be recognized as distinct no matter how close they are. In a physical experiment, however, this assumption is not realistic. Values of $\mathbf{x}$ that differ slightly cannot always be treated as distinct (noise considerations or round-off errors, for example). The presence of the term $\ln\delta$ in (15-78) is, in a sense, a recognition of this ambiguity.

2. As in the case of arbitrary partitions, the entropy of a discrete-type RV $\mathbf{x}$ is positive and it is used as a measure of uncertainty about $\mathbf{x}$. This is not so, however, for continuous-type RVs. Their entropy can take any value from $-\infty$ to ∞ and it is used to measure only changes in uncertainty. The various properties of partitions also apply to continuous-type RVs if, as is generally the case, they involve only differences of entropies.

Entropy as expected value. The integral in (15-79) is the expected value of the RV $\mathbf{y} = -\ln f(\mathbf{x})$ obtained through the transformation $g(x) = -\ln f(x)$:

$$H(\mathbf{x}) = E\{-\ln f(\mathbf{x})\} = -\int_{-\infty}^{\infty} f(x) \ln f(x)\, dx \tag{15-81}$$

Similarly, the sum in (15-75) can be written as the expected value of the RV $-\ln p(\mathbf{x})$:

$$H(\mathbf{x}) = E\{-\ln p(\mathbf{x})\} = -\sum_i p_i \ln p_i \tag{15-82}$$

where now $p(x)$ is a function defined only for $x = x_i$ and such that $p(x_i) = p_i$.

Example 5-11. If

$$f(x) = ce^{-cx}U(x) \qquad \text{then} \qquad E\{-\ln f(\mathbf{x})\} = E\{c\mathbf{x} - \ln c\}$$

Since $E\{c\mathbf{x}\} = 1$, this yields

$$H(\mathbf{x}) = 1 - \ln c = \ln \frac{e}{c} \tag{15-83}$$

Example 15-12. If

$$f(x) = \frac{1}{\sigma\sqrt{2\pi}} e^{-(x-\eta)^2/2\sigma^2}$$

then

$$E\{-\ln f(\mathbf{x})\} = \ln \sigma\sqrt{2\pi} + E\left\{\frac{(\mathbf{x}-\eta)^2}{2\sigma^2}\right\} = \ln \sigma\sqrt{2\pi} + \frac{\sigma^2}{2\sigma^2}$$

Hence the entropy of a *normal* RV equals

$$H(\mathbf{x}) = \ln \sigma\sqrt{2\pi e} \tag{15-84}$$

Joint entropy. Suppose that $\mathbf{x}$ and $\mathbf{y}$ are two discrete-type RVs taking the values x_i and y_j respectively with

$$P\{\mathbf{x} = x_i, \mathbf{y} = y_j\} = p_{ij}$$

Their joint entropy, denoted by $H(\mathbf{x}, \mathbf{y})$, is by definition the entropy of the product of their respective partitions. Clearly, the elements of $\mathfrak{A}_x \cdot \mathfrak{A}_y$ are the events $\{\mathbf{x} = x_i, \mathbf{y} = y_j\}$. Hence

$$H(\mathbf{x}, \mathbf{y}) = H(\mathfrak{A}_x \cdot \mathfrak{A}_y) = -\sum_{i,j} p_{ij} \ln p_{ij}$$

The above can be written as an expected value

$$H(\mathbf{x}, \mathbf{y}) = E\{-\ln p(\mathbf{x}, \mathbf{y})\}$$

where $p(x, y)$ is a function defined only for $x = x_i$ and $y = y_j$ and it is such that $p(x_i, y_j) = p_{ij}$.

The joint entropy $H(\mathbf{x}, \mathbf{y})$ of two continuous-type RVs $\mathbf{x}$ and $\mathbf{y}$ is defined as the limit of the sum

$$H(\mathbf{x}_\delta, \mathbf{y}_\delta) + 2 \ln \delta$$

where $\mathbf{x}_\delta$ and $\mathbf{y}_\delta$ are their staircase approximation. Reasoning as in (15-78), we obtain

$$H(\mathbf{x}, \mathbf{y}) = -\int_{-\infty}^{\infty}\int_{-\infty}^{\infty} f(x, y) \ln f(x, y)\, dx\, dy = E\{-\ln f(\mathbf{x}, \mathbf{y})\} \quad (15\text{-}85)$$

Example 15-13. If the RVs $\mathbf{x}$ and $\mathbf{y}$ are *jointly normal* as in (6-15), then

$$\ln f(x, y) = \frac{-1}{2(1 - r^2)}\left[\frac{(x - \eta_1)^2}{\sigma_1^2} - 2r\frac{(x - \eta_1)(y - \eta_2)}{\sigma_1\sigma_2} + \frac{(y - \eta y_2)^2}{\sigma_2^2}\right]$$
$$- \ln 2\pi\sigma_1\sigma_2\sqrt{1 - r^2}$$

In this case,

$$E\left\{\frac{(\mathbf{x} - \eta_1)^2}{\sigma_1^2} - 2r\frac{(\mathbf{x} - \eta_1)(\mathbf{y} - \eta_2)}{\sigma_1\sigma_2} + \frac{(\mathbf{y} - \eta_2)^2}{\sigma_2^2}\right\} = 1 - 2r^2 + 1$$

Hence

$$E\{-\ln f(\mathbf{x}, \mathbf{y})\} = 1 + \ln 2\pi\sigma_1\sigma_2\sqrt{1 - r^2}$$

From the above and (15-85) it follows that the joint entropy of two jointly normal RVs equals

$$H(\mathbf{x}, \mathbf{y}) = \ln 2\pi e\sqrt{\Delta} \quad (15\text{-}86)$$

where

$$\Delta = \mu_{11}\mu_{22} - \mu_{12}^2 \qquad \mu_{11} = \sigma_1^2 \qquad \mu_{22} = \sigma_2^2 \qquad \mu_{12} = r\sigma_1\sigma_2$$

Conditional entropy. Consider two discrete-type RVs $\mathbf{x}$ and $\mathbf{y}$ taking the values x_i and y_j with

$$P\{\mathbf{x} = x_i | \mathbf{y} = y_j\} = \pi_{ji} = p_{ji}/p_j$$

The conditional entropy $H(\mathbf{x}|y_j)$ of $\mathbf{x}$ assuming $\mathbf{y} = y_j$ is by definition the conditional entropy of the partition $\mathfrak{A}_x$ of $\mathbf{x}$ assuming $\{\mathbf{y} = y_j\}$. From the above and (15-39) it follows that

$$H(\mathbf{x}|y_j) = -\sum_i \pi_{ji} \ln \pi_{ji} \quad (15\text{-}87)$$

The conditional entropy $H(\mathbf{x}|\mathbf{y})$ of $\mathbf{x}$ assuming $\mathbf{y}$ is the conditional entropy of $\mathfrak{A}_x$ assuming $\mathfrak{A}_y$. Thus [see (15-41)]

$$H(\mathbf{x}|\mathbf{y}) = -\sum_j p_j H(\mathbf{x}|y_j) = -\sum_{i,j} p_{ji} \ln \pi_{ji} \quad (15\text{-}88)$$

For continuous-type RVs the corresponding concepts are defined similarly

$$H(\mathbf{x}|y) = -\int_{-\infty}^{\infty} f(x|y)\ln f(x|y)\,dx \tag{15-89}$$

$$H(\mathbf{x}|\mathbf{y}) = -\int_{-\infty}^{\infty} f(y)H(\mathbf{x}|y)\,dy = \int_{-\infty}^{\infty}\int_{-\infty}^{\infty} f(x,y)\ln f(x|y)\,dx\,dy \tag{15-90}$$

The above integrals can be written as expected values [see also (7-66)]

$$H(\mathbf{x}|y) = E\{-\ln f(\mathbf{x}|\mathbf{y})\,|\mathbf{y} = y\} \tag{15-91}$$

$$H(\mathbf{x}|\mathbf{y}) = E\{-\ln f(\mathbf{x}|\mathbf{y})\} = E\{E\{-\ln f(\mathbf{x}|\mathbf{y})\,|\mathbf{y}\}\} \tag{15-92}$$

The discrete case leads to similar expressions.

Mutual information. Guided by (15-53), we shall call the function

$$I(\mathbf{x},\mathbf{y}) = H(\mathbf{x}) + H(\mathbf{y}) - H(\mathbf{x},\mathbf{y}) \tag{15-93}$$

the mutual information of the RVs **x** and **y**.

From (15-81) and (15-85) it follows that $I(\mathbf{x},\mathbf{y})$ can be written as an expected value

$$I(\mathbf{x},\mathbf{y}) = E\left[\ln\frac{f(\mathbf{x},\mathbf{y})}{f(\mathbf{x})f(\mathbf{y})}\right] \tag{15-94}$$

Since $f(x, y) = f(x|y)f(y)$ it follows from the above and (15-92) that

$$I(\mathbf{x},\mathbf{y}) = H(\mathbf{x}) - H(\mathbf{x}|\mathbf{y}) = H(\mathbf{y}) - H(\mathbf{y}|\mathbf{x}) \tag{15-95}$$

Example 15-14. If two RVs **x** and **y** are *jointly normal* with zero mean, then [see (7-42)] the conditional density $f(x|y)$ is normal with mean $r\sigma_x/\sigma_y$ and variance $\sigma_x^2(1 - r^2)$. From this and (15-84) it follows that

$$H(\mathbf{x}|y) = E\{-\ln f(\mathbf{x}|y)\} = \ln \sigma_x\sqrt{2\pi e(1 - r^2)} \tag{15-96}$$

Since this is independent of y, it follows from (15-92) that

$$H(\mathbf{x}|\mathbf{y}) = H(\mathbf{x}|y) \tag{15-97}$$

This yields [see (15-95)]

$$I(\mathbf{x},\mathbf{y}) = H(\mathbf{x}) - H(\mathbf{x}|\mathbf{y}) = -0.5\ln(1 - r^2) \tag{15-98}$$

We note finally that [see (15-86)]

$$H(\mathbf{x}|\mathbf{y}) + H(\mathbf{y}) = \ln 2\pi e\sqrt{\Delta} = H(\mathbf{x},\mathbf{y})$$

Special Case. Suppose that $\mathbf{y} = \mathbf{x} + \mathbf{n}$ where **n** is independent of **x** and $E\{\mathbf{n}^2\} = N$. In this case,

$$E\{\mathbf{xy}\} = \sigma_x^2 \qquad E\{\mathbf{y}^2\} = \sigma_x^2 + N \qquad r^2 = \frac{\sigma_x^2}{\sigma_x^2 + N}$$

Inserting into (15-98), we obtain

$$I(\mathbf{x},\mathbf{y}) = 0.5\ln\left(1 + \frac{\sigma_x^2}{N}\right) \tag{15-99}$$

PROPERTIES. The properties of entropy, developed in Sec. 15-2 for arbitrary partitions, are obviously true for the entropy of discrete-type RVs and can be simply established as appropriate limits for continuous-type RVs. It might be of interest however, to prove directly theorems (15-45) and (15-49) using the representation of entropy as expected value. The proofs are based on the following version of inequality (15-38): If $\mathbf{x}$ and $\mathbf{y}$ are two RVs with respective densities $a(x)$ and $b(y)$, then

$$E\{\ln a(\mathbf{x})\} \geq E\{\ln b(\mathbf{x})\} \tag{15-100}$$

Equality holds iff $a(x) = b(x)$.

Proof. Applying the inequality $\ln z \leq z - 1$ to the function $z = b(x)/a(x)$, we obtain

$$\ln b(x) - \ln a(x) = \ln\frac{b(x)}{a(x)} \leq \frac{b(x)}{a(x)} - 1$$

Multiplying by $a(x)$ and integrating, we obtain

$$\int_{-\infty}^{\infty} a(x)[\ln b(x) - \ln a(x)]\, dx \leq \int_{-\infty}^{\infty} [b(x) - a(x)]\, dx = 0$$

and (15-100) results. The right side is 0 because the functions $a(x)$ and $b(x)$ are densities by assumption.

Inequality (15-100) can be readily extended to n-dimensional densities. For example, if $a(x, y)$ and $b(z, w)$ are the joint densities of the RVs $\mathbf{x}$, $\mathbf{y}$ and $\mathbf{z}, \mathbf{w}$ respectively, then

$$E\{\ln a(\mathbf{x}, \mathbf{y})\} \geq E\{\ln b(\mathbf{x}, \mathbf{y})\} \tag{15-101}$$

THEOREM 1.

$$H(\mathbf{x}, \mathbf{y}) \leq H(\mathbf{x}) + H(\mathbf{y}) \tag{15-102}$$

Proof. Suppose that $f_{xy}(x, y)$ is the joint density of the RVs $\mathbf{x}$ and $\mathbf{y}$ and $f_x(x)$ and $f_y(y)$ their marginal densities. Clearly, the product $f_x(z)f_y(w)$ is the joint density of two independent RVs $\mathbf{z}$ and $\mathbf{w}$. Applying (15-101), we conclude that

$$E\{\ln f_{xy}(\mathbf{x}, \mathbf{y})\} \geq E\{\ln[f_x(\mathbf{x})f_y(\mathbf{y})]\} = E\{\ln f_x(\mathbf{x})\} + E\{\ln f_y(\mathbf{y})\}$$

and (15-102) results.

THEOREM 2.

$$H(\mathbf{x}, \mathbf{y}) = H(\mathbf{x}|\mathbf{y}) + H(\mathbf{y}) = H(\mathbf{y}|\mathbf{x}) + H(\mathbf{x}) \tag{15-103}$$

Proof. Inserting the identity $f(x, y) = f(x|y)f(y)$ into (15-85), we obtain

$$H(\mathbf{x}, \mathbf{y}) = E\{-\ln f(\mathbf{x}, \mathbf{y})\} = E\{-\ln f(\mathbf{x}|\mathbf{y})\} + E\{-\ln f(\mathbf{y})\}$$

and the first equality in (15-103) results. The second follows because $H(\mathbf{x}, \mathbf{y}) = H(\mathbf{y}, \mathbf{x})$.

COROLLARY. Comparing (15-102) with (15-103), we conclude that

$$H(\mathbf{x}|\mathbf{y}) \leq H(\mathbf{x}) \tag{15-104}$$

Note If the RV $\mathbf{y}$ is of discrete type, then $H(\mathbf{y}|\mathbf{x}) \geq 0$ and (15-103) yields $H(\mathbf{x}) \leq H(\mathbf{x}, \mathbf{y})$. This is not, however, true in general if $\mathbf{y}$ is of continuous type.

Generalizations. The preceding results can be readily generalized to an arbitrary number of RVs: Suppose that $\mathbf{x}_1, \ldots, \mathbf{x}_n$ are n RVs with joint density $f(x_1, \ldots, x_n)$. Extending (15-85), we define their *joint entropy* as an expected value

$$H(\mathbf{x}_1, \ldots, \mathbf{x}_n) = E\{-\ln f(\mathbf{x}_1, \ldots, \mathbf{x}_n)\} \tag{15-105}$$

If the RVs $\mathbf{x}_i$ are *independent*, then

$$f(x_1, \ldots, x_n) = f(x_1) \cdots f(x_n)$$

and (15-105) yields

$$H(\mathbf{x}_1, \ldots, \mathbf{x}_n) = H(\mathbf{x}_1) + \cdots + H(\mathbf{x}_n) \tag{15-106}$$

Conditional entropies are defined similarly. For example [see (15-92)]

$$H(\mathbf{x}_n|\mathbf{x}_{n-1}, \ldots, \mathbf{x}_1) = E\{-\ln f(\mathbf{x}_n|\mathbf{x}_{n-1}, \ldots, \mathbf{x}_1)\} \tag{15-107}$$

Chain rule From the identity [see (8-37)]

$$f(x_1, \ldots, x_n) = f(x_n|x_{n-1}, \ldots, x_1) \cdots f(x_2|x_1) f(x_1)$$

and (15-107) it follows that

$$H(\mathbf{x}_1, \ldots, \mathbf{x}_n) = H(\mathbf{x}_n|\mathbf{x}_{n-1}, \ldots, \mathbf{x}_1) + \cdots + H(\mathbf{x}_2|\mathbf{x}_1) + H(\mathbf{x}_1) \tag{15-108}$$

The following relationships are simple extensions of (15-102) and (15-103):

$$\begin{aligned} H(\mathbf{x}, \mathbf{y}|\mathbf{z}) &\leq H(\mathbf{x}|\mathbf{z}) + H(\mathbf{y}|\mathbf{z}) \\ H(\mathbf{x}, \mathbf{y}|\mathbf{z}) &= H(\mathbf{x}|\mathbf{z}) + H(\mathbf{y}|\mathbf{x}, \mathbf{z}) \\ H(\mathbf{x}_1, \ldots, \mathbf{x}_n) &\leq H(\mathbf{x}_1) + \cdots + H(\mathbf{x}_n) \end{aligned} \tag{15-109}$$

Example 15-15. If the RVs $\mathbf{x}_i$ are jointly normal with covariance matrix C as in (8-58), then

$$E\{-\ln f(\mathbf{x}_1, \ldots, \mathbf{x}_n)\} = \ln\sqrt{(2\pi)^n \Delta} + \tfrac{1}{2} E\{\mathbf{X} C^{-1} \mathbf{X}'\} \tag{15-110}$$

This yields (see Prob. 8-23)

$$H(\mathbf{x}_1, \ldots, \mathbf{x}_n) = \ln\sqrt{(2\pi e)^n \Delta} \tag{15-111}$$

Transformations of RVs

We shall compare the entropy of the RVs $\mathbf{x}$ and $\mathbf{y} = g(\mathbf{x})$.

Discrete type. If the RV $\mathbf{x}$ is of discrete type, then

$$H(\mathbf{y}) \le H(\mathbf{x}) \tag{15-112}$$

with equality iff the transformation $y = g(x)$ has a unique inverse.

Proof. Suppose that $\mathbf{x}$ takes the values x_i with probability p_i and $g(x)$ has a unique inverse. In this case,

$$P\{\mathbf{y} = y_i\} = P\{\mathbf{x} = x_i\} = p_i \qquad y_i = g(x_i)$$

hence $H(\mathbf{y}) = H(\mathbf{x})$. If the transformation is not one-to-one, then $\mathbf{y} = y_i$ for more than one value of $\mathbf{x}$. This results in a reduction of $H(\mathbf{x})$ [see (15-31)].

Continuous type. If the RV $\mathbf{x}$ is of continuous type, then

$$H(\mathbf{y}) \le H(\mathbf{x}) + E\{\ln|g'(\mathbf{x})|\} \tag{15-113}$$

with equality iff the transformation $y = g(x)$ has a unique inverse.

Proof. As we know [see (5-5)] if $y = g(x)$ has a unique inverse $x = g^{(-1)}(y)$, then

$$f_y(y) = \frac{f_x(x)}{|g'(x)|} \qquad dy = g'(x)\,dx$$

Hence

$$\begin{aligned} H(\mathbf{y}) &= -\int_{-\infty}^{\infty} f_y(y)\ln f_y(y)\,dy = -\int_{-\infty}^{\infty} f_x(x)\ln\frac{f_x(x)}{|g'(x)|}\,dx \\ &= -\int_{-\infty}^{\infty} f_x(x)\ln f_x(x)\,dx + \int_{-\infty}^{\infty} f_x(x)\ln|g'(x)|\,dx \end{aligned}$$

and (15-113) results.

Several RVs. Reasoning as in (15-113), we can similarly show that if

$$\mathbf{y}_i = g_i(\mathbf{x}_1, \dots, \mathbf{x}_n) \qquad i = 1, \dots, n$$

are n functions of the RVs $\mathbf{x}_i$, then

$$H(\mathbf{y}_1, \dots, \mathbf{y}_n) \le H(\mathbf{x}_1, \dots, \mathbf{x}_n) + E\{\ln|J(\mathbf{x}_1, \dots, \mathbf{x}_n)|\} \tag{15-114}$$

where $J(x_1, \dots, x_n)$ is the jacobian of the above transformation [see (8-9)]. Equality holds iff the transformation has a unique inverse.

Linear transformations Suppose that

$$\mathbf{y}_i = a_{i1}\mathbf{x}_1 + \cdots + a_{in}\mathbf{x}_n$$

Denoting by Δ the determinant of the coefficients, we conclude from (15-114) that if $\Delta \ne 0$ then

$$H(\mathbf{y}_1, \dots, \mathbf{y}_n) = H(\mathbf{x}_1, \dots, \mathbf{x}_n) + \ln|\Delta| \tag{15-115}$$

because the transformation has a unique inverse and Δ does not depend on $\mathbf{x}_i$.

Stochastic Processes and Entropy Rate

As we know, the statistics of most stochastic processes are determined in terms of the joint density $f(x_1, \ldots, x_m)$ of the RVs $\mathbf{x}(t_1), \ldots, \mathbf{x}(t_m)$. The joint entropy

$$H(\mathbf{x}_1, \ldots, \mathbf{x}_m) = E\{-\ln f(\mathbf{x}_1, \ldots, \mathbf{x}_m)\} \tag{15-116}$$

of these RVs is the *mth-order entropy* of the process $\mathbf{x}(t)$. This function equals the uncertainty about the above RVs and it equals the information gained when they are observed.

In general, the uncertainty about the values of $\mathbf{x}(t)$ on the entire t axis or even on a finite interval, no matter how small, is infinite. However, if $\mathbf{x}(t)$ can be expressed in terms of its values on a countable set of points, as is the case for bandlimited processes, then a rate of uncertainty can be introduced. It suffices, therefore, to consider only discrete-time processes.

The mth-order entropy of a discrete-time process $\mathbf{x}_n$ is the joint entropy $H(\mathbf{x}_1, \ldots, \mathbf{x}_m)$ of the m RVs

$$\mathbf{x}_n, \mathbf{x}_{n-1}, \ldots, \mathbf{x}_{n-m+1} \tag{15-117}$$

defined as in (15-116). We shall assume throughout that the process $\mathbf{x}_n$ is SSS. In this case, $H(\mathbf{x}_1 \cdots \mathbf{x}_m)$ is the uncertainty about any m consecutive values of the process $\mathbf{x}_n$. The first-order entropy will be denoted by $H(\mathbf{x})$. Thus $H(\mathbf{x})$ equals the uncertainty about $\mathbf{x}_n$ for a specific n.

Clearly [see (15-109)]

$$H(\mathbf{x}_1, \ldots, \mathbf{x}_m) \le H(\mathbf{x}_1) + \cdots + H(\mathbf{x}_m) = mH(\mathbf{x}) \tag{15-118}$$

Special cases (a) If the process $\mathbf{x}_n$ is *strictly white*, that is, if the RVs $\mathbf{x}_n, \mathbf{x}_{n-1}, \ldots$ are independent, then [see (15-106)]

$$H(\mathbf{x}_1, \ldots, \mathbf{x}_m) = mH(\mathbf{x}) \tag{15-119}$$

(b) If the process $\mathbf{x}_n$ is *Markoff*, then [see (16-99)]

$$f(x_1, \ldots, x_m) = f(x_m|x_{m-1}) \cdots f(x_2|x_1) f(x_1) \tag{15-120}$$

This yields

$$H(\mathbf{x}_1, \ldots, \mathbf{x}_m) = H(\mathbf{x}_m|\mathbf{x}_{m-1}) + \cdots + H(\mathbf{x}_2|\mathbf{x}_1) + H(\mathbf{x}_1) \tag{15-121}$$

From (15-103) and the stationarity of $\mathbf{x}_n$ it follows, therefore, that

$$H(\mathbf{x}_1, \ldots, \mathbf{x}_m) = (m-1)H(\mathbf{x}_1, \mathbf{x}_2) - (m-2)H(\mathbf{x}) \tag{15-122}$$

We have thus expressed the mth-order entropy of a Markoff process in terms of its first- and second-order entropies.

CONDITIONAL ENTROPY. The conditional entropy of order m:

$$H(\mathbf{x}_n|\mathbf{x}_{n-1}, \ldots, \mathbf{x}_{n-m})$$

of a process $\mathbf{x}_n$ is the uncertainty about its present under the assumption that its m most recent values have been observed. Extending (15-104), we can readily show that

$$H(\mathbf{x}_n|\mathbf{x}_{n-1},\ldots,\mathbf{x}_{n-m}) \le H(\mathbf{x}_n|\mathbf{x}_{n-1},\ldots,\mathbf{x}_{n-m+1}) \qquad (15\text{-}123)$$

Thus the above conditional entropy is a decreasing function of m. If, therefore, it is bounded from below, it tends to a limit. This is certainly the case if the RVs $\mathbf{x}_n$ are of discrete type because then all entropies are positive. The limit will be denoted by $H_c(\mathbf{x})$ and will be called the *conditional entropy* of the process $\mathbf{x}_n$:

$$H_c(\mathbf{x}) = \lim_{m\to\infty} H(\mathbf{x}_n|\mathbf{x}_{n-1},\ldots,\mathbf{x}_{n-m}) \qquad (15\text{-}124)$$

The function $H_c(\mathbf{x})$ is a measure of our uncertainty about the present of $\mathbf{x}_n$ under the assumption that its entire past is observed.

Special cases (a) If $\mathbf{x}_n$ is *strictly white*, then

$$H_c(\mathbf{x}) = H(\mathbf{x})$$

(b) If $\mathbf{x}_n$ is a *Markoff* process, then

$$H(\mathbf{x}_n|\mathbf{x}_{n-1},\ldots,\mathbf{x}_{n-m}) = H(\mathbf{x}_n|\mathbf{x}_{n-1})$$

Since $\mathbf{x}_n$ is a stationary process, the above equals $H(\mathbf{x}_2|\mathbf{x}_1)$. Hence

$$H_c(\mathbf{x}) = H(\mathbf{x}_2|\mathbf{x}_1) = H(\mathbf{x}_1,\mathbf{x}_2) - H(\mathbf{x}) \qquad (15\text{-}125)$$

This shows that if $\mathbf{x}_{n-1}$ is observed, then the past has no effect on the uncertainty of the present.

ENTROPY RATE. The ratio $H(\mathbf{x}_1 \cdots \mathbf{x}_m)/m$ is the average uncertainty per sample in a block of m consecutive samples. The limit of this average as $m \to \infty$ will be denoted by $\overline{H}(\mathbf{x})$ and will be called the *entropy rate* of the process $\mathbf{x}_n$:

$$\overline{H}(\mathbf{x}) = \lim_{m\to\infty} \frac{1}{m} H(\mathbf{x}_1,\ldots,\mathbf{x}_m) \qquad (15\text{-}126)$$

If $\mathbf{x}_n$ is *strictly white*, then

$$\overline{H}(\mathbf{x}) = H(\mathbf{x}) = H_c(\mathbf{x})$$

If $\mathbf{x}_n$ is *Markoff*, then [see (15-122)]

$$\overline{H}(\mathbf{x}) = H(\mathbf{x}_1,\mathbf{x}_2) - H(\mathbf{x}) = H_c(\mathbf{x}) \qquad (15\text{-}127)$$

Thus, in both cases, the limit in (15-126) exists and it equals $H_c(\mathbf{x})$. We show next that this is true in general.

THEOREM. The entropy rate of a process $\mathbf{x}_n$ equals its conditional entropy

$$\overline{H}(\mathbf{x}) = H_c(\mathbf{x}) \qquad (15\text{-}128)$$

Proof. This is a consequence of the following simple property of convergent sequences: If

$$a_k \to a \qquad \text{then} \qquad \frac{1}{m}\sum_{k=1}^{m} a_k \to a \qquad (15\text{-}129)$$

Since $\mathbf{x}_n$ is stationary we conclude, as in (15-108), that

$$H(\mathbf{x}_1, \ldots, \mathbf{x}_m) = H(\mathbf{x}) + \sum_{k=1}^{m} H(\mathbf{x}_n | \mathbf{x}_{n-1}, \ldots, \mathbf{x}_{n-k})$$

Dividing by m and using (15-129), we obtain (15-128) because

$$H(\mathbf{x}_n | \mathbf{x}_{n-1}, \ldots, \mathbf{x}_{n-k})$$

tends to $H_c(\mathbf{x})$ as $k \to \infty$.

Note If $\mathbf{x}_n$ equals the samples $\mathbf{x}(nT)$ of $\mathbf{x}(t)$, then the entropy rate is measured in bits per T seconds. If we wish to measure it in bits per second, we must divide by T.

Normal processes. We shall show that if $\mathbf{x}_n$ is a normal process with power spectrum $S(\omega)$, then

$$\overline{H}(\mathbf{x}) = \ln\sqrt{2\pi e} + \frac{1}{4\pi}\int_{-\pi}^{\pi} \ln S(\omega)\, d\omega \tag{15-130}$$

Proof. As we know, the function $f(x_{m+1} | x_m, \ldots, x_1)$ is a one-dimensional normal density with variance Δ_{m+1}/Δ_m [see (8-85) and (14-66)]. Hence

$$H(\mathbf{x}_n | \mathbf{x}_{n-1}, \ldots, \mathbf{x}_{n-m}) = \ln\sqrt{\frac{2\pi e\, \Delta_{m+1}}{\Delta_m}} \tag{15-131}$$

as in (15-84). This leads to the conclusion that

$$H_c(\mathbf{x}) = \ln\sqrt{2\pi e} + \frac{1}{2}\lim_{m\to\infty} \ln\frac{\Delta_{m+1}}{\Delta_m} \tag{15-132}$$

and (15-130) follows from (14-70) and Prob. 14-15.

ENTROPY RATE OF SYSTEM RESPONSE. We shall show that the entropy rate $\overline{H}(\mathbf{y})$ of the output $\mathbf{y}_n$ of a linear system $\mathbf{L}(z)$ is given by

$$\overline{H}(\mathbf{y}) = \overline{H}(\mathbf{x}) + \frac{1}{2\pi}\int_{-\pi}^{\pi} \ln\left|\mathbf{L}(e^{j\omega})\right| d\omega \tag{15-133}$$

where $\overline{H}(\mathbf{x})$ is the entropy rate of the input $\mathbf{x}_n$ (Fig. 15-15).

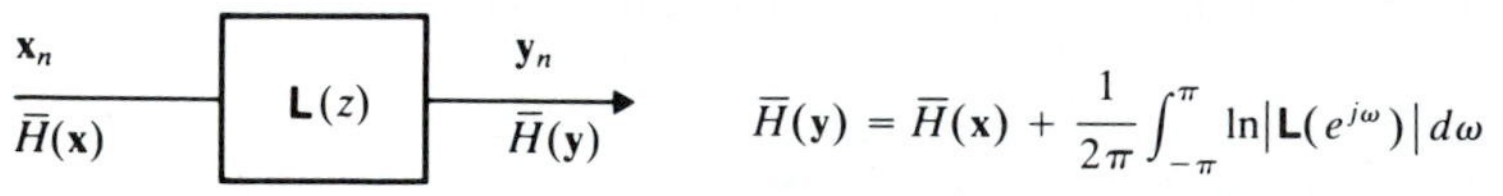

$\mathbf{x}_n$, $\overline{H}(\mathbf{x})$ → $\mathbf{L}(z)$ → $\mathbf{y}_n$, $\overline{H}(\mathbf{y})$

$$\overline{H}(\mathbf{y}) = \overline{H}(\mathbf{x}) + \frac{1}{2\pi}\int_{-\pi}^{\pi} \ln\left|\mathbf{L}(e^{j\omega})\right| d\omega$$

FIGURE 15-15

Suppose, first, that $\mathbf{x}_n$ is a normal process. In this case $\mathbf{y}_n$ is also normal and its entropy rate is given by (15-130) where

$$S(\omega) = S_y(\omega) = S_x(\omega)|\mathbf{L}(e^{j\omega})|^2 \tag{15-134}$$

This yields

$$\overline{H}(\mathbf{y}) = \ln\sqrt{2\pi e} + \frac{1}{4\pi}\int_{-\pi}^{\pi}\left[\ln S_x(\omega) + \ln|\mathbf{L}(e^{j\omega})|^2\right]d\omega \tag{15-135}$$

and (15-133) follows.

The proof for arbitrary processes is involved. We shall sketch a justification based on (15-115): If the RVs $\mathbf{y}_1, \ldots, \mathbf{y}_m$ depend linearly on the RVs $\mathbf{x}_1, \ldots, \mathbf{x}_m$, then

$$H(\mathbf{y}_1, \ldots, \mathbf{y}_m) = H(\mathbf{x}_1, \ldots, \mathbf{x}_m) + K_o \tag{15-136}$$

where $K_o = \ln|\Delta|$ is a constant that depends only on the coefficients of the transformation. The process $\mathbf{y}_n$ depends linearly on $\mathbf{x}_n$:

$$\mathbf{y}_n = \sum_{k=0}^{\infty} l_k \mathbf{x}_{n-k} \qquad n = -\infty, \ldots, \infty \tag{15-137}$$

where now the transformation matrix is of infinity order. Extending (15-136) to infinitely many variables, we conclude with (15-126) that

$$\overline{H}(\mathbf{y}) = \overline{H}(\mathbf{x}) + K \tag{15-138}$$

where again K is a constant that depends only on the parameters of the system $\mathbf{L}(z)$. As we have seen, if $\mathbf{x}_n$ is normal, then K equals the integral in (15-133). And since K is independent of $\mathbf{x}_n$, it must equal that integral for any $\mathbf{x}_n$.

15-4 THE MAXIMUM ENTROPY METHOD

The MEM is used to determine various parameters of a probability space subject to given constraints. The resulting problem can be solved, in general, only numerically and it involves the evaluation of the maximum of a function of several variables. In a number of important cases, however, the solution can be found analytically or it can be reduced to a system of algebraic equations. In this section, we consider certain special cases, concentrating on constraints in the form of expected values. The results can be obtained with the familiar variational techniques involving Lagrange multipliers or Euler's equations. For most problems under consideration, however, it suffices to use the following form of (15-100).

If $f(x)$ and $\varphi(x)$ are two arbitrary densities, then

$$-\int_{-\infty}^{\infty}\varphi(x)\ln\varphi(x)\,dx \le -\int_{-\infty}^{\infty}\varphi(x)\ln f(x)\,dx \tag{15-139}$$

Example 15-16. In the coin experiment, the probability of heads is often viewed as an RV $\mathbf{p}$ (see bayesian estimation, Sec. 9-2). We shall show that if no prior information about $\mathbf{p}$ is available, then, according to the ME principle, its density $f(p)$ is uniform in the interval $(0,1)$. In this problem we must maximize $H(\mathbf{p})$ subject to the constraint (dictated by the meaning of $\mathbf{p}$) that $f(p)=0$ outside the interval $(0,1)$. The corresponding entropy is, therefore, given by

$$H(\mathbf{p}) = -\int_0^1 f(p)\ln f(p)\,dp$$

and our problem is to find $f(p)$ such as to maximize the above integral.

We maintain that $H(\mathbf{p})$ is maximum if

$$f(p) = 1 \qquad H(\mathbf{p}) = 0$$

Indeed, if $\varphi(p)$ is any other density such that $\varphi(p)=0$ outside the interval $(0,1)$, then [see (15-139)]

$$-\int_0^1 \varphi(p)\ln\varphi(p) \le -\int_0^1 \varphi(p)\ln f(p)\,dp = 0 = H(\mathbf{p})$$

Example 15-17. Suppose that $\mathbf{x}$ is an RV vanishing outside the interval $(-\pi,\pi)$. Using the MEM, we shall determine the density $f(x)$ of $\mathbf{x}$ under the assumption that the coefficients c_n of its Fourier series expansion

$$f(x) = \sum_{n=-\infty}^{\infty} c_n e^{jnx} \qquad -\pi \le x \le \pi$$

are known for $|n| \le N$. Our problem now is to maximize the integral

$$H(\mathbf{x}) = -\int_{-\pi}^{\pi} f(x)\ln f(x)\,dx$$

subject to the constraints

$$c_n = \frac{1}{2\pi}\int_{-\pi}^{\pi} f(x)e^{-jnx}\,dx \qquad |n| \le N \tag{15-140}$$

Clearly, $H(\mathbf{x})$ depends on the unknown coefficients c_n and it is maximum iff

$$\frac{\partial H}{\partial c_n} = \frac{\partial H}{\partial f}\frac{\partial f}{\partial c_n} = -\int_{-\pi}^{\pi}[\ln f(x)+1]e^{jnx}\,dx = 0 \qquad |n| > N$$

This shows that the coefficients γ_n of the Fourier series expansion of the function $\ln f(x) + 1$ in the interval $(-\pi,\pi)$ are 0 for $|n| > N$. Hence

$$\ln f(x) + 1 = \sum_{k=-N}^{N} \gamma_k e^{jkx}$$

From the above it follows that

$$f(x) = \exp\left\{-1 + \sum_{k=-N}^{N} \gamma_k e^{jkx}\right\} \qquad -\pi \le x \le \pi \tag{15-141}$$

We have thus shown that the unknown function is given by an exponential involving the parameters γ_k. These parameters can be determined from (15-140). The resulting system is nonlinear and can only be solved numerically.

Constraints as Expected Values

We shall consider now a class of problems involving constraints in the form of expected values. Such problems are common in statistical mechanics. We start with the one-dimension case.

We wish to determine the density $f(x)$ of an RV $\mathbf{x}$ subject to the condition that the expected values η_i of n known functions $g_i(\mathbf{x})$ of $\mathbf{x}$ are given

$$E\{g_i(\mathbf{x})\} = \int_{-\infty}^{\infty} g_i(x) f(x)\, dx = \eta_i \qquad i = 1, \ldots, n \tag{15-142}$$

Using (15-139), we shall show that the MEM leads to the conclusion that $f(x)$ must be an exponential

$$f(x) = A \exp\{-\lambda_1 g_1(x) - \cdots - \lambda_n g_n(x)\} \tag{15-143}$$

where λ_i are n constants determined from (15-142) and A is such as to satisfy the density condition

$$A \int_{-\infty}^{\infty} \exp\{-\lambda_1 g_1(x) - \cdots - \lambda_n g_n(x)\}\, dx = 1 \tag{15-144}$$

Proof. Suppose that $f(x)$ is given by (15-143). In this case,

$$\int_{-\infty}^{\infty} f(x) \ln f(x)\, dx = \int_{-\infty}^{\infty} f(x)[\ln A - \lambda_1 g_1(x) - \cdots - \lambda_n g_n(x)]\, dx$$

Hence

$$H(\mathbf{x}) = \lambda_1 \eta_1 + \cdots + \lambda_n \eta_n - \ln A \tag{15-145}$$

To prove (15-143), it suffices, therefore, to show that, if $\varphi(x)$ is any other density satisfying the constraints (15-142), then its entropy cannot exceed the right side of (15-145). This follows readily from (15-139):

$$\begin{aligned} -\int_{-\infty}^{\infty} \varphi(x) \ln \varphi(x)\, dx &\le -\int_{-\infty}^{\infty} \varphi(x) \ln f(x)\, dx \\ &= \int_{-\infty}^{\infty} \varphi(x)[\lambda_1 g_1(x) + \cdots + \lambda_n g_n(x) - \ln A]\, dx \\ &= \lambda_1 \eta_1 + \cdots + \lambda_n \eta_n - \ln A \end{aligned}$$

We note that, if $f(x) = 0$ outside a certain set R, then $f(x)$ is again given by (15-143) for every x in R and the region of integration in (15-144) is the set R.

Example 15-18. We shall determine $f(x)$ assuming that $\mathbf{x}$ is a positive RV with known mean η. With $g(x) = x$, it follows from (15-143) that

$$f(x) = \begin{cases} Ae^{-\lambda x} & x > 0 \\ 0 & x < 0 \end{cases}$$

We have thus shown that if an RV is positive with specified mean, then its density obtained with the MEM, is an exponential.

THE PARTITION FUNCTION. In certain problems, it is more convenient to express the given constraints in terms of the partition function (Zustandsumme)

$$Z(\lambda_1,\ldots,\lambda_n) = \frac{1}{A} = \int_{-\infty}^{\infty} \exp\{-\lambda_1 g_1(x) - \cdots - \lambda_n g_n(x)\}\, dx \quad (15\text{-}146)$$

Indeed, differentiating with respect to λ_i, we obtain

$$-\frac{\partial Z}{\partial \lambda_i} = \int_{-\infty}^{\infty} g_i(x)\exp\left\{-\sum_{k=1}^{n} \lambda_k g_k(x)\right\} dx = Z\int_{-\infty}^{\infty} g_i(x) f(x)\, dx$$

This yields

$$-\frac{1}{Z}\frac{\partial Z}{\partial \lambda_i} = -\frac{\partial}{\partial \lambda_i}\ln Z = \eta_i \qquad i = 1,\ldots,n \quad (15\text{-}147)$$

The above is a system of n equations equivalent to (15-142) and can be used to determine the n parameters λ_i.

Example 15-19. In the coin experiment of Example 15-16, we assume that $\mathbf{p}$ is an RV with known mean η. Since $f(p) = 0$ outside the interval (0, 1), (15-143) yields

$$f(p) = \begin{cases} Ae^{-\lambda p} & 0 \le p \le 1 \\ 0 & \text{otherwise} \end{cases} \qquad Z = \int_0^1 e^{-\lambda p}\, dp = \frac{1 - e^{-\lambda}}{\lambda}$$

The constant λ is determined from (15-147):

$$-\frac{1}{Z}\frac{\partial Z}{\partial \lambda} = \frac{1 - e^{-\lambda} - \lambda e^{-\lambda}}{\lambda(1 - e^{-\lambda})} = \eta$$

In Fig. 15-16, we plot λ and $f(p)$ for various values of η. Note that if $\eta = 0.5$, then $\lambda = 0$ and $f(p) = 1$.

Example 15-20. A collection of particles moves in a conservative field whose potential equals $V(x)$. For a specific t, the x component of the position of a particle is an RV $\mathbf{x}$ with density $f(x)$ independent of t (stationary state). Thus the

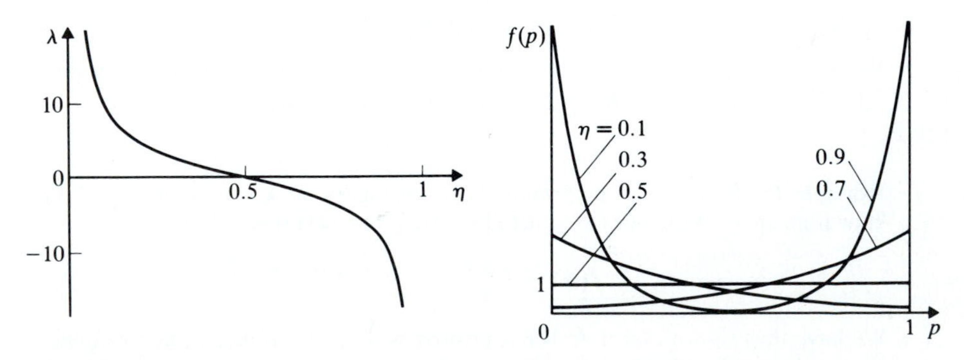

FIGURE 15-16

probability that the particle is between x and $x + dx$ equals $f(x)\,dx$ and the total energy per unit mass of the ensemble equals

$$I = \int_{-\infty}^{\infty} V(x) f(x)\,dx = E\{V(\mathbf{x})\}$$

We shall find $f(x)$ under the assumption that the function $g(x) = V(x)$ and the mean I of $V(\mathbf{x})$ are given. Inserting into (15-143), we obtain

$$f(x) = \frac{1}{Z} e^{-\lambda V(x)} \qquad (15\text{-}148)$$

where

$$Z = \int_{-\infty}^{\infty} e^{-\lambda V(x)}\,dx \qquad \frac{1}{Z}\int_{-\infty}^{\infty} V(x) e^{-\lambda V(x)}\,dx = I$$

Special Case. In a gravitational field, the potential $V(x) = Mgx$ is proportional to the distance x from the ground. Since $f(x) = 0$ for $x < 0$, it follows from (15-148) that

$$f(x) = \frac{Mg}{I} e^{-Mgx/I} U(x)$$

The resulting atmospheric pressure is proportional to $1 - F(x)$.

Example 15-21. We shall find $f(x)$ such that $E\{\mathbf{x}^2\} = m_2$. With $g_1(x) = x^2$, (15-143) yields

$$f(x) = Ae^{-\lambda x^2} \qquad (15\text{-}149)$$

Thus, if the second moment m_2 of an RV $\mathbf{x}$ is specified, then $\mathbf{x}$ is $N(0, m_2)$. We can show similarly that if the variance σ^2 of $\mathbf{x}$ is specified, then $\mathbf{x}$ is $N(\eta, \sigma^2)$ where η is an arbitrary constant.

Special Case. We consider again a collection of particles in stationary motion and we denote by $\mathbf{v}_x$ the x component of their velocity. We shall determine the density $f(v_x)$ of $\mathbf{v}_x$ under the constraint that the corresponding average kinetic energy $K_x = E\{M\mathbf{v}_x^2/2\}$ is specified. This is a special case of (15-149) with $m_2 = 2K_x/M$. Hence

$$f(v_x) = \sqrt{\frac{M}{4\pi K_x}}\, e^{-Mv^2/4K_x}$$

Discrete type RVs. Suppose that an RV $\mathbf{x}$ takes the values x_k with probability p_k. We shall use the MEM to determine p_k under the assumption that the expected values

$$E\{g_i(\mathbf{x})\} = \sum_k g_i(x_k) p_k = \eta_i \qquad (15\text{-}150)$$

of the n known functions $g_i(\mathbf{x})$ are given.

Using (15-37), we can show as in (15-143) that the unknown probabilities equal

$$p_k = A \exp\{-\lambda_1 g_1(x_k) - \cdots - \lambda_n g_n(x_k)\} \tag{15-151}$$

where

$$\frac{1}{A} = Z = \sum_k \exp\{-[\lambda_1 g_1(x_k) + \cdots + \lambda_n g_n(x_k)]\} \tag{15-152}$$

The n constants λ_i are determined either from (15-150) or from the equivalent system

$$-\frac{1}{Z}\frac{\partial Z}{\partial \lambda_i} = \eta_i \qquad i = 1, \ldots, n \tag{15-153}$$

Example 15-22. A die is rolled a large number of times and the average number of dots up equals η. Assuming that η is known, we shall determine the probabilities p_k of the six faces f_k using the MEM. For this purpose, we form an RV $\mathbf{x}$ such that $\mathbf{x}(f_k) = k$. Clearly,

$$E\{\mathbf{x}\} = p_1 + 2p_2 + \cdots + 6p_6 = \eta$$

With $g(x) = x$, it follows from (15-151) that

$$p_k = \frac{1}{Z}e^{-k\lambda} \qquad Z = w + w^2 + \cdots + w^6$$

where $w = e^{-\lambda}$. Hence

$$p_k = \frac{w^k}{w + w^2 + \cdots + w^6} \qquad \frac{w + 2w^2 + \cdots + 6w^6}{w + w^2 + \cdots + w^6} = \eta$$

as in Fig. 15-17. We note that if $\eta = 3.5$, then $p_k = \frac{1}{6}$.

Joint density. The MEM can be used to determine the density $f(X)$ of the random vector $\mathbf{X}$: $[\mathbf{x}_1, \ldots, \mathbf{x}_M]$ subject to the n constraints

$$E\{g_i(\mathbf{X})\} = \eta_i \qquad i = 1, \ldots, n \tag{15-154}$$

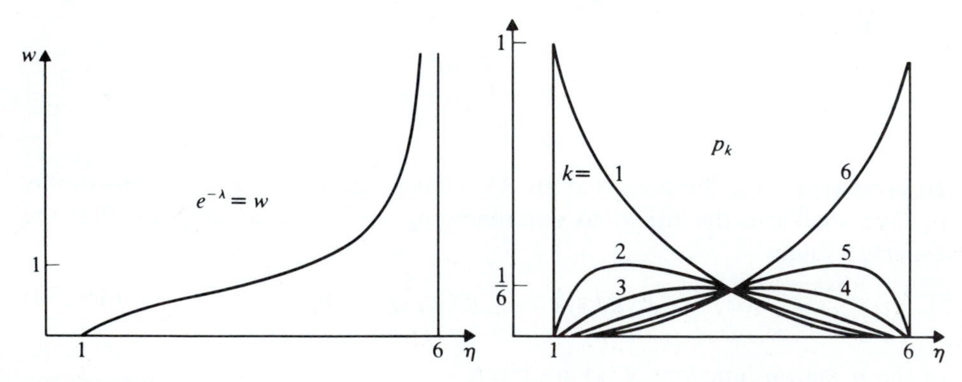

FIGURE 15-17

Reasoning as in the scalar case, we conclude that

$$f(X) = A\exp\{-\lambda_1 g_1(X) - \cdots - \lambda_n g_n(X)\} \tag{15-155}$$

Second-Order Moments and Normality

We are given the correlation matrix

$$R = E\{\mathbf{X}'\mathbf{X}\} \tag{15-156}$$

of the random vector $\mathbf{X}$ and we wish to find its density using the MEM. We maintain that $f(X)$ is normal with zero mean as in (8-58)

$$f(X) = \frac{1}{\sqrt{(2\pi)^M \Delta}} \exp\{-\tfrac{1}{2}XR^{-1}X^t\} \tag{15-157}$$

Proof. The elements $R_{jk} = E\{\mathbf{x}_j\mathbf{x}_k\}$ of R are the expected values of the M^2 RVs $g_{jk}(\mathbf{X}) = \mathbf{x}_j\mathbf{x}_k$. Changing the subscript i in (15-154) to double subscript, we conclude from (15-155) that

$$f(X) = A\exp\left\{-\sum_{j,k}\lambda_{jk}x_j x_k\right\} \tag{15-158}$$

This shows that $f(X)$ is normal. The M^2 coefficients λ_{jk} can be determined from the M^2 constraints in (15-156). As we know [see (8-58)], these coefficients equal the elements of the matrix $R^{-1}/2$ as in (15-157).

The preceding results are acceptable only if the matrix R is p.d. Otherwise, the function $f(X)$ in (15-157) is not a density. The p.d. condition is, of course, satisfied if the given R is a true correlation matrix. However, even then (15-157) might not be acceptable if only a subset of the elements of R is specified. In such cases, it might be necessary, as we shall presently see, to introduce the unspecified elements of R as auxiliary constraints.

Suppose, first, that we are given only the diagonal elements of R:

$$E\{\mathbf{x}_i^2\} = R_{ii} \qquad i = 1, \ldots, M \tag{15-159}$$

Inserting the functions $g_{ii}(x) = x_i^2$ into (15-155), we obtain

$$f(X) = A\exp\{-\lambda_{11}x_1^2 - \cdots - \lambda_{MM}x_M^2\} \tag{15-160}$$

This shows that the RVs $\mathbf{x}_i$ are normal, independent, with zero mean and variance $R_{ii} = 1/2\lambda_{ii}$.

The above solution is acceptable because $R_{ii} > 0$. If, however, we are given $N < M^2$ arbitrary joint moments, then the corresponding quadratic in (15-158) will contain only the terms $x_j x_k$ corresponding to the given moments. The resulting $f(X)$ might not then be a density. To find the ME solution for this case, we proceed as follows: We introduce as constraints the M^2 joint moments R_{jk} where now only N of these moments are given and the other $M^2 - N$ moments are unknown parameters. Applying the MEM, we obtain

(15-157). The corresponding entropy equals [see (15-111)].

$$H(\mathbf{x}_1,\ldots,\mathbf{x}_M) = \ln\sqrt{(2\pi e)^M \Delta} \qquad \Delta = |R| \tag{15-161}$$

This entropy depends on the unspecified parameters of R and it is maximum if its determinant Δ is maximum. Thus the RVs $\mathbf{x}_i$ are again normal with density as in (15-157) where the unspecified parameters of R are such as to maximize Δ.

Note From the above it follows that the determinant Δ of a correlation matrix R is such that

$$\Delta \le R_{11} \cdots R_{MM}$$

with equality iff R is diagonal. Indeed, (15-159) is a restricted moment set; hence the ME solution (15-160) maximizes Δ.

Stochastic processes. The MEM can be used to determine the statistics of a stochastic process subject to given constraints. We shall discuss the following case.

Suppose that $\mathbf{x}_n$ is a WSS process with autocorrelation

$$R[m] = E\{\mathbf{x}_{n+m}\mathbf{x}_n\}$$

We wish to find its various densities assuming that $R[m]$ is specified either for some or for all values of m. As we know [see (15-158)] the MEM leads to the conclusion that, in both cases, $\mathbf{x}_n$ must be a normal process with zero mean. This completes the statistical description of $\mathbf{x}_n$ if $R[m]$ is known for all m. If, however, we know $R[m]$ only partially, then we must find its unspecified values. For finite-order densities, this involves the maximization of the corresponding entropy with respect to the unknown values of $R[m]$ and it is equivalent to the maximization of the correlation determinant Δ [see (15-161)]. An important special case is the MEM solution to the extrapolation problem considered in Sec. 13-3. We shall reexamine this problem in the context of the entropy rate.

We start with the simplest case: Given the average power $E\{\mathbf{x}_n^2\} = R[0]$ of $\mathbf{x}_n$, we wish to find its power spectrum. In this case, the entropy of the RVs

$$\mathbf{x}_n,\ldots,\mathbf{x}_{n+M}$$

is maximum if these RVs are normal and independent for any M [see (15-160)], that is, if the process $\mathbf{x}_n$ is normal white noise with $R[m] = R[0]\delta[m]$.

Suppose now that we are given the $N + 1$ values (data)

$$R[0],\ldots,R[N]$$

of $R[m]$ and we wish to find the density $f(X)$ of the $M + 1$ RVs $\mathbf{x}_n,\ldots,\mathbf{x}_{n+M}$. If $M \le N$, then the correlation matrix of $\mathbf{X}$ is specified in terms of the data and $f(X)$ is given by (15-157). This is not the case, however, if $M > N$ because then only the center diagonal and the N upper and lower diagonals of the correlation matrix

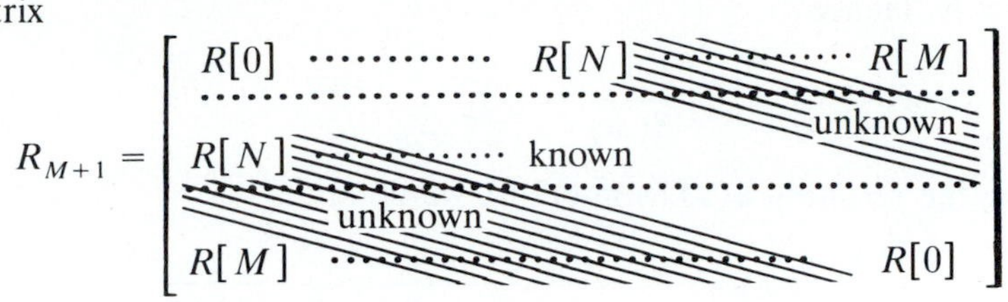

$$R_{M+1} = \begin{bmatrix} R[0] & \cdots\cdots & R[N] & \cdots\cdots & R[M] \\ \cdots & \cdots & \cdots & \cdots & \text{unknown} \\ R[N] & \cdots\cdots & \text{known} & & \\ \cdots & \text{unknown} & \cdots & \cdots & \cdots \\ R[M] & \cdots\cdots & & \cdots & R[0] \end{bmatrix}$$

are known. To complete the specification of R_{M+1}, we maximize the determinant Δ_{M+1} with respect to the unknown values of $R[m]$.

Example 15-23. Given $R[0]$ and $R[1]$, we shall find $R[2]$ using the maximum determinant method. In this case,

$$\Delta = \begin{vmatrix} R[0] & R[1] & R[2] \\ R[1] & R[0] & R[1] \\ R[2] & R[1] & R[0] \end{vmatrix}$$

Hence

$$\frac{\partial \Delta}{\partial R[2]} = -2R[0]R[2] + 2R^2[1] = 0 \qquad R[2] = \frac{R^2[1]}{R[0]}$$

THE MEM IN SPECTRAL ESTIMATION. We are given again $R[m]$ for $|m| \le N$. The power spectrum

$$S(\omega) = R[0] + 2\sum_{m=1}^{\infty} R[m]\cos m\omega$$

of $\mathbf{x}_n$ involves the values of $R[m]$ for every m. To find its unspecified values, we maximize the correlation determinant Δ_M and examine the form of the resulting $R[m]$ as $M \to \infty$. This is equivalent to the maximization of the *entropy rate* $\overline{H}(\mathbf{x})$ of the process $\mathbf{x}_n$. Using this equivalence, we shall develop a more direct method for determining $S(\omega)$.

As we know, the MEM leads to the conclusion that under the given constraints (second-order moments), the process $\mathbf{x}_n$ must be normal with zero mean. From this and (15-130) it follows that

$$\overline{H}(\mathbf{x}) = \ln\sqrt{2\pi e} + \frac{1}{4\pi}\int_{-\pi}^{\pi} \ln S(\omega)\, d\omega$$

The entropy rate $\overline{H}(\mathbf{x})$ depends on the unspecified values of $R[m]$ and it is maximum if

$$\frac{\partial \overline{H}}{\partial R[m]} = \frac{1}{2\pi}\int_{-\pi}^{\pi} \frac{1}{S(\omega)} e^{-jm\omega}\, d\omega = 0 \qquad |m| > N \tag{15-162}$$

This shows that the coefficients of the Fourier series expansion of $1/S(\omega)$ are 0 for $|m| > N$. Hence

$$\frac{1}{S(\omega)} = \sum_{k=-N}^{N} c_k e^{-jk\omega}$$

Factoring the resulting $\mathbf{S}(z)$ as in (12-6), we obtain

$$S(\omega) = \frac{1}{|b_0 + b_1 e^{-j\omega} + \cdots + b_N e^{-jN\omega}|^2} \tag{15-163}$$

This is the spectrum obtained in Sec. 13-3 [see (13-141)] and it shows that the MEM leads to an AR model. The coefficients b_k can be obtained either from the Yule–Walker equations or from Levinson's algorithm.

Note The MEM also has applications in nonprobabilistic problems involving the determination of unknown parameters from insufficient data. In such cases, probabilistic models are created where the unknown parameters take the form of statistical variables that are determined with the MEM. We should point out, however, that the results obtained are not unique because more than one model can be used in the same problem. In the following, we illustrate this approach using as an example the one-dimensional form of an important problem in *crystallography*.

A deterministic application of the MEM. We wish to find a nonnegative periodic function $f(x)$ with period 2π:

$$0 < f(x) = \sum_{n=-\infty}^{\infty} c_n e^{jnx}$$

having access only to partial information about its Fourier series coefficients

$$c_n = r_n e^{j\varphi_n}$$

The truncation problem We assume that c_n is known only for $|n| \le N$.

Solution 1. We create the following probabilistic model: In the interval $(-\pi, \pi)$, the unknown function $f(x)$ is the density of an RV $\mathbf{x}$ taking values between $-\pi$ and π. We determine $f(x)$ so as to maximize the entropy

$$I = -\int_{-\pi}^{\pi} f(x) \ln f(x)\, dx$$

of $\mathbf{x}$. This yields [see (15-141)]

$$f(x) = \exp\left\{-1 + \sum_{n=-N}^{N} \gamma_n e^{jnx}\right\}$$

The constants γ_n are determined in terms of the known values of c_n.

Solution 2. We assume that $f(x)$ is the power spectrum of a stochastic process $\mathbf{x}_n$ and we determine $f(x)$ so as to maximize the entropy rate (we omit incidental constants)

$$I = \int_{-\pi}^{\pi} \ln f(x)\, dx$$

of $\mathbf{x}_n$. In this case, $f(x)$ is given by [see (15-163)]

$$f(x) = \frac{1}{\sum_{n=-N}^{N} d_n e^{jnx}}$$

The constants d_n are again determined in terms of the known values of c_n. (Levinson's algorithm).

The phase problem We assume that we know only the amplitudes r_n of c_n for $|n| \leq N$.

To solve the problem, we form again the integral I, either as the entropy or as the entropy rate, and we maximize it with respect to the unknown parameters which are now the coefficients c_n (amplitudes and phases) for $|n| > N$, and the phase φ_n for $|n| \leq N$. An equivalent approach involves the determination of $f(x)$ as in the truncation problem, treating the phases φ_n as parameters, and the maximization of the resulting I with respect to these parameters. In either case, the required computations are not simple.

15-5 CODING

Coding belongs to a class of problems involving the efficient search and identification of an object ζ_i from a set $\mathscr{S}$ of N objects. This topic is extensive and it has many applications. We shall present here merely certain aspects related to entropy and probability, limiting the discussion to binary instantly decodable codes. The underlying ideas can be readily generalized.

Binary coding can be also described in terms of the familiar game of 20 questions: A person selects an object ζ_i from a set $\mathscr{S}$. Another person wants to identify the object by asking "yes" or "no" questions. The purpose of the game is to find ζ_i using the smallest possible number of questions.

The various search techniques can be described in three equivalent forms: (a) as chains of dichotomies of the set $\mathscr{S}$; (b) in the form of a binary tree; (c) as binary codes (Fig. 15-18). We start with an explanation of these approaches, ignoring for the moment optimality considerations. The criteria for selecting the "best" search method will be developed later.

Set dichotomies. We subdivide the set $\mathscr{S}$ into two nonempty sets $\mathscr{A}_0$ and $\mathscr{A}_1$ (first-generation sets). We subdivide each of the sets $\mathscr{A}_0$ and $\mathscr{A}_1$ into two nonempty sets $\mathscr{A}_{00}, \mathscr{A}_{01}$ and $\mathscr{A}_{10}, \mathscr{A}_{11}$ (second-generation sets). We continue with such dichotomies until the final sets consist of a single element each.

The indices of the sets of each generation are binary numbers formed by attaching 0 or 1 to the indices of the preceding generation sets.

In Fig. 15-18, we illustrate the above with a set consisting of nine elements. We shall use the chain of sets so formed to identify the element ζ_7 by a sequence of appropriate questions (set dichotomies): Is it in $\mathscr{A}_0$? No. Is it in $\mathscr{A}_{10}$? No. Is it in $\mathscr{A}_{110}$? Yes. Is it in $\mathscr{A}_{1100}$? Yes. Hence the unknown element is ζ_7 because $\mathscr{A}_{1100} = \{\zeta_7\}$.

Binary trees. A tree is a simply connected graph consisting of line segments called *branches*. In a binary tree, each branch splits into *two* other branches or

Binary code

x_i	000	0010	0011	010	011	10	1100	1101	111
l_i	3	4	4	3	3	2	4	4	3

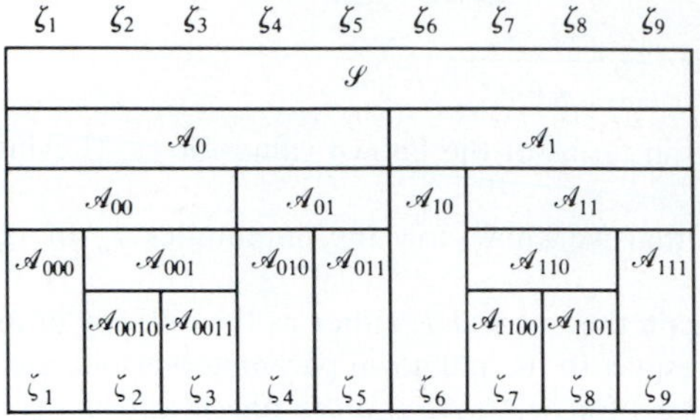

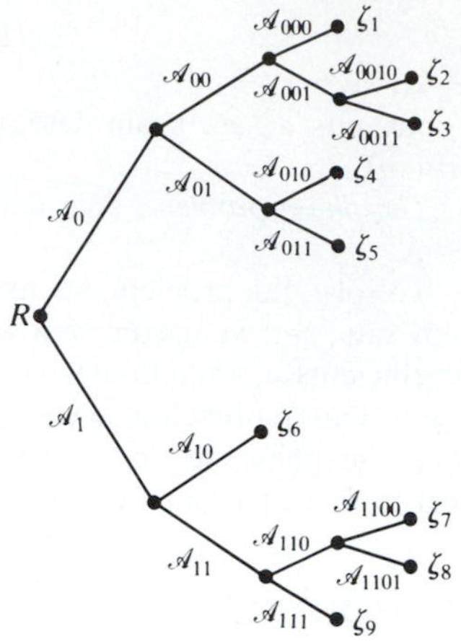

FIGURE 15-18

it terminates. The points of termination are the *endpoints* of the tree and the starting point R is its *root* (Fig. 15-18). A *path* is a part of the tree from R to an endpoint. The two branches closest to the root are the first-generation branches. They split into two branches each, forming the second generation. Since each branch splits into two or it terminates, the number of branches in each generation is always *even*. The *length* of a path is the total number of its branches.

There is one-to-one correspondence between set dichotomies and trees. The kth-generation sets correspond to the kth-generation branches and each set dichotomy to the splitting of the corresponding branch. The terminal sets $\{\zeta_i\}$ correspond to the terminal branches and the elements ζ_i to the endpoints of the tree. The indices of the sets are also used to identify the corresponding branches where we use the following convention: When a branch splits, 0 is assigned to the left new branch and 1 to the right. The index of a terminal branch is also used to identify the corresponding endpoint ζ_i. Thus each element ζ_i of $\mathscr{S}$ is identified by a binary number x_i (Fig. 15-18). The number of digits l_i of x_i equals the length of the path ending at ζ_i. This number also equals the number of questions (dichotomies) required to identify ζ_i.

Binary codes. A binary code is a one-to-one correspondence between the elements ζ_i of a set $\mathscr{S}$ and the elements x_i of a set $X = \{x_1, x_2, \dots\}$ of binary numbers. *Encoding* is the process of constructing such a correspondence.

The set $\mathscr{S}$ will be called the *source* and its elements ζ_i the *source words*. The corresponding binary numbers x_i will be called the *code words*. The binary digits 0 and 1 form the *code alphabet*. The *length* l_i of a code word x_i is the total number of its binary digits.

A *message* is a sequence of source words

$$\zeta_{i_1} \cdots \zeta_{i_k} \cdots \zeta_{i_n} \qquad \zeta_{i_k} \in \mathscr{S} \tag{15-164}$$

The sequence of the corresponding code words

$$x_{i_1} \cdots x_{i_k} \cdots x_{i_n} \tag{15-165}$$

is a *coded message*.

The indices of the terminal elements of a tree, or, equivalently, of a chain of set dichotomies, specify a code. Codes can, of course, be formed in other ways; however, other codes will not be considered here. The term *code* will mean a *binary code* specified by a tree as above.

In Fig. 15-18, we show the code words x_i of a source $\mathscr{S}$ consisting of $N = 9$ elements, and the corresponding word lengths l_i.

THEOREM. If a source $\mathscr{S}$ has N words and the lengths of the corresponding code words equal l_i, then

$$\sum_{i=1}^{N} \frac{1}{2^{l_i}} = 1 \tag{15-166}$$

Proof. The last-generation branches of the tree are terminal and they form pairs. The two branches of one such pair are the ends of two paths of length l_r (Fig. 15-19). If they are removed, the tree *contracts* into a tree with $N - 1$ endpoints. In this operation, the two paths are replaced with one path of length $l_r - 1$ and the two terms 2^{-l_r} in (15-166) are replaced with the term $2^{-(l_r-1)}$.

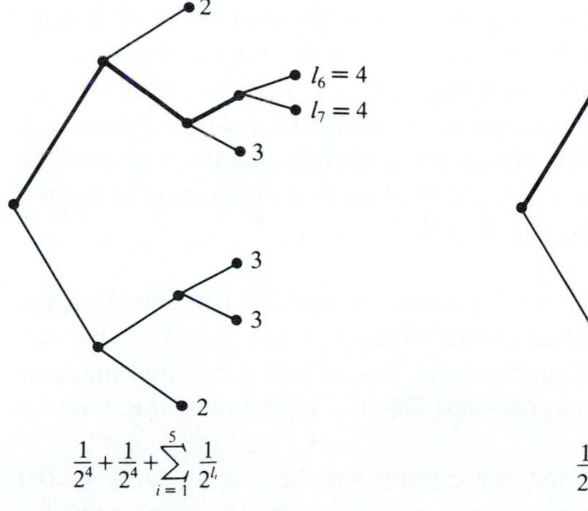

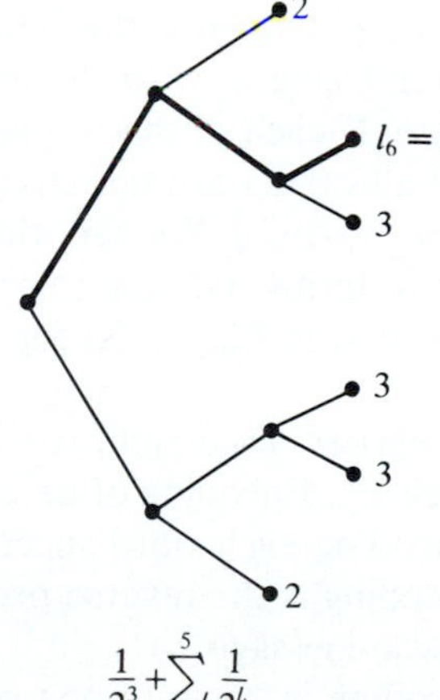

FIGURE 15-19

Tree construction

l	1	2	3	4	5	6	7	8
l_i	2	2	3	3	4	4	4	4
	2	2	3	3	3		3	
	2	2	2		2			
	1		1					
x_i	11	10	011	010	0011	0010	0001	0000

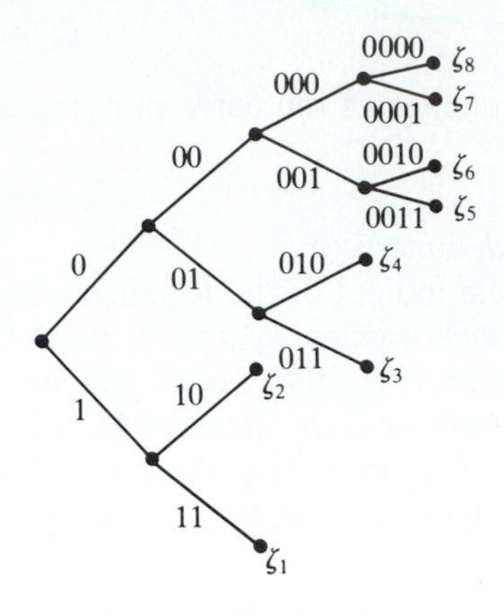

FIGURE 15-20

Since

$$2^{-l_r} + 2^{-l_r} = 2^{-(l_r - 1)} \tag{15-167}$$

the sum does not change. Thus the binary length sum in (15-166) is invariant to a contraction. Repeating the process until we are left with only two first-generation branches, we obtain (15-166) because $2^{-1} + 2^{-1} = 1$.

CONVERSE THEOREM. Given N integers l_i satisfying (15-166), we can construct a code with lengths l_i.

Proof. It suffices to construct a binary tree with path lengths l_i. From (15-166) it follows that if l_r is the largest of the integers l_i, then the number n of lengths that equal l_r is even. Using $n = 2m$ segments, we form the rth (last) generation branches of our tree. If each of the m pairs of integers l_r is replaced by a single integer $l_r - 1$ and all others are not changed, the resulting set of numbers will satisfy (15-166) [see (15-167)]. We can, therefore, continue this process until we are left with only two terms. These terms yield the two first-generation branches. The above is illustrated in Fig. 15-20 for $N = 8$.

Decoding. In the earlier discussion, we presented a method for encoding the words ζ_i of a source $\mathscr{S}$. Encoding of an entire message of the form (15-164) can be obtained by encoding each word successively. The result is a coded message as in (15-165). Decoding is the reverse process: Given a coded message, find the corresponding source message.

Since word coding is a one-to-one correspondence between ζ_i and x_i, the decoding of each word of a message is unique. However, an entire message cannot always be so decoded because there is no space separating the code words (this would require an additional letter in the code alphabet). The problem of separation does not exist for codes constructed through dichotomies (they are, we repeat, the only codes considered here) because such codes have the following property: *No code word is the beginning of another code word*. This property is a consequence of the fact that in any tree, each path terminates at an endpoint; therefore, it cannot be part of another path. Codes with this property are called "instantaneous" because they are instantly decodable; that is, if we start from the beginning of a message, we can identify in real time the end of each word without any reference to the future.

Example 15-24. We wish to decode the message

$$1011010000101000101111000000010$$

formed with the code shown in Fig. 15-18. Starting from the beginning, we identify the code words by underlying them with the help of the table of Fig. 15-18:

$$\underline{10} \quad \underline{1101} \quad \underline{000} \quad \underline{010} \quad \underline{10} \quad \underline{0010} \quad \underline{111} \quad \underline{1100} \quad \underline{000} \quad \underline{0010}$$

The corresponding source message is the sequence

$$\zeta_6\zeta_8\zeta_1\zeta_4\zeta_6\zeta_2\zeta_9\zeta_7\zeta_1\zeta_2$$

Note We have identified each source word with a single symbol ζ_i. It is possible, however, that ζ_i might be a grouping of other symbols. For example, the source $\mathscr{S}$ might consist of: All the *letters* of the English alphabet, certain frequently used words (for instance, the word *the*) and even a number of common phrases like *happy birthday*. Such sources are equivalent to single-symbol sources if each word is viewed as a single element.

Optimum Codes

In the absence of prior information, the two subsets of each set dichotomy are so chosen as to have nearly equal elements. The resulting code lengths are then nearly equal to log N. If, however, prior information is available, then more efficient codes can be constructed. The information is usually given in terms of relative frequencies and it is used to form codes with minimum average length. Since relative frequencies are best described in terms of probabilities, we shall assume from now on that the source $\mathscr{S}$ is a probability space.

DEFINITIONS. A *random code* is a process of assigning to every source word ζ_i a binary number x_i.

Since ζ_i is an element of the probability space $\mathscr{S}$, a random code defines an RV $\mathbf{x}$ such that

$$\mathbf{x}(\zeta_i) = x_i$$

The *length* of a random code is an RV $\mathbf{L}$ such that

$$\mathbf{L}(\zeta_i) = l_i \tag{15-168}$$

where l_i is the length of the code word x_i assigned to the element ζ_i.

The expected value of $\mathbf{L}$ is denoted by L and it is called the *average length* of the random code $\mathbf{x}$. Thus

$$L = E\{\mathbf{L}\} = \sum_i p_i l_i \tag{15-169}$$

where $p_i = P\{\mathbf{x} = x_i\} = P\{\zeta_i\}$.

Optimum code. An *optimum code* is a code whose average length does not exceed the average length of any other code. A basic objective of coding theory is the determination of such a code. Optimum codes have the following properties:

1. Suppose that ζ_a and ζ_b are two elements of $\mathscr{S}$ such that

$$p_a = P\{\zeta_a\} \qquad p_b = P\{\zeta_b\} \qquad l(\zeta_a) = l_a \qquad l(\zeta_b) = l_b$$

We maintain that if the code is optimum and

$$p_a > p_b \qquad \text{then} \quad l_a \le l_b \tag{15-170}$$

Proof. Suppose that $l_a > l_b$. Interchanging the codes assigned to the elements ζ_a and ζ_b, we obtain a new code with average length

$$L_1 = L - (p_a l_a + p_b l_b) + (p_a l_b + p_b l_a) = L - (p_a - p_b)(l_a - l_b)$$

And since $(p_a - p_b)(l_a - l_b) > 0$, we conclude that $L_1 < L$. This, however, is impossible because L is the optimum code length; hence $l_a \le l_b$.

Repeated application of (15-170) leads to the conclusion that if

$$p_1 \ge p_2 \ge \cdots \ge p_N \qquad \text{then} \quad l_1 \le l_2 \le \cdots \le l_N \tag{15-171}$$

2. The elements (source words) with the two smallest probabilities p_{N-1} and p_N are in the last generation of the tree; that is, their code lengths are l_{N-1} and l_N.

Proof. This is a consequence of (15-171) and the fact that the number of branches in each generation is even.

The following basic theorem shows the relationship between the entropy

$$H(\mathfrak{S}) = -\sum_{i=1}^{N} p_i \log p_i$$

of the source word partition $\mathfrak{S}$ and the average length L of an arbitrary random code $\mathbf{x}$.

THEOREM.

$$H(\mathfrak{S}) \le L \tag{15-172}$$

Proof. As we have seen from (15-166), if l_i are the lengths of the code words of $\mathbf{x}$ and $q_i = 1/2^{l_i}$, then the sum of the q_i's equals 1. With $a_i = p_i$ and $b_i = q_i$ it follows, therefore, from (15-37) that

$$-\sum_i p_i \log p_i \le -\sum_i p_i \log q_i = \sum_i p_i l_i = L \qquad (15\text{-}173)$$

and (15-172) results.

In general, $H(\mathfrak{S}) < L$. We maintain, however, that $H(\mathfrak{S}) = L$ iff the probabilities p_i are binary decimals, that is, iff $p_i = 1/2^{n_i}$.

Proof. If $H(\mathfrak{S}) = L$, then (15-173) is an equality; hence $p_i = q_i = 1/2^{l_i}$ [see (15-37)] and our assertion is true because the lengths l_i are integers.

Conversely, if $p_i = 1/2^{n_i}$ and n_i are integers, then we can construct a code with lengths $l_i = n_i$ because the sum of the p_i's equals 1. The length L of this code equals $H(\mathfrak{S})$. In other words, if all p_i's are binary decimals, then the code with lengths $l_i = n_i$ is optimum.

Shannon, Fano, and Huffman Codes

The preceding theorem gives us a low bound for the average code length L but it does not say how close we can come to this bound. At the end of the section we show that, if we encode not each word but an entire message, then we can construct codes with average length per word less than $H(\mathfrak{S}) + \varepsilon$ for any $\varepsilon > 0$.

In the following, we present three well-known codes including the optimum code (Huffman). The description of these codes is clarified in Example 15-25.

The Shannon code. As we noted, if all probabilities p_i are binary decimals, then the code with lengths $l_i = -\log p_i$ is optimum. Guided by this, we shall construct a code for all other cases.

Each p_i specifies an integer n_i such that

$$\frac{1}{2^{n_i}} \le p_i < \frac{1}{2^{n_i - 1}} \qquad (15\text{-}174)$$

where $p_i > 1/2^{n_i}$ for at least one p_i (assumption). With n_m the largest of the integers n_i, it follows from the above that

$$\sum_{i=1}^{N} \frac{1}{2^{n_i}} \le 1 - \frac{1}{2^{n_m}} \qquad (15\text{-}175)$$

because the left side is a binary integer smaller than 1. If, therefore, n_m is changed to $n_m - 1$, the resulting value of the sum in (15-175) will not exceed 1. We continue the process of reducing the largest integer by 1 until we reach a set of integers l_i such that

$$\sum_{i=1}^{N} \frac{1}{2^{l_i}} = 1 \qquad l_i \le n_i \qquad (15\text{-}176)$$

With this set of integers we construct a code and we denote by L^a its average length. Thus

$$L^a = \sum_{i=1}^{N} p_i l_i \le \sum_{i=1}^{N} p_i n_i$$

We maintain that

$$H(\mathfrak{S}) \le L^a < H(\mathfrak{S}) + 1 \tag{15-177}$$

Proof. From (15-174) it follows that $n_i < -\log p_i + 1$. Multiplying by p_i and adding, we obtain

$$\sum_{i=1}^{N} p_i n_i < \sum_{i=1}^{N} p_i(-\log p_i + 1) = H(\mathfrak{S}) + 1$$

and (15-177) results [see (15-172)].

The Fano code. We shall describe this code in terms of set dichotomies based on the following rule of subdivision. We number the probabilities p_i in descending order

$$p_1 \ge p_2 \ge \cdots \ge p_N \tag{15-178}$$

and we select the sets $\mathscr{A}_0$ and $\mathscr{A}_1$ of the first generation so as to have equal or nearly equal probabilities. To do so, we determine k such that

$$p_1 + \cdots + p_k \le 0.5 \le p_{k+1} + \cdots + p_N$$

and we set $\mathscr{A}_0$ equal to $\{\zeta_1, \ldots, \zeta_k\}$ or to $\{\zeta_1, \ldots, \zeta_{k+1}\}$. The same rule is used in all subsequent subdivisions. As we see in Example 15-25, the length L^b of the resulting code is close to the Shannon code length L^a.

We note that, since there is an ambiguity in the choice of the subsets in each dichotomy, the Fano code is not unique.

The Huffman code. We denote by $\mathbf{x}_N^0$ the optimum N-element code and by L_N^0 its average length. We shall determine $\mathbf{x}_N^0$ using the following operation: We arrange the probabilities p_i of the elements ζ_i of $\mathscr{S}$ in descending order as in (15-178) and we number the corresponding elements ζ_i accordingly. We then replace the last two elements ζ_{N-1} and ζ_N with a new element and we assign to this element the probability $p_{N-1} + p_N$. A new source results with $N - 1$ elements. This operation will be called *Huffman contraction*.

In the table of Example 15-25, the new element is identified by a box in which the replaced elements are shown.

Rearranging the probabilities of the new source in descending order, we repeat the above operation until we reach a set with only two elements.

To each element ζ_i of the source $\mathscr{S}$, we shall assign a code word x_i starting from the last digit: We assign the numbers 0 and 1 respectively to the last digits of the code words of the elements ζ_{N-1} and ζ_N. At each subsequent contraction, we assign the numbers 0 and 1 to the *left* of the partially completed code words of all elements that are included in the last two boxes.

The code so formed (Huffman) will be denoted by $\mathbf{x}_N^c$ and its average length by L_N^c. We shall show that this code is optimal.

Proof. The proof of the optimality is based on the following observation. We can readily see that the last two code words x_{N-1} and x_N have the same length l_r. In Example 15-25,

$$N = 9 \qquad x_8 = 00000 \qquad x_9 = 00001 \qquad l_r = 5$$

If we replace these two words with a single word consisting of their common part, we obtain the Huffman code $\mathbf{x}_{N-1}^c$ for the set of $N-1$ elements and the code length of the new element equals l_{r-1}. This leads to the conclusion that

$$L_N^c - (p_{N-1} + p_N)l_r = L_{N-1}^c - (p_{N-1} + p_N)(l_r - 1)$$

Hence

$$L_N^c = L_{N-1}^c + p_{N-1} + p_N \tag{15-179}$$

In the example

$$L_9^c = \sum_{i=1}^{7} p_i l_i + 5p_8 + 5p_9 \qquad L_8^c = \sum_{i=1}^{7} p_i l_i + 4(p_8 + p_9)$$

Induction The Huffman code is optimum for $N = 2$ because there is only one code with two words. We assume that it is optimum for every source with $k \le N - 1$ elements and we shall show that it is optimum for $k = N$. Suppose that there is an N-element source $\mathscr{S}$ for which this is not true, that is, suppose that

$$L_N^0 < L_N^c \tag{15-180}$$

As we know, the two elements ζ_{N-1} and ζ_N with the smallest probabilities are in the last-generation branches of the optimum code tree. If they are removed, the contracted tree specifies a new code with length L_{N-1}. Reasoning as in (15-179), we conclude with (15-180) that

$$L_{N-1} + p_{N-1} + p_N = L_N^0 < L_N^c = L_{N-1}^c + p_{N-1} + p_N$$

hence $L_{N-1} < L_{N-1}^c$. But this is impossible because the Huffman code of order $N-1$ is optimum by assumption.

Example 15-25. We shall describe the above codes using as source a set $\mathscr{S}$ with nine elements. Their probabilities are shown in the table below:

i	1	2	3	4	5	6	7	8	9
p_i	0.22	0.19	0.15	0.12	0.08	0.07	0.07	0.06	0.04

The resulting entropy equals

$$H(\mathfrak{S}) = -\sum_{i=1}^{9} p_i \log p_i = 2.703$$

Arbitrary code We form a code using a chain of dichotomies chosen arbitrarily as in Fig. 15-19. In the table below we show the code words and their lengths.

i	1	2	3	4	5	6	7	8	9	
x_i	000	0010	0011	010	011	10	1100	1101	111	$L = \sum_{i=1}^{9} p_i l_i = 3.40$
l_i	3	4	4	3	3	2	4	4	3	

Shannon code In the table below we show the integers n_i determined from (15-174) and the required reductions until the final lengths l_i are reached. The corresponding code tree is shown in Fig. 15-20.

p_i	0.22	0.19	0.15	0.12	0.08	0.07	0.07	0.06	0.04	
	$\frac{1}{2^3} \le p_i < \frac{1}{2^2}$			$\frac{1}{2^4} \le p_i < \frac{1}{2^3}$				$\frac{1}{2^5} \le p_i < \frac{1}{2^4}$		$\sum_{i=1}^{N} \frac{1}{2^{n_i}}$
n_i	3	3	3	4	4	4	4	5	5	12/16
	3	3	3	3	3	4	4	4	4	14/16
l_i	3	3	3	3	3	3	3	4	4	1
x_i	000	001	010	011	100	101	110	1110	1111	$L^a = 3.1$

Fano code In the table below we show the subsets obtained with the Fano dichotomies, and their probabilities. The last-generation sets are the elements ζ_i of $\mathscr{S}$; their probabilities are shown on the first row of the table. The dichotomies start with

$$\mathscr{A}_0 = \{\zeta_1, \zeta_2, \zeta_3\} \qquad P(\mathscr{A}_0) = 0.22 + 0.19 + 0.15 = 0.56$$

p_i	0.22	0.19	0.15	0.12	0.08	0.07	0.07	0.06	0.04	
	$\mathscr{A}_0$	0.56		$\mathscr{A}_1$				0.44		
	$\mathscr{A}_{00}$	$\mathscr{A}_{01}$	0.34	$\mathscr{A}_{10}$	0.20	$\mathscr{A}_{11}$		0.24		
		$\mathscr{A}_{010}$	$\mathscr{A}_{011}$	$\mathscr{A}_{100}$	$\mathscr{A}_{101}$	$\mathscr{A}_{110}$	0.14	$\mathscr{A}_{111}$	0.10	
	ζ_1	ζ_2	ζ_3	ζ_4	ζ_5	$\mathscr{A}_{1100}$ ζ_6	$\mathscr{A}_{1101}$ ζ_7	$\mathscr{A}_{1110}$ ζ_8	$\mathscr{A}_{1111}$ ζ_9	
x_i	00	010	011	100	101	1100	1101	1110	1111	
l_i	2	3	3	3	3	4	4	4	4	$L^b = 3.02$

Optimum code In the table below we show the original set $\mathscr{S}_9$ consisting of nine elements and the sets obtained with each Huffman contraction. The elements ζ_i are identified by their indices and the combined elements by boxes. Each box contains all elements ζ_i of the original source involved in each contraction, and the evolution of their code words x_i starting with the last digit. The rows below each $\mathscr{S}_i$ line show the probabilities of the various elements of $\mathscr{S}_i$. For example, the number 0.10 in the line below $\mathscr{S}_7$ is the probability of the box (element of $\mathscr{S}_7$) that contains the elements ζ_8 and ζ_9.

The column at the extreme right shows the sum of the two smallest probabilities of the elements in $\mathscr{S}_i$. This number is used to form the row $\mathscr{S}_{i+1}$ by reordering the elements of $\mathscr{S}_i$.

Evolution of Huffman code

$\mathscr{S}_9$	1	2	3	4	5	6	7	8	9	
$p_{i,9}$	0.22	0.19	0.15	0.12	0.08	0.07	0.07	0.06	0.04	0.10
$\mathscr{S}_8$	1	2	3	4	8 0	9 1	5	6	7	
$p_{i,8}$	0.22	0.19	0.15	0.12	0.10		0.08	0.07	0.07	0.14
$\mathscr{S}_7$	1	2	3	6 0	7 1	4	8 0	9 1	5	
$p_{i,7}$	0.22	0.19	0.15	0.14	0.12		0.10		0.08	0.18
$\mathscr{S}_6$	1	2	8 00	9 01	5 1	3	6 0	7 1	4	
$p_{i,6}$	0.22	0.19	0.18			0.15	0.14		0.12	0.26
$\mathscr{S}_5$	6 00	7 01	4 1	1	2	8 00	9 01	5 1	3	
$p_{i,5}$	0.26			0.22	0.19	0.18			0.15	0.33
$\mathscr{S}_4$	8 000	9 001	5 01	3 1	6 100	7 01	4 1	1	2	
$p_{i,4}$	0.33				0.26			0.22	0.19	0.41
$\mathscr{S}_3$	1 0	2 1	8 000	9 001	5 01	3 1	6 00	7 01	4 1	
$p_{i,3}$	0.41		0.33				0.26			0.59
$\mathscr{S}_2$	8 0000	9 001	5 001	3 01	6 100	7 101	4 11	1 0	2 1	
$p_{i,2}$	0.59							0.41		1
$\mathscr{S}_1$	8 00000	9 00001	5 0001	3 001	6 0100	7 0101	4 011	1 10	2 11	

The completed code words x_i taken from the last line of the table and their code lengths l_i are listed below.

	1	2	3	4	5	6	7	8	9	
x_i	10	11	001	011	0001	0100	0101	00000	00001	$L^0 = 3.01$
l_i	2	2	3	3	4	4	4	5	5	

The Shannon Coding Theorem

In the earlier discussion, we considered only codes of the elements ζ_i of a set $\mathscr{S}$ and we showed that the optimum code is between $H(\mathfrak{S})$ and $H(\mathfrak{S}) + 1$:

$$H(\mathfrak{S}) \le L^0 \le H(\mathfrak{S}) + 1 \tag{15-181}$$

This follows from (15-172) and (15-177). We show next that if we encode not merely single words but entire messages, then the code length per word can be reduced to less than $H(\mathfrak{S}) + \varepsilon$ for any $\varepsilon > 0$.

A message of length n is any element of the product space $\mathscr{S}^n$. The number of such messages is N^n and a code of the space $\mathscr{S}^n$ is a correspondence between its elements and a set of N^n binary numbers. This correspondence defines the RV $\mathbf{x}_n$ (random code) on the space $\mathscr{S}^n$ and the lengths of the code words form another RV $\mathbf{L}_n$ (random code length). The expected value L_n of $\mathbf{L}_n$ is the average code length. From the definition it follows that L_n is the average number of digits required to encode the elements of $\mathscr{S}^n$. The ratio

$$\overline{L} = \frac{L_n}{n} \tag{15-182}$$

is the *average code length* per word. The term *word* means, of course, an element of $\mathscr{S}$.

We shall assume that $\mathscr{S}^n$ is the space of n *independent* trials.

THEOREM. We can construct a code of the space $\mathscr{S}^n$ such that

$$H(\mathfrak{S}) \le \overline{L} \le H(\mathfrak{S}) + \frac{1}{n} \tag{15-183}$$

Proof. We shall give two proofs. The first is a direct consequence of (15-181). The second is based on the concept of typical sequences.

1. Applying the earlier results to the source $\mathscr{S}^n$, we construct a code L_n such that

$$H(\mathfrak{S}^n) \le L_n < H(\mathfrak{S}^n) + 1 \tag{15-184}$$

This yields (15-183) because $L_n = n\overline{L}$ and $H(\mathfrak{S}^n) = nH(\mathfrak{S})$ [see (15-67)].

2. As we know the space $\mathscr{S}^n$ can be divided into two sets: the set $\mathbb{T}$ of all typical sequences and the set $\overline{\mathbb{T}}$ of all rare sequences. To prove (15-183), we construct a code tree consisting of $2^{nH(\mathfrak{S})} - 1$ short paths of length $l_t = nH(\mathfrak{S})$ and 2^l paths of length $l_t + l$. The short paths are used as the code words of the typical sequences and the long paths for the long sequences (Fig. 15-21). Since $P(\mathbb{T}) \simeq 1$ and $P(\overline{\mathbb{T}}) \simeq 0$, we conclude that the average length of the resulting code equals

$$L_n = l_t P(\mathbb{T}) + (l + l_t) P(\overline{\mathbb{T}}) \simeq l_t = nH(\mathfrak{S})$$

Thus $\overline{L} \simeq H(\mathfrak{S})$ and (15-183) results.

We note that (15-184) holds even if the trials are not independent. In this case, the theorem is true if $H(\mathfrak{S})$ is replaced by $H(\mathfrak{S}^n)/n$.

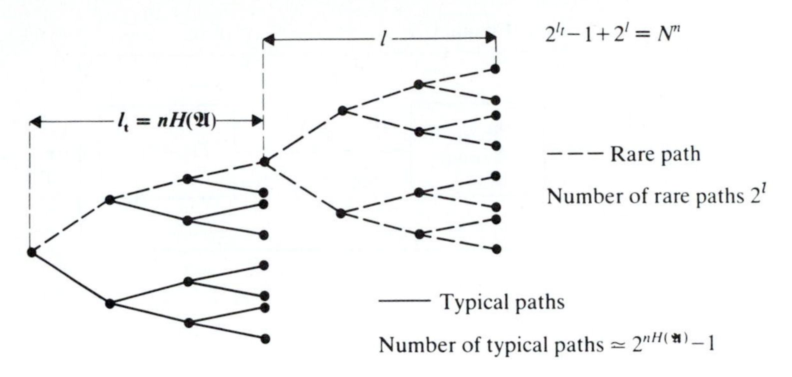

FIGURE 15-21

15-6 CHANNEL CAPACITY

We wish to transmit a message from point A to point B by means of a communications channel (a telephone cable, for example). The message to be transmitted is a stationary process $\mathbf{x}_n$ generating at the receiving end another process $\mathbf{y}_n$. The output $\mathbf{y}_n$ depends not only on the input $\mathbf{x}_n$ but also on the nature of the channel. Our objective is to determine the maximum rate of information that can be transmitted through the channel. To simplify the discussion, we make the following assumptions:

1. The channel is *binary*; that is, the input $\mathbf{x}_n$ and the output $\mathbf{y}_n$ take only the values 0 and 1.
2. The channel is *memoryless*; that is, the present value of $\mathbf{y}_n$ depends only on the present value of $\mathbf{x}_n$.
3. The input $\mathbf{x}_n$ is *strictly white noise*.

From assumptions 2 and 3 it follows that $\mathbf{y}_n$ is also white noise.

4. The messages are transmitted at the rate of *one word per second*.

This is a mere normalization stating that the duration T of each transmitted state equals one second.

Example 15-26. In Fig. 15-22 we show a simple realization of a channel as a system with input $\mathbf{x}_n$ and output $\mathbf{y}_n$. The input to the physical channel is a time signal $\mathbf{x}(t)$ taking the values E and $-E$ (binary transmission). These values correspond to the two states 1 and 0 of $\mathbf{x}_n$. The received signal $\mathbf{y}(t)$ is a distorted version of $\mathbf{x}(t)$ contaminated possibly by noise. The system output $\mathbf{y}_n$ is obtained by some decision rule (detector) translating the time signal $\mathbf{y}(t)$ into a discrete-time signal consisting of 0's and 1's.

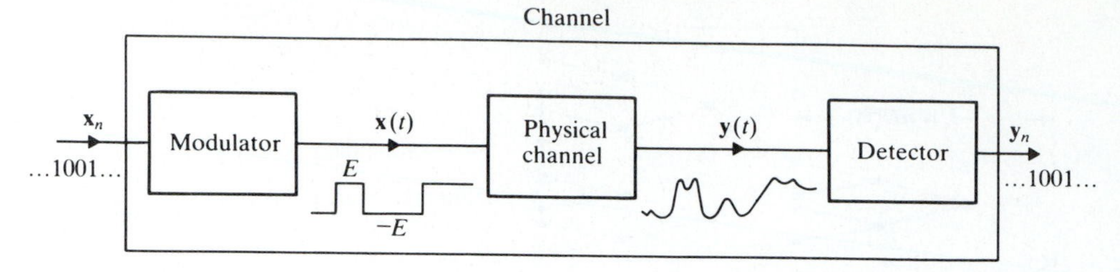

FIGURE 15-22

Noiseless Channel

We shall say that a channel is noiseless† if there is a one-to-one correspondence between the input $\mathbf{x}_n$ and the output $\mathbf{y}_n$. For a binary channel this means that if $\mathbf{x}_n = 0$, then $\mathbf{y}_n = 0$; if $\mathbf{x}_n = 1$, then $\mathbf{y}_n = 1$.

In a given channel, the uncertainty per transmitted word equals the entropy rate $\overline{H}(\mathbf{x}) = H(\mathbf{x})$ of the input $\mathbf{x}_n$. If the channel is noiseless, then the observed output $\mathbf{y}_n$ determines $\mathbf{x}_n$ uniquely; hence it removes this uncertainty. Thus the rate of transmitted information equals $H(\mathbf{x})$.

Definition of channel capacity. The maximum value of $H(\mathbf{x})$, as $\mathbf{x}$ ranges over all possible inputs, is denoted by C and is called the *channel capacity*

$$C = \max_{\mathbf{x}_n} H(\mathbf{x}) \tag{15-185}$$

It appears that C does not depend on the channel but that is not so because the channel determines the number of the input states. If it is binary, then $\mathbf{x}_n$ has two possible states with probabilities p and $q = 1 - p$, respectively; hence

$$H(\mathbf{x}) = -p \log p - (1 - p)\log(1 - p) = \boldsymbol{r}(p) \tag{15-186}$$

where $\boldsymbol{r}(p)$ is the function of Fig. 15-2. Since $\boldsymbol{r}(p)$ is maximum for $p = 0.5$ and $\boldsymbol{r}(0.5) = 1$, we conclude that the capacity of a binary noiseless channel equals 1 bit/s.

Similarly, if the channel accepts N input states, then its capacity equals $\log N$ bit/s.

Rate of information transmission. We repeat: The channel transmits messages at the rate of 1 word/s. It transmits information at the rate $H(\mathbf{x})$ bits/s. This rate depends on the source and it is maximum if the two states of the source are equally likely.

†This definition does not lead to any conclusion about the actual presence of noise in the channel.

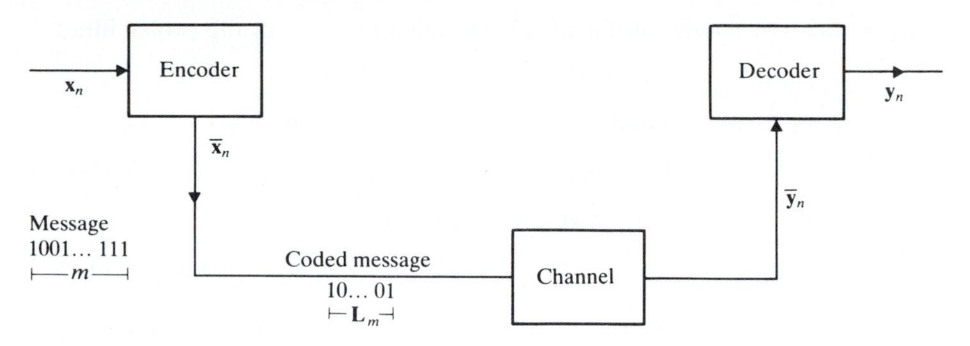

FIGURE 15-23

THEOREM. The maximum rate of 1 bit/s can be reached even if the input $\mathbf{x}_n$ is arbitrary, provided that it is properly encoded prior to transmission.

Proof. 1. An m-word message is a binary number with m digits. There are 2^m such messages forming the space $\mathscr{S}_x^m$ and every realization of the input $\mathbf{x}_n$ is a sequence of such messages. We encode optimally the space $\mathscr{S}_x^m$ into a set of binary numbers $\bar{\mathbf{x}}_n$ using the techniques of the last section (Fig. 15-23). The number of digits (code length) of each $\bar{\mathbf{x}}_n$ is an RV $\mathbf{L}_m$ with mean $L_m = E\{\mathbf{L}_m\}$. As we know,

$$mH(\mathbf{x}) \le L_m < mH(\mathbf{x}) + 1 \tag{15-187}$$

Hence $\bar{L}_m \simeq H(\mathbf{x})$ for large m. A code word $\bar{\mathbf{x}}_n$ requires $\mathbf{L}_m$ seconds to be transmitted because it consists of $\mathbf{L}_m$ binary digits. Hence the average time required to transmit the m-word messages of $\mathbf{x}_n$ in code form equals $L_m \simeq mH(\mathbf{x})$ seconds. And since the information contained in each message equals $mH(\mathbf{x})$ bits, we conclude that the average rate of information transmission equals $mH(\mathbf{x})/mH(\mathbf{x}) = 1$ bit/s.

Proof. 2. We have 2^m messages of length m. In a direct transmission (not encoded), each message requires the same transmission time: m seconds. However, of all these messages, only $2^{mH(\mathbf{x})}$ are likely to occur (typical sequences). To reduce the time of transmission, we encode all typical sequences into words of length $l_t \simeq mH(\mathbf{x})$ as in Fig. 15-21. The rare sequences require longer codes; however, the probability of their occurrence is negligible. Hence the average time of transmission of each message is reduced from m seconds to $mH(\mathbf{x})$ seconds.

Noisy Channel

Due to a variety of factors, a physical channel establishes not a functional but a statistical relationship between the input $\mathbf{x}_n$ and the output $\mathbf{y}_n$. For a binary channel, this relationship is completely specified in terms of the probabilities

$$P\{\mathbf{x}_n = 0\} = p \qquad P\{\mathbf{x}_n = 1\} = q$$

of the two states of the input, and the conditional probabilities

$$P\{\mathbf{y}_n = j | \mathbf{x}_n = i\} = \pi_{ij} \qquad i, j = 0, 1 \tag{15-188}$$

The probabilities of the output states are given by

$$P\{\mathbf{y}_n = 0\} = \pi_{00} p + \pi_{10} q \qquad P\{\mathbf{y}_n = 1\} = \pi_{01} p + \pi_{11} q \tag{15-189}$$

DEFINITION. A noisy channel is a random system establishing a statistical relationship between the input $\mathbf{x}_n$ and the output $\mathbf{y}_n$.

For a memoryless channel, this relationship is completely specified in terms of the *channel matrix* Π whose elements π_{ij} are the conditional probabilities between the input states and the output states. For a binary channel

$$\Pi = \begin{bmatrix} \pi_{00} & \pi_{01} \\ \pi_{10} & \pi_{11} \end{bmatrix} \qquad \text{where} \qquad \begin{array}{l} \pi_{00} + \pi_{01} = 1 \\ \pi_{10} + \pi_{11} = 1 \end{array} \tag{15-190}$$

The channel is called *symmetrical* if $\pi_{10} = \pi_{01} = \beta$. In a symmetrical channel, $\pi_{00} = \pi_{11} = 1 - \beta$ and

$$\Pi = \begin{bmatrix} 1-\beta & \beta \\ \beta & 1-\beta \end{bmatrix} \tag{15-191}$$

Example 15-27. To give some idea of the nature of the channel matrix, we show in Fig. 15-24 a simple version of a symmetrical channel. The input $\mathbf{x}(t)$ is a time signal as in Example 15-26, and the resulting output $\mathbf{y}(t)$ is the sum

$$\mathbf{y}(t) = \mathbf{x}(t) + \boldsymbol{\nu}_n \qquad nT \le t < nT + T \tag{15-192}$$

where $\boldsymbol{\nu}_n$ is a sequence of independent RVs with density the even function $f(\upsilon)$. The output states are determined as follows:

$$\mathbf{y}_n = \begin{cases} 1 & \text{if } \mathbf{y}(t) \ge 0 \\ 0 & \text{if } \mathbf{y}(t) < 0 \end{cases}$$

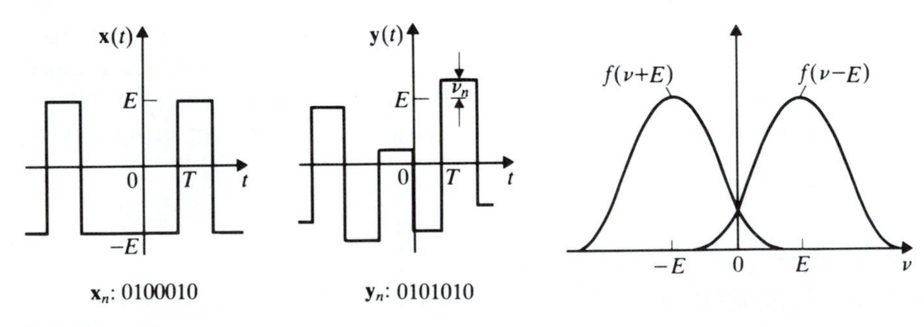

FIGURE 15-24

From this we conclude that the channel is symmetrical and

$$\beta = P\{\mathbf{y}_n = 1 | \mathbf{x}_n = 0\} = \int_0^\infty f(\nu + E)\, d\boldsymbol{\nu} = P\{\boldsymbol{\nu} > E\}$$

Channel capacity. Prior to transmission, the uncertainty about the input $\mathbf{x}_n$ equals $H(\mathbf{x})$ per word. In a noiseless channel, the observed output $\mathbf{y}_n$ reduces the uncertainty to 0. This is not so, however, for a noisy channel because $\mathbf{y}_n$ does not determine $\mathbf{x}_n$ uniquely. Knowledge of $\mathbf{y}_n$ reduces the uncertainty about $\mathbf{x}_n$ from $H(\mathbf{x})$ to $H(\mathbf{x}|\mathbf{y})$ and the difference

$$I(\mathbf{x}, \mathbf{y}) = H(\mathbf{x}) - H(\mathbf{x}|\mathbf{y}) \tag{15-193}$$

is the *rate of information transmission*.†

If the channel is noiseless, then $H(\mathbf{x}|\mathbf{y}) = 0$; hence $I(\mathbf{x}, \mathbf{y}) = H(\mathbf{x})$. If the output $\mathbf{y}_n$ is independent of the input, then $H(\mathbf{x}|\mathbf{y}) = H(\mathbf{x})$; hence $I(\mathbf{x}, \mathbf{y}) = 0$. In other words, such a channel is useless (it does not transmit any information).

DEFINITION. The function $I(\mathbf{x}, \mathbf{y})$ depends on the matrix Π and on the input $\mathbf{x}_n$. The capacity C of a noisy channel is the maximum value of $I(\mathbf{x}, \mathbf{y})$ as $\mathbf{x}_n$ ranges over all possible inputs

$$C = \max_{\mathbf{x}_n} I(\mathbf{x}, \mathbf{y}) \tag{15-194}$$

This is consistent with (15-185) because, for noiseless channels, $I(\mathbf{x}, \mathbf{y}) = H(\mathbf{x})$.

Example 15-28. We shall show that the capacity of a *binary symmetrical channel* with channel matrix as in (15-191) (Fig. 15-25) equals

$$C = 1 - \mathfrak{r}(\beta) \qquad \text{where} \quad \mathfrak{r}(p) = -p \log p - q \log q \tag{15-195}$$

Proof. The entropy of a two-state partition equals $\mathfrak{r}(p)$ where p is the probability of one of the states. Thus the entropy $H(\mathbf{x})$ of the input to the channel equals $\mathfrak{r}(p)$ and the entropy of the output equals

$$H(\mathbf{y}) = \mathfrak{r}(\gamma) \qquad \gamma = (1 - 2\beta)p + \beta \tag{15-196}$$

because [see (15-189)]

$$P\{\mathbf{y}_n = 0\} = (1 - \beta)p + \beta(1 - p) = \gamma$$

The above holds also for conditional entropies. Thus, since

$$P\{\mathbf{y}_n = 0 | \mathbf{x}_n = 0\} = P\{\mathbf{y}_n = 1 | \mathbf{x}_n = 1\} = 1 - \beta$$

we conclude that

$$H(\mathbf{y}|\mathbf{x}_n = 0) = H(\mathbf{y}|\mathbf{x}_n = 1) = \mathfrak{r}(1 - \beta)$$

†The conditional entropy $H(\mathbf{x}|\mathbf{y})$ is Shannon's *equivocation*.

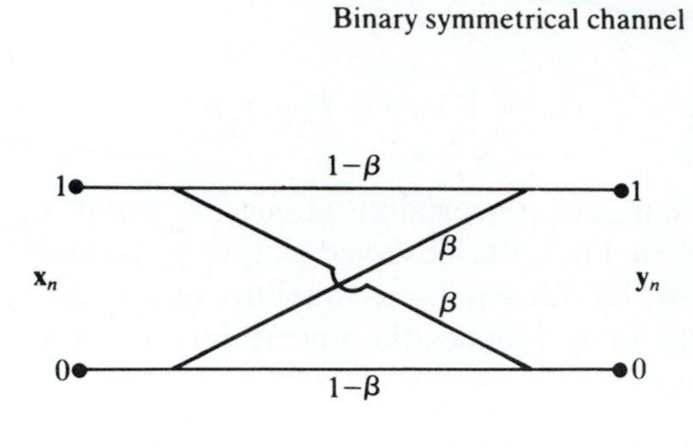

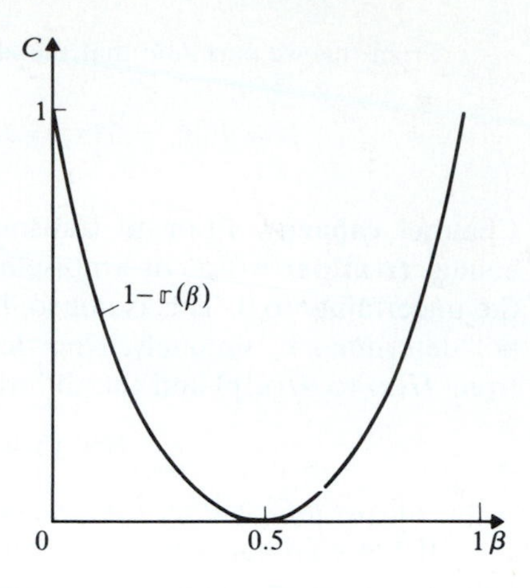

FIGURE 15-25

Inserting into (15-41) and using the fact that $r(\beta) = r(1-\beta)$, we obtain

$$H(\mathbf{x}|\mathbf{y}) = H(\mathbf{y}|\mathbf{x}) = p r(\beta) + q r(\beta) = r(\beta)$$

From the above it follows that $I(\mathbf{x}, \mathbf{y}) = r(\gamma) - r(\beta)$. This yields (15-195) because $r(\beta)$ does not depend on p and $r(\gamma)$ is maximum if $\gamma = 0.5$.

Redundant and random codes Consider a set $\mathscr{A}$ (source) with N elements and a set $\mathscr{B}$ (code) with M elements where $N < M$. A redundant code is a one-to-one correspondence between the elements of $\mathscr{A}$ and the elements of a subset $\mathscr{B}_1$ of $\mathscr{B}$.

The subset $\mathscr{B}_1$ consists of N elements that can be selected in many ways. If the elements of $\mathscr{B}_1$ are chosen at random from the M elements of $\mathscr{B}$, the resulting code is called *random*.† From the definition it follows that the probability that a specific element of $\mathscr{B}$ is in the randomly selected set $\mathscr{B}_1$ equals N/M.

In the next example we show that redundant encoding can be used to reduce the probability of error in transmission.

Example 15-29. In a symmetrical channel, the probability of error equals β. To reduce this error, we encode the input set $\mathscr{A} = \{0, 1\}$ into the subset $\mathscr{B}_1 = \{000, 111\}$ of the set $\mathscr{B}$ of all three-digit binary numbers. In the earlier notation, $N = 2$ and $M = 8$.

The input $\mathbf{x}_n$ is thus encoded into a signal $\bar{\mathbf{x}}_n$ consisting of triplets of 0's and 1's yielding as output a signal $\bar{\mathbf{y}}_n$ (Fig. 15-26). The decoding scheme is the *majority rule*: If the received triplet consists of at least two 0's, then $\mathbf{y}_n = 0$, otherwise $\mathbf{y}_n = 1$.

---

†This definition of a random code is not the definition given on page 583.

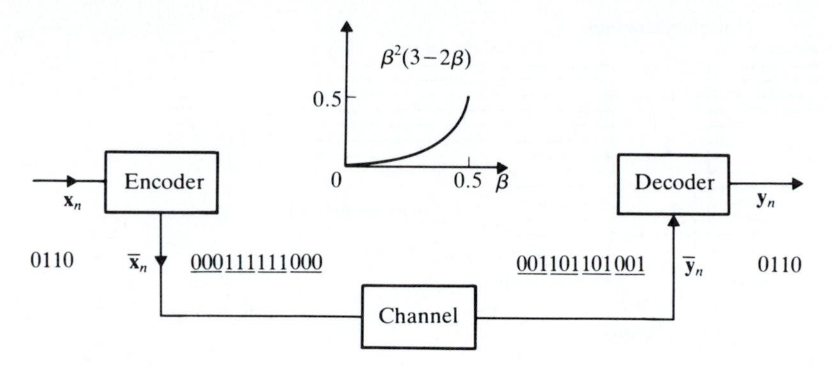

FIGURE 15-26

It can be readily seen that (Prob. 15-23) the probability that a transmitted word will be detected incorrectly equals $\beta^2(3 - 2\beta)$. This is less than β if $\beta < 0.5$. However, the rate of transmission is also reduced from 1 word per second to 1 word per three seconds.

It appears from the above that reduction of the probability of error by redundant encoding must result in transmission rates that tend to 0 as the error tends to 0. This, however, is not so. As the following remarkable theorem shows, it is possible to achieve arbitrarily small error probabilities while maintaining the rate of information transmission close to the channel capacity.

The Channel Capacity Theorem

Information can be transmitted through a noisy channel at a rate nearly equal to the channel capacity C with negligible probability of error.

Proof. *Preliminary remarks* From the definition of channel capacity, it follows that the maximum of $H(\mathbf{x})$ is at least equal to C because

$$H(\mathbf{x}) = I(\mathbf{x}, \mathbf{y}) + H(\mathbf{x}|\mathbf{y}) \geq I(\mathbf{x}, \mathbf{y}) \tag{15-197}$$

This shows that we can find a source with entropy rate as close to C as we want. We shall show that if $\mathbf{x}_n$ is a source with entropy rate

$$H(\mathbf{x}) < C \tag{15-198}$$

then it can be transmitted at the rate of 1 word per second with probability of error less than α for any $\alpha > 0$. This will prove the theorem because the information per word equals $H(\mathbf{x})$.

As in the noiseless case, the proof is based on proper encoding of the space $\mathscr{S}_x^m$ consisting of all possible segments of $\mathbf{x}_n$ of length m. However, as the following remarks show, the objectives are different.

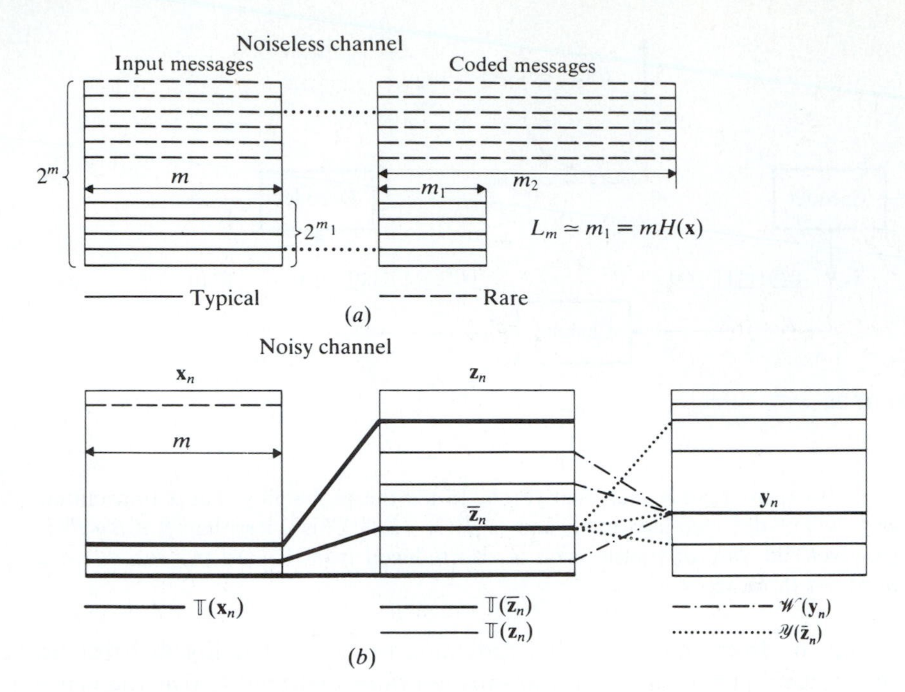

FIGURE 15-27

Noiseless channel The code set consists of two groups of binary numbers (Fig. 15-27*a*). The first group has 2^{m_1} elements of length $m_1 = mH(\mathbf{x})$ and it is used to encode the 2^{m_1} typical sequences of the input space $\mathscr{S}_x^m$. The second group is used to encode the rare sequences of $\mathscr{S}_x^m$. Since the set of all rare sequences has negligible probability, the average length of the code equals m_1.

Thus, in the noiseless case, the purpose of coding is reduction of the time of transmission of m-word messages from m seconds to m_1 seconds. This results in an increase of the rate of information transmission from $mH(\mathbf{x})$ bits per m seconds to $mH(\mathbf{x})$ bits per $m_1 = mH(\mathbf{x})$ seconds.

Noisy channel Reasoning as in (15-197), we conclude that, given $\varepsilon > 0$, we can find a process $\mathbf{z}_n$ such that

$$H(\mathbf{z}) - H(\mathbf{z}|\mathbf{y}) > C - \varepsilon \tag{15-199}$$

Choosing $\varepsilon < C - H(\mathbf{x})$, we obtain

$$H(\mathbf{z}) > H(\mathbf{x}) + H(\mathbf{z}|\mathbf{y}) \geq H(\mathbf{x}) \tag{15-200}$$

because $H(\mathbf{z}|\mathbf{y}) > 0$.

All sequences of $\mathbf{z}_n$ of length m form a space $\mathscr{S}_z^m$ consisting of 2^m elements. We can, therefore, encode the input set $\mathscr{S}_x^m$ into the set $\mathscr{S}_z^m$. The resulting code is one-to-one (Fig. 15-27*b*). The code can, however, be viewed as redundant if we consider only the mapping of the subset $\mathbb{T}(\mathbf{x}_n)$ of all typical sequences of $\mathscr{S}_x^m$ into the subset $\mathbb{T}(\mathbf{z}_m)$ of all typical sequences of $\mathscr{S}_z^m$. Indeed, $\mathbb{T}(\mathbf{x}_n)$ has $N = 2^{mH(\mathbf{x})}$ elements and $\mathbb{T}(\mathbf{z}_n)$ has $M = 2^{mH(\mathbf{z})}$ elements where

$$N = 2^{mH(\mathbf{x})} \ll 2^{mH(\mathbf{z})} = M \tag{15-201}$$

because $H(\mathbf{x}) < H(\mathbf{z})$ and $m \gg 1$. We denote by $\bar{\mathbf{z}}_n$ the code word of a typical $\mathbf{x}_n$ message and by $\mathbb{T}(\bar{\mathbf{z}}_n)$ the set of all such code words. Clearly, $\mathbb{T}(\bar{\mathbf{z}}_n)$ is a subset of the set $\mathbb{T}(\mathbf{z}_n)$ consisting of $N \ll M$ elements.

The purpose of the coding is to select the set $\mathbb{T}(\bar{\mathbf{z}}_n)$ such that its elements are at a "large distance" from each other in the following sense: Since the channel is noisy, the output due to a specific element $\bar{\mathbf{z}}_n$ is not unique. We denote by $\mathscr{Y}(\bar{\mathbf{z}}_n)$ the set of all output sequences due to this element, and we attempt to design the code such that the probability of the intersection of the output sets $\mathscr{Y}(\bar{\mathbf{z}}_n)$ as $\bar{\mathbf{z}}_n$ ranges over every element of the set $\mathbb{T}(\bar{\mathbf{z}}_n)$ is negligible. This will ensure the unique determination of $\bar{\mathbf{z}}_n$ in terms of the observed output $\mathbf{y}_n$.

Random code To complete the proof, we shall show that among all N-element subsets of the set $\mathbb{T}(\mathbf{z}_n)$ there exists at least one that meets our requirements. In fact, we shall prove a stronger statement: If we select *at random* N elements $\bar{\mathbf{z}}_n$ from the M elements of $\mathbb{T}(\mathbf{z}_n)$ and use the resulting set $\mathbb{T}(\bar{\mathbf{z}}_n)$ to encode the set $\mathbb{T}(\mathbf{x}_n)$ then, almost certainly, the probability of error in transmission will be negligible.

We note that, once the *code set* $\mathbb{T}(\bar{\mathbf{z}}_n)$ has been selected, the probability that an element of $\mathbb{T}(\mathbf{z}_n)$ is in $\mathbb{T}(\bar{\mathbf{z}}_n)$ equals N/M. From this it follows that, if $\mathscr{W}$ is a randomly selected subset of $\mathbb{T}(\mathbf{z}_n)$ consisting of N_w elements, then the probability P_w that it will intersect the set $\mathbb{T}(\bar{\mathbf{z}}_n)$ equals

$$P_w = 1 - \left(1 - \frac{N}{M}\right)^{N_w} \simeq \frac{NN_w}{M} \tag{15-202}$$

because $N \ll M$.

Suppose that we transmit the selected m-word message $\mathbf{z}_n$ through the channel and we observe at the output the m-word message $\mathbf{y}_n$. Since the channel is noisy, the same $\mathbf{y}_n$ might result from many other input messages. We denote by $\mathscr{W}(\mathbf{y}_n)$ the set consisting of all elements of $\mathbb{T}(\mathbf{z}_n)$ that will produce the same output $\mathbf{y}_n$, excluding the actually transmitted message $\bar{\mathbf{z}}_n$ (Fig. 15-27*b*). If the set $\mathscr{W}(\mathbf{y}_n)$ does not intersect the code set $\mathbb{T}(\bar{\mathbf{z}}_n)$, there is no error because the observed signal $\mathbf{y}_n$ determines uniquely the transmitted signal $\bar{\mathbf{z}}_n$. The error probability equals, therefore, the probability P_w that the sets $\mathscr{W}(\mathbf{y}_n)$ and $\mathbb{T}(\bar{\mathbf{z}}_n)$ intersect. As we know [see (15-74)] the number N_w of typical elements in $\mathscr{W}(\mathbf{y}_n)$ equals $2^{mH(\mathbf{z}|\mathbf{y})}$. Neglecting all others, we conclude from (15-202) that

$$P_w \simeq \frac{NN_w}{M} = 2^{mH(\mathbf{z}|\mathbf{y})}2^{m[H(\mathbf{x})-H(\mathbf{z})]}$$

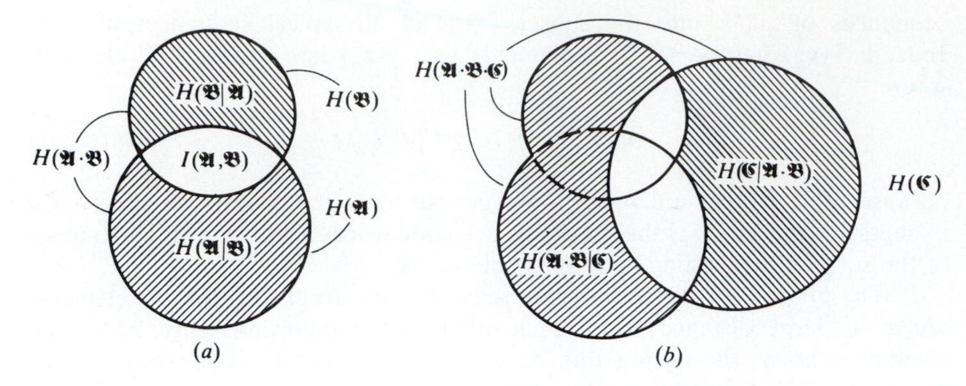

FIGURE P15-3

This shows that

$$P_w \to 0 \qquad \text{as} \quad m \to \infty$$

because $H(\mathbf{z}|\mathbf{y}) + H(\mathbf{x}) - H(\mathbf{z}) < 0$, and the proof is complete.

We note, finally, that the maximum rate of information transmission cannot exceed C bits per second.

Indeed, to achieve a rate higher than C, we would need to transmit a signal $\mathbf{z}_n$ such that $H(\mathbf{z}) - H(\mathbf{z}|\mathbf{y}) > C$. This, however, is impossible [see (15-194)].

PROBLEMS

15-1. Show that $H(\mathfrak{A} \cdot \mathfrak{B}|\mathfrak{B}) = H(\mathfrak{A}|\mathfrak{B})$.

15-2. Show that if $\varphi(p) = -p \log p$ and $p_1 < p_1 + \varepsilon < p_2 - \varepsilon < p_2$, then

$$\varphi(p_1 + p_2) < \varphi(p_1) + \varphi(p_2) < \varphi(p_1 + \varepsilon) + \varphi(p_2 - \varepsilon)$$

15-3. In Fig. P15-3*a*, we give a schematic representation of the identities

$$H(\mathfrak{A} \cdot \mathfrak{B}) = H(\mathfrak{A}) + H(\mathfrak{B}|\mathfrak{A}) = H(\mathfrak{A}) + H(\mathfrak{B}) - I(\mathfrak{A}, \mathfrak{B})$$

where each quantity equals the area of the corresponding region. Extending formally this representation to three partitions (Fig. P15-3*b*), we obtain the identities

$$H(\mathfrak{A} \cdot \mathfrak{B} \cdot \mathfrak{C}) = H(\mathfrak{A}) + H(\mathfrak{B} \cdot \mathfrak{C}|\mathfrak{A}) = H(\mathfrak{A} \cdot \mathfrak{B}) + H(\mathfrak{C}|\mathfrak{A} \cdot \mathfrak{B})$$

$$H(\mathfrak{A} \cdot \mathfrak{B} \cdot \mathfrak{C}) = H(\mathfrak{A}) + H(\mathfrak{B}|\mathfrak{A}) + H(\mathfrak{C}|\mathfrak{A} \cdot \mathfrak{B})$$

$$H(\mathfrak{B} \cdot \mathfrak{C}|\mathfrak{A}) = H(\mathfrak{B}|\mathfrak{A}) + H(\mathfrak{C}|\mathfrak{A} \cdot \mathfrak{B})$$

Show that these identities are correct.

15-4. Show that

$$I(\mathfrak{A} \cdot \mathfrak{B}, \mathfrak{C}) + I(\mathfrak{A}, \mathfrak{B}) = I(\mathfrak{A} \cdot \mathfrak{C}, \mathfrak{B}) + I(\mathfrak{A}, \mathfrak{C})$$

and identify each quantity in the representation of Fig. P15-3*b*.

15-5. The conditional mutual information of two partitions $\mathfrak{A}$ and $\mathfrak{B}$ assuming $\mathfrak{C}$ is by definition

$$I(\mathfrak{A},\mathfrak{B}|\mathfrak{C}) = H(\mathfrak{A}|\mathfrak{C}) + H(\mathfrak{B}|\mathfrak{C}) - H(\mathfrak{A}\cdot\mathfrak{B}|\mathfrak{C})$$

(*a*) Show that

$$I(\mathfrak{A},\mathfrak{B}|\mathfrak{C}) = I(\mathfrak{A},\mathfrak{B}\cdot\mathfrak{C}) - I(\mathfrak{A},\mathfrak{C}) \qquad \text{(i)}$$

and identify each quantity in the representation of Fig. P15-3*b*.
(*b*) From (i) it follows that $I(\mathfrak{A},\mathfrak{B}\cdot\mathfrak{C}) \geq I(\mathfrak{A},\mathfrak{C})$. Interpret this inequality in terms of the subjective notion of mutual information.

15-6. In an experiment $\mathscr{S}$, the entropy of the binary partition $\mathfrak{A} = [\mathscr{A}, \overline{\mathscr{A}}]$ equals $\mathfrak{r}(p)$ where $p = P(\mathscr{A})$. Show that in the experiment $\mathscr{S}^3 = \mathscr{S} \times \mathscr{S} \times \mathscr{S}$, the entropy of the eight-element partition $\mathfrak{A}^3 = \mathfrak{A}\cdot\mathfrak{A}\cdot\mathfrak{A}$ equals $3\mathfrak{r}(p)$ as in (15-67).

15-7. Show that

$$H(\mathbf{x}+a) = H(\mathbf{x}) \qquad H(\mathbf{x}+\mathbf{y}|\mathbf{x}) = H(\mathbf{y}|\mathbf{x})$$

In the above, $H(\mathbf{x}+a)$ is the entropy of the RV $\mathbf{s}+a$ and $H(\mathbf{x}+\mathbf{y}|\mathbf{x})$ is the conditional entropy of the Rv $\mathbf{x}+\mathbf{y}$.

15-8. The RVs $\mathbf{x},\mathbf{y}$ are of discrete type and independent. Show that if $\mathbf{z} = \mathbf{x}+\mathbf{y}$ and the line $x + y = z_i$ contains no more than one mass point, then

$$H(\mathbf{z}|\mathbf{x}) = H(\mathbf{y}) \leq H(\mathbf{z})$$

Hint: Show that $\mathfrak{A}_z = \mathfrak{A}_x \cdot \mathfrak{A}_y$.

15-9. The RV $\mathbf{x}$ is uniform in the interval $(0, a)$ and the RV $\mathbf{y}$ equals the value of $\mathbf{x}$ rounded off to the nearest multiple of δ. Show that $I(\mathbf{x},\mathbf{y}) = \ln a/\delta$.

15-10. Show that, if the transformation $\mathbf{y} = g(\mathbf{x})$ is one-to-one and $\mathbf{x}$ is of discrete type, then

$$H(\mathbf{x},\mathbf{y}) = H(\mathbf{x})$$

Hint: $p_{ij} = P\{\mathbf{x} = x_i\}\delta[i-j]$.

15-11. Show that for discrete-type RVs

$$H(\mathbf{x},\mathbf{x}) = H(\mathbf{x}) \qquad H(\mathbf{x}|\mathbf{x}) = 0 \qquad H(\mathbf{y}|\mathbf{x}) = H(\mathbf{y},\mathbf{x}|\mathbf{x})$$

$$H(\mathbf{y}|\mathbf{x}_1,\ldots,\mathbf{x}_n) = H\left(\mathbf{y}, \sum_{k=1}^{n} a_k\mathbf{x}_k|\mathbf{x}_1,\ldots,\mathbf{x}_n\right)$$

For continuous-type RVs, the relevant densities are singular. The above holds, however, if we set $H(\mathbf{x},\mathbf{x}) = H(\mathbf{x})$ and use theorem (15-103) and its extensions to several variables to define recursively all conditional entropies.

15-12. The process $\mathbf{x}_n$ is normal white noise with $E\{\mathbf{x}_n^2\} = 5$, and

$$\mathbf{y}_n = \sum_{k=0}^{\infty} 2^{-k}\mathbf{x}_{n-k}$$

(*a*) Find the mutual information of RVs $\mathbf{x}_n$ and $\mathbf{y}_n$. (*b*) Find the entropy rate of the process $\mathbf{y}_n$.

15-13. The RVs $\mathbf{x}_n$ are independent and each is uniform in the interval (4, 6). Find the entropy rate of the process

$$\mathbf{y}_n = 5 \sum_{k=0}^{\infty} 2^{-k} \mathbf{x}_{n-k}$$

15-14. Find the ME density of an RV $\mathbf{x}$ if $f(x) = 0$ for $|x| > 1$ and $E\{\mathbf{x}\} = 0.31$.

15-15. It is observed that the duration of the telephone calls is a number $\mathbf{x}$ between 1 and 5 minutes and its mean is 3 min 37 sec. Find its ME density.

15-16. We are given a die with $P\{\text{even}\} = 0.5$ and are told that the mean of the number $\mathbf{x}$ of faces up equals 4.44. Find the ME values of $p_i = P\{\mathbf{x} = i\}$.

15-17. Suppose that $\mathbf{x}$ is an RV with entropy $H(\mathbf{x})$ and $\mathbf{y} = 3\mathbf{x}$. Express the entropy $H(\mathbf{y})$ of $\mathbf{y}$ in terms of $H(\mathbf{x})$. (*a*) if $\mathbf{x}$ is of discrete type, (*b*) if $\mathbf{x}$ is of continuous type.

15-18. In the experiment of two fair dice. $\mathfrak{A}$ is a partition consisting of the events $\mathscr{A}_1 = \{\text{seven}\}$, $\mathscr{A}_2 = \{\text{eleven}\}$, and $\mathscr{A}_3 = \overline{\mathscr{A}_1 \cup \mathscr{A}_2}$. (*a*) Find its entropy. (*b*) The dice were rolled 100 times. Find the number of typical and atypical sequences formed with the events $\mathscr{A}_1$, $\mathscr{A}_2$, and $\mathscr{A}_3$.

15-19. The process $\mathbf{x}[n]$ is SSS with entropy rate $\overline{H}(\mathbf{x})$. Show that, if

$$\mathbf{w}_n = \sum_{k=0}^{n} \mathbf{x}_{n-k} h_k$$

then

$$\lim_{n \to \infty} \frac{1}{n+1} H(\mathbf{w}_0, \ldots, \mathbf{w}_n) = \overline{H}(\mathbf{x}) + \ln h_0$$

15-20. In the coin experiment, the probability of "heads" is an RV $\mathbf{p}$ with $E\{\mathbf{p}\} = 0.6$. Using the MEM, find its density $f(p)$.

15-21. (The Brandeis dice problem†) In a die experiment, the average number of dots up equals 4.5. Using the MEM, find $p_i = P\{f_i\}$.

15-22. Using the MEM, find the joint density $f(x_1, x_2, x_3)$ of the RVs $\mathbf{x}_1, \mathbf{x}_2, \mathbf{x}_3$ if

$$E\{\mathbf{x}_1^2\} = E\{\mathbf{x}_2^2\} = E\{\mathbf{x}_3^2\} = 4 \qquad E\{\mathbf{x}_1\mathbf{x}_2\} = E\{\mathbf{x}_1\mathbf{x}_3\} = 1$$

15-23. A source has seven elements with probabilities

0.3 0.2 0.15 0.15 0.1 0.06 0.04

respectively. Construct a Shannon, a Fano, and a Huffman code and find their average code lengths.

15-24. Show that in the redundant coding of Example 15-29, the probability of error equals $\beta^2(3 - 2\beta)$.

Hint: $P\{\mathbf{y}_n = 1 | \mathbf{x}_n = 0\} = \beta^3 + 3\beta^2(1 - \beta)$.

15-25. Find the channel capacity of a symmetrical binary channel if the received information is always wrong.

---

†E. T. Jaynes: Brandeis lectures, 1962.

CHAPTER 16

SELECTED TOPICS

16-1 THE LEVEL-CROSSING PROBLEM

Given a random process $\mathbf{x}(t)$ and a constant a, we denote by $\boldsymbol{\tau}_i$ the time instances when $\mathbf{x}(t)$ crosses the line L_a shown in Fig. 16-1. This line is parallel to the time axis and

$$\mathbf{x}(\boldsymbol{\tau}_i) = a$$

The level-crossing problem is the determination of the statistical properties of the point process $\boldsymbol{\tau}_i$ so formed. A special case is the zero-crossing problem ($a = 0$) when L_a coincides with the t axis.

The level-crossing problem is, in general, complicated. We discuss next only certain aspects that lead to simple results.

EXPECTED NUMBER OF CROSSINGS†. We assume that the process $\mathbf{x}(t)$ is stationary and we denote by $\mathbf{n}_a(T)$ the number of points $\boldsymbol{\tau}_i$ in an interval of length T. The following basic theorem expresses the mean of $\mathbf{n}_a(T)$ in terms of the first-order density $f_x(x)$ of $\mathbf{x}(t)$ and the conditional mean of its derivative.

THEOREM. If $\mathbf{x}'(t)$ exists, then

$$E\{\mathbf{n}_a(T)\} = Tf_x(a)E\{|\mathbf{x}'(t)||\mathbf{x}(t) = a\} \tag{16-1}$$

---

†A. Blanc-Lapierre: *Modèles Statistiques pour l'étude de phenomenes de Fluctuations*, Masson et cie, Paris, 1963.

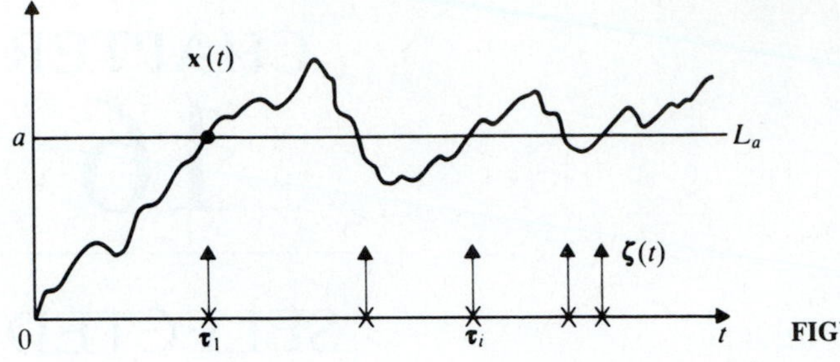

FIGURE 16-1

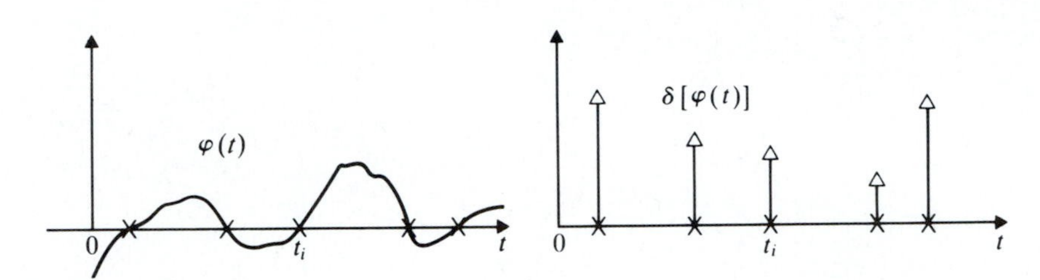

FIGURE 16-2

Proof. We shall prove the theorem using the following property of the impulse function (see Papoulis, 1968): If t_i are all the real 0's of a function $\varphi(t)$, then (Fig. 16-2)

$$\delta[\varphi(t)] = \sum_i \frac{\delta(t - t_i)}{|\varphi'(t_i)|} = \frac{1}{|\varphi'(t)|} \sum_i \delta(t - t_i) \tag{16-2}$$

The 0's of the function $\varphi(t) = \mathbf{x}(t) - a$ are the L_a crossings of $\mathbf{x}(t)$. Thus

$$\varphi(\boldsymbol{\tau}_i) = \mathbf{x}(\boldsymbol{\tau}_i) - a = 0 \qquad \varphi'(t) = \mathbf{x}'(t)$$

Inserting into (16-2), we obtain

$$\sum_i \delta(t - \boldsymbol{\tau}_i) = |\mathbf{x}'(t)| \delta[\mathbf{x}(t) - a] \tag{16-3}$$

The sum

$$\boldsymbol{\zeta}(t) = \sum_i \delta(t - \boldsymbol{\tau}_i)$$

is a stationary process consisting of a sequence of impulses at the points $\boldsymbol{\tau}_i$. The area of each impulse equals 1 and the number of impulses in the interval $(t, t + T)$ equals $\mathbf{n}_a(T)$. Hence

$$\mathbf{n}_a(T) = \int_t^{t+T} \boldsymbol{\zeta}(\alpha)\, d\alpha \qquad E\{\mathbf{n}_a(T)\} = TE\{\boldsymbol{\zeta}(t)\}$$

To prove (16-1), it suffices, therefore, to find the mean of $\boldsymbol{\zeta}(t)$. As we see from (16-3), the RV $\boldsymbol{\zeta}(t)$ is a function of the RVs $\mathbf{x}(t)$ and $\mathbf{x}'(t)$. Denoting by $f(x, x')$ the joint density of $\mathbf{x}(t)$ and $\mathbf{x}'(t)$, we conclude from (16-3) and (7-2) that

$$E\{\boldsymbol{\zeta}(t)\} = \int_{-\infty}^{\infty}\int_{-\infty}^{\infty} |x'|\delta(x-a)f(x,x')\,dx\,dx' = \int_{-\infty}^{\infty} |x'|f(a,x')\,dx'$$

This yields (16-1) because $f(a, x') = f_x(a)f(x'|a)$.

Note that the conditional mean $E\{|\mathbf{x}'(t)|\,|\mathbf{x}(t) = a\}$ is the average of the slopes $|\mathbf{x}'(t)|$ of all processes that cross the line L_a at time t.

LEVEL-CROSSING DENSITY. We denote by $p_a(\tau)$ the probability that in an interval I_τ of length τ there is one and only one crossing. If $\tau \to 0$ then $p_a(\tau) \to 0$. Furthermore, with the exception of unusual cases, the probability that there is more than one crossing in a small interval τ is small compared to $p_a(\tau)$. Assuming that this is the case for the process $\mathbf{x}(t)$, we introduce the limit

$$\lambda_a = \lim_{\tau \to 0} \frac{1}{\tau} p_a(\tau)$$

If this limit exists, it is the density of the L_a crossings. Thus $p_a(\Delta\tau) \simeq \lambda_a\,\Delta\tau$ for small $\Delta\tau$.

We maintain that†

$$\lambda_a = \frac{1}{T}E\{\mathbf{n}_a(T)\} \tag{16-4}$$

Proof. If $\Delta\tau$ is small, then

$$P\{\mathbf{n}_a(\Delta\tau) = 1\} = p_a(\Delta\tau) \qquad P\{\mathbf{n}_a(\Delta\tau) = 0\} = 1 - p_a(\Delta\tau)$$

Hence

$$E\{\mathbf{n}_a(\Delta\tau)\} = 1 \times P\{\mathbf{n}_a(\Delta\tau) = 1\} = p_a(\Delta\tau) \simeq \lambda_a\,\Delta\tau$$

From this it follows that (16-4) is true for T small. And since

$$\mathbf{n}_a(T_1 + T_2) = \mathbf{n}_a(T_1) + \mathbf{n}_a(T_2)$$

we conclude that it is true for any T.

We have thus shown that if $\mathbf{x}(t)$ is differentiable, then the level-crossing density λ_a exists and it equals

$$\lambda_a = f_x(a)E\{|\mathbf{x}'(t)|\,|\mathbf{x}(t) = a\} \tag{16-5}$$

Density of maxima. The maxima and minima (extrema) of a process $\mathbf{x}(t)$ are the zero crossings of the process $\mathbf{y}(t) = \mathbf{x}'(t)$. From this it follows that the density λ_m of the extrema of $\mathbf{x}(t)$ is obtained from (16-5) if we set $a = 0$ and replace

†S. O. Rice: "Mathematical Analysis of Random Noise," in *Selected Papers on Noise and Stochastic Processes*, Dover, N.Y., 1954.

$\mathbf{x}(t)$ by $\mathbf{x}'(t)$. This yields

$$\lambda_m = f_{x'}(0)E\{|\mathbf{x}''(t)|\,|\mathbf{x}'(t) = 0\} \tag{16-6}$$

The density of maxima equals $\lambda_m/2$.

Normal Processes

We shall apply the preceding results to normal processes under the assumption that their mean η_x is 0. This assumption is not essential because the a-level crossings of the process $\mathbf{x}(t)$ are identical to the $(a - \eta_x)$-level crossings of the centered process $\mathbf{x}(t) - \eta_x$.

THEOREM. If $\mathbf{x}(t)$ is a differentiable process with autocorrelation $R(\tau)$, then

$$\lambda_a = f_x(a)E\{|\mathbf{x}'(t)|\} = \frac{1}{\pi}\sqrt{\frac{-R''(0)}{R(0)}}\,e^{-a^2/2R(0)} \tag{16-7}$$

Proof. As we know

$$R_{xx'}(\tau) = -R'(\tau) \qquad R_{x'x'}(\tau) = -R''(\tau)$$

From the existence of $\mathbf{x}'(t)$ it follows that $R''(\tau)$ exists. Therefore, $R'(\tau)$ also exists and $R'(0) = 0$ because $R(\tau)$ is even. Hence

$$E\{\mathbf{x}(t)\mathbf{x}'(t)\} = -R'(0) = 0$$

This shows that the RVs $\mathbf{x}(t)$ and $\mathbf{x}'(t)$ are orthogonal. And since they are normal with zero mean, they are independent. The first equality in (16-7) follows, therefore, from (16-5).

To prove the second equality, we observe that the variance of $\mathbf{x}(t)$ equals $R(0)$ and the variance of $\mathbf{x}'(t)$ equals $-R''(0)$. This yields [see (5-45)]

$$f_x(a) = \frac{1}{\sqrt{2\pi R(0)}}e^{-a^2/2R(0)} \qquad E\{|\mathbf{x}'(t)|\} = \sqrt{\frac{-2R''(0)}{\pi}}$$

and (16-7) results.

Zero-crossing density. Denoting by λ_0 the density of the 0's of $\mathbf{x}(t)$, we conclude from (16-7) with $a = 0$ that

$$\lambda_0^2 = \frac{-R''(0)}{\pi^2 R(0)} = \frac{\int_{-\infty}^{\infty}\omega^2 S(\omega)\,d\omega}{\pi^2\int_{-\infty}^{\infty}S(\omega)\,d\omega} \tag{16-8}$$

Example 16-1. If $S(\omega) = 0$ for $|\omega| > \sigma$, then (16-8) yields $\lambda_0 \le \sigma/\pi$ because

$$\int_{-\sigma}^{\sigma}\omega^2 S(\omega)\,d\omega \le \sigma^2\int_{-\sigma}^{\sigma}S(\omega)\,d\omega$$

If also $S(\omega) = S_0$ for $|\omega| < \sigma$ (ideal low-pass), then

$$\lambda_0 = \frac{\sigma}{\pi\sqrt{3}}$$

Example 16-2. (a) If $\mathbf{a}$ and $\mathbf{b}$ are two independent RVs with variance σ^2 and

$$\mathbf{x}(t) = \mathbf{a}\cos\omega_0 t + \mathbf{b}\sin\omega_0 t$$

as in Example 10-13, then

$$R_x(\tau) = \sigma^2 \cos\omega_0\tau \qquad R_x(0) = \sigma^2 \qquad R_x''(0) = -\omega_0^2\sigma^2$$

and (16-8) yields

$$\lambda_0 = \frac{\omega_0}{\pi}$$

(b) If the RVs $\mathbf{a}$ and $\mathbf{b}$ are independent of the process $\boldsymbol{\nu}(t)$ and

$$\mathbf{x}(t) = \mathbf{a}\cos\omega_0 t + \mathbf{b}\sin\omega_0 t + \boldsymbol{\nu}(t)$$

then

$$R_x(\tau) = \sigma^2\cos\omega_0\tau + R_\nu(\tau)$$

and (16-8) yields

$$\lambda_0 = \frac{1}{\pi}\sqrt{\frac{\omega_0^2\sigma^2 - R_\nu''(0)}{\sigma^2 + R_\nu(0)}}$$

Nondifferentiable processes. We have shown that if $\mathbf{x}'(t)$ exists, then the probability $p_0(\tau)$ that there is a zero crossing in a small interval τ equals $\lambda_0\tau$. If $\mathbf{x}'(t)$ does not exist, then $p_0(\tau)$ is no longer proportional to τ. In the following, we examine the asymptotic form of $p_0(\tau)$ as $\tau \to 0$ under the assumption that $R(\tau)$ has a corner at the origin as in the Ornstein–Uhlenbeck process [see (11-15)].

Suppose that $R'(\tau)$ is discontinuous at $\tau = 0$ but $R'(0^+)$ exists. In this case

$$R(\tau) = R(0) + R'(0^+)\tau + O(\tau^2) \qquad \tau > 0 \tag{16-9}$$

We shall show that $p_0(\tau)$ is proportional to $\sqrt{\tau}$:

$$p_0(\tau) \simeq \frac{1}{\pi}\sqrt{-\frac{2R'(0^+)\tau}{R(0)}} \tag{16-10}$$

Proof. If τ is small, then we can neglect more than one zero crossing in the interval $(t, t+\tau)$. With this assumption, we have one crossing iff $\mathbf{x}(t+\tau)\mathbf{x}(t) < 0$ (Fig. 16-3). Hence

$$p_0(\tau) = P\{\mathbf{x}(t+\tau)\mathbf{x}(t) < 0\}$$

The RVs $\mathbf{x}(t+\tau)$ and $\mathbf{x}(t)$ are jointly normal with correlation coefficient $r(\tau) = R(\tau)/R(0)$. Applying (6-46), we conclude that

$$p_0(\tau) \simeq \frac{\beta}{\pi} \qquad \cos\beta = r(\tau)$$

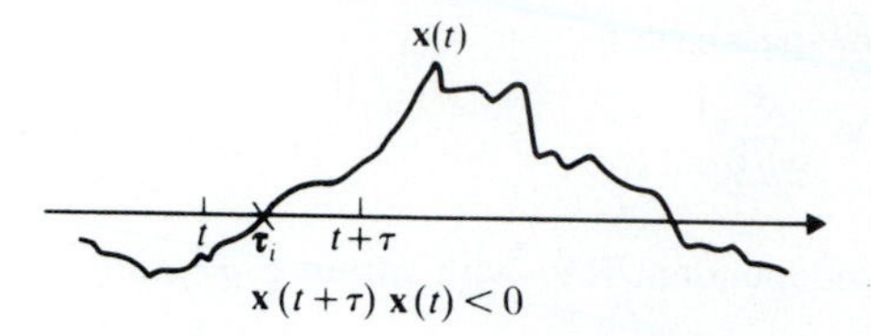

FIGURE 16-3

This yields

$$r(\tau) = \cos\beta = \cos\pi p_0(\tau) \simeq 1 - \frac{\pi^2 p_0^2(\tau)}{2}$$

for $p_0(\tau) \ll 1$. Thus, for small τ,

$$\pi p_0(\tau) \simeq \sqrt{2[1 - r(\tau)]} \qquad r(\tau) = \frac{R(\tau)}{R(0)} \tag{16-11}$$

Inserting (16-9) into the above, we obtain (16-10).

Note Using (16-11), we can reestablish (16-10). Indeed, in this case, $R'(0) = 0$ and $R''(0)$ exists. Hence

$$R(\tau) = R(0) + \tfrac{1}{2}R''(0)\tau^2 + O(\tau^3)$$

Inserting into (16-11), we obtain

$$p_0(\tau) \simeq \frac{\tau}{\pi}\sqrt{\frac{-R''(0)}{R(0)}} \tag{16-12}$$

This yields (16-8) because $p_0(\tau) \simeq \lambda_0 \tau$ for small τ.

Example 16-3. If $\mathbf{a}(t)$ and $\mathbf{b}(t)$ are two normal independent processes and

$$\mathbf{x}(t) = \mathbf{a}(t)\cos\omega_0 t - \mathbf{b}(t)\sin\omega_0 t$$

as in (11-62), then $\mathbf{x}(t)$ is normal, stationary, with autocorrelation [see (11-65)]

$$R_x(\tau) = R_a(\tau)\cos\omega_0\tau$$

(a) We shall show that, if $\mathbf{x}(t)$ is differentiable, then its zero-crossing density equals

$$\lambda_0 = \sqrt{\bar{\lambda}_0^2 + \frac{\omega_0^2}{\pi^2}} \qquad \text{where} \quad \bar{\lambda}_0 = \frac{1}{\pi}\sqrt{\frac{-R_a''(0)}{R(0)}} \tag{16-13}$$

is the zero-crossing density of $\mathbf{a}(t)$. Indeed, in this case, $R_a'(0) = 0$; hence

$$R_x(0) = R_a(0) \qquad R_x'(0) = 0 \qquad R_x''(0) = R_a''(0) - \omega_0^2 R_a(0)$$

and (16-13) follows from (16-8).

(b) In the nondifferentiable case, we have

$$R_x(0) = R_a(0) \qquad R_x'(0^+) = R_a'(0^+) \neq 0$$

This shows that the zero-crossing probabilities $p_0(\tau)$ and $\bar{p}_0(\tau)$ of the processes $\mathbf{x}(t)$ and $\mathbf{a}(t)$ are equal [see (16-10)].

Example 16-4. In the preceding discussion we assumed that all processes are stationary. The results can be readily extended to nonstationary processes. We illustrate the nondifferentiable case using as an example the Wiener process $\mathbf{w}(t)$.

We shall show that for small τ, the probability $p_0(t, t+\tau)$ that $\mathbf{w}(t)$ crosses the t axis in the interval $(t, t+\tau)$ equals

$$p_0(t, t+\tau) \simeq \frac{1}{\pi}\sqrt{\frac{\tau}{t}} \tag{16-14}$$

Proof. Reasoning as in the stationary case, we conclude that $p_0(t, t+\tau)$ is given by (16-11) where $r(\tau)$ is the correlation coefficient of the RVs $\mathbf{x}(t+\tau)$ and $\mathbf{x}(t)$. Thus [see (11-5)]

$$r^2(\tau) = \frac{R^2(t+\tau, t)}{R(t+\tau, t+\tau)R(t,t)} = \frac{\alpha^2 t^2}{\alpha(t+\tau)\alpha t} = \frac{t}{t+\tau}$$

Inserting into (16-11), we obtain (16-14) because $\sqrt{t/(t+\tau)} \simeq 1 - \tau/2t$ for small τ.

Density of maxima. The extrema of $\mathbf{x}(t)$ are the 0's of $\mathbf{x}'(t)$. Hence their density λ_m is given by (16-8) if we replace the autocorrelation $R(\tau)$ of $\mathbf{x}(t)$ by the autocorrelation $-R''(\tau)$ of $\mathbf{x}'(t)$. From this it follows that

$$\lambda_m^2 = \frac{-R^{(4)}(0)}{\pi^2 R''(0)} = \frac{\int_{-\infty}^{\infty} \omega^4 S(\omega)\, d\omega}{\pi^2 \int_{-\infty}^{\infty} \omega^2 S(\omega)\, d\omega} \tag{16-15}$$

provided that $\mathbf{x}'(t)$ exists.

Example 16-5. (a) If $S(\omega) = 0$ for $|\omega| > \sigma$, then the above yields $\lambda_m \le \sigma/\pi$ because

$$\int_{-\sigma}^{\sigma} \omega^4 S(\omega)\, d\omega \le \sigma^2 \int_{-\sigma}^{\sigma} \omega^2 S(\omega)\, d\omega$$

If also $S(\omega) = S_0$ for $|\omega| \le \sigma$, then

$$\lambda_m = \frac{\sigma}{\pi}\sqrt{\frac{3}{5}}$$

(b) If $S(\omega) = S_0$ for $\omega_1 < |\omega| < \omega_2$ and 0 otherwise (ideal bandpass), then

$$\lambda_m = \frac{1}{\pi}\sqrt{\frac{3(\omega_2^5 - \omega_1^5)}{5(\omega_2^3 - \omega_1^3)}}$$

FIRST-PASSAGE TIME. We denote by $\boldsymbol{\tau}_1$ the first a-level crossing to the right of the origin (Fig. 16-4a). The first-passage problem is the determination of the distribution function $F_\tau(\tau, a)$ of the RV $\boldsymbol{\tau}_1$. We shall solve the problem under the assumption that $\mathbf{x}(t)$ is the Wiener process (11-20). We should note, however, that in the solution we make use only of the fact that the increment

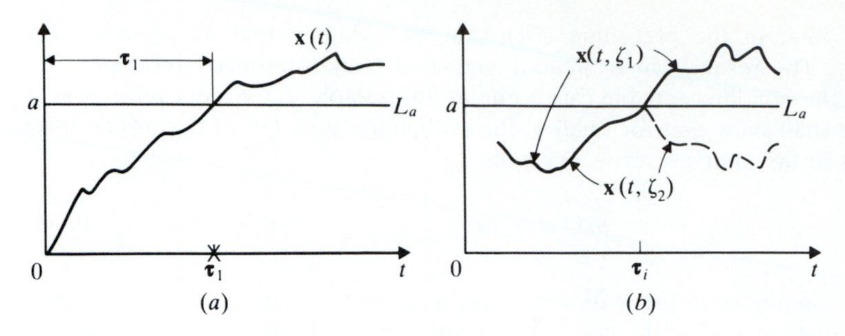

FIGURE 16-4

$\mathbf{x}(t_2) - \mathbf{x}(t_1)$ is *independent* of $\mathbf{x}(t_1)$ and its density is *even*:

$$P\{\mathbf{x}(t_2) - \mathbf{x}(t_1) \le w\} = P\{\mathbf{x}(t_2) - \mathbf{x}(t_1) \ge -w\} \tag{16-16}$$

The reflection principle. We shall show that the samples $\mathbf{x}(t, \zeta)$ of $\mathbf{x}(t)$ that cross the line L_a continue on symmetrical paths. In other words, if a sample $\mathbf{x}(t, \zeta_1)$ crosses the line L_a at $t = \tau_i$, then there exists another sample $\mathbf{x}(t, \zeta_2)$ that coincides with $\mathbf{x}(t, \zeta_1)$ for $t \le \boldsymbol{\tau}_i$ and for $t > \boldsymbol{\tau}_i$ is the reflection of $\mathbf{x}(t, \zeta_1)$ on the line L_a (Fig. 16-4*b*)

$$\mathbf{x}(t, \zeta_1) = \mathbf{x}(t, \zeta_2) \qquad t \le \boldsymbol{\tau}_i$$

$$a - \mathbf{x}(t, \zeta_1) = \mathbf{x}(t, \zeta_2) - a \qquad t > \boldsymbol{\tau}_i$$

This result, known as the *reflection principle*, can be stated as follows:

THEOREM. For any x (less than or greater than a)

$$P\{\mathbf{x}(t) \le x | \boldsymbol{\tau}_1 \le t\} = P\{\mathbf{x}(t) \ge 2a - x | \boldsymbol{\tau}_1 \le t\} \tag{16-17}$$

Proof. It suffices to show that (see Prob. 4-13)

$$P\{\mathbf{x}(t) \le x | \boldsymbol{\tau}_1 = \tau\} = P\{\mathbf{x}(t) \ge 2a - x | \boldsymbol{\tau}_1 = \tau\} \tag{16-18}$$

for every $\tau \le t$. From (16-16) it follows that

$$P\{\mathbf{x}(t) - \mathbf{x}(\tau) \le x - a\} = P\{\mathbf{x}(t) - \mathbf{x}(\tau) \ge a - x\} \tag{16-19}$$

Since $\mathbf{x}(t) - \mathbf{x}(\tau)$ is independent of $\mathbf{x}(\tau)$, we can write (16-19) in the form

$$P\{\mathbf{x}(t) - \mathbf{x}(\tau) \le x - a | \mathbf{x}(\tau) = a\} = P\{\mathbf{x}(t) - \mathbf{x}(\tau) \ge a - x | \mathbf{x}(\tau) = a\}$$

This is true for any t and τ; hence it is true for $\tau = \boldsymbol{\tau}_1$. In this case $\{\mathbf{x}(\tau) = a\} = \{\boldsymbol{\tau}_1 = \tau\}$ and (16-18) results because $\mathbf{x}(\boldsymbol{\tau}_1) = a$.

COROLLARY. If $x \le a$, then

$$P\{\mathbf{x}(t) \le x, \boldsymbol{\tau}_1 \le t\} = 1 - F_x(2a - x, t) \tag{16-20}$$

$$P\{\mathbf{x}(t) \le x, \boldsymbol{\tau}_1 > t\} = F_x(x, t) + F_x(2a - x, t) - 1 \tag{16-21}$$

where

$$F_x(x,t) = \mathbb{G}\left(\frac{x}{\sqrt{2\alpha t}}\right)$$

is the first-order distribution of the Wiener process $\mathbf{x}(t)$.

Proof. Multiplying both sides of (16-17) by $P\{\boldsymbol{\tau}_1 \le t\}$, we obtain

$$P\{\mathbf{x}(t) \le x, \boldsymbol{\tau}_1 \le t\} = P\{\mathbf{x}(t) \ge 2a - x, \boldsymbol{\tau}_1 \le t\} \tag{16-22}$$

for every x and t. If $x \le a$, then $\mathbf{x}(t) > 2a - x$ iff $\boldsymbol{\tau}_1 \le t$; hence the right side of (16-22) equals $P\{\mathbf{x}(t) \ge 2a - x\}$ and (16-20) results. The second equation follows because the sum of the left sides of (16-20) and (16-21) equals $P\{\mathbf{x}(t) \le x\}$.

First-passage distribution. The distribution $F_\tau(t,a)$ of the RV $\boldsymbol{\tau}_1$ equals

$$P\{\boldsymbol{\tau}_1 \le t\} = P\{\boldsymbol{\tau}_1 \le t, \mathbf{x}(t) \le a\} + P\{\boldsymbol{\tau}_1 \le t, \mathbf{x}(t) > a\} \tag{16-23}$$

Clearly, if $\mathbf{x}(t) > a$, then there must be a crossing prior to time t; hence $\boldsymbol{\tau}_1 \le t$. From this it follows that

$$P\{\boldsymbol{\tau}_1 \le t, \mathbf{x}(t) > a\} = P\{\mathbf{x}(t) > a\}$$

Setting $x = a$ in (16-20), we obtain

$$P\{\boldsymbol{\tau}_1 \le t, \mathbf{x}(t) \le a\} = 1 - F_x(a,t) = P\{\mathbf{x}(t) > a\}$$

Thus the two terms on the right of (16-23) equal $P\{\mathbf{x}(t) > a\}$. Therefore,

$$P\{\boldsymbol{\tau}_1 \le t\} = F_\tau(t,a) = 2P\{\mathbf{x}(t) > a\} = 2 - 2F_x(a,t) \tag{16-24}$$

Absorbing wall We replace the line L_a by an absorbing barrier. This means that the resulting process $\mathbf{y}(t)$ equals a for every $t > \boldsymbol{\tau}_1$ (Fig. 16-5). We

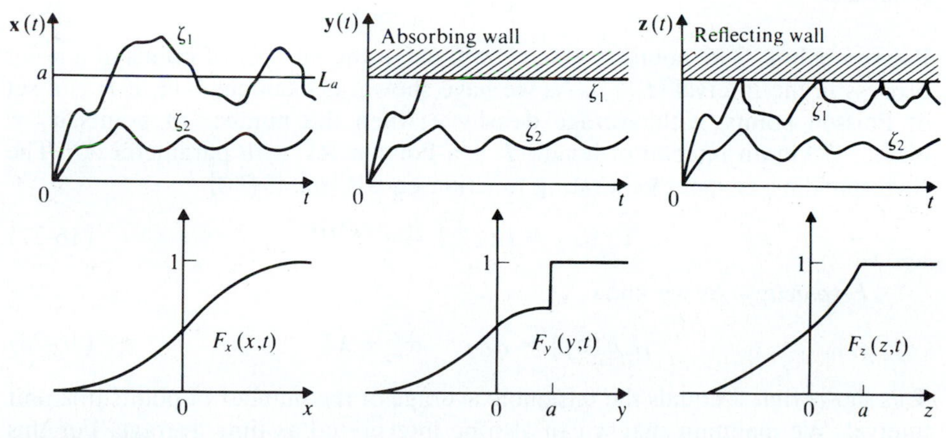

FIGURE 16-5

shall show that the distribution function of $\mathbf{y}(t)$ equals

$$F_y(y,t) = F_x(y,t) + F_x(2a - y, t) - 1 \qquad y < a \tag{16-25}$$

and, of course, $F_y(y,t) = 1$ for $y \geq a$.

Proof. If $y < a$, then $\mathbf{y}(t, \zeta_i) \leq y$ for some outcome ζ_i iff $\mathbf{x}(t, \zeta_i) \leq y$ *and* $\mathbf{x}(t, \zeta_i)$ does not reach the line L_a prior to time t. Hence

$$\{\mathbf{y}(t) \leq y\} = \{\mathbf{x}(t) \leq y, \boldsymbol{\tau}_1 > t\}$$

and (16-25) follows from (16-21).

Reflecting wall We replace now the line L_a by a reflecting barrier. This means that the resulting process $\mathbf{z}(t)$ equals $\mathbf{x}(t)$ if $\mathbf{x}(t) < a$ and it equals $2a - \mathbf{x}(t)$ if $\mathbf{x}(t) > a$ (Fig. 16-5). We shall show that in this case

$$F_z(z,t) = F_x(z,t) + 1 - F_x(2a - z, t) \qquad z < a \tag{16-26}$$

and $F_z(z,t) = 1$ for $z > a$.

Proof. If $z < a$, then $\mathbf{z}(t, \zeta_i) \leq z$ iff either $\mathbf{x}(t, \zeta_i) \leq a$ or $\mathbf{x}(t, \zeta_i) > 2a - z$. Hence

$$\{\mathbf{z}(t) \leq z\} = \{\mathbf{x}(t) \leq z\} + \{\mathbf{x}(t) > 2a - z\}$$

and (16-26) results.

16-2 QUEUEING THEORY

Queueing theory deals with point processes (arrivals and departures) and random intervals (waiting and servicing). This involves the statistical properties of the number of random points in intervals of random length. As a preparation, we introduce the underlying concepts in the context of Poisson points and renewals.

Poisson points. The notation $\mathbf{n}(t_1, t_2)$ will mean the number of points of a point process in the interval (t_1, t_2). As we have shown in Example 4-11, if $\mathbf{t}_i$ is a set of Poisson points, with average density λ, then the number of points $\mathbf{n}_T = \mathbf{n}(t, t+T)$ in an interval of length T is a Poisson RV with parameter λT. The corresponding moment-generating function equals [see (5-79)]

$$\Gamma_{n_T}(z) = E\{z^{\mathbf{n}_T}\} = e^{(z-1)\lambda T} \tag{16-27}$$

Ergodicity As we know,

$$E\{\mathbf{n}_T\} = \lambda T \qquad \sigma_{n_T}^2 = \lambda T \tag{16-28}$$

This shows that λ equals the ensemble average of the number of points in a unit interval. We maintain that λ can also be interpreted as time average. For this purpose, we form the time average $\boldsymbol{\eta}_T = \mathbf{n}_T / T$. As we see from (16-28), $\boldsymbol{\eta}_T$ is an RV such that

$$E\{\boldsymbol{\eta}_T\} = \lambda \qquad \sigma^2_{\eta_T} = \frac{\lambda}{T}$$

Hence $\sigma^2_{\eta_T} \to 0$ as $T \to \infty$. From this it follows that

$$\boldsymbol{\eta}_T = \frac{\mathbf{n}_T}{T} \xrightarrow[T\to\infty]{} \lambda \tag{16-29}$$

and $\mathbf{n}_T \simeq \lambda T$ for sufficiently large T.

Poisson Points in Random Intervals

Suppose now that $\mathbf{c}$ is a positive RV and

$$\mathbf{n}_c = \mathbf{n}(t, t + \mathbf{c})$$

is the number of points in an interval of length $\mathbf{c}$. We shall determine its statistics.

If $\mathbf{c} = c$ is a constant, then $\mathbf{n}_c$ is a Poisson RV with parameter λc. Hence

$$E\{\mathbf{n}_c | \mathbf{c} = c\} = \lambda c$$

From this and (7-66) it follows that

$$E\{\mathbf{n}_c\} = E\{E\{\mathbf{n}_c|\mathbf{c}\}\} = E\{\lambda \mathbf{c}\} = \lambda \eta_c \tag{16-30}$$

Denoting by ρ_c the average number of points in the random interval $\mathbf{c}$, we conclude that $\rho_c = \lambda \eta_c$.

Reasoning similarly, we can find all moments of $\mathbf{n}_c$. In the following, we determine directly the moment function

$$\Gamma_{n_c}(z) = E\{z^{\mathbf{n}_c}\}$$

of the discrete-type RV $\mathbf{n}_c$ in terms of the moment function

$$\boldsymbol{\Phi}_c(s) = E\{e^{s\mathbf{c}}\} \tag{16-31}$$

of the continuous-type RV $\mathbf{c}$.

THEOREM.

$$\Gamma_{n_c}(z) = \boldsymbol{\Phi}_c(\lambda z - \lambda) \tag{16-32}$$

Proof. If $\mathbf{c} = c$, then $\mathbf{n}_c$ is a Poisson RV with parameter λc. Hence [see (16-27)]

$$E\{z^{\mathbf{n}_c} | \mathbf{c} = c\} = e^{(z-1)\lambda c}$$

This yields

$$E\{z^{\mathbf{n}_c}\} = E\{E\{z^{\mathbf{n}_c}|\mathbf{c}\}\} = E\{e^{(z-1)\lambda \mathbf{c}}\}$$

and (16-32) follows from (16-31) with $s = (z - 1)\lambda$.

Using (16-32) and the moment theorems (5-67) and (5-77), we can express the moments of $\mathbf{n}_c$ in terms of the moments of $\mathbf{c}$. We note, in particular, that

$$E\{\mathbf{n}_c\} = \Gamma'_{n_c}(1) = \lambda\Phi'_c(0) = \lambda E\{\mathbf{c}\}$$

in agreement with (16-30). Similarly,

$$E\{\mathbf{n}_c(\mathbf{n}_c - 1)\} = \Gamma''_{n_c}(1) = \lambda^2\Phi''_c(0) = \lambda^2 E\{\mathbf{c}^2\} \tag{16-33}$$

Hence

$$E\{\mathbf{n}_c^2\} = \lambda^2 E\{\mathbf{c}^2\} + \rho_c \qquad \sigma_{n_c}^2 = \lambda^2\sigma_c^2 + \rho_c$$

In the above, $\rho_c = \lambda\eta_c$ is the average number of points in the random interval $\mathbf{c}$, and σ_c^2 is the variance of $\mathbf{c}$.

Example 16-6. If†

$$f_c(c) = \mu e^{-\mu c}$$

then $E\{\mathbf{c}\} = 1/\mu$, $E\{\mathbf{c}^2\} = 2/\mu^2$, and

$$\Phi_c(s) = \frac{\mu}{\mu - s}$$

Hence

$$E\{\mathbf{n}_c\} = \frac{\lambda}{\mu} = \rho_c \qquad E\{\mathbf{n}_c^2\} = \frac{2\lambda^2}{\mu^2} + \frac{\lambda}{\mu}$$

and

$$\Gamma_{\mathbf{n}_c}(z) = \frac{\mu}{\mu + \lambda - \lambda z}$$

From this it follows that

$$P\{\mathbf{n}_c = n\} = \frac{\mu}{\mu + \lambda}\left(\frac{\lambda}{\mu + \lambda}\right)^n \qquad n = 0, 1, \ldots$$

We have thus shown that the number $\mathbf{n}_c$ of Poisson points in an exponentially distributed random interval $\mathbf{c}$ has a *geometric distribution* with ratio $\lambda/(\mu + \lambda)$.

Example 16-7. If $\mathbf{c} = c$ is a constant, then

$$\Phi_c(s) = e^{cs} \qquad \Gamma_{n_c}(z) = e^{\lambda c(z-1)}$$

Thus $\mathbf{n}_c$ is a Poisson RV with parameter λc, as it should be.

RENEWAL PROCESSES. Consider a stationary point process $\mathbf{t}_i$ such that the RVs

$$\mathbf{c}_i = \mathbf{t}_i - \mathbf{t}_{i-1} \tag{16-34}$$

†Since we deal only with positive RVs, we shall assume tacitly that all densities are 0 on the negative axis.

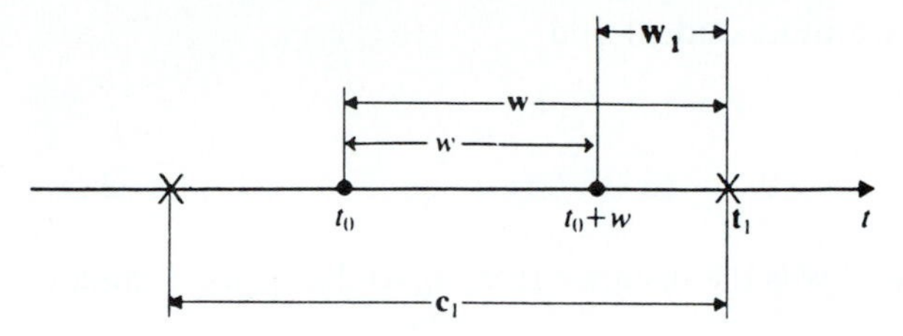

FIGURE 16-6

are i.i.d. with distribution $F_c(c)$. With t_0 a fixed point, we form the RV

$$\mathbf{w} = \mathbf{t}_1 - t_0$$

where $\mathbf{t}_1$ is the first random point to the right of t_0 (Fig. 16-6). We shall express the density $f_w(w)$ of this RV in terms of $F_c(c)$.

THEOREM. We maintain that

$$f_w(w) = \frac{1}{\eta_c}[1 - F_c(w)] \tag{16-35}$$

where

$$\eta_c = E\{\mathbf{c}_i\} = \int_0^\infty [1 - F_c(c)]\, dc \tag{16-36}$$

is the mean of $\mathbf{c}_i$ [see (5-27)].

Proof. Given a number w, we define the Rv $\mathbf{w}_1$ as the distance from the point $t_w = t_0 + w$ to the next random point $\mathbf{t}_i$ to its right (Fig. 16-6). From the stationarity of the points $\mathbf{t}_i$, it follows that the RVs $\mathbf{w}$ and $\mathbf{w}_1$ have the same density $f_w(w)$. Suppose that the first random point $\mathbf{t}_1$ to the right of t_0 is also to the right of t_w. In this case, there is no random point between t_0 and t_w and

$$\mathbf{w} = \mathbf{t}_1 - t_0 > w \qquad \mathbf{w}_1 = \mathbf{t}_1 - t_w \qquad \mathbf{c}_1 > \mathbf{w}_1 + w$$

From the above it follows that the events

$$\{\mathbf{w} > w\} \qquad \text{and} \qquad \{\mathbf{c}_1 > \mathbf{w}_1 + w\}$$

are equal provided that in the second set we consider only the outcomes such that $\mathbf{c}_1 > \mathbf{w}_1$. Hence

$$P\{\mathbf{w} > w\} = P\{\mathbf{c}_1 > \mathbf{w}_1 + w | \mathbf{c}_1 > \mathbf{w}_1\}$$

Reasoning similarly, we conclude that

$$P\{w < \mathbf{w} < w + dw\} = P\{0 < \mathbf{w}_1 < dw, \mathbf{c}_1 > w\}$$

The above yields

$$f_w(w)\, dw = f_w(0)\, dw[1 - F_c(w)]$$

because the RVs $\mathbf{w}_1$ and $\mathbf{c}_1$ are independent and

$$P\{\mathbf{c}_1 > w\} = 1 - F_c(w)$$

This completes the proof because the area of $f_w(w)$ equals one; it also shows that $f_w(0) = 1/\eta_c$.

The theorem is also true if $\mathbf{w}$ is the distance from t_0 to the nearest random point on its left.

COROLLARY. The moment function $\mathbf{\Phi}_w(s)$ of $\mathbf{w}$ equals

$$\mathbf{\Phi}_w(s) = \frac{1}{\eta_c s}[\mathbf{\Phi}_c(s) - 1] \tag{16-37}$$

Proof. Since $F_c'(w) = f_c(w)$ and $F_c(w) = 0$ for $w \le 0$, we conclude, integrating by parts, that for Re $s < 0$:

$$-\int_0^\infty F_c(w)e^{sw}\,dw = \frac{1}{s}\mathbf{\Phi}_c(s) \qquad -\int_0^\infty e^{sw}\,dw = \frac{1}{s}$$

and (16-37) follows from (16-35).

Differentiating (16-37), we obtain

$$\mathbf{\Phi}_w'(0) = \frac{\mathbf{\Phi}_c''(0)}{2\eta_c}$$

because $\mathbf{\Phi}_c(0) = 1$. Hence [see (5-67)]

$$\eta_w = \frac{E\{\mathbf{c}^2\}}{2\eta_c} \tag{16-38}$$

Example 16-8. If $f_c(c) = \lambda e^{-\lambda c}$, then

$$F_c(c) = 1 - e^{-\lambda c} \qquad \eta_c = \frac{1}{\lambda}$$

and (16-35) yields $f_w(w) = \lambda e^{-\lambda w} = f_c(w)$. This is the case iff $\mathbf{t}_i$ is a set of Poisson points.

Note With $\mathbf{c}_i$ as in (16-34), we form the point process

$$\boldsymbol{\tau}_i = \mathbf{c}_1 + \mathbf{c}_2 + \cdots + \mathbf{c}_i$$

starting from $t = 0$. In general, this process does not have the same statistics as the original process $\mathbf{t}_i$ (it is not even stationary). It is, however, asymptotically equivalent to $\mathbf{t}_i$. The process $\mathbf{t}_i$ is given by

$$\mathbf{t}_i = \mathbf{w} + \mathbf{c}_2 + \cdots + \mathbf{c}_i$$

It can, therefore, be constructed in terms of the sequence $\mathbf{c}_i$ and the RV $\mathbf{w}$ specified by (16-35).

Arrivals and Departures

The term *queueing* is used to describe a large class of phenomena involving arrivals, waiting, servicing, and departures. In Fig. 16-7 we show a typical model (*queueing system*) whose inputs are certain objects (*units*) identified in terms of their *arrival times* $\mathbf{t}_i$. Each unit stays in the system $\mathbf{a}_i$ seconds (*total system time*) and it departs at the *departure time* $\boldsymbol{\tau}_i$:

$$\boldsymbol{\tau}_i = \mathbf{t}_i + \mathbf{a}_i$$

The number of units in the system (*state of the system*) at time t will be denoted by $\mathbf{N}(t)$. Thus $\mathbf{N}(t)$ is a discrete-state process increasing by 1 at $\mathbf{t}_i$ and decreasing by 1 at $\boldsymbol{\tau}_i$.

Our objective in this section is not to develop this involved topic in detail. We plan merely to introduce the main ideas in the context of earlier concepts. We start with a general theorem that does not rely on any special conditions about the interarrival times, the nature of the system, or the properties of $\mathbf{a}_n$. It assumes merely that all processes are SSS with finite second moments.

LITTLE'S THEOREM. Suppose that the processes $\mathbf{t}_i$ and $\mathbf{a}_i$ are mean-ergodic:

$$\frac{\mathbf{n}_T}{T} \xrightarrow[T\to\infty]{} \lambda \qquad \frac{1}{n}\sum_{k=1}^{n} \mathbf{a}_k \xrightarrow[n\to\infty]{} E\{\mathbf{a}_n\} \tag{16-39}$$

In the above, $\mathbf{n}_T$ is the number of points $\mathbf{t}_i$ in the interval $(0, T)$ and $\lambda = E\{\mathbf{n}_T\}/T$ is the mean density of these points.

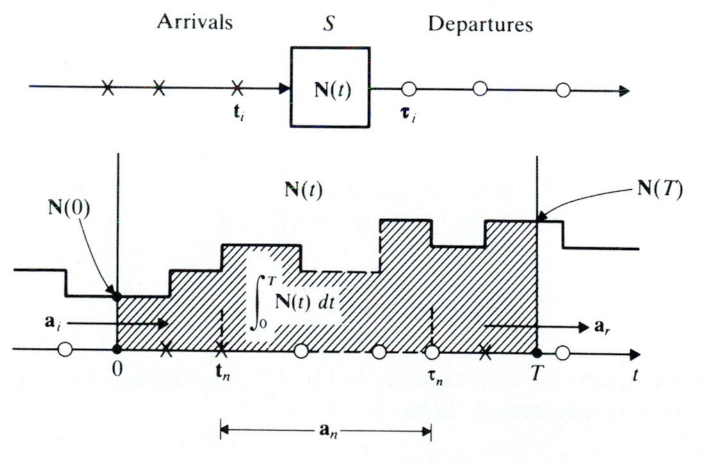

FIGURE 16-7

We maintain that†

$$E\{\mathbf{N}(t)\} = \lambda E\{\mathbf{a}_i\} \tag{16-40}$$

In fact, we shall establish the stronger statement that $\mathbf{N}(t)$ is also mean-ergodic:

$$\lim_{T\to\infty} \frac{1}{T}\int_0^T \mathbf{N}(t)\,dt = \lambda E\{\mathbf{a}_i\} = E\{\mathbf{N}(t)\} \tag{16-41}$$

Equation (16-40) seems reasonable: The mean $E\{\mathbf{N}(t)\}$ of the number of units in the system equals the mean number λ of arrivals per second multiplied by the mean time $E\{\mathbf{a}_i\}$ that each unit remains in the system. It is not, however, always true, although it holds under general conditions.

Proof. We start with the observation that

$$-\sum_{r=1}^{\mathbf{N}(T)} \mathbf{a}_r \le \int_0^T \mathbf{N}(t)\,dt - \sum_{n=1}^{\mathbf{n}_T} \mathbf{a}_n \le \sum_{i=1}^{\mathbf{N}(0)} \mathbf{a}_i \tag{16-42}$$

In the above, the terms $\mathbf{a}_n$ of the second sum are due to the $\mathbf{n}_T$ units that arrived in the interval $(0, T)$; the terms $\mathbf{a}_i$ of the last sum are due to the $\mathbf{N}(0)$ units that are in the system at $t = 0$; the terms $\mathbf{a}_r$ of the first sum are due to the $\mathbf{N}(T)$ units that are still in the system at $t = T$. The details of the reasoning that establishes (16-42) are omitted.

As we know (see Prob. 8-9)

$$E\left\{\left(\sum_{k=1}^{\mathbf{N}(t)} \mathbf{a}_k\right)^2\right\} \le E\{\mathbf{N}^2(t)\}E\{\mathbf{a}_k^2\} < \infty \tag{16-43}$$

Dividing (16-42) by T, we conclude that, if T is sufficiently large, then

$$\frac{1}{T}\int_0^T \mathbf{N}(t)\,dt \simeq \frac{1}{T}\sum_{n=1}^{\mathbf{n}_T} \mathbf{a}_n \tag{16-44}$$

because the left and right sides of (16-42) tend to 0 after the division by T [see (16-43)]. Furthermore, assumption (16-39) yields $\mathbf{n}_T \simeq \lambda T$ and

$$\frac{1}{T}\sum_{n=1}^{\mathbf{n}_T} \mathbf{a}_n \simeq \frac{\lambda}{\mathbf{n}_T}\sum_{n=1}^{\mathbf{n}_T} \mathbf{a}_n \simeq \lambda E\{\mathbf{a}_n\}$$

Inserting into (16-44), we obtain the first equality in (16-41). The second follows because the mean of the left side equals $E\{\mathbf{N}(t)\}$.

†F. J. Beutler: "Mean Sojourn Times...," *IEEE Transactions Information Theory*, March, 1983.

Immediate Service ($M|G|\infty$)

In general, the system time $\mathbf{a}_n$ is a sum

$$\mathbf{a}_n = \mathbf{b}_n + \mathbf{c}_n$$

where $\mathbf{b}_n$ is the *waiting time* (or *queueing time*) and $\mathbf{c}_n$ is the *service time* of the nth unit. In many applications, $\mathbf{b}_n = 0$ and $\mathbf{a}_n = \mathbf{c}_n$. This is the case if the number of servers is infinite or when no servers are involved (visits to a park, for example). We consider this problem next.

We shall assume that the arrival times $\mathbf{t}_n$ are Poisson points with mean density λ and the service times $\mathbf{c}_n$ are i.i.d. with distribution an arbitrary function $F_c(c)$. In queueing theory this is written in the following form (Kendall)

$$M|G|\infty$$

The first position in this notation refers to arrivals, the second to service times, and the third to the number of servers in the system. The letter M (for *Markoff* or *memoryless*) in the first position means Poisson arrivals; in the second position it means exponentially distributed service times. The letter G (general) indicates that the arrivals or the service times are arbitrary. The letter D (deterministic) indicates that they are constant.

THEOREM. The state $\mathbf{N}(t)$ of an $M|G|\infty$ system is Poisson distributed

$$P\{\mathbf{N}(t) = k\} = e^{-\rho}\frac{\rho^k}{k!} \tag{16-45}$$

with parameter

$$\rho = \lambda E\{\mathbf{c}_n\} = \lambda \eta_c \tag{16-46}$$

This parameter is called *traffic intensity* or *offered load*.

Proof. Using the point t as the origin (Fig. 16-8), we divide the time axis into consecutive intervals I_i: (α_i, α_{i+1}) of length $\Delta\alpha = \alpha_{i+1} - \alpha_i$. Denoting by $\Delta\mathbf{n}_i$ the number of arrivals in the interval I_i and by $\Delta\mathbf{N}(t, \alpha_i)$ the contribution to the state $\mathbf{N}(t)$ of the system at time t due to these arrivals, we have

$$\mathbf{N}(t) = \sum_i \Delta\mathbf{N}(t, \alpha_i) \tag{16-47}$$

If $\Delta\alpha$ is small, then, within probabilities of order $\Delta\alpha$, the RV $\Delta\mathbf{n}_i$ takes the

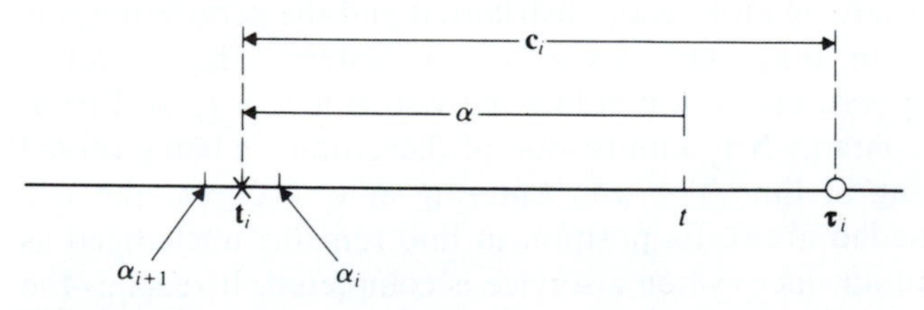

FIGURE 16-8

values 0 or 1 and

$$P\{\Delta\mathbf{n}_i = 1\} = \lambda\,\Delta\alpha \tag{16-48}$$

If $\Delta\mathbf{n}_i = 0$, then $\Delta\mathbf{N}(t, \alpha_i) = 0$; if $\Delta\mathbf{n}_i = 1$, then

$$\Delta\mathbf{N}(t, \alpha_i) = \begin{cases} 1 & \text{if } \mathbf{c}_i > \alpha_i \\ 0 & \text{if } \mathbf{c}_i < \alpha_i \end{cases}$$

where $\mathbf{c}_i$ is the service time of the single unit that enters the system in the interval I_i. Hence

$$P\{\Delta\mathbf{N}(t, \alpha_i) = 1 | \Delta\mathbf{n}_i = 1\} = P\{\mathbf{c}_i > \alpha_i\} = 1 - F_c(\alpha_i) \tag{16-49}$$

Multiplying (16-48) and (16-49), we obtain

$$P\{\Delta\mathbf{N}(t, \alpha_i) = 1\} = [1 - F_c(\alpha_i)]\lambda\,\Delta\alpha$$

The RV $\Delta\mathbf{N}(t, \alpha_i)$ takes the values 0 or 1 and they are independent because the RVs $\Delta\mathbf{n}_i$ are independent (Poisson points in nonoverlapping intervals.) Hence (see Example 8-8*b*) the sum in (16-47) tends to a Poisson RV with parameter

$$\lambda\int_0^\infty [1 - F_c(\alpha)]\,d\alpha = \lim_{\Delta\alpha \to 0} \sum_{i=1}^{\infty} \lambda\,\Delta\alpha[1 - F_c(\alpha_i)]$$

and (16-45) results because the sum in (16-47) equals $\mathbf{N}(t)$ and the above integral equals $E\{\mathbf{c}_n\}$ [see (16-36)].

COROLLARY. The traffic intensity ρ equals the average number of units in the system

$$E\{\mathbf{N}(t)\} = \rho = \lambda\eta_c \tag{16-50}$$

Proof. It follows from (16-45) and (5-36).

Example 16-9. (*a*) $(M|M|\infty)$ If $\mathbf{c}$ is an exponential with parameter μ, then $\eta_c = 1/\mu$; hence $\rho = \lambda/\mu$.

(*b*) $(M|D|\infty)$ If $\mathbf{c} = c$ is a constant, then $\eta_c = c$; hence $\rho = \lambda c$.

Single-Server Queue $(M|G|1)$

We conclude the section with a detailed treatment of the $M|G|1$ system. In this system, the arrival times $\mathbf{t}_i$ are again Poisson distributed and the service times $\mathbf{c}_i$ are i.i.d. However, there is only one server in the system. The model is described in Fig. 16-9: Suppose that a unit enters the system at $t = \mathbf{t}_i$. Just prior to its arrival, the system contains $\mathbf{N}(\mathbf{t}_i^-)$ units; one of these units is being served and the others are waiting in line. The unit entering at $\mathbf{t}_i$ occupies the last position in the queue (shaded area). Its position in line remains unchanged as other units arrive; the unit advances when a service is completed. It reaches the server (position 1 in line when service starts) at $t = \boldsymbol{\tau}_{i-1}$ and it leaves the system

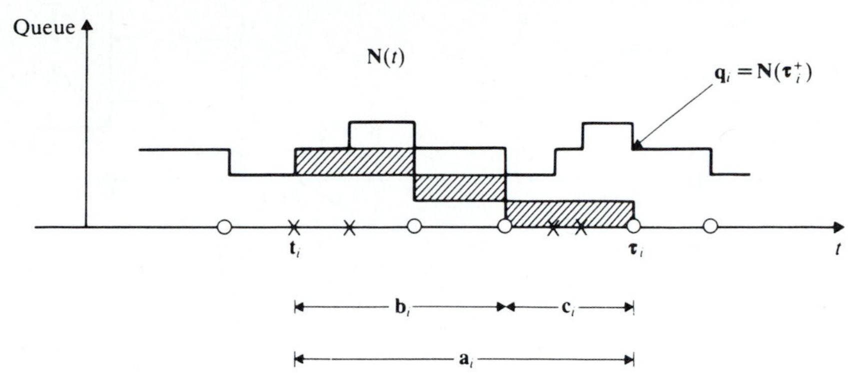

FIGURE 16-9

at time $t = \boldsymbol{\tau}_i$:

$$\boldsymbol{\tau}_i = \boldsymbol{\tau}_{i-1} + \mathbf{c}_i$$

where $\mathbf{c}_i$ is the *service time*. Denoting by $\mathbf{b}_i$ the *waiting time* and by $\mathbf{a}_i$ the *system time*, we conclude that

$$\mathbf{b}_i = \boldsymbol{\tau}_{i-1} - \mathbf{t}_i \qquad \mathbf{a}_i = \boldsymbol{\tau}_i - \mathbf{t}_i = \mathbf{b}_i + \mathbf{c}_i$$

This completes the description of the $M|G|1$ model.

Our objective is to express the statistics of the various parameters of the system in terms of the distribution $F_c(c)$ of the service times $\mathbf{c}_i$ and the average density λ of the arrival times $\mathbf{t}_i$ under the assumption that $\mathbf{N}(t)$ is *stationary*. As we shall see, the system reaches stationarity only if the mean η_c of $\mathbf{c}_i$ (mean service time) is less than the *average interarrival time* $1/\lambda$.

Note If the state $\mathbf{N}(t)$ of a system is a stationary $M|G|1$ process, then $\mathbf{N}(-t)$ is a $G|M|1$ process obtained from $\mathbf{N}(t)$ by interchanging arrivals and departures. This shows that the properties of a $G|M|1$ process follow from the properties of the corresponding $M|G|1$ process.

THE IMBEDDED MARKOFF CHAIN. The ith unit arrives at $t = \mathbf{t}_i$ and departs at $t = \boldsymbol{\tau}_i$ remaining in the system $\mathbf{a}_i = \boldsymbol{\tau}_i - \mathbf{t}_i$ seconds. During the interval $(\mathbf{t}_i, \boldsymbol{\tau}_i)$, $\mathbf{n}_{a_i}$ new units arrive and all other units depart. Hence $\mathbf{n}_{a_i}$ equals the state $\mathbf{N}(\boldsymbol{\tau}_i^+)$ of the system at $t = \boldsymbol{\tau}_i^+$. This number is denoted by $\mathbf{q}_i$ and is called the Markoff chain imbedded in the process $\mathbf{N}(t)$ (see Sec. 16-4). Thus

$$\mathbf{q}_i = \mathbf{n}_{a_i} = N(\boldsymbol{\tau}_i^+) \tag{16-51}$$

We wish to relate $\mathbf{q}_i$ to $\mathbf{q}_{i-1}$. Suppose, first, that $\mathbf{q}_{i-1} \geq 1$ (Fig. 16-10a). In this case, at least one unit is waiting at $t = \boldsymbol{\tau}_{i-1}$; hence $\mathbf{c}_i = \boldsymbol{\tau}_i - \boldsymbol{\tau}_{i-1}$ is the ith service interval. During this internal, $\mathbf{n}_{c_i}$ units arrive and none depart. And

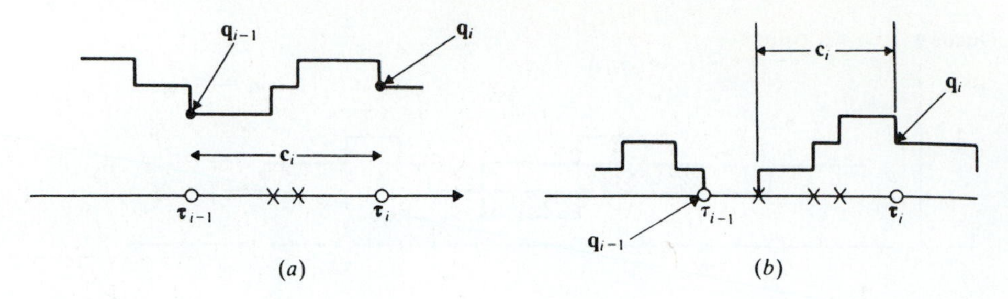

FIGURE 16-10

since at $t = \boldsymbol{\tau}_i$ the ith unit leaves, we conclude that

$$\mathbf{q}_i = \mathbf{q}_{i-1} + \mathbf{n}_{c_i} - 1 \qquad \mathbf{q}_{i-1} \geq 1$$

If $\mathbf{q}_{i-1} = 0$, then the ith service starts not at $t = \boldsymbol{\tau}_{i-1}$ (the system is then empty) but at the time of the arrival of the next unit (Fig. 16-10b). Hence

$$\mathbf{q}_i = \mathbf{n}_{c_i} \qquad \mathbf{q}_{i-1} = 0$$

where $\mathbf{n}_{c_i}$ equals the arrivals in the *interior* of the service interval.

From the above it follows that $\mathbf{q}_i$ satisfies the recursion equation

$$\mathbf{q}_i = \bar{\mathbf{q}}_{i-1} + \mathbf{n}_{c_i} \tag{16-52}$$

where

$$\bar{\mathbf{q}}_i = \begin{cases} \mathbf{q}_i - 1 & \mathbf{q}_i \geq 1 \\ 0 & \mathbf{q}_i = 0 \end{cases} \tag{16-53}$$

Using (16-52), we shall determine the state probabilities $p_k = P\{\mathbf{q}_i = k\}$ at $\boldsymbol{\tau}_i^+$.

Zero state We maintain that

$$p_0 = 1 - \rho \qquad \rho = \lambda \eta_c \tag{16-54}$$

Proof. Clearly, $\mathbf{n}_{c_i}$ is the number of Poisson points in the random interval $\mathbf{c}_i$; hence [see (16-30)]

$$E\{\mathbf{n}_{c_i}\} = \lambda \eta_c = \rho$$

With

$$\eta_q = \sum_{k=0}^{\infty} k p_k = \sum_{k=1}^{\infty} k p_k$$

the mean of $\mathbf{q}_i$, it follows from (16-53) that

$$E\{\bar{\mathbf{q}}_i\} = \sum_{k=1}^{\infty} (k-1) p_k = \eta_q - \sum_{k=1}^{\infty} p_k = \eta_q - (1 - p_0)$$

And since (stationarity assumption) $E\{\bar{\mathbf{q}}_i\} = E\{\bar{\mathbf{q}}_{i-1}\}$, (16-52) yields

$$\eta_q = \eta_q - (1 - p_0) + \lambda \eta_c$$

and (16-54) results.

Proceeding similarly, we can find the moments of $\mathbf{q}_i$ (see Prob. 16-11). It is simpler, however, to determine directly the moment function

$$\Gamma_q(z) = E\{z^{\mathbf{q}_i}\} = p_0 + \sum_{k=1}^{\infty} p_i z^k \tag{16-55}$$

of the sequence $\mathbf{q}_i$. Using (16-52), we shall show that $\Gamma_q(z)$ can be expressed in terms of the moment function

$$\Gamma_{n_c}(z) = \mathbf{\Phi}_c(\lambda z - \lambda)$$

In the above, $\mathbf{n}_{c_i}$ is the number of arrivals in the random interval $\mathbf{c}_i$, $\Gamma_{n_c}(z)$ is the moment function of $\mathbf{n}_{c_i}$, and $\mathbf{\Phi}_c(s)$ is the moment function of $\mathbf{c}_i$ [see (16-32)].

THEOREM.

$$\Gamma_q(z) = \frac{p_0(1-z)}{1 - z/\mathbf{\Phi}_c(\lambda z - \lambda)} \tag{16-56}$$

Proof. The moment function of $\bar{\mathbf{q}}_i$ equals

$$\Gamma_{\bar{q}}(z) = p_0 + \sum_{k=1}^{\infty} p_k z^{k-1} = p_0 + z^{-1}\left[\Gamma_q(z) - p_0\right] \tag{16-57}$$

From the independence of $\mathbf{q}_i$ and $\mathbf{c}_i$ it follows that $\mathbf{q}_i$ and $\mathbf{n}_{c_i}$ are also independent. Hence [see (16-52)]

$$\Gamma_q(z) = \Gamma_{\bar{q}}(z)\Gamma_{n_c}(z) \tag{16-58}$$

Inserting (16-57) into (16-58), we obtain

$$z\Gamma_q(z) = \left[\Gamma_q(z) + p_0 z - p_0\right]\Gamma_{n_c}(z) \tag{16-59}$$

and (16-56) results.

Equation (16-59) can be used to obtain all moments of $\mathbf{q}_i$. We shall use it to determine η_q. As a preparation, however, we reestablish (16-54): Since $\Gamma_q(1) = 1$, we conclude differentiating (16-59) that

$$1 + \Gamma'_q(1) = \Gamma'_q(1) + p_0 + \Gamma'_{n_c}(1)$$

This yields (16-54) because $\Gamma'_{n_c}(1) = E\{\mathbf{n}_c\} = \lambda \eta_c$ [see (16-30)].

Pollaczek–Khinchin formula We maintain that the mean of $\mathbf{q}_i$ equals

$$\eta_q = \rho + \frac{\lambda^2 E\{\mathbf{c}_i^2\}}{2(1-\rho)} \tag{16-60}$$

Proof. Differentiating (16-59) twice and setting $z = 1$, we obtain

$$2\Gamma_q'(1) = 2\left[\Gamma_q'(1) + p_0\right]\Gamma_{n_c}'(1) + \Gamma_{n_c}''(1)$$

But [see (16-33)]$\Gamma_{n_c}''(1) = \lambda^2 E\{\mathbf{c}_i^2\}$; hence

$$2\eta_q = 2(\eta_q + p_0)\lambda\eta_c + \lambda^2 E\{\mathbf{c}_i^2\}$$

and (16-60) results because $\rho = 1 - p_0 = \lambda\eta_c$.

Mean system time and waiting time The process $\mathbf{q}_i$ equals the number of arrivals $\mathbf{n}_{a_i}$ during the system time $\mathbf{a}_i$. Hence [see (16-30)]

$$E\{\mathbf{q}_i\} = E\{\mathbf{n}_{a_i}\} = \lambda E\{\mathbf{a}_i\}$$

and (16-60) yields

$$E\{\mathbf{a}_i\} = \eta_c + \frac{\lambda E\{\mathbf{c}_i^2\}}{2p_0} \tag{16-61}$$

This is the *mean system time*. Since $\mathbf{b}_i = \mathbf{a}_i - \mathbf{c}_i$, the *mean waiting time* equals

$$E\{\mathbf{b}_i\} = \frac{\lambda E\{\mathbf{c}_i^2\}}{2p_0} \tag{16-62}$$

Other moments can be found using moment functions. Denoting by $\mathbf{\Phi}_a(s)$ the moment function of $\mathbf{a}_i$, we conclude from (16-32) that $\Gamma_q(z) = \mathbf{\Phi}_a(\lambda z - \lambda)$. Hence

$$\mathbf{\Phi}_a(s) = \Gamma_q\left(1 + \frac{s}{\lambda}\right) \tag{16-63}$$

where $\Gamma_q(z)$ is given by (16-56).

Note As we know (Little's theorem) $E\{\mathbf{N}(t)\} = \lambda E\{\mathbf{a}_i\}$. Comparing with (16-61) and (16-60), we obtain

$$E\{\mathbf{N}(t)\} = E\{\mathbf{N}(\boldsymbol{\tau}_i^+)\} = \lambda\eta_c + \frac{\lambda^2(\eta_c^2 + \sigma_c^2)}{2(1 - \lambda\eta_c)} \tag{16-64}$$

Thus the mean of $\mathbf{N}(\tau_i^+)$ equals the mean of $\mathbf{N}(t)$ for any t. At the end of this section we shall prove that the processes $\mathbf{N}(t)$ and $\mathbf{N}(\tau_i^+)$ have not only equal means but also identical distributions.

Equation (16-64) shows that, if the mean service time η_c is specified, then the mean $E\{\mathbf{N}(t)\}$ of the number of units in the system is minimum if $\sigma_c = 0$. This is so iff the service time $\mathbf{c}_i$ is constant.

Example 16-10. $(M|M|1)$ If $F_c(c) = 1 - e^{-\mu c}$, then

$$\mathbf{\Phi}_c(s) = \frac{\mu}{\mu - s} \qquad E\{\mathbf{c}_i\} = \frac{1}{\mu} \qquad E\{\mathbf{c}_i^2\} = \frac{2}{\mu^2}$$

Thus $\rho = \lambda/\mu$, $p_0 = 1 - \lambda/\mu$, and (16-56) yields

$$\Gamma_q(z) = \frac{\mu - \lambda}{\mu - \lambda z} = \frac{p_0}{1 - \rho z}$$

This shows that $\mathbf{q}_i$ has a geometric distribution with ratio the traffic intensity ρ:

$$P\{\mathbf{q}_i = k\} = p_0 \rho^k \qquad E\{\mathbf{q}_i\} = \frac{\rho}{1-\rho} = \frac{\lambda}{\mu - \lambda}$$

We note, finally, that [see (16-63)]

$$\Phi_a(s) = \frac{\mu p_0}{\mu p_0 - s} \qquad F_a(a) = 1 - e^{-\mu p_0 a}$$

Thus the system time $\mathbf{a}_i$ is exponential with parameter μp_0 and the mean system time equals

$$E\{\mathbf{a}_i\} = \frac{1}{\mu p_0} = \frac{1}{\mu - \lambda}$$

Example 16-11. $(M|D|1)$ If $\mathbf{c} = c$ is a constant then

$$\Phi_c(s) = e^{cs} \qquad E\{\mathbf{c}\} = c \qquad E\{\mathbf{c}^2\} = c^2$$

$\rho = \lambda c$, $p_0 = 1 - \rho$, and (16-60) yields

$$E\{\mathbf{q}_i\} = \frac{\rho(2-\rho)}{2(1-\rho)}$$

Finally [see (16-56) and (16-63)]

$$\Gamma_q(z) = \frac{p_0(1-z)}{1 - ze^{-\lambda c(z-1)}} \qquad \Phi_a(s) = \frac{p_0 s/\lambda}{(1 + s/\lambda)e^{-cs} - 1}$$

BUSY PERIOD. The state of a queuing system is characterized by a succession of *idle periods*, when $\mathbf{N}(t) = 0$, and *busy periods*, when there is at least one unit in the system. We denote by $\mathbf{x}$ the length of an idle period and by $\mathbf{y}$ the length of a busy period. Clearly, $\mathbf{x}$ is exponentially distributed as in (11-31) because there are no arrivals during an idle period. Our objective is to determine the properties of $\mathbf{y}$.

The busy period starts with the arrival of a unit at $t = \mathbf{t}_0$ into an empty system. The unit is served instantly and it departs at $t = \boldsymbol{\tau}_0$. The difference $\mathbf{c} = \boldsymbol{\tau}_0 - \mathbf{t}_0$ is its service time. Denoting by $\mathbf{n}_c$ the number of arrivals in the first service interval $(\mathbf{t}_0, \boldsymbol{\tau}_0)$, we conclude that

$$\mathbf{N}(\mathbf{t}_0^+) = 1 \qquad \mathbf{N}(\boldsymbol{\tau}_0^+) = \mathbf{n}_c$$

The busy period $\mathbf{y}$ equals, thus, the interval of time from $t = \mathbf{t}_0$ when $\mathbf{N}(t_0^+) = 1$ until the moment $t = t_0 + \mathbf{y}$ when $\mathbf{N}(t) = \mathbf{N}(t_0^+) - 1 = 0$ *for the first time*. But the variations of $\mathbf{N}(t)$ during that interval do not depend on its initial value $\mathbf{N}(t_0^+)$. Hence, a busy period $\mathbf{y}_i$ can be characterized statistically as follows:

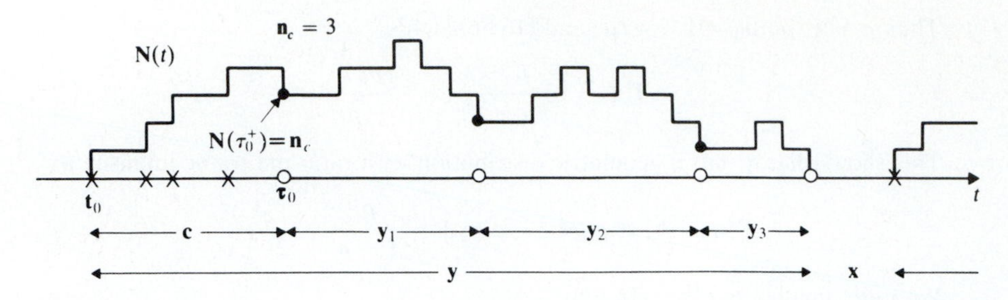

FIGURE 16-11

It is an RV equal to the time interval from $t = \tau_i$, when a service period begins, to $t = \tau_i + y_i$ when $N(t) = N(\tau_i^+) - 1$ *for the first time.*

From the above it follows that the RVs y_1, y_2, y_3 in Fig. 16-11 are independent and each is a busy period. Furthermore, the total number of these RVs equals n_c because, at the start of y_1, the state of the system equals $N(\tau_0^+) = n_c$ and at the end of the total busy period y, $N(t)$ equals 0. Hence

$$y = c + y_1 + \cdots + y_{n_c} \tag{16-65}$$

Mean busy period. Using (16-65), we shall show that

$$\eta_y = \frac{\eta_c}{p_0} \qquad p_0 = 1 - \lambda\eta_c \tag{16-66}$$

Proof. Since $E\{n_c\} = \lambda\eta_c$, it follows from (8-47) that

$$E\left\{\sum_{i=1}^{n_c} y_i\right\} = E\{n_c\}E\{y_i\} = \lambda\eta_c\eta_y$$

Inserting into (16-65), we obtain $\eta_y = \eta_c + \lambda\eta_c\eta_y$ and (16-66) results.

Mean number of units served. The number of units served in a busy period equals the number of arrivals n_y in the random interval y. From (16-30) and (16-66) it follows that the mean of this number equals

$$E\{n_y\} = \lambda\eta_y = \frac{\rho}{1 - \rho} \tag{16-67}$$

Moment function. We shall show that the moment function $\Phi_y(s)$ of the busy period y satisfies the functional equation

$$\Phi_y(s) = \Phi_c\left[s + \lambda\Phi_y(s) - \lambda\right] \tag{16-68}$$

Proof. If $\mathbf{c} = c$, then $\mathbf{n}_c$ is a Poisson RV with parameter λc. And since the RVs $\mathbf{y}$ and $\mathbf{y}_i$ have the same statistics, we conclude that (see Prob. 8-13)

$$E\{\exp[s(\mathbf{y}_1 + \cdots + \mathbf{y}_{n_c})]\} = e^{\lambda c \Phi_y(s) - \lambda c}$$

From this and (16-65) it follows that

$$E\{e^{s\mathbf{y}}\} = E\{E\{e^{s\mathbf{y}}|\mathbf{c}\}\} = E\{e^{[s + \lambda\Phi_y(s) - \lambda]\mathbf{c}}\}$$

and (16-68) results.

Note that (16-68) does not give $\Phi_y(s)$ explicitly. It can, however, be used to determine the moments of $\mathbf{y}$. For example, its derivative at $s = 0$ yields

$$\Phi_y'(0) = [1 + \lambda\Phi_y'(0)]\Phi_c'(0)$$

in agreement with (16-66).

Ergodicity. The process $\mathbf{N}(t)$ and the sequences $\mathbf{N}(t_i)$ and $\mathbf{N}(\boldsymbol{\tau}_i)$ are ergodic. This means that statistical averages can be expressed as time averages. The proof is a consequence of the fact that the state of the system is a succession of idle and busy periods and its behavior in each period does not depend on what happens elsewhere in time. The details, however, are omitted. We note the following consequences:

1. The state of the system $\mathbf{N}(t_i^-)$ just before a unit arrives is an ergodic sequence of RVs; hence the probability that $\mathbf{N}(\mathbf{t}_i^-) = k$ equals the number of times this occurs in a large interval T divided by the number of points $\mathbf{t}_i$ in this interval. This number also equals the number of times $\mathbf{N}(\boldsymbol{\tau}_i^+)$ equals k because every time $\mathbf{N}(t)$ crosses the line $\mathbf{N} = k$ increasing, it crosses it also decreasing. Hence

$$P\{\mathbf{N}(\mathbf{t}_i^-) = k\} = P\{\mathbf{N}(\tau_i^+) = k\}$$

In other words, the processes $\mathbf{N}(\mathbf{t}_i^-)$ and $\mathbf{N}(\boldsymbol{\tau}_i^+)$ have the same statistics.

2. The expected values $\eta_x = E\{\mathbf{x}_i\}$, $\eta_y = E\{\mathbf{y}_i\}$ of the idle and busy periods, and their total number n in the interval $T = T_x + T_y$ are such that

$$\eta_x \simeq \frac{T_x}{n} \qquad \eta_y \simeq \frac{T_y}{n}$$

where T_x and T_y are the total times the system is idle or busy. Furthermore, $P\{\mathbf{N}(t) = 0\} \simeq T_x/T$. This leads to the conclusion that

$$P\{\mathbf{N}(t) = 0\} = \frac{\eta_x}{\eta_x + \eta_y} = \frac{1/\lambda}{1/\lambda + \eta_c/p_0} = \frac{p_0}{p_0 + \lambda\eta_c}$$

because $\eta_x = 1/\lambda$ and $\eta_y = \eta_c/p_0$ [see (6-66)]. And since $p_0 = 1 - \lambda\eta_c = P\{N(\boldsymbol{\tau}_i^+) = 0\}$, we conclude that

$$P\{\mathbf{N}(t) = 0\} = P\{N(\boldsymbol{\tau}_i^+) = 0\} = p_0 \qquad P\{\mathbf{N}(t) > 0\} = \rho$$

State statistics. We shall, finally, show that the process $\mathbf{N}(t)$ and the sequence $\mathbf{q}_i = \mathbf{N}(\tau_i^+)$ have the same distribution

$$P\{\mathbf{N}(t) = k\} = P\{\mathbf{q}_i = k\} = p_k \qquad k \geq 0 \tag{16-69}$$

Proof. If we eliminate all idle periods and connect all busy periods together, contracting the t axis, we obtain a renewal process specified in terms of the service times $\mathbf{c}_i = \boldsymbol{\tau}_i - \boldsymbol{\tau}_{i-1}$. With t an arbitrary constant, we denote by $\mathbf{w}$ the distance from t to the nearest point $\boldsymbol{\tau}_{i-1}$ on its left and by $\mathbf{n}_w$ the number of arrivals in the interval $(\boldsymbol{\tau}_{i-1}, t)$. From (16-37) and (16-32) it follows that

$$\begin{aligned} \mathbf{\Phi}_w(s) &= \frac{1}{\eta_c s}\left[\mathbf{\Phi}_c(s) - 1\right] \\ \mathbf{\Gamma}_{n_w}(z) &= \mathbf{\Phi}_w(\lambda z - \lambda) = \frac{\mathbf{\Gamma}_{n_c}(z) - 1}{\rho(z-1)} \end{aligned} \tag{16-70}$$

This holds for every t in a busy period, that is, when $\mathbf{N} = \mathbf{N}(t) \neq 0$. Returning to the original time (Fig. 16-12), we observe that, if $\mathbf{N} \neq 0$, then

$$\mathbf{N} = \tilde{\mathbf{q}}_{i-1} + \mathbf{n}_w \qquad \text{where} \quad \tilde{\mathbf{q}}_i = \begin{cases} \mathbf{q}_i & \mathbf{q}_i > 0 \\ 1 & \mathbf{q}_i = 0 \end{cases}$$

Since $\mathbf{n}_w$ is independent of $\tilde{\mathbf{q}}_i$, the above yields

$$E\{z^{\mathbf{N}} | \mathbf{N} \neq 0\} = \mathbf{\Gamma}_{\tilde{q}}(z)\mathbf{\Gamma}_{n_w}(z)$$

where

$$\mathbf{\Gamma}_{\tilde{q}}(z) = p_0 z + \sum_{k=1}^{\infty} p_k z^k = p_0 z + \mathbf{\Gamma}_q(z) - p_0 \tag{16-71}$$

As we know, $P\{\mathbf{N} = 0\} = p_0$, $P\{\mathbf{N} > 0\} = \rho$. Hence

$$\mathbf{\Gamma}_N(z) = E\{z^{\mathbf{N}}\} = E\{z^{\mathbf{N}} | \mathbf{N} = 0\}p_0 + E\{z^{\mathbf{N}} | \mathbf{N} > 0\}\rho$$

From this and (16-59) it follows, after some manipulations, that $\mathbf{\Gamma}_N(z) = \mathbf{\Gamma}_q(z)$ and (16-69) results.

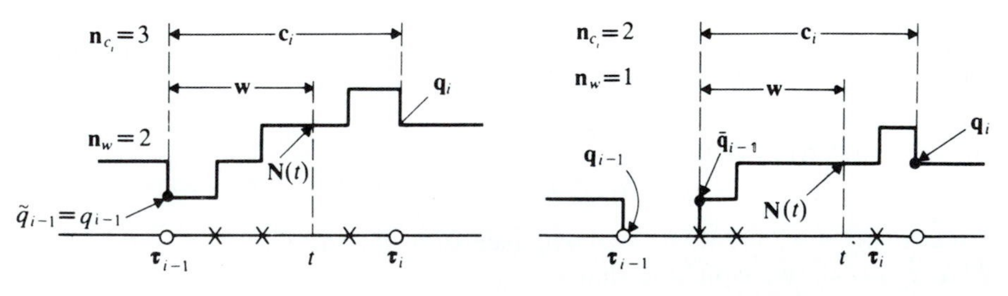

FIGURE 16-12

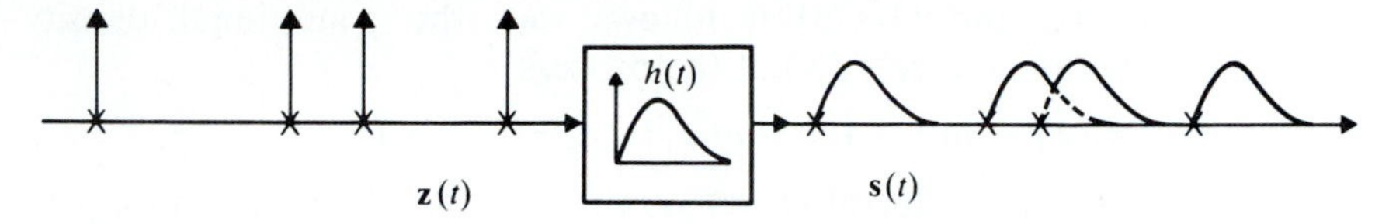

FIGURE 16-13

16-3 SHOT NOISE

Shot noise is the process

$$\mathbf{s}(t) = \sum_i h(t - \mathbf{t}_i) \tag{16-72}$$

where $h(t)$ is a given function and $\mathbf{t}_i$ is a set of Poisson points. The process $\mathbf{s}(t)$ is the output of a system with impulse response $h(t)$ and input a Poisson impulse train $\mathbf{z}(t)$ (Fig. 16-13). If $h(t) = U(t)$, then $\mathbf{s}(t)$ is a Poisson process (Fig. 10-3*a*). If $h(t)$ is a pulse of width c, then $\mathbf{s}(t)$ is the queueing process $M|D|\infty$ of Example 16-9*b*. In Sec. 11-2, we determined the second-order properties of the shot noise. In the following, we evaluate its general statistics.

DENSITY FUNCTION. We start with the determination of the density $f_s(x)$ of $\mathbf{s}(t)$:

$$f_s(x)\,dx = P\{x < \mathbf{s}(t) \le x + dx\}$$

under the assumption that $h(t)$ is of finite duration

$$h(t) = 0 \qquad \text{for} \quad t < 0 \quad \text{and} \quad t > T \tag{16-73}$$

Denoting by $\mathbf{n}_T$ the number of Poisson points in the interval $(t - T, t)$, we have [see (4-48)]

$$f_s(x) = \sum_{k=0}^{\infty} f_s(x|\mathbf{n}_T = k)P\{\mathbf{n}_T = k\} \tag{16-74}$$

In the above

$$P\{\mathbf{n}_T = k\} = \frac{e^{-\lambda T}(\lambda T)^k}{k!}$$

To find $f_s(x)$, it suffices, therefore, to find the conditional density $f_s(x|\mathbf{n}_T = k)$ assuming that there are k points in the interval $(t - T, t)$. The evaluation of this density is based on the following property of Poisson points [see (3-51)]:

If it is known that there are exactly k points in an interval (t_1, t_2), then these points have the same statistics as k arbitrary points placed at random in this interval. In other words, the k points can be assumed to be k independent RVs uniform in the interval (t_1, t_2).

From the above and (16-73) it follows that the conditional density $f_s(x|\mathbf{n}_T = 1)$ equals the density $g_1(x)$ of the process

$$\mathbf{x}_1(t) = h(t - \mathbf{t}_1) = h(\boldsymbol{\tau}_1) \qquad \boldsymbol{\tau}_1 = t - \mathbf{t}_1$$

where $\mathbf{t}_1$ is uniform in the interval $(t - T, t)$ or, equivalently, $\boldsymbol{\tau}_1$ is uniform in the interval $(0, T)$. The function $g_1(x)$ is independent of t and can be found with the techniques of Sec. 5-2.

Similarly, $f_s(x|\mathbf{n}_T = 2)$ equals the density of the process

$$\mathbf{x}_2(t) = h(t - \mathbf{t}_1) + h(t - \mathbf{t}_2) = h(\boldsymbol{\tau}_1) + h(\boldsymbol{\tau}_2)$$

where $\boldsymbol{\tau}_1$ and $\boldsymbol{\tau}_2$ are two independent RVs uniform in the interval $(0, T)$. Hence the RVs $h(\boldsymbol{\tau}_1)$ and $h(\boldsymbol{\tau}_2)$ have the same density $g_1(x)$ and they are independent because they are functions of the independent RVs $\boldsymbol{\tau}_1$ and $\boldsymbol{\tau}_2$. From this it follows that [see (6-39)] the density $g_2(x)$ of $\mathbf{x}_2(t)$ is the convolution

$$g_2(x) = g_1(x) * g_1(x)$$

Reasoning similarly, we conclude that

$$f_s(x|\mathbf{n}_T = k) = g_k(x) = g_1(x) * \cdots * g_1(x) \tag{16-75}$$

We note, finally, that if $\mathbf{n}_T = 0$, then $\mathbf{s}(t) = 0$; therefore,

$$f_s(x|\mathbf{n}_T = 0) = g_0(x) = \delta(x)$$

Inserting into (16-74), we obtain

$$f_s(x) = e^{-\lambda T} \sum_{k=0}^{\infty} \frac{g_k(x)(\lambda T)^k}{k!} \tag{16-76}$$

This formula is useful mainly for "*low density*" shot noise, that is, when λT is of the order of 1.

Example 16-12. Suppose that $h(t)$ is a trapezoid as in Fig. 16-14. In this case

$$P\{\mathbf{x}_1(t) \le x\} = P\{\boldsymbol{\tau} \ge (1.5 - x)T\} = x - 0.5$$

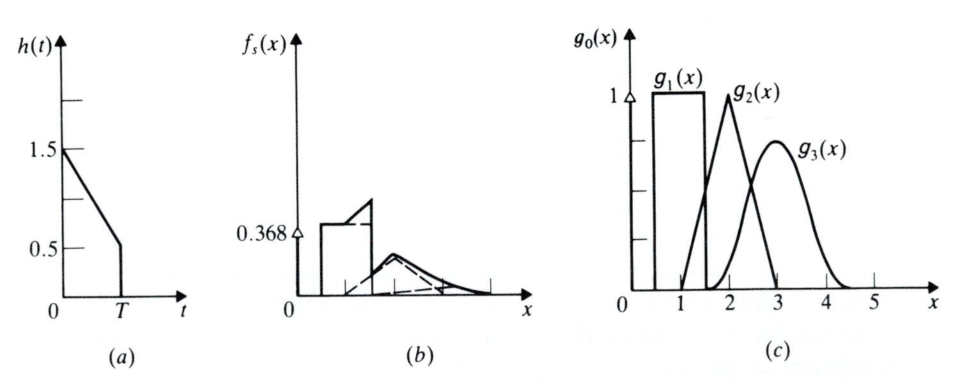

FIGURE 16-14

for $0.5 \le x \le 1.5$ and 0 otherwise. Hence $g_1(x)$ is uniform in the interval $(0.5, 1.5)$, $g_2(x)$ is a triangle in the interval $(1, 3)$, and $g_3(x)$ consists of three parabola pieces in the interval $(1.5, 4.5)$. Assuming that $\lambda = 1/T$, we have

$$P\{\mathbf{n}_T = 0\} = \frac{1}{e} \qquad P\{\mathbf{n}_T = 1\} = \frac{1}{e}$$

$$P\{\mathbf{n}_T = 2\} = \frac{1}{2e} \qquad P\{\mathbf{n}_T = 3\} = \frac{1}{6e}$$

Inserting into (16-76) and neglecting higher-order terms, we obtain the density $f_s(x)$ shown in Fig. 16-14.

Moment function. The moment function of $\mathbf{s}(t)$ equals

$$\Phi(s) = \int_{-\infty}^{\infty} e^{sx} f_s(x)\, dx = \exp\left\{\lambda \int_0^T [e^{sh(\alpha)} - 1]\, d\alpha\right\} \qquad (16\text{-}77)$$

Proof. This is a special case of (16-80) but can be established directly: As we have seen, $\mathbf{s}(t)$ can be written as a sum

$$\mathbf{s}(t) = \sum_{i=0}^{\mathbf{n}T} h(\boldsymbol{\tau}_i) \qquad (16\text{-}78)$$

where $\mathbf{n}_T$ is a Poisson RV with parameter λT and $\boldsymbol{\tau}_i$ are independent RVs uniform in the interval $(0, T)$. Reasoning as in Prob. 8-13, we obtain (16-77) because

$$E\{e^{sh(\boldsymbol{\tau}_i)}\} = \frac{1}{T}\int_0^T e^{sh(\alpha)}\, d\alpha \qquad (16\text{-}79)$$

General Properties

Suppose now that the points $\mathbf{t}_i$ are *nonuniform* with mean density $\lambda(t)$ and that the function $h(t)$ is arbitrary. We shall show that the second moment function of the resulting *nonhomogeneous* shot-noise process $\mathbf{s}(t)$ equals

$$\Psi(s) = \ln \Phi(s) = \int_{-\infty}^{\infty} \lambda(\alpha)[e^{h(t-\alpha)s} - 1]\, d\alpha \qquad (16\text{-}80)$$

Proof. We divide the time axis into consecutive intervals I_i: (α_i, α_{i+1}) of length $\Delta\alpha = \alpha_{i+1} - \alpha_i$ as in Fig. 16-15 and we denote by $\Delta\mathbf{n}_i$ the number of points $\mathbf{t}_i$

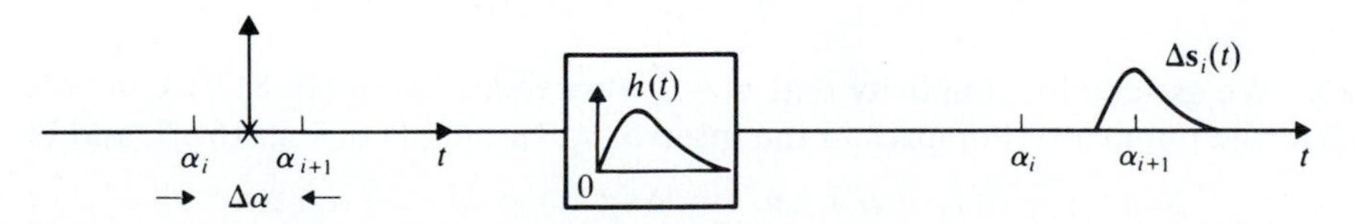

FIGURE 16-15

in I_i. If $\Delta\alpha$ is sufficiently small, then the contribution $\Delta\mathbf{s}_i(t)$ to $\mathbf{s}(t)$ due to these points is given by [see (16-72)]

$$\Delta\mathbf{s}_i(t) \simeq h(t-\alpha_i)\,\Delta\mathbf{n}_i \tag{16-81}$$

As we know, $\Delta\mathbf{n}_i$ is a Poisson RV with parameter

$$\int_{\alpha_i}^{\alpha_i+\Delta\alpha} \lambda(\alpha)\,d\alpha \simeq \lambda(\alpha_i)\,\Delta\alpha$$

Hence the moment function of $\Delta\mathbf{s}_i(t)$ equals

$$\Delta\Phi_i(s) \simeq E\{e^{sh(t-\alpha_i)\Delta\mathbf{n}_i}\} = \exp\{\lambda(\alpha_i)\,\Delta\alpha[e^{h(t-\alpha_i)s}-1]\} \tag{16-82}$$

Furthermore,

$$\mathbf{s}(t) = \sum_i \Delta\mathbf{s}_i(t) \simeq \sum_i h(t-\alpha_i)\,\Delta\mathbf{n}_i$$

And since the RVs $\Delta\mathbf{n}_i$ are independent, we conclude that

$$\Psi(s) = \sum_i \ln\Delta\Phi_i(s) \simeq \sum_i \lambda(\alpha_i)\,\Delta\alpha[e^{h(t-\alpha_i)s}-1] \tag{16-83}$$

As $\Delta\alpha \to 0$, the last sum tends to the integral in (16-80).

Generalization of Campbells' theorem. The mean η_s and the variance σ_s^2 of the nonhomogeneous shot-noise process $\mathbf{s}(t)$ are given by

$$\eta_s = \int_{-\infty}^{\infty} \lambda(\alpha)h(t-\alpha)\,d\alpha \qquad \sigma_s^2 = \int_{-\infty}^{\infty} \lambda(\alpha)h^2(t-\alpha)\,d\alpha \tag{16-84}$$

Proof. Expanding the exponential in (16-80) into a series and integrating termwise, we obtain (16-84) because [see (5-73)]

$$\eta_s = \Psi'(0) \qquad \sigma_s^2 = \Psi''(0)$$

JOINT CHARACTERISTIC FUNCTIONS. The joint second moment function Ψ of the n RVs

$$\mathbf{s}(t_1),\ldots,\mathbf{s}(t_n)$$

equals

$$\Psi(s_1,\ldots,s_n) = \int_{-\infty}^{\infty} \lambda(\alpha)(e^{\beta}-1)\,d\alpha \tag{16-85}$$

where

$$\beta = s_1h(t_1-\alpha) + \cdots + s_nh(t_n-\alpha)$$

Proof. We assume for simplicity that $n=2$. Proceeding as in (16-81), we denote by $\Delta\mathbf{n}_i$ the number of impulses in the interval I_i: (α_i, α_{i+1}) of Fig. 16-15, and by

$$\Delta\mathbf{s}_i(t_1) \simeq h(t_1-\alpha_i)\,\Delta\mathbf{n}_i \qquad \Delta\mathbf{s}_i(t_2) \simeq h(t_2-\alpha_i)\,\Delta\mathbf{n}_i$$

the response of the system at $t=t_1$ and $t=t_2$ respectively due to these impulses. The joint moment function $\mathbf{\Phi}$ of the RVs $\Delta\mathbf{s}_i(t_1)$ and $\Delta\mathbf{s}_i(t_2)$ equals

$$E\{e^{[s_1h(t_1-\alpha_i)+s_2h(t_2-\alpha_i)]\Delta\mathbf{n}_i}\}$$

This is of the same form as the second term in (16-82) if we replace the exponent $sh(t-\alpha_i)$ by the sum $s_1h(t_1-\alpha_i)+s_2h(t_2-\alpha_i)$. Hence, as in (16-83),

$$\mathbf{\Psi}(s_1,s_2) \simeq \sum_i \lambda(\alpha_i)[e^{s_1h(t_1-\alpha_i)+s_2h(t_2-\alpha_i)}-1]\,\Delta\alpha$$

and with $\Delta\alpha \to 0$, (16-85) results.

Covariance. We shall use (16-85) to determine the autocovariance $C(t_1,t_2)$ of $\mathbf{s}(t)$. As we know [see (page 160)], $C(t_1,t_2)$ is the coefficient of s_1s_2 in the series expansion of $\mathbf{\Psi}(s_1,s_2)$ about the origin. Expanding the term e^{β} in (16-85), we conclude that

$$C(t_1,t_2) = \int_{-\infty}^{\infty}\lambda(\alpha)h(t_1-\alpha)h(t_2-\alpha)\,d\alpha \tag{16-86}$$

If $\lambda(t)=\lambda=$ constant, then, with $t_1-t_2=\tau$, the above yields

$$C(\tau) = \lambda\int_{-\infty}^{\infty}h(\tau+\alpha)h(\alpha)\,d\alpha \tag{16-87}$$

in agreement with (11-50).

High density and normality. We shall show that if the density λ of stationary shot noise is large compared with the time constants of $h(t)$, then $\mathbf{s}(t)$ is approximately normal. For this purpose, we introduce the normalizations

$$\lambda_0 = \frac{\lambda}{k} \qquad h_0(t) = \sqrt{k}\,h(t) \qquad \beta_0 = \sqrt{k}\,\beta$$

and examine the form of $\mathbf{\Psi}(s_1,s_2)$ as $k\to\infty$. Clearly,

$$\lambda(e^{\beta}-1) = \lambda_0\left(\sqrt{k}\,\beta_0 + \frac{\beta_0^2}{2!} + \frac{\beta_0^3}{3!\sqrt{k}} + \cdots\right)$$

Neglecting negative powers of k, we conclude from (16-85) that

$$\mathbf{\Psi}(s_1,\ldots,s_2) \simeq \lambda_0\int_{-\infty}^{\infty}\left(\sqrt{k}\,\beta_0 + \frac{\beta_0^2}{2}\right)d\alpha \tag{16-88}$$

This shows that $\mathbf{\Psi}$ is a quadratic function of s_i because β_0 is linear in s_i. Hence $\mathbf{s}(t)$ is nearly normal. This result is exact in the limit as $k\to\infty$ if the linear term is omitted (centering).

INTENSITY OF SHOT NOISE. The square

$$\mathbf{I}(t) = \mathbf{s}^2(t)$$

of the shot noise $\mathbf{s}(t)$ is a stationary process with mean

$$\eta_I = E\{\mathbf{s}^2(t)\} = \lambda \int_{-\infty}^{\infty} h^2(t)\,dt + \lambda^2 \left[\int_{-\infty}^{\infty} h(t)\,dt\right]^2 \qquad (16\text{-}89)$$

We shall determine its autocorrelation $R_I(\tau)$ under the assumption that

$$H(0) = \int_{-\infty}^{\infty} h(t)\,dt = 0 \qquad (16\text{-}90)$$

With this assumption, the mean and autocorrelation of $\mathbf{s}(t)$ are given by

$$\eta_s = 0 \qquad R_s(\tau) = \lambda \int_{-\infty}^{\infty} h(\tau + \alpha) h(\alpha)\,d\alpha \qquad (16\text{-}91)$$

We maintain that

$$R_I(\tau) = \lambda^2 E^2 + 2R_s^2(\tau) + \lambda \int_{-\infty}^{\infty} h^2(\tau + \alpha) h^2(\alpha)\,d\alpha \qquad (16\text{-}92)$$

where

$$E = \int_{-\infty}^{\infty} h^2(t)\,dt$$

is the energy of $h(t)$.

Proof. The second moment function $\Psi(s_1, s_2)$ of the RVs $\mathbf{s}(t_1)$ and $\mathbf{s}(t_2)$ equals the integral in (16-85) where

$$\beta = s_1 h(t_1 - \alpha) + s_2 h(t_2 - \alpha) \qquad \lambda(t) = \lambda$$

As we know [see (7-34)] the coefficient of $s_1^2 s_2^2$ in the expansion of the moment function $\Phi(s_1, s_2)$ equals

$$\frac{E\{\mathbf{s}^2(t_1)\mathbf{s}^2(t_2)\}}{4}$$

We introduce the functions

$$\gamma_n = \frac{\lambda}{n!} \int_{-\infty}^{\infty} \beta^n\,d\alpha$$

Since

$$e^{\beta} - 1 = \beta + \frac{\beta^2}{2} + \cdots$$

and $\gamma_1 = 0$ [see (16-90)], we conclude that

$$\Psi(s_1, s_2) = \gamma_2 + \gamma_3 + \gamma_4 + \cdots$$

$$\Phi(s_1, s_2) = 1 + \gamma_2 + \gamma_3 + \gamma_4 + \cdots + \frac{(\gamma_2 + \gamma_3 + \gamma_4 + \cdots)^2}{2} + \cdots$$

In this expansion, only the sum $\gamma_4 + \gamma_2^2/2$ will have terms in $s_1^2 s_2^2$. Furthermore,

$$\gamma_4 = \cdots + \gamma \frac{s_1^2 s_2^2}{4!} \int_{-\infty}^{\infty} \binom{4}{2} h^2(t_1 - \alpha) h^2(t_2 - \alpha)\, d\alpha + \cdots$$

$$\gamma_2^2 = \cdots + \lambda^2 \frac{s_1^2 s_2^2}{2} \int_{-\infty}^{\infty} h^2(t_1 - \alpha)\, d\alpha \int_{-\infty}^{\infty} h^2(t_2 - \alpha)\, d\alpha$$

$$+ \lambda^2 s_1^2 s_2^2 \left[\int_{-\infty}^{\infty} h(t_1 - \alpha) h(t_2 - \alpha)\, d\alpha \right]^2$$

and with $t_1 = t_2 + \tau$, (16-92) results.

Power spectrum. The power spectrum $S_s(\omega)$ of $\mathbf{s}(t)$ equals [see (16-91)]

$$S_s(\omega) = \lambda |H(\omega)|^2 \tag{16-93}$$

Furthermore,

$$\lambda^2 E^2 \leftrightarrow 2\pi\lambda^2 E^2 \delta(\omega) \qquad 2\pi R_s^2(\tau) \leftrightarrow S_s(\omega) * S_s(\omega)$$

The integral in (16-92) equals $h^2(\tau) * h^2(-t)$. And since

$$2\pi h^2(t) \leftrightarrow H(\omega) * H(\omega)$$

we conclude from (16-92) that the power spectrum of the intensity $\mathbf{I}(t)$ of $\mathbf{s}(t)$ equals

$$2\pi\lambda^2 E^2 \delta(\omega) + \frac{\lambda^2}{\pi} |H(\omega)|^2 * |H(\omega)|^2 + \frac{\lambda}{4\pi^2} |H(\omega) * H(\omega)|^2 \tag{16-94}$$

High density If λ is sufficiently large, then the third term above can be neglected. In this case, $\mathbf{s}(t)$ is nearly normal and the power spectrum of $\mathbf{s}^2(t)$ equals the sum of the first two terms in (16-94) [see also (10-68)].

Low density If λ is small, then the first two terms in (16-94) can be neglected. In this case,

$$\mathbf{I}(t) = \mathbf{s}^2(t) \simeq \sum_i h^2(t - \mathbf{t}_i) \tag{16-95}$$

because the probability that the terms $h(t - \mathbf{t}_i)$ and $h(t - \mathbf{t}_k)$ have a significant overlap is negligible. This means that the square of $\mathbf{s}(t)$ is approximately a shot-noise process generated by $h^2(t)$.

16-4 MARKOFF PROCESSES

A Markoff process is a stochastic process whose past has no influence on the future if its present is specified. This means the following:

If $t_{n-1} < t_n$, then

$$P\{\mathbf{x}(t_n) \le x_n | \mathbf{x}(t), t \le t_{n-1}\} = P\{\mathbf{x}(t_n) \le x_n | \mathbf{x}(t_{n-1})\} \tag{16-96}$$

From this it follows that if

$$t_1 < t_2 < \cdots < t_n$$

then

$$P\{\mathbf{x}(t_n) \le x_n | \mathbf{x}(t_{n-1}), \ldots, \mathbf{x}(t_1)\} = P\{\mathbf{x}(t_n) \le x_n | \mathbf{x}(t_{n-1})\} \quad (16\text{-}97)$$

The above definition holds also for discrete-time processes if $\mathbf{x}(t_n)$ is replaced by $\mathbf{x}_n$.

In this section, we develop various properties of Markoff processes, concentrating on three classes: (*a*) discrete-time, discrete-state; (*b*) continuous-time, discrete-state; (*c*) continuous-time, continuous-state. We start with a brief discussion of certain general properties phrasing the results in terms of discrete-time, continuous-state processes.

1. From (16-97) it follows that

$$f(x_n | x_{n-1}, \ldots, x_1) = f(x_n | x_{n-1}) \quad (16\text{-}98)$$

Applying the chain rule (8-37) to the above, we obtain

$$f(x_1, \ldots, x_n) = f(x_n | x_{n-1}) f(x_{n-1} | x_{n-2}) \cdots f(x_2 | x_1) f(x_1) \quad (16\text{-}99)$$

Conversely, if (16-99) is true for all n, then the process $\mathbf{x}_n$ is Markoff because, in this case,

$$f(x_n | x_{n-1}, \ldots, x_1) = \frac{f(x_1, \ldots, x_{n-1}, x_n)}{f(x_1, \ldots, x_{n-1})} = f(x_n | x_{n-1}) \quad (16\text{-}100)$$

2. From (16-98) it follows that

$$E\{\mathbf{x}_n | \mathbf{x}_{n-1}, \ldots, \mathbf{x}_1\} = E\{\mathbf{x}_n | \mathbf{x}_{n-1}\} \quad (16\text{-}101)$$

3. A Markoff process is also Markoff if time is reversed:

$$f(x_n | x_{n+1}, \ldots, x_{n+k}) = f(x_n | x_{n+1}) \quad (16\text{-}102)$$

Proof. The left side of (16-102) equals

$$\frac{f(x_n, x_{n+1}, \ldots, x_{n+k})}{f(x_{n+1}, \ldots, x_{n+k})} = \frac{f(x_{n+1} | x_n)}{f(x_{n+1})} f(x_n)$$

And since

$$f(x_{n+1} | x_n) f(x_n) = f(x_n, x_{n+1}) = f(x_n | x_{n+1}) f(x_{n+1})$$

(16-102) results.

4. If the present is specified, then the past is independent of the future in the following sense: If $k < m < n$, then

$$f(x_n, x_k | x_m) = f(x_n | x_m) f(x_k | x_m) \quad (16\text{-}103)$$

Proof. From (16-99) it follows that

$$f(x_n, x_k | x_m) = \frac{f(x_n, x_m, x_k)}{f(x_m)} = \frac{f(x_n|x_m)f(x_m|x_k)}{f(x_m)} f(x_k)$$

and (16-103) results.

The above relationship can be used to express conditional densities involving the past and the future, in terms of the conditional (*transition*) densities $f(x_k | x_{k+1})$. For example, if $k < m < n$, then

$$f(x_m | x_n, x_k) = \frac{f(x_k|x_m)}{f(x_k|x_n)} f(x_m|x_n) \tag{16-104}$$

Example 16-13. If $\mathbf{x}_n$ satisfies the recursion equation

$$\mathbf{x}_{n+1} - a(\mathbf{x}_n, n) = \boldsymbol{\nu}_n \tag{16-105}$$

and $\boldsymbol{\nu}_n$ is strictly white noise, then $\mathbf{x}_n$ is Markoff.

Proof. The process $\mathbf{x}_{n+1}$ is determined in terms of $\mathbf{x}_n$ and $\boldsymbol{\nu}_n$; hence it is independent of $\mathbf{x}_k$ for $k < n$, assuming $\mathbf{x}_n = x_n$.

Special case (generalized random walk). If $\mathbf{x}_0 = 0$ and $a(\mathbf{x}_n, n) = \mathbf{x}_n$, then

$$\mathbf{x}_n = \mathbf{x}_{n-1} + \boldsymbol{\nu}_{n-1} = \boldsymbol{\nu}_1 + \boldsymbol{\nu}_2 + \cdots + \boldsymbol{\nu}_{n-1}$$

Thus the sum of independent RVs is a Markoff sequence.

Homogeneous processes. From the chain rule (16-99) it follows that the statistics of any order of a Markoff process can be determined in terms of the conditional densities $f(x_n | x_{n-1})$ and the first-order density $f(x_n)$. If the process $\mathbf{x}_n$ is *stationary*, then the functions $f(x_n)$ and $f(x_n | x_{n-1})$ are invariant to a shift of the origin. In this case, the statistics of $\mathbf{x}_n$ are completely determined in terms of the second-order density

$$f(x_1, x_2) = f(x_2 | x_1) f(x_1)$$

A Markoff process $\mathbf{x}_n$ is called *homogeneous* if the conditional density $f(x_n | x_{n-1})$ is invariant to a shift of the origin but the first-order density $f(x_n)$ might depend on n. In general, a homogeneous process is not stationary. However, in many cases, it tends to a stationary process as $n \to \infty$.

The Chapman–Kolmogoroff equation The conditional density $f(x_n | x_k)$ can be expressed in terms of $f(x_n | x_m)$ and $f(x_m | x_k)$ for any $n > m > k$:

$$f(x_n | x_k) = \int_{-\infty}^{\infty} f(x_n | x_m) f(x_m | x_k)\, dx_m \tag{16-106}$$

This follows from (8-39) because $f(x_n | x_m, x_k) = f(x_n | x_m)$.

Discrete-Time Markoff Chains

A discrete-time Markoff chain is a Markoff process $\mathbf{x}_n$ having a countable number of states a_i. A Markoff chain is specified in terms of its *state probabilities*

$$p_i[n] = P\{\mathbf{x}_n = a_i\} \qquad i = 1, 2, \ldots \tag{16-107}$$

and the *transition probabilities*

$$\pi_{ij}[n_1, n_2] = P\{\mathbf{x}_{n2} = a_j | \mathbf{x}_{n_1} = a_i\} \tag{16-108}$$

As we know [see (7-48)]

$$\sum_j \pi_{ij}[n_1, n_2] = 1 \qquad \sum_i p_i[k]\pi_{ij}[k, n] = p_j[n] \tag{16-109}$$

Furthermore, if $n_1 < n_2 < n_3$, then

$$\pi_{ij}[n_1, n_3] = \sum_r \pi_{ir}[n_1, n_2]\pi_{rj}[n_2, n_3] \tag{16-110}$$

This is the discrete form of the Chapman–Kolmogoroff equation and it follows readily from (8-40).

HOMOGENEOUS CHAINS. If the process $\mathbf{x}_n$ is homogeneous, then the transition probabilities in (16-108) depend only on the difference $m = n_2 - n_1$. Thus

$$\pi_{ij}[m] = P\{\mathbf{x}_{n+m} = a_j | \mathbf{x}_n = a_i\} \tag{16-111}$$

Setting $n_2 - n_1 = k$, $n_3 - n_2 = n$ in (16-110), we obtain

$$\pi_{ij}[n + k] = \sum_r \pi_{ir}[k]\pi_{rj}[n] \tag{16-112}$$

For a finite-state Markoff chain, the above can be written in vector form:

$$\Pi[n + k] = \Pi[n]\Pi[k] \tag{16-113}$$

where $\Pi[n]$ is a Markoff matrix with elements $\pi_{ij}[n]$. This yields

$$\Pi[n] = \Pi^n \qquad \text{where} \quad \Pi = \Pi[1] \tag{16-114}$$

is the one-step transition matrix with elements $\pi_{ij} = \pi_{ij}[1]$. The above is the solution of the first-order recursion equation [see (16-113)]

$$\Pi[n + 1] = \Pi[n]\Pi \tag{16-115}$$

The matrix Π is shown schematically in Fig. 16-16. The circles in the diagram represent the states a_i of the process and the number on each segment, from a_i to a_j, the transition probabilities π_{ij}. The number on the loop from a_i to a_i equals π_{ii}. This loop (dashed line) can be omitted because the sum of the row elements of Π (segments leaving a state, including the loop) equals 1 [see (16-109)].

$$\Pi = \begin{bmatrix} \pi_{11} & \pi_{12} & \cdots & \pi_{1N} \\ \pi_{21} & \pi_{22} & \cdots & \pi_{2N} \\ \cdots & \cdots & \cdots & \cdots \\ \pi_{N1} & \pi_{N2} & \cdots & \pi_{NN} \end{bmatrix}$$

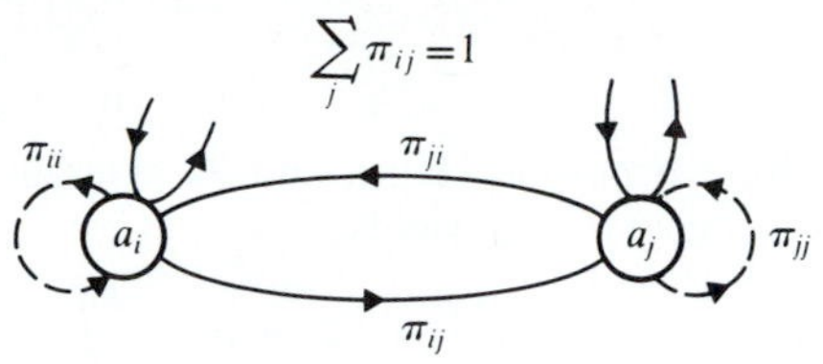

$\sum_j \pi_{ij} = 1$ $\quad \pi_{ii} \quad \pi_{ji} \quad a_i \quad a_j \quad \pi_{jj} \quad \pi_{ij}$

FIGURE 16-16

Writing (16-109) in vector form and using (16-114), we conclude that

$$P[n] = \cdots = P[n-k]\Pi^k = \cdots = P[0]\Pi^n \tag{16-116}$$

where $P[n]$ is a vector whose elements are the state probabilities $p_i[n]$.

In general, $P[n]$ depends on n. However, if the initial state vector

$$P[1] = P = [p_1, \ldots, p_N] \tag{16-117}$$

is such that $P[2] = P$, then $P[n] = P$ for all n. In this case, the homogeneous process $\mathbf{x}_n$ is also stationary and its state vector P is the solution of the system

$$P\Pi = P \qquad \sum_i p_i = 1 \tag{16-118}$$

The state probability vector P of a stationary Markoff chain is thus an eigenvector of its transition matrix Π and the corresponding eigenvalue equals 1 [see (16-109)].

If the initial state $P[1]$ of a homogeneous chain $\mathbf{x}_n$ does not equal P, then $\mathbf{x}_n$ is not stationary. In this case, certain of its states might never be reached. The details of the underlying theory will not be discussed. We note only that, if Π^n tends to a limit as $n \to \infty$, then $\mathbf{x}_n$ is asymptotically stationary. This is the case if all the elements π_{ij} of Π are strictly positive (Prob. 16-19).

Example 16-14. In the random walk experiment (Sec. 10-1), we place two reflecting walls at $x = 2s$ and $x = -2s$ as in Fig. 16-17. The resulting motion $\mathbf{x}(t)$ between the two walls generates a homogeneous Markoff chain $\mathbf{x}_n = \mathbf{x}(nT)$ taking the

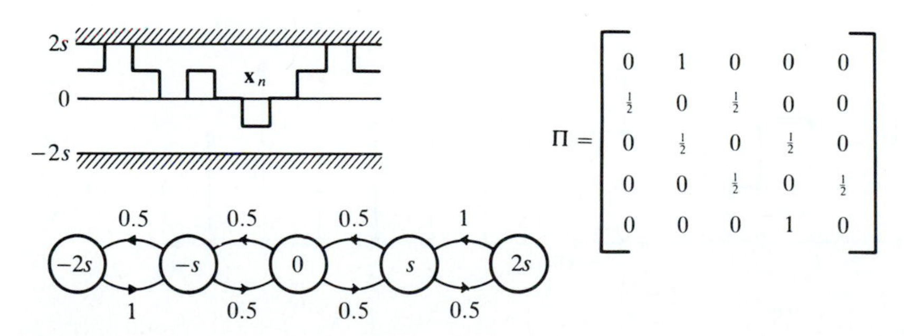

FIGURE 16-17

values

$$-2s \qquad -s \qquad 0 \qquad s \qquad 2s$$

The resulting transition matrix and its diagram are shown in the figure. In this case, the system (16-118) yields

$$p_1 = \frac{p_2}{2} \qquad p_2 = p_1 + \frac{p_3}{2} \qquad p_3 = \frac{p_2 + p_4}{2}$$

$$p_4 = p_5 + \frac{p_3}{2} \qquad p_1 + p_2 + p_3 + p_4 + p_5 = 1$$

Solving, we obtain

$$p_1 = p_5 = \tfrac{1}{8} \qquad p_2 = p_3 = p_4 = \tfrac{1}{4}$$

These are the initial state probabilities that generate a stationary process.

Continuous-Time Markoff Chains

A continuous-time Markoff chain is a Markoff process $\mathbf{x}(t)$ consisting of a family of staircase functions (discrete states) with discontinuities at the random points $\mathbf{t}_n$ (Fig. 16-18a). The values

$$\mathbf{q}_n = \mathbf{x}(t_n^+) \tag{16-119}$$

of $\mathbf{x}(t)$ at these points (Fig. 16-18b) form a discrete-state Markoff sequence called the *Markoff chain imbedded* in the process $\mathbf{x}(t)$.

A discrete-state stochastic process is called *semi-Markoff* if it is not Markoff but the imbedded sequence $\mathbf{q}_n$ is a Markoff chain. An example is the queueing process $\mathbf{N}(t)$ of Sec. 16-2.

A Markoff chain $\mathbf{x}(t)$ is specified in terms of the underlying point process $\mathbf{t}_n$ and the imbedded Markoff chain $\mathbf{q}_n$.

We denote by

$$p_i(t) = P\{\mathbf{x}(t) = a_i\} \tag{16-120}$$

the *state probabilities* of $\mathbf{x}(t)$ and by

$$\pi_{ij}(t_1, t_2) = P\{\mathbf{x}(t_2) = a_j | \mathbf{x}(t_1) = a_i\} \tag{16-121}$$

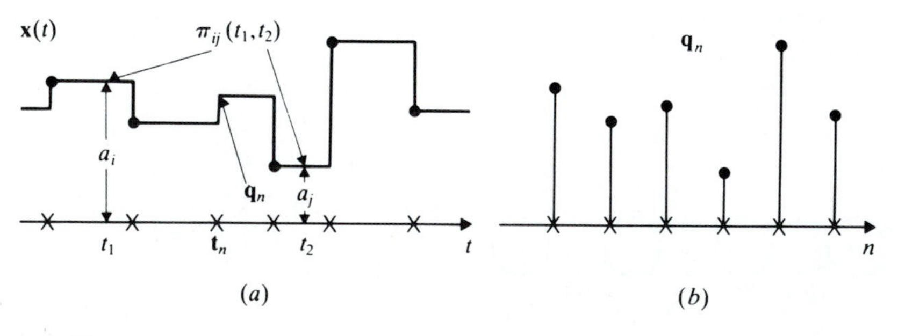

FIGURE 16-18

its *transition probabilities*. These functions are such that

$$\sum_j \pi_{ij}(t_1, t_2) = 1 \qquad \sum_i p_i(t_1)\pi_{ij}(t_1, t_2) = p_j(t_2) \tag{16-122}$$

and they satisfy the Chapman–Kolmogoroff equation

$$\pi_{ij}(t_1, t_3) = \sum_r \pi_{ir}(t_1, t_2)\pi_{rj}(t_2, t_3) \qquad t_1 < t_2 < t_3 \tag{16-123}$$

In specific problems the functions $\pi_{ij}(t_1, t_2)$ are not given directly. As we shall presently see, however, they can be determined in terms of the transition probability rates to be presently defined. For simplicity, we shall consider only *homogeneous processes*.

A Markoff process $\mathbf{x}(t)$ is homogeneous if its transition probabilities depend on the difference $\tau = t_2 - t_1$:

$$\pi_{ij}(\tau) = P\{\mathbf{x}(t+\tau) = a_j | \mathbf{x}(t) = a_i\} \qquad \tau \geq 0 \tag{16-124}$$

From the above and (16-123) it follows with $\alpha = t_3 - t_2$ that

$$\pi_{ij}(\tau + \alpha) = \sum_r \pi_{ir}(\tau)\pi_{rj}(\alpha) \tag{16-125}$$

This is the Chapman–Kolmogoroff equation for continuous-time Markoff chains and it can be written in vector form:

$$\Pi(\tau + \alpha) = \Pi(\tau)\Pi(\alpha) \qquad \tau, \alpha \geq 0 \tag{16-126}$$

where $\Pi(\tau)$ is a matrix with elements $\pi_{ij}(\tau)$.

Probability rates. In the discrete-time case, we showed that the matrix $\Pi[n]$ satisfies the recursion equation (16-115) and can be determined in terms of the one-step transition matrix Π. We show next that the transition matrix $\Pi(\tau)$ of a continuous-time chain $\mathbf{x}(t)$ satisfies a differential equation and can be determined in terms of the matrix

$$\Pi'(0^+) = \Lambda \equiv \begin{bmatrix} \lambda_{11}, \ldots, \lambda_{1n} \\ \cdots\cdots\cdots \\ \lambda_{n1}, \ldots, \lambda_{nn} \end{bmatrix} \tag{16-127}$$

whose elements $\lambda_{ij} = \pi'_{ij}(0^+)$ are the derivatives from the right of the elements $\pi_{ij}(\tau)$ of $\Pi(\tau)$. These derivatives will be called the *transition probability rates* of $\mathbf{x}(t)$. Clearly,

$$\sum_j \lambda_{ij} = 0 \qquad \text{because} \qquad \sum_j \pi_{ij}(\tau) = 1$$

and since

$$\pi_{ij}(0) = \delta[i-j] = \begin{cases} 1 & i = j \\ 0 & i \neq j \end{cases} \tag{16-128}$$

we conclude with $\mu_i = -\lambda_{ii}$ that

$$\mu_i = \sum_j{}' \lambda_{ij} \geq 0 \qquad \lambda_{ij} \geq 0 \qquad i \neq j \tag{16-129}$$

The prime indicates summation for every $j \neq i$.

In the above, we have assumed that $\pi_{ij}(\tau)$ is differentiable at $\tau = 0^+$. This is so only if the probability that there is one discontinuity point in the interval $(t, t + \Delta t)$ is of the order of Δt:

$$P\{\mathbf{x}(t + \Delta t) = a_j | \mathbf{x}(t) = a_i\} = \begin{cases} 1 - \mu_i \Delta t & i = j \\ \lambda_{ij} \Delta t & i \neq j \end{cases} \tag{16-130}$$

The Kolmogoroff equations. Differentiating (16-126) with respect to α and setting $\alpha = 0$, we obtain

$$\Pi'(\tau) = \Pi(\tau)\Lambda \qquad \Pi(0) = 1 \tag{16-131}$$

This is a system of linear differential equations with constant coefficients and its initial condition $\Pi(0)$ is the identity matrix [see (16-128)]. Solving, we obtain

$$\Pi(\tau) = e^{\Lambda\tau} \tag{16-132}$$

We have thus expressed $\Pi(\tau)$ in terms of the transition rate matrix Λ.

The state probabilities $p_i(t)$ satisfy a similar system: Denoting by $P(t)$ a vector with elements $p_i(t)$, we conclude from (16-122) that

$$P(t + \tau) = P(t)\Pi(\tau) \tag{16-133}$$

Differentiating with respect to τ and setting $\tau = 0$, we obtain

$$P'(t) = P(t)\Lambda \tag{16-134}$$

This is a system of N equations of the form

$$p_i'(t) = -\mu_i p_i(t) + \sum_j{}' \lambda_{ji} p_j(t) \tag{16-135}$$

Its formal solution is a vector exponential

$$P(t) = P(0)e^{\Lambda t} \tag{16-136}$$

We have thus expressed $P(t)$ in terms of Λ and the initial state probabilities $p_i(0)$.

If $\mathbf{x}(t)$ is stationary, then $p_i(t) = p_i$ = constant. Hence [see (16-135)]

$$\mu_i p_i = \sum_j{}' \lambda_{ji} p_j \qquad \sum_i p_i = 1 \tag{16-137}$$

This is a system expressing the state probabilities of a stationary process in terms of the transition rates λ_{ij}.

Example 16-15 Generalized telegraph signal. Suppose that $\mathbf{x}(t)$ takes two values (Fig. 16-19)

$$a_1 = A \qquad a_2 = -A$$

and

$$\begin{aligned} P\{\mathbf{x}(t+\Delta t) = A \mid \mathbf{x}(t) = A\} &= 1 - \mu_1 \Delta t = \pi_{11}(\Delta t) \\ P\{\mathbf{x}(t+\Delta t) = -A \mid \mathbf{x}(t) = -A\} &= 1 - \mu_2 \Delta t = \pi_{22}(\Delta t) \end{aligned} \tag{16-138}$$

In this case, $\lambda_{12} = \mu_1$, $\lambda_{21} = \mu_2$. Inserting into (16-135), we obtain

$$p_1'(t) + \mu_1 p_1(t) = \mu_2 p_2(t)$$

And since $p_2(t) = 1 - p_1(t)$, we conclude that

$$p_1(t) = \frac{\mu_2}{\mu_1 + \mu_2}\left[1 - e^{-(\mu_1+\mu_2)t}\right] + p_1(0)e^{-(\mu_1+\mu_2)t}$$

Note that

$$p_1(t) \xrightarrow[t\to\infty]{} \frac{\mu_2}{\mu_1+\mu_2} = p_1 \qquad p_2(t) \xrightarrow[t\to\infty]{} \frac{\mu_1}{\mu_1+\mu_2} = p_2$$

The transition probabilities are determined from (16-131)

$$\pi_{11}'(\tau) + \mu_1 \pi_{11}(\tau) = \mu_2 \pi_{12}(\tau) \qquad \pi_{11}(0) = 1$$

$$\pi_{22}'(\tau) + \mu_2 \pi_{22}(\tau) = \mu_1 \pi_{21}(\tau) \qquad \pi_{22}(0) = 1$$

where

$$\pi_{12}(\tau) = 1 - \pi_{11}(\tau) \qquad \pi_{21}(\tau) = 1 - \pi_{22}(\tau)$$

The above yields

$$\begin{aligned} \pi_{11}(\tau) &= p_1 + p_2 e^{-(\mu_1+\mu_2)\tau} \\ \pi_{21}(\tau) &= p_2 + p_1 e^{-(\mu_1+\mu_2)\tau} \end{aligned} \tag{16-139}$$

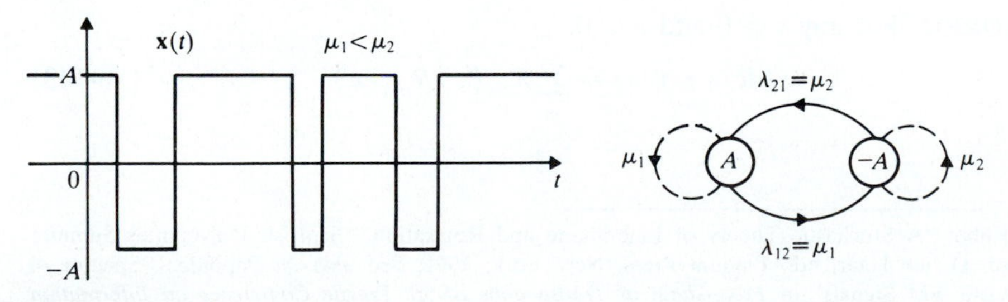

FIGURE 16-19

Mean and autocorrelation The process $\mathbf{x}(t)$ is asymptotically stationary with

$$P\{\mathbf{x}(t) = a_i\} = p_i \qquad a_1 = A \qquad a_2 = -A$$

and

$$P\{\mathbf{x}(t+\tau) = a_j, \mathbf{x}(t) = a_i\} = p_i \pi_{ij}(\tau) \qquad i, j = 1, 2$$

From this it follows that

$$E\{\mathbf{x}(t)\} = \eta = p_1 A - p_2 A$$
$$R(\tau) = \eta^2 + 4A^2 p_1 p_2 e^{-(\mu_1+\mu_2)|\tau|}$$

If $\mu_1 = \mu_2 = \lambda$, then $\eta = 0$ and $R(\tau) = e^{-2\lambda|\tau|}$ as in (10-19). In this case only, the discontinuity points $\mathbf{t}_i$ of $\mathbf{x}(t)$ are Poisson.

SPECTRA OF STOCHASTIC FM SIGNALS†. We shall determine the power spectrum of the FM signal

$$\mathbf{w}(t) = e^{j\boldsymbol{\varphi}(t)} \qquad \boldsymbol{\varphi}(t) = \int_0^t \mathbf{x}(\alpha)\,d\alpha \tag{16-140}$$

(see also Sec. 11-3) where we assume that the instantaneous frequency $\mathbf{x}(t)$ is a stationary Markoff chain as in (16-120). Clearly,

$$R(\tau) = E\left\{\exp\left[j\int_t^{t+\tau} \mathbf{x}(\alpha)\,d\alpha\right]\right\} = E\left\{\exp\left[j\int_0^{\tau} \mathbf{x}(\alpha)\,d\alpha\right]\right\} = E\{\mathbf{w}(\tau)\}$$

We introduce the conditional correlations

$$R_{ik}(\tau) = E\{\mathbf{w}(\tau) | \mathbf{x}(0) = a_i, \mathbf{x}(\tau) = a_k\} \pi_{ik}(\tau) \tag{16-141}$$

where $\pi_{ik}(\tau)$ are the transition probabilities defined in (16-124). Clearly,

$$P\{\mathbf{x}(0) = a_i, \mathbf{x}(\tau) = a_k\} = p_i \pi_{ik}(\tau)$$

hence

$$R(\tau) = \sum_{i,k} p_i R_{ik}(\tau) \tag{16-142}$$

To determine $R(\tau)$ it suffices, therefore, to find $R_{ik}(\tau)$.

THEOREM. For any $\tau > 0$ and $\nu > 0$:

$$R_{ik}(\tau + \nu) = \sum_m R_{im}(\tau) R_{mk}(\nu) \tag{16-143}$$

†R. Kubo: "A Stochastic Theory of Line-Shape and Relaxation," Scottish Universities Summer School, D. ter Haar, ed., Plenum Press, New York, 1961. See also A. Papoulis: "Spectra of Stochastic FM Signals" in *Proceedings of Transactions of 9th Prague Conference on Information Theory*, 1982.

Proof. In the following, the conditions

$$\mathbf{x}(0) = a_i \qquad \mathbf{x}(\tau) = a_m \qquad \mathbf{x}(\tau + \nu) = a_k$$

will be abbreviated as a_i, a_m, and a_k respectively. Reasoning as in (16-104), we obtain

$$P\{\mathbf{x}(\tau) = a_m | a_i, a_k\} = \frac{\pi_{im}(\tau)\pi_{mk}(\nu)}{\pi_{ik}(\tau + \nu)} \tag{16-144}$$

Furthermore [see (8-43)]

$$E\{\mathbf{w}(\tau + \nu) | a_i, a_k\} = \sum_m E\{\mathbf{w}(\tau + \nu) | a_i, a_m, a_k\} P\{\mathbf{x}(\tau) = a_m | a_i, a_k\} \tag{16-145}$$

If $\mathbf{x}(\tau) = a_m$ is specified, then the integrals

$$\int_0^\tau \mathbf{x}(\alpha)\, d\alpha \qquad \int_\tau^{\tau+\nu} \mathbf{x}(\alpha)\, d\alpha$$

are conditionally independent. Hence

$$E\left\{\exp\left[j\int_0^{\tau+\nu}\mathbf{x}(\alpha)\, d\alpha\bigg|_{a_i, a_m, a_k}\right]\right\}$$
$$= E\left\{\exp\left[j\int_0^{\tau}\mathbf{x}(\alpha)\, d\alpha\bigg|_{a_i, a_m}\right]\right\} E\left\{\exp\left[j\int_\tau^{\tau+\nu}\mathbf{x}(\alpha)\, d\alpha\bigg|_{a_m, a_k}\right]\right\}$$

From the stationarity of $\mathbf{x}(t)$ it follows that the last term equals

$$E\left\{\exp\left[j\int_0^{\nu}\mathbf{x}(\alpha)\, d\alpha\bigg|_{\mathbf{x}(0) = a_m,\, \mathbf{x}(\nu) = a_k}\right]\right\}$$

Inserting into (16-145) and using (16-144), we obtain (16-143).

The Kolmogoroff equations. Differentiating (16-143) with respect to ν and setting $\nu = 0$, we obtain

$$R'_{ik}(\tau) = \sum_m R_{im}(\tau) R'_{mk}(0^+) \qquad \tau > 0 \tag{16-146}$$

Initial conditions To determine $R_{ik}(\tau)$ it suffices, therefore, to find the values of $R_{ik}(\tau)$ and its derivative at $\tau = 0^+$. We maintain that

$$R_{ik}(0^+) = \begin{cases} 1 & i = k \\ 0 & i \neq k \end{cases} \qquad R'_{ik}(0^+) = \begin{cases} ja_i - \mu_i & i = k \\ \lambda_{ik} & i \neq k \end{cases} \tag{16-147}$$

Proof. For small τ,

$$\int_0^\tau \mathbf{x}(\alpha)\, d\alpha \simeq \mathbf{x}(0)\tau \qquad \pi_{ik}(\tau) \simeq \begin{cases} 1 - \mu_i\tau & i = k \\ \lambda_{ik}\tau & i \neq k \end{cases}$$

Neglecting terms of the order of τ^2, we conclude that

$$R_{ii}(\tau) \simeq E\{\exp[j\mathbf{x}(0)\tau|a_i]\}\pi_{ii}(\tau) \simeq e^{ja_i\tau}(1-\mu_i\tau) \simeq 1 + ja_i\tau - \mu_i\tau$$

Furthermore, for $i \neq k$:

$$R_{ik}(\tau) \simeq e^{ja_i\tau}\lambda_{ik}\tau \simeq \lambda_{ik}\tau$$

and (16-147) follows.

Combining (16-147) and (16-146), we obtain

$$R'_{ik}(\tau) = (ja_k - \mu_k)R_{ik}(\tau) + \sum_m{}' \lambda_{mk}R_{im}(\tau) \qquad (16\text{-}148)$$

This yields $R_{ik}(\tau)$. The autocorrelation $R(\tau)$ of $\mathbf{w}(t)$ is determined from (16-142). Since the coefficients of (16-148) are constant, we conclude that the power spectrum of an FM signal, whose instantaneous frequency is a finite-state Markoff chain, is rational.

Example 16-16. Suppose that $\mathbf{x}(t)$ is a symmetrical telegraph signal (Fig. 16-20a). In this case,

$$a_1 = -a_2 = A \qquad \mu_1 = \mu_2 = \lambda$$

and (16-148) yields

$$R'_{11}(\tau) = (jA - \lambda)R_{11}(\tau) + \lambda R_{12}(\tau) \qquad R_{11}(0) = 1$$

$$R'_{12}(\tau) = \lambda R_{11}(\tau) - (jA + \lambda)R_{12}(\tau) \qquad R_{12}(0) = 0$$

Denoting by $\mathbf{S}^+_{ik}(s)$ the Laplace transform of $R^+_{ik}(\tau)$, we conclude from the above that

$$s\mathbf{S}^+_{11}(s) - 1 = (jA - \lambda)\mathbf{S}^+_{11}(s) + \lambda\mathbf{S}^+_{12}(s)$$

$$s\mathbf{S}^+_{12}(s) = \lambda\mathbf{S}^+_{11}(s) - (jA + \lambda)\mathbf{S}^+_{12}(s)$$

Hence

$$\mathbf{S}^+_{11}(s) = \frac{s + jA + \lambda}{D(s)} \qquad \mathbf{S}^+_{12}(s) = \frac{\lambda}{D(s)}$$

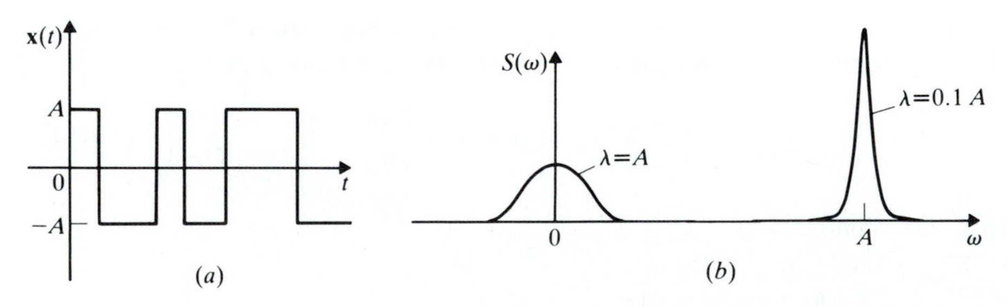

FIGURE 16-20

where

$$D(s) = s^2 + 2\lambda s + A^2$$

Reasoning similarly, we find

$$\mathbf{S}_{21}^{+}(s) = \frac{\lambda}{D(s)} \qquad \mathbf{S}_{22}^{+}(s) = \frac{s - jA + \lambda}{D(s)}$$

And since $p_1 = p_2 = 0.5$, (16-142) yields

$$\mathbf{S}^{+}(s) = \frac{s + 2\lambda}{D(s)} \qquad S(\omega) = 2\operatorname{Re}\mathbf{S}^{+}(j\omega)$$

In Fig. 16-20b, we plot $S(\omega)$ for $\lambda = A$ and $\lambda = 0.1A$. Note that the discontinuity points (zero crossings) of $\mathbf{x}(t)$ are Poisson distributed and their average density equals λ.

BIRTH PROCESSES. A birth process is Markoff chain $\mathbf{x}(t)$ consisting of a family of increasing staircase functions (Fig. 16-21). The process $\mathbf{x}(t)$ (population size) takes the values $1, 2, 3, \ldots$ and increases by 1 at the discontinuity points $\mathbf{t}_i$ (birth times). From the definition it follows that the transition rates λ_{ij} are different from 0 only if $i = j$ or $i = j - 1$. Thus

$$-\lambda_{ii} = \mu_i \qquad \lambda_{i(i+1)} = \mu_i \qquad \lambda_{ij} = 0 \qquad \text{otherwise}$$

The above shows that a birth process is specified in terms of the parameter μ_i. Clearly [see (16-130)]

$$\begin{aligned} P\{\mathbf{x}(t + \Delta t) = n \mid \mathbf{x}(t) = n\} &= 1 - \mu_n \Delta t \quad n \geq 1 \\ P\{\mathbf{x}(t + \Delta t) = n \mid \mathbf{x}(t) = n - 1\} &= \mu_{n-1} \Delta t \quad n > 1 \end{aligned} \tag{16-149}$$

Hence

$$p_1(t + \Delta t) = p_1(t)(1 - \mu_1 \Delta t)$$

$$p_n(t + \Delta t) = p_n(t)(1 - \mu_n \Delta t) + p_{n-1}(t)\mu_{n-1} \Delta t \qquad n > 1$$

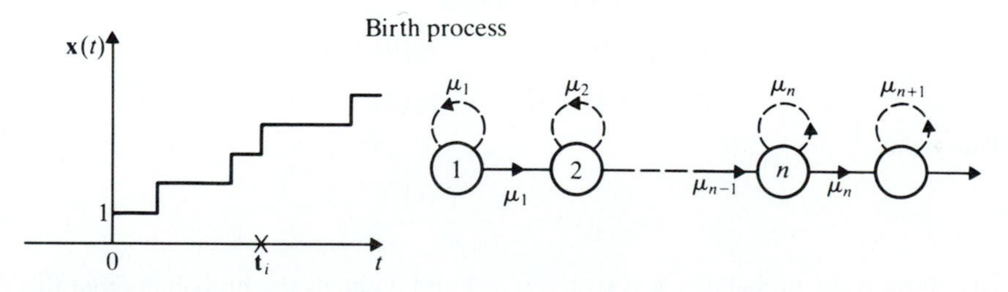

FIGURE 16-21

This yields

$$p_1'(t) = -\mu_1 p_1(t)$$
$$p_n'(t) = -\mu_n p_n(t) + \mu_{n-1} p_{n-1}(t) \qquad n > 1 \tag{16-150}$$

in agreement with (16-135).

Note The difference $\mathbf{x}(t_2) - \mathbf{x}(t_1)$ equals the number of discontinuity points $\mathbf{t}_i$ in the interval (t_1, t_2). This shows that a birth process is completely specified in terms of the point process $\mathbf{t}_i$.

Example 16-17. If the birthrate is proportional to the population size n:

$$\mu_n = nc \tag{16-151}$$

then $\mathbf{x}(t)$ is called the *simple birth process* (the constant c is the birthrate per person). We shall determine $p_n(t)$ under the (unrealistic!) assumption that (16-151) holds for every $n \geq 1$ and that $\mathbf{x}(0) = 1$. In this case,

$$p_1(0) = 1 \qquad p_n(0) = 0 \qquad n > 1 \tag{16-152}$$

Setting $\mu_n = nc$ in (16-150), we obtain

$$p_1'(t) + cp_1(t) = 0$$
$$p_n'(t) + ncp_n(t) = (n-1)cp_{n-1}(t) \qquad n > 1 \tag{16-153}$$

The above yields

$$p_1(t) = p_1(0)e^{-ct} = e^{-ct}$$

and with a simple recursion,

$$p_n(t) = e^{-ct}(1 - e^{-ct})^{n-1} \tag{16-154}$$

This function is called the *Yule–Furry* density. Thus, for a specific t, the RV $\mathbf{x}(t) - 1$ has a geometric distribution with ratio $(1 - e^{-ct})$. Hence (see Prob. 5-35)

$$E\{\mathbf{x}(t)\} = e^{ct} \qquad E\{\mathbf{x}^2(t)\} = 2e^{2ct} - e^{ct}$$

Example 16-18. We now assume that the rate of increase of $\mathbf{x}(t)$ is independent of its present state

$$\mu_n = \lambda = \text{constant}$$

As we shall see, the resulting $\mathbf{x}(t)$ is a *Poisson process*. We assume again that $\mathbf{x}(0) = 1$. Setting $\mu_n = \lambda$ in (16-150), we obtain

$$p_1'(t) + \lambda p_1(t) = 0 \qquad p_1(0) = 1$$
$$p_n'(t) + \lambda p_n(t) = \lambda p_{n-1}(t) \qquad p_n(0) = 0 \qquad n > 1$$

This yields

$$p_{n+1}(t) = \frac{e^{-\lambda t}(\lambda t)^n}{n!}$$

The above is the probability that $\mathbf{x}(t) = n + 1$ and it equals the probability that the number of points $\mathbf{x}(t) - \mathbf{x}(0)$ in the interval $(0, t)$ equals n.

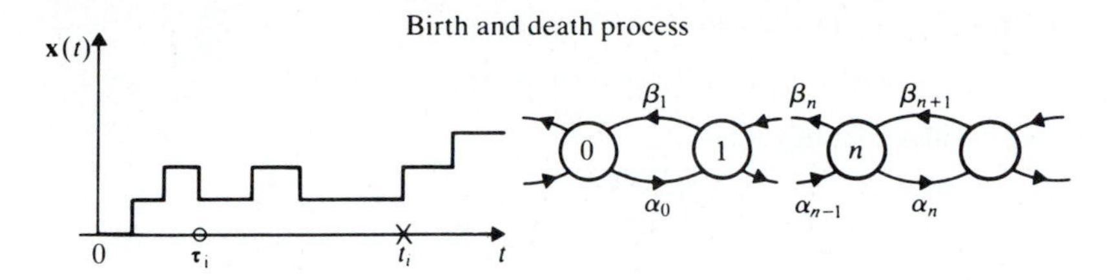

FIGURE 16-22

BIRTH–DEATH PROCESSES. Suppose now that a Markoff chain takes the values $0, 1, 2, \ldots$ and its discontinuities equal $+1$ or -1. (Fig. 16-22). We then say that $\mathbf{x}(t)$ is a birth–death process. In this case, λ_{ij} is different from 0 only if $i = j$ or $j-1$ or $j+1$. Hence $\mathbf{x}(t)$ is specified in terms of the two parameters

$$\alpha_i = \lambda_{i(i+1)} \qquad \beta_i = \lambda_{i(i-1)}$$

Thus $-\lambda_{ii} = \mu_i = \alpha_i + \beta_i$ and

$$P\{\mathbf{x}(t+\Delta t) = n \,|\, \mathbf{x}(t) = n-1\} = \alpha_{n-1}\,\Delta t$$

$$P\{\mathbf{x}(t+\Delta t) = n \,|\, \mathbf{x}(t) = n\} = [1 - (\alpha_n + \beta_n)\,\Delta t]$$

$$P\{\mathbf{x}(t+\Delta t) = n \,|\, \mathbf{x}(t) = n+1\} = \beta_{n+1}\,\Delta t$$

From (16-135) or, directly from the above, it follows that

$$\begin{aligned} p_0'(t) + \alpha_0 p_0(t) &= \beta_1 p_1(t) \\ p_n'(t) + (\alpha_n + \beta_n) p_n(t) &= \alpha_{n-1} p_{n-1}(t) + \beta_{n+1} p_{n+1}(t) \qquad n > 0 \end{aligned} \tag{16-155}$$

Example 16-19 ***M*|*M*|1 queue.** A queueing process $\mathbf{N}(t)$ is not, in general, Markoff. The $M|M|1$ case, however, is an exception. In this case, the arrival times $\mathbf{t}_i$ are Poisson with average density λ and the service time $\mathbf{c}_i$ is an RV with density $\mu e^{-\mu c}$ where $\mu > \lambda$. We maintain that the resulting $\mathbf{N}(t)$ is a birth–death process.

The probability that a unit will arrive in the interval $(t, t+\Delta t)$ equals $\lambda\,\Delta t$ (property P_1 of Poisson points). We shall show that the probability that a unit will depart in the interval $(t, t+\Delta t)$ equals $\mu\,\Delta t$. Indeed, denoting by $\boldsymbol{\tau}_i$ the first departure point to the right of t and by $\mathbf{c}_i$ the corresponding service time, we conclude that (see Example 7-10)

$$P\{t < \mathbf{c}_i \le t + \Delta t\} \simeq f_c(0)\,\Delta t = \mu\,\Delta t$$

no matter when this service started.

We have thus shown that

$$\begin{aligned} P\{\mathbf{N}(t+\Delta t) = n \,|\, \mathbf{N}(t) = n-1\} &= \lambda\,\Delta t \\ P\{\mathbf{N}(t+\Delta t) = n-1 \,|\, \mathbf{N}(t) = n\} &= \mu\,\Delta t \end{aligned} \tag{16-156}$$

Thus $\mathbf{N}(t)$ is a birth–death process with $\alpha_n = \lambda$ and $\beta_n = \mu$. We shall determine its state probabilities p_n for the stationary case.

In this case, $p_n'(t) = 0$ and (16-155) yields

$$\lambda p_0 = \mu p_1 \qquad (\lambda + \mu) p_n = \lambda p_{n-1} + \mu p_{n+1}$$

From this it follows readily that

$$p_n = p_0 \left(\frac{\lambda}{\mu} \right)^n \qquad p_0 = 1 - \frac{\lambda}{\mu}$$

as in Example 16-10.

Continuous-State Processes

A continuous-state Markoff process $\mathbf{x}(t)$ is specified in terms of its first-order density

$$p(x,t) = \frac{\partial P(x,t)}{\partial x} \qquad P(x,t) = P\{\mathbf{x}(t) \leq x\}$$

and the conditional (transition) density

$$\pi(x, x_0; t, t_0) = f_{\mathbf{x}(t)}(x | \mathbf{x}(t_0) = x_0) \qquad t > t_0$$

These functions are such that, if $t_0 < t < t_1$, then

$$\begin{aligned} &\int_{-\infty}^{\infty} p(x,t)\, dx = 1 \\ &p(x,t) = \int_{-\infty}^{\infty} p(x_0, t_0) \pi(x, x_0; t, t_0)\, dx_0 \end{aligned} \tag{16-157}$$

and

$$\begin{aligned} &\int_{-\infty}^{\infty} \pi(x, x_0; t, t_0)\, dx = 1 \\ &\pi(x, x_0; t, t_0) = \int_{-\infty}^{\infty} \pi(x, x_1; t, t_1) \pi(x_1, x_0; t_1, t_0)\, dx_1 \end{aligned} \tag{16-158}$$

as in (16-122) and (16-123).

Furthermore,

$$\pi(x, x_0; t, t_0) \xrightarrow[t \to t_0]{} \delta(x - x_0) \tag{16-159}$$

We shall show that the function $\pi(x, x_0; t, t_0)$ can be determined in terms of the slopes η and σ^2 of the *conditional mean* $a(x_0; t, t_0)$ and the *conditional variance* $b(x_0; t, t_0)$ of $\mathbf{x}(t)$ assuming $\mathbf{x}(t_0) = x_0$, defined as follows:

$$\begin{aligned} a(x_0; t, t_0) &= \int_{-\infty}^{\infty} x \pi(x, x_0; t, t_0)\, dx \\ b(x_0; t, t_0) &= \int_{-\infty}^{\infty} (x - a)^2 \pi(x, x_0; t, t_0)\, dx \end{aligned} \tag{16-160}$$

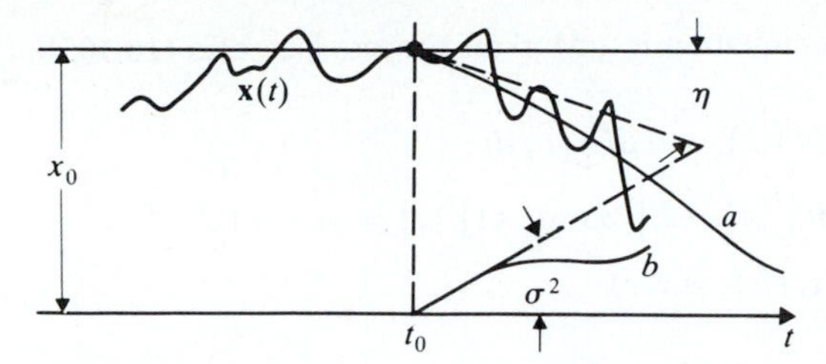

FIGURE 16-23

We assume that these functions are differentiable from the right and we denote by $\eta(x_0, t_0)$ and $\sigma^2(x_0, t_0)$ respectively their slopes at $t = t_0$ (Fig. 16-23)

$$\begin{aligned} \eta(x_0, t_0) &= \frac{\partial}{\partial t} a(x_0; t, t_0)\bigg|_{t=t_{0^+}} \\ \sigma^2(x_0; t_0) &= \frac{\partial}{\partial t} b(x_0; t, t_0)\bigg|_{t=t_{0^+}} \end{aligned} \tag{16-161}$$

Clearly [see (16-159) and (16-160)]

$$a(x_0; t_0, t_0) = x_0 \qquad b(x_0; t_0, t_0) = 0$$

Hence, for $\Delta t > 0$,

$$\begin{aligned} a(x_0; t_0 + \Delta t, t_0) &\simeq x_0 + \eta(x_0, t_0)\,\Delta t \\ b(x_0; t_0 + \Delta t, t_0) &\simeq \sigma^2(x_0, t_0)\,\Delta t \end{aligned} \tag{16-162}$$

If the process $\mathbf{x}(t)$ is homogeneous, then the function $\pi(x, x_0; t, t_0)$ depends on $\tau = t - t_0$. In this case, the slopes $\eta(x_0)$ and $\sigma^2(x_0)$ of a and b are independent of t_0.

From (16-162) and the definition of the functions a and b it follows that

$$\begin{aligned} E\{d\mathbf{x}(t)\,|\,\mathbf{x}(t) = x\} &= \eta(x, t)\,dt \\ E\{[d\mathbf{x}(t) - \eta(\mathbf{x}, t)\,dt]^2|x\} &= \sigma^2(x, t)\,dt \end{aligned} \tag{16-163}$$

As we show in the next example, these equations can often be used to determine $\eta(x, t)$ and $\sigma^2(x, t)$ directly in terms of the specifications of the process $\mathbf{x}(t)$.

Example 16-20. Consider the nonlinear stochastic differential equation

$$\frac{d\mathbf{x}(t)}{dt} + \beta(\mathbf{x}, t) = \frac{d\mathbf{w}(\mathbf{x}, t)}{dt}$$

where $\mathbf{w}(x, t)$ is a process with independent increments and such that

$$E\{d\mathbf{w}(x, t)\} = 0 \qquad E\{[d\mathbf{w}(x, t)]^2\} = \gamma(x, t)\,dt$$

The solution of the above equation is a Markoff process [see also (16-105)]. Clearly,

$$E\{d\mathbf{x}(t)|x\} = -\beta(x,t)\,dt$$

$$E\{[d\mathbf{x}(t) + \beta(\mathbf{x},t)\,dt]^2|x\} = E\{[d\mathbf{w}(x,t)]^2|x\} = \gamma(x,t)\,dt$$

Hence $\eta(x,t) = -\beta(x,t)$, $\sigma^2(x,t) = \gamma(x,t)$.

THE DIFFUSION EQUATIONS. We shall show that the conditional density $\pi = \pi(x, x_0; t, t_0)$ satisfies the *diffusion equations*

$$\begin{gathered}\frac{\partial\pi}{\partial t} + \frac{\partial}{\partial x}[\eta(x,t)\pi] - \frac{1}{2}\frac{\partial^2}{\partial x^2}[\sigma^2(x,t)\pi] = 0 \\ \frac{\partial\pi}{\partial t_0} + \eta(x_0,t_0)\frac{\partial\pi}{\partial x_0} + \frac{1}{2}\sigma^2(x_0,t_0)\frac{\partial^2\pi}{\partial x_0^2} = 0\end{gathered} \qquad (16\text{-}164)$$

The first is called *forward* (or *Fokker–Planck*) and the second *backward*.

Proof. If $f(x)$ is a density with mean η and "small" variance, then [see (5-55)]

$$\int_{-\infty}^{\infty} g(\xi)f(\xi)\,d\xi \simeq g(\eta) + \frac{\sigma^2}{2}g''(\eta) \qquad (16\text{-}165)$$

From (16-158) it follows with $x_1 = \xi$ and $t_1 = t_0 + \varepsilon$ that

$$\pi(x, x_0; t, t_0) = \int_{-\infty}^{\infty} \pi(x, \xi; t, t_0 + \varepsilon)\pi(\xi, x_0; t_0 + \varepsilon, t_0)\,d\xi$$

In the above, $\pi(\xi, x_0; t_0 + \varepsilon, t_0)$ is a density in the variable ξ and its mean and variance equal [see (16-160) and (16-162)]

$$a(x_0; t_0 + \varepsilon, t_0) \simeq x_0 + \varepsilon\eta(x_0, t_0) \qquad b(x; t_0 + \varepsilon, t_0) \simeq \varepsilon\sigma^2(x_0, t_0)$$

Therefore, with

$$g(\xi) = \pi(x, \xi; t, t_0 + \varepsilon) \qquad f(\xi) = \pi(\xi, x_0; t_0 + \varepsilon, t_0)$$

(16-165) yields (Fig. 16-24)

$$\pi(x, x_0; t, t_0) \simeq \pi(x, x_0 + \varepsilon\eta; t, t_0 + \varepsilon) + \frac{\varepsilon\sigma^2}{2}\frac{\partial^2}{\partial x_0^2}\pi(x, x_0 + \varepsilon; t, t_0 + \varepsilon)$$

within $O(\varepsilon^2)$. Expanding the right side into a power series in ε and retaining only linear terms, we obtain the second equation in (16-164). The proof of the first is similar.

COROLLARY. The first-order density $p = p(x,t)$ of the process $\mathbf{x}(t)$ satisfies the Fokker–Planck equation

$$\frac{\partial p}{\partial t} + \frac{\partial}{\partial x}[\eta(x,t)p] - \frac{1}{2}\frac{\partial^2}{\partial x^2}[\sigma^2(x,t)p] = 0 \qquad (16\text{-}166)$$

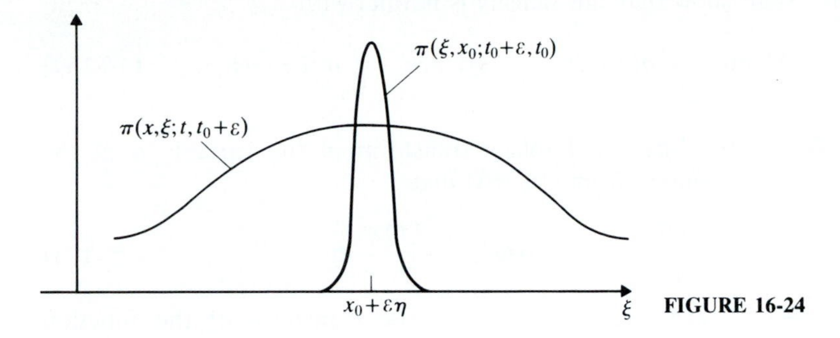

FIGURE 16-24

Proof. It follows if we express the function $p(x, t)$ in terms of the integral in (16-157) and use (16-164).

Example 16-21. The velocity $\mathbf{v}(t)$ of a particle in brownian motion satisfies Langevin's equation

$$\mathbf{v}'(t) + \beta\mathbf{v}(t) = \mathbf{w}'(t)$$

where $\mathbf{w}(t)$ is a process with orthogonal increments and such that

$$E\{d\mathbf{w}(t)\} = 0 \qquad E\{[d\mathbf{w}(t)]^2\} = \gamma\, dt$$

Hence (see Example 16-20) $\mathbf{v}(t)$ is a Markoff process with $\eta(x, t) = -\beta v$, $\sigma^2(x, t) = \gamma$, and (16-166) yields

$$\frac{\partial p}{\partial t} = \beta\frac{\partial(vp)}{\partial v} + \frac{\gamma}{2}\frac{\partial^2 p}{\partial v^2} \tag{16-167}$$

where $p = p(v, t)$ is the density of $\mathbf{v}(t)$.

SOLUTION OF THE FOKKER–PLANCK EQUATION. We shall solve the forward equation in (16-164) under the assumption that the conditional density $\pi(x, x_0; t, t_0)$ does not depend explicitly on the state $\mathbf{x}(t_0) = x_0$ of $\mathbf{x}(t)$ but only on the increment $u = x - x_0$. In this case [see (16-160)]

$$a(x_0; t, t_0) = \int_{-\infty}^{\infty}(u + x_0)\pi(u; t, t_0)\, du = a_1(t, t_0) + x_0$$

Inserting into the second integral in (16-160), we conclude that the conditional variance $b = b(t, t_0)$ does not depend on x_0. From the above it follows that the slopes $\eta(t_0)$ and $\sigma^2(t_0)$ of a and b are independent of x_0. This simplifies the form of the forward equation

$$\frac{\partial\pi}{\partial t} + \eta(t)\frac{\partial\pi}{\partial x} - \frac{1}{2}\sigma^2(t)\frac{\partial^2\pi}{\partial x^2} = 0 \tag{16-168}$$

The solution $\pi(u; t, t_0)$ of (16-168) is a density in the variable u. In fact, it is the density of the increment $\mathbf{x}(t) - \mathbf{x}(t_0)$ of $\mathbf{x}(t)$ under the assumption that

$\mathbf{x}(t_0) = x_0$. We shall show that this density is normal with

$$\text{Mean: } \int_{t_0}^{t} \eta(\tau)\,d\tau \qquad \text{Variance: } \int_{t_0}^{t} \sigma^2(\tau)\,d\tau \tag{16-169}$$

Proof. With $\Phi(s,t)$ the bilateral Laplace transform in the variable u of the function $\pi(u; t, t_0)$, it follows from (16-168) that

$$\frac{\partial \Phi}{\partial t} = -\eta(t)s\Phi + \frac{\sigma^2(t)s^2}{2}\Phi \tag{16-170}$$

The function $\Phi(s,t)$, evaluated at $t = t_0$, is the transform of the function $\pi(u; t_0, t_0) = \delta(u)$ [see (16-159)]. Hence $\Phi(s, t_0) = 1$ and (16-170) yields

$$\ln \Phi(s,t) = -s\int_{t_0}^{t} \eta(\tau)\,d\tau + \frac{s^2}{2}\int_{t_0}^{t} \sigma^2(\tau)\,d\tau \tag{16-171}$$

This shows that $\Phi(s,t)$ is the moment function of a normal density with mean and variance as in (16-169).

PROBLEMS

16-1. Show that the probability that the Wiener process $\mathbf{w}(t)$ does not cross the line L_a in the interval $(0, t)$ equals $2\mathbb{G}(a/\sqrt{\alpha t}) - 1$.

16-2. We denote by $P_0(\tau)$ the conditional probability that the number of 0's of a normal process $\mathbf{x}(t)$ in the interval $(t, t+\tau)$, assuming $\mathbf{x}(t) = 0$, is odd. Show that if $E\{\mathbf{x}(t)\} = 0$, then

$$\cos \pi P_0(\tau) = -R'(\tau)\left[-R''(0)R(0) + \frac{R''(0)R^2(\tau)}{R(0)}\right]^{-1/2}$$

16-3. Show that if $f_w(w, t)$ is the first-order density of the Wiener process $\mathbf{w}(t)$ and $f_\tau(\tau, a)$ is the density of the first passage time $\boldsymbol{\tau}_1$ (Fig. 16-4a), then†

$$f_\tau(\tau, a) = \frac{a}{\tau} f_w(a, \tau)$$

16-4. Passengers arrive at a terminal boarding the next bus. The times of their arrival are Poisson with density $\lambda = 1$ per minute. The times of departure of each bus are Poisson with density $\mu = 2$ per hour. (a) Find the mean number of passengers in each bus. (b) Find the mean number of passengers in the first bus that leaves after 9 A.M.

Answer: (a) 30; (b) 60.

16-5. Passengers arrive at a terminal after 9 A.M. The times of their arrival are Poisson with mean density $\lambda = 1$ per minute. The time interval from 9 A.M. to the

†A. A. Borovkov: "On the First Passage Time . . . ," *Theory of Probability and Its Applications*, vol. X, no. 2, 1965.

departure of the next bus is an RV **c**. Find the mean number of passengers in this bus (*a*) if **c** has an exponential density with mean $\eta_c = 30$ min, (*b*) if **c** is uniform between 0 and 60 min.

Answer: (*a*) 30; (*b*) 30.

16-6. The point process $\mathbf{t}_i$ is stationary and the RVs $\mathbf{c}_i = \mathbf{t}_i - \mathbf{t}_{i-1}$ are uniform in the interval $(0, a)$. Show that if $\mathbf{t}_1$ is the first point to the right of a fixed point t_0, then $E\{\mathbf{t}_1 - t_0\} = a/3$.

16-7. (*a*) The RVs $\mathbf{c}_i$ are i.i.d. and $E\{e^{j\omega \mathbf{c}_i}\} = \Phi_c(\omega)$. The process $\mathbf{n}(t)$ is Poisson with parameter λt and independent of $\mathbf{c}_i$. Show that, if (Fig. P16-7*a*)

$$\mathbf{x}(t) = \sum_{i=1}^{\mathbf{n}(t)} \mathbf{c}_i \qquad \text{then} \quad E\{e^{j\omega \mathbf{x}(t)}\} = \exp\{\lambda t[\Phi_c(\omega) - 1]\}$$

Special case: If the RV $\mathbf{c}_i$ takes the values 1 and 0 (Fig. P16-7*b*) and $P\{\mathbf{c}_i = 1\} = p$, then $\mathbf{x}(t)$ is a Poisson process with parameter λpt (see also Prob. 8-11).

Hint: $E\{e^{j\omega \mathbf{x}(t)} | \mathbf{n}(t) = n\} = \Phi_c^n(\omega)$.

(*b*) Using the above, show that, if $\mathbf{t}_i$ is a Poisson point process with mean density λ and $\boldsymbol{\tau}_i$ is a process obtained by eliminating at random a subset of $\mathbf{t}_i$, then $\boldsymbol{\tau}_i$ is a Poisson point process with mean density λp where p is the probability that a point of $\mathbf{t}_i$ is not eliminated.

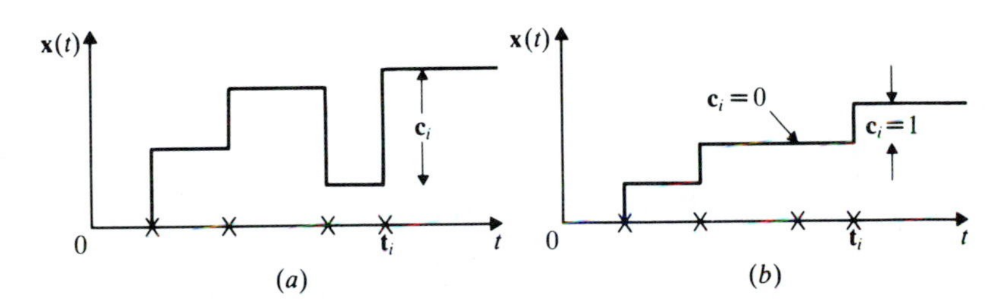

FIGURE P16-7

16-8. (*a*) Show that if $\mathbf{t}_n$ is a Poisson point process, then the process $\mathbf{t}_{2n}$ consisting of every other point of $\mathbf{t}_n$ is not Poisson. (*b*) Show that if $\boldsymbol{\alpha}_n$ and $\boldsymbol{\beta}_n$ are two independent Poisson point processes with densities λ_α and λ_β respectively, then the process $\mathbf{t}_n$ consisting of all the points of $\boldsymbol{\alpha}_n$ and $\boldsymbol{\beta}_n$ is Poisson with density $\lambda_\alpha + \lambda_\beta$.

16-9. Visitors enter a park at Poisson times with mean density $\lambda = 2$ per minute. Each visitor stays in the park **c** minutes where **c** is an RV uniform between 30 and 90 min. Find the mean and the variance of the number $\mathbf{N}(t)$ of visitors in the park.

16-10. In the $M|M|1$ queue (Example 16-10), **y** is the busy period and $\boldsymbol{\Phi}_y(s)$ is its moment function. Show that

$$\lambda \boldsymbol{\Phi}_y^2(s) - (\lambda + \mu - s)\boldsymbol{\Phi}_y(s) + \mu = 0 \qquad \boldsymbol{\Phi}_y(s) \xrightarrow[S \to \infty]{} 0$$

16-11. (*a*) With $\bar{q}_i$ as in (16-53), show that

$$E\{\bar{\mathbf{q}}_i^2\} = E\{\mathbf{q}_i^2\} - 2\eta_q + \rho \tag{i}$$

(*b*) Prove the Pollaczek–Khinchin formula (16-60), using (i) and the identity

$$E\{\mathbf{q}_i^2\} = E\{\bar{\mathbf{q}}_i^2\} + E\{\mathbf{n}_{c_i}^2\} + 2E\{\bar{\mathbf{q}}_i\}E\{\mathbf{n}_{c_i}\}$$

16-12. In a single-server queueing system, the arrival times $\mathbf{t}_i$ are Poisson with mean density $\lambda = 9$ per hour. Find the mean of the following: the service time **c**, the waiting time **b**, the system time **a**, the idle period **x**, the busy period **y**, and the number $\mathbf{n}_y$ of units served during a busy period. Consider two cases: the density of the service time **c** is (*a*) uniform between 4 and 8 min; (*b*) it equals $\mu^2 ce^{-\mu c}$ where $\mu = 1/3$.

16-13. In an $M|M|1$ queue, the service time density equals $\mu e^{-\mu c}$. Find the density of the distance from a fixed point t_0 to the next departure point.

Hint: The probability that t_0 is a point of a busy period equals λ/μ.

16-14. Find the probability $P\{\mathbf{s}(t) \geq 2\}$ of the shot-noise process

$$\mathbf{s}(t) = \sum_i h(t - t_i) \qquad h(t) = 4U(t) - 3U(t-1) - U(t-2)$$

where $\mathbf{t}_i$ are Poisson points with $\lambda = 2$.

16-15. The shot-noise process $\mathbf{s}(t)$ is a train of triangles

$$\mathbf{s}(t) = \sum_i h(t - \mathbf{t}_i) \qquad h(t) = \begin{cases} 5(2 - |t|) & |t| < 2 \\ 0 & |t| > 2 \end{cases}$$

and the points $\mathbf{t}_i$ are Poisson with $\lambda = 0.01$. (*a*) Find its power spectrum $S_s(\omega)$. (*b*) Find its first-order density. (Note that $2\lambda \ll 1$.)

16-16. The points $\mathbf{t}_i$ are Poisson with density λ and

$$\mathbf{s}(t) = \sum_i h(t - \mathbf{t}_i) \qquad h(t) = e^{-\alpha t}U(t)$$

(*a*) Find the mean, the variance, and the power spectrum of $\mathbf{s}(t)$. (*b*) Find the power spectrum of the process $\mathbf{y}(t) = \mathbf{s}^2(t)$ for $\lambda \gg \alpha$ and for $\lambda \ll \alpha$.

16-17. The RVs $\mathbf{x}_n$ are i.i.d. taking the values $+1$ and -1 with $P\{\mathbf{x}_n = 1\} = 0.6$ and $P\{\mathbf{x}_n = -1\} = 0.4$. Show that the process $\mathbf{y}_n = \mathbf{x}_n + \mathbf{x}_{n-1} + \cdots + \mathbf{x}_1$ is a Markoff chain and find its state probabilities $p_i[n]$ and transition probabilities $\pi_{ij}[m]$.

16-18. Show that if $\mathbf{x}(0) = 0$ and $\mathbf{x}(t)$ is a process with independent increments, then it is Markoff.

16-19. Given a two-state Markoff chain $\mathbf{x}_n$ taking the values 1 and 0 with state probability vector $P[n]$ and transition matrix Π. Show that, if

$$\Pi = \begin{bmatrix} \frac{2}{3} & \frac{1}{3} \\ \frac{1}{3} & \frac{2}{3} \end{bmatrix} \qquad \text{then} \quad \Pi^n \xrightarrow[n\to\infty]{} \begin{bmatrix} \frac{1}{2} & \frac{1}{2} \\ \frac{1}{2} & \frac{1}{2} \end{bmatrix} \qquad P[n] \xrightarrow[n\to\infty]{} [\tfrac{1}{2}, \tfrac{1}{2}]$$

Find $P[2]$ and $P[3]$ if $\mathbf{x}_1 = 0$.

16-20. Show that, if $\mathbf{x}(t)$ is a discrete-state Markoff process taking the values a_i and

$$P\{\mathbf{x}(t) = a_i\} = p_i(t) \qquad P\{\mathbf{x}(t_2) = a_j | \mathbf{x}(t_1) = a_i\} = \pi_{ij}(t_1, t_2)$$

then its autocorrelation equals

$$R_x(t_1, t_2) = \sum_{i,j} a_i a_j \pi_{ij}(t_1, t_2) p_i(t_1)$$

16-21. Show that if $\pi_{ij}(t_1, t_2)$ are the transition probabilities of a Markoff chain $\mathbf{x}(t)$ and

$$P\{\mathbf{x}(t + \Delta\tau) = a_i | \mathbf{x}(t) = a_i\} \simeq 1 - \mu(t)\,\Delta t$$

$$P\{\mathbf{x}(t + \Delta\tau) = a_j | \mathbf{x}(t) = a_i\} = \lambda_{ij}\,\Delta t$$

then

$$\frac{\partial \pi_{ji}(t, t_0)}{\partial t} = -\mu_i(t)\pi_{ji}(t, t_0) + \sum_k{}' \lambda_{ki}(t)\pi_{jk}(t, t_0)$$

$$\frac{\partial \pi_{ji}(t, t_0)}{\partial t_0} = \pi_{ji}(t, t_0)\mu_j(t_0) + \sum_k{}' \pi_{ki}(t, t_0)\lambda_{jk}(t_0)$$

16-22. The telegraph signal $\mathbf{x}(t)$ of Example 16-15 is stationary with $\mu_2 = 3\mu_1 = 6$ and $A = 100$. (*a*) Find its mean η_x and autocorrelation $R_x(\tau)$. (*b*) Find the power spectrum $S_w(\omega)$ of the FM signal

$$\mathbf{w}(t) = e^{j\varphi(t)} \qquad \varphi(t) = \int_0^t \mathbf{x}(\alpha)\,d\alpha$$

(*c*) Show that $\mathbf{w}(t)$ satisfies the time-varying stochastic differential equation

$$\mathbf{w}'(t) + j\mathbf{x}(t)\mathbf{w}(t) = 0 \qquad \mathbf{w}(0) = 1$$

Find $E\{\mathbf{w}(t)\}$ and $R_w(t_1, t_2)$.

16-23. Show that the distribution function

$$F(x, x_0; t, t_0) = P\{\mathbf{x}(t) < x | \mathbf{x}(t_0) = x_0\} = \int_{-\infty}^{\infty} \pi(\xi, x_0; t, t_0)\,d\xi$$

of a Markoff process satisfies the backward diffusion equation

$$\frac{\partial F}{\partial t_0} + \eta(x_0, t_0)\frac{\partial F}{\partial x_0} + \frac{1}{2}\sigma^2(x_0, t_0)\frac{\partial^2 F}{\partial x_0^2} = 0$$

BIBLIOGRAPHY

Abramson, N. M. (1963): *Information Theory and Coding*, McGraw-Hill, New York.
Antoniou, A. (1979): *Digital Filters: Analysis and Design*, McGraw-Hill, New York.
Ash, R. (1965): *Information Theory*, Interscience, New York.
Bharucha-Reid, A. T. (1960): *Elements of the Theory of Markov Processes and Their Applications*, McGraw-Hill, New York.
Blackman, R. B., and J. W. Tukey (1959): *The Measurement of Power Spectra*, Dover, New York.
Blanc-Lapierre, A. and R. Fortet (1953): *Theorie des Fonctions Aleatoires*, Masson et Cie, Paris.
Childers, D. G., ed. (1978): *Modern Spectrum Analysis*, Wiley, New York.
Cooper, R. B. (1981): *Introduction to Queuing Theory*, North-Holland, New York.
Cramer, H. (1946): *Mathematical Methods of Statistics*, Princeton University Press, Princeton, NJ.
Davenport, W. B., Jr. and W. L. Root (1958): *An Introduction to the Theory of Random Signals and Noise*, McGraw-Hill, New York.
Doob, J. L. (1953): *Stochastic Processes*, Wiley, New York.
Feinstein, A. (1958): *Foundations of Information Theory*, McGraw-Hill, New York.
Feller, W. (1957 and 1967): *An Introduction to Probability Theory and Its Applications*, Vols. I and II, Wiley, New York.
Franks, L. E. (1979): *Signal Theory*, Prentice-Hall, Englewood Cliffs, NJ.
Gardner, W. A. (1987): *Statistical Spectral Analysis: A Non-Probabilistic Theory*, Prentice-Hall, Englewood Cliffs, NJ.
Helstrom, C. W. (1968): *Statistical Theory of Signal Detection*, 2d ed., Pergamon Press, New York.
Jenkins, G. M. and D. G. Watts (1968): *Spectral Analysis and Its Applications*, Holden-Day, San Francisco, CA.
Kleinrock, L. (1975–1976): *Queuing Systems*, 2 vols., Wiley, New York.
Laning, J. H. and R. H. Battin (1956): *Random Processes in Automatic Control*, McGraw-Hill, New York.
Lebedev, V. L.: "Random Processes in Electric and Mechanical Systems," NSF and NASA Technical Translations, Washington, DC.
Marple, S. L. (1987): *Digital Spectral Analysis*, Prentice-Hall, Englewood Cliffs, NJ.
Nahi, N. E. (1969): *Estimation Theory and Applications*, Wiley, New York.
Oppenheim, A. V. and R. W. Schafer (1975): *Digital Signal Processing*, Prentice-Hall, Englewood Cliffs, NJ.
Papoulis, A. (1962): *The Fourier Integral and Its Applications*, McGraw-Hill, New York.

Papoulis, A. (1968): *Systems and Transforms with Applications in Optics*, McGraw-Hill, New York. Reprinted (1981) by Krieger Publishing Company, Melbourne, FL.
Papoulis, A. (1977): *Signal Analysis*, McGraw-Hill, New York.
Papoulis, A. (1980): *Circuits and Systems: A Modern Approach*, Holt, Rinehart and Winston, New York.
Papoulis, A. (1990): *Probability and Statistics*, Prentice-Hall, Englewood Cliffs, NJ.
Parzen, E. (1960): *Modern Probability Theory and Its Applications*, Wiley, New York.
Priestley, M. (1981): *Spectral Analysis and Time Series*, 2 vols., Academic, London.
Proakis, J. (1983): *Introduction to Digital Communications*, McGraw-Hill, New York.
Schwartz, M. (1977): *Computer-Communication Network Design and Analysis*, Prentice-Hall, Englewood Cliffs, NJ.
Schwartz, M. and L. Shaw (1975): *Signal Processing*, McGraw-Hill, New York.
Wainstein, L. A. and V. D. Zubakov (1962): *Extraction of Signals from Noise* (translated from Russian), Prentice-Hall, Englewood Cliffs, NJ.
Wiener, N. (1949): *Extrapolation, Interpolation, and Smoothing of Stationary Time Series*, MIT Press, Cambridge, MA.
Woodward, P. (1953): *Probability and Information Theory with Applications to Radar*, Pergamon, New York.
Yaglom, A. M. (1962): *Stationary Random Functions* (translated from Russian), Prentice-Hall, Englewood Cliffs, NJ.
Yaglom, A. M. (1987): *Correlation Theory of Stationary and Related Random Functions*, 2 vols., Springer, New York.

INDEX